W0036865

Crustal and Upper Mantle Structure in Europe
Monograph No. 1
European Seismological Commission

Publication of the
Deutsche Geophysikalische Gesellschaft

Explosion Seismology in Central Europe

Data and Results

Edited by
P. Giese C. Prodehl A. Stein

With 284 Figures and 3 Maps

Springer-Verlag
Berlin Heidelberg New York 1976

Prof. Dr. PETER GIESE
Institut für Geophysikalische Wissenschaften
Freie Universität Berlin
Rheinbabenallee 49, 1000 Berlin 33, FRG

Dr. CLAUS PRODEHL
Geophysikalisches Institut
Technische Universität Karlsruhe
Hertzstraße 16, 7500 Karlsruhe, FRG

Dr. ALBERT STEIN
Niedersächsisches Landesamt für Bodenforschung
Alfred Bentz Haus, Postfach 230153
3000 Hannover-Buchholz, FRG

ISBN-13: 978-3-642-66405-2 e-ISBN-13: 978-3-642-66403-8
DOI: 10.1007/978-3-642-66403-8

Library of Congress Cataloging in Publication Data. Main entry under title: Explosion seismology in Central Europe. Sponsored by the German Geophysical Society. Bibliography: p. Includes index. 1. Geology - Central Europe. I. Giese, Peter. II. Prodehl, C., 1936 -. III. Stein, Albert, Dr. IV. Deutsche Geophysikalische Gesellschaft. QE260.E95. 551.1'3'0943. 7621047.

Softcover reprint of the hardcover 1st edition 1976

Offsetprinting: Beltz Offsetdruck, Hemsbach/Bergstr.

Foreword

The determination of crustal structure by means of explosion seismology has been one of the major objectives of the European Seismological Commission (ESC) over the past twenty-five years. It was decided some time ago to publish the results of regional crustal investigations in Europe in a series of monographs. This publication entitled "Explosion Seismology in Central Europe - Data and Results" is Volume 1 in a sequence of publications dealing with the crustal structure in Europe.

The European seismologists are indebted to the German Geophysical Society (Deutsche Geophysikalische Gesellschaft) for taking the initiative to publish this book. Thanks are due to the German Research Society (Deutsche Forschungsgemeinschaft) for providing generous financial support of the field measurements and data evaluation. It is hoped that this publication will stimulate a continuation of investigations of the earth's lithosphere in order to elucidate the details which are still not fully understood.

STEPHAN MUELLER
(President of the
European Seismological Commission)

Preface

The investigation of the earth's crust started in 1910, when A. MOHOROVIČIĆ deduced the existence of the crust/mantle boundary from near-earthquake records. At about the same time, L. MINTROP in Göttingen started experiments to investigate the uppermost sedimentary layers with portable seismographs and a dropping weight. The use of large explosions for the study of the deeper crust was introduced in Germany by E. WIECHERT, director of the geophysical institute at Göttingen, and the first field experiments were carried out as early as 1923. G. ANGENHEISTER, sen., B. BROCKAMP, H. REICH, W. SCHWEYDAR, and K. WÖLCKEN continued this fundamental explosion work.

The second stage of seismic crustal studies in Germany started with the big explosions on Helgoland in 1947 and near Haslach in the Black Forest, in 1948. Since the beginning of the fifties, commercial quarry blasts have been used increasingly for the investigation of crustal structure, as they proved to be a powerful and a low-cost tool. These studies were carried out first by the Geophysical Department of the Geological Survey at Hannover and the Institute of Applied Geophysics of the University of München. Other geophysical institutes co-operated occasionally in major experiments, for instance by the recording of a deep-borehole shot at Tölz, Southern Bavaria, in 1954. Here, the idea of a continuous co-operation for seismic crustal studies was initiated.

The third stage of seismic crustal research started in 1957 with the project "Geophysikalische Erforschung des tieferen Untergrundes Mitteleuropas" (Geophysical Investigation of Crustal Structure in Central Europe), sponsored by the Deutsche Forschungsgemeinschaft (German Research Society) and carried out by the geophysical institutes of the universities and the geophysical departments of the Geological Surveys. In order to enable effective co-operation between all participating institutions, some working groups were founded in order to develop new instruments, to plan and organize field work, and to interpret and compile the data.

In the investigation of crust and upper mantle, seismic-refraction and reflection methods have been applied. In 1950, H. REICH, using a quarry blast, observed for the first time clear reflections from the Mohorovičić discontinuity in southern Germany. Using oil-prospecting shots in the following years, an increasing number of deep reflections were recorded. The main activity of crustal studies was, however, within the domain of seismic-refraction investigations.

When, in 1960, the International Union of Geodesy and Geophysics suggested a world-wide project for the investigation of the "Upper Mantle", much work had already been done in this field so that the fourth stage of crustal research was easily introduced with programs extending beyond the borders of western Germany, especially into the Alpine region. In order to improve the methods applied and to exchange data and results, special annual meetings were organized by the German Research Society between 1960 and 1964, and corresponding reports were distributed to all participants in this research program.

The joint geophysical research of crust and upper mantle demonstrated the advantage of close co-operation between all German geophysical research institutions. For this reason, the "Forschungskollegium Physik des Erdkörpers" (Research Council of Physics of the Earth) was established in 1964 as a permanent council in which all geophysical research institutions of the Federal Republic of Germany are represented. For the branch of explosion seismology, a special working group was set up in the same year, called "Arbeitsgruppe Seismische Feldmessungen und Auswertung (ASFA)" (Working group for seismic field experiments and interpretation). Planning, organization, and data evaluation of the seismic measurements within the "Upper Mantle Project" are the tasks of this working group. Since 1964, it has been compiling internal activity reports, including the data observed in the form of tables and record sections. This material was the basis for the presentation of data in Section 3.

Since the beginning of the seventies, the number of quarry blasts suitable for seismic research work has decreased continuously. At the same time, special explosions for deep-seismic sounding purposes have been arranged with increasing frequency in western Europe, initiating the fifth stage of German explosion seismology. Most of these experiments are not confined to central Europe and require therefore close international cooperation. Since their interpretation is mostly not yet completed, a description of this last stage of explosion seismology is beyond the scope of this monograph. Only two reports concerning the area of the Alps and the Rhine Graben are included.

The following reports give an outline of the present state of knowledge on crustal structure in central Europe. The progress which has been achieved since the beginning of intensive investigation in 1957 is quite remarkable. Many details and problems, however, are not yet understood because of the deeper location of the sources. As proposed for the International Geodynamics Project the more recent experiments of explosion seismology have been extended to the investigation of the whole lithosphere.

H. CLOSS
F. GOERLICH
M. SIEBERT

Contents

Contributors

ANGENHEISTER, G. Institut für Allgemeine und Angewandte Geophysik, Universität München, Theresienstraße 41/IV,(Block C), 8000 München 2, FRG

ANSORGE, J. Institut für Geophysik, Eidgenössische Technische Hochschule, Hönggerberg, Postfach 266, CH-8049 Zürich, Switzerland

BAMFORD, D. Department of Geophysics, University of Edinburgh, 6 South Oswald Road, Edinburgh EH9 2HX, Great Britain

BARTELSEN, H. Institut für Geophysik, Universität Kiel, Neue Universität, Haus B2, Olshausenstraße 40-60, 2300 Kiel, FRG

BERCKHEMER, H. Institut für Meteorologie und Geophysik, Universität Frankfurt, Feldbergstraße 47, 6000 Frankfurt 1, FRG

BLOHM, E.-K. Niedersächsisches Landesamt für Bodenforschung, Alfred-Bentz-Haus, Postfach 510153, 3000 Hannover-Buchholz, FRG

BURKHARDT, H. Institut für Geophysik, Technische Universität Clausthal, Adolf-Römer-Straße 2A, 3392 Clausthal-Zellerfeld, FRG

BUTTKUS, B. Bundesanstalt für Geowissenschaften und Rohstoffe, Alfred-Bentz-Haus, Postfach 510153, 3000 Hannover-Buchholz, FRG

CLOSS, H. Bundesanstalt für Geowissenschaften und Rohstoffe, Alfred-Bentz-Haus, Postfach 510153, 3000 Hannover-Buchholz, FRG

DOHR, G. Preussag A.G., Erdöl und Erdgas, Arndtstraße 1, 3000 Hannover 1, FRG

EDEL, J.B. Institut de Physique du Globe, 5, rue René Descartes, 67084 Strasbourg CEDEX, France

EMTER, D. Geowiss. Gemeinschaftsobservatorium Schiltach, Heubach 206, 7620 Wolfach, FRG

FIELITZ, K. Bundesanstalt für Geowissenschaften und Rohstoffe, Alfred-Bentz-Haus, Postfach 510153, 3000 Hannover-Buchholz, FRG

FLATHE, H. Bundesanstalt für Geowissenschaften und Rohstoffe, Alfred-Bentz-Haus, Postfach 510153, 3000 Hannover-Buchholz, FRG

FUCHS, K. Geophysikalisches Institut, Universität Karlsruhe, Hertzstraße 16, Bau 42, 7500 Karlsruhe-West, FRG

GEBRANDE, H. Institut für Allgemeine und Angewandte Geophysik, Geophysikalische Wissenschaften, Universität München, Theresienstraße 41/IV, 8000 München 2, FRG

GIESE, P. Institut für Geophysikalische Wissenschaften, Freie Universität Berlin, Rheinbabenallee 49, 1000 Berlin 33, FRG

GLOCKE, A. PRAKLA-SEISMOS GmbH, Haarstraße 5, 3000 Hannover, FRG

GOERLICH, F. Deutsche Forschungsgemeinschaft, Kennedyallee 40, 5300 Bonn-Bad Godesberg, FRG

GRUBBE, K. Niedersächsisches Landesamt für Bodenforschung, Alfred-Bentz-Haus, Postfach 510153, 3000 Hannover-Buchholz, FRG

HÄNEL, R. Niedersächsiches Landesamt für Bodenforschung, Alfred-Bentz-Haus, Postfach 510153 3000 Hannover-Buchholz, FRG

HAHN, A. Niedersächsisches Landesamt für Bodenforschung, Alfred-Bentz-Haus, Postfach 510153 3000 Hannover-Buchholz, FRG

HINZ, E. Niedersächsisches Landesamt für Bodenforschung, Alfred-Bentz-Haus, Postfach 510153 3000 Hannover-Buchholz, FRG

JACOBSHAGEN, V. Institut für Geologie, Freie Universität Berlin, Altensteinstraße 34A, 1000 Berlin 33, FRG

KAMINSKI, W. Geophysikalisches Institut, Universität Karlsruhe, Hertzstraße 16, Bau 42, 7500 Karlsruhe-West, FRG

KELLER, F. Institut für Geophysik, Technische Universität Clausthal, Adolf-Römer-Straße 2A, 3392 Clausthal-Zellerfeld, FRG

KOSCHYK, K. Niedersächsisches Landesamt für Bodenforschung, Alfred-Bentz-Haus, Postfach 510153 3000 Hannover-Buchholz, FRG

MEISSNER, R. Institut für Geophysik, Universität Kiel, Neue Universität, Haus B 2, Olshausenstraße 40-60, 2300 Kiel, FRG

MENZEL, H. Institut für Geophysik, Universität Hamburg, Bundesstraße 55, 2000 Hamburg 13, FRG

MILLER, H. Institut für Allgemeine und Angewandte Geophysik, Universität München, Theresienstraße 41/IV, (Block C), 8000 München 2, FRG

MÜLLER, G. Geophysikalisches Institut, Universität Karlsruhe, Hertzstraße 16, Bau 42, 7500 Karlsruhe-West, FRG

MUELLER, S. Institut für Geophysik, Eidgenössische Technische Hochschule, Hönggerberg, Postfach 266, CH-8049 Zürich, Switzerland

PETERSCHMITT, E. Institut de Physique du Globe, 5, rue René Descartes, F 67084 Strasbourg CEDEX, France

PLAUMANN, S. Niedersächsisches Landesamt für Bodenforschung, Alfred-Bentz-Haus, Postfach 510153 3000 Hannover-Buchholz, FRG

POHL, J. Institut für Allgemeine und Angewandte Geophysik, Universität München, Theresienstraße 41/IV (Block C), 8000 München 2, FRG

PRODEHL, C. Geophysikalisches Institut, Universität Karlsruhe, Hertzstraße 16, Bau 42, 7500 Karlsruhe-West, FRG

SCHELIGA, G. Patentamt München, 8000 München, FRG

SCHRÖDER, H. Bundesanstalt für Geowissenschaften und Rohstoffe, Alfred-Bentz-Haus, Postfach 510153, 3000 Hannover-Buchholz, FRG

SCHULT, A. Institut für Allgemeine und Angewandte Geophysik, Universität München, Theresienstraße 41/IV (Block C), 8000 München 2, FRG

STEIN, A. Niedersächsisches Landesamt für Bodenforschung, Alfred-Bentz-Haus, Postfach 510153 3000 Hannover-Buchholz, FRG

STEINWACHS, M. Niedersächsisches Landesamt für Bodenforschung, Alfred-Bentz-Haus, Postfach 510153 3000 Hannover-Buchholz, FRG

VEES, R. Institut für Geophysik, Technische Universität Clausthal, Adolf-Römer-Straße 2A, 3392 Clausthal-Zellerfeld, FRG

VETTER, U. Institut für Geophysik, Universität Kiel, Neue Universität, Haus B 2, Olshausenstraße 40-60, 2300 Kiel, FRG

WILL, M. Institut für Geophysik, Schwingungs- und Schalltechnik der Westfälischen Berggewerkschaftskasse, Herner Straße 45, 4630 Bochum, FRG

ZSCHAU, J. Institut für Geophysik, Universität Kiel, Neue Universität, Haus B 2, Olshausenstraße 40-60, 2300 Kiel, FRG

Introduction

P. GIESE, C. PRODEHL, and A. STEIN

In close cooperation between geophysical institutes of universities and the geophysical departments of government agencies, the earth's crust and upper mantle in western Germany, the area between the North Sea and the Alps, have been investigated by explosion seismology. These investigations were performed within two priority programs of the German Research Society: The Deep Structure of Central Europe (1958-1964) and Upper Mantle Project (1965-1974). These priority programs also included the participation of the German group in the international program of the investigation of crustal structure in the Alps. Further support was given by a part of the priority program Geodynamics of the Mediterranean Area (1971-1975). After more than one decade the intensive and successful research has reached a state where it is adequate to present a compilation of the data and the essential results. The monograph on seismic research in the Western Alps, edited by CLOSS and LABROUSTE (1963) was the first monograph containing all usable seismograms in a large uniform time scale and other necessary data allowing later reinterpretation, and in this respect, differed from monographs published for other areas also dealing with explosion seismology (see STEINHART and MEYER, 1961; STEINHART and SMITH, 1966; HART, 1969; HEACOCK, 1971; KOSMINSKAYA, 1971; SOLLOGUB et al., 1972).

The suggestion to publish a further monograph on explosion seismology for western Germany was supported by all scientists participating in this common research. Numerous papers were submitted, the review of which showed that the edition of a major publication was justified. For various reasons, the compilation of this monograph took several years. Some investigations were completed several years ago, and where necessary other investigations were supplemented by the most recently obtained results.

The emphasis of this monograph is on the presentation and evaluation of seismic-refraction data obtained in western Germany. Additionally the results of seismic investigations of the Alps and Rhine Graben are reviewed. As mentioned, the presentation of data is regarded as of equal importance as the presentation of interpretations. The study of velocity distribution resulting from seismic measurements must be related to other geophysical and geological data. For this reason this publication also contains some brief reviews from closely related geosciences.

The first two chapters deal with the mean geological features of western Germany and with some results of other geophysical methods.

In Chapter 3, the data of shotpoints and profiles are compiled for the period 1958 to 1970. In an associated appendix, all record sections which were obtained during these investigations are presented. A description of the instruments used completes this section.

Chapter 4 deals with the problems of evaluation of seismic-refraction data. Here different steps of the interpretation procedure such as correlation, presentation of some basic data, time cross-sections are analyzed. Aspects of synthetic record sections are discussed later.

Chapter 5 describes the results of seismic-refraction studies in western Germany. The first three sections deal with the results of generalized interpretations of the data. Section 4 gives a review of the whole area under study and compares the seismic results with the main features of geological structure.

Chapter 6 deals with the results of regional studies from different parts of western Germany including the entire Rhine Graben area. On one hand it shows the complexity of crustal structure of the area under investigation and conversely the application of different methods of data evaluation.

The main results of seismic crustal studies in the Alps are reviewed in Chapter 7 and complete the picture of crustal structure in central Europe.

Chapter 8 contains comparisons of crustal and upper-mantle structures in central Europe and some other regions without, however, claiming completeness. Seismic studies and gravimetric data are also included.

1. Main Geologic Features of the Federal Republic of Germany

V. JACOBSHAGEN

ABSTRACT

The Federal Republic of Germany consists of four orographic regions: the lowlands of northern Germany, the central European uplands, the rather flat foreland of the Alps, and the northernmost chains of the Alps. Within the extra-Alpine regions three geological tiers can be distinguished: (1) an orogenic deformed substratum - in the north of Precambrian and Caledonian age respectively, in the uplands mainly Hercynian - is regionally covered by (2) hardrocks of the Upper Permian and the Mesozoic and (3) Cenozoic softrocks.

In the lowlands of northern Germany the substratum is situated very deep. The Permo-Mesozoic tier is intensively structured by Zechstein salt domes and diapirs, and covered by very thick sediments of the Tertiary and the Pleistocene. In the uplands the substratum is uplifted in blocks (e.g. Rhenish Massif, Harz, Bohemian Massif, Oberrhein Massifs). To the north the substratum vanishes under the germanotype structures of the Permo-Mesozoic (Saxonicum); to the south it is covered by the mainly flat Mesozoic layers of the main block of southern Germany (Süddeutsche Großscholle). From SSW to NNE western Germany is crossed by a big lineament, the Mediterranean-Mjösen-Zone of STILLE, marked by the Rhine Graben or the Hessen Depression. The Cenozoic volcanism follows to this zone and to EW strips.

The Bavarian Alps and their foreland are part of the Eastern Alps. The mountain chains can be subdivided into the Northern Calcareous Alps, the Flysch Zone and the Helvetic Zone with thick miogeosynclinal and flysch sediments of Mesozoic and Paleogene age. These zones are extremely deformed (naps, schuppen, and fold structures) and on the whole overthrusted on the thick Tertiary layers of their marginal trough (Molasse) now being the northern foreland of the Alps.

ZUSAMMENFASSUNG

Die Bundesrepublik Deutschland gliedert sich in vier Landschaftstypen: Norddeutsches Flachland, deutsche Mittelgebirge, Alpenvorland und Bayerische Alpen.

Am geologischen Aufbau des außeralpinen Bereiches sind drei Krusten-Stockwerke beteiligt: Ein orogen deformiertes Grundgebirgsstockwerk (1) - im Norden präkambrischen bzw. kaledonischen Alters, im Mittelgebirgsbereich fast ausschließlich variskisch - wird gebietsweise diskordant von Festgesteinen des Zechsteins und des Mesozoikums (2) überlagert, die ihrerseits von Lockersedimenten des Känozoikums (3) bedeckt sein können. Im Norddeutschen Flachland liegt das Grundgebirge tief. Das permomesozoische Deckgebirge ist durch mächtiges Zechsteinsalz halokinetisch verformt. Darüber liegen mächtige Sedimente des Tertiärs und des Eiszeitalters. In den Mittelgebirgen ist das Grundgebirgsstockwerk in einigen großen Blöcken (vor allem Rheinische Masse, Harz, Böhmische Masse und Oberrheinische Massive) bis an die heutige Erdoberfläche herausgehoben. Nach N sinkt das Grundgebirge in Staffeln unter das intensiv germanotyp strukturierte Mesozoikum von Süd-Niedersachsen und Nordwestfalen (Saxonikum) ab, im S ist es unter dem meist ruhiger gelagerten Mesozoikum der Süddeutschen Großscholle verborgen. Flachland und Mittelgebirge werden von einem NNE-streichenden Lineament, der Mittelmeer-Mjösen-Zone, durchzogen, die durch eine Flucht von Gräben und tektonischen Senken (z.B. Oberrheintal-Graben, Hessische Senke) markiert ist.

An sie sowie an zwei etwa EW-verlaufende Zonen war auch der känozoische Vulkanismus gebunden.

Die Bayrischen Alpen und ihr Vorland stellen einen nördlichen Ausschnitt der Ostalpen dar: Der Gebirgsbereich umfaßt die Nördlichen Kalkalpen, die ostalpine Flyschzone und die Helvetische Zone mit mächtigen Miogeosynklinal- bzw. Flysch-Ablagerungen des Mesozoikums und des Alttertiärs, durch Decken-, Schuppen- und Faltenstukturen extrem verformt und insgesamt auf das mächtige Tertiär ihrer Randsenke (Molasse), die das heutige Alpenvorland aufbauen, überschoben.

1.1 Introduction

The structure and geological history of central Europe are extremely complicated. Therefore, a short review

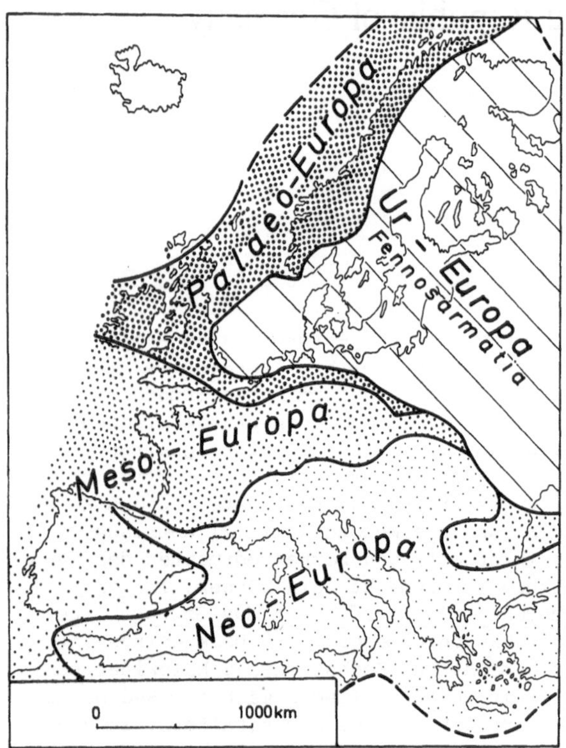

Fig. 1. Main geotectonic units of Europe. (After STILLE, 1924; LOTZE, 1971; BARTENSTEIN et al., 1968)

of this topic involves the risk of false generalizations. A more detailed study would be necessary and could be based on LOTZE (1971) and KNETSCH (1963) and, with certain restrictions, on RUTTEN (1969); these descriptions, however, cannot replace the extensive number of recent special publications.

The region dealt with in this study is not limited to the Federal Republic of Germany, but includes the western parts of the GDR and the ČSSR as well as the northern parts of the Eastern Alps and the adjoining regions of eastern France and the Benelux countries. Roughly seen, four orographic regions can be distinguished:

a) the Cenozoic mountain chains of the Alps in the south

b) the northern foreland of the Alps, a rather flat country consisting of sediments of their late-orogenic marginal trough (Molasse)

c) The central European uplands with their rather complex geological structure

d) the lowlands of northern Germany.

Referring to its geological history, STILLE (1924) has divided Europe into four main regions (Fig. 1):

1. The primeval Ur-Europa with its Precambrian basement and nearly undisturbed sediments of the Phanerozoic,
2. Palaeo-Europa, originated by the Caledonian revolution,
3. Meso-Europa with the Hercynian orogeny being the main structural event,
4. Neo-Europa, the region of Alpine mountain belts surrounding the Mediterranean.

Based on today's experiences, slight changes of limitation and interpretation of these regions are necessary. Ur-Europa did not simply increase by

---

Fig. 2. Main geologic features of western central Europe ▶

Structural limits — fractures — overthrusts

folds — Astroblemes (meteor impacts)

a) extra-Alpine region

- volcanic rocks of the Cenozoic
- Tertiary (sediments)
- southern border of the Quaternary cover of the northern lowlands
- Zechstein (Upper Permian) and Mesozoic
- salt diapirs
- Permian volcanic rocks
- Upper Carboniferous and Rotliegendes
- sediments and volcanic rocks older than Upper Carboniferous
- crystalline rocks of Hercynian age and older

b) Alpine region

- Helvetic zone and Préalpes
- autochthonous massifs
- Penninic zone, flysch of the Eastern Alps
- Northern Calcareous Alps
- central zone of the Eastern Alps

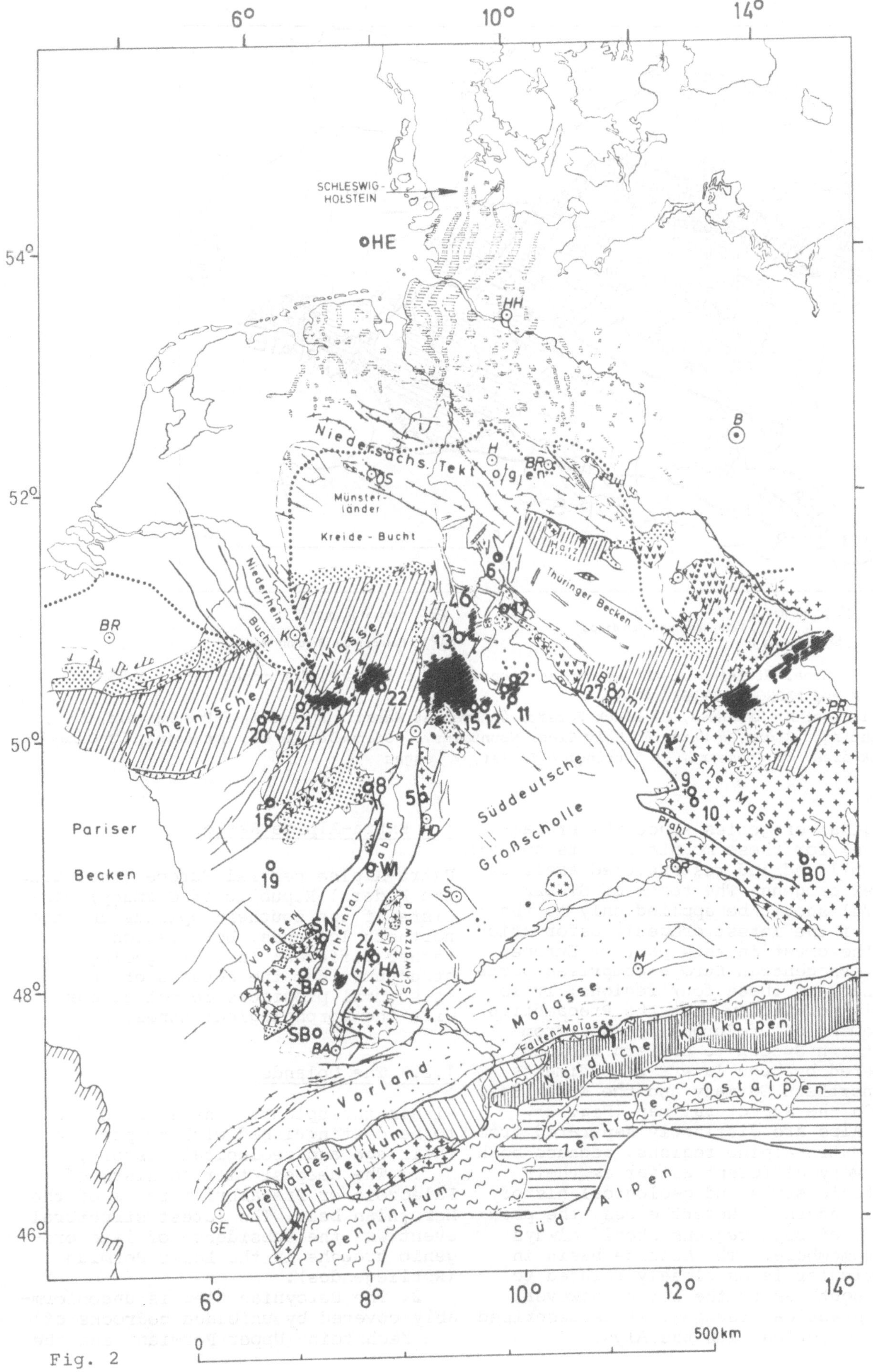

Fig. 2

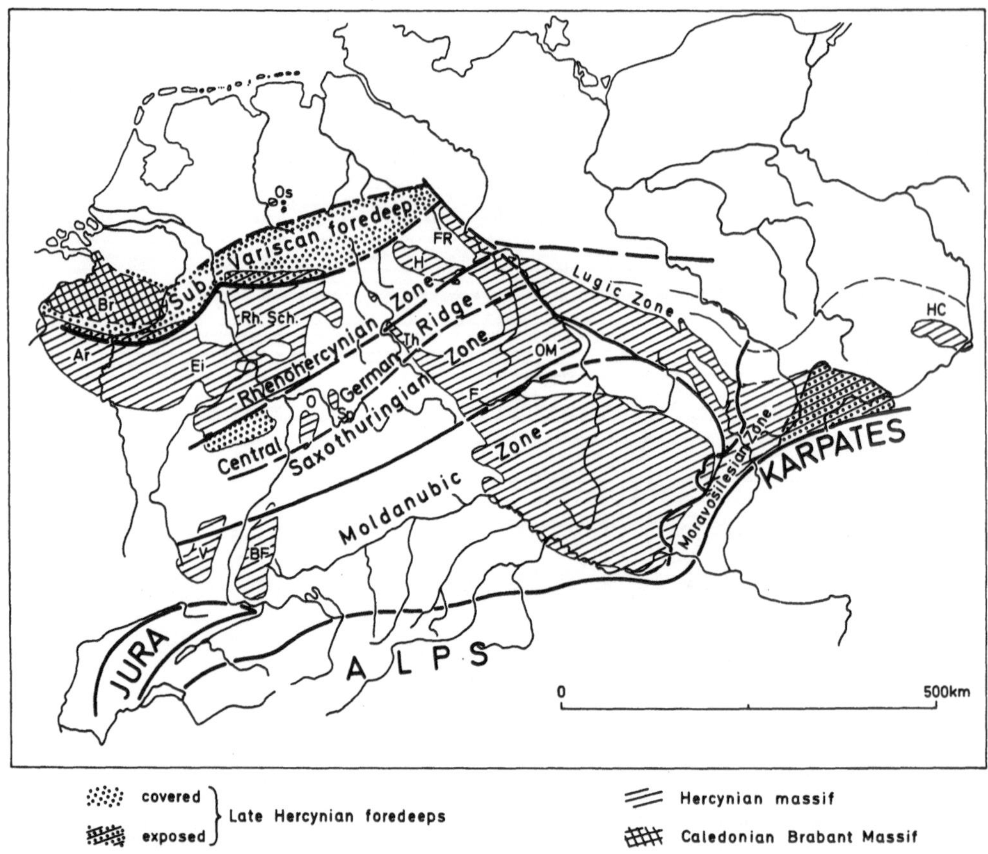

Fig. 3. Zonal division of the Hercynian mountain belt in Central Europe. (After KOSSMAT; DVORAK and PARROTH; FRASL et al.)
Localities: *Ar* Ardennes, *BF* Black Forest, *Br* Brabant Massif, *E* Eifel, *F* Frankenwald, *FR* Flechtingen Ridge, *H* Harz, *HC* Holy Cross Mountains, *O* Odenwald, *OM* Ore Mountains, *Os* Osnabrück area, *Sp* Spessart, *Th* Thuringian Forest, *V* Vosges

marginal orogenies since the Precambrian; for these mountain belts contain remobilized and restructured areas of older orogens. The terms of STILLE (1924) should be applied only to the age of the latest orogenic deformation of the crust in this part of Europe. Western central Europe comprises parts of each of these four regions; it is well known today that the Precambrian basement of Ur-Europa reaches from southern Sweden to eastern England, covered by huge layers of sediments in northern Germany and the North Sea.

On the structural map (Fig. 2), the Alps are distinctly separated from the extra-Alpine regions, because both are very different as far as their tectonic style and geological history are concerned. Nevertheless, the proximity of both regions should always be remembered. The Molasse basin in particular is as closely related to the south as to the north; however, for practical reasons, it is described in connection with the Alps.

1.2 Extra-Alpine Region

Extra-Alpine central Europe (including the Federal Republic of Germany) consists of the southern uplands and the northern lowlands. Both regions are rather different in their geological structure, but the borders of the structural provinces do not concur with the morphological zones.

1.2.1 The Uplands

Within the uplands, three tiers are to be distinguished with respect to their age and geological history.

1. The deepest tier consists of Paleozoic or Precambrian rocks of the Hercynian belt. The latest structural event was the subsidence of late orogenic troughs in the Lower Permian (Rotliegendes).

2. The Hercynian tier is unconformably covered by unfolded bedrocks of the Zechstein (Upper Permian) and the

Mesozoic. These strata are deformed by fractures or domes in some districts.

3. The uppermost tier, consisting of Tertiary and Quaternary soft rocks, is separated from the adjacent strata by a large stratigraphic hiatus. The Cenozoic accumulations reach a certain extention and thickness only in some basins and grabens. The fractured tectonics of the Tertiary follow the Mesozoic pattern in various places.

In addition to these crustal tiers, volcanic masses of the Cenozoic are very important regionally.

1.2.2 The Hercynian Substratum

The large units of the Hercynian substratum of central Europe are shown in Fig. 3: the Rhenish Massif with the Flechtingen ridge (F), the Harz mountains (H), and the Thuringian Forest (Th) in the east and some crystalline massifs near the Rhine river, the so-called Oberrhein Massifs such as the Black Forest (BF), the Vosges (V), the Odenwald (O), and the Spessart (Sp) mountains. Some smaller uplifts, situated mainly between the Bohemian and the Rhenish Massifs, could not be drawn to scale.

The geosynclinal phase of the Hercynian orogen comprises the Devonian and the Lower Carboniferous. Sedimentation has been accompanied by ophiolitic initial magmatism (e.g. Lahn-Dill district, Sauerland, Harz, southern parts of the Thuringian Forest). The orogenic movements began within the inner zones of the Hercynian belt during the Upper Devonian and culminated at the end of the Lower Carboniferous (Sudetic phase of STILLE); during the Upper Carboniferous, the folding activity diminished and ended within the Lower Permian. Together with the orogenic deformation, flysch accumulation began. The flysch trough along the front of folding migrated continually north, followed by the folding movements; this process can be noticed particularly well within the Kulm facies of the Lower Carboniferous. In relation with the diminishing folding activities, Molasse-type sediments were deposited in a northern foredeep (Subvariskische Saumtiefe) and in intramontaneous basins since the lowermost period of the Upper Carboniferous. These troughs contain the well-known deposits of coal (e.g. Ruhr district, Saar basin, small basins within the Black Forest and the Vosges). During the Lower Permian, the intramontaneous basins expanded, and new ones appeared (Rotliegendes), the marginal trough having already been filled up. For the Hercynian units north of the Alps, E. SUESS and F. KOSSMAT developed the well-known sequence of NE ("erzgebirgisch") striking zones, which are characterized in the following for the western part of central Europe from the south to the north:

1. The Moldanubic zone is exposed in the Vosges, the Black Forest, the Bavarian, and the Bohemian Forests. Volcanic ejections as well as deep drillings have proved the connections between these Massifs now covered by the Mesozoic of southern Germany. In this zone, highly metamorphized schists (e.g. widely spread in the Bohemian Massif cordierite gneisses) are interspersed with late- and post-kinematic intrusions mainly of granitic character. The composition of the metamorphics varies widely. They are derived from Precambrian or sometimes from Paleozoic rocks, the former being polymetamorphic in some districts. Locally (e.g. in the Black Forest), the metamorphosis has been intensified to anatexis and palingenesis. Sediments of the Devonian and the Lower Carboniferous forming the non-metamorphic cover of the crystalline areas only exist in relics (Vosges, Black Forest, vicinity of Prague in the ČSSR).

2. The typical Saxothuringian zone in the region of the Bohemian Massif can be divided into three sections: the main element, a crystalline ridge, is exposed in small outcrops in the north of NW-Saxony and northern Thuringia and, to a larger extent, in the Thuringian Forest; in SW-Germany, the Spessart and the Odenwald also belong to this ridge. Acid to mafic intrusions have penetrated into highly metamorphized schists. This so-called Central German Ridge (Mitteldeutsche Schwelle; BRINKMANN, 1948) emerged out of the geosyncline after the Upper Devonian and provided weathering detritus for the sedimentary basins in the NW and SE.

To the SE, there follows a large area of Paleozoic and even Precambrian sediments (Thuringian Schiefergebirge) which, further toward the SE, become gradually metamorphized including autochthonous massifs of igneous and metamorphic rocks.

Finally, a broad anticline with Precambrian crystalline cores (e.g. Fichtelgebirge = F and Oberpfälzer

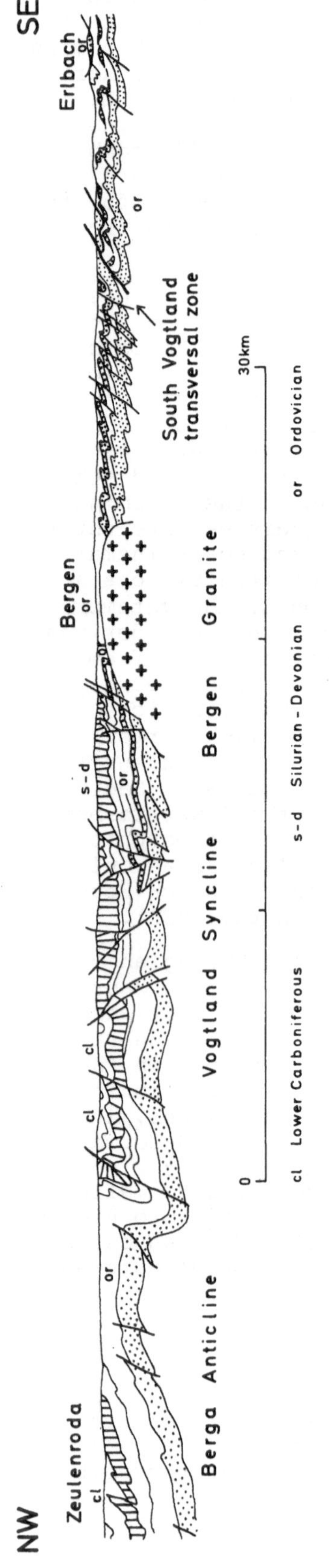

Fig. 4. Geologic cross section through the southern part of the Saxothuringian Zone. (Simplified from GAERTNER)

Wald, Ore Mountains = OM) borders the Moldanubic zone.

In SW Germany, the central section of the Saxothuringicum is completely covered by Mesozoic strata. Its southernmost parts are exposed in the northern Vosges consisting mainly of Paleozoic sediments and crystalline areas. Small outcrops exist also in the northern Black Forest.

The main tectonic features of the Saxothuringian zone are complicated fold and schuppen structures surrounding the old crystalline cores and being overturned into various directions (Fig. 4).

In the Rhenish Massif the northern border of the Saxothuringian zone is marked by a deep fracture (Hunsrück fault).

3. The main blocks of the Rhenohercynian zone are the Rhenish Massif, the Harz, and the Flechtingen ridge, consisting of Devonian and Carboniferous bedrocks (mainly graywackes, sandstones, and shales, including some limestones, cherts, and others). This series (nowhere completely exposed) reaches 5-10,000 m. The occurrence of pre-Devonian rocks is limited to a few anticlines.

The Rhenohercynian zone has been deformed relatively close to the surface of the Hercynian orogen, its southernmost part alone being slightly metamorphized. Only in the Harz Mountains, intrusions have reached this level (granites of the Brocken and the Ramberg, Harzburg gabbro). The tectonic feature of this zone is characterized by folds and schuppen structures overturned mainly to NW (Fig. 5), and by slaty cleavage.

4. The northern foredeep filled up by very thick molasse-type sediments of the Upper Carboniferous is called the Subvariscan Zone. In the Federal Republic of Germany, it is visible only in the Ruhr district and in the vicinity of Aachen at the Belgian border, while it is covered everywhere else by Cretaceous and Cenozoic sediments. Late Upper Carboniferous movements - the Asturic phase of STILLE - effected very intensive fold structures in the southern part of this zone, which flatten towards the north (Fig. 6). North of a line from Xanten to Osnabrück, the unfolded foreland of the Hercynian belt begins (HOYER et al., 1969). Especially in the Subvariscan Zone, deep transversal faults and fissures developed during the orogenetic processes, although some of

them were reactivated at a later period.

5. Intramontaneous basins. Together with the Subvariscan foredeep, some basins sank in the inner parts of the Hercynian mountain belt. During the Upper Carboniferous, these basins stretched parallel to the Hercynian fold axes, whereas the Lower Permian basins of the Rotliegendes were of a more irregular shape, signalizing the end of the orogenic movements.

In these Rotliegendes troughs, continental sediments - usually red-colored except for the lowermost parts - have been accumulated to a maximum thickness of more than 1500 m in the Saar-Nahe Basin.

Lower Permian fracturing was followed by volcanic events; this so-called subsequent volcanism produced mainly intermediate to acid rocks, but also basic types (e.g. tholeiites).

The eastern part of the Rhenish Massif (Rheinisches Schiefergebirge), has been especially carefully studied (Fig. 7). This area once was the Devonian geosyncline. Roughly seen, it consists of NE-striking fold systems, the cores of the anticlinoria being built up by huge masses of Lower Devonian clastics (maximum thickness within the Siegen anticlinorium and the Taunus several thousands of meters). Pre-Devonian rocks of uncertain age are exposed at the southern metamorphic border of the Rhenish Massif; Ordovician and Silurian crop out within the northern anticlinoria (Remscheid-Altenaer "Sattel", Ebbe-"Sattel") and, with some tectonic lenses, in some places near the eastern border of the massif. Since, in wide districts, the fold-axes are dipping slightly towards NE, the synclinoria become deeper and broader in this direction containing plenty of Upper Devonian and Lower Carboniferous sediments and ophiolites, mainly in the eastern part of the Rhenish Massif (Mosel-Lahn synclinorium, Dill "syncline", Waldeck Kulm "syncline").

In the south, the Rhenish Massif is divided from the Tertiary filling the Rhine Graben by an enormous fault.

1.2.3 Post-Hercynian Areas

After orogenic movements ceased, the Hercynian belt was largely flattened by degradation at the end of the Lower Permian. While Upper Carboniferous fracturing roughly followed the NE-striking of the mountain belt, the geotectonic pattern and with it the paleogeographical situation were completely reversed: the main part of the Rhenish Massif on the one hand and the core of the Bohemian Massif on the other remained from that time almost entirely solid and areas of degradation; the Harz, the Thuringian Forest, and the smaller blocks in central Germany were lifted but at a later period.

Between the Rhenish and the Bohemian Massifs, within the pattern of the Zechstein Basin, the Rhenish furrow with its "Rhenish" NNE-strike began to show its first outlines, belonging to a lineament, STILLE's Mediterranean-Mjösen-Zone, which had been active as early as from the Paleozoic and probably even before that time. From the Rhenanian furrow, the Hassian depression (Hessische Senke), at times also a marine channel, developed later and connected the North German depositional basin with the South German basin.

Most of the fractures that originated during the Mesozoic and later periods and which also decomposed the Hercynian substratum strike in WNW to NW direction (Hercynic) or - especially near the Mediterranean-Mjösen Zone - in the "Rhenish" (NNE) direction (cf. KNETSCH, 1969); besides, disturbances with NNW-"eggisch") or NE- to ENE-striking ("erzgebirgisch") are frequent in some parts.

The major part of South Germany was still an area of denudation in the Upper Permian. Only in the Lower Triassic, the region north of the Danube and the regions around the Black Forest and the Vosges subsided and were included in the Mesozoic depositional basin. Towards the east, the latter stretched beyond the present western border of the Bohemian Massif only in some places and then not very far. In the south, it is bordered by a crystalline ridge which branched off from the Bohemian Massif near Regensburg and can be followed down to central Switzerland. During the Mesozoic, this "Vindelician Ridge" as well as the Bohemian Isle were mainly elevated areas supplying sediments (such as the sandstones of the Keuper). In the west, the South German basin was, during the Triassic, bordered by a Gallic Land most of which, however, sank below sea-level in the Jurassic. South Germany's present structure developed during the Tertiary. In the older Paleogene, a dome

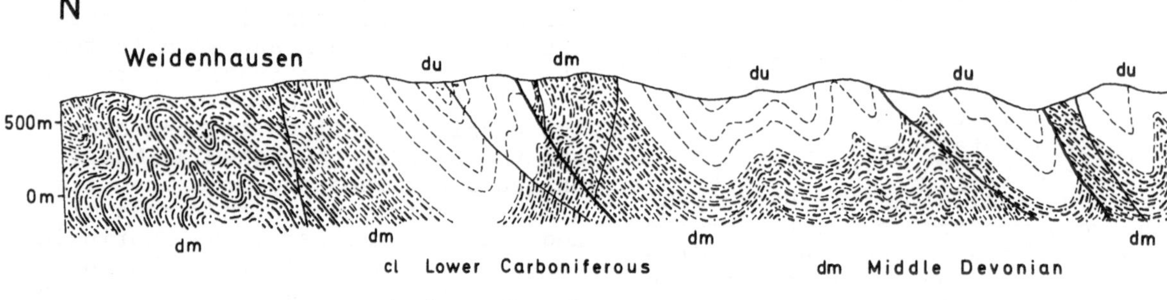

Fig. 5. Typical cross section of the central Rhenish Massif, southeastern border of the

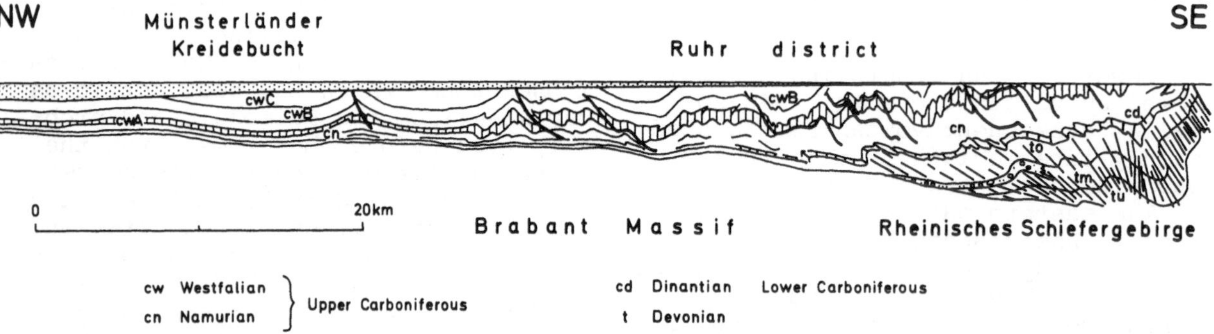

Fig. 6. Schematic cross section of the Subvariscan zone of Westphalia. (After TEICHMÜLLER)

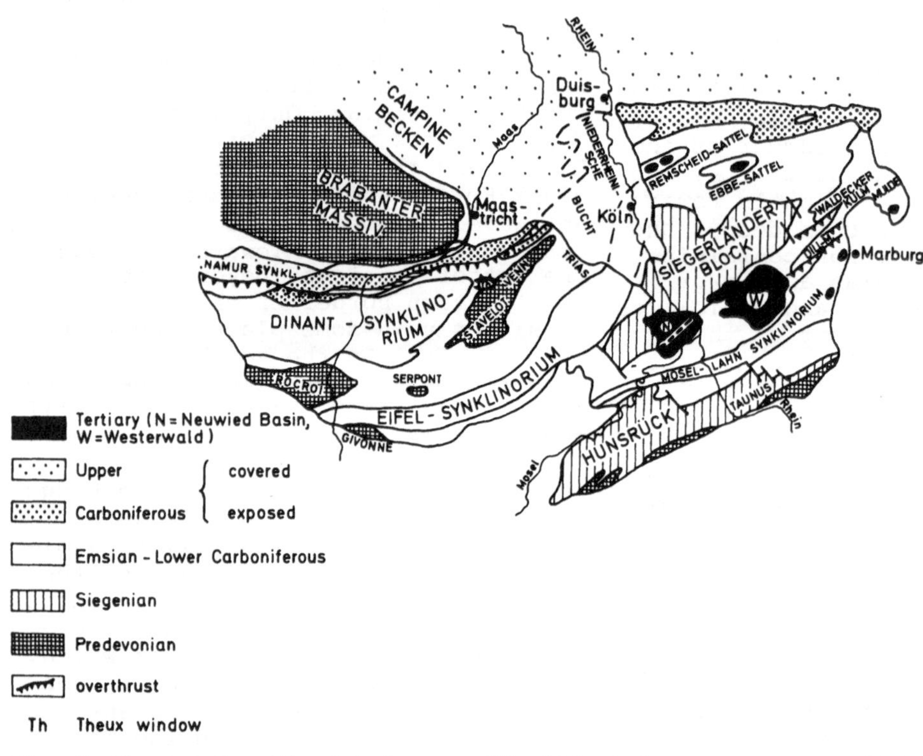

Fig. 7. Rhenish Massif and Ardennes. (Simplified from v. BUBNOFF, 1930)

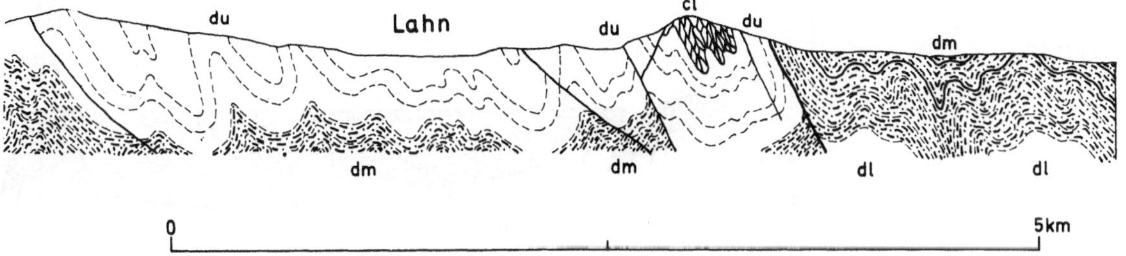

Siegerländer Block. (After REICH and SCHMIERER)

vaulted along the present Rhine Valley between Basel and Frankfurt in the crest of which the Oberrhein Graben subsided (Fig. 8); this uplift is marked by the crystalline blocks of the Oberrhein Massifs (Vosges, Black Forest, Vorspessart).

The Oberrhein Graben has a complex history (ILLIES, 1965). Its Cenozoic sediments partly reach a total thickness of 4000 m, the highest values being reached by the Paleogene in the South, the Neogene and the Quaternary in the north.

Thus, the main block of southern Germany had come into existence between the Oberrhein Graben and the Bohemian Massif, on the whole a triangularly shaped plate, consisting of beds of the Triassic and the Jurassic up to 1500 m, slightly dipping from SE to E (Fig. 9). Not only its western but also its eastern border is slightly swept up since the Bohemian Massif is somewhat upthrust towards the SW. The southern part of the main block of South Germany is covered by Molasse sediments of the Alpine marginal trough. In the interior, the block is subdivided by zones of fracturing which, near the Oberrhein Graben, strike NNE as this Graben does and NW (Hercynian) in almost all other regions, especially in the group of Franconian fractures which lead to the Bohemian uplift. Moreover, there are flat elevations and tectonic depressions with NE-striking and tracings of the Hercynian substratum, especially in the northern part or the block.

The Hessen depression, i.e. the region between the Hercynian blocks of the Rhenish Massif and the Thuringian Forest, is nearly a prolongation of the Oberrhein Graben somewhat shifted towards the east. It consists of flatly deposited sediments of Zechstein and the Triassic (several hundred meters thick) into which narrow grabens with NNW-, NNE-, and SE-striking have been downthrust. These grabens originated partly before the Lower Cretaceous. On the other hand, in a few small horsts, the Hercynian substratum can be noticed. The central part of the depression is filled with Tertiary sediments of varying thickness (locally over 200 m) which often cover the Mesozoic fractures but are themselves cut by younger faults in the same directions.

The Thuringian Forest, similar to the Harz, may probably have been raised as a block only towards the end of the Mesozoic. During this process, the Thuringian basin (a flat plate of Triassic strata) subsided between these two uplifts and the Hercynian mountain range, cut by a few mainly NW-striking fracture zones.

North of the Rhenish Massif expands the Cretaceous basin of Münster. The eastern border of the Cretaceous area is bended up increasingly steeply from south to north, the northern border of the Teutoburger Forest even overturned to the south. Here lies the southern border of the Saxonian "Bruchfalten-Gebirge" (mountain range of germanotype orogenesis), so famous since the days of STILLE. Today it is called Tectogen of Lower Saxony (BOIGK, 1968).

This WNW-striking (Hercynian) crust stripe consists of fractured blocks and wide folds in about the same direction (Fig. 10). Although halokinetic movements of the Zechstein salt deposits played their part, forming even diapir-type structures, the pattern of the tectogen is due to deep-reaching tectonic movements within the crust by which the pre-Permian substratum has been faulted (i.e. upthrust of the Harz block towards NE, and of the Flechtingen Ridge or the Carboniferous-bearing block near Osnabrück).

Fig. 8. Geological cross section of the southern Oberrhein Graben. (After SITTLER)

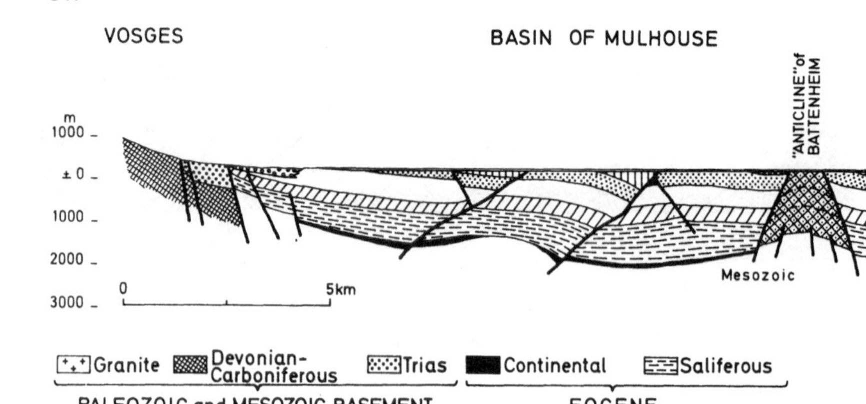

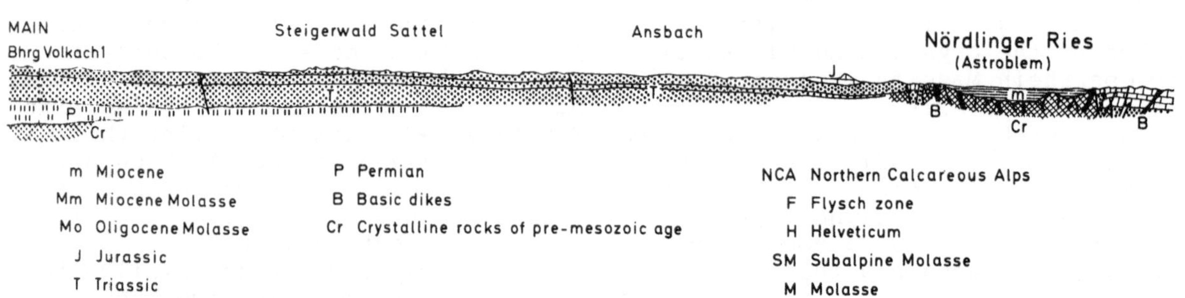

Fig. 9. Schematic cross section of the main block of southern Germany. (Simplified from

West of the Weser, the Tectogen of Lower Saxony has been upthrust to the neighboring blocks in the north as well as in the south; the Osning overthrust onto the Münster basin is considerable. Further to the east, i.e. in the region of Hannover-Braunschweig the mainly Hercynian-structured tectogen is traversed by the NNE-striking Leinetal Graben, another part of the Mediterranean-Mjösen Zone. In the field of interference of both directions, a very complex block mosaic exists.

The thickness of the Mesozoic tiers in the Lower Saxonian tectogen amounts to several thousand meters, but nevertheless, as mentioned above, the Hercynian substratum is visible in some regions.

East of the Weser river, the northern half of the Lower Saxonian Tectogen belongs, from an orographic point of view, to the North German Lowland; its tectonic structures are completely buried under softrocks of the Cenozoic.

The northwest parts of Germany - the Münster Basin as well as the Rhenish Massif - are cut by a Tertiary graben structure, the Lower Rhine depression (Niederrheinische Bucht), which is the southern prolongation of the North Sea Graben. This wedge-

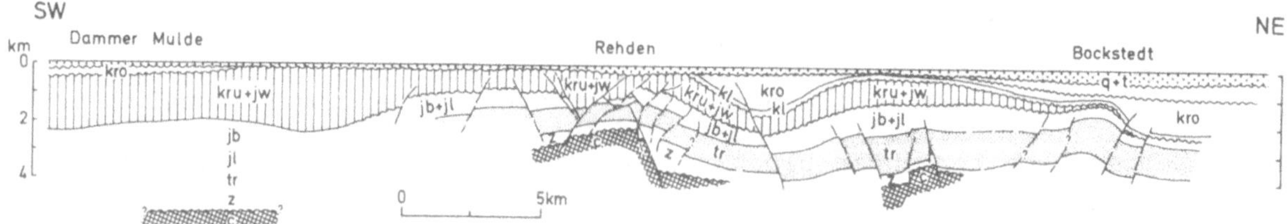

Fig. 10. Cross section of the northwestern part of the Lower Saxonian Tectogen. (After BOIGK, 1968).
Localities: *c* Carboniferous, *z* Zechstein, *tr* Triassic, *jl* Liassic, *jb* Dogger, *jw* Malm, *kru* Lower Cretaceous, *kro* Upper Cretaceous, *q+t* Qaternary + Tertiary

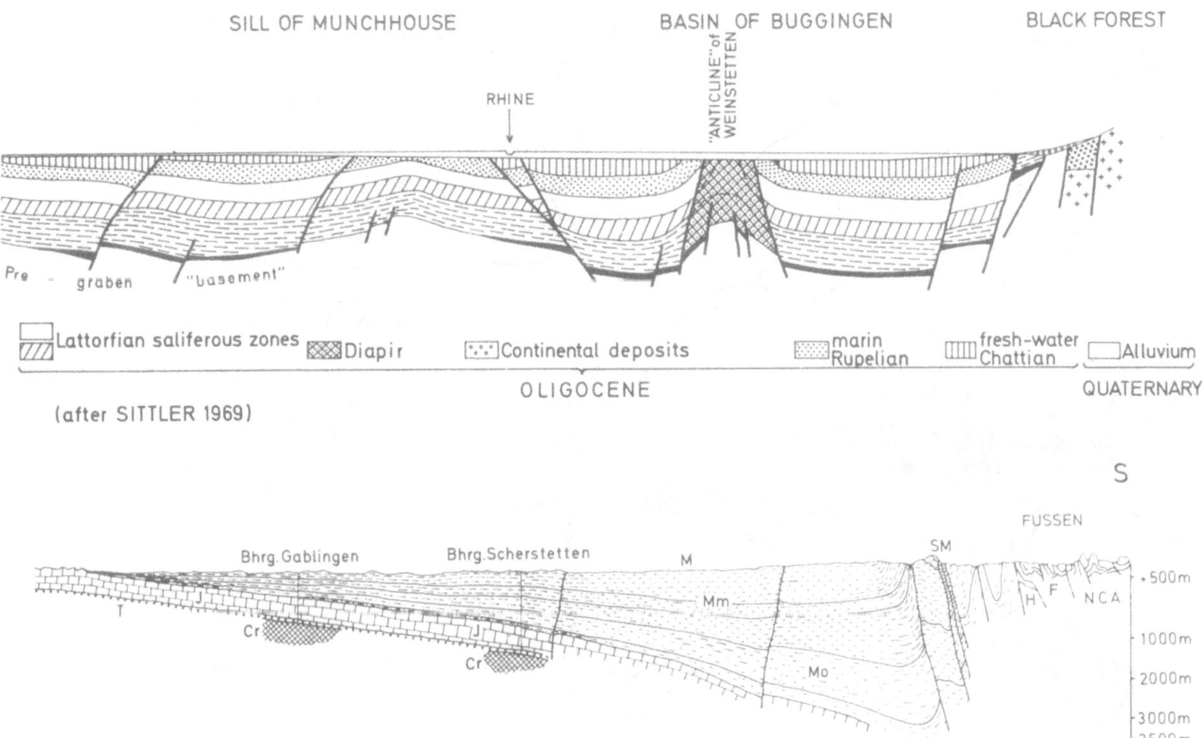

Geol. Map of Bavaria 1 : 500,000)

shaped block is limited by still active faults up to the city of Bonn. In its interior, drillings have proved the maximum thickness of the Cenozoic softrocks of approx. 800 m. In the south these sediments cover Paleozoic rocks, whereas from Duisburg towards the north, Zechstein and Triassic sediments have been found at the base of the Tertiary.

1.2.4 Cenozoic Volcanism

With the Tertiary, a period of strong volcanic activity began in Germany; it reached its peak in the Neogene and alackened in the Pleistocene. The eruption centers and the distribution of vocanic rocks (mainly basaltic laves and ashes; of secondary importance also intermediate the acid rocks, partly with alcaline character) are to be seen in Fig. 11. Most of them are restricted to narrow NNE- or ENE-striking zones where fracturing has been active during the Tertiary. In the Mediterranean-Mjösen Zone e.g., the Kaiserstuhl, the Vogelsberg, and the Lower Hassian Basalts are situated.

An ENE-trend is shown by the following volcanic areas:

1. Eifel-Middle Rhenish centers (Siebengebirge, Neuwied Basin), Westerwald, Vogelsberg, Rhön. The Vogelsberg, as the largest center of Cenozoic volcanism in central Europe, is situated at the crossing of this zone with the Hessen depression (Mediterranean-Mjösen Zone).

2. North Bohemian centers; Eger, Doupov Mountains, Bohemian "Mittelgebirge".

3. Kaiserstuhl-Urach area.

Some other centers are more isolated: the Hegau volcanoes at the east end of the Bonndorf Graben near Lake Constance in the south and in northern Germany a basic pluton, indicated by the gravity maximum at Bramsche which will probably be of Tertiary age.

Within the Eifel and the Neuwied Basin, volcanic activity has been known since the Pleistocene. The latest eruptions there took place at the end of the Upper Pleistocene (pumice ashes of the crater of the Laacher See in the Neuwied basin). Today, mofettes and high geothermal gradients are the last traces of volcanic activity in the mentioned areas.

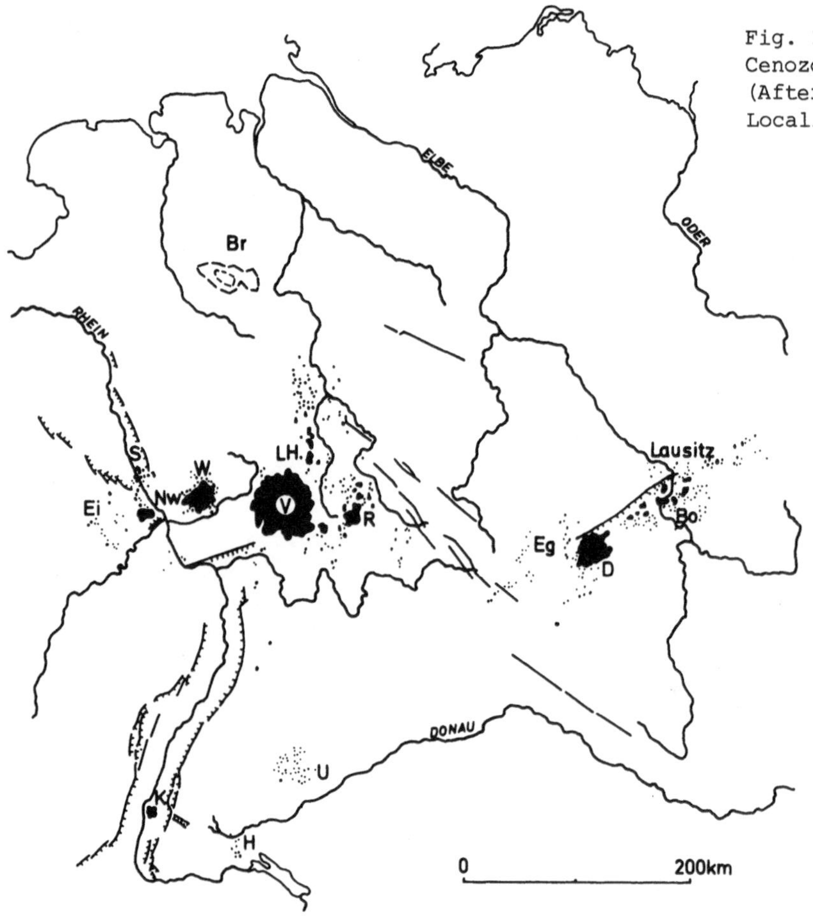

Fig. 11. Eruption centers of the Cenozoic volcanism in central Europe. (After KNETSCH, 1963, altered)
Localities:

Bo Bohemian Massif
Br Bramsche Massif
D Doupov Mountains
Eg Eger Basin
Ei Eifel
H Hegau
K Kaiserstuhl
LH Lower Hassia
Nw Neuwied Basin
R Röhn
S Siebengebirge
U Urach area
V Vogelsberg

1.3 The North German Lowland

Although the geological structures of the North German Lowland are almost completely buried by deposits of the Pleistocene ice age, they are now well-known owing to intensive petroleum exploration (cf. e.g. GRIPP, 1964).

In the north, the basement consists of Precambrian rocks adjacent to Fennosarmatia (see BARTENSTEIN, 1968; 4 of this volume, Fig. 5). In its southern part, it belonged to a Caledonian mountain belt which crops out e.g. in the Brabant Massif in Belgium: During the Hercynian orogenesis the basement of North Germany belonged to the foreland of the mountain range

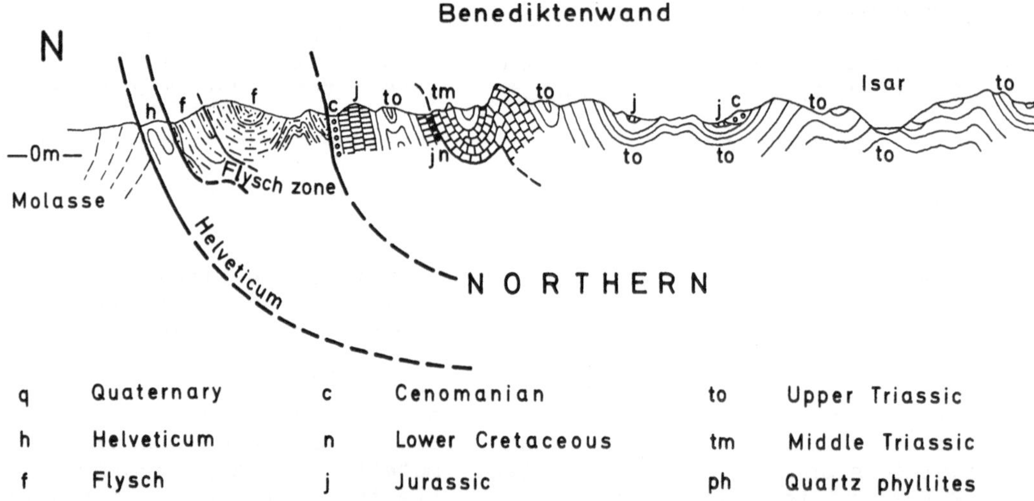

Fig. 12. Cross section through northern regions of the Eastern Alps, Bavaria/Tyrol.

but was, with the exception of Schleswig-Holstein, flooded by a shelf sea in the Carboniferous. Since the Lower Permian, the entire North German Lowland was a sedimentation area. The crust subsided with different velocity, broken into some larger blocks. Permian and Mesozoic together can reach a thickness of over 5000 m. General data, however, cannot be given since the structures of these sediments were formed by halokinetic movements: the very thick salt layers of the Upper Permian were affected by a large number of diapirs which began to ascend in the Upper Mesozoic and still continue in some cases.

In the Tertiary, the North German Lowland belonged to the sea. The thicknesses of the Tertiary sediments increase from south to north up to 2000 m in some places (Schleswig-Holstein). In addition, there are the sediments of the Quaternary (mainly basal moraine of the continental ice-sheet of northern Europe), reaching locally a thickness of 200 m or more.

1.3.1 The Alps and their Northern Foreland

From a geological point of view, southern Germany south of the Danube river belongs to the Eastern Alps. The Erläuterungen zur Geologischen Karte von Bayern 1 : 500,000 (1964) offer a good survey of this region. Only the southernmost border of these mountains belongs to the folded mountain range; the largest part, the Alpine foreland, is built up by unfolded sediments of the Alpine marginal trough, the so-called Molasse basin.

Because of their different general structure, the Eastern Alps are marked off against the Western Alps by a line running south from Lake Constance. A deep-reaching lineament zone runs through both parts of the Alps striking from Torino up to southeastern Austria: it is called Insubric Line in Switzerland, and Tonale, Pustertal, or Gailtal Line in the Eastern Alps. The tectonic character and the importance of this zone has often been discussed (BÖGEL, 1975). It is agreed that it marks the geotectonic border between the moderately deformed Southern Alps with their mainly southern overturn and the Eastern Alps in a geological sense which are heavily tectonized and almost completely inclined to the north. The latter consist, in rough outlines, of a metamorphic zone in the south (Central Alps) and some sedimentary zones in the north. In the following, only the mountain range north of the large line of lineaments will be dealt with.

The history of the Alpine geosyncline began in the Triassic and lasted into the Paleogene. With the strike of the present mountain range, long troughs developed at that time with sequences of sediments that differ considerably in facies and thickness and that - according to the rock structure - were differently deformed during the orogeny (orogenic paroxysms mainly between the younger part of the Lower Cretaceous and the beginning of the Neogene). Generally, the trough fillings were disconnected at large overthrusts and transported in naps

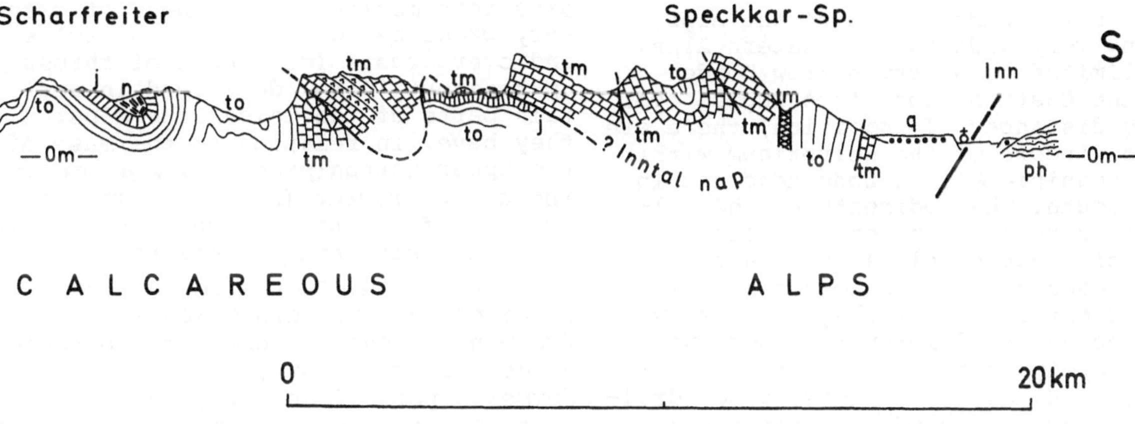

(Simplified from SCHMIDT-THOME)

towards the north; for this reason, the sediments deposited farthest south now build up the uppermost naps. Thus, the Alps may be subdivided into several zones which may be characterized according to their sequences of strata as well as to their tectonic style. These are from N to S:

Molasse Zone
Helvetic Zone
Penninic Zone
Austroalpine Zone

These zones themselves are again subdivided, but we shall not go into details in this respect. Fig. 12 shows a cross-section from the Molasse to the Austroalpine Zone.

1.3.2 Molasse Zone

Its sediments, mainly conglomerates, psammitic and pelite rocks of the Oligocene to Pliocene, originated from the late orogenic erosion rubble of the Alps. From minor thicknesses at the Danube, they swell up to 6500 m at the Alpine border (Fig. 9). These strata are on the whole undeformed; only near the front of the Alps, are some wide-spanned folds and schuppen structures to be found. Seismic reflection research and drilling profiles in eastern Austria suggest, however, that this main fault at the northern border of the Alps becomes more shallow in depth and that the Alps were overthrust to the Molasse at least 10-15 km.

1.3.3 Helvetic Zone

This northernmost zone of the Alps, being very wide in the Western Alps, is limited to a very narrow border in the Eastern Alps, interrupted over long distances. In addition, there are some windows of the Helveticum within the Penninic Flysch Zone adjacent in the south. The sediments of the Helvetic zone (Carboniferous-Paleogene) are of miogeosynclinal type and do not reach great thicknesses before the Upper Jurassic. They are mostly dislodged in close-knit slices and moderately folded only in the western part of Bavaria; but here, also, drillings have revealed an intensive upthrow structure. There is no evidence how far the Helvetic zone continues southward under the cover of higher naps; in western Bavaria, at any rate, the Penninic Flysch Zone shows an overthrust of at least 15 km over the Helceticum.

1.3.4 Flysch Zone of the Eastern Alps

The Flysch Zone forms the northern border of the Penninicum in the Eastern Alps. The southern areas of the Penninicum are buried under Eastern Alpine naps and crop out only in windows. (see 1.3.5).

In the Flysch Zone we find sediments from the Lower Cretaceous into the Eocene, rising up to more than 2000 m. The typical flysch facies appears towards the end of the Lower Cretaceous, simultaneously with the most important folding period of the Eastern Alps (Austrian Phase according to STILLE). Wide folds which originated at the end of the Paleogene are the typical tectonic structures of the Flysch Zone. An internal nap structure is believed to exist only near the western border of Bavaria and in West Austria.

1.3.5 Austroalpine Zone

The naps of the Austroalpine zone were thrust over wide distances of the Flysch Zone. They build the Northern Calcareous Alps, the Graywacke Zone and, in the main, the Central Alps.

Northern Calcareous Alps. These consist of a series of mainly carbonatic rocks several thousand meters thick which were sedimentated in the period from the Permo-Triassic to the Oligocene. Middle Triassic to Lower Cretaceous layers are pronouncedly miogeosynclinal. Owing to several phases of compressive deformation during the Cretaceous, very complicated structures of folds and overthrusts (distances of thrust up to about 10 km) developed in certain parts of the Calcareous Alps; they have, in relation to the base of the Upper Astroalpine nappe, a parautochthonous character (Fig. 12). The existence of an internal nappe structure is uncontested only in eastern Bavaria.

Graywacke Zone. In the south, the rocks of the Northern Calcareous Alps frequently cover slightly metamorphic Paleozoic of the Graywacke Zone - formerly part of an Hercynian mountain belt. The original sedimentary contact between the two mountain regions is, however, to a large extent cut up by fault zones. The tectonic structure of the Graywacke Zone situated outside the territory of the Federal Republic of Germany is, so far, not very well known, but internal deformation is, at any rate, very strong. In parts, the Graywacke Zone is interrupted; in these cases the Central Alps, otherwise adjacent in the south, border on the Northern Calcareous Alps directly.

Central Alps. The Central Alps consist mainly of metamorphic rocks. Apart from a few disputed places, their northern border is determined by the overthrust structure of the orogen with total fractures leading to overlaps in some places. The Central Alps do not consist of uniform material. In addition to Hercynian and, in some areas, also older, possibly even Precambrian series, more or less subjected to alpine metamorphism, there are, in places, Mesozoic metasediments. Especially widely spread, however, are Pre-Alpine crystalline rock. Relatively thin sheets of (1) slightly metamorphic Paleozoic, or (2) mainly with carbonate rocks in a facies that slightly differs from the Northern Calcareous Alps in some parts lie on this "old crystalline rock" with sedimentary or tectonic contact. Metamorphic series of the Penninic zone underlie the "old crystalline rock", outcropping in several large windows (e.g., Engadin window and Tauern window). The metamorphic complex of the Central Alps is again subdivided into naps. A detailed description of the tectonic structure must be omitted here for brevity's sake.

2. Contributions by Other Geophysical Methods

A. HAHN

The results of refraction and reflection seismics and their interpretation with respect to the structure of the crust and upper mantle ought to be consistent with the appropriate results obtained by other geophysical methods. The papers in this chapter are presented to illustrate this point. They deal with the following subjects:

Seismic noise
Magnetics, Gravimetry
Geoelectric deep-sounding (DC)
Geothermic investigations
Laboratory experiments with high pressure and temperature.

1. Natural seismic noise is of great importance for the installation of seismic-recording systems with respect to the signal/noise ratio. The contribution presented in this chapter gives some quantitative information on the noise level for the different geological units in the area of the seismic-refraction investigations.

The other methods are concerned with the transformation into geological and general geophysical terms:

2. The distribution of rock bodies with higher magnetization is found primarily by magnetics. Up to now it is not yet possible to give a clear correlation between magnetization and the petrological composition of rock complex. But the interpretation of a map of the local magnetic anomalies reflects the arrangement of crustal regions which differ in magnetic characteristics at a depth of some km below the transition zone between nonmetamorphosed and metamorphosed sediments. Thus from the magnetic results, usually obtained on a grid system over a large area, one can try to obtain some hints about the shape and situation of the environmental area where seismic results - usually obtained along profiles - can be expected to be representative.

3. The electrical DC-deep sounding (Schlumberger array) is at present carried out only in a few places in the Federal Republic of Germany, yielding resistivity profiles down to approximately 12 km. If the main problems of interpretation - deviation of the interfaces between layers of different resistivity from horizontal layering as well as anisotropy of the sedimentary series - can be overcome, then this method yields depth values for boundaries between layers of different petrological character with an accuracy which is comparable to that obtained in seismics. Thus, apart from the fact that a contrast in seismic velocity and in electrical resistivity does not need to be present at the same depth in every case, the results of the two methods may check each other at the depth mentioned above.

4. The results of geothermal investigations usually give information about the terrestrial heat flow and the subsurface temperature field. We can expect, particularly in the areas of unusual velocity distribution, to find corresponding anomalies in the geothermal parameters. A clear correlation should, for example, be possible in the case of a layer with extremely low velocity at a high level. Here the thermal conditions should allow the assumption of an at least partially molten state of the corresponding layer if no other explanation for the low velocity is possible.

5. The possibility of measuring temperatures directly is restricted by the depth of the boreholes. A reliable extrapolation downwards on the basis of the present knowledge of the terrestrial heat flow is limited by the lack of accurate information about the distribution of heat sources at greater depths.

6. Laboratory investigation of seismic velocity and electrical conductivity as a function of pressure and temperature, combined with the results of seismology and magnetic deep-sounding, enable us to make some estimations about temperature values

in certain depth intervals depending on the rock types present. On the other hand these considerations contribute essentially to an interpretation of seismic results in petrological terms.

2.1 Investigation of Seismic Noise in the Federal Republic of Germany

M. STEINWACHS

Seismic noise in the frequency range 0.3 to 8.3 Hz was recorded on magnetic tape by a mobile seismic station over a period of four years. The recordings generally extend over a period of at least 24 h. At some stations the recordings were repeated after half a year or more. The stations were selected in such a way to allow adequate distance from visible artifical noise sources (towns, industry planes, traffic) or evident natural noise sources (trees and bushes moved by wind, running water etc.).

The spectral density functions, correlation functions, directional characteristics and ground-particle movements were computed from the recordings. The total number of interpreted intervals from all stations was more than 300.

Fig. 1 shows the stations and the average spectral density functions. Most of the curves show a downward trend with increasing frequency. The spectra with low noise level show a sharp peak near 2 Hz. The investigation of the daily variation of the noise shows that clear differences between day- and night-level can be recognized in the range from 3-8 Hz.

For the analysis of the time and space variations of the seismic noise it was possible to distinguish on a profile through the Molasse (southern Germany), the influence of the geological subsurface on the spectral density function. The result is a monotonic increase of the spectral density with increasing thickness of the Molasse.

The complete result of this investigation was published by STEINWACHS (1974).

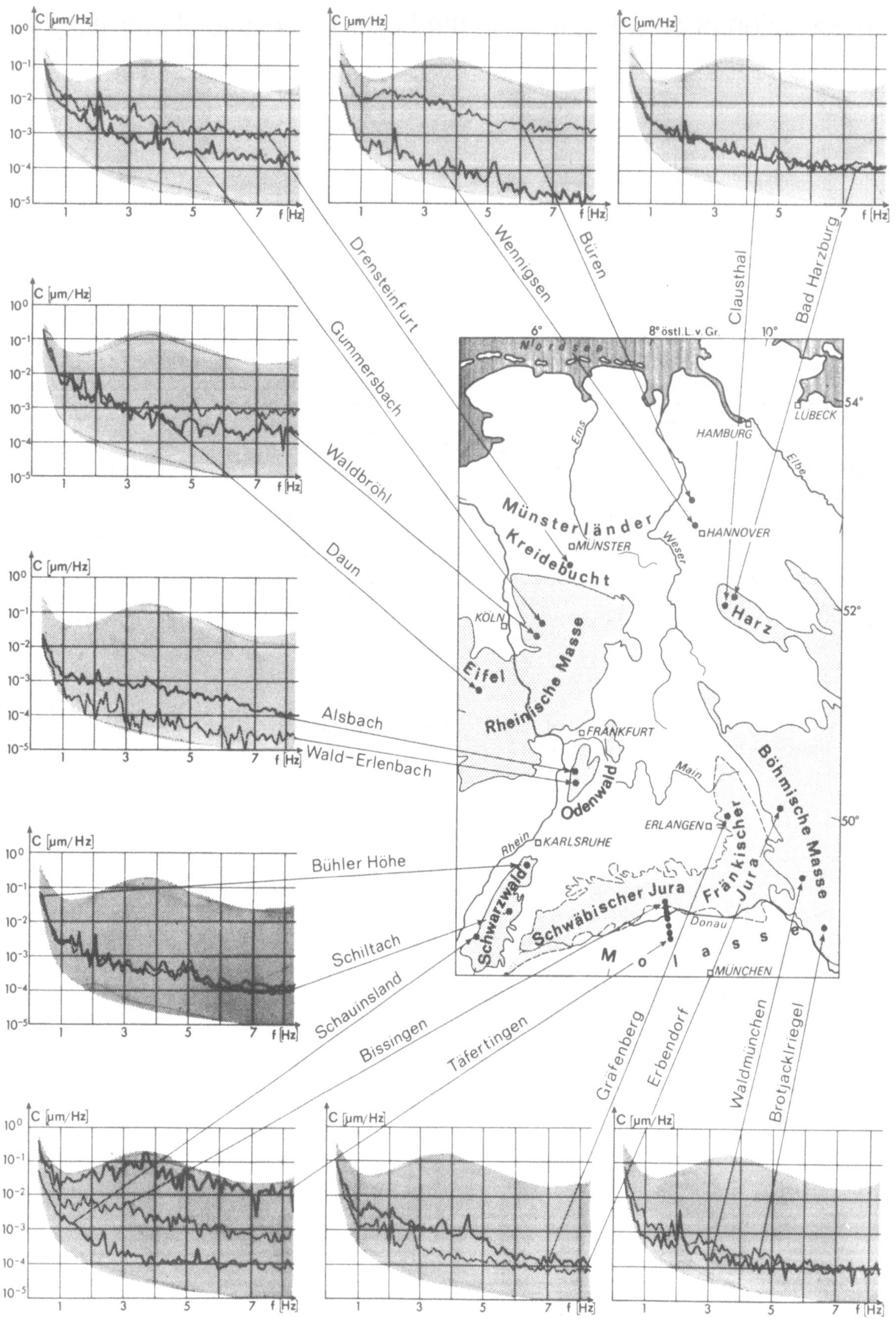

Fig. 1. Stations and the average spectral density functions of the short-period seismic noise in the Federal Republic of Germany

2.2 Magnetics

A. HAHN

ABSTRACT

In order to compare the results of the interpretation of the aeromagnetic map of the Federal Republic of Germany with the results from explosion seismology, bodies at 0 to 0.3 km, 0.3 to 3 km, and 3 to 15 km depth are considered here. For the deepest interval, which in many ways is the most interesting, only a relatively small part of the map area has been interpreted both seismically and magnetically. For this small part (Hessen depression - Leine Graben), the results of the two methods correlate well; for the two shallower intervals, the correlation is also good.

ZUSAMMENFASSUNG

Um die Ergebnisse, die aus der Interpretation der aeromagnetischen Karte der Bunderepublik Deutschland einerseits und aus der Tiefenseismik andererseits erhalten wurden, miteinander zu vergleichen, werden drei Tiefenintervalle betrachtet: 0-0,3 km, 0,3-3 km, 3-15 km. Im tiefsten Intervall, das offensichtlich von besonderem Interesse ist, gibt es bisher nur in einem relativ kleinen Gebiet (Hessische Senke - Leinegraben) Ergebnisse beider Methoden. Diese stimmen gut überein. In den flacheren Tiefenintervallen findet man ebenfalls gute Übereinstimmung.

A generalized version (1 : 1,000,000 scales) of the aeromagnetic map of the Federal Republic of Germany was published in 1973 (EBERLE, 1973). It shows anomalies which result from magnetic sources at various depths. In the following, these sources are discussed according to depth intervals of 0-0.3 km, 0.3-3 km, and 3-15 km. The locations mentioned in the following can be found in the geological map of the Federal Republic of Germany (JACOBSHAGEN, Chapter 1 of this volume). The aeromagnetic map is shown in Fig. 1.

2.2.1 Sources 0-0.3 km Deep

The magnetic geological bodies in this depth interval correlate poorly with seismic data. The bodies are mostly magnetic extrusives or shallow intrusives of late Paleozoic or Tertiary age. The only exceptions are in the northwestern Odenwald (49.8° N, 8.7° E), northwestern Spessart (50° N, 9.2° E), and southeastern Oberpfälzer Forest (49.2° N, 12.9° E). Gabbro and serpentinite, basic crystalline rocks of several types, and amphibolite crop out here. The fact that in these regions the crystalline basement crops out is doubtlessly interesting for the interpretation of seismic results, but this exposure can be seen in much more detail from the geological map than from the magnetic map.

2.2.2 Sources 0.3-3 km Deep

Anomalies of this type are found mainly in southern Germany and are concentrated in two linear zones. One zone strikes northeast across the northern half of the Rhine Graben and the central part of the Main (the axis of this zone connects the points 49° N, 8° E and 50° N, 10° E). The zone is about 100 km wide. BOSUM and ULRICH (1969) suggested that these anomalies resulted from magnetized bodies whose tops are at about sea level at the northwestern boundary of the zone and are at increasingly greater depths to the southeast.

The second zone of anomalies follows outcropping Jurassic rocks from 9.4° E to 11.6° E, 48.5° N to 48.8° N. EBERLE (1975) interpreted the anomalies as due to magnetized bodies whose tops lie at sea level or a little deeper. These bodies would then correspond to the Vindelician crystalline rocks.

There is a significant relationship between the magnetic data and the map of the surface of Variscan Mountain System German Research Group (1964, Fig. 8). A high part of this surface trends northeastward from a

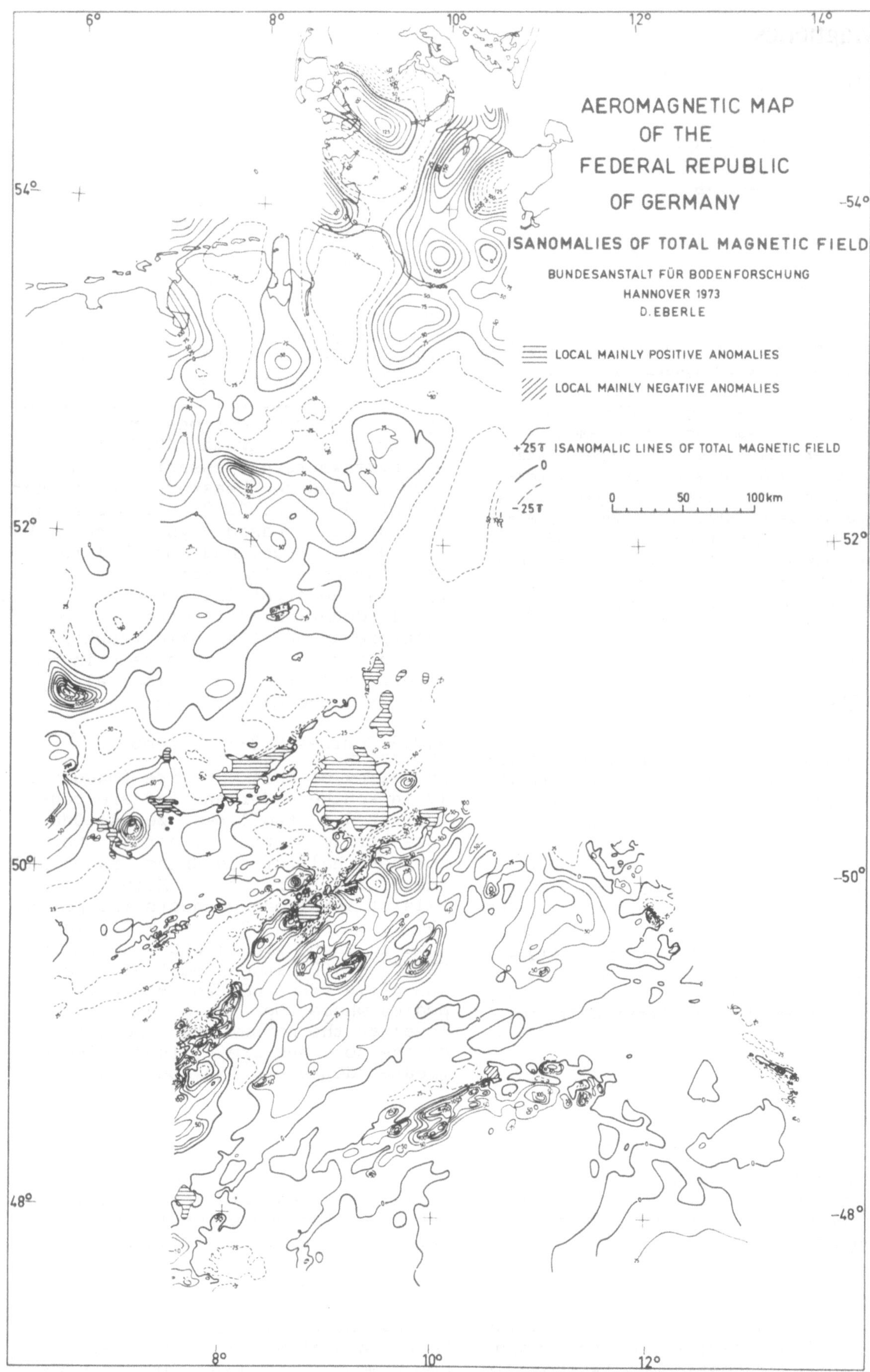

Fig. 1. Aeromagnetic map of the Federal Republic of Germany. (After EBERLE, 1973)

point 20 km south of Frankfurt. This high is then just below the northern anomaly chain of the Rhine-Main zone. The gentle southeastward slope of the surface also corresponds to the increasing depth to the magnetized bodies.

Similarly, the Variscan surface forms a narrow high about 0.2 to 0.6 km deep beneath the western half of the anomaly zone of the Vindelician crystalline rocks. Beneath the eastern half, the narrow high merges with a larger high having its crest about 30-50 km to the north but the anomalies continue to follow the 0.2-0.6 km contours along the southeastern flank of the larger high.

In conclusion, it is apparent that there is rather good correspondence between the seismic and magnetic results. The seismically determined depths seem to be a little greater than is to be expected from the interpretation of the magnetic anomalies. This small discrepancy can be explained by the empirical data that granite, for example, does not reach its full velocity of 6.2 km/s at depths less than 2 km.

2.2.3 Sources 3-15 km Deep

With respect to these sources, only the part of the aeromagnetic map from its northern boundary to 50.7° N has been interpreted. Most seismic data, however, is from the more southern part of the Federal Republic of Germany. Hence, we can compare only one feature, located in a zone extending north-south between 52° N to 50° N and located between 9° E and 10° E. The Contour Map Depth $z(x_c)$ of GIESE and STEIN (1971, Fig. 12) shows in this area a long depression with its axis striking 350°. The depth varies from 24 km at the borders to more than 28 km in the center of the depression.

A more or less parallel trending depression occurs on the aeromagnetic map in the same area. If this depression of ΔT-values is interpreted to result from the relief of the upper surface of a magnetized layer, the layer would have a general depth of about 10 km in which there is a depression 2 to 3 km deep (depending on the magnetization value chosen for this layer).

It should be added that this depression follows the general trend of the so-called Mediterranean-Mjösen-Zone in the section Hessen depression - Leine Graben whose most prominent part is the Rhine rift valley where subsidence has been about 3 km.

2.3 Geoelectric Ultra-Deep Sounding

E. K. Blohm and H. Flathe

ABSTRACT

Some problems connected with the application of direct-current resistivity sounding for depths down to 10 km are discussed.

Several measurements with maximal electrode spacing between 20 and 150 km were carried out in various regions of Germany. The problem of anisotropy is of importance in the interpretation of the ultra-deep sounding data and can be studied from the data presently available, but is complicated by the presence of sedimentary basins.

ZUSAMMENFASSUNG

Einige Probleme der geoelektrischen Widerstandsmethode bei Eindringtiefen bis zu 10 km Tiefe werden diskutiert. Messungen mit maximalen Elektrodenentfernungen zwischen 20 und 150 km wurden in verschiedenen Gegenden Deutschlands durchgeführt. Das Problem der Anisotropie, das anhand aller vorliegender Messungen studiert werden konnte, kommt bei der Interpretation ultratiefer Sondierungen eine einschneidende Bedeutung zu. Entsprechende Untersuchungen werden durch die Gegenwart von Sedimentbecken bedeutend erschwert.

The geophysicist who attempts to see deep subsurface structure by means of geophysical measurements at the earth's surface, is confronted with the uncomfortable fact that the physical character of near-surface bodies can strongly influence his measurements. If we consider, for example, electrical conductivity as a physical parameter, it is obvious that a layer at the surface with good conductivity will have a shielding effect and will give the geophysicist trouble when he tries to "look through" it. In any case, it is necessary to find a means to determine the effect of the shielding layer quantitatively in order to make at all possible an interpretation of what lies at greater depth.

This problem appears in the study of large-scale electrical conductivity anomalies using magnetic deep-sounding when the study area has a thick, sedimentary cover with good electrical conductivity as is the case in the Northwest German Lowland. A quantitative approach to data about greater depths is made possible by first knowing the thickness and conductivity of this sedimentary cover. Useful in determining this surficial influence are geophysical methods which use natural fields (magnetotellurics) or artificial fields (direct-current resistance method) to subdivide the horizontally stratified subsurface according to its electrical conductivity.

The application of artificially produced fields is described in this chapter. These methods have the advantage that the initial form of the field is known and one has them therefore mathematically better "in hand". The classical procedure for determining the conductivity of a horizontally stratified subsurface is the geoelectric resistance method. Direct current is applied to the earth through two electrodes, A and B. The various subsurface conductivities influence the artificially produced field. For the so-called deep sounding, various electrode spacings $\overline{AB}$ are used and the "apparent" resistivity relative to the electrode spacing is determined (this yields the sounding curve). The true resistivities of the individual strata are then determined from the apparent resistivity.

For many years, this method has been successfully applied world-wide in the study of shallow objects (up to a few hundred m deep), for example, in the clarification of hydrogeologic problems (groundwater exploitation). The technical problems are comparatively small but increase rapidly when one wishes to probe greater depths. For this, greater electrode spacings $\overline{AB}$ are necessary and that requires longer cables for transmit-

Fig. 1. Map of western Germany showing location of geoelectric ultra-deep-sounding resistivity measurements
1 Hannover-Hildesheim
2 Schwetzingen
3 Schopfheim
4 Heide
5 Rhine Graben
6 Wamberger anticline
7 Fohrenbühl
8 Reinhardswald
9 Markdorf
10 Nördlinger Ries

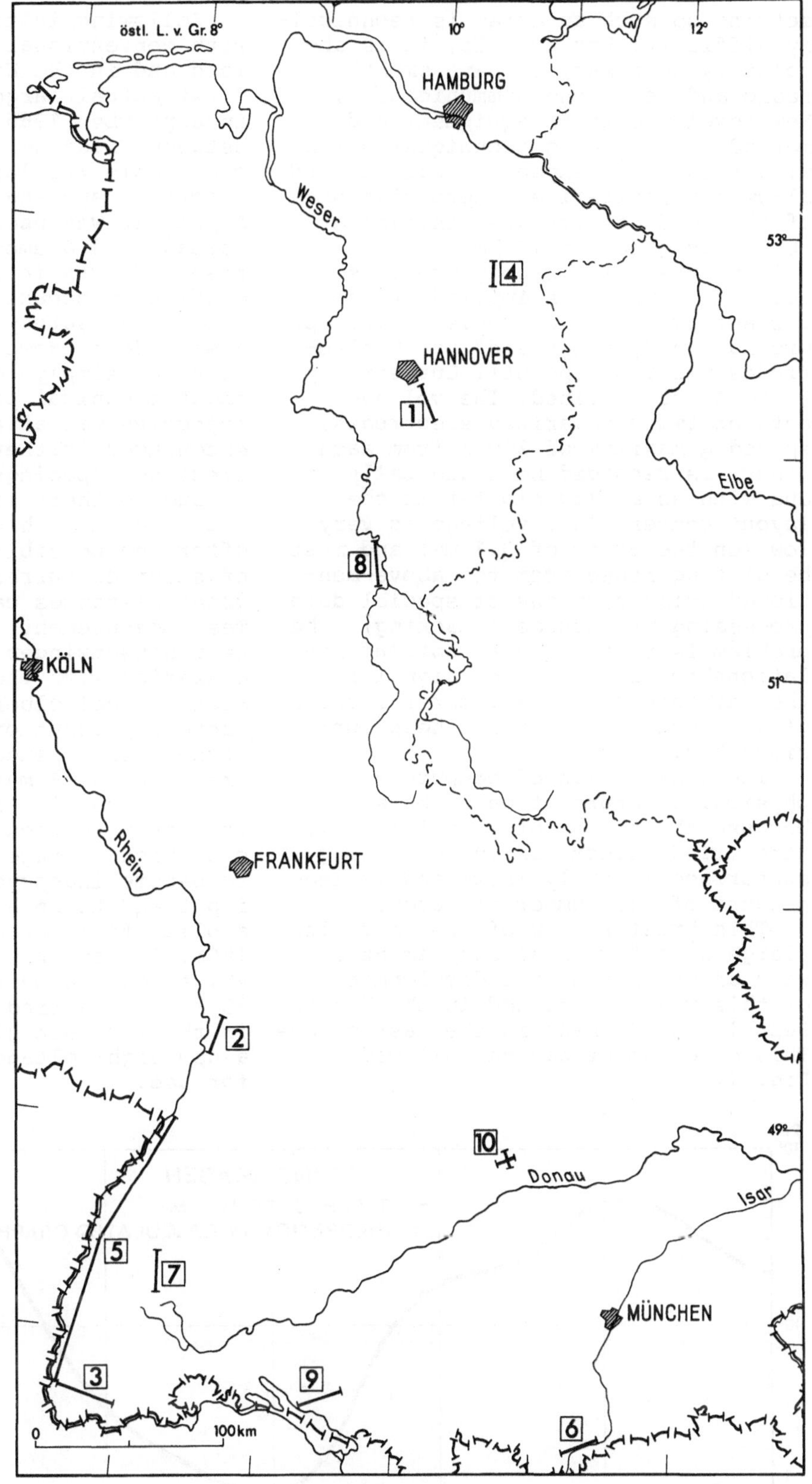

ting the current. The power loss in the cables requires a large power supply. Sensitive equipment which can measure low voltages is necessary. Of course, this equipment can be considerably influenced by natural and random currents (noise).

Theoretically, there is nothing to hinder deeper penetration by means of enlarging the survey array, but setting up such an array is technically difficult. For example, it is absolutely necessary to have smooth radio and telephone communication. New developments in equipment and methods have led to techniques which, with reasonable expense, indicate and allow a quantitative interpretation of the conductivity distribution at depths as great as 10 km.

A 60 kVA power supply provides current. Current is applied at the surface over special cable to multiple grounds (a "groundfield" of 50 electrodes per input point). Currents up to 80 A are obtained. The voltage between two unpolarized electrodes, spaced a maximum of 200 m from each other, is recorded both digitally and with an analog plotter in the layout center. This voltage is very low (on the order of 0.1 mV) and must be distinguished from the above-mentioned noise by means of special data processing procedures (stacking). The problem is to achieve an optimum relationship of the "true" signal to the "disturbed" signal; careful choice of the recording site may help considerably.

The application of computers for physical interpretation is possible because the theory of a horizontally stratified, electrically conducting subsurface is fully known and is independent of the number of layers.

This brief review of the so-called "large array" will be supplemented with an outline of the development of this research method in the Federal Republic of Germany in the last decade. This development can be followed in Fig. 1.

Following initial test surveys with conventional equipment in Münsterland and in the Harz Mountains, the first actual large array was laid out in September 1965 along the autobahn between Hildesheim and Hannover. Special cable was laid with a maximum electrode spacing ($\overline{AB}$ max) of 20 km. A plotter was used to record the probe voltage and a smooth sounding curve resulted. Its interpretation presented difficulty because, as was previously known, the survey line passed over a salt dome. However, the aim of this test was simply to obtain information about technical possibilities and this objective was attained. The results encouraged further tests with greater electrode spacings ($\overline{AB}$).

Due to their larger conductivity cross section, high voltage cables offer the possibility of the input of stronger currents. The loss over great distances can be kept small. Test measurements were carried out near Schwetzingen in March 1966 in cooperation with the 400 KV Research Society (Heidelberg). This test with "foreign" input over "foreign" cable technically opened a way to large arrays with $\overline{AB}$ max greater than 100 km.

Subsequently, self-contained generators were used to apply current over high voltage conductors in order to become independent of the "foreign" input and to be able to control the applied field at will. In November 1966, in cooperation with Badenwerk AG, a deep sounding with AB max = 30 km was carried out in this manner in the southern Black Forest where a new high voltage line was available for use.

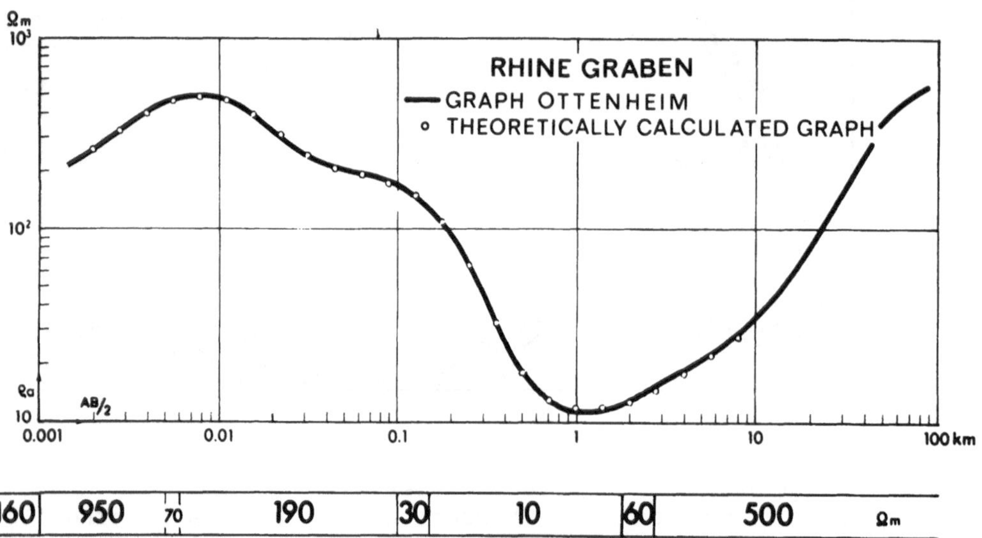

Fig. 2. Ultra-deep sounding curve of the Rhine Graben

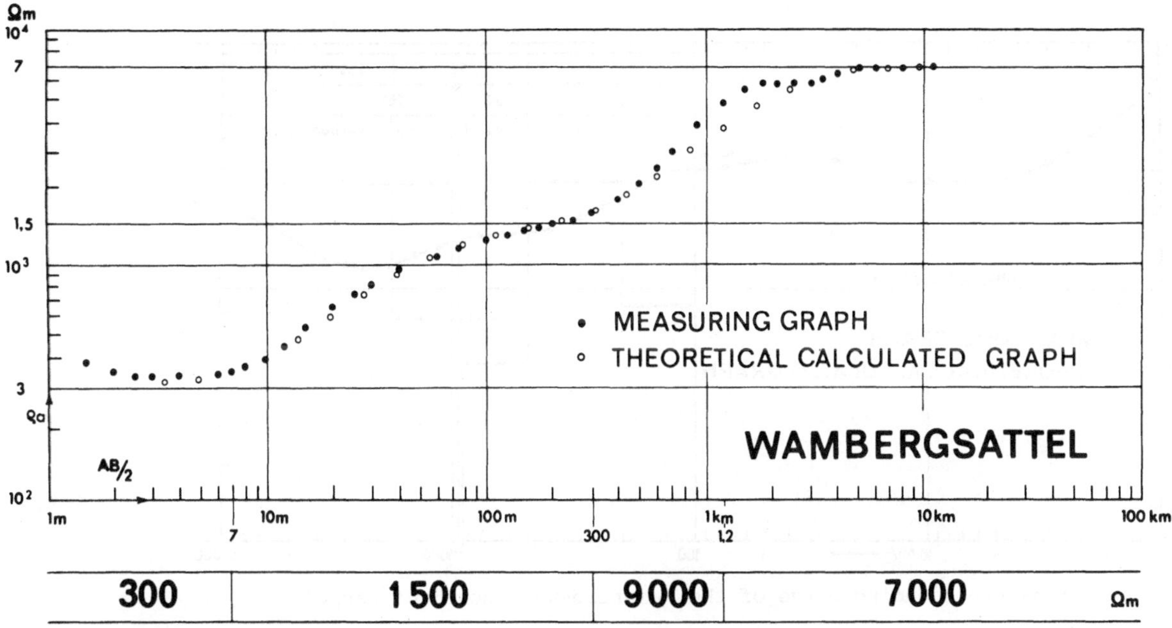

Fig. 3. Ultra-deep sounding curve of the Wamberger anticline

The results of the above test were used for the preparation of the large-scale test with $\overline{AB}$ max = 150 km along the Rhine Graben between Karlsruhe and Badenweiler in Autumn 1967. Previously, experience was gained by carrying out grounding tests in the heath north-east of Hannover where dry sand makes grounding of the electrodes very difficult.

The large array ($\overline{AB}$ max = 150 km) in the Rhine Graben was the largest array in Europe up to that time. The resulting smooth sounding curve (Fig. 2) has an over-steepened rear part of the curve which allows several essentially different interpretations (BLOHM and FLATHE, 1970; BLOHM, 1972). The large array in the Rhine Graben led to additional studies. These projects used their own cable and power supply and the aim of independence in the selection of the survey area was achieved at the so-called "Wamberger anticline" south of München. Here it was possible to study an overthrust of the Calcareous Alps over the Molasse. Considering the conclusions which may be of consequence to the oil industry, this survey symbolized the independence of the large array from "foreign" equipment. With the equipment used, maximum electrode spacings of 40 km could be reached relatively easily (Fig. 3).

Using this technical knowledge, the Rhine Graben curve was supplemented by parallel measurements in the high Black Forest (near Fohrenbühl). These supplementary measurements aided in the interpretation of the Rhine Graben curve relative to the problems of anisotropy and the variation from horizontal stratification (HOMILIUS and BLOHM, 1973).

Additional ultradeep soundings have been carried out in the Reinhardswald north of Kassel (Fig. 4), near Markdorf on Lake Constance (Fig. 5), and in the Nördlinger Ries.

By far the largest ultra-deep sounding is presently underway in the south African shield area and is making use of the Cabora Bassa DC-high power line. This deep sounding is subdivided into several overlapping sections and, in 1974, reached an electrode spacing $\overline{AB}$ of 960 km. In 1975, the spacing is to be extended to 1200 km with an electrode current of 3300 Amp. This project is being carried out in close cooperation with Brown Boverie and Cie (Switzerland), the C.S.I.R. in Pretoria, the University of Rhodesia, and the Universities of Clausthal, Göttingen, München, and Münster.

The problem of anisotropy, which is of importance in the interpretation of ultra-deep sounding, can be studied from the data presently available. In the following concluding remarks, this problem will be briefly discussed.

It is known that quasi-horizontally stratified sediments show better electrical conductivity parallel to bedding than perpendicular to bedding. Although borehole measurements of resistivity only indicate the conduc-

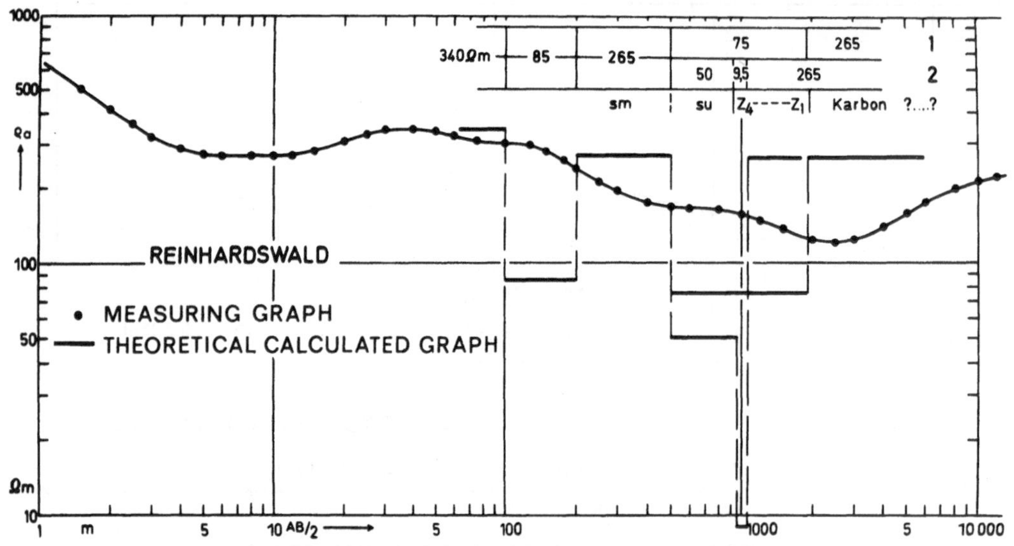

Fig. 4. Ultra-deep sounding curve of the Reinhardswald north of Kassel

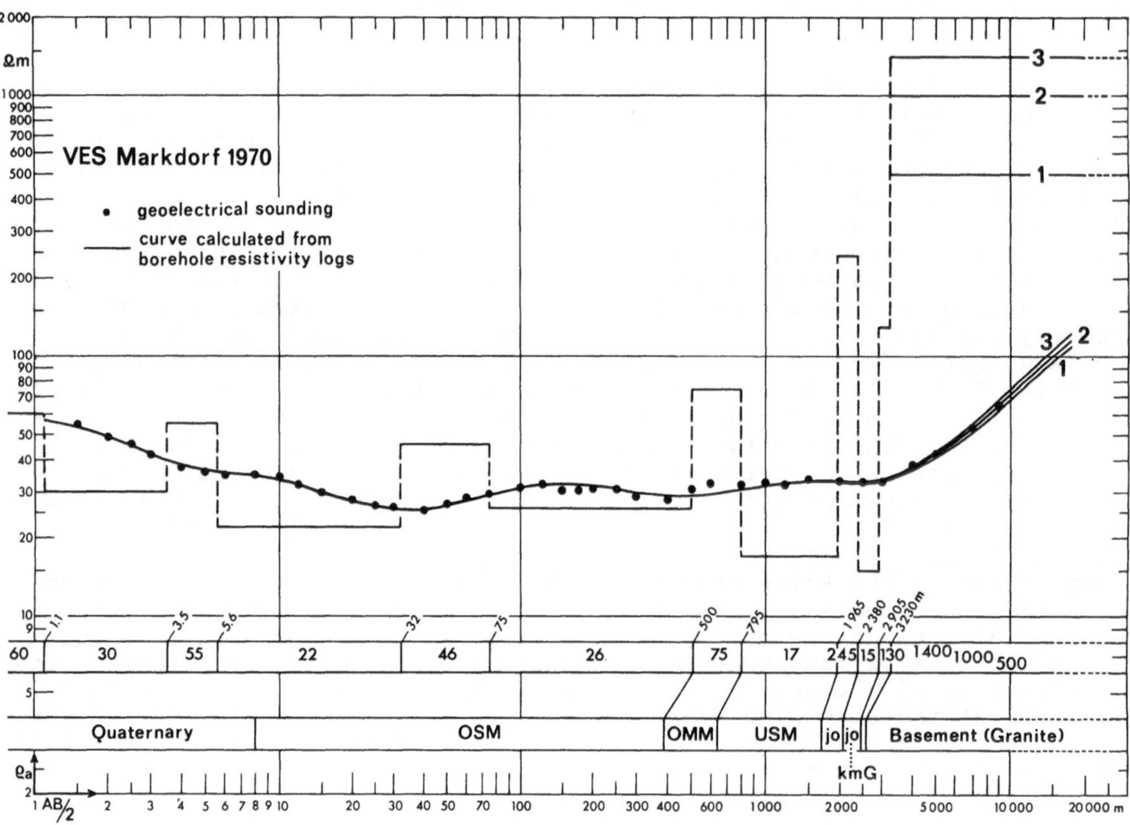

Fig. 5. Ultra-deep sounding curve of Markdorf on Lake Constance

tivity parallel to bedding (RÜLKE, 1967; FLATHE, 1967; FLATHE and HOMILIUS, 1973), measurements of the resistivity at the surface yield so-called "pseudo-resistivities" which are somewhere intermediate between the resistivities parallel and normal to bedding. Related to the pseudo-resistivity of a bed or group of beds is a pseudo-thickness. This thickness is that of an isotropic, homogeneous layer which would have the same effect at the earth's surface as the true, anisotropic layer (with higher conductivity in horizontal direction) would have.

The pseudo-thickness is greater than the true thickness with the difference varying according to a coefficient of anisotropy λ which is greater than 1. This coefficient can become very large for ultradeep soundings across a sedimentary basin (values up to $\lambda = 3$ are not unusual). This is a serious handicap for the quantitative interpretation of the sounding curves. For example, the top of crystalline rocks beneath thick sediments can only be determined if there is information about λ. In principle, such information cannot be obtained from the sounding curve; other means (such as drilling or seismic and magnetotelluric surveys) must be used to obtain it.

Nevertheless, in the case of a sediment "plate" with good electrical conductivity above poorly conducting crystalline rocks, the ratio of thickness to resistance remains invariant whether either a pseudo-value or true value (in the horizontal direction) is concerned. This means that geoelectric deep sounding can indicate quantitatively the influence of sediments on subsurface conductivities to depths extending into the upper mantle. Thus, a 10-year development has reached the initially stated objective.

2.4 Geothermic Investigations

R. HÄNEL

ABSTRACT

The importance of geothermics for the investigation of the deeper subsurface is increasing. At the present time geothermics is in the data-gathering stage. Up to now the various thermal properties determined (thermal conductivity, specific heat, density, thermal diffusivity) have been discussed under different point of views. In addition, the terrestrial heat flow in Germany shows a marked trend, increasing from NW to SE. The present mean is 1.71 μcal/cm^2 s. Finally a heat flow anomaly is interpreted in the Rhine Graben and a connection is shown with the low velocity in the subsurface.

ZUSAMMENFASSUNG

Die Bedeutung der Geothermik nimmt bei der Erforschung des tieferen Untergrundes ständig zu. Doch vorerst befindet sie sich noch im Stadium des Wertesammelns. Die bisher gesammelten thermischen Stoffwerte der Gesteine (Wärmeleitfähigkeit, spezifische Wärme, Dichte, Temperaturleitfähigkeit) werden nach verschiedenen Kriterien untersucht und diskutiert. Darüber hinaus wird ein Anstieg der terrestrischen Wärmestromdichte in Deutschland von NW nach SE nachgewiesen. Der derzeitige Mittelwert beträgt 1,71 μcal/cm^2 s. Schließlich wird noch die Wärmestromdichte-Anomalie im Rheintalgraben interpretiert und ein Zusammenhang zur Geschwindigkeitserniedrigung im Untergrund gegeben.

2.4.1 Introduction

Since all physical and chemical processes which take place within the solid earth are connected with a production or resorption of heat, the knowledge of the distribution in time and space of temperature provides criteria for a series of hypothesis concerning these terrestrial processes. The temperature distribution in great depths can be determined with the help of heat-flow measurements, thermal conductivity and heat production on rocks.

Since only a small number of records are available in geothermics as compared with seismics, the main object at present is to collect data. The first interpretations of the existing material have therefore always been considered with some reservation.

2.4.2 Thermal Properties

Up to now the thermal properties of about 200 rock samples from Germany

Table 1. Thermal properties at 50°C; λ = thermal conductivity, c = specific heat, ρ = density, κ = diffusivity

Rock	Quantity	λ (10^{-3} cal/cm/s/°C)	c (cal/g/°C)	ρ (g/cm^3)	κ = λ/ρc (10^{-3}/cm^2/s)
Limestone	9	5.52 ± 0.90	0.209 ± 0.011	2.55 ± 0.08	10.34 ± 1.57
Massive Limestone	16	7.08 ± 1.12	0.217 ± 0.010	2.67 ± 0.08	12.21 ± 1.54
Dolomite	9	7.43 ± 1.06	0.223 ± 0.009	2.62 ± 0.14	12.70 ± 1.34
Marl, limy and clayey marl	19	5.08 ± 0.93	0.214 ± 0.018	2.54 ± 0.28	9.34 ± 1.48
Clay, Clay marl	19	6.64 ± 1.54	0.211 ± 0.010	2.60 ± 0.12	10.84 ± 3.15
Slate	11	6.28 ± 1.76	0.218 ± 0.016	2.59 ± 0.08	11.17 ± 3.32
Sandstone	54(28)	7.75 ± 1.64	0.197 ± 0.010	2.59 ± 0.10	16.45 ± 3.17
Basalt	3	4.01	0.211	2.81	6.58
Granite	16	7.37 ± 1.08	0.205 ± 0.014	2.64 ± 0.07	13.78 ± 2.12
Gneiss	4	6.46 ± 0.39	0.193 ± 0.011	2.71 ± 0.01	12.24 ± 1.25

Table 2. Thermal properties as a function of the formation at 50°C; λ = thermal conductivity, c = specific heat, ρ = density, κ = diffusivity

Formation	Quantity	λ (10^{-3} cal/cm/s/°C)	c (cal/g/°C)	ρ (g/cm^3)	$\kappa = \lambda/\rho c$ (10^{-3}/cm^2/s
Tertiary	8	4.84 ± 1.89	0.195 ± 0.010	2.48 ± 0.06	10.01 ± 2.46
Cretaceous	5	5.16 ± 0.75	0.225 ± 0.010	2.56 ± 0.06	9.01 ± 1.54
Jurassic	42	6.54 ± 1.37	0.219 ± 0.012	2.62 ± 0.09	11.39 ± 2.06
Triassic	23	6.64 ± 1.06	0.211 ± 0.025	2.38 ± 0.17	13.46 ± 2.82
Permian	2	8.24		2.56	
Carboniferous	67(31)	7.38 ± 2.04	0.208 ± 0.013	2.68 ± 0.10	13.99 ± 4.46

have been studied. In this context, measurements were made of the thermal conductivity λ, the specific heat c, and the density ρ. Moreover, the thermal diffusivity κ was calculated. The properties determined can be taken from Table 1, in which the mean value of several measurements and the standard deviation are given. These values can at the same time be taken in the case of lacking property values.

In Table 2, the results were classified according to stratigraphy. It is obvious that, apart from the specific heat, the values increase with increasing age of the rocks (at increasing density).

The rock samples are from depths between 0 m and 5700 m; and the mean depth is calculated to be 1033 m, or 1000 m roughly. The corresponding mean values of the thermal properties (λ and c measured at ca. 50°C) are:

Thermal conductivity λ_m = (6.69 ± 1.89) 10^{-3} cal/cm/s/deg
Specific heat c_m = (0.212 ± 0.014) cal/g/deg
Density ρ_m = (2.60 ± 0.17) g/cm
Thermal diffusivity κ_m = (12.14 ± 3.69) 10^{-3}cm^2/s

The thermal conductivity is moreover a function of pressure and temperature. The dependence upon pressure can be approximated by:

$$\lambda = \lambda_o \ (1 + \alpha \ P) \tag{1}$$

where P(kg/cm^2) the pressure and α (cm^2/kg) is the pressure coefficient (cf. also Table 3). As at a depth of 30 km about 8 · 10^3 kg/cm^2 may be expected, the increase of λ in the case of basalt, for instance, amounts to less than 1% in this depth.

As to the thermal conductivity between 0°C and 400°C an equation developed by BIRCH and CLARK (1940) is available:

$$\lambda = \left(\frac{600}{300 + T} + 4\right) \cdot 10^{-3} \text{ cal/cm/s/deg}$$

$$T = \text{temperature (deg)} \tag{2}$$

For 50°C, $\lambda = 5.71 \cdot 10^{-3}$cal/cm/s/deg is obtained from it. This value is not too different from the mean value given above.

Table 3. The pressure coefficient for different rocks after BIRCH and CLARK (1940) and BRIDGMAN (1924)

Rock	α (10^{-6} cm^2/kg)
Limestone	1 - 95
Marble	28 -185
Dolomite	30 -160
Sandstone	57 -255
Basalt	2.2- 4.7
Gabbro	28

2.4.3 Terrestrial Heat Flow

The determination of the terrestrial heat flow q is possible on the basis of the well-known relation:

$$q = \lambda \cdot \text{grad } T \ (\text{cal/cm}^2/\text{s}) \tag{3}$$

where λ (cal/cm/s/deg) = thermal conductivity and grad T (deg/cm) = temperature gradient.

Many values of heat flow are now available from Germany. With the help of the criteria applied in statistics the data were examined with respect to any significant information (HÄNEL, 1971a,b). The following results were obtained:

1. In Fig. 1 the result of the trent analysis of the first order is given.

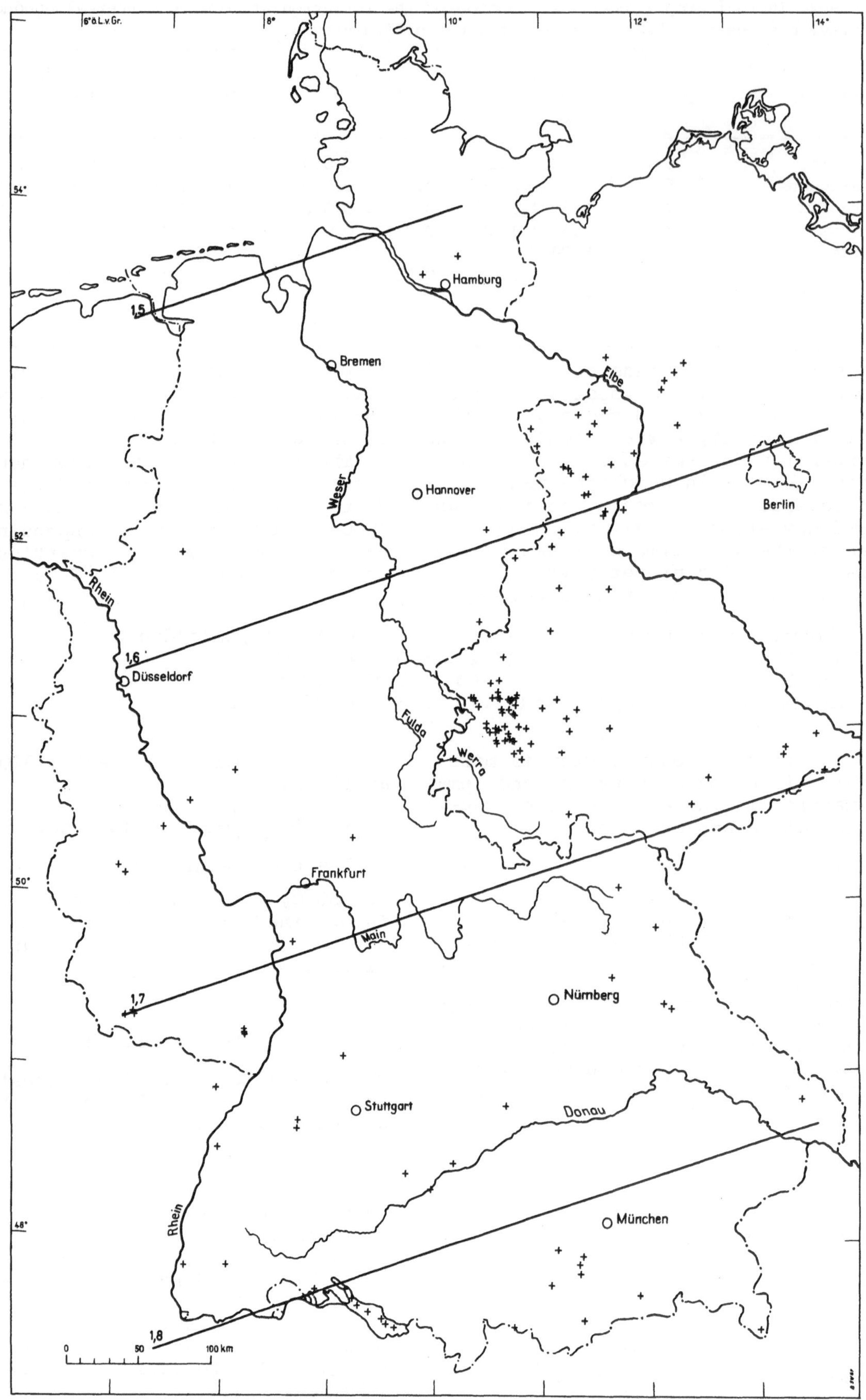

Fig. 1. Distribution of heat flow measuring points and isolines of a trend analysis of the 1sr order in μ cal/cm²/s of Germany

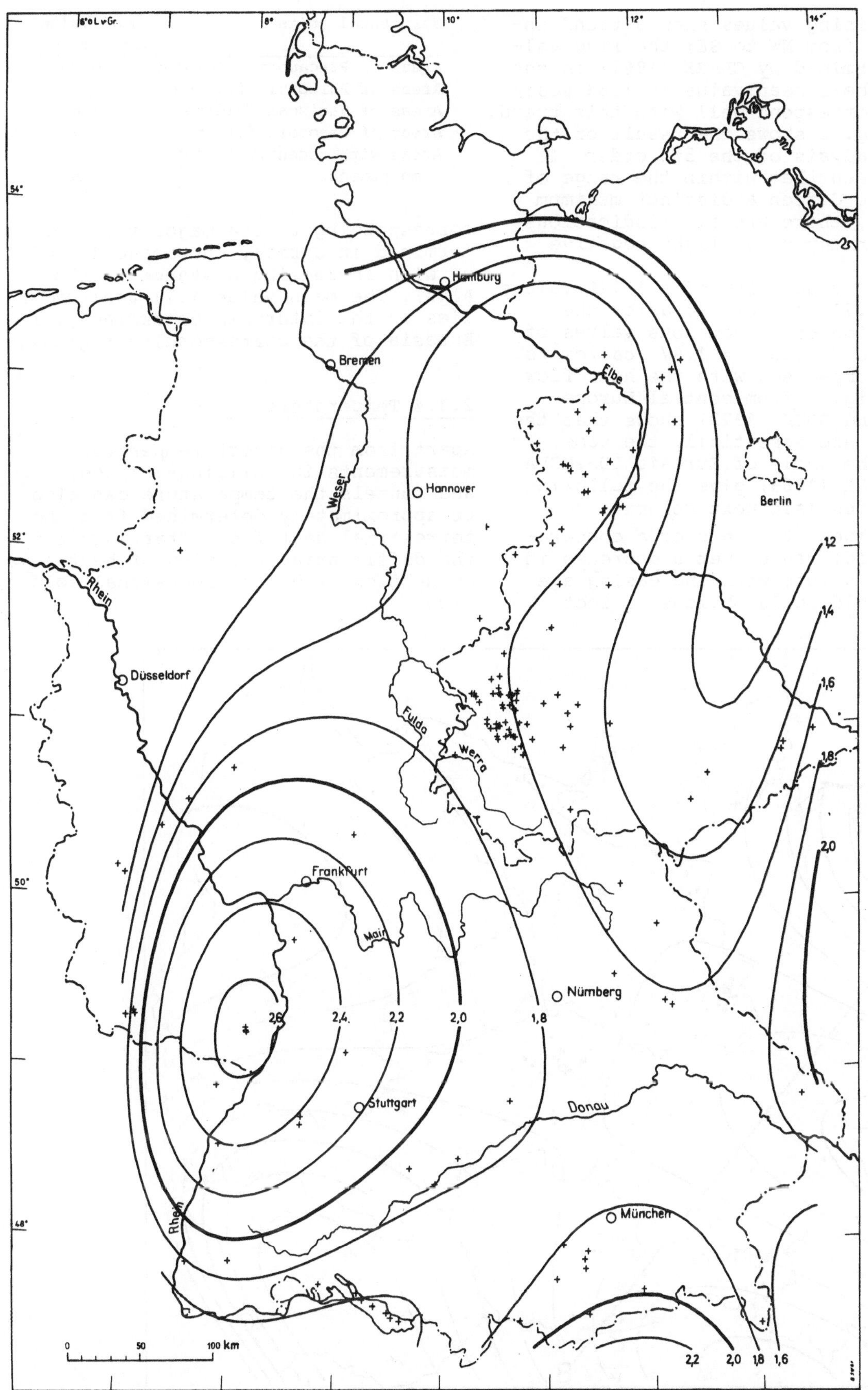

Fig. 2. Distribution of heat flow measuring points and isolines of a trend analysis of the 5th order in μ cal/cm^2/s of Germany

The remaining values show a trend increasing from NW to SE; the five values determined by CLARK (1961) in the Alps - their mean value is 1.83 μcal/cm/s - correspond well with this trend.

2. Fig. 2 shows the result of the trend analysis of the 5th order. It can be seen that within the range of the Rhine Graben a distinct maximum is found, there are also indications in NE Germany of a slight negative anomaly.

3. The mean value of all 146 values is $\overline{q}$ = 1.71 μcal/cm²/s; after the elimination of 6 anomalous values of the Rhine Graben $\overline{q}$ = 1.67 μcal/cm²/s.

The comparison with the heat flow map in Fig. 3 from central Europe (HÄNEL and ZOTH, 1973) shows that the contours are essentially the same.

For the whole of Eurasia LUBIMOVA and POLYAK (1969) give the following mean values (see next column).

Structural units	Mean value (μcal/cm²/s)
Areas of Precambrian folding	0.96
Areas of Paleozoic folding	1.6
Areas of Variscan folding	1.6
Areas of Cenozoic folding	1.7
Areas with recently folded mountains	1.8

These values, too, are of a comparatively wide range, yet a decrease in the mean values with increasing age of the folding is obvious, a fact corresponding to the major trend determined in Germany (decrease in value with increasing distance to the Alps). The mean value 1.71 μcal/cm²/s lies in the intervals determined for Eurasia of the corresponding regions.

2.4.4 Temperature Field

Apart from the direct temperature measurements in drillings, mines, and tunnels the temperature can also be approximately determined from the terrestrial heat flow. Starting from the diffirential equation of heat conduction, the one-dimensional solution

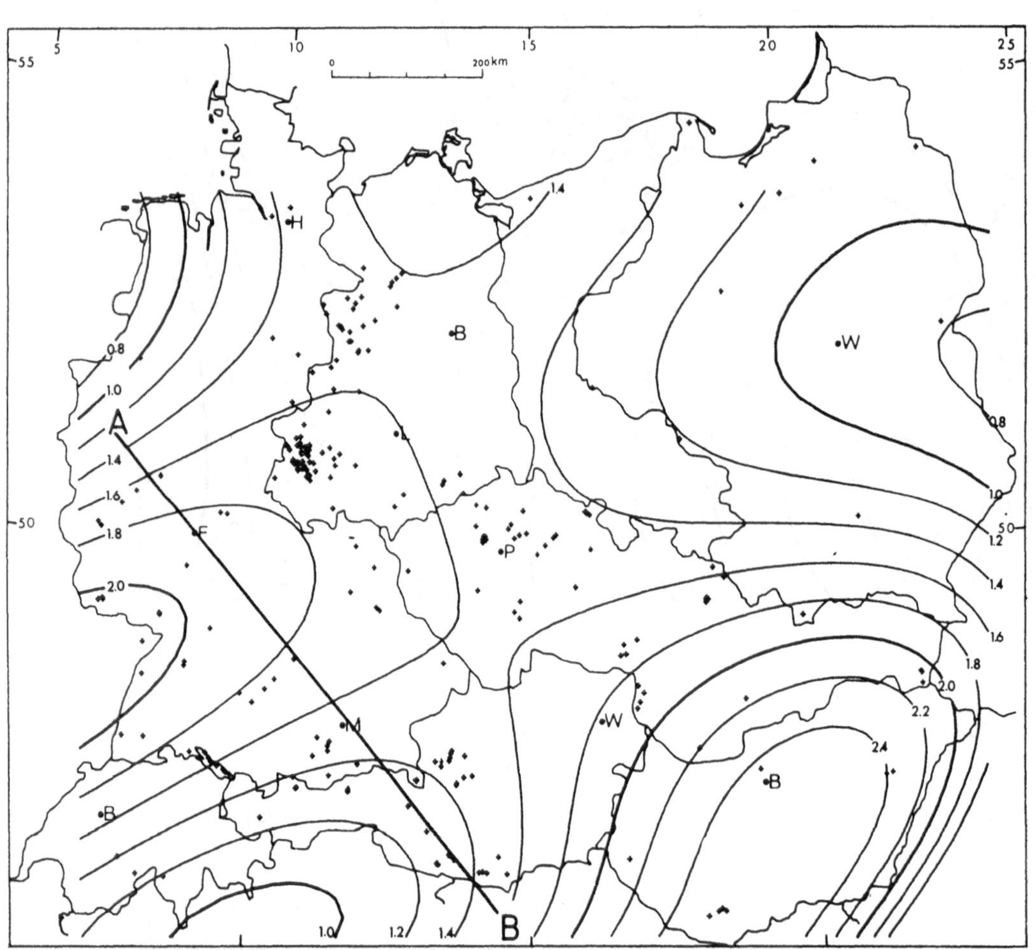

Fig. 3. Distribution of heat flow values and isolines of a trend analysis of the 5th order in μcal/cm²/s of Central Europe

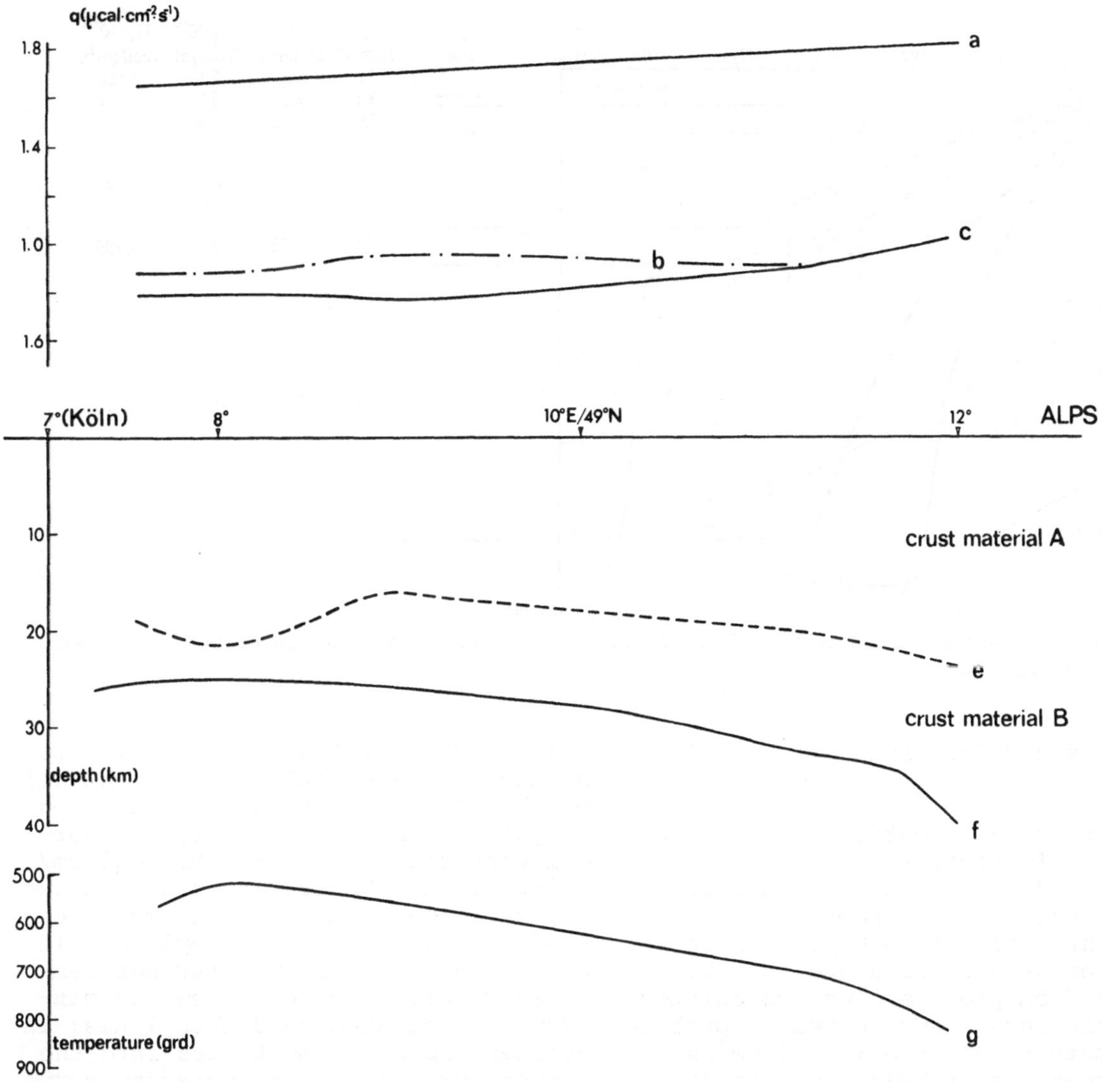

Fig. 4. Profile Köln-München with: *a* distribution of heat flow according to Fig. 1, *b* difference between heat flow on profiles *a* and *c*, *c* heat flow above the Mahorovičić discontinuity, *d* earth surface, *e* transition from acid (crust material *A*) to basic (crust material *B*) material, *f* Mohorovičić discontinuity, *g* temperature at the Mohorovičić discontinuity referring to a mean percentage of 33.3% in the crust material *A*

$$T_{i+1} = T_i + \frac{1}{\lambda_i}\left(q_i \cdot z - \frac{H_i \cdot z^2}{2}\right), \qquad (4)$$

$$q_{i+1} = q_i - H_i z$$

is obtained by double integration, provided the thermal conductivity and the heat production H are independent of time and place. In this case, i = 1, 2, 3 .... number of layers assumed, T_{i+1} = required temperature at the lower boundary of each layer, $T_{i=o}$ = mean annual temperature at the interface of air/earth, z = depth, $z_i \leq z \leq z_{i+1}$.

The difficulty of determining T_{i+1} lies in finding plausible assumptions for λ_i and H_i.

Fig. 4 shows a profile of the earth's crust approximately from Köln to München, with the Mohorovičić discontinuity representing its lower boundary according to GIESE and STEIN (1971). The structures within the earth's crust are not, in general, so early identified as the Mohorovičić discontinuity. At some points of the profile GIESE indicated a transition zone of dominantly acidic rocks to dominantly basic material (personal communication). It is represented here, however, as a continuous boundary, in order to simplify the calculations.

The temperature calculation was carried out with thermal conductivity $\lambda = 6 \cdot 10^{-3}$ cal/cm/s/deg for lack of accurate values; and the heat production for crust material A (sediment with 33.3% granite) is $H = 0.35 \cdot 10^{-12}$

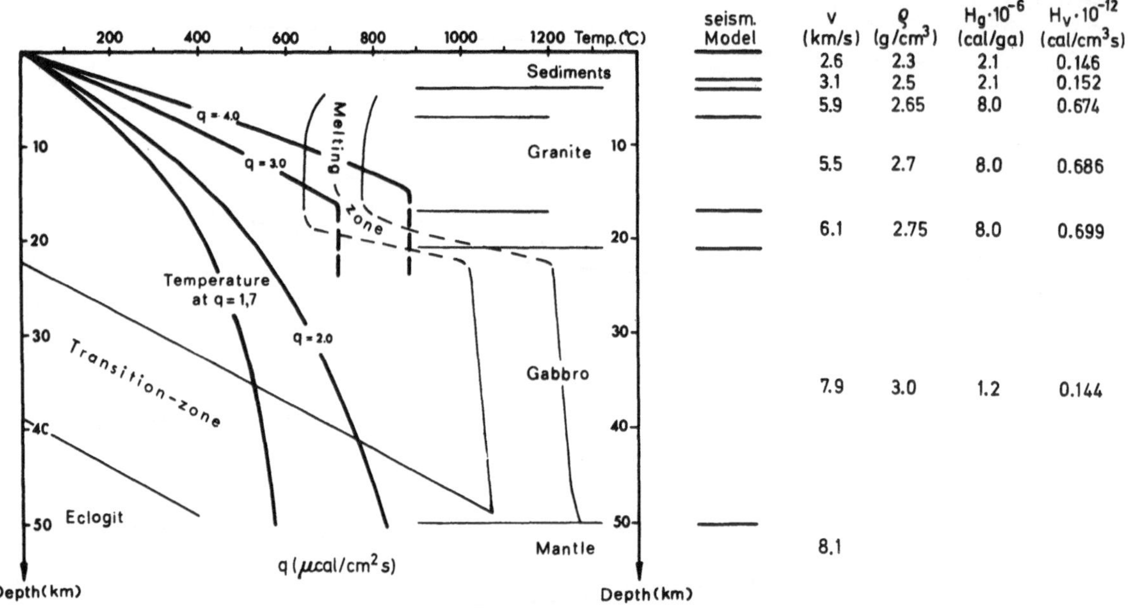

Fig. 5. Temperature distribution in the case of various heat flows (Rhine Graben), plotted in a fusion diagram

cal/cm³/s and for crust material B (basic rocks) $H = 0.16 \cdot 10^{-12}$ cal/cm³/s.

The resultant contribution to the heat flow is represented as profile c in Fig. 4. It amounts, on average, to 0.8 μcal/cm²/s. Moreover, the pertinent heat flow from Fig. 1 is represented as profile a and the difference from profile c was calculated. The difference, represented in profile b, amounts to about 0.9 μcal/cm²/s; the percentage of heat flow rising from below the Mohorovičić discontinuity is concerned here. At last the temperatures at the Mohorovičić discontinuity is shown in profile g of Fig. 4.

In Fig. 5 the temperature field of the anomaly in the Upper Rhine Graben (cf. Fig. 2) is examined (HÄNEL, 1970). In the same diagram the fusion zone of acid rocks (WINKLER, 1967), as well as the fusion and alteration zones of basic rocks (YODER and TILLEY, 1962) are marked. The temperature were calculated with the help of the values of the heat production H_V given in Fig. 5 and uniform thermal conductivity $\lambda = 6 \cdot 10^{-3}$ cal/cm/s/deg. It can be seen that within the depth of 10-20 km the rocks must be fused either partly or completely if q = 3-4 μcal/cm²/s. The further temperature distribution below the maximum is indicated by dashed lines; in greater depths it must pass into the temperatures of the temperature curve of q = 1.7-2.0 μcal/cm²/s. The non-molten rocks are influenced by the partly or completely molten rocks to the effect that the seismic velocity is reduced, as has been proved by seismic measurements (MUELLER et al., 1967).

Another possible interpretation of the positive heat flow anomaly in the Rhine Graben is given. If there is a granitic layer embedded in the sediments, a higher heat flow can be

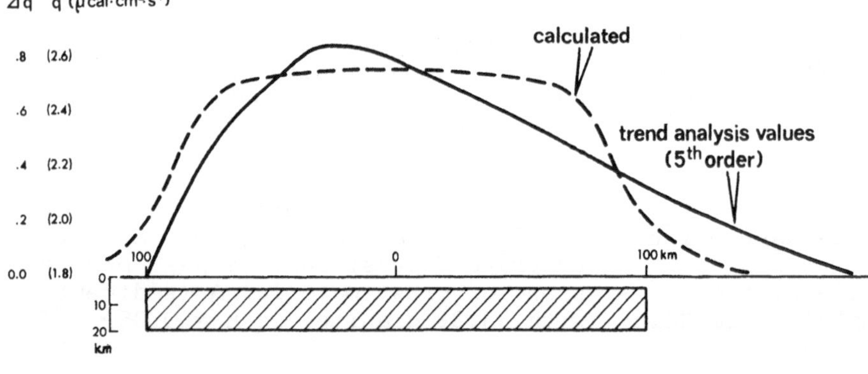

Fig. 6. A granite plate as a possible explanation of the measured increased heat flow in the upper Rhine Graben

expected. The calculation is carried out after a method by SIMMONS (1967) with a plate of radius R = 100 km, thickness h = 16 km, depth of the plate center z = 12 km, and a heat production of $H_1 = 0.69 \cdot 10^{-12}$ cal/cm^3/s was taken. From an average heat production of the surrounding rock $H_2 = 0.18 \cdot 10^{-12}$ cal/cm^3/s, it follows that $H = H_1 - H_2 = 0.51 \cdot 10^{-12}$ cal/cm^3/s is the heat production difference between plate and surrounding. The calculated heat flow Δq is shown in Fig. 6 (dash-lined curve). If it is anomalous, when greater than 1.8 µcal/cm^2/s (see also Fig. 1), the heat flow distribution is obtained, as given in Fig. 6 (solid curve) for a profile across the center of the anomaly in Fig. 2 parallel to the circle latitude.

There are indications that the curves correspond with each other; this correspondence would probably be even better if the plate were slightly inclined. It must, however, be emphasized that the "measured curve" is taken from the trend surface and that, at its maximum anomaly value, it only reached about half the maximum anomaly value which was measured there.

In summary, it can be stated that the geothermal data within the range of the Rhine Graben anomaly are not in contradiction with the seismic data. By geothermal methods both a granitic "anomalous body" and a low velocity channel can be found there.

2.5 Compressional and Shear Wave Velocities as a Function of Temperature in Rocks at High Pressure[1]

K. Fielitz

ABSTRACT

The results of wave-velocity measurements in rock samples at pressures in the region of 4 kb and temperatures up to 750°C are reported. The different relative variations of v_p and v_s with temperature in peridotite, eclogite, gabbro, granite, and quartzite are compared, and the velocity decrease in granite at the α/β-quartz inversion temperature is discussed with regard to low-velocity layers in the earth's crust.

ZUSAMMENFASSUNG

Es wird über Ergebnisse von Geschwindigkeitsmessungen an Gesteinsproben bei Drucken nahe 4 kb und Temperaturen bis zu 750°C berichtet. Die relativen Änderungen von v_p und v_s mit der Temperatur im Peridotit, Eklogit, Gabbro, Granit und Quarzit werden miteinander verglichen. Die Geschwindigkeitsabnahme in Granit bei der α/β-Quarz-Inversionstemperatur wird im Hinblick auf Geschwindigkeitsinversionen in der Erdkruste diskutiert.

2.5.1 Introduction

For a more detailed interpretation of the variation of seismic velocities with depth, the effect of temperature and pressure on the elasticity of rocks has to be taken into consideration.

The influence of pressure on sound velocities in rock specimens at room temperature has already been studied rather intensively (e.g. BIRCH, 1960, 1961; SIMMONS, 1964a, b), but data on the influence of temperature at high pressure are scanty.

HUGHES and MAURETTE (1957a, b) measured compressional and shear wave velocities in some rocks at temperatures up to 300°C and at variable pressures up to 6 kb.

Their results were questioned by BIRCH (1958), who among others pointed out that velocity variations observed at low pressures and high temperatures are partly irreversible in general because of the formation of microcracks, and are not representative for the conditions in the earth's crust. He showed that more representative reversible changes of velocities can be observed, if the temperature is varied at a constant pressure of some kilobars.

BIRCH (1943, 1958) measured shear wave velocities as a function of temperature at high pressure, but no compressional wave velocities. There is no evident relation between his results on the relative change of shear wave velocities with temperature and the mineral content of the rock specimens used.

In the following, some new measurements of compressional and shear wave velocities in rocks at temperatures of up to 750°C and pressures of 2 to 4.5 kb are reported.

2.5.2 Apparatus

For pressure generation a triaxial press was used, the six pistons of which acted upon the plane surfaces of a cube-shaped rock specimen with dimensions of 40 × 40 × 40 mm^3.

The contacting surfaces of the pistons and the cube were ground, polished and coated with a thin film of a high temperature lubricant to reduce friction.

With equal press forces applied in the three perpendicular directions, approximate hydrostatic pressure conditions could be generated.

The variations of specimen length were measured by means of displacement transducers. Heating coils wound around the pistons near to the specimen and thermoelements within the pistons

[1] The publication is an excerpt of a dissertation completed at the Institut für Geophysik of the Technische Universität Clausthal.

served to control and measure the temperature. The opposite end of each piston was water-cooled.

For the determination of elastic wave velocities, ultrasonic pulses were generated by a piezoelectric transducer fixed to the cooled end of one piston, and were transmitted to another transducer fixed to the opposite piston.

As the travel time of such a pulse includes time delays due to the path through the specimen plus the two pistons, the travel time belonging to the path through the pistons alone had to be determined separately by reflection measurements, and to be subtracted from the travel time of the transmitted pulse. The phase shifts produced by reflection and by the transducer-piston bond were determined by calibration measurements, during which the specimen was removed.

By the use of different transducers vibrating in thickness and shear modes, both compressional and shear wave travel times could be measured alternatively.

The rock specimens were subjected to a constant pressure of approximately 4 kb in general, while the temperature was varied step-wise within the range of $50^{\circ}C$ to $750^{\circ}C$.

2.5.3 Results

Fig. 1 shows the variations of relative compressional wave velocities v_p/v^o_p and relative shear wave velocities v_s/v^o_s with temperature in peridotite at 4.1 kb. For v^o_p and v^o_s the velocities at $100^{\circ}C$ were chosen for experimental reasons. Different symbols stand for measurements with increasing or decreasing temperature respectively. The short vertical lines indicate the accuracy of velocity measurements.

The relative change of v_s is slightly greater than the relative change of v_p. The variations of both are close to those calculated from forsterite single crystal data (GRAHAM and BARSCH, 1969) by the Voigt-Reuss-Hill method. As GRAHAM and BARSCH did not

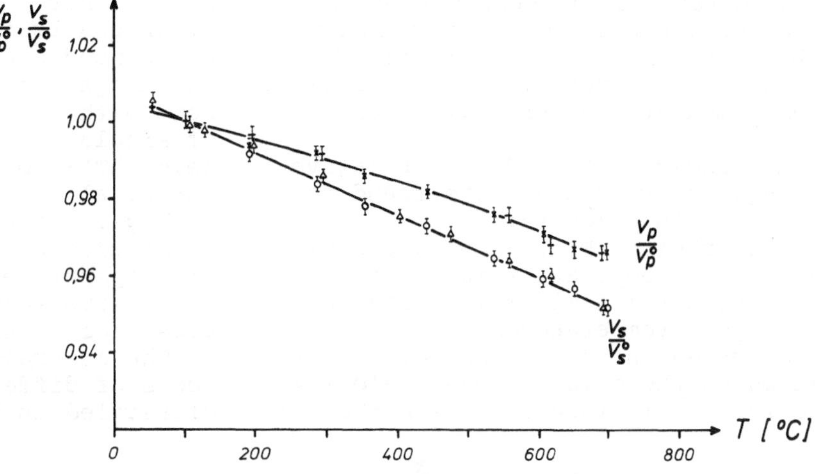

Fig. 1. Relative compressional wave velocity v_p/v_p^o and relative shear-wave velocity v_s/v_s^o as a function of temperature in peridotite at 4.1 kb. $v_p^o = v_p(100^{\circ}C) = 7.8$ km/s, $v_s^o = v_s(100^{\circ}C) = 4.5$ km/s. Circles and diagonal crosses mark values measured at rising temperature, triangles and vertical crosses those measured at falling temperature

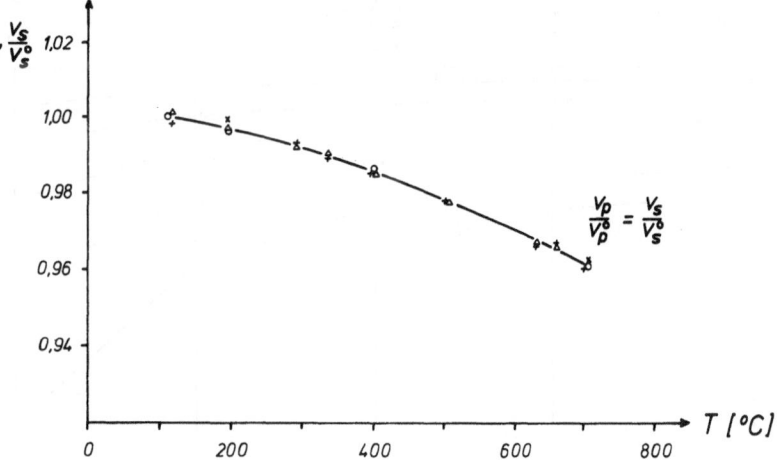

Fig. 2. Relative compressional wave velocity v_p/v_p^o and relative shear-wave velocity v_s/v_s^o as a function of temperature in eclogite at 4.1 kb. $v_p^o = v_p(100^{\circ}C) = 7.9$ km/s, $v_s^o = v_s(100^{\circ}C) = 4.6$ km/s

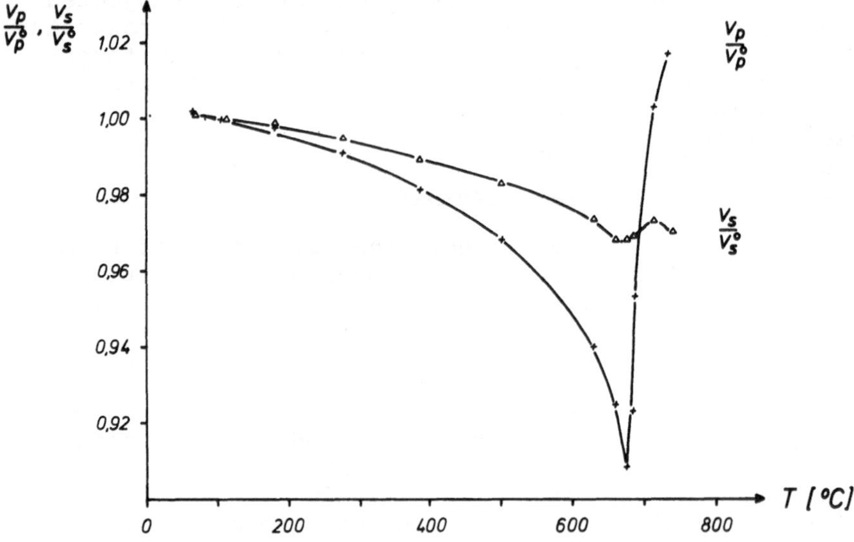

Fig. 3. Relative compressional wave velocity v_p/v_p^o and relative shear-wave velocity v_s/v_s^o as a function of temperature in granite at 4.2 kb. $v_p^o = v_p(100^oC) = 6.1$ km/s, $v_s^o = v_s(100^oC) = 3.5$ km/s

use high pressures, it may be concluded that at least in this case for ideally close-packed aggregates the influence of temperature on elasticity can be taken as independent of pressure to a first degree of approximation. The same conclusion could be drawn by a comparison of measurements performed at pressures of 4 kb and 2 kb.

As shown in Fig. 2, the relative changes of v_p and v_s were found to be equal in eclogite.

In granite (Fig. 3) v_p decreases at rising temperature more sharply than v_s until a minimum is reached. At higher temperatures it shows a pronounced increase. The temperature at which the compressional velocity is lowest, is identical with the α-β inversion temperature of quartz. In general this temperature is close to 573°C at a pressure of 1 bar, and increases with pressure by about 26°C per kilobar.

In quartzite similar but even more pronounced changes of v_p were observed, which again showed rather good agreement with Voigt-Reuss-Hill averages of single crystal data for quartz (e.g. ATANASOFF and HART, 1941; KAMMER et al., 1948).

Figs. 4 and 5 show a compilation of results concerning the influence of temperature on compressional and shear wave velocities in rocks at pressures close to 4 kb.

The diagrams contain data of 17 rocks of different origin, which are classified in 5 groups. Each point

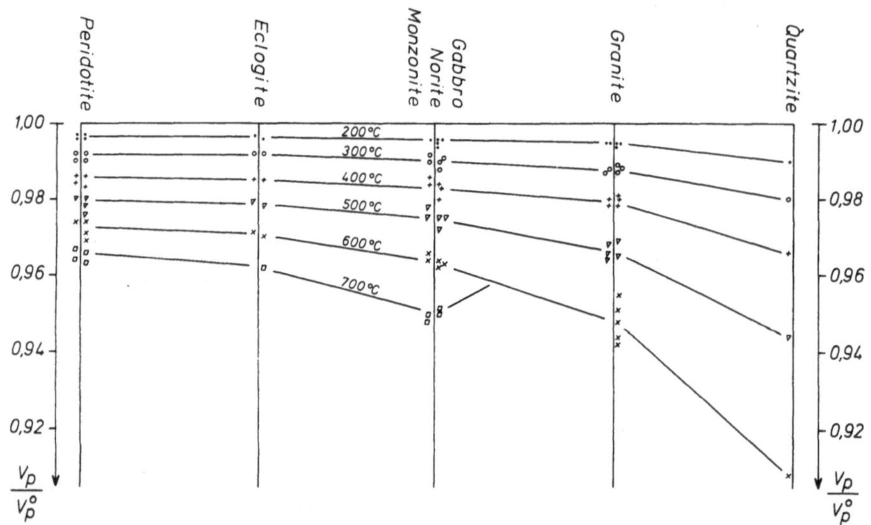

Fig. 4. Compilation of results concerning the relative change of the compressional wave velocity v_p with temperature in different samples of the indicated types of rocks at pressures of about 4 kb. $v_p^o = v_p(100^oC)$

Fig. 5. Compilation of results concerning the relative change of the shear-wave velocity v_s with temperature in different samples of the indicated types of rocks at pressures of about 4 kb. $v_s^o = v_s$ (100°C)

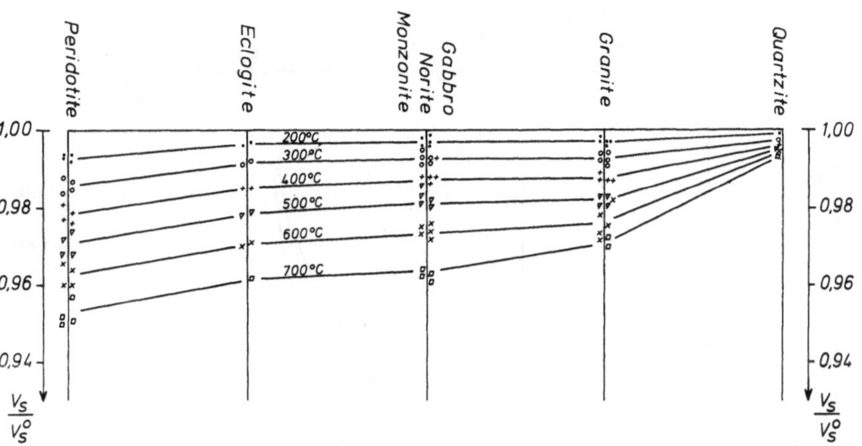

symbol represents relative velocities measured in the different types of rocks at the same temperature.

Although there are differences in mineral content and structure of the rocks collected in one group, the scattering of points is not too large, and a common tendency may be observed. There is an increasing influence of temperature on v_p and a decreasing influence on v_s, if the types of rocks are listed in the sequence peridotite, eclogite, gabbro, granite, quartzite, as is done in the diagram.

Taking the main mineral content of the rock samples into consideration this arrangement corresponds with the common sequence of silicates starting with orthosilicates and ending with framework silicates and quartz.

The difference in the behavior of v_p and v_s in relation to temperature is noteworthy. It implies that Poisson's ratio σ is not independent of temperature. With rising temperature σ increases slightly in peridotite and decreases more strongly in granite and quartzite below the α-β quartz inversion temperature.

2.5.4 Geophysical Implications

In several seismological investigations of the earth's crust (e.g. MÜLLER and LANDISMAN, 1966; GIESE, 1968a) a low-velocity layer characterized by a decrease of velocities of up to 10% was found in a range of depth of 8 to 20 km. It has been supposed by different authors that this velocity reversal is caused by the influence of temperature on seismic velocities in solid rocks or by the effect of partial melting.

The results shown above may serve for discussing this question more quantitatively. As dry rock samples were used for the measurements, the data are suitable for describing the conditions in homogeneous layers of rocks, which do not contain compressed water. In the temperature range up to 750°C this means that the rocks are not affected by partial melting. Therefore, if for instance a homogeneous layer of the composition of an average granite is assumed for the range of depth of 8-20 km, it may be concluded from the results of the measurements, whether the influence of temperature on seismic velocities in the solid material is a sufficient explanation for the velocity reversal, or whether an additional effect of partial melting has to be taken into consideration.

For such an estimation the influence of both temperature and pressure on the velocities can be determined approximately by adding the relative change with pressure at 20°C (e.g. BIRCH, 1960) and the relative change with temperature at 4 kb given in Figs. 3, 4 and 5. The shift of the α-β quartz inversion temperature by 26°C per kb can be taken into account by a small correction.

Fig. 6 shows the variation of v_p and v_s with depth for granite determined in this way. A pressure gradient of 0.26 kb/km and two different temperature distributions marked (1) and (2) according to LUBIMOVA (1967) and BORCHERT (1960) respectively were assumed for the calculation. The variations of v_p and v_s are different mainly as a consequence of the strong influence of temperature on Poisson's ratio in quartzite. In the range 10 to 20 km a maximal decrease of 2.5% for v_p and of 0.5% for v_s result from the temperature distribution (2).

In a similar estimation BIRCH (1958) did not obtain a velocity reversal for v_p. There are mainly two reasons for this: first, experimental

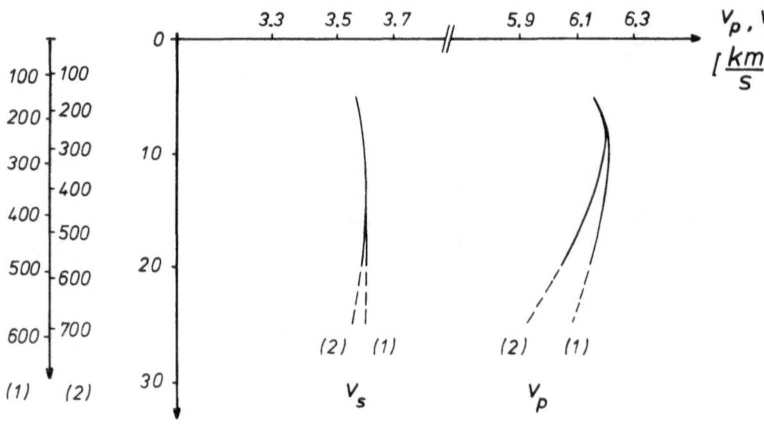

Fig. 6. Distribution of v_p and v_s in an assumed layer of homogeneous dry granite in a range of depth of 5 to 25 km, calculated from laboratory data. The curves marked (*1*) and (*2*) result from the temperature distributions (*1*) and (*2*) respectively

data for the relative change of v_s with temperature were used to calculate changes of v_p, supposing Poisson's ratio to be independent of temperature; secondly, a smaller temperature gradient was assumed.

From the data given above, a low-velocity layer with a mean compressional wave velocity about 1.2% lower than the velocity on top of the layer can be constructed. In the seismological investigations cited, more pronounced low velocity layers were found. Therefore, in these cases the assumption of temperature distribution (2) with no fluid components present is not sufficient to be in correspondence with the seismic observations.

Even if higher temperatures are assumed, a compressional wave velocity reduced by more than 4% within a range of depth of more than 5 km seems to be explainable only by the existence of compressed water in pore spaces and possibly the additional effect of partial melting.

Results of seismological investigations (ZSCHAU and KOSCHYK, 6.10 of this volume) showing a small increase of Poisson's ratio in the low velocity layer may also suggest the existence of a fluid component in addition to thermal effects, for in dry granitic material (see Fig. 6) Poisson's ratio should decrease with increasing temperature.

Acknowledgments. This study has been sponsored by the Deutsche Forschungsgemeinschaft. I am indebted to Prof. Dr. O. ROSENBACH for his encouragement and help during the investigation.

2.6 Some Petrophysical Aspects of Deep Structure

A. SCHULT

ABSTRACT

Combining laboratory measurements on olivines or other "mantle materials" with resistivity profiles of the earth's interior, one can obtain information on temperature distribution in the earth, as temperature is one of the main factors controlling resistivity. There are indications that the temperature is relatively low in the upper mantle. Because of the large scatter of the results the method is subjected to some uncertainty at the present stage.

ZUSAMMENFASSUNG

Der Vergleich der elektrischen Leitfähigkeit von Olivin und anderen "Mantel-Materialien" mit der Leitfähigkeit im Erdinnern gibt Hinweise auf die Temperaturverteilung in der Erde. Viele Ergebnisse deuten auf eine verhältnismäßig geringe Temperatur im oberen Mantel. Allerdings ist die Methode wegen der großen Streuung bei den Labormessungen mit einiger Unsicherheit behaftet.

Combined information from seismic properties of the earth with laboratory measurements on rocks and minerals has provided the greatest part of the information of crustal and upper mantle structure. The study of the electrical conductivity of minerals and rocks, together with the conductivity distribution in the earth by magnetotelluric methods can yield information on the temperature distribution, a variable not clearly determined until now.

Considerable experimental data exist on the electrical conductivity of minerals and rocks under pressure and temperature conditions of the upper mantle for plausible materials. The temperature dependence of the electrical conductivity of minerals and rocks can be described in the relevant range by

$$\sigma = \sum_i \sigma_i \exp(-E_i/kT).$$

E_i are activation energies, T is the absolute temperature, k Boltzmann's constant and σ_i constants. In most cases i takes three values or less depending on the temperature range investigated.

The parameters σ_i and E_i are in most cases pressure-dependent but relatively weakly. Thus the conductivity is mainly temperature-controlled. Fig. 1 summarizes conductivity data for the olivine system (solid solution between forsterite Mg_2SiO_4 and fayalite Fe_2SiO_4), Fig. 2 for some other minerals, and Fig. 3 for some rocks with pressure as parameter. In nearly all cases pressure tends to increase the conductivity in the temperature ranges shown. When the pressure is raised from near atmospheric to 50 Kbar for instance, the conductivity may be increased by an order of magnitude or less. If the pressure becomes essentially higher than 50 kbar its effect on the conductivity is not known. There are indications that the conductivity may decrease in this range (SCHOBER, 1971).

The important factors controlling the conductivity of the olivine system can be seen in Fig. 1. The single-crystal forsterite (very pure, synthetic) is the most resistive at any temperature. The conductivity is largely increased by polycrystallinity (see Fig. 1). SHANKLAND (1969) attributed this increase to the grain boundaries which can be both a source of carriers and a relatively high mobility channel for the movement of ions. Single and polycrystalline samples of MgO show the same behavior (see Fig. 2). For natural olivines, however, SCHOBER (1971) found practically no difference in conductivity for single and polycrystalline samples.

Recently it was shown that the oxidation state of iron in olivine and also pyroxene is very important (e.g. DUBA and NICHOLLS, 1973). Single crystals of natural olivine (about 10% fayalite) with no Fe^{3+} content can have very low conductivities similar

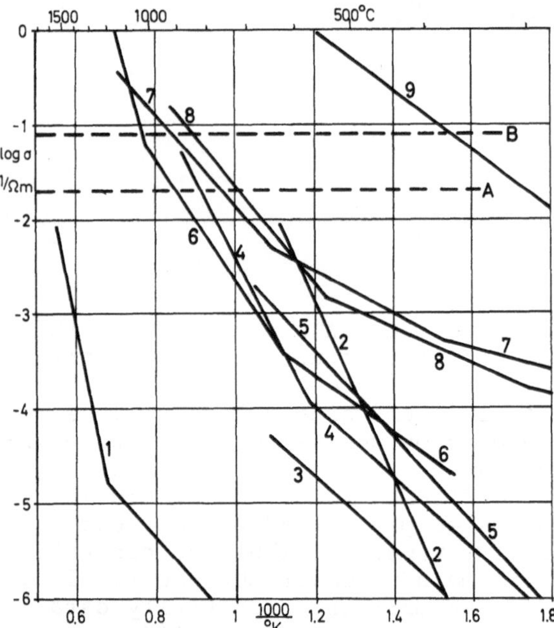

Fig. 1. The electrical conductivity for the olivine system.
Curve *1* forsterite single-crystal, synthetic, 0 kbar (SHANKLAND, 1969). *2* forsterite, poly-crystal (as all following samples) synthetic, 23 kbar (BRADLEY et al., 1964). *3* olivine 10.4% fayalite, natural, 22 kbar (HAMILTON, 1965). *4* olivine 8.6% fayalite, natural, 24 kbar (SCHULT and SCHOBER, 1969). *5* olivine, 10% fayalite, synthetic, 23 kbar (BRADLEY et al., 1964). *6* olivine, natural, 0 kbar (from PARKHOMENKO, 1967). *7* olivine, 8,6% fayalite, natural 35 kbar (SCHOBER, 1971). *8* olivine, 11% fayalite natural, single crystal, 30 kbar (SCHOBER, 1971). *9* fayalite, synthetic, 23 kbar (BRADLEY et al., 1964). Curve *A* global conductivity in about 100 km depth (SCHMUCKER, 1970). Curve *B* upper limit of conductivity in the depth range 0-400 km (BANKS, 1969)

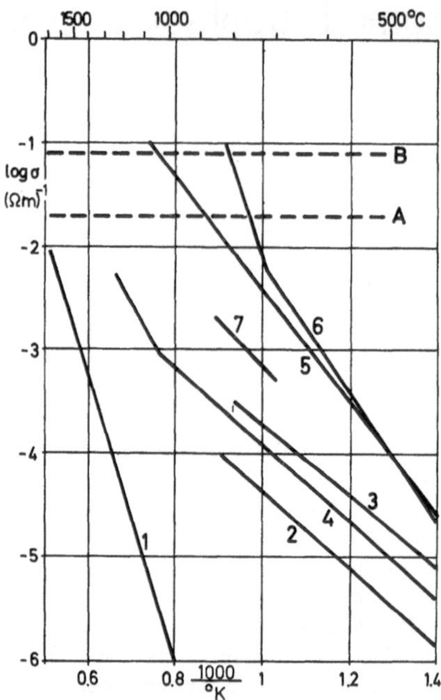

Fig. 2. The electical conductivity of several minerals. *1* MgO, single-crystal synthetic, 0 kbar (MITOFF, 1964). *2* MgO, poly-crystal as all following samples, synthetic 0 kbar (HAMILTON, 1965). *3* MgO poly-crystal, synthetic, 42 kbar (HAMILTON, 1965). *4* enstatite, natural, 0 kbar (UHRI, 1961). *5* granat, natural, 0 kbar (UHRI, 1961). *6* granat, natural, 20 kbar (STAUDACHER, 1968). *7* enstatite, natural, 48 kbar (STAUDACHER, 1968)

as given by curve 1 in Fig. 1. The samples in Fig. 1 represented by curve 2-8 have a small Fe^{3+} content (so far determined) and have therefore a high oxidation state (SCHULT, 1974). The oxidation state is controlled by oxygen fugacity which is not known in the Upper Mantle. It has been mentioned however by NITSAN (1973) that the oxidation state in olivines from peridotite bombs and other presumed samples from the Upper Mantle is usually high. This would be consistent with a high oxygen fugacity in the Upper Mantle.

The conductivity mechanism in olivines is probably an electronic one. By the aid of measurements of the Hall mobility and the thermoelectric power, SCHOBER (1970) concluded that the conduction can well be described by the electronic "hopping mechanism" for temperatures up to 1200°C. Electronic hopping is due to electron transfer after the relation $Fe^{3+} + e \rightleftharpoons Fe^{2+}$. The changes in the slopes in the log σ versus $^1/T$-diagram indicate, however, that the conductivity mechanisms are different in certain temperature ranges up to 1200°C (see Fig. 1). In the higher temperature ranges (> 1400°C) ionic conductivity probably dominates (e.g. PLUSCHKELL and ENGELL, 1968; SCHANKLAND, 1969).

The scatter of the results is large but it seems to be smaller in the temperature range above 800°C and some useful information about temperature can be drawn from the data. Magnetotelluric investigations yield a conductivity of about $1 \cdot 10^{-2}/\Omega m$ in 100 km depth for the global distribution (e.g. SCHMUCKER, 1970). The conductivity does not increase very much down to about 400 km depth. An upper limit for the conductivity down to 400 km is $8 \cdot 10^{-2}/\Omega m$ after BANKS (1969).

By a temperature calculation based on results for olivines with approximate 10% fayalite content one gets about 800°C for 100 km depth (Fig. 1). The upper limit down to 400 km depth lies between 900°C and 1000°C. The corresponding (average) values for eclogites, peridotites, gabbros, or basalts are 850°C and 1100°C (Fig. 3). Similar values one gets for several other - not olivine - minerals (Fig. 2). Temperatures of about 1600°C in 250 km depth often proposed (e.g. GREEN and RINGWOOD, 1967) would yield only for very high resistive materials, such as olivines in a low oxidation state (no Fe^{3+} content) the above conductivity values in the upper mantle (see Figs. 1 and 2). In a recent paper TOZER (1970) showed that there are also indications from models of a thermally convecting earth that the temperature may generally be below 1000°C down to depths of several hundred kilometers. Convection in the upper mantle (heat conduction essentially takes place by mass transport) implies a nearly adiabatic thermal gradient (about 0.5°C/km) consistent with the weak increase of the electrical conductivity below 100 km depth down to depths between 400 and 600 km.

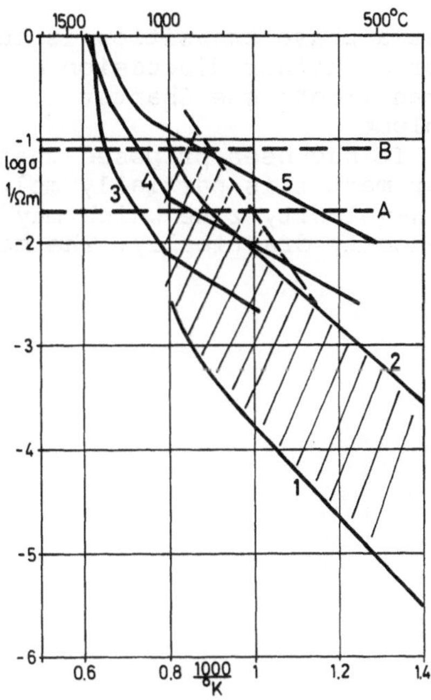

Fig. 3. The electrical conductivity of several rocks. Range between curves *1* and *2* peridotite, eclogite, gabbro, various basalts from various sources, 0 kbar (from PARKHOMENKO, 1967). *3* Al-rich tholeiite, *4* quartz tholeiite, *5* olivine tholeiite, 28 kbar (KHITAROV et al., 1970)

Many regional anomalies of greater conductivities than the mean value in shallower depth have been observed. The anomalies indicate temperatures in the uppermost mantle which are several hundred degrees greater than the mean. An example is the conductivity anomaly beneath the Rhine Graben of about $2 \cdot 10^{-1}/\Omega m$ in about 30-40 km depth (HAAK, 1970) yielding a temperature of 1100°C or more in that depth. This may be an explanation for the relatively high temperatures one can deduce locally from volcanism.

The increase of conductivity between 400 and 600 km to about $1/\Omega m$ is often attributed to phase transition of olivine to a spinel-like phase or to a post-spinel phase. This increase in conductivity by the phase transition has not been verified experimentally until now for olivines with low fayalite content.

Several attempts have been made to calculate temperature gradients and the temperature from seismic data. As the velocity of seismic waves increases with increasing pressure and decreases with increasing temperature in the low-velocity zone in the upper mantle or in the low-velocity channel in the crust, a high temperature gradient can cancel out the effect of pressure thus producing negative velocity gradients. The critical temperature gradient is in the order of 6 to 10°C/km for P-waves and 2 to 4°C/km for shear waves (ANDERSON et al., 1968). As the velocity decreases in the low-velocity layer larger temperature gradients than the critical gradients are required, thus yielding relatively high temperature gradients. Recent calculations (e.g. BIRCH, 1970; ANDERSON and SAMMIS, 1970) gave temperatures of at least 1500°C in 100 km depth, even taking into account a possible change in mineralogy.

This temperature is much too high in this depth and not consistent with temperatures as derived above. (There are several other objections against high-temperature gradients in the low-velocity zone as discussed by ANDERSON and SAMMIS (1970).) Also the velocity inversion in the crust (low-velocity channel) can hardly be attributed to high-temperature gradient alone. As the P-velocity is decreased in the order of 5% the temperature increase should be about 1000°C (see e.g. BIRCH, 1970), if not a phase transition is considered. For a further discussion of phase transitions see Chapter 2.5 of this volume.

It has been proposed that the upper mantle is partially molten in the low-velocity zone which may drop the velocity drastically. The melting point of dry "mantle-material" basic rocks varies from about 1300° to 1500°C in 100 km depth (quoted in ANDERSON and SAMMIS, 1970). It is known that the melting point is lowered several hundred degrees by a small water content under upper mantle pressures.

2.7 Gravity

S. PLAUMANN

ABSTRACT

A 1 : 1,000,000 map of Bouguer gravity anomalies of western Germany was published by GERKE (1957) and in a smaller scale by GERKE and WATERMANN (1960). For the purposes of deep-seismic sounding only the regional anomalies are of interest. This map reflects the areas covered by young sediments as well as regions showing a decrease in thickness of the upper crust.

ZUSAMMENFASSUNG

Eine Karte der Bouguerschwere-Anomalien von Westdeutschland wurde von GERKE (1957) im Maßstab 1 : 1.000.000 und in kleinerem Maßstab von GERKE und WATERMANN (1960) publiziert. Für Zwecke der seismischen Tiefensondierung sind lediglich die großräumigen Anomalien von Interesse. Diese Karte spiegelt sowohl die Areale junger Sedimente wider als auch Regionen mit einer dünneren Oberkruste.

This chapter is intended to help the reader of the articles on deep-seismic sounding to relate them to the gravity field of the Federal Republic of Germany. For this reason, instead of a detailed map, a small-scale Bouguer-anomaly map is presented which does not confuse by short wave local anomalies but allows quick comprehension of the general features. It is greatly appreciated that the publishers of the Institute of Applied Geodesy, Frankfurt, gave permission to reproduce the 1 : 4,000,000 gravity map (Fig. 1) of GERKE and WATERMANN (1960). This map was produced by generalizing the isogam contours of the 1 : 1,000,000 Bouguer gravity map of West Germany (GERKE , 1957), which in turn represents the state of the regional gravity survey of the Federal Republic of Germany at that time. It was primarily the result of the "territorial geophysical reconnaissance".

The Territorial Geophysical Reconnaissance was initiated by governmental resolution in 1934; up to 1945 most parts of West Germany had been covered. Thyssen gravimeters were first used and later Askania gravimeters. Work stopped in 1945 but began again in the fifties with the establishment of a new gravity base net. This net allowed a better connection of all local and regional surveys to the Potsdam datum than was possible with the older pendulum base net. The gaps left by the territorial geophysical reconnaissance can be recognized in Fig. 1 by the dashed isogam lines. Here only the new gravity base net stations were available at the time of preparation of the 1 : 1,000,000 gravity map by GERKE (1957). Subsequently the Geological Survey of Lower Saxony (Niedersachsen) in Hannover undertook completion of the regional gravity survey of West Germany. In a few years all gaps will be filled and a new series of maps will be published.

To a certain extent gravity anomalies provide an independent control on the results derived from seismic experiments. In several places qualitative agreement will be found, such as gravity highs above relatively thin crust and gravity lows above greater crustal thicknesses. But sometimes quantitative analysis lead to apparent contradictions. Density variations in the upper mantle have been postulated to overcome those difficulties. A large percentage of the anomalies are of course related to structures in the top few km of the earth's crust; they generally produce the high frequency portion of the gravity field, though not exclusively. Likewise the extended province of negative Bouguer anomalies in the northwest German lowland is evidently caused by the thick sequence of young, unconsolidated sediments. Similar conditions hold for the Upper Rhine Graben, the gravity anomaly of which (-40 to -50 mgals) can be explained mostly by the Tertiary graben fill

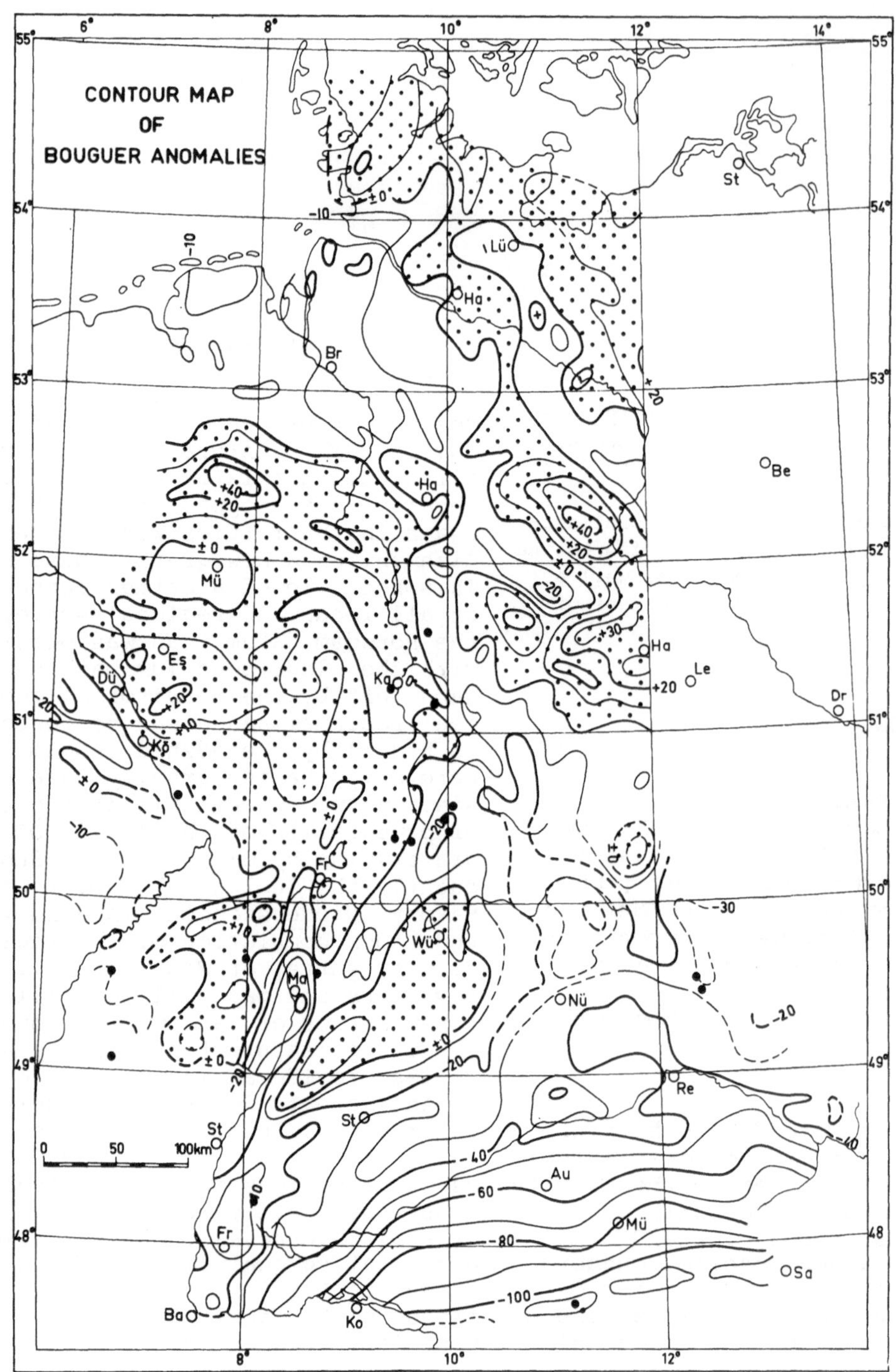

Fig. 1. Map of western Germany showing Bouguer anomalies. (After GERKE and WATERMANN, 1960)

(MUELLER et al., 1967; CLOSS and PLAUMANN, 1968). On the other hand, the most striking feature of the gravity map, the broad, pronounced decrease of the anomaly values just north of the Alps, is clearly the consequence of crustal thickening in the area of the Alpine orogeny. In the same way the general increase of Bouguer anomaly values from west to east in the Rhenish Massif seems to have its origin in decreasing thickness of the upper crust. Deep seismic soundings (GIESE and STEIN, 1971) revealed an 8 km rise of the crust-mantle boundary from SW to NE in this area.

It is not intended that each satisfactorily interpreted anomaly be discussed in detail. However, at least one of the prominent anomalies in the Bouguer anomaly maps of West

Germany is worth mentioning, namely the Bramsche anomaly situated at approximately 52° 30' N and 7° 40' E. It has an amplitude of more than +40 mgals and corresponds with a magnetic anomaly. It is attributed to a body of higher density and higher magnetization resting upon the basement which in that region should be at a depth of 10 km or more (HAHN and KIND, 1971).

After greater parts of Germany had been covered by the Territorial Geophysical Reconnaissance, several scientists made studies of relations between gravity and known or hypothetical subsurface structures. The articles of CLOSS (1944) and v. ZWERGER (1948) may be mentioned in this respect. At that time as well as at present the large variety of density contrasts encountered gives rise to difficulties in analysing the relationships. However, today it is possible to add to the solution "from above", which starts with the known superficial structural elements, a solution "from below" using deep seismic sounding, so that future interpreters may be able to shed more light on the medium depth range of the earth's crust.

3. Basic Data of the Deep-Seismic Sounding Observations

P. GIESE and C. PRODEHL

This section contains the seismic-refraction data obtained since 1958 within the scope of joint programs of the geophysical institutions of the Federal Republic of Germany. As mentioned in the introduction, one of the most important aims of this monograph is to publish all data in such a form that they can serve as basic material for future detailed interpretation.

It further includes contributions on properties of seismic signals observed from quarry blasts.

In Chapter 3.1, the shotpoint data are described and collected. BURKHARDT and VEES (Chap. 3.2) discuss some characteristic dynamic properties of signals derived from measurements carried out next to the source as well as from recordings at greater distances with regard to the reproducibility of the signal source, the question whether a scaling law can be defined, the frequency contents or the pulse shape of the signal etc. The next Chapter (3.3) contains technical details on profiles, recording points and instrumentation. Furtheron all references are listed corresponding to each profile. All available record sections are presented in a unique time-scale as appendix to 3.4. The magnetic-tape recording system used since 1966 is described in 3.5. Some technical data are also included. A special investigation of seismic observations of quarry blasts at distances from 300 to 670 km is described by BURKHARDT et al. (3.6). The authors discuss thoroughly a number of conditions under which quarry blasts can also be used satisfactorily for long-distance observations, aiming for upper-mantle investigations. Finally, some general remarks on the cooperation between the geophysical institutes, the geological survey and the oil industry are given.

3.1 Shotpoint Data of Quarry Blasts 1958-1972

A. STEIN and H. SCHRÖDER

ABSTRACT

This section gives information on the used quarry blasts as to geology, coordinates, and shot-time.

ZUSAMMENFASSUNG

Dieser Abschnitt enthält Informationen über die benutzten Steinbruchsprengungen, wie Geologie, Koordinaten und Schußzeit.

3.1.1 Introduction

Commercial quarry blasts offer a convenient and economical energy source for seismic crustal and upper mantle investigations. In the twenties, shortly after the development of portable seismic-refraction equipment, initiated by E. WIECHERT, the geophysical institute of Göttingen started a research program using such commercial quarry blasts (e.g. WIECHERT, 1926, 1929; SCHWEYDAR and REICH, 1927; BROCKAMP and WOELCKEN, 1929; BROCKAMP, 1931; for details see also SCHULZE, 1974). After the second world war, quarry blasts were used in increasing numbers in different parts of the world for deep-seismic sounding investigations (e.g. REINHARDT, 1954; STEINHART and MEYER, 1961; PAPKE, 1967).

The data and the results presented in this monograph are based mainly on a systematic use of all major and suitable quarry blasts in the Federal Republic of Germany from 1958 to 1972.

The quarry blasts used for this research program were mainly arranged as explosions in T-shaped galleries, only in a few cases were large drill-hole shots observed. The latter are generally fired with delays in the order of some ten ms. This delay technique reduces the suitability for seismic purposes. As the number of explosions in galleries has decreased within the last five years in favor of borehole shots, the chances of long-range seismic refraction observation with the aid of quarry blasts have been drastically reduced.

Details on the shooting technique in quarries are given, e.g. by BOMMERT (1953), KOCHANOWSKY (1955), and LAMPE (1959).

3.1.2 Determination of Shot-Data

Shotpoint coordinates cannot be determined with an accuracy better than 10-20 m because of the linear distribution of charges along some 10 m. The coordinates shown in Tables 1 and 2 refer to the center of the array.

Table 1. List of the quarries used for deep-seismic sounding

Q	LOCALITY	SMAP
01	ESCHENLOHE	8332
02	HILDERS	5426
03	GERSFELD/SCHWARZER ACKER	5525
04	GROSSENRITTE	4722
05	BIRKENAU	6418
06	ADELEBSEN	4324
07	GERSFELD/NALLENBERG	5525
08	KIRCHHEIMBOLANDEN	6313
09	BOEHMISCHBRUCK	6440
10	VOGGENDORF	6540
11	BISCHOFSHEIM	5526
12	ROMSTHAL	5622
13	DORHEIM	5021
14	MEHRBERG	5309
15	BUEDINGEN	5621
16	TABEN-RODT	6405
17	BRANSRODE	4725
18	LAHR	5414
19	MERLENBACH	6806
20	BIRRESBORN	5805
21	BERMEL	5708
22	DORNDORF	5414
23	WILSENROTH	5414
24	STEINACH	7714
25	VILS	8429
26	LOHNE	4821
27	MAUTHAUS KRONACH	5634

The resulting position is marked on a topographic map with a scale of 1 : 25,000. The positions are given in the rectangular Gauß-Krüger system as well as in geographic coordinates. The obtained accuracy of ± 10 m is comparable to accuracy obtained for the recording points.

The blasts are fired electrically except in the quarry Bransrode (17 in Table 1). The shot-time is usually recorded by electromagnetic induction from the current which fires the charge. In addition, a short seismic refraction line is installed some 10 m away from the shooting area. The difference between both time-break determinations is within 5 ms. The shot-times in Table 2 are rounded off to hundredths of a second.

The time-signal at the shotpoints is recorded in a similar way as at the field stations (see 3.2).

3.1.3 Tables of Shot-Data

The shot-data for the period from 1958 to 1972 are compiled in Tables 1 and 2. Table 1 starts with a short list of the quarries, giving code (Q) name of locality and code number of the corresponding topographic map 1 : 25,000 (SMAP).

This map system is divided by geographic coordinates, each map extending over 6' in latitude and 10' in longitude. The map number consists of four digits and refers to the geographic coordinates of the upper left corner. The first two digits multiplied by 6' give a latitude in minutes referring to the "zero-latitude" $\phi = 54^o$, being positive to the south. The last two digits multiplied by 10' give a longitude in minutes, referring to the "zero-longitude" $\lambda = 6^o40'$, being positive to the east. This system facilitates the finding of the corresponding map if the geographic coordinates are given.

The following example explains the procedure. The assumed point shall have the coordinates:
$\lambda = 11^o24'$ and $\phi = 50^o40'$.
In this case, the map number can be found as follows:
$54^o - 50^o40' = 3^o20' = 200' = 33{,}3 \cdot 6'$
$11^o24' - 6^o40' = 4^o44' = 284' = 28{,}4 \cdot 10$
The resulting code is 3328.

The column B of the main list (Table 2) shows the current number of the seismically used explosions in the corresponding quarry.

Shot-times are presented in hours of Central European Time (MEZ), minutes, seconds, and hundredths of a second.

The following column (TO) gives the total amount of explosives in to (= 1000 kg).

Geographic coordinates are listed in the columns LONGIT (longitude) and LATIT (latitude). Here, the respective first two digits stand for degrees, and the following five digits, including the decimal point, for minutes with an accuracy of hundredths of a minute.

The next group of columns shows the rectangular Gauß-Krüger coordinates Y and X, given here in kilometers. The column ALT indicates the altitude in meters above sea level.

Table 2. Shotpoint data of quarry blasts 1958-1972

Q B	DATUM	SHOT-TIME H M	S	TO	COORDINATES LONGIT.	LATIT.	Y(LONG)	X(LAT)	ALT
0101	15.FEB 58	11.11	01.63	12.0	1109.46	4737.92	4436.70	5277.33	
02	15.FEB 58	11.13	01.76	12.5	1108.91	4737.89	4436.01	5277.26	
03	27.JUN 59	11.13	29.61	16.0	1108.94	4737.93	4436.05	5277.33	
04	17.OCT 59	11.13	30.26	8.8	1109.88	4737.99	4436.60	5277.46	
05	17.OCT 59	11.18	00.	5.6	1108.94	4737.91	4436.06	5277.32	
06	30.JAN 60	11.13	30.29	8.3	1108.80	4737.89	4435.90	5277.29	
07	30.APR 60	11.13	29.36	11.2	1108.80	4737.89	4435.90	5277.29	
08	02.JUL 60	11.13	31.76	7.3	1109.47	4738.00	4436.73	5277.48	
09	03.SEP 60	11.13	30.05	7.2	1108.80	4737.89	4435.90	5277.29	
10	19.NOV 60	11.13	16.53	12.5	1108.85	4737.91	4435.95	5277.31	
11	19.NOV 60	11.14	31.49	7.6	1109.42	4738.00	4436.65	5277.48	
12	04.MAR 61	11.13	13.69	9.7	1108.78	4737.89	4435.87	5277.28	
13	04.MAR 61	11.14	29.02	6.9	1109.42	4738.00	4436.65	5277.48	
14	10.MAY 61	17.56	30.28	9.2	1108.81	4737.87	4435.89	5277.23	
15	24.NOV 61	16.25	00.44	7.9	1109.50	4738.00	4436.75	5277.48	
16	02.DEC 61	11.13	30.09	19.5	1108.82	4737.88	4435.95	5277.28	

Table 2 (continued)

Q B	DATUM	SHOT-TIME H M	S	TO	COORDINATES LONGIT.	LATIT.	Y(LONG)	X(LAT)	ALT
17	07.APR 62	11.16	30.28	6.0	1109.17	4737.95	4436.35	5277.39	
18	07.APR 62	11.18	00.04	11.0	1109.40	4738.00	4436.63	5277.48	
19	07.JUL 62	11.13	29.33	8.5	1108.78	4737.89	4435.87	5277.28	
20	21.SEP 62	17.10	29.67	13.0	1108.92	4737.92	4436.04	5277.34	
21	21.SEP 62	17.11	58.51	7.0	1109.41	4738.00	4436.64	5277.48	
22	24.MAY 63	18.50	30.10	11.0	1108.95	4737.92	4436.06	5277.33	
23	31.AUG 63	11.13	30.29	18.0	1108.83	4737.90	4435.91	5277.29	
24	06.DEC 63	15.30	49.80	11.0	1108.80	4737.89	4435.89	5277.28	670
25	06.DEC 63	15.32	52.62	9.2	1109.48	4737.99	4436.73	5277.47	650
26	17.APR 64	18.00	07.65	7.6	1109.47	4738.00	4436.73	5277.48	670
27	15.MAY 64	17.59	45.20	14.0	1108.94	4737.92	4436.06	5277.32	670
28	18.SEP 64	17.53	30.19	11.4	1108.73	4737.90	4435.81	5277.31	700
29	05.DEC 64	11.13	30.04	12.8	1108.88	4737.92	4435.98	5277.33	650
30	02.JUL 65	17.58	30.63	14.3	1108.86	4737.91	4435.97	5277.32	700
31	04.DEC 65	11.13	31.35	10.5	1108.85	4737.91	4435.93	5277.30	700
32	27.MAY 66	15.04	59.70	7.0	1108.87	4737.92	4435.98	5277.33	670
33	07.OCT 66	15.05	01.10	12.4	1108.86	4737.91	4435.99	5277.31	700
34	30.JUN 67	15.04	59.19	11.0	1108.83	4737.92	4435.92	5277.31	700
35	06.OCT 67	15.04	59.61	7.0	1108.84	4737.92	4435.96	5277.34	720
36	22.DEC 67	14.05	00.09	16.3	1108.72	4737.90	4435.78	5277.30	700
37	26.APR 68	15.05	00.40	10.2	1108.80	4737.90	4435.88	5277.28	700
38	26.JUL 68	15.05	00.03	9.1	1108.80	4737.90	4435.88	5277.28	700
39	06.DEC 68	15.04	59.38	7.1	1108.88	4737.91	4435.98	5277.33	700
40	14.MAR 69	15.05	00.59	7.7	1108.68	4737.90	4435.74	5277.28	700
41	13.JUN 69	15.05	00.51	5.7	1108.87	4737.92	4435.93	5277.33	700
42	18.JUL 69	15.05	00.59	12.8	1108.80	4737.91	4435.88	5277.31	700
43	24.OCT 69	15.04	59.86	21.7	1108.74	4737.90	4435.82	5277.31	700
44	27.FEB 70	15.05	01.40	8.8	1108.68	4737.91	4435.74	5277.30	700
45	10.JUL 70	15.05	01.13	16.4	1108.76	4737.91	4435.83	5277.31	700
46	06.NOV 70	15.05	00.54	8.1	1108.64	4737.91	4435.73	5277.32	700
47	22.JAN 71	15.04	59.38	6.5	1109.18	4737.97	4436.35	5277.41	700
48	02.JUL 71	15.05	01.16	14.0	1108.81	4737.91	4435.90	5277.31	700
0201	28.MAR 58	16.	.	10.0	1002.40	5032.50	3573.71	5601.03	730
02	28.SEP 58	16.00	.	10.0	1002.43	5032.46	3573.75	5600.95	730
03	11.MAR 60	16.	.	7.6	1002.48	5032.47	3573.71	5600.99	730
04	20.MAY 60	16.	.	8.6	1002.34	5032.46	3573.64	5600.96	730
05	07.DEC 61	15.30	31.02	14.6	1002.45	5032.47	3573.79	5600.98	730
06	16.NOV 62	16.05	36.21	14.0	1002.38	5032.48	3573.71	5600.99	730
07	07.JUN 63	16.05	40.81	9.8	1002.47	5032.45	3573.80	5600.98	730
08	11.OCT 63	16.05	01.34	8.1	1002.34	5032.49	3573.64	5601.02	730
09	16.DEC 63	15.30	10.51	13.0	1002.38	5032.50	3573.68	5601.03	730
10	28.AUG 64	16.30	00.64	16.0	1002.37	5032.49	3573.68	5601.04	750
11	26.MAR 65	16.00	00.70	16.0	1002.42	5032.50	3573.75	5601.06	730
12	07.MAY 65	16.05	01.04	6.8	1002.32	5032.50	3573.62	5601.04	730
13	08.OCT 65	16.30	01.35	18.0	1002.42	5032.52	3573.73	5601.08	730
14	22.APR 66	16.05	00.67	10.8	1002.36	5032.52	3573.66	5601.08	730
15	29.JUL 66	16.05	01.11	14.9	1002.51	5032.50	3573.85	5601.04	730
16	07.OCT 66	16.05	01.02	8.4	1002.41	5032.53	3573.71	5601.09	730
17	14.APR 67	16.05	01.09	12.3	1002.48	5032.54	3573.80	5601.10	730
18	14.JUL 67	16.05	00.93	9.0	1002.28	5032.47	3573.57	5601.00	730
19	25.AUG 67	16.05	00.74	4.1	1002.52	5032.44	3573.85	5600.95	750
20	06.OCT 67	16.05	00.80	11.3	1002.42	5032.52	3573.73	5601.09	730
21	24.NOV 67	16.05	00.65	5.4	1002.53	5032.43	3573.87	5600.93	700
22	22.MAR 68	16.05	00.80	10.0	1002.53	5032.49	3573.86	5601.04	730
23	26.APR 68	16.05	00.74	9.2	1002.39	5032.50	3573.70	5601.06	730
24	23.AUG 68	16.05	00.90	11.8	1002.48	5032.56	3573.79	5601.13	730
25	07.MAR 69	15.05	00.89	10.8	1002.51	5032.51	3573.84	5601.07	730
26	11.APR 69	16.05	00.73	9.4	1002.42	5032.51	3573.74	5601.07	730
27	27.JUN 69	16.05	00.91	12.0	1002.36	5032.50	3573.67	5601.06	730

Table 2 (continued)

Q B	DATUM	SHOT-TIME H M	S	TO	COORDINATES LONGIT.	LATIT.	Y(LONG)	X(LAT)	ALT
28	13.MAR 70	16.05	00.60	21.8	1002.44	5032.55	3573.76	5601.09	730
29	16.JUL 70	16.05	00.74	11.7	1002.33	5032.49	3573.63	5601.04	730
30	02.OCT 70	15.45	00.87	21.4	1002.39	5032.52	3573.70	5601.09	730
31	14.MAY 71	16.05	00.38	8.4	1002.30	5032.48	3573.59	5601.02	730
32	20.OCT 71	16.05	00.54	9.2	1002.32	5032.50	3573.62	5601.06	730
33	14.JUL 72	16.05	00.95	5.5	1002.31	5032.53	3573.61	5601.12	730
34	01.SEP 72	17.05	01.25	12.2	1002.38	5032.56	3573.68	5601.16	730
35	24.NOV 72	16.05	00.83	9.8	1002.28	5032.53	3573.57	5601.11	730
0301	12.JUN 58	16.00	.	4.6	0957.79	5027.78	3568.39	5592.23	760
02	15.MAR 62	16.46	01.85	5.0	0957.72	5027.81	3568.33	5592.28	730
03	18.MAY 63	15.00	00.48	5.4	0957.77	5027.83	3568.35	5592.30	730
0401	27.FEB 59	16.29	.	5.0	0921.93	5114.80	3525.52	5679.01	310
0501	03.APR 59	13.59	.	7.3	0843.11	4933.22	3479.63	5490.67	220
0601	12.JUN 59	17.01	.	8.2	0944.76	5136.44	3551.67	5719.32	350
02	16.JUN 61	17.30	.	9.0	0944.65	5136.53	3551.55	5719.50	250
03	26.JUL 63	16.05	02.68	6.3	0944.74	5136.48	3551.64	5719.41	350
04	26.APR 65	16.15	01.32	6.5	0944.64	5136.55	3551.54	5719.52	350
05	10.JUN 66	15.45	01.08	6.4	0944.62	5136.55	3551.52	5719.53	350
06	23.JUN 67	15.45	00.80	6.0	0944.80	5136.48	3551.72	5719.40	350
07	14.MAR 68	15.45	01.20	5.4	0944.68	5136.53	3551.58	5719.50	350
08	18.SEP 68	15.45	00.82	3.0	0944.67	5136.59	3551.57	5719.60	400
09	03.OCT 68	15.45	01.00	5.9	0944.61	5136.51	3551.51	5719.46	350
10	13.JUN 69	15.45	00.93	6.1	0944.67	5136.55	3551.56	5719.54	350
11	17.JUL 69	15.45	00.36	6.4	0944.80	5136.50	3551.73	5719.44	350
12	22.MAY 70	15.45	00.84	7.5	0944.78	5136.51	3551.70	5719.46	350
13	30.APR 71	15.45	00.53	7.0	0944.82	5136.50	3551.75	5719.44	350
14	11.JUN 71	15.45	00.52	6.2	0944.78	5136.53	3551.70	5719.50	350
15	24.MAR 72	15.45	00.80	8.5	0944.80	5136.53	3551.72	5719.52	350
16	02.JUN 72	15.45	00.72	8.4	0944.77	5136.56	3551.68	5719.55	350
0701	03.JUL 59	16.01	.	15.5	0953.04	5026.69	3562.79	5590.14	720
0801	13.APR 60	16.00	.	2.6	0757.49	4940.54	3424.81	5504.75	310
02	23.SEP 70	15.05	00.56	10.5	0757.59	4940.51	3424.93	5504.67	380
0901	09.JUL 60	11.13	31.69	2.7	1221.34	4934.09	4525.73	5492.33	
02	29.APR 61	11.13	29.27	5.3	1221.32	4934.10	4525.70	5492.33	
03	08.JUN 62	17.59	30.38	5.3	1221.36	4934.12	4525.75	5492.38	510
04	24.AUG 63	11.30	29.48	6.0	1221.36	4934.12	4525.75	5492.38	510
05	26.JUN 64	16.59	49.37	5.0	1221.36	4934.11	4525.75	5492.38	510
06	28.MAY 65	11.13	30.02	5.0	1221.36	4934.12	4525.75	5492.38	510
07	07.APR 66	11.13	30.13	6.5	1221.36	4934.12	4525.75	5492.38	525
08	07.APR 67	10.54	59.19	5.0	1221.36	4934.12	4525.75	5492.38	525
09	20.APR 68	11.05	00.41	6.0	1221.36	4934.11	4525.75	5492.36	510
10	25.APR 69	11.05	01.19	3.5	1221.36	4934.11	4525.75	5492.36	510
11	09.JUL 70	18.05	00.64	5.0	1221.31	4934.15	4525.75	5492.47	520
12	26.FEB 71	11.05	00.07	6.7	1221.42	4934.18	4525.76	5492.48	550
1001	29.APR 61	13.59	48.50		1222.57	4927.91	4527.20	5480.88	
02	30.APR 61	14.00	.		1222.57	4927.91	4527.20	5480.88	
03	26.APR 63	11.59	30.99	7.0	1222.32	4927.93	4526.95	5480.92	500
04	24.APR 64	16.59	46.04	5.0	1222.32	4927.93	4526.95	5480.92	500
05	16.JUL 65	11.14	42.17	3.0	1222.32	4927.93	4526.95	5480.92	500
06	26.NOV 65	11.13	30.13	10.0	1222.32	4927.93	4526.95	5480.92	
07	19.APR 68	17.05	00.67	2.7	1222.32	4927.93	4526.95	5480.92	500

Table 2 (continued)

Q B	DATUM	SHOT-TIME H M	S	TO	COORDINATES LONGIT.	LATIT.	Y(LONG)	X(LAT)	ALT
1101	10.JUN 61	11.13	30.90	6.5	1000.13	5025.90	3571.19	5588.76	680
02	29.AUG 70	17.00	01.51	5.1	1001.71	5025.29	3573.08	5587.66	620
03	15.JUL 72	14.05	00.12	4.6	1001.62	5025.31	3572.98	5587.70	620
1201	30.JUN 61	16.00	.	3.8	0922.88	5019.87	3527.15	5577.18	240
02	29.NOV 63	15.59	58.49	1.4	0922.87	5019.83	3527.13	5577.12	250
03	30.OCT 64	14.59	11.35	4.6	0922.85	5019.88	3527.13	5577.19	240
04	18.MAR 66	17.00	02.09	4.2	0922.79	5019.83	3527.05	5577.10	270
05	03.JUN 66	16.05	00.97	2.4	0922.83	5019.84	3527.07	5577.13	250
06	10.MAY 68	16.05	00.48	6.8	0922.83	5019.86	3527.09	5577.16	230
1301	25.OCT 62	16.05	30.82	6.3	0913.10	5057.89	3515.34	5647.65	330
02	12.MAY 66	16.05	00.98	4.5	0913.13	5057.92	3515.38	5647.67	320
03	15.FEB 67	16.05	01.15	5.3	0913.09	5057.89	3515.33	5647.62	320
04	01.MAR 68	15.05	01.09	10.0	0913.14	5057.90	3515.38	5647.65	320
05	20.MAR 69	16.05	01.33	7.8	0913.12	5057.89	3515.36	5647.63	320
06	05.MAR 70	16.05	01.11	7.0	0913.11	5057.87	3515.35	5647.59	320
07	28.MAY 71	16.05	00.84	6.7	0913.16	5057.88	3515.40	5647.61	320
08	21.OCT 71	16.05	00.79	3.5	0913.10	5057.87	3515.34	5647.58	320
09	29.JUN 72	16.05	00.87	3.5	0913.15	5057.88	3515.39	5647.61	320
10	27.OCT 72	16.05	01.01	5.6	0913.12	5057.87	3515.35	5647.59	320
1401	12.DEC 64	10.59	08.04	6.0	0717.87	5036.60	3379.52	5609.52	420
02	09.APR 65	18.00	01.47	2.3	0717.96	5036.58	2591.97	5608.90	420
03	19.SEP 67	17.02	00.63	6.0	0717.85	5036.55	2591.83	5608.83	400
04	04.OCT 68	17.05	01.63	4.2	0717.85	5036.54	2591.83	5608.81	400
05	06.FEB 69	17.05	01.38	3.9	0717.84	5036.53	2591.82	5608.80	400
06	28.MAR 69	17.05	00.76	3.9	0717.85	5036.53	2591.84	5608.80	400
07	16.MAY 69	17.05	01.10	3.1	0717.82	5036.54	2591.79	5608.82	400
08	03.MAY 72	18.25	01.66	2.6	0717.80	5036.55	2591.77	5608.83	370
1501	09.MAR 66	16.00	01.67	2.6	0911.38	5018.12	3513.51	5573.89	280
02	31.AUG 66	17.00	00.79	9.2	0911.28	5018.12	3513.40	5573.88	280
1601	27.APR 66	15.45	00.43	2.1	0636.74	4932.96	2544.30	5490.33	310
02	30.MAR 67	15.45	00.68	6.3	0636.69	4932.93	2544.25	5490.29	270
03	04.OCT 67	15.45	00.96	3.3	0636.64	4932.94	2544.19	5490.31	185
04	24.APR 68	15.50	01.11	4.6	0636.68	4932.95	2544.23	5490.32	230
05	11.JUN 69	15.45	00.79	4.5	0636.65	4932.96	2544.19	5490.33	185
06	08.OCT 69	15.45	01.13	7.7	0636.72	4932.95	2544.28	5490.32	230
07	04.JUN 70	14.50	00.88	4.5	0636.64	4932.96	2544.17	5490.34	185
08	29.OCT 70	14.50	00.64	5.2	0636.69	4932.96	2544.24	5490.34	240
09	25.MAR 71	14.50	01.39	8.4	0636.76	4932.95	2544.32	5490.32	280
10	03.JUN 71	14.50	00.66	5.1	0636.68	4932.90	2544.23	5490.23	200
11	08.JUN 72	14.25	01.06	4.4	0636.72	4932.88	2544.28	5490.18	240
1701	29.JUN 66	16.04	58.36	16.0	0951.55	5113.96	3560.01	5677.74	690
02	28.JUL 67	16.05	11.09	24.0	0951.54	5113.95	3559.99	5677.72	690
03	20.JUN 68	17.04	58.47	19.5	0951.44	5113.90	3559.87	5677.63	690
04	02.OCT 68	15.05	27.09	14.0	0951.37	5113.92	3559.79	5677.66	690
05	18.JUL 69	17.05	02.97	22.0	0951.59	5113.93	3560.05	5677.69	690
06	12.JUN 70	16.05	01.16	23.0	0951.51	5113.90	3559.96	5677.63	690
07	23.OCT 70	16.05	00.63	20.3	0951.63	5114.00	3560.09	5677.80	690
1801	07.JUL 66	16.05	01.80	3.4	0809.57	5031.40	3440.38	5598.84	340
1901	25.OCT 66	08.46	56.62	0.9	0647.93	4909.73	2558.25	5447.40	
02	25.OCT 66	09.06	57.69	5.3	0647.95	4909.70	2558.28	5447.35	
03	25.OCT 66	11.07	01.69	2.8	0647.97	4909.67	2558.30	5447.29	
04	25.OCT 66	11.37	02.40	1.6	0647.98	4909.65	2558.32	5447.25	

Table 2 (continued)

Q B	DATUM	SHOT-TIME H M	S	TO	COORDINATES LONGIT.	LATIT.	Y(LONG)	X(LAT)	ALT
05	18.NOV 66	08.57	00.08	4.5	0647.27	4909.81	2557.45	5447.54	
06	18.NOV 66	10.56	59.88	5.0	0647.31	4909.83	2557.50	5447.58	
07	18.NOV 66	12.36	59.32	3.4	0647.35	4909.85	2557.55	2447.62	
2001	17.MAY 68	16.30	00.39	5.7	0637.82	5011.76	2545.02	5562.26	425
02	13.JUL 68	14.30	01.39	4.3	0637.78	5012.19	2544.95	5563.06	380
03	09.APR 69	16.45	00.57	2.9	0638.18	5011.39	2545.45	5561.58	420
04	19.JUN 70	16.30	00.55	4.2	0637.81	5011.69	2545.00	5562.13	430
05	30.JUN 70	16.30	00.61	7.3	0638.19	5011.39	2545.46	5561.58	420
06	26.MAY 72	15.30	00.55	3.6	0638.07	5011.44	2545.31	5561.67	420
2201	22.AUG 70	13.00	01.06	2.9	0800.57	5028.63	3429.71	5593.85	250
02	05.DEC 70	12.30	00.71	2.9	0800.57	5028.62	3429.71	5593.81	250
03	08.APR 71	17.05	00.76	2.8	0800.54	5028.61	3429.66	5593.74	250
2401	19.JUL 68	16.50	03.89	2.9	0803.64	4817.53	3430.44	5350.96	260
02	27.JUN 69	16.40	36.25	1.8	0803.73	4817.64	3430.42	5350.98	260
03	18.JUL 69	16.47	47.61	2.4	0803.68	4817.70	3430.35	5351.09	260
04	03.JUL 70	16.46	46.88	2.0	0803.68	4817.70	3430.35	5351.09	230
05	10.NOV 70	16.47	18.28	2.0	0804.12	4817.68	3430.40	5351.00	270
06	16.DEC 71	15.30	15.10	2.0	0803.69	4817.69	3430.37	5351.08	270
2501	29.MAY 69	16.05	03.21	1.4	1037.00	4732.70	4395.88	5268.22	1060
2601	04.SEP 70	16.30	00.78	8.4	0914.73	5111.31	3517.17	5672.50	345
02	27.MAY 71	16.05	00.24	6.6	0914.73	5111.33	3517.16	5672.53	345
2701	16.DEC 70	15.30	02.43	9.8	1129.34	5019.87	4463.61	5577.22	420
02	01.SEP 71	15.01	00.73	4.0	1129.45	5019.92	4463.76	5577.33	406
03	24.SEP 71	11.30	59.21	1.0	1129.43	5019.92	4463.79	5577.32	406

3.1.4 The System of Gauß-Krüger Coordinates

The system of the rectangular Gauß-Krüger coordinates is divided into stripes, each covering 3° degrees longitude. The code of each stripe is identical with the first digit of the coordinate Y. The Federal Republic of Germany is covered by three stripes as follows:

Code	Central longitude	Width of the stripe
2	$\lambda = 6^{\circ}00'$	$4^{\circ}30' \leq \lambda < 7^{\circ}30'$
3	$\lambda = 9^{\circ}00'$	$7^{\circ}30' \leq \lambda < 10^{\circ}30'$
4	$\lambda = 12^{\circ}00'$	$10^{\circ}30' \leq \lambda < 13^{\circ}30'$

The seismic refraction profiles in western Germany are in general not longer than 300 km. Therefore, the earth curvature can be neglected when calculating distances. If a profile crosses several stripes, it is necessary to transform coordinates from one stripe to the other. Formulas and tables for such a transformation have been published by HRISTOW (1943).

3.1.5 Position and Code of the Quarries

Fig. 1 shows the position of the quarries used for crustal studies in a geological map of Central Europe prepared by JACOBSHAGEN (1 of this volume). Details of the shotpoints are presented in Tables 1 and 2.

Igneous rocks are quarried in Böhmischbruck, 09 (granite) and Voggendorf, 10 (gabbro), both located in the Oberpfälzer Forest. The quarry Steinach, 24, situated in the Black Forest, wins granitic rocks, too. Quartz porphyrites are mined in the quarries Birkenau, 05, (Odenwald) and Kirchheimbolanden, 08, (Pfälzer Forest).

Most of the quarries used for the studies described here are located in areas with Tertiary basalts: in the Rhön are situated the quarries Hilders, 02, Gersfeld/Schwarzer Acker, 03, Gers-

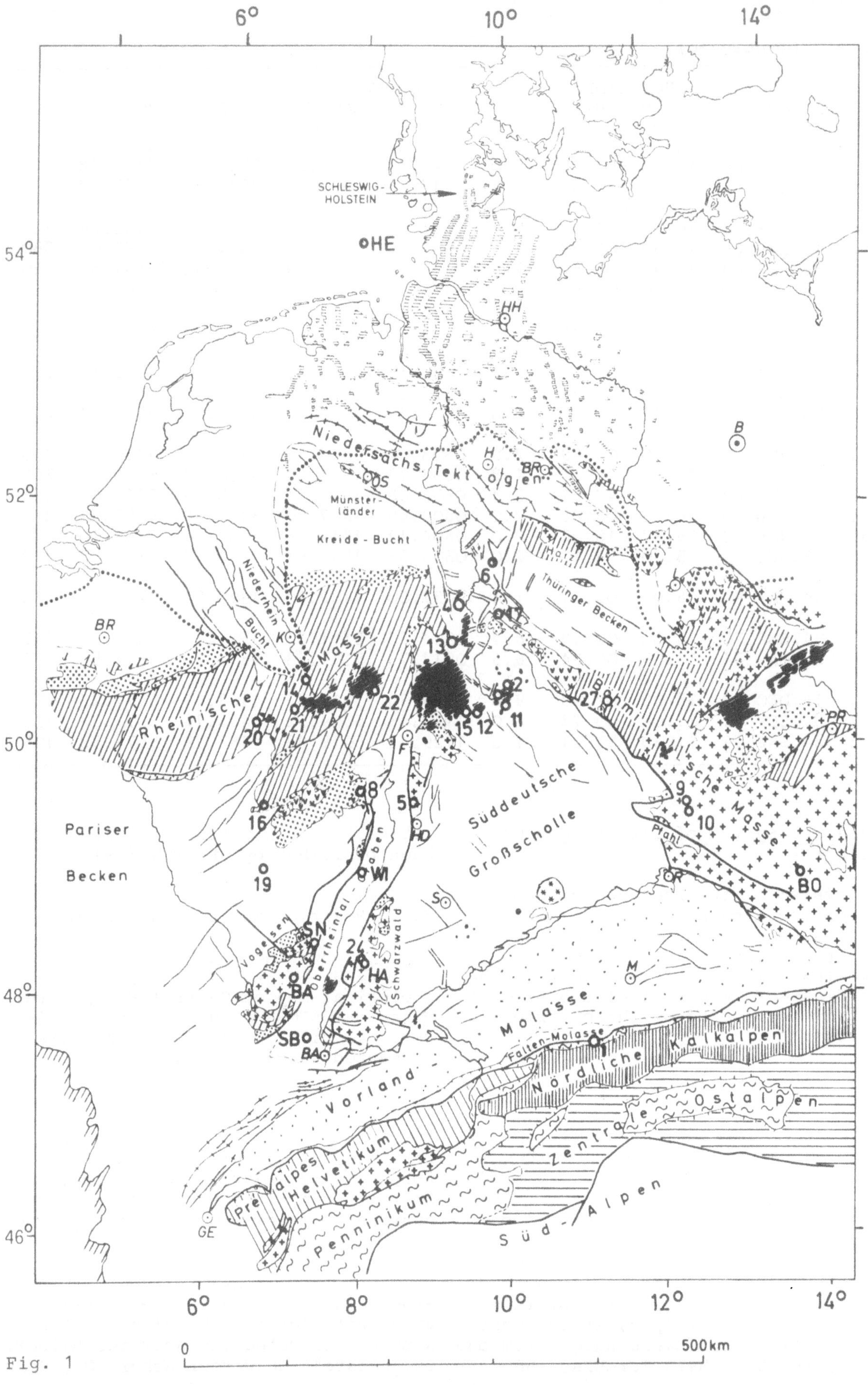

SCHLESWIG-HOLSTEIN
Niedersachs. Tektogen
Münsterländer
Kreide-Bucht
Niederrhein Bucht
Rheinische Masse
Thüringer Becken
Harz
Pariser Becken
Süddeutsche Großscholle
Oberrheintal-Graben
Schwarzwald
Vogesen
Vorland-Molasse
Falten-Molasse
Nördliche Kalkalpen
Zentrale Ostalpen
Helvetikum
Penninikum
Süd-Alpen
Prealpes
Prahl
500km

Fig. 1

◀ Fig. 1. Geological map of central Europe with the position of shotpoints. (From JACOBSHAGEN, see 1 of this volume, Fig. 2). Code of shotpoints see Table 1

Structural limits — fractures — overthrusts

folds — Astroblemes (meteor impacts)

a) extra-Alpine region

- volcanic rocks of the Cenozoic
- Tertiary (sediments)
- southern border of the Quaternary cover of the northern lowlands
- Zechstein (Upper Permian) and Mesozoic
- salt diapirs
- Permian volcanic rocks
- Upper Carboniferous and Rotliegendes
- sediments and volcanic rocks older than Upper Carboniferous
- crystalline rocks of Hercynian age and older

b) Alpine region

- Helvetic zone and Préalpes
- autochthonous massifs
- Penninic zone, flysch of the Eastern Alps
- Northern Calcareous Alps
- central zone of the Eastern Alps

feld/Nallenberg, 07, and Bischofsheim, 11; in the area of Kassel, the quarries Großenritte, 04, Adelebsen, 06, Dorheim, 13, Bransrode, 17, and Lohne, 26; in the Vogelsberg, the quarries Lahr, 18, Dorndorf, 22, and Wilsenroth, 23; and in the Eifel the quarries Birresborn, 20, and Bermel, 21.

In the very active quarry of Eschenlohe, 01, situated at the northern margin of the Eastern Alps in the Helvetic zone, quartzitic sandstone is being mined. In the quarry of Taben Rodt (16), quartzites of lower Devonian age are won.

3.2 Seismic Signals from Quarry Blasts

H. BURKHARDT and R. VEES

ABSTRACT

The properties of the seismic signals, radiated from quarry blasts are investigated, based on the Eschenlohe quarry. The seismic signal is supposed to be generated predominantly by the secondary effect of natural vibrations in close vicinity of the shotpoint. Quarry blasts are approximately reproducible seismic signal sources. There are no uniform scaling laws for the amplitude-charge relation. A correlation between signal shape and wave path is indicated, stressing the importance of control measurements of the source signal for quantitative seismogram interpretation.

ZUSAMMENFASSUNG

Die Eigenschaften der bei Steinbruchsprengungen abgestrahlten seismischen Signale werden am Beispiel des Steinbruchs Eschenlohe untersucht. Es wird angenommen, daß die Hauptquelle für das Signal in Sekundäreffekten besteht, in Form von Eigenschwingungen im Nahbereich der Sprengstelle. Steinbruchsprengungen stellen eine näherungsweise reproduzierbare seismische Signalquelle dar. Es gibt dabei keine einheitliche Beziehung zwischen Amplitude und Ladungsmenge. Eine Korrelation zwischen Signalform und Wellenweg deutet sich an, wodurch die Bedeutung von Kontrollmessungen des Quellsignals im Hinblick auf quantitative Seismogrammauswertung unterstrichen wird.

3.2.1 Introduction

For deep seismic soundings in West Germany industrial quarry blasts have been the main signal source. With the introduction of more detailed interpretation methods of the seismograms, extending the usual travel-time analysis to other signal parameters, such as amplitude and pulse shape, the properties of the signal source and the source pulse are gaining increasing importance.

Whereas for underwater explosions many experimental and theoretical investigations have clarified the process of the generation of seismic signals with this type of explosion to a certain and normally sufficient extent (MÜLLER et al., 1962; BURKHARDT, 1964; VEES, 1965; O'BRIEN, 1965) the situation is rather different for quarry blasts. In contrast to underwater explosions the mechanism of signal generation with quarry blasts is much more complex and the necessary systematic investigations are difficult and expensive. Thus there are only a few reports dealing quantitatively with quarry blast signals (see e.g. MÜLLER et al., 1962; VEES, 1965; BERCKHEMER, 1970 for a summary of the results).

As a first step towards a better knowledge of the generation process the following remarks describe some characteristic properties of seismic signals, generated by different quarries and used for deep-seismic soundings.

From the extensive data material existing (e.g. ASFA, 1969, 1970, 1971; GIESE and STEIN, 1971) a selection was made and the positions of the used quarries and recording sites are given in Fig. 1 and Table 1. Special emphasis is given to the quarry Eschenlohe (01), its properties as a seismic signal source being to some extent representative for this type of quarry blast.

3.2.2 Seismic Signals from the Quarry Eschenlohe (01)

Due to the favorable situation for seismic profiling, the quarry Eschenlohe has been used extensively as signal source for deep-seismic soundings in central Europe (e.g. GIESE and STEIN, 1971; GIESE et al. 3.3 of this volume). There are thus numerous recordings of Eschenlohe shots with

Fig. 1. Location map of shot- and registration points. Shotpoints refer to Table 1. Exact positions of recording points are defined in the respective figures by distance and azimuth

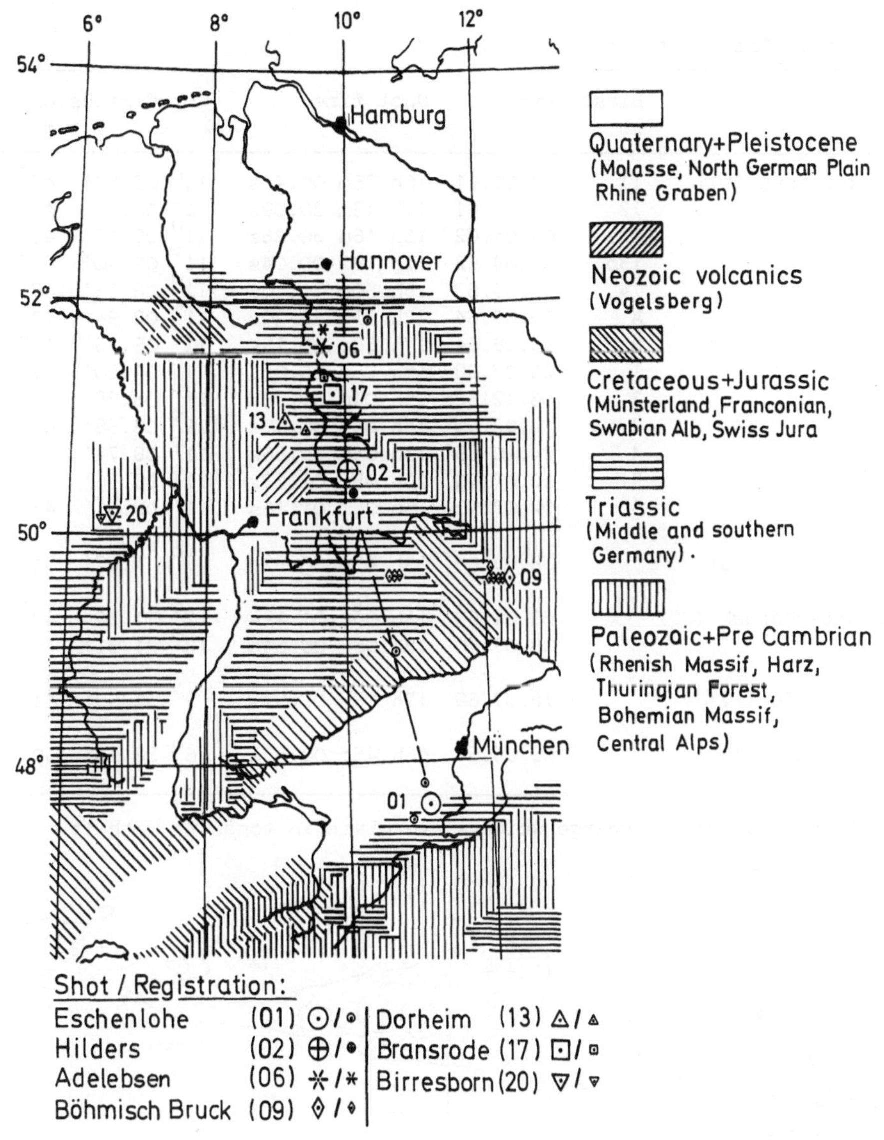

different charge weights in different recording distances. In the following, a selection of these recordings is utilized for interpretation.

The quarry Eschenlohe is situated at the southern border of the Bavarian Molasse Basin (Fig. 1), the exploitable rock consisting of quartzitic sandstone. The geometrical shape of the rock body is shown in Fig. 2 together with the positions of the explosion chambers as used in this investigation. From a geological point of view the quarry can be approximately described as a massive block of rock surrounded by sediments.

All explosions were momentary layer blasts; the characteristic configuration of the explosion chambers is given in Fig. 2 for two shots.

3.2.2.1 Nearby Measurements within the Quarry

In order to obtain some information about the radiated seismic source signal, quantitative measurements were made within the quarry on the floor in the distance interval 128-560 m. The positions of the respective shot and recording points are given in Fig. 2. Some examples of the seismic signals are shown in Fig. 3 for the distances 297 m and 560 m, featuring the characteristic properties of the nearby seismic signal.

Using displacement transducers the transfer function of the calibrated recording equipment is frequency-independent from 3 to 120 cps. Qualitatively, a direct inspection of the

Table 1. Shotpoint data

	Blast no.	Date	Shot time	Coordinates of shotpoint λ	φ	Chargeweight (t)[a]
Eschenlohe (01)	11	24.11.61	16h 25m 00.44s	11° 09.50'	47° 38.00'	7.9
	12	02.12.61	11h 13m 30.09s	11° 08.82'	47° 37.88'	19.5
	13a	07.04.62	11h 16m 30.28s	11° 09.16'	47° 37.95'	6.0
	13b	07.04.62	11h 18m 00.04s	11° 09.40'	47° 38.00'	11.0
	14	24.05.63	18h 50m 30.10s	11° 08.95'	47° 37.92'	11.0
	A	15.05.64	17h 59m 45.20s	11° 08.94'	47° 37.92'	14.0
	B	18.09.64	17h 53m 30.19s	11° 08.73'	47° 37.90'	11.4
	1	26.07.68	15h 05m 00.03s	11° 08.80'	47° 37.90'	9.1
	2	06.12.68	15h 04m 59.38s	11° 08.88'	47° 37.91'	7.1
	3	14.03.69	15h 05m 00.59s	11° 08.68'	47° 37.90'	7.7
	4	24.10.69	15h 04m 59.86s	11° 08.74'	47° 37.90'	21.7
	5	10.07.70	15h 05m 01.13s	11° 08.76'	47° 37.91'	16.4
	6	02.07.71	15h 05m 01.16s	11° 08.81'	47° 37.91'	14.0
Adelebsen (06)		18.09.68	15h 45m 00.82s	9° 44.67'	51° 36.59'	3.0
Böhmischbruck (09)		25.04.69	11h 05m 01.19s	12° 21.36'	49° 34.11'	3.5
		09.07.70	18h 05m 00.64s	12° 21.31'	49° 34.15'	5.0
Bransrode (17)		18.07.69	17h 05m 02.97s	9° 51.59'	51° 13.93'	22.0
Birresborn (20)		09.04.69	16h 45m 00.57s	6° 38.18'	50° 11.39'	2.9

[a] In this paper all charge weights are given in tons (t) with 1t = 1000 kg.

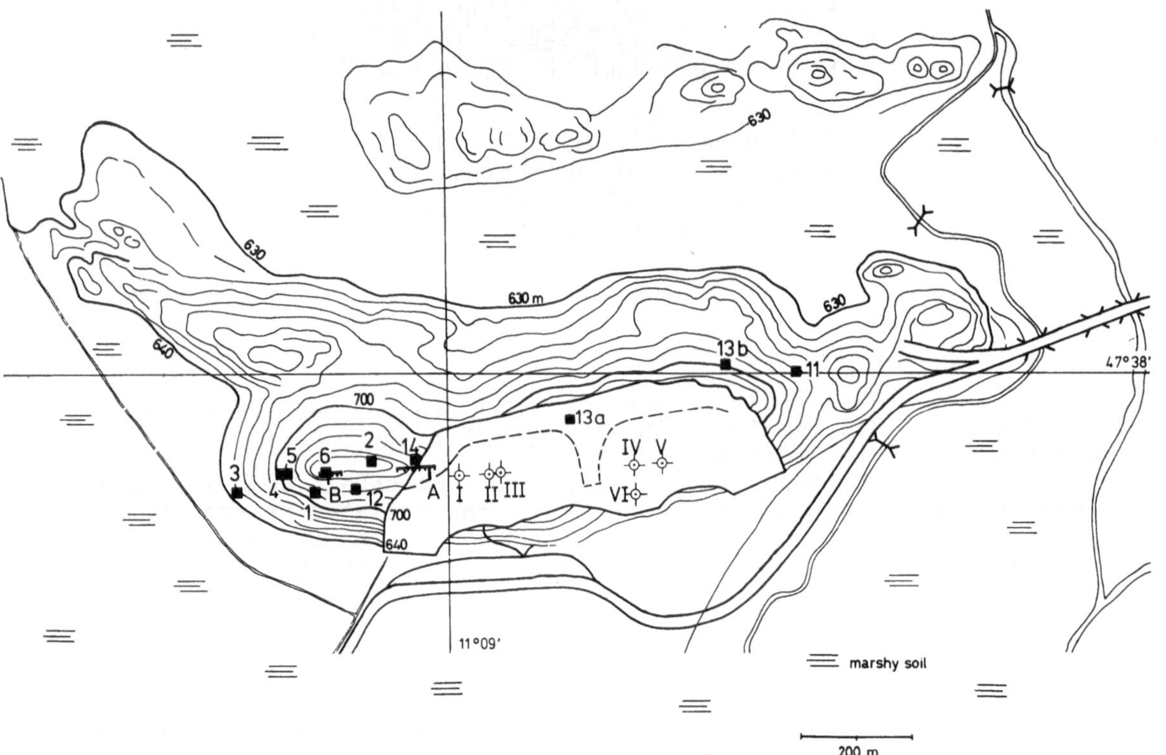

Fig. 2. Location map of Eschenlohe quarry (01). (After survey map 8332 Unterammergau, ed. 1959)
A, B, 1-14 mark the center of the blasts given in Table 1. (For blasts *A* and *B* the complete arrangement of explosion chambers is indicated. *III, IV* recording sites for blast *A*. *I, II, IV, V, VI* recording sites for blast *B*. Dashed line indicates the approximate course of the lower quarry floor at the time of shot *A*

Fig. 3. Example for seismic near-by measurements in the Eschenlohe quarry (O1) (shot no. *B*, see Table 1): Seismic signals in two distances at sites *II* and *IV* (see Fig. 2) and ground particle movement in two planes at site *IV*. $Z, H_{\parallel}, H_{\perp}$ three-component seismometer arrangement (displacement transducer), Z vertical component (velocity transducer)

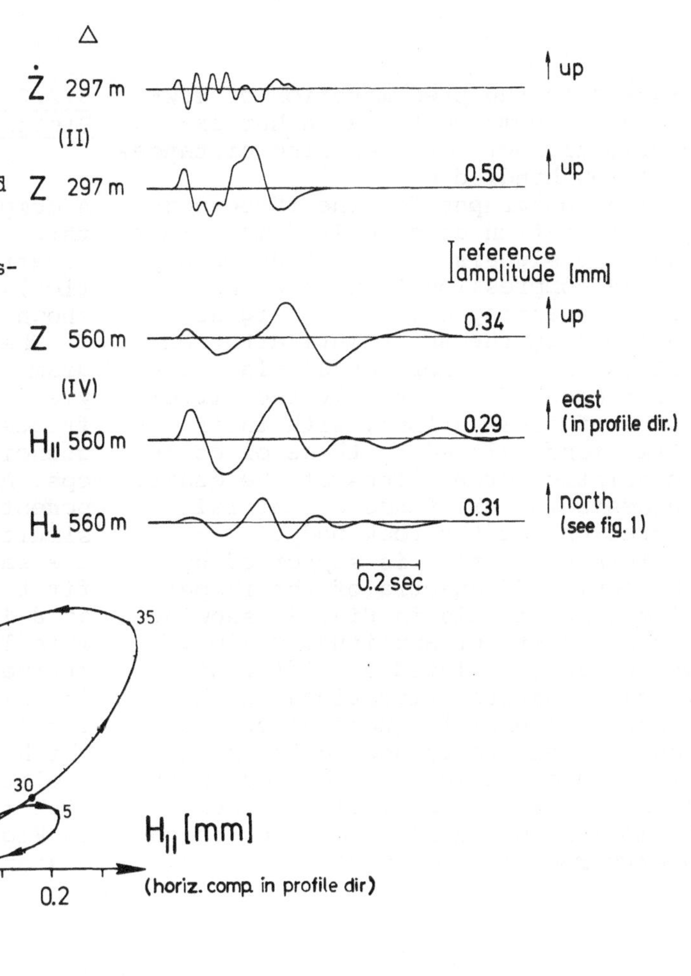

HORIZONTAL PLANE

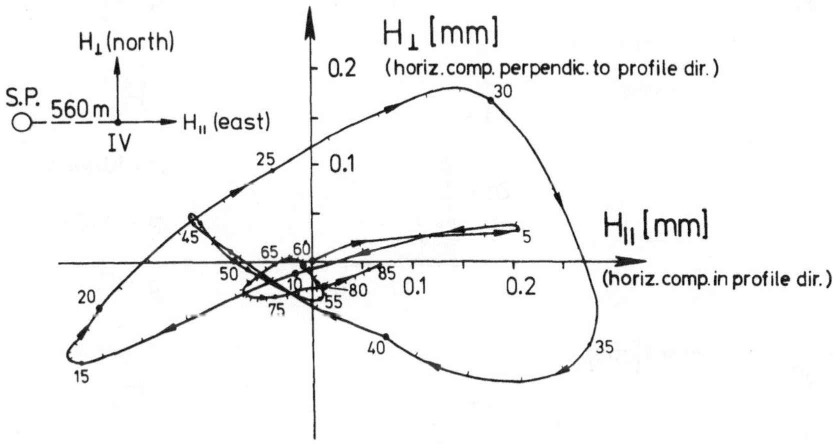

time signals at site II (Δ = 297 m) yields two distinct frequency components of ca 4 cps and 20 cps, the latter being preferably obvious for the additional velocity-transducer trace. At site IV only the lower-frequency components dominate, suggesting either a strong absorption of higher frequencies or the influence of interferences and sharp-focused radiation of these components. The time duration of the signals increases from 500 ms at site II to ca 1 s at site IV. A remarkable feature of the signals is the predomination of low-frequency components, a rather exceptional fact for recording distances of a few hundred m.

The hodographs for the three-component station at site IV demonstrate that the signals are built up mainly by one compressional and one shear event, starting with the P signal, followed by the shear wave after ca. 300 ms. For the horizontal plane the occurrence of essentially two directions of oscillations, with their directions related to those of characteristic border lines of the quarry, suggests the existance of natural vibrations of the rock body.

This suggestion is supported by the amplitude spectra of the respective time signals in Fig. 4, showing several distinct amplitude maxima, which can be related to different modes of natural vibrations of the quarry. Several frequencies can be found which are connected to characteristic dimensions of the body with a special emphasis of first-order harmonics (measured P velocity of the quarry rock is 5000 m/s).

3.2.2.2 The Seismic Signal in Medium Distances

A compilation of seismic records in three recording distances, 0.3 km (quarry floor), 22 km (reference station), and 151 km (profile 01-345) is shown in Fig. 5.

The first arrivals of the seismograms in all three distances again yield a clear distinction of different frequency groups, with central frequencies at 4 cps, 8-10 cps and 20 cps. At Δ = 0.3 km the frequency components at 4 cps and 20 cps occur simultaneously. At Δ = 22 km it is the same during the first s of the first arrival, with a frequency group at 8-10 cps following after 1.5 s. This latter group appears as the first arrival with small amplitudes at Δ = 151 km and is clearly separated against the low-frequency group of 4 cps which follows after 1.4 s with large amplitudes. Comparing this evidence with the spectra of the nearby signals in Fig. 4 it can be seen that the distant signals are built up mainly

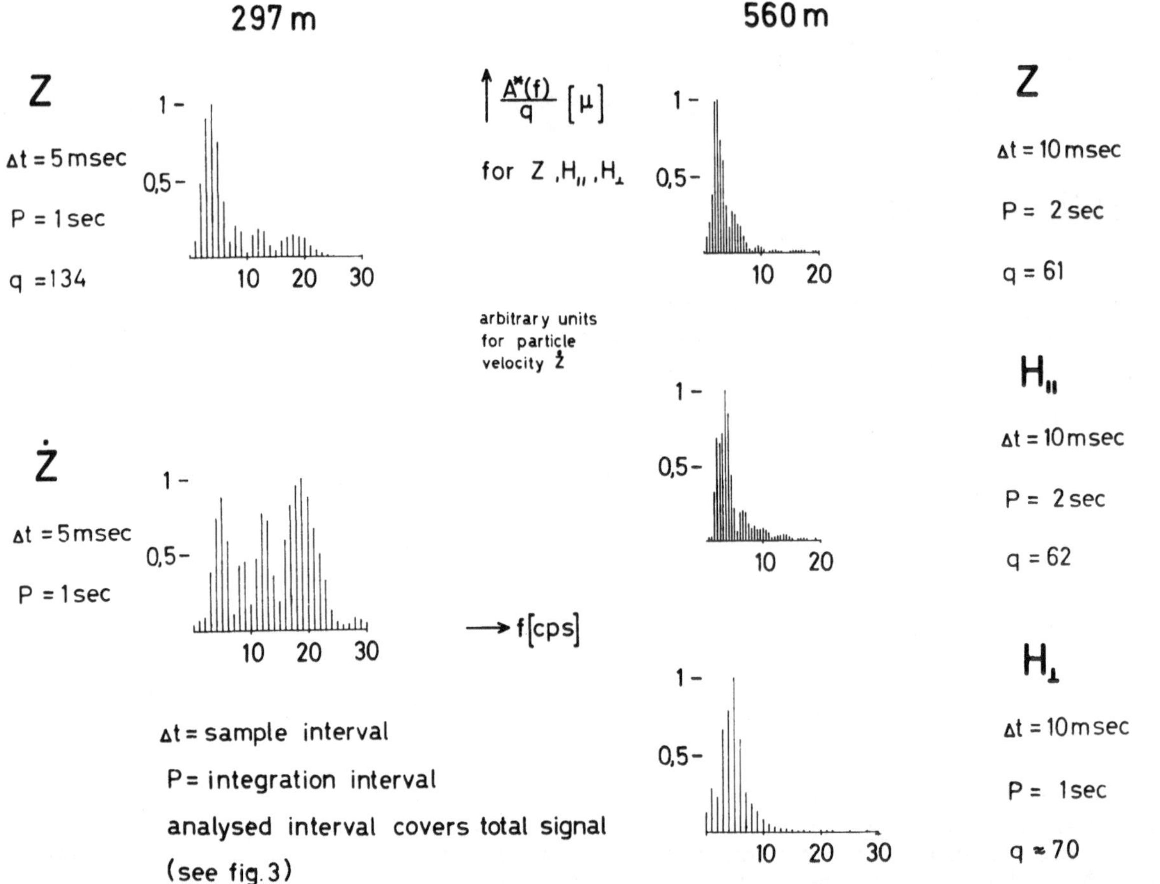

Fig. 4. Frequency spectra for signals shown in Fig. 3

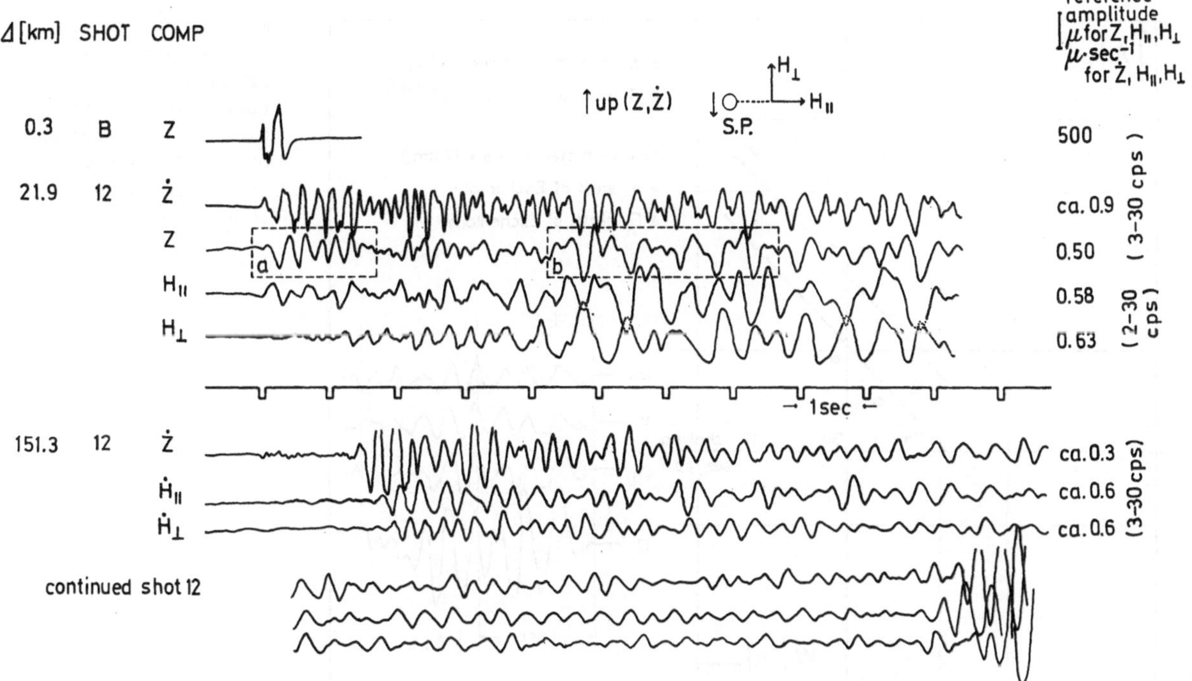

Fig. 5. Seismogram samples in three distances for shotpoint Eschenlohe (01). Sections *a* and *b* of third trace refer to Fig. 6. Frequency range for reference amplitude is given in the last column

with the same dominant frequency components as the nearby signal.

A third group is characterized by large amplitudes and low frequencies with about 2 cps, demonstrating the well-known effect that later arrivals are often lower frequency signals. These later arrivals have the largest amplitudes in the seismogram and appear 4 s at 22 km and 21 s at 151 km after the first arrival.

Several recordings at the reference station at 22 km were used to investigate the relation between seismic signal amplitudes and charge weight for the Eschenlohe quarry. Amplitudes of two different groups - first arrival (a) with mean frequency 4 cps and the later arrival (b) with ca 2 cps referring to Fig. 5 - are shown in Fig. 6 for four different charge weights. Although the number of recordings is not sufficient to establish a scaling law, the tendency of the relation is quite obvious and similar for both amplitude groups. For comparison two dashed straight lines are drawn corresponding to a power law with amplitude linearly proportional to charge weight and showing an approximate agreement with the experimental values. This kind of approximate linear relation between seismic amplitude and charge weight corresponds to the normally assumed scaling law for quarry blasts.

For the wave group (a) the used seismogram sections are also given in Fig. 6, demonstrating that the wave shape does not change with charge weight, although the positions of the explosion chambers for the different shots in the quarry were different (see Fig. 2). This fact again suggests the existence of some secondary effect, probably natural vibrations of the rock body, as the main source for seismic waves radiated with quarry blasts.

3.2.2.3 Long-Distance Recordings

During test measurements on the profile Eschenlohe N to investigate the possibility of extensions of existing seismic-refraction profiles (3.6 of this volume), a reference station in Clausthal (Δ = 471 km) was used, which could record several Eschenlohe shots with different charge weights. Some results of these reference recordings are shown in Fig. 7.

In the upper part a complete signal (vertical component) for one shot is shown in three different bandpass-filter settings, yielding three characteristic signal sections, which are the only seismogram parts with essential amplitudes. Sections (a) and (b)

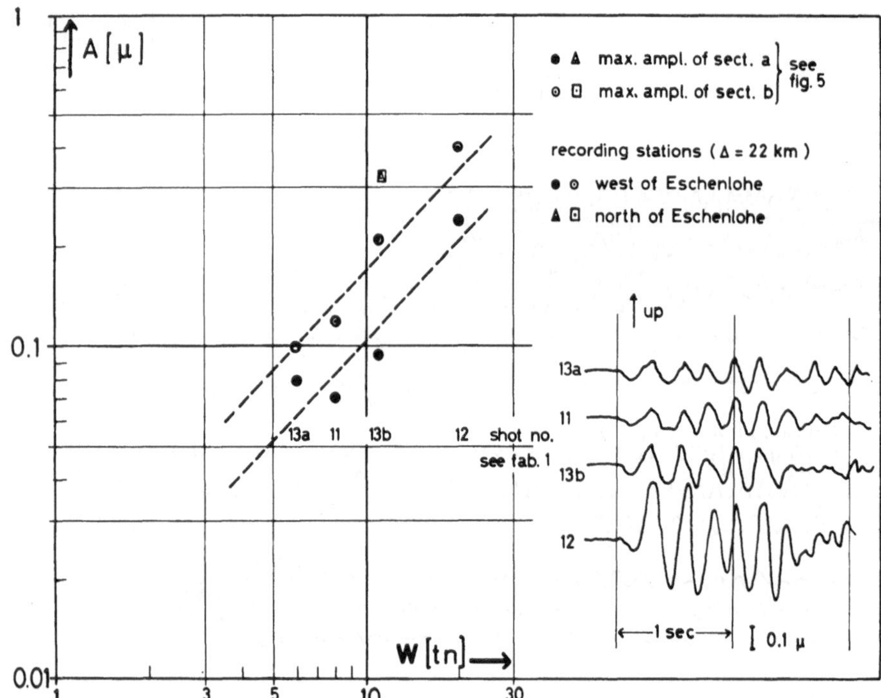

Fig. 6. Seismogram section of first arrival and amplitude - charge relation in a distance of ca 22 km for shotpoint Eschenlohe (01)

are characterized by distinctively different signal shapes. The applied analogue filter settings are chosen such as to describe the frequency band of the signals, thus depicting clearly the signal bandwiths. It can be seen further that the first arrival is the only signal part with unexpectedly high frequencies above 10 cps in the whole seismogram.

The verification for the reality of this strange high-frequency arrival at such a large recording distance is given by the traces in the left side of the lower part of the figure, where the seismogram section (a) is shown with expanded time scale for different Eschenlohe shots. For both arrival time and signal shape, these samples demonstrate the excellent reproducibility of the first arrival.

A comparison of this arrival with the high-frequency signal components in smaller distances (see Fig. 5) depicts the similarity of these phenomena, showing that the different frequency components of the signal can be traced from the immediate vicinity of the explosion up to recording distances of 470 km (see also 3.1 of this volume).

An amplitude-charge relation for the maximum amplitudes of signal sections (a) and (b) (Fig. 8) gives no uniform dependence for the two signals. Whereas the higher frequency-amplitudes of section (a) show a tendency to decrease or remain constant with increasing charge weight, in section (b) the lower frequency amplitudes increase or remain constant in contrast to the result in a recording distance of 22 km (Fig. 6). The dashed lines are auxiliary ones.

Comparing the signal amplitudes (a) for different shots with the shot position in the quarry (Fig. 2) it can be seen, that the amplitude decreases relatively with the shotpoint approaching the western border of the quarry. So the amplitude decrease with increasing charge weight might be produced by different excitation of amplitudes of different modes of natural vibrations of the quarry for different shot locations, taking into account the remarks in the foregoing sections.

Fig. 7. Seismogram samples (vertical components): Shotpoint Eschenlohe (01) - reference station Clausthal/Harz (Δ = 471.4 km, α = 353°). Upper part: complete signal, *3* bandpass-filter settings, shot no. *5* (see Table 1), original time scale. Lower part: section *a* and *b* with expanded time scale for 6 different shots (see Table 1, shots *1-6*), filter setting 0.5-20 cps. Recording equipment: MARS 66 (BERCKHEMER, 1970). Reference amplitude is valid for all traces with filter setting 0.5-20 cps. *W* charge weight

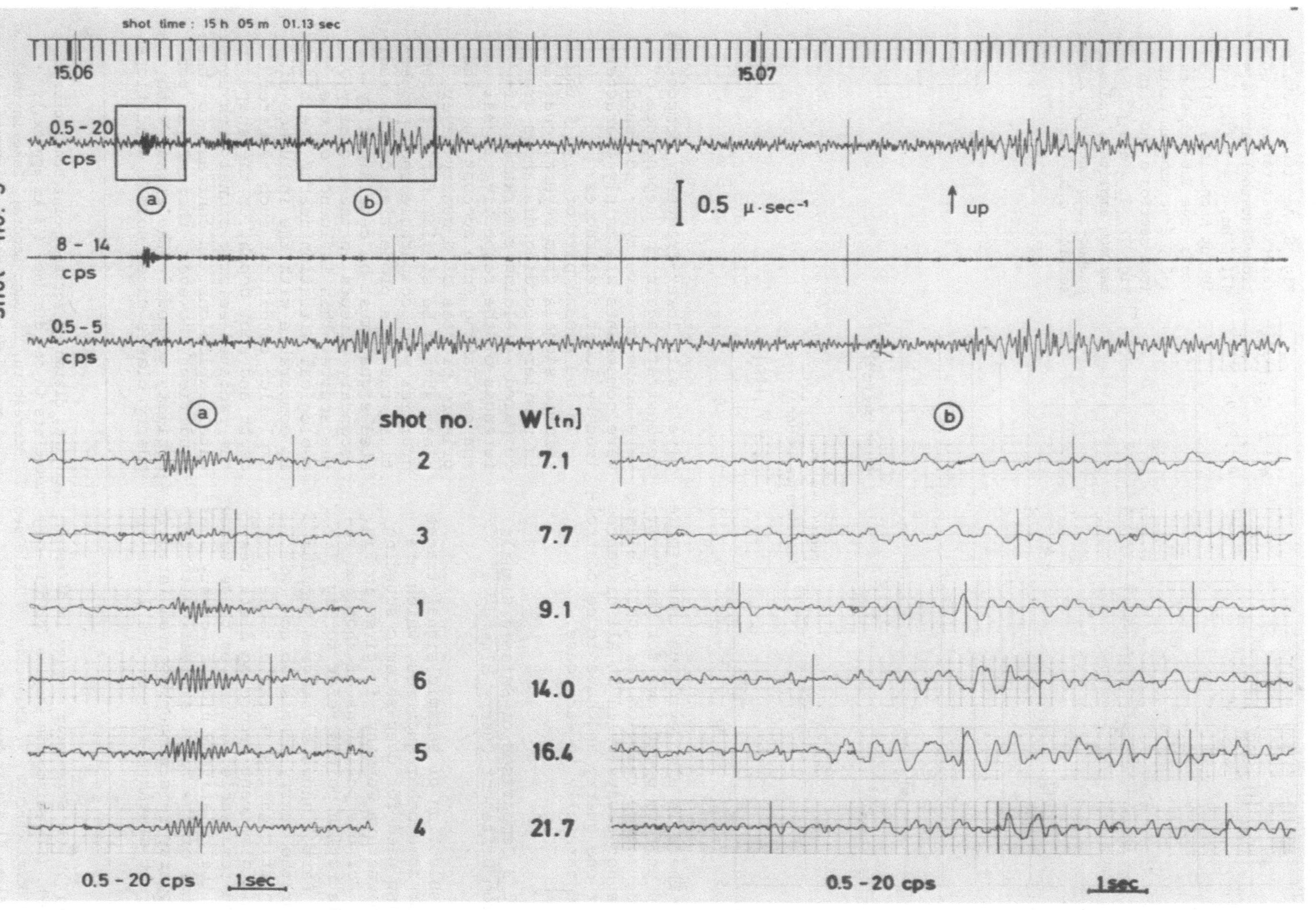
shot time : 15 h 05 m 01.13 sec
15.06
15.07
shot no. 5
0.5 - 20 cps
a
b
0.5 μ·sec⁻¹
up
8 - 14 cps
0.5 - 5 cps
a
shot no.
W[tn]
b
2
7.1
3
7.7
1
9.1
6
14.0
5
16.4
4
21.7
0.5 - 20 cps
1sec
0.5 - 20 cps
1sec

Fig. 7

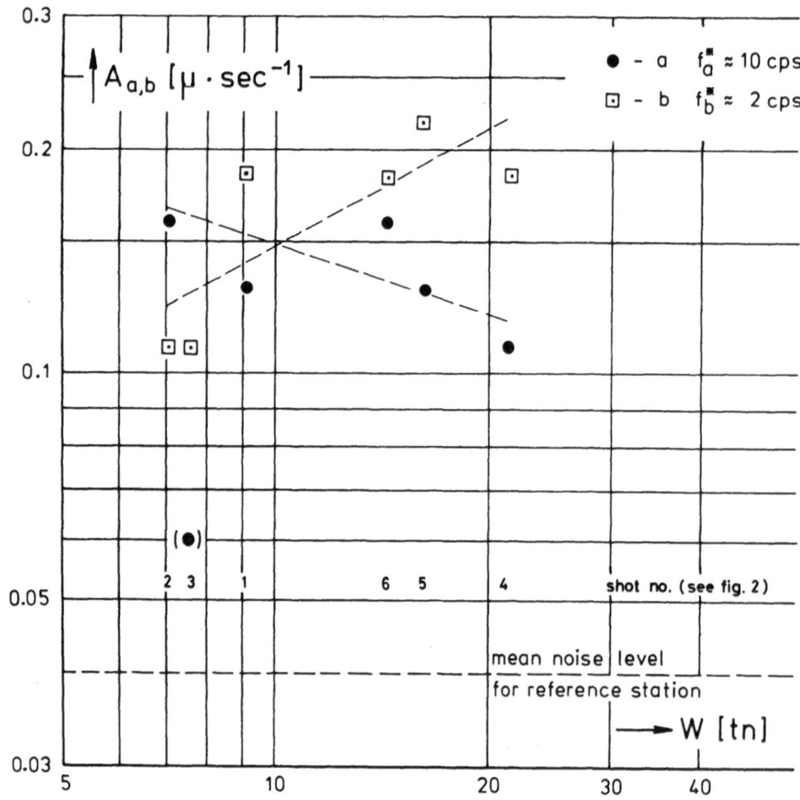

Fig. 8. Amplitude - charge relation at reference station Clausthal/Harz (Δ = 471.4 km, α = 353°), 6 different Eschenlohe (01) - shots (see Table 1, shots 1-6), 2 seismogram parts (see section a and b in Fig. 7). $A_{a,b}$ and $f^*_{a,b}$: maximum amplitude and mean pseudo-frequency for signal sections a, b, resp.

There is also evidence that simple and uniform power laws for the amplitude-charge relation for quarry blasts are not always valid in all distances for one quarry and have to be controlled carefully in each case.

3.2.3 Comparison of Seismic Signals from Different Quarry Blasts

A selection of seismic signal traces for some other quarries, which are also used for deep-seismic sounding in western Germany, is given in Fig. 9. For the positions of shot and registration points refer to Fig. 1.

Part I shows seismogram samples for the quarry Böhmischbruck (09) - granitic gneiss as exploitable rock - in different recording distances. The remarkable feature of these traces is the relatively high frequency character of the first arrival, with this property remaining preserved for different distance ranges. The later arrivals with lower-frequency components demonstrate again the dependence of wave shape on wave type; so, shear wave components are mainly characterized by lower frequencies.

In the second part of Fig. 9 the seismic signals from several quarries in similar recording distances are compared, yielding remarkable differences of the respective signal shapes. Taking into account that the primary process of a quarry layer blast should be similar for different quarries with comparable rock properties, the main properties of the radiated signal must be determined by secondary effects in the vicinity of the source. This influence of the geological situation nearby the blast is comparable with the influence of explosion conditions such as water depth and shot depth, on the shape of the radiated seismic signal with underwater explosions. In contrast to those conditions, however, there is no possibility to modify the signal with quarry blasts under normal conditions.

▶

Fig. 9. (I) Seismogram samples in different recording distances for shotpoint Böhmischbruck (09). (II) Seismogram samples for different shotpoints in one distance (15 km approx.). *09* Böhmischbruck, *17* Bransrode, *06* Adelebsen, *20* Birresborn, *LL* Lago Lagorai, Lake in the Italian Alps near Bolzano (not indicated in location map of Fig. 1). Values of ground velocity (μ/s) refer to indicated reference amplitude. Recording equipment: MARS 66

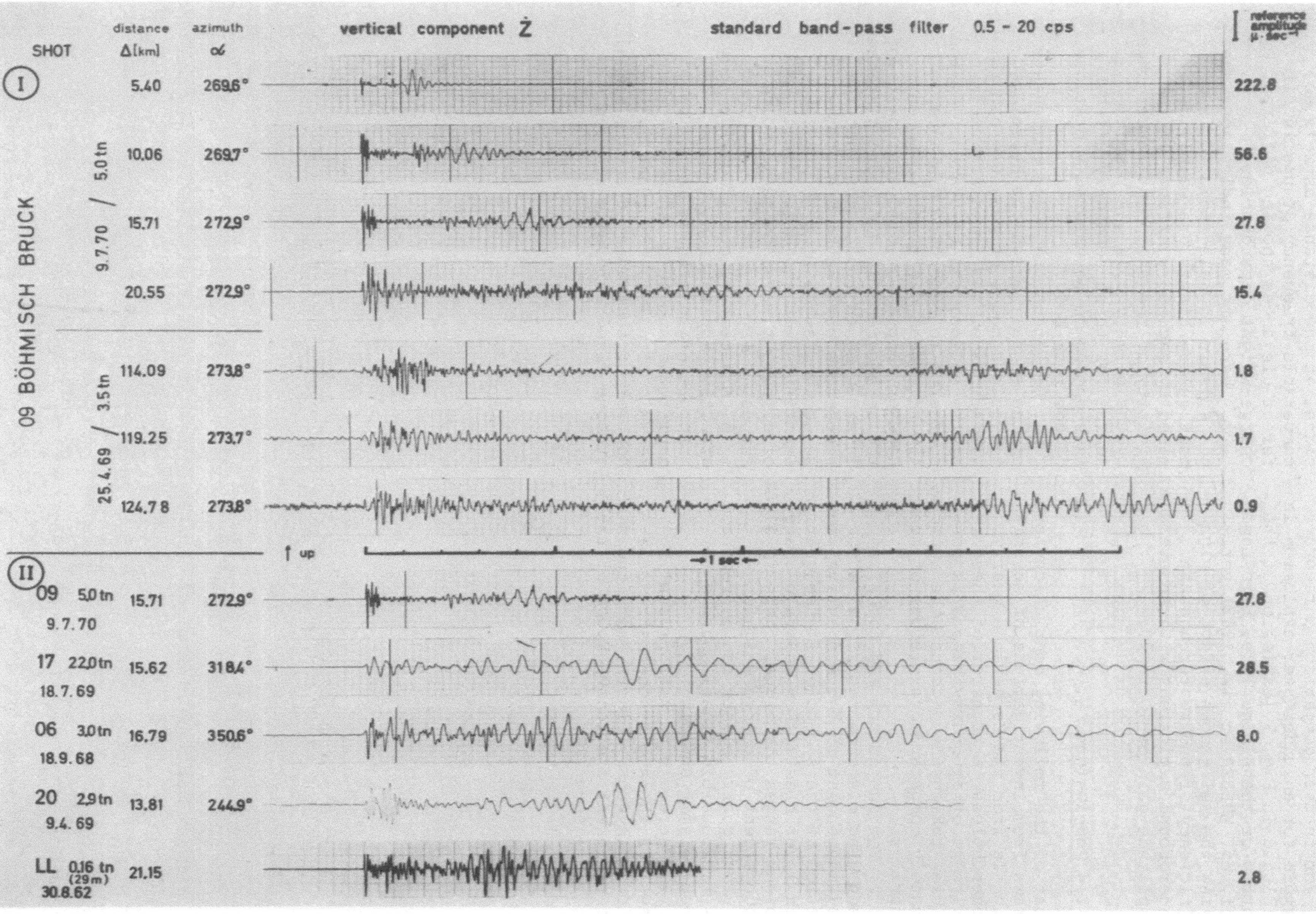
vertical component Ż
standard band-pass filter 0.5 - 20 cps
reference amplitude μ·sec⁻¹
SHOT
distance Δ[km]
azimuth α
I
09 BÖHMISCH BRUCK
9.7.70 / 5.0 tn
25.4.69 / 3.5 tn
5.40
269.6°
222.8
10.06
269.7°
56.6
15.71
272.9°
27.8
20.55
272.9°
15.4
114.09
273.8°
1.8
119.25
273.7°
1.7
124.78
273.8°
0.9
up
1 sec
II
09 5.0 tn 9.7.70
15.71
272.9°
27.8
17 22.0 tn 18.7.69
15.62
318.4°
28.5
06 3.0 tn 18.9.68
16.79
350.6°
8.0
20 2.9 tn 9.4.69
13.81
244.9°
LL 0.16 tn (29m) 30.8.62
21.15
2.8

Fig. 9

For further comparison a record sample for an underwater explosion is included in Fig. 9 (explosion at the bottom of the Alpine lake, Lago Lagorai, cf. MÜLLER et al., 1962), verifying the well-known effect of high efficiency of underwater against underground explosions with an average amplitude gain of 10 to 20.

3.2.4 Conclusions

Most of the following conclusions are drawn from investigations of seismic signals from the Eschenlohe quarry. Nevertheless, a transfer of these results to other quarry blasts is possible, taking into account the approximately comparable conditions of most quarries.

1. For one quarry and for fixed recording distance the signal shape is independent of the respective position of the shotpoint in the quarry for layer blasts. For moderate charge weight ranges the signal shape is also independent of charge weight.

Thus quarry blasts can be regarded as reproducible signal sources for deep-seismic sounding.

2. There are no uniform scaling laws for the amplitude-charge relation with quarry blasts, amplitudes of different frequency groups of the signal at one distance as well as comparable frequency groups in different distances showing different charge-weight dependencies.

3. There are many indications that the seismic signals radiated from most quarry blasts are mainly determined by secondary effects in the vicinity of the shotpoint rather than by the primary pulse. A possible source for these secondary effects is supposed to be natural vibrations of the rock body.

4. Different frequency components of the signal, which can be related to various modes of possible natural vibrations of the immediate vicinity of the source, can be traced from the source up to recording distances of 450 km. These different components, however, appear at different parts of the seismograms in different distances. This may be an indication that the respective wave paths are frequency-selective, suggesting the possible identification and discrimination of different waves by their pulse shapes.

5. A necessary condition for the effective application of detailed interpretation methods including amplitudes and pulse shapes of the seismic signal is the quantitative control of the radiated source signal at some fixed reference station.

3.3 Description of Profiles

P. Giese, E. Hinz, C. Prodehl, H. Schröder, and A. Stein

ABSTRACT

This chapter describes the network of seismic-refraction profiles observed in western Germany and adjacent areas. Most of the seismograms were obtained by the recording of quarry blasts. The arrangement of the profiles is mainly determined by the distribution of these quarry blasts. Additionally some special seismic-refraction investigations were performed. The data of the recording points are presented in tables.

ZUSAMMENFASSUNG

Dieses Kapitel beschreibt das Profilnetz der refraktionsseismischen Profile in Westdeutschland und unmittelbar benachbarter Gebiete. Der größte Teil der Registrierungen stützt sich auf Steinbruchsprechungen. Die Anordnung der Profile ist durch die Lage der Steinbrüche bestimmt. Zusätzlich wurden einige spezielle Refraktionsprofile beobachtet. Die Daten der Registrierpunkte sind in Tabellenform aufgeführt.

3.3.1 Choice of Shotpoints

When the extensive research program for the investigation of crustal and upper-mantle structure in central Europe by explosion seismology started at the end of the fifties, the use of quarry blasts offered the most economic seismic energy source.

This special circumstance explains the unequal distribution of usable shotpoints (Fig. 1). Most of them are located in the Rhenish Massif and Hessisches Bergland where Paleozoic rocks and Tertiary basalts are outcropping. Some shotpoints are also situated in the crystalline regions of Vosges, Black Forest, and Bohemian Massif.

In the southernmost part of Germany only the shotpoint Eschenlohe (01) situated at the northern margin of the Alps offers the possibility of recording long seismic-refraction lines up to more than 300 km. Further details of the shotpoints are given in 3.1 of this volume.

3.3.2 Location of Profiles

When planning the observation lines the following two conditions were followed as closely as possible: each profile should be reversed and profiles from several shotpoints should form an overlapping system. It must be confessed that these conditions could not be realized adequately. Consequently, in general geological aspects could not be taken into account as much as would have been advisable.

In the course of the program some special projects were undertaken. Using especially arranged borehole shots, a wide-angle profile was obtained both in the Bavarian Molasse Basin (FR) and in the Rhenish Massif (LO). In another case, borehole shots of a commercial seismic-refraction program in northern Germany enabled the recording of profiles useful for crustal studies.

Borehole shots (BO) fired on the International Profile VII traversing Poland and the Bohemian Massif in the ČSSR in NE-SW direction, were recorded on a profile in SE Bavaria. Thus it was possible to connect the existing seismic-refraction network in western and eastern Europe.

The network of seismic-refraction profiles in the area under investigation is shown in Fig. 1.

In Table 1 all refraction profiles are listed showing the number of recordings and the corresponding references. The first column gives the code of each profile: code of shotpoint - azimuth - reverse shotpoint. The code of the reverse shotpoint is indicated only if reversed observations exist or are planned. Sometimes only very few recordings are available for testing purposes here designed by the letter V. From some shotpoints, indicated by the letter F, fan observations exist. In this case the second number means the recording distance. The next three columns show number and type of recordings up to January 1971. N1 means the number of explosions which were used for the corresponding profile, N2 describes the number of photographic records whereas N3 is the number of magnetic tape records.

The last column gives publications dealing with the corresponding profile.

3.3.3 Instrumentation

The explosions of Helgoland and Haslach (Black Forest) were observed with recording systems based on a mechanical-optical amplification of ground movements (system of SCHULZE and FÖRTSCH, amplification 40,000-fold; REICH et al., 1948). At the beginning of the systematic investigation of crustal structure it was necessary to develop new recording systems with higher magnification combined with more flexibility in field operations. Two systems were constructed and applied each in connection with high-sensitive electromagnetic and electrodynamic seismographs with a natural frequency of 1-2 cps: one system was based on electronic amplifiers of quite different types with low-sensitive galvanometers whereas the other used only high-sensitive galvanometers of the type Kipp and Zonen. Both systems reach a maximum magnification of 10^6 in the frequency range of about 5 to 20 cps. After a few years the use of seismometers of type FS-60 after BERCKHEMER (Stuttgart, Frankfurt) became more and more common.

At the beginning of the sixties between 35 and 40 pieces of equipment were available. Because of the limited number of stations and the rarity of large explosions it was not suitable to apply the method of continuous profiling, as used in eastern Europe in

---

Fig. 1. Location map of seismic-refraction observations in central Europe.
Code of observations: (a) Profiles: 16-080-02 shotpoint code - azimuth - reversed shotpoint
BO-215 shotpoint code - azimuth (no reversed shot), 06-280-V test measurements,
—·—●—·—common-depth-point profiles with center near *FR* Friedberg, *LO* Loreley
═════ X-350-WE moving shotpoint (X) - azimuth - fixed stations near Weißenburg (WE)
KMgW-profile: Profile recorded continuously, based on the correlation method of refracted waves (Korrelations-Methode gebrochener Wellen) (KNOTHE and WALTHER, 1968).
(b) Fan observations: 01-220-F shotpoint code - distance - F

● Shotpoints: (see also Table 1, 3.1)

01 Eschenlohe	*11* Bischofsheim	*21* Bermel
02 Hilders	*12* Romsthal	*22* Dorndorf
03 Gersfeld Schw.A.	*13* Dorheim	*23* Wilsenroth
04 Grossenritte	*14* Mehrberg	*24* Steinach
05 Birkenau	*15* Büdingen	*25* Vils
06 Adelebsen	*16* Taben Rodt	*26* Lohne
07 Gersfeld Nall.	*17* Bransrode	*27* Mauthaus Kronach
08 Kirchheimbolanden	*18* Lahr	*28* Ueffeln
09 Böhmischbruck	*19* Merlebach	*29* Suhl
10 Voggendorf	*20* Birresborn	
AL Altengönner	*HE* Helgoland	*SN* Saint Nabor
BA Col des Bagenelles	*MB* Merlebach	*VS* Všetaty
BO Boubin	*NE* Nepomyšl	*WI* Wissembourg
EI Eineborn	*SB* Steinbrunn	
HA Haslach	*SC* Schernberg	

o Cities:

AU Augsburg	*HH* Hamburg	*NÜ* Nürnberg
BA Basel	*IN* Innsbruck	*SA* Salzburg
BE Berlin	*KA* Karlsruhe	*ST* Strasbourg
BR Bremen	*KÖ* Köln	*SU* Stuttgart
ER Erfurt	*KS* Kassel	*UL* Ulm
FR Frankfurt	*LE* Leipzig	*WÜ* Würzburg
HA Hannover	*MS* Münster	
HE Heidelberg	*MÜ* München	

For more details in the southwest corner of the map (Rhine Graben area) see Fig. 3, Chapter 6.8 of this volume

deep-seismic sounding. Therefore, the stations were spaced on the average every 3-5 km; occasionally the spacing was smaller.

In the mid-sixties a third generation of equipment for explosion seismology was developed using magnetic-tape recording on the principle of multiplex-frequency modulation. This system is described in detail by BERCKHEMER (1970, 3.5 of this volume). In the beginning of the seventies the German geophysical institutions possessed some 60 homogeneous recording systems of this type, called MARS 66 (Magnetband-Apparatur für Refraktions-Seismik, Baujahr 1966). Consequently the density of data on profiles recorded more recently is higher. Furthermore the use of magnetic-tape recordings offers new possibilities of data evaluation.

The equipment of the second and third generation usually worked with three seismic channels. In most cases the three channels were connected with

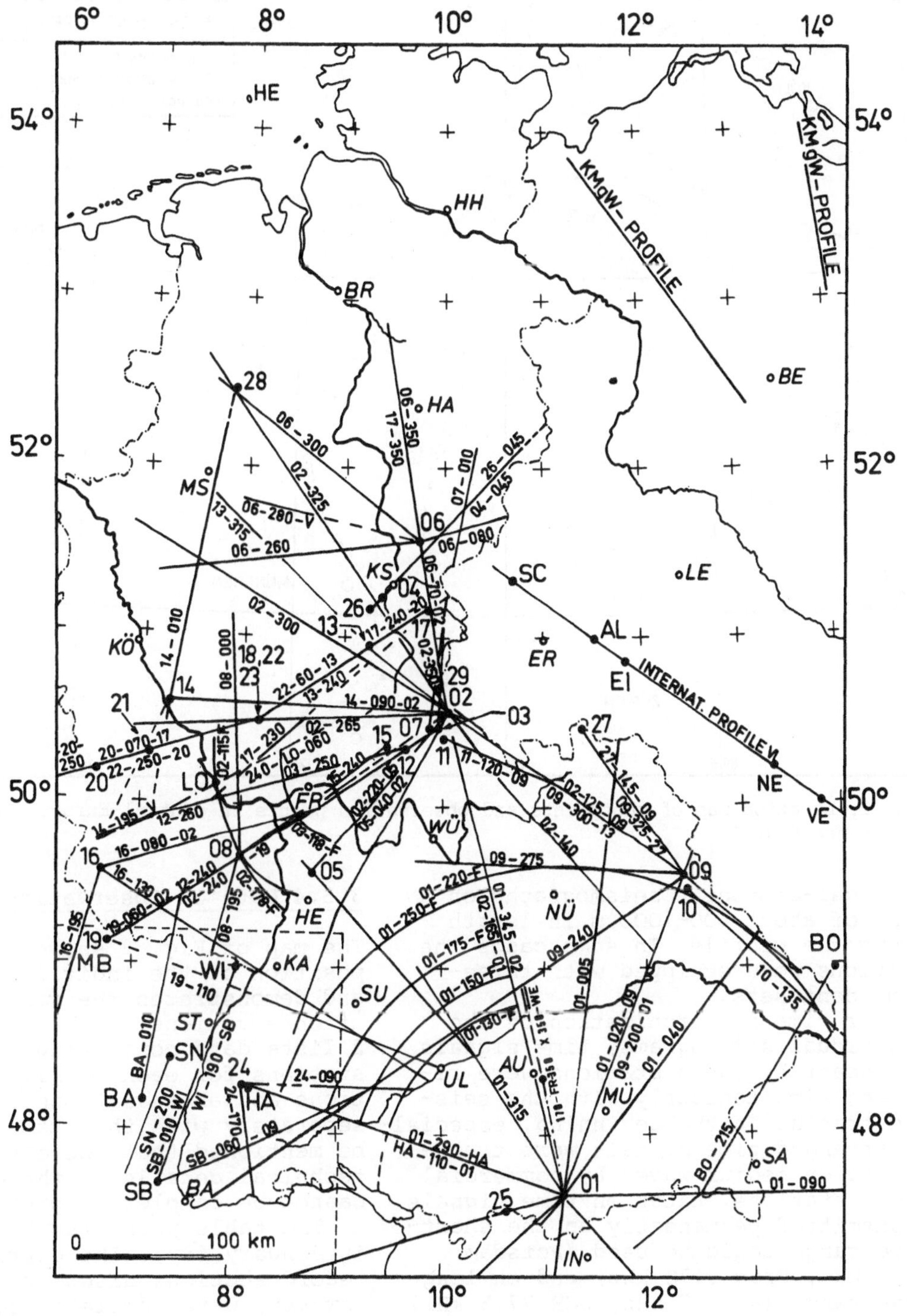

Fig. 1

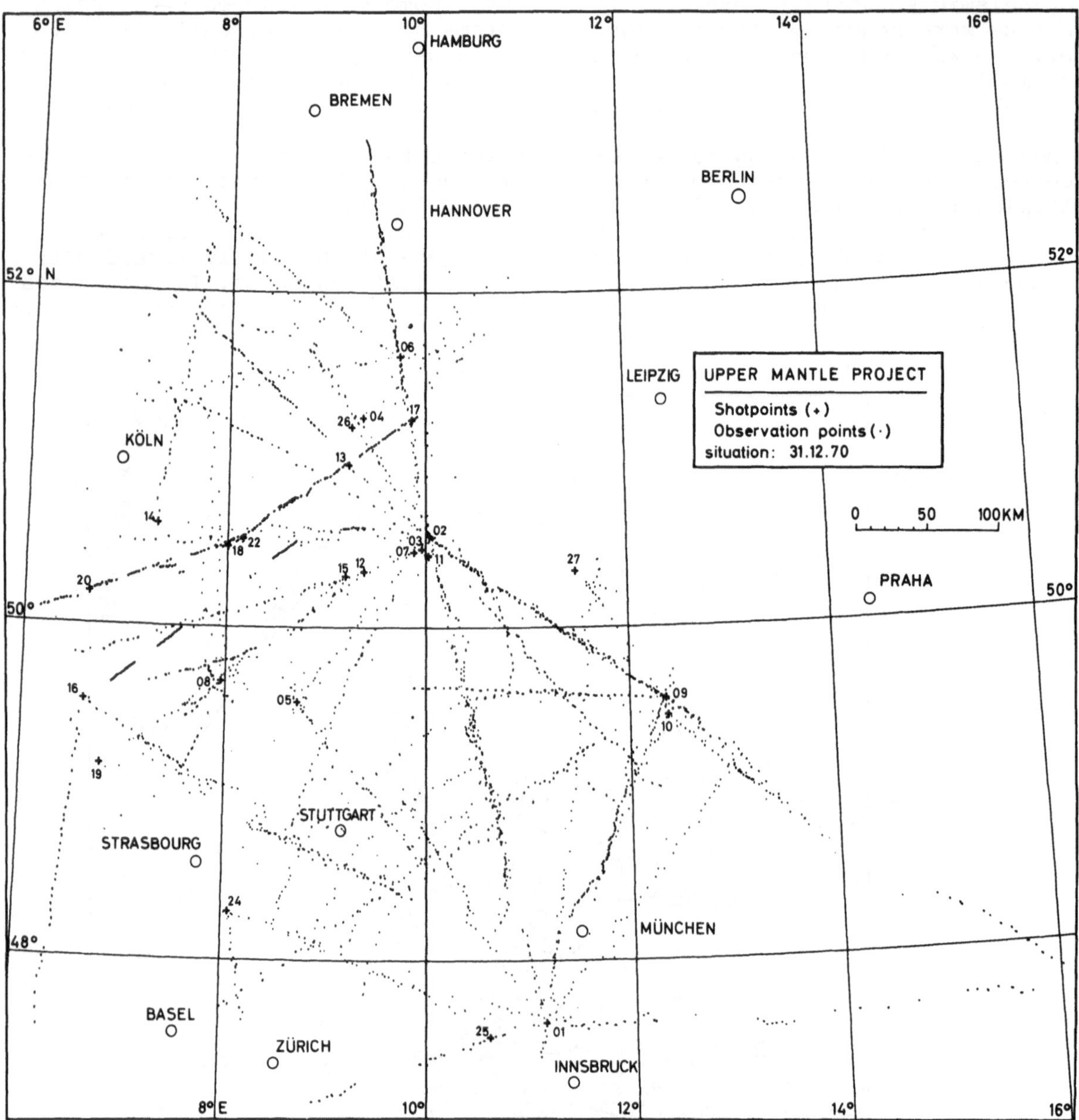

Fig. 2. Location map of shotpoints and observation points in central Europe. Code of shotpoints see Fig. 1

vertical-component seismographs on a line of about 500-1000 m in length along the profile. In some cases the stations were equipped with three-component sets.

For time synchronization of the individual seismograms, time signals broadcast by radio stations were recorded simultaneously with the seismic traces. In the beginning, especially arranged time signals were transmitted on medium waves by commercial radio stations. Later on time signals transmitted permanently on the short-wave range could be used (Potsdam 4525 kHz, Prag 3170 kHz) and on long-wave range (HBG 75 kHz, DCF 77.5 kHz).

3.3.4 Data of Observation Points

The map of Fig. 2 shows the actual position of the individual stations and demonstrates the varying density of observations on the profiles. Table 2 lists data concerning the recording stations for each profile. Explanations and abbreviations are given separately with this table. It must be mentioned that the data concerning technical details of the equipment cannot be completely presented.

The table presents the situation in January 1971. The corresponding record sections discussed in 3.4, however, were completed some time be-

Fig. 3. Examples of registration cards

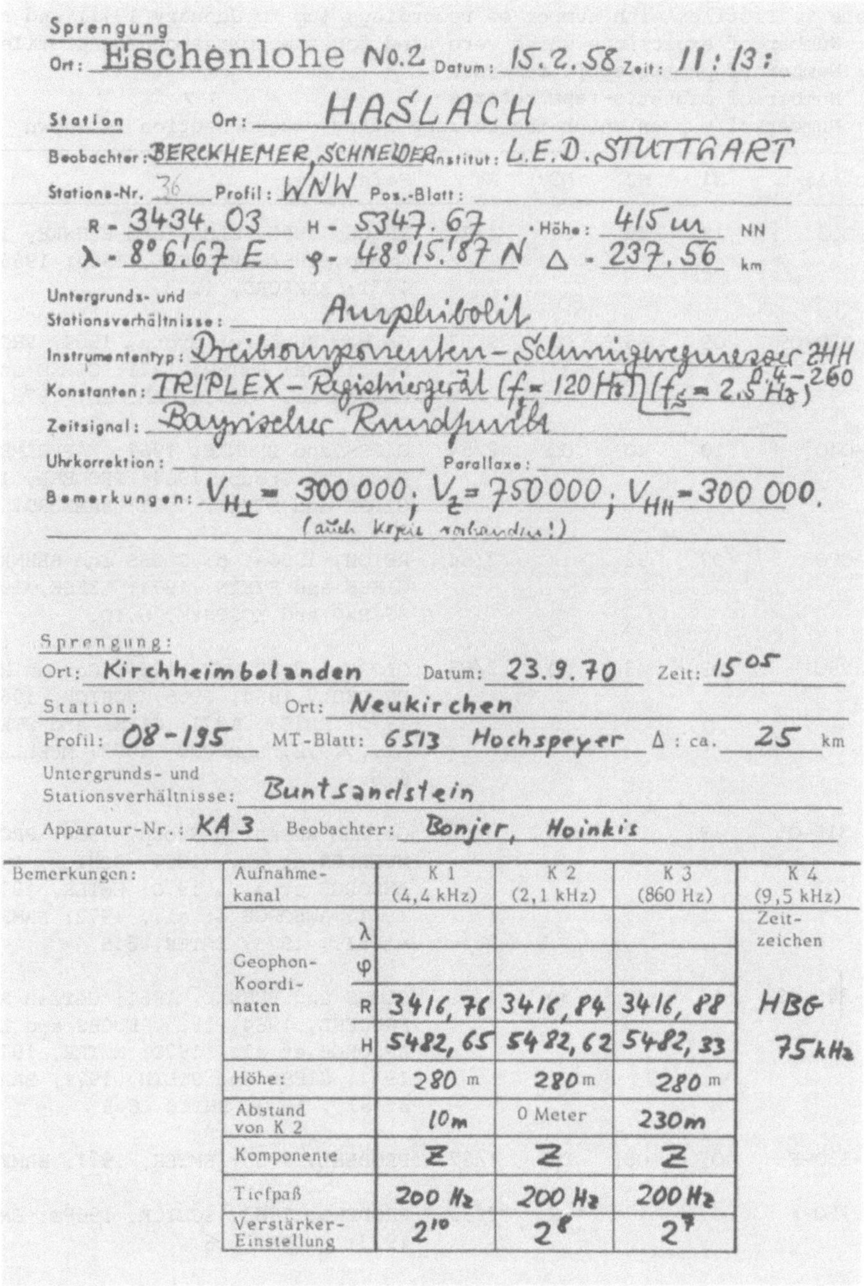

Sprengung

Ort: Eschenlohe No.2 Datum: 15.2.58 Zeit: 11:13:

Station Ort: HASLACH

Beobachter: BERCKHEMER, SCHNEIDER Institut: L.E.D. STUTTGART

Stations-Nr. 36 Profil: WNW Pos.-Blatt:

R - 3434,03 H - 5347,67 Höhe: 415 m NN

λ - 8°6,67 E φ - 48°15,87 N Δ - 237,56 km

Untergrunds- und Stationsverhältnisse: Amphibolit

Instrumententyp: Dreikomponenten-Schwingwegmesser 2HH

Konstanten: TRIPLEX-Registriergerät (f_f = 120 Hz) (f_s = 2,5 Hz) 0,4–260

Zeitsignal: Bayrischer Rundfunk

Uhrkorrektion: Parallaxe:

Bemerkungen: $V_{H\perp}$ = 300 000; V_z = 750 000; V_{HII} = 300 000.
(auch Kopie vorhanden!)

Sprengung:

Ort: Kirchheimbolanden Datum: 23.9.70 Zeit: 15^{05}

Station: Ort: Neukirchen

Profil: 08-195 MT-Blatt: 6513 Hochspeyer Δ : ca. 25 km

Untergrunds- und Stationsverhältnisse: Buntsandstein

Apparatur-Nr.: KA 3 Beobachter: Bonjer, Hoinkis

Bemerkungen:	Aufnahmekanal		K 1 (4,4 kHz)	K 2 (2,1 kHz)	K 3 (860 Hz)	K 4 (9,5 kHz)
	Geophon-Koordinaten	λ				Zeitzeichen
		φ				
		R	3416,76	3416,84	3416,88	HBG
		H	5482,65	5482,62	5482,33	75 kHz
	Höhe:		280 m	280 m	280 m	
	Abstand von K 2		10 m	0 Meter	230 m	
	Komponente		Z	Z	Z	
	Tiefpaß		200 Hz	200 Hz	200 Hz	
	Verstärker-Einstellung		2^{10}	2^{8}	2^{7}	

fore or after this date. Therefore, some stations are not listed or some records are not shown. In addition, for many stations the table contains the data for several traces, but in the corresponding record section only one trace could be plotted, as will be discussed in detail in 3.4.

For each field record a registration card was completed which is stored in the DFG-Zentralarchiv in the Geophysical Institute, Stuttgart (BINDER, 1969). Examples of these cards are presented in Fig. 3. Most could be included in the tables presented here, but additional details concerning equipment and environment of seismometers may be found on these cards.

In the DFG-Zentralarchiv all recordings except those obtained by magnetic-tape recording system are stored, whereas copies of the magnetic tapes are collected in the Institut für Meteorologie und Geophysik, Frankfurt. Copies can be obtained from both.

Table 1. Profiles with number of recordings (up to January 1971) and references
N1: Number of explosions which were used for the corresponding profile
N2: Number of photographic records
N3: Number of magnetic-tape records
N4: Number of map on which the corresponding record section is shown

Profile	N1	N2	N3	N4	References
01-005	18	48	01	1/30	REICH, 1952; CLOSS and BEHNKE, 1961; German Research Group, 1964; PRODEHL, 1964, 1965; GIESE and STEIN, 1971; BAMFORD, 1973.
01-020-09	09	65	00	2/57	German Research Group, 1964; PRODEHL, 1964, 1965; GIESE, 1968a; EMTER, 1971; GIESE and STEIN, 1971; LANDISMAN et al., 1971; BAMFORD, 1973.
01-040	10	40	00	2/54	CLOSS and BEHNKE, 1961; KAPPELMEYER, 1961; German Research Group, 1964; PRODEHL, 1964, 1965; GIESE, 1968a; GIESE and STEIN, 1971; BAMFORD, 1973.
01-090	09	52	14	3/64	REICH, 1960a, b; CLOSS and BEHNKE, 1961; KOSCHYK, 1969; GIESE and STEIN, 1971; GIESE, 1972; BAMFORD, 1973; ZSCHAU and KOSCHYK, 6.10.
01-290	09	43	00	2/55	CLOSS and BEHNKE, 1961; German Research Group, 1964; PRODEHL, 1964, 1965; SCHICK, 1968a; ANSORGE et al., 1970; EMTER, 1971; GIESE and STEIN, 1971; ANSORGE et al., 1972; BAMFORD, 1973; MUELLER et al., 1973; EMTER, 6.5.
01-315-05	19	87	15	3/63	German Research Group, 1964; PRODEHL, 1964, 1965; MUELLER et al., 1967; SCHICK, 1968a; EMTER, 1969; ANSORGE et al., 1970; EMTER, 1971; GIESE and STEIN, 1971; ANSORGE et al., 1972; BAMFORD, 1973; MUELLER et al., 1973; EMTER, 6.5.
01-345-02	13	39	30	1/31	CLOSS and BEHNKE, 1961; German Research Group, 1964; PRODEHL, 1964, 1965; FUCHS and LANDISMAN, 1966a, b; ANSORGE et al., 1970; EMTER, 1971; FUCHS and MÜLLER, 1971; GIESE and STEIN, 1971; BAMFORD, 1973; MUELLER et al., 1973; EMTER, 6.5.
01-130-F	07	08	07	1/37	PRODEHL, 1965; EMTER, 1971; BAMFORD, 1973; EMTER, 6.5.
01-150-F	07	17	02	2/53	PRODEHL, 1965; SCHICK, 1968a; EMTER, 1971; BAMFORD, 1973; EMTER, 6.5.
01-175-F	03	24	00	1/35	SCHICK, 1968a; EMTER, 1971; BAMFORD, 1973; EMTER, 6.5.
01-210-F	07	03	22	2/51	EMTER, 1971; BAMFORD, 1973; EMTER, 6.5.
01-220-F	04	11	15	1/12	PRODEHL, 1965; BAMFORD, 1973.
02-125-09	11	32	49	1/19	GIESE and STEIN, 1971; BAMFORD, 1973.
02-140	04	00	38	2/49	GIESE and STEIN, 1971; BAMFORD, 1973.
02-165-01	10	52	21	2/44	German Research Group, 1964; FUCHS and LANDISMAN, 1966a, b; ANSORGE et al., 1970; WANGEMANN, 1970; FUCHS and MÜLLER, 1971; GIESE and STEIN, 1971; BAMFORD, 1973; MUELLER et al., 1973; MÜLLER and FUCHS, 4.6.
02-215	04	33	16	2/47	MEISSNER and BERCKHEMER, 1967; MUELLER et al., 1969; WILDE, 1969; MEISSNER et al., 1970; GIESE and STEIN, 1971; BAMFORD, 1973; MUELLER et al., 1973; MEISSNER

Table 1 (continued)

Profile	N1	N2	N3	N4	References
					and VETTER, 1974; Rhine Graben Research Group, 1974; MEISSNER et al., 6.7.
02-220-05	02	17	00	2/39	STROBACH, 1963; German Research Group, 1964; GIESE and STEIN, 1971; BAMFORD, 1973; MEISSNER and VETTER, 1974; Rhine Graben Research Group, 1974; MEISSNER et al., 6.7; PRODEHL et al., 6.8.
02-240-19	03	27	08	2/48	German Research Group, 1964; HÄNEL, 1964; MEISSNER and BERCKHEMER, 1967; MEISSNER et al., 1970; GIESE and STEIN, 1971; BAMFORD, 1973; MEISSNER and VETTER, 1974; MEISSNER et al., 6.7.
02-265	05	39	00	1/10	BEHNKE, 1961a; CLOSS and BEHNKE, 1961; German Research Group, 1964; HÄNEL, 1964; GIESE and STEIN, 1971; BAMFORD, 1973.
02-300	04	14	24	1/16	THYSSEN et al., 1971; BAMFORD, 1973.
02-325	11	55	10	1/34	BROCKAMP, 1967; THYSSEN et al., 1971; BAMFORD, 1973.
02-350-06	07	19	22	1/9	German Research Group, 1964; FUCHS and LANDISMAN, 1966a, b; WANGEMANN, 1970; GIESE and STEIN, 1971; BAMFORD, 1973.
03-250	02	37	00	1/5	HÄNEL, 1963; German Research Group, 1964; HÄNEL, 1964; GIESE and STEIN, 1971; BAMFORD, 1973; MEISSNER and VETTER, 1974; MEISSNER et al., 6.7.
03-118-F	01	08	00	1/15	German Research Group, 1964; HÄNEL, 1964.
04-045	01	11	00	1/21	BEHNKE, 1961b; CLOSS and BEHNKE, 1961; German Research Group, 1964.
05-040-02	01	12	00	2/41	STROBACH, 1963; German Research Group, 1964; GIESE and STEIN, 1971; BAMFORD, 1973; MEISSNER and VETTER, 1974; Rhine Graben Research Group, 1974; MEISSNER et al., 6.7; PRODEHL et al., 6.8.
06-080	02	10	00	1/20	CLOSS and BEHNKE, 1961; PLAUMANN, 1961a; German Research Group, 1964.
06-170-02	04	45	08	1/25	German Research Group, 1964; FUCHS and LANDISMAN, 1966a, b; WANGEMANN, 1970; GIESE and STEIN, 1971; BAMFORD, 1973.
06-260	04	23	11	1/32	CLOSS and BEHNKE, 1961; PLAUMANN, 1961a; German Research Group, 1964; GIESE and STEIN, 1971; BAMFORD, 1973.
06-260-V	01	05	00	1/6a	-
06-280-V	01	00	05	1/6b	-
06-300	03	37	00	1/7	BROCKAMP, 1967; GIESE and STEIN, 1971; THYSSEN et al., 1971; BAMFORD, 1973.
06-350	03	04	46	1/2	HINZ et al., 6.3.
07-010	01	17	00	-	German Research Group, 1964; BAMFORD, 1973.
08-000	01	20	00	1/23	PLAUMANN, 1961b; HÄNEL, 1963; German Research Group, 1964; GIESE and STEIN, 1971; PRODEHL et al., 6.8.

Table 1 (continued)

Profile	N1	N2	N3	N4	References
08-195	01	00	12	1/36	Rhine Graben Research Group, 1974; EDEL et al., 1975; PRODEHL et al., 6.8.
09-135	02	19	09	-	-
09-200-01	05	80	00	3/61	GIESE, 1963; German Research Group, 1964; PRODEHL, 1964, 1965; GIESE, 1968a; GIESE and STEIN, 1971; LANDISMAN et al., 1971; BAMFORD, 1973.
09-240	07	44	22	2/50a,b	PRODEHL, 1964, 1965; EMTER, 1969; ANSORGE et al., 1970; EMTER, 1971; GIESE and STEIN, 1971; LANDISMAN et al., 1971; ANSORGE et al., 1972; EMTER, 6.5.
09-275-08	02	00	38	2/40	GIESE and STEIN, 1971; BAMFORD, 1973.
09-300-13	06	47	19	1/17	German Research Group, 1964; GIESE and STEIN, 1971; BAMFORD, 1973; STEIN, 1973.
09-325-27	02	00	14	1/26	PETERS, 1974.
10-135	04	55	00	2/56	GIESE, 1968a; WOLBER, 1968; GIESE and STEIN, 1971; BAMFORD, 1973; MÜLLER and FUCHS, 4.6.
10-305	01	01	04	1/4	-
11-120-09	01	16	00	1/18	STEIN, 1963; German Research Group, 1964; GIESE and STEIN, 1971; BAMFORD, 1973.
12-260	03	30	06	1/29	HÄNEL, 1963; German Research Group, 1964; HÄNEL, 1964; GIESE and STEIN, 1971; BAMFORD, 1973; MEISSNER and VETTER, 1974; MEISSNER et al., 6.7.
12-115-F	01	10	00	1/14	German Research Group, 1964.
13-060-17	02	00	11	-	-
13-120-09	05	53	20	1/13	STEIN, 1963; German Research Group, 1964; GIESE and STEIN, 1971; BAMFORD, 1973.
13-240-20	03	00	55	2/43a,b	GIESE and STEIN, 1971; BAMFORD, 1973.
14-010	05	13	34	1/3	GIESE and STEIN, 1971; THYSSEN et al., 1971.
14-090-02	02	36	00	1/28	German Research Group, 1964; GIESE and STEIN, 1971; BAMFORD, 1973.
14-195-V	01	00	13	1/38	-
15-240-08	02	15	01	2/42	MEISSNER and BERCKHEMER, 1967; MEISSNER et al., 1970; GIESE and STEIN, 1971; BAMFORD, 1973; MEISSNER and VETTER, 1974; MEISSNER et al., 6.7.
16-080-02	02	19	09	2/45	MEISSNER and BERCKHEMER, 1967; MEISSNER et al., 1970; GIESE and STEIN, 1971; BAMFORD, 1973; MEISSNER and VETTER, 1974; Rhine Graben Research Group, 1974; MEISSNER et al., 6.7; PRODEHL et al., 6.8.
16-130	08	12	58	3/59	MUELLER et al., 1967, 1969; ANSORGE et al., 1970; EMTER, 1971; GIESE and STEIN, 1971; ANSORGE et al., 1972; MUELLER et al., 1973; Rhine Graben Research Group, 1974; EDEL et al., 1975; EMTER, 6.5; PRODEHL et al., 6.8.

Table 1 (continued)

Profile	N1	N2	N3	N4	References
16-195	03	00	15	2/52	ANSORGE et al., 1972; MUELLER et al., 1973; Rhine Graben Research Group, 1974; EDEL et al., 1975; PRODEHL et al., 6.8.
17-170-01	03	05	25	1/22	WANGEMANN, 1970.
17-230	01	00	20	1/24	-
17-240-20	05	00	95	-	GIESE and STEIN, 1971.
17-230-V	01	05	03	1/1	THYSSEN et al., 1971.
17-350-06	04	14	38	1/8	WANGEMANN, 1970; GIESE and STEIN, 1971; HINZ et al., 6.3.
19-110	07	27	08	2/46	MUELLER et al., 1967, 1969; ANSORGE et al., 1970; EMTER, 1971; GIESE and STEIN, 1971; ANSORGE et al., 1972; MUELLER et al., 1973; Rhine Graben Research Group, 1974; EDEL et al., 1975; EMTER, 6.5; PRODEHL et al., 6.8.
20-070-17	05	00	59	-	GIESE and STEIN, 1971.
20-250	04	00	13	-	-
20-060-13	02	00	13	-	-
22-250-20	02	00	06	-	-
23-180-08	27	48	00	-	-
24-090	02	00	15	1/33	EMTER, 1971; ANSORGE et al., 1972; MUELLER et al., 1973; Rhine Graben Research Group, 1974; EDEL et al., 1975; EMTER, 6.5; PRODEHL et al., 6.8.
24-170	03	00	15	1/11	ANSORGE et al., 1970; EMTER, 1971; ANSORGE et al., 1972; MUELLER et al., 1973; Rhine Graben Research Group, 1974; EDEL et al., 1975; EMTER, 6.5; PRODEHL et al., 6.8.
27-145-09	01	00	08	1/27	PETERS, 1974.
HA-120-01	2	18	-	3/60	REICH et al., 1948; ROTHÉ and PETERSCHMITT, 1950; FÖRTSCH, 1951; ROTHÉ, 1958; CLOSS and BEHNKE, 1961; German Research Group, 1964; PRODEHL, 1965; LANDISMAN and MUELLER, 1966; MUELLER et al., 1967; ANSORGE et al., 1970; EMTER, 1971; GIESE and STEIN, 1971; ANSORGE et al., 1972; MUELLER et al., 1973; Rhine Graben Research Group, 1974; EDEL et al., 1975; PRODEHL et al., 6.8.
HE-160	1	22	-	2/58	CHARLIER, 1947; WILLMORE, 1949a, b; REICH, 1950; SCHULZE and FÖRTSCH, 1950; REICH et al., 1951; CLOSS and BEHNKE, 1961; German Research Group, 1964; GIESE and STEIN, 1971; ANSORGE, 1975.
240-LO-60	8	-	173	3/62	BARTELSEN, 1970; GLOCKE, 1970; GIESE and STEIN, 1971; MEISSNER and VETTER, 1974; GLOCKE and MEISSNER, 6.2; MEISSNER et al., 6.1, 6.7.
X-350-WE	18	108	-	3/65a,b	BRAM, 1967; BRAM and GIESE, 1968; MEISSNER, 1970; GIESE and STEIN, 1971.

Table 1 (continued)

Profile	N1	N2	N3	N4	References
350-FR-170	18	72	18	-	German Research Group, 1966; MEISSNER, 1966, 1967a, b, 1970.
BO-215	9	-	201	3/66a, b,c,d,e	HOLUB, 1972; BERANEK et al., 1973, 1974; HOLUB, 1974; MILLER and GEBRANDE, 6.11.

Explanation of Table 2

Head of each profile

QUARRY NO: code of quarry and name (STEIN and SCHRÖDER 3.1 of this volume, Table 1)

SMAP: number of topographic map 1:25,000

PROFILE: code of profile: shotpoint code - azimuth - code of reversed shotpoint

or code of fan: shotpoint code - recording distance - F

COORDINATES rectangular coordinates related to the Gauss-Krüger system, the first value refers to the eastern direction (X, km), the second one to the northern direction (Y, km). G indicates the geographic coordinates: longitude, latitude. Full degree's longitude (latitude) are designed by the first (first two) digit(s), the other digits give the minutes (735.24 means 7°35.24' longitude; 4926.13 means 49°26.13' latitude).

Columns

B number of explosion in the corresponding quarry (STEIN and SCHRÖDER 3.1 of this volume, Table 2)

IN abbreviation of the institute setting up the station

- B Berlin
- BO Bochum
- CL Clausthal-Zellerfeld
- F Frankfurt
- FF Fürstenfeldbruck
- GO Göttingen
- H Hannover
- HH Hamburg
- K Köln
- KA Karlsruhe
- KI Kiel
- KR Krefeld
- M München
- MS Münster
- MZ Mainz
- PR Prakla, Hannover
- S Stuttgart
- SE Seismos, Hannover
- ST Strasbourg

A equipment no.

C letter or number refers to the trace of the corresponding record

ALT altitude above sea level in m

X(KM) distance from the shotpoint in km

SMAP number of the topographic map 1 : 25,000 on which the station is located

LOCALITY village or another geographic point closest to the station

G type of equipment and/or seismometer

- HT seismometer type Hottinger
- MP magnetic-tape recording system after PETÖFALVI, Clausthal, Hamburg
- BA, B4, B5 seismometer after BAULE, Bochum
- HA seismometer type Hall-Sears
- MH seismometer after SCHULZE and FÖRTSCH, also called type Helgoland
- no indication: seismometer type FS 60

C

- X horizontal component parallel to the profile
- Y horizontal component perpendicular to the profile
- no indication = vertical component

DATA OF THE SHOTPOINTS

Q	LOCALITY	SMAP
01	ESCHENLOHE	8332
02	HILDERS	5426
03	GERSFELD/SCHWARZER ACKER	5525
04	GROSSENRITTE	4722
05	BIRKENAU	6418
06	ADELEBSEN	4324
07	GERSFELD/WALLENBERG	5525
08	KIRCHHEIMBOLANDEN	6313
09	BOEHMISCHBRUCK	6440
10	VOGGENDORF	6540
11	BISCHOFSHEIM	5526
12	ROMSTHAL	5622
13	DORHEIM	5021
14	MEHRBERG	5309
15	BUEDINGEN	5621
16	TABEN-RODT	6405
17	BRANSRODE	4725
18	LAHR	5414
19	MERLENBACH	6806
20	BIRRESBORN	5805
21	BERMEL	5708
22	DORNDORF	5414
23	HILSENROTH	5414
24	STEINACH	7714
25	VILS	8429
26	LOHNE	4821
27	MAUTHAUS KRONACH	5634

Q B	DATUM	SHOT-TIME H M	S	TO	COORDINATES LONGIT.	LATIT.	Y(LONG)	X(LAT)	ALT
0101	15.FEB 58	11.11	01.63	12.0	1109.46	4737.92	4436.70	5277.33	
02	15.FEB 58	11.13	01.76	12.5	1108.91	4737.89	4436.01	5277.26	
03	27.JUN 59	11.13	29.61	16.0	1108.94	4737.93	4436.05	5277.33	
04	17.OCT 59	11.13	30.26	8.8	1109.88	4737.99	4436.60	5277.46	
05	17.OCT 59	11.18	00.	5.6	1108.94	4737.91	4436.06	5277.32	
06	30.JAN 60	11.13	30.29	8.3	1108.80	4737.89	4435.90	5277.29	
07	30.APR 60	11.13	29.36	11.2	1108.80	4737.89	4435.90	5277.29	
08	02.JUL 60	11.13	31.76	7.3	1109.47	4738.00	4436.73	5277.48	
09	03.SEP 60	11.13	30.05	7.2	1108.80	4737.89	4435.90	5277.29	
10	19.NOV 60	11.13	16.53	12.5	1108.85	4737.91	4435.95	5277.31	
11	19.NOV 60	11.14	31.49	7.6	1109.42	4738.00	4436.65	5277.48	
12	04.MAR 61	11.13	13.69	9.7	1108.78	4737.89	4435.87	5277.28	
13	04.MAR 61	11.14	29.02	6.9	1109.42	4738.00	4436.65	5277.48	
14	10.MAY 61	17.56	30.28	9.2	1108.81	4737.87	4435.89	5277.23	
15	24.NOV 61	16.25	00.44	7.9	1109.50	4738.00	4436.75	5277.48	
16	02.DEC 61	11.13	30.09	19.5	1108.82	4737.88	4435.95	5277.28	
17	07.APR 62	11.16	30.28	6.0	1109.17	4737.95	4436.35	5277.39	
18	07.APR 62	11.18	00.04	11.0	1109.40	4738.00	4436.63	5277.48	
19	07.JUL 62	11.13	29.33	8.5	1108.78	4737.89	4435.87	5277.28	
20	21.SEP 62	17.10	29.67	13.0	1108.92	4737.92	4436.04	5277.34	
21	21.SEP 62	17.11	58.51	7.0	1109.41	4738.00	4436.64	5277.48	
22	24.MAY 63	18.50	30.10	11.0	1108.95	4737.92	4436.06	5277.33	
23	31.AUG 63	11.13	30.29	18.0	1108.83	4737.90	4435.91	5277.29	
24	06.DEC 63	15.30	49.80	11.0	1108.80	4737.89	4435.89	5277.28	670
25	06.DEC 63	15.32	52.62	9.2	1109.48	4737.99	4436.73	5277.47	650
26	17.APR 64	18.00	07.65	7.6	1109.47	4738.00	4436.73	5277.48	670
27	15.MAY 64	17.59	45.20	14.0	1108.94	4737.92	4436.06	5277.32	670
28	18.SEP 64	17.53	30.19	11.4	1108.73	4737.90	4435.81	5277.31	700
29	05.DEC 64	11.13	30.04	12.8	1108.88	4737.92	4435.98	5277.33	650
30	02.JUL 65	17.58	30.63	14.3	1108.86	4737.91	4435.97	5277.32	700
31	04.DEC 65	11.13	31.35	10.5	1108.85	4737.91	4435.93	5277.30	700
32	27.MAY 66	15.04	59.70	7.0	1108.87	4737.92	4435.98	5277.33	670
33	07.OCT 66	15.05	01.10	12.4	1108.86	4737.91	4435.99	5277.31	700
34	30.JUN 67	15.04	59.19	11.0	1108.83	4737.92	4435.92	5277.31	700
35	06.OCT 67	15.04	59.61	7.0	1108.84	4737.92	4435.96	5277.34	720
36	22.DEC 67	14.05	00.09	16.3	1108.72	4737.90	4435.78	5277.30	700
37	26.APR 68	15.05	00.40	10.2	1108.80	4737.90	4435.88	5277.28	700
38	26.JUL 68	15.05	00.03	9.1	1108.80	4737.90	4435.88	5277.28	700
39	06.DEC 68	15.04	59.38	7.1	1108.88	4737.91	4435.98	5277.33	700
40	14.MAR 69	15.05	00.59	7.7	1108.68	4737.90	4435.74	5277.28	700
41	13.JUN 69	15.05	00.51	5.7	1108.87	4737.92	4435.93	5277.33	700
42	18.JUL 69	15.05	00.59	12.8	1108.80	4737.91	4435.88	5277.31	700
43	24.OCT 69	15.04	59.86	21.7	1108.74	4737.90	4435.82	5277.31	700
44	27.FEB 70	15.05	01.40	8.8	1108.68	4737.91	4435.74	5277.30	700
45	10.JUL 70	15.05	01.13	16.4	1108.76	4737.91	4435.83	5277.31	700
46	06.NOV 70	15.05	00.54	8.1	1108.64	4737.91	4435.73	5277.32	700
0201	28.MAR 58	16.	.	10.0	1002.40	5032.50	3573.71	5601.03	730
02	18.SEP 58	16.00	.	10.0	1002.43	5032.46	3573.75	5600.95	730
03	11.MAR 60	16.	.	7.6	1002.48	5032.47	3573.71	5600.99	730
04	20.MAY 60	16.	.	8.6	1002.34	5032.46	3573.64	5600.96	730
05	07.DEC 61	15.30	31.02	14.6	1002.45	5032.47	3573.79	5600.98	730
06	16.NOV 62	16.05	36.21	14.0	1002.38	5032.48	3573.71	5600.99	730
07	07.JUN 63	16.05	40.81	9.8	1002.47	5032.45	3573.80	5600.98	730
08	11.OCT 63	16.05	01.34	8.1	1002.34	5032.49	3573.64	5601.02	730
09	16.DEC 63	15.30	10.51	13.0	1002.38	5032.50	3573.68	5601.03	730
10	28.AUG 64	16.30	00.64	16.0	1002.37	5032.49	3573.68	5601.04	750
11	26.MAR 65	16.00	00.70	16.0	1002.42	5032.50	3573.75	5601.06	730
12	07.MAY 65	16.05	01.04	6.8	1002.32	5032.50	3573.62	5601.04	730
13	08.OCT 65	16.30	01.35	18.0	1002.42	5032.52	3573.73	5601.08	730
14	22.APR 66	16.05	00.67	10.8	1002.36	5032.52	3573.66	5601.08	730
15	29.JUL 66	16.05	01.11	14.9	1002.51	5032.50	3573.85	5601.04	730
16	07.OCT 66	16.05	01.02	8.4	1002.41	5032.53	3573.71	5601.09	730
17	14.APR 67	16.05	01.09	12.3	1002.48	5032.54	3573.80	5601.10	730
18	14.JUL 67	16.05	00.93	9.0	1002.28	5032.47	3573.57	5601.00	730
19	25.AUG 67	16.05	00.74	4.1	1002.52	5032.44	3573.85	5600.95	750
20	06.OCT 67	16.05	00.80	11.3	1002.42	5032.52	3573.73	5601.09	730
21	24.NOV 67	16.05	00.65	5.4	1002.53	5032.43	3573.87	5600.93	700
22	22.MAR 68	16.05	00.80	10.0	1002.53	5032.49	3573.86	5601.04	730
23	26.APR 68	16.05	00.74	9.2	1002.39	5032.50	3573.70	5601.06	730
24	23.AUG 68	16.05	00.90	11.8	1002.48	5032.56	3573.79	5601.13	730
25	07.MAR 69	15.05	00.89	10.8	1002.51	5032.51	3573.84	5601.07	730
26	11.APR 69	16.05	00.73	9.4	1002.42	5032.51	3573.74	5601.07	730
27	27.JUN 69	16.05	00.91	12.0	1002.36	5032.50	3573.67	5601.06	730
28	13.MAR 70	16.05	00.60	21.8	1002.44	5032.55	3573.76	5601.09	730
29	16.JUL 70	16.05	00.74	11.7	1002.33	5032.49	3573.63	5601.04	730
30	02.OCT 70	15.45	00.87	21.4	1002.39	5032.52	3573.70	5601.09	730
0301	12.JUN 58	16.00	.	4.6	0957.79	5027.78	3568.39	5592.23	760
02	15.MAR 62	16.46	01.85	5.0	0957.72	5027.81	3568.33	5592.28	730
03	18.MAY 63	15.00	00.48	5.4	0957.77	5027.83	3568.35	5592.30	730
0401	27.FEB 59	16.29	.	5.0	0921.93	5114.80	3525.52	5679.01	310
0501	03.APR 59	13.59	.	7.3	0843.11	4933.22	3479.63	5490.67	220
0601	12.JUN 59	17.01	.	8.2	0944.76	5136.44	3551.67	5719.32	350
02	16.JUN 61	17.30	.	9.0	0944.65	5136.53	3551.55	5719.50	250
03	26.JUL 63	16.05	02.68	6.3	0944.74	5136.48	3551.64	5719.41	350
04	26.APR 65	16.15	01.32	6.5	0944.64	5136.55	3551.54	5719.52	350
05	10.JUN 66	15.45	01.08	6.4	0944.62	5136.55	3551.52	5719.53	350
06	23.JUN 67	15.45	00.80	6.0	0944.80	5136.48	3551.72	5719.40	350
07	14.MAR 68	15.45	01.20	5.4	0944.68	5136.53	3551.58	5719.50	350
08	18.SEP 68	15.45	00.82	3.0	0944.67	5136.59	3551.57	5719.60	400
09	03.OCT 68	15.45	01.00	5.9	0944.61	5136.51	3551.51	5719.46	350
10	13.JUN 69	15.45	00.93	6.1	0944.67	5136.55	3551.56	5719.54	350
11	17.JUL 69	15.45	00.36	6.4	0944.80	5136.50	3551.73	5719.44	350
12	22.MAY 70	15.45	00.84	7.5	0944.78	5136.51	3551.70	5719.46	350
0701	03.JUL 59	16.01	.	15.5	0953.04	5026.69	3562.79	5590.14	720
0801	13.APR 60	16.00	.	2.6	0757.49	4940.54	3424.81	5504.75	310
02	23.SEP 70	15.05	00.56	10.5	0757.59	4940.51	3424.93	5504.67	380
0901	09.JUL 60	11.13	31.69	2.7	1221.34	4934.09	4525.73	5492.33	
02	29.APR 61	11.13	29.27	5.3	1221.32	4934.10	4525.70	5492.33	
03	08.JUN 62	17.59	30.38	5.3	1221.36	4934.12	4525.75	5492.38	510
04	24.AUG 63	11.30	29.48	6.0	1221.36	4934.12	4525.75	5492.38	510
05	26.JUN 64	16.59	49.37	5.0	1221.36	4934.11	4525.75	5492.38	510
06	28.MAY 65	11.13	30.02	5.0	1221.36	4934.12	4525.75	5492.38	510
07	07.APR 66	11.13	30.13	6.5	1221.36	4934.12	4525.75	5492.38	525
08	07.APR 67	10.54	59.19	5.0	1221.36	4934.12	4525.75	5492.38	525
09	20.APR 68	11.05	00.41	6.0	1221.36	4934.11	4525.75	5492.36	510
10	25.APR 69	11.05	01.19	3.5	1221.36	4934.11	4525.75	5492.36	510
11	09.JUL 70	18.05	00.64	5.0	1221.31	4934.15	4525.75	5492.47	520
1001	29.APR 61	13.59	48.50		1222.57	4927.91	4527.20	5480.88	
02	30.APR 61	14.00	.		1222.57	4927.91	4527.20	5480.88	
03	26.APR 63	11.59	30.99	7.0	1222.32	4927.93	4526.95	5480.92	500
04	24.APR 64	16.59	46.04	5.0	1222.32	4927.93	4526.95	5480.92	500
05	16.JUL 65	11.14	42.17	3.0	1222.32	4927.93	4526.95	5480.92	500
06	26.NOV 65	11.13	30.13	10.0	1222.32	4927.93	4526.95	5480.92	
07	19.APR 68	17.05	00.67	2.7	1222.32	4927.93	4526.95	5480.92	500
1101	10.JUN 61	11.13	30.90	6.5	1000.13	5025.90	3571.19	5588.76	680
02	29.AUG 70	17.00	01.51	5.1	1001.71	5025.29	3573.08	5587.66	620
1201	30.JUN 61	16.00	.	3.8	0922.88	5019.87	3527.15	5577.18	240
02	29.NOV 63	15.59	58.49	1.4	0922.87	5019.83	3527.13	5577.12	250
03	30.OCT 64	14.59	11.35	4.6	0922.35	5019.88	3527.13	5577.19	240
04	18.MAR 66	17.00	02.09	4.2	0922.79	5019.83	3527.05	5577.10	270
05	03.JUN 66	16.05	00.97	2.4	0922.83	5019.84	3527.07	5577.13	250
06	10.MAY 68	16.05	00.48	6.8	0922.83	5019.86	3527.09	5577.16	230
1301	25.OCT 62	16.05	30.82	6.3	0913.10	5057.89	3515.34	5647.65	330
02	12.MAY 66	16.05	00.98	4.5	0913.13	5057.92	3515.38	5647.67	320
03	15.FEB 67	16.05	01.15	5.3	0913.09	5057.89	3515.33	5647.62	320
04	01.MAR 68	15.05	01.09	10.0	0913.14	5057.90	3515.38	5647.65	320
05	20.MAR 69	16.05	01.33	7.8	0913.12	5057.89	3515.36	5647.63	320
06	05.MAR 70	16.05	01.11	7.0	0913.11	5057.87	3515.35	5647.59	320
1401	12.DEC 64	10.59	08.04	6.0	0717.87	5036.60	3379.52	5609.52	420
02	09.APR 65	18.00	01.47	2.3	0717.96	5036.58	2591.97	5608.90	420
03	19.SEP 67	17.02	00.63	6.0	0717.85	5036.55	2591.83	5608.83	400
04	04.OCT 68	17.05	01.63	4.2	0717.85	5036.54	2591.83	5608.81	400
05	06.FEB 69	17.05	01.38	3.9	0717.84	5036.53	2591.82	5608.80	400
06	28.MAR 69	17.05	00.76	3.9	0717.85	5036.53	2591.84	5608.80	400
07	16.MAY 69	17.05	01.10	3.1	0717.82	5036.54	2591.79	5608.82	400
1501	09.MAR 66	16.00	01.67	2.6	0911.38	5018.12	3513.51	5573.89	280
02	31.AUG 66	17.00	00.79	9.2	0911.28	5018.12	3513.40	5573.88	280
1601	27.APR 66	15.45	00.43	2.1	0636.74	4932.96	2544.30	5490.33	310
02	30.MAR 67	15.45	00.68	6.3	0636.69	4932.93	2544.25	5490.29	270
03	04.OCT 67	15.45	00.96	3.3	0636.64	4932.94	2544.19	5490.31	185
04	24.APR 68	15.50	01.11	4.6	0636.68	4932.95	2544.23	5490.32	230
05	11.JUN 69	15.45	00.79	4.5	0636.65	4932.96	2544.19	5490.33	185
06	08.OCT 69	15.45	01.13	7.7	0636.72	4932.95	2544.28	5490.32	230
07	04.JUN 70	14.50	00.88	4.5	0636.64	4932.96	2544.17	5490.34	185
08	29.OCT 70	14.50	00.64	5.2	0636.69	4932.96	2544.24	5490.34	240
1701	29.JUN 66	16.04	58.36	16.0	0951.55	5113.96	3560.01	5677.74	690
02	28.JUL 67	16.05	11.09	24.0	0951.54	5113.95	3559.99	5677.72	690
03	20.JUN 68	17.04	58.47	19.5	0951.44	5113.90	3559.87	5677.63	690
04	02.OCT 68	15.05	27.09	14.0	0951.37	5113.92	3559.79	5677.66	690
05	18.JUL 69	17.05	02.97	22.0	0951.59	5113.93	3560.05	5677.69	690
06	12.JUN 70	16.05	01.16	23.0	0951.51	5113.90	3559.96	5677.63	690
07	23.OCT 70	16.05	00.63	20.3	0951.63	5114.00	3560.09	5677.80	690
1801	07.JUL 66	16.05	01.80	3.4	0809.57	5031.40	3440.38	5598.84	340
1901	25.OCT 66	08.46	56.62	0.9	0647.93	4909.73	2558.25	5447.40	
02	25.OCT 66	09.06	57.69	5.3	0647.95	4909.70	2558.28	5447.35	
03	25.OCT 66	11.07	01.69	2.8	0647.97	4909.67	2558.30	5447.29	
04	25.OCT 66	11.37	02.40	1.6	0647.98	4909.65	2558.32	5447.25	
05	18.NOV 66	08.57	00.08	4.5	0647.27	4909.81	2557.45	5447.54	
06	18.NOV 66	10.56	59.88	5.0	0647.31	4909.83	2557.50	5447.58	
07	18.NOV 66	12.36	59.32	3.4	0647.35	4909.85	2557.55	5447.62	
2001	17.MAY 68	16.30	00.39	5.7	0637.82	5011.76	2545.02	5562.26	425
02	13.JUL 68	14.30	01.39	4.3	0637.78	5012.19	2544.95	5563.06	380
03	09.APR 69	16.45	00.57	2.9	0638.18	5011.39	2545.45	5561.58	420
04	19.JUN 70	16.30	00.55	4.2	0637.81	5011.69	2545.00	5562.13	430
05	30.JUN 70	16.30	00.61	7.3	0638.19	5011.39	2545.46	5561.58	420
2201	22.AUG 70	13.00	01.06	2.9	0800.57	5028.63	3429.71	5593.85	250
02	05.DEC 70	12.30	00.71	2.9	0800.57	5028.62	3429.71	5593.81	250
2401	19.JUL 68	16.50	03.89	2.9	0803.64	4817.53	3430.44	5350.96	260
02	27.JUN 69	16.40	36.25	1.8	0803.73	4817.64	3430.42	5350.98	260
03	18.JUL 69	16.47	47.61	2.4	0803.68	4817.70	3430.35	5351.09	260
04	03.JUL 70	16.46	46.88	2.0	0803.68	4817.70	3430.35	5351.09	230
05	10.NOV 70	16.47	18.28	2.0	0804.12	4817.68	3430.40	5351.00	270
2501	29.MAY 69	16.05	03.21	1.4	1037.00	4732.70	4395.88	5268.22	1060
2601	04.SEP 70	16.30	00.78	8.4	0914.73	5111.31	3517.17	5672.50	345
2701	16.DEC 70	15.30	02.43	9.8	1129.34	5019.87	4463.61	5577.22	420

DATA OF THE OBSERVATION POINTS

QUARRY-NO. 01 ESCHENLOHE SMAP 8332
PROFILE 01-005 COMPILED BY AICHELE.FUCHS.KAMINSKI

B	IN	AC	COORDINATES	ALT	X(KM)	SMAP	LOCALITY	G C
06	M		4442.69 5278.35	660	6.87	8333	OHLSTADT	MAY
12	M		4442.69 5278.35	660	6.90	8333	OHLSTADT	MA
12	M		4442.69 5278.35	660	6.90	8333	OHLSTADT	MAX
12	M		4442.69 5278.35	660	6.90	8333	OHLSTADT	MAY
07	M		4442.53 5283.00	680	8.75	8333	RIEGSEE-S	
07	M		4443.10 5283.83	680	9.72	8333	RIEGSEE-N	
06	CL		4439.11 5293.00	623	16.03	8233	GUTLINDEN	
07	M		4440.78 5295.71	650	19.06	8233	EBERFING	
13	M		4439.88 5301.32	632	24.06	8133	STEINBERG	MA
12	M		4439.88 5301.32	632	24.37	8133	STEINBERG	MA
13	M		4440.36 5304.52	610	27.30	8133	WILZHOFEN	MA
12	M		4440.36 5304.52	610	27.61	8133	WILZHOFEN	MA
06	M		4442.42 5307.37	670	30.78	8033	MONATSHAUSEN	
15	M	B	4440.65 5309.80	730	32.56	8033	KERSCHLACH	
15	M	C	4440.65 5309.80	730	32.56	8033	KERSCHLACH	MA
13	M		4442.13 5312.95	700	35.89	8033	MACHTLFING	MA
12	M		4442.13 5312.95	700	36.21	8033	MACHTLFING	MA
19	M		4442.17 5315.12	695	38.37	8033	ROTHENFELD	
13	M		4442.79 5316.35	690	39.35	8033	LANDSTETTEN	MA
12	M		4442.79 5316.35	690	39.68	8033	LANDSTETTEN	MA
15	M		4442.78 5317.33	665	40.31	8033	LANDSTETTEN	MA
15	M		4442.75 5319.24	665	42.19	7933	FRIEDING-B	MA
22	M		4441.62 5319.50	660	42.54	7933	FRIEDING	MA
06	M		4441.49 5319.54	664	42.62	7933	FRIEDING	MH
13	M		4442.93 5321.87	630	44.83	7933	HACKEN	MA
12	M		4442.93 5321.87	630	45.14	7933	HACKEN	MA
04	M		4444.24 5325.22	595	48.37	7933	ETTENHOFEN	MH
05	M		4444.24 5325.22	595	48.59	7933	ETTENHOFEN	MH
13	M		4445.52 5329.12	572	52.39	7933	ST.GILDEN/H.	MA
12	M		4445.52 5329.12	572	52.73	7933	ST.GILDEN/H.	MA
04	M		4445.70 5330.58	565	53.89	7833	THALHOF	MH
05	M		4445.70 5330.58	565	54.12	7833	THALHOF	MH
06	M		4445.68 5330.59	565	54.19	7833	THALHOF	MH
21	M	A	4446.12 5336.52	570	59.80	7833	FUERSTENFELDBR.	Y
21	M	B	4446.12 5336.52	570	59.80	7833	FUERSTENFELDBR.	
21	M	C	4446.12 5336.52	570	59.80	7833	FUERSTENFELDBR.	X
11	M		4446.24 5336.56	565	59.86	7833	FUERSTENFELDBR.	MH
11	M		4446.24 5336.56	565	59.86	7833	FUERSTENFELDBR.	
20	M	A	4446.12 5336.52	570	60.03	7833	FUERSTENFELDBR.	Y
20	M	B	4446.12 5336.52	570	60.03	7833	FUERSTENFELDBR.	
20	M	C	4446.12 5336.52	570	60.03	7833	FUERSTENFELDBR.	X
07	M		4446.15 5336.53	570	60.12	7833	FUERSTENFELDBR.	
10	M		4446.24 5336.56	565	60.14	7833	FUERSTENFELDBR.	MH
10	M		4446.24 5336.56	565	60.14	7833	FUERSTENFELDBR.	
06	M		4446.24 5336.56	565	60.16	7833	FUERSTENFELDBR.	
09	M		4446.24 5336.56	565	60.16	7833	FUERSTENFELDBR.	
16	M		4467.64 5334.48	516	65.39	7835	MUENCHEN	
13	M		4448.23 5342.52	508	66.06	7733	GERLINDEN	MA
12	M		4448.23 5342.52	508	66.40	7733	GERLINDEN	MA
06	M		4445.02 5346.98	520	70.28	7733	ZOETZELHOFEN	
06	M		4445.02 5346.98	520	70.28	7733	ZOETZELHOFEN	X
06	M		4445.02 5346.98	520	70.28	7733	ZOETZELHOFEN	Y
07	M		4445.02 5346.98	520	70.28	7733	ZOETZELHOFEN	
07	M		4445.02 5346.98	520	70.28	7733	ZOETZELHOFEN	X
07	M		4445.02 5346.98	520	70.28	7733	ZOETZELHOFEN	Y
13	M		4447.77 5350.15	508	73.52	7733	ANGER SE	MA
13	M		4447.77 5350.15	508	73.52	7733	ANGER SE	
12	M		4447.77 5350.15	508	73.83	7733	ANGER SE	MA
12	M		4447.77 5350.15	508	73.83	7733	ANGER SE	
04	M		4449.39 5353.15	505	76.76	7633	GROSS-BERGHOFEN	
21	M		4448.72 5358.70	520	82.11	7633	NEUSREUTH	
20	M		4448.72 5358.70	520	82.34	7633	NEUSREUTH	
06	M		4448.75 5358.72	515	82.44	7633	NEUSREUTH	MA
04	K		4449.95 5359.32	480	82.95	7633	EICHHOFEN	
05	K		4449.95 5359.32	480	83.17	7633	EICHHOFEN	
13	M		4450.42 5365.05	500	88.64	7533	MICHELSKIRCHEN	MA
12	M		4450.42 5365.05	500	88.97	7533	MICHELSKIRCHEN	MA
04	CL		4450.10 5367.52	510	91.07	7533	UNT.DUNKELHAUSEN	
04	CL		4450.10 5367.52	510	91.07	7533	UNT.DUNKELHAUSEN	X
04	CL		4450.10 5367.52	510	91.07	7533	UNT.DUNKELHAUSEN	Y
05	CL		4450.10 5367.52	510	91.29	7533	UNT.DUNKELHAUSEN	
05	CL		4450.10 5367.52	510	91.29	7533	UNT.DUNKELHAUSEN	X
05	CL		4450.10 5367.52	510	91.29	7533	UNT.DUNKELHAUSEN	Y
13	M	A	4450.22 5371.75	504	95.24	7533	DUCKENRIED	MA
13	M	B	4450.22 5371.75	504	95.24	7533	DUCKENRIED	
12	M	A	4450.22 5371.75	504	95.55	7533	DUCKENRIED	MA
12	M	B	4450.22 5371.75	504	95.55	7533	DUCKENRIED	
06	M		4451.78 5375.46	495	99.45	7434	KREUTH	SI
06	K		4450.72 5377.92	460	101.71	7433	OBER-LAUTERBACH	
06	M		4455.20 5389.80	415	114.15	7334	WINTERSOLN	BS
06	CL		4456.85 5410.05	391	134.40	7134	WETTSTETTEN	BS
06	M		4456.90 5416.64	480	140.92	7134	SCHELLDORF	
23	CL		4458.20 5422.70	500	147.11	7034	GEBELSEE	Y
23	CL		4458.20 5422.70	500	147.11	7034	GEBELSEE	X
06	M		4460.52 5431.35	410	156.01	6934	KIRCHANHAUSEN	
06	M		4461.14 5441.13	525	165.77	6834	WINTERSZHOFEN	
06	M		4461.27 5449.64	460	174.21	6834	HAPPERSDORF	
06	M		4464.32 5459.02	525	183.94	6735	HELENA-E	
06	M		4464.32 5459.02	525	183.94	6735	HELENA-E	
06	M		4464.32 5459.02	525	183.94	6735	HELENA-E	
06	M		4464.44 5468.30	520	193.14	6635	LITZLOHE	
06	M		4464.44 5468.30	520	193.14	6635	LITZLOHE	
06	M		4464.44 5468.30	520	193.14	6635	LITZLOHE	
07	M		4464.70 5476.69	545	201.47	6535	LIERITZHOFEN	
07	M		4466.50 5482.06	580	207.04	6535	MITTELBURG	
07	M		4468.10 5490.85	550	215.97	6435	RUINE HAUSECK	
33	MM		4458.30 5494.56	627	218.40	6434	ALGERSDORF	
34	MM	3	4465.14 5494.24	476	218.89	6435	HARTENSTEIN	
34	MM	1	4464.95 5495.16	493	219.77	6435	HARTENSTEIN	
34	MM	2	4464.95 5495.16	493	219.77	6435	HARTENSTEIN	
07	M		4470.77 5500.10	430	225.52	6335	KROTTENSEE	
07	MM		4471.00 5514.09	540	239.39	6235	TROSCHENREUTH	
07	MM		4472.23 5522.13	440	247.52	6135	CREUSSEN	
26	MZ		4469.13 5528.89	453	253.49	6135	ROEDENDORF	
26	F		4469.75 5537.30	445	261.91	5935	BAYREUTH	
26	F		4468.25 5546.17	420	270.53	5935	MICHELSREUTH	
27	F		4473.06 5550.16	503	275.34	5935	B.MARKTSCHORGAST	
12	M		4473.97 5554.02	530	279.35	5835	WEISSENBACH	
23	CL		4477.85 5556.19	553	282.03	5836	STAMMBACH	
27	F		4478.02 5560.20	625	285.98	5836	FOERSTENREUTH	
23	CL		4477.42 5565.51	660	291.19	5736	WUESTENSELBITZ	
27	CL		4476.72 5568.39	675	293.90	5736	SUTTENBACH	
27	CL		4476.72 5568.44	675	293.95	5736	SUTTENBACH	
27	CL		4479.15 5572.17	565	297.99	5736	HAIDENGRUEN	
27	CL		4479.19 5572.18	565	298.01	5736	HAIDENGRUEN	
33	MM		4481.21 5577.19	585	303.27	5636	NAILA	
33	MM		4479.87 5583.90	500	309.72	5636	ISSIGAU	

QUARRY-NO. 01 ESCHENLOHE SMAP 8332
PROFILE 01-020-09 COMPILED BY GIESE.PRODEHL

B	IN	AC	COORDINATES	ALT	X(KM)	SMAP	LOCALITY	G C
09	M		4453.55 5324.78	565	50.66	7934	GAUTING	
15	M		4467.64 5334.48	516	64.83	7835	MUENCHEN	
17	M		4467.64 5334.48	516	65.10	7835	MUENCHEN	
24	M	A	4455.45 5347.10	510	72.51	7734	GUENDING	
25	M	B	4455.44 5347.58	510	72.56	7734	GUENDING	
24	M	B	4455.44 5347.58	510	72.97	7734	GUENDING	
25	M	C	4455.51 5348.03	500	73.02	7734	GUENDING	
24	M	C	4455.51 5348.03	500	73.42	7734	GUENDING	
25	M	A	4456.08 5350.00	510	75.07	7734	PELHEIM	
24	M	A	4456.08 5350.00	510	75.47	7734	PELHEIM	
25	M	B	4456.26 5350.46	510	75.56	7734	PELHEIM	
24	M	B	4456.26 5350.46	510	75.96	7734	PELHEIM	
25	M	C	4456.31 5350.94	505	76.03	7734	PELHEIM	
24	M	C	4456.31 5350.94	505	76.44	7734	PELHEIM	
25	M	A	4457.55 5354.19	490	79.49	7634	SIGMERTSHS/N	
24	M	A	4457.55 5354.19	490	79.90	7634	SIGMERTSHS/N	
25	M	B	4457.79 5354.73	490	80.08	7634	SIGMERTSHS/N	
24	M	B	4457.79 5354.73	490	80.49	7634	SIGMERTSHS/N	
13	M		4486.00 5341.83	517	81.09	7736	FINSING	
12	M		4486.00 5341.83	517	81.73	7736	FINSING	
25	M	A	4459.95 5356.00	490	81.89	7634	BIBERBACH	
24	M	A	4459.95 5356.00	490	82.31	7634	BIBERBACH	
25	M	B	4460.29 5356.69	492	82.65	7634	BIBERBACH	
24	M	B	4460.29 5356.69	492	83.07	7634	BIBERBACH	
25	M	C	4460.61 5357.32	488	83.34	7634	BIBERBACH	
24	M	C	4460.61 5357.32	488	83.77	7634	BIBERBACH	
25	M	A	4462.65 5359.29	475	85.83	7634	GIEBING	
24	M	A	4462.65 5359.29	475	86.27	7634	GIEBING	
25	M	B	4463.00 5359.70	480	86.32	7634	GIEBING	
25	M	C	4463.11 5359.90	475	86.55	7634	GIEBING	
24	M	B	4463.00 5359.70	480	86.76	7634	GIEBING	
24	M	C	4463.11 5359.90	475	86.99	7634	GIEBING	
25	M	A	4464.19 5361.25	500	88.17	7635	LAUTERBACH	
24	M	A	4464.19 5361.25	500	88.61	7635	LAUTERBACH	
25	M	B	4464.60 5361.72	505	88.74	7635	LAUTERBACH	
25	M	C	4464.79 5362.03	500	89.09	7635	LAUTERBACH	
24	M	B	4464.60 5361.72	505	89.19	7635	LAUTERBACH	
24	M	C	4464.79 5362.03	500	89.54	7635	LAUTERBACH	
25	M	A	4466.45 5362.81	495	90.37	7535	LAIMBACH	
24	M	A	4466.45 5362.81	495	90.83	7535	LAIMBACH	
25	M	B	4466.64 5363.40	500	90.99	7535	LAIMBACH	
25	M	C	4466.85 5363.77	485	91.41	7535	LAIMBACH	
24	M	B	4466.64 5363.40	500	91.45	7535	LAIMBACH	
24	M	C	4466.85 5363.77	485	91.86	7535	LAIMBACH	
25	M	A	4467.21 5364.17	490	91.90	7535	EGLHAUSEN	
24	M	A	4467.21 5364.17	480	92.36	7535	EGLHAUSEN	
25	M	B	4467.45 5364.67	495	92.45	7535	EGLHAUSEN	
25	M	C	4467.60 5364.93	485	92.75	7535	EGLHAUSEN	
24	M	B	4467.45 5364.67	495	92.91	7535	EGLHAUSEN	
24	M	C	4467.60 5364.67	485	92.97	7535	EGLHAUSEN	
25	M	D	4467.76 5365.27	465	93.12	7535	EGLHAUSEN	
24	M	D	4467.76 5365.27	465	93.58	7535	EGLHAUSEN	
25	M	A	4468.54 5368.02	490	95.97	7535	UNTERKIENBG.	
24	M	A	4468.54 5368.02	490	96.44	7535	UNTERKIENBG.	
25	M	B	4468.67 5368.38	490	96.36	7535	UNTERKIENBG.	
25	M	C	4468.51 5368.80	495	96.70	7535	UNTERKIENBG.	
24	M	B	4468.67 5368.38	490	96.82	7535	UNTERKIENBG.	
24	M	C	4468.51 5368.80	495	97.16	7535	UNTERKIENBG.	
25	M	A	4470.88 5369.94	470	98.57	7535	AUFHAM	
25	M	B	4470.96 5370.36	490	99.00	7535	AUFHAM	
24	M	A	4470.88 5369.94	470	99.05	7535	AUFHAM	
24	M	B	4470.96 5370.36	490	99.47	7535	AUFHAM	
26	M	A	4474.20 5372.40	470	102.05	7535	AMPERTSHN.	
26	M	D	4474.52 5372.58	500	102.33	7535	AMPERTSHN.	
26	M	E	4474.83 5372.66	480	102.52	7353	AMPERTSHN.	
26	M	A	4478.52 5375.64	515	106.69	7436	SILLERTSHN/S	
26	M	B	4478.85 5375.83	515	106.99	7436	SILLERTSHN/S	
26	M	C	4479.01 5376.04	505	107.25	7436	SILLERTSHN/S	
26	M	A	4477.97 5379.04	480	109.61	7436	GUENZENHSN.	
26	M	B	4478.14 5379.35	480	109.96	7436	GUENZENHSN.	
26	M	C	4478.28 5379.83	470	110.46	7436	GUENZENHSN.	
26	M	C	4477.95 5381.75	505	112.12	7436	RUDERTSHSN.	
26	M	B	4478.10 5382.10	500	112.50	7436	RUDERTSHSN.	
26	M	A	4478.34 5382.81	500	113.25	7436	RUDERTSHSN.	
30	M	A	4480.59 5383.66	486	115.32	7436	ENZELHAUSEN	
30	M	C	4481.25 5384.15	445	116.03	7436	ENZELHAUSEN	
30	M	B	4481.43 5384.99	495	116.87	7336	STEINBACH	
30	M	A	4481.55 5385.69	480	117.57	7336	STEINBACH	
30	M	C	4481.55 5385.69	480	117.57	7336	STEINBACH	
26	M	A	4482.03 5387.39	475	118.88	7336	UNTEREMPFENB.	
26	M	B	4482.35 5387.90	460	119.47	7336	UNTEREMPFENB.	
26	M	B	4482.35 5387.90	460	119.47	7336	UNTEREMPFENB.	
26	M	C	4482.73 5388.41	430	120.09	7336	UNTEREMPFENB.	
26	M	A	4482.70 5390.72	455	122.22	7336	SCHLEISSBACH	
26	M	B	4483.08 5391.26	445	122.86	7336	SCHLEISSBACH	
26	M	A	4482.95 5392.68	455	124.13	7336	LINDKIRCHEN	
26	M	B	4483.26 5393.23	445	124.75	7336	LINDKIRCHEN	
26	M	C	4483.49 5393.62	455	125.20	7336	LINDKIRCHEN	
26	M	A	4484.29 5394.59	445	126.40	7336	RATZENHOFEN	
26	M	B	4484.43 5395.00	425	126.83	7336	RATZENHOFEN	
26	M	C	4484.53 5395.40	425	127.24	7336	RATZENHOFEN	
26	M	A	4483.63 5397.39	440	128.76	7236	ELSENDORF	
26	M	B	4483.78 5398.00	405	129.38	7236	ELSENDORF	
27	M	A	4483.78 5398.00	405	129.77	7236	ELSENDORF	
26	M	C	4483.93 5398.48	420	129.88	7236	ELSENDORF	
27	M	A	4485.14 5399.03	405	131.23	7236	ST.JOHANN	
27	M	B	4485.66 5399.45	415	131.82	7236	ST.JOHANN	

B	IN	AC	COORDINATES	ALT	X(KM)	SMAP	LOCALITY
27	M	A	4486.21 5400.74	400	133.22	7236	NEUKIRCHEN
27	M	A	4486.76 5402.54	400	135.09	7236	FUCHS-BERG
27	M	B	4486.97 5402.81	410	135.42	7236	FUCHS-BERG
27	M	C	4487.37 5403.18	390	135.92	7236	FUCHS-BERG
27	M	A	4487.31 5404.31	420	136.94	7236	FORSTDURNB.
27	M	B	4487.57 5405.02	400	137.70	7236	FORSTDURNB.
27	M	A	4486.93 5406.03	400	138.40	7236	ULRAIN/S
27	M	B	4487.10 5406.72	390	139.10	7236	ULRAIN/S
27	M	A	4486.45 5408.02	360	140.08	7136	ULRAIN/N
28	M	B	4486.65 5407.89	370	140.13	7136	ULRAIN/N
28	M	A	4486.45 5408.02	360	140.18	7136	ULRAIN/N
28	M	A	4488.12 5409.81	365	141.73	7137	ABENSBERG/N
28	M	B	4488.43 5410.05	375	142.79	7137	ABENSBERG/N
27	M	A	4488.53 5410.69	375	143.32	7137	ARNHOFEN
28	M	C	4488.76 5410.76	390	143.57	7137	ABENSBERG/N
27	M	B	4488.56 5413.22	430	145.69	7137	PULLACH
27	M	A	4488.62 5413.87	430	146.32	7137	PULLACH
27	M	A	4489.34 5414.45	430	147.12	7137	HOLZHARLANDN
27	M	B	4489.74 5415.09	430	147.86	7137	HOLZHARLANDN
28	M	B	4489.23 5416.12	410	148.73	7137	THALDORF
28	M	A	4489.28 5416.21	400	148.84	7137	THALDORF
28	M	A	4490.03 5418.36	455	151.11	7037	KELHEIM/S
28	M	B	4490.03 5418.50	470	151.24	7037	KELHEIM/S
28	M	C	4490.24 5418.68	465	151.49	7037	KELHEIM/S
28	M	A	4492.49 5421.53	450	154.96	7037	KELHEIM/N
28	M	B	4492.51 5421.73	460	155.15	7037	KELHEIM/N
28	M	A	4494.22 5424.89	440	158.72	7037	SPARBENECK
28	M	B	4494.53 5425.16	440	159.08	7037	SPARBENECK
28	M	C	4494.64 5425.46	455	159.40	7037	SPARBENECK
28	M	C	4495.24 5427.27	445	161.31	7037	VIEHHAUSEN
28	M	B	4495.21 5427.63	465	161.63	7037	VIEHHAUSEN
28	M	A	4495.20 5427.92	460	161.90	7037	VIEHHAUSEN
28	M	C	4495.51 5429.03	435	163.04	6937	THUMHAUSEN
28	M	B	4495.45 5429.38	400	163.35	6937	THUMHAUSEN
28	M	A	4495.64 5429.72	400	163.73	6937	THUMHAUSEN
28	M	A	4496.08 5431.76	460	165.79	6937	UNDORF
28	M	B	4496.26 5431.80	440	165.90	6937	UNDORF
30	M	A	4497.53 5435.90	375	170.11	6937	DISTELHAUSEN
30	M	B	4497.48 5436.17	375	170.34	6937	DISTELHAUSEN
30	M	C	4497.64 5436.32	385	170.54	6937	DISTELHAUSEN
30	M	A	4405.00 5430.37	345	172.81	6937	OBER FREIUNG
30	M	B	4496.00 5439.65	345	173.07	6937	OBER-FREIUNG
30	M	C	4496.01 5439.96	345	173.37	6937	OBER-FREIUNG
30	M	A	4495.87 5444.39	370	177.48	6837	WEICHSELDORF
30	M	A	4497.37 5448.41	375	181.77	6837	ZAAR B.KALLM.
30	M	B	4497.37 5448.41	375	181.77	6837	ZAAR B.KALLM.
30	M	C	4497.53 5448.65	405	182.05	6837	ZAAR B.KALLM.
30	M	A	4499.86 5451.02	370	185.08	6837	SEE
30	M	B	4499.99 5451.30	360	185.39	6837	SEE
30	M	A	4501.43 5454.38	417	188.77	6738	SAASS
30	M	B	4501.41 5454.71	415	189.08	6738	SAASS
30	M	C	4501.35 5454.94	420	189.27	6738	SAASS
30	M	A	4503.94 5457.64	475	192.71	6738	BUBACH
30	M	B	4504.04 5457.86	478	192.95	6738	BUBACH
30	M	A	4500.36 5462.37	470	195.93	6638	NEUKIRCHEN
30	M	B	4500.62 5462.82	486	196.44	6638	NEUKIRCHEN
30	M	A	4519.07 5462.78	391	203.23	6639	TAXOELDERN
30	M	B	4519.29 5463.01	388	203.53	6639	TAXOELDERN
30	M	A	4520.47 5465.96	500	206.70	6639	FUHRN
30	M	B	4520.42 5466.38	550	207.06	6639	FUHRN
30	M	A	4513.10 5471.78	448	209.20	6639	ZILCHENRICHT
30	M	A	4515.82 5475.26	405	213.44	6539	SCHWARZACH
30	M	B	4515.82 5475.26	405	213.44	6539	SCHWARZACH
29	M	C	4520.15 5477.99	400	217.60	6539	GUTENECK
29	M	A	4520.15 5477.99	400	217.60	6539	GUTENECK
29	M	B	4520.18 5478.48	420	218.06	6539	GUTENECK
29	M	B	4520.46 5481.45	510	220.91	6539	WEIDENTHAL
29	M	A	4520.52 5481.84	525	221.29	6539	WEIDENTHAL
29	M	A	4523.88 5489.08	615	229.27	6439	TAENNESBERG
29	M	C	4523.88 5489.08	615	229.27	6439	TAENNESBERG
37	CL	13	4531.42 5488.08	704	231.44	6440	PULLENRIED
37	CL	12	4531.05 5488.32	728	231.51	6440	PULLENRIED
37	CL	11	4530.75 5488.54	726	231.58	6440	PULLENRIED
29	M	A	4525.81 5491.40	555	232.15	6440	BOEHM-BRUCK
29	M	B	4525.81 5491.40	555	232.15	6440	BOEHM-BRUCK
37	CL	23	4519.71 5494.68	510	233.00	6439	LEUCHTENBERG
37	CL	22	4519.32 5494.86	470	233.03	6439	LEUCHTENBERG
37	CL	21	4518.93 5495.07	470	233.09	6439	LEUCHTENBERG
29	M	A	4525.60 5495.44	560	235.80	6440	UNTERE HILLE
29	M	B	4525.60 5495.44	560	235.80	6440	UNTERE HILLE
29	M	C	4525.86 5495.65	550	236.10	6440	UNTERE HILLE
29	M	A	4528.87 5497.87	570	239.30	6340	BRAUNETSV.
29	M	B	4528.87 5497.87	570	239.30	6340	BRAUNETSV.
29	M	C	4528.95 5498.29	560	239.72	6340	BRAUNETSV.
29	M	A	4531.28 5499.93	540	242.14	6340	SPIELHOF
29	M	B	4531.45 5500.29	555	242.54	6340	SPIELHOF
29	M	A	4528.73 5504.77	550	245.32	6340	HAGENMUEHLE
29	M	B	4528.73 5504.77	550	245.62	6340	HAGENMUEHLE
29	M	C	4528.85 5505.05	570	245.93	6340	HAGENMUEHLE
29	M	A	4531.04 5509.28	690	250.67	6240	GRENZE
29	M	B	4531.04 5509.28	690	250.67	6240	GRENZE

QUARRY-NO. 01 ESCHENLOHE SMAP 8332
PROFILE 01-040 COMPILED BY PRODEHL.GEBRANDE.KAPPELMEYER

B	IN	AC	COORDINATES	ALT	X(KM)	SMAP	LOCALITY	G C
07	M		4442.69 5278.35	660	6.87	8333	OHLSTADT	
03	K		4444.12 5288.04	680	13.41	8233	HABACHING	
15	M	B	4448.44 5288.96	690	16.38	8233	FRAUENRAIN	
00	S		4448.86 5292.84	595	19.90	8233	IFFELDORF (2.4.55)	
15	M	B	4451.90 5295.21	612	23.33	8234	UNTER-EURACH	
04	M		4454.07 5297.65	656	26.70	8234	MAIERWALD	
05	M		4454.07 5297.65	656	27.16	8234	MAIERWALD	
03	M		4455.22 5302.27	655	31.47	8134	EURASBURG	
00	M		4461.28 5308.99	665	40.30	8034	NEUFAHRN (2.4.55)	
15	M	A	4464.08 5310.72	635	43.04	8035	AUFHOFEN	
00	M		4464.95 5315.96	640	49.66	8035	OBERBIBERG (2.4.55)	
04	M		4476.52 5318.43	593	57.20	7936	BRUNNTHAL	
05	M		4476.52 5318.43	593	57.68	7936	BRUNNTHAL	
00	M		4484.68 5318.08	620	62.90	7936	EGMATING (2.4.55)	
04	M		4484.37 5325.32	565	67.62	7936	MOESCHENFELD	
05	M		4484.37 5325.32	565	68.10	7936	MOESCHENFELD	
13	M		4484.36 5334.63	524	74.45	7836	GRUB	
12	M		4484.36 5334.63	524	75.10	7836	GRUB	
04	M		4491.13 5332.04	530	77.15	7837	ANZINGER SAUSCHUETT	
05	M		4491.13 5332.04	530	77.63	7837	ANZINGER SAUSCHUETT	
03	M		4489.95 5342.69	525	84.82	7737	HOF STOCKER	
08	M		4493.96 5344.25	500	87.94	7737	KIRCHOETTING	
04	M		4502.63 5352.86	520	100.22	7638	UEBERMIETHING	
05	M		4502.63 5352.86	520	100.69	7638	UEBERMIETHING	
03	M		4504.13 5357.70	500	105.33	7638	OBERBIERBACH	
03	M		4509.16 5363.93	510	113.33	7538	DIEMATING	
03	MZ		4515.05 5371.30	505	122.76	7539	NARRENSTETTEN	
03	MM		4516.88 5376.77	480	128.15	7439	ATTENKOFEN	
03	CL		4521.36 5376.68	500	130.95	7439	WOLFSBACH	
31	KI		4528.16 5386.56	390	142.98	7340	NIEDERVIEHBACH	
04	M		4530.34 5392.24	400	148.20	7340	RIMBACH	
05	M		4530.34 5392.24	400	148.64	7340	RIMBACH	
04	M		4536.43 5399.12	455	157.38	7240	WUNDER	
05	M		4536.43 5399.12	455	157.83	7240	WUNDER	
06	M		4530.65 5403.23	414	163.17	7241	I FIBELFING	
06	M		4547.14 5406.78	360	170.71	7241	OBERSCHNEIDING	
06	M		4547.14 5406.78	360	171.36	7241	OBERSCHNEIDING	
07	M		4548.56 5414.20	326	177.30	7141	FRUHSTORF	
06	M		4551.53 5418.44	412	182.47	7042	BOGENBERG	
03	M		4554.83 5422.73	450	187.75	7042	WINDBERG	
11	M		4558.17 5426.67	442	192.42	7042	OBERMUEHLBACH	
10	M		4558.17 5426.67	442	192.99	7042	OBERMUEHLBACH	
08	M		4559.62 5431.14	880	196.76	6942	ENGLMAR	
11	M		4561.50 5430.37	895	197.39	6943	MARKBUCHEN	
10	M		4561.50 5430.37	895	197.97	6943	MARKBUCHEN	
03	M		4563.00 5434.83	650	202.29	6943	KOLLNBURG	
03	SE		4569.27 5439.56	660	209.92	6943	NEU-NUSSBERG	
07	M		4568.44 5443.64	690	212.69	6843	BAERENLOCH	
07	M		4572.39 5447.80	843	218.41	6843	ECK	
07	M		4572.65 5447.80	875	218.57	6843	ECK	
03	M		4578.62 5451.88	640	225.38	6744	LAM	

QUARRY-NO. 01 ESCHENLOHE SMAP 8332
PROFILE 01-080 COMPILED BY REICH.KOSCHYK

B	IN	AC	COORDINATES	ALT	X(KM)	SMAP	LOCALITY	G C
08	M		4442.69 5278.35	660	6.02	8333	OHLSTADT	
08	BO		4447.70 5279.43	646	11.14	8333	KOCHEL	
08	BO		4452.80 5279.15	650	16.16	8334	KOCHEL	

QUARRY-NO. 01 ESCHENLOHE SMAP 8332
PROFILE 01-090 COMPILED BY GIESE.KOSCHYK

B	IN	AC	COORDINATES	ALT	X(KM)	SMAP	LOCALITY	G C
11	BO		4439.48 5276.24	632	3.09	8333	ESCHENLOHE	
23	M		4442.70 5278.42	660	6.88	*8333	OHLSTADT	
40	M	A	G 1118.77 4738.31	650	12.66	*8333	FELSENKELLER	
40	M	B	G 1119.08 4738.36	630	13.05	8333	FELSENKELLER	
40	M	C	G 1119.36 4738.34	640	13.40	8333	FELSENKELLER	
08	BO		4452.80 5279.15	650	16.16	*8334	KOCHEL	
41	M	A	G 1122.81 4736.49	840	17.66	8334	SACHENBACH	
41	M	B	G 1123.12 4736.53	830	18.04	*8334	SACHENBACH	
41	M	C	G 1123.38 4736.55	860	18.35	8334	SACHENBACH	
40	M	A	G 1125.24 4736.37	810	20.94	8334	JACHENAU	
40	M	B	G 1125.45 4736.36	790	21.19	*8334	JACHENAU	
41	M	A	G 1128.45 4736.42	760	24.68	8334	BAECKER	
41	M	C	G 1128.83 4736.56	780	25.14	*8334	BAECKER	
40	M	A	G 1129.54 4736.91	800	26.17	8334	HINTERBICHL	
40	M	B	G 1129.69 4736.93	750	26.35	8334	HINTERBICHL	
40	M	C	G 1129.82 4736.97	740	26.52	*8334	HINTERBICHL	
39	M	B	G 1132.30 4735.39	810	29.72	8435	HOHER ZWIESL.	
39	M	C	G 1132.58 4735.34	800	30.08	*8435	HOHER ZWIESL.	
41	M	1	G 1135.21 4737.79	800	32.99	8335	RAUCHENBERG	
41	M	2	G 1135.36 4737.77	710	33.18	*8335	RAUCHENBERG	
41	M	3	G 1135.39 4737.82	710	33.22	8335	RAUCHENBERG	
40	M	1	G 1138.31 4736.78	1010	37.15	8335	GLASHUETTE	
40	M	2	G 1138.46 4736.68	940	37.52	8335	GLASHUETTE	
40	M	3	G 1138.59 4736.70	910	37.75	*8335	GLASHUETTE	
39	M	A	G 1142.52 4737.12	960	42.17	8336	AM SCHILDEN	
39	M	B	G 1142.74 4737.21	940	42.44	*8336	AM SCHILDEN	
39	M	C	G 1143.01 4737.22	900	42.77	8336	AM SCHILDEN	
39	M	A	G 1147.62 4736.57	870	48.60	8336	STEIN.KREUZ	
39	M	B	G 1147.93 4736.46	880	49.00	*8336	STEIN.KREUZ	
39	M	C	G 1147.27 4736.50	900	49.42	8336	STEIN.KREUZ	
39	M	A	G 1152.24 4737.70	920	54.32	*8337	VALEPP	
39	M	B	G 1152.55 4737.74	940	54.70	8337	VALEPP	
39	M	C	G 1152.76 4737.67	950	54.97	8337	VALEPP	
39	M	A	G 1158.10 4737.28	850	61.67	8337	KLOASCHAU	
39	M	B	G 1158.10 4737.28	850	61.67	8337	KLOASCHAU	
39	M	C	G 1158.42 4737.39	820	62.07	*8337	KLOASCHAU	
32	M	A	4500.62 5277.16	960	64.64	8338	SILLBERGHAUS	
32	M	B	4500.62 5277.16	960	64.64	*8338	SILLBERGHAUS	
40	M	1	G 1203.22 4739.55	1160	68.37	8338	ROSENGASSE	
40	M	3	G 1203.38 4739.61	1090	68.57	8338	ROSENGASSE	
40	M		G 1203.40 4739.65	1070	68.60	*8338	ROSENGASSE	
32	M	A	4506.46 5281.88	880	70.63	8338	TATZELWURM	
32	M	B	4506.46 5281.88	880	70.63	8338	TATZELWURM	
40	M	1	G 1205.62 4740.28	790	71.43	8338	TATZELWURM 2	
40	M	2	G 1205.82 4740.22	760	71.68	*8338	TATZELWURM 2	
40	M	3	G 1206.00 4740.28	780	71.91	8338	TATZELWURM 2	
39	M	A	G 1209.05 4737.85	660	75.37	8338	MUELAU	
32	M	A	4511.40 5276.97	620	75.42	8338	OBERAUDORF	
39	M	B	G 1209.10 4737.92	640	75.43	8338	MUELAU	
32	M	B	4511.76 5277.06	610	75.78	8338	OBERAUDORF	
39	M	C	G 1209.43 4737.93	630	75.85	*8338	MUELAU	
40	M	1	G 1210.54 4737.49	560	77.49	8339	RIED	
40	M	2	G 1210.64 4737.61	510	77.62	8339	RIED	
40	M	3	G 1210.69 4737.72	510	77.68	8339	RIED	
32	M	A	4517.38 5279.74	670	81.44	*8339	NIEDERNDORF	
32	M	B	4517.38 5279.74	670	81.44	8339	NIEDERNDORF	
32	M	C	4517.59 5279.63	660	81.64	8339	NIEDERNDORF	
32	M	A	4520.84 5278.85	700	84.87	*8339	MIESBERG	

B	IN	AC	COORDINATES	ALT	X(KM)	SMAP	LOCALITY	G C
32	M	B	4521.34 5278.91	700	85.37	8339	MIESBERG	
32	M	A	4525.41 5280.60	880	89.49	8340	KRANZACH	
32	M	B	4525.59 5280.71	800	89.67	8340	KRANZACH	
32	M	C	4525.59 5280.71	800	89.67	*8340	KRANZACH	
32	M	A	4530.82 5278.04	720	94.84	8340	KOESSEN	
32	M	B	4530.97 5278.27	720	94.99	*8340	KOESSEN	
32	M	C	4530.97 5278.27	720	94.99	8340	KOESSEN	
32	M	A	4533.05 5279.68	700	97.10	8340	KALTENBACH	
32	M	B	4533.18 5279.70	700	97.23	*8340	KALTENBACH	
32	M	C	4533.48 5279.81	700	97.53	8340	KALTENBACH	
32	M	A	4538.39 5278.23	1240	102.41	8341	OB.HEMMERS.	
32	M	B	4538.73 5278.00	1250	102.75	*8341	OB.HEMMERS.	
32	M	A	G 1232.91 4739.25	880	105.25	8341	SEEGATTERL	
32	M	B	G 1232.91 4739.25	880	105.25	8341	SEEGATTERL	
32	M	C	G 1232.91 4739.25	880	105.25	*8341	SEEGATTERL	
32	M	A	4549.74 5274.83	1325	113.79	*8341	LOFERER ALPE	
32	M	B	4549.74 5274.83	1325	113.79	8341	LOFERER ALPE	
32	M	A	4551.73 5273.53	810	115.81	*8442	LOFER	
32	M	B	4551.73 5273.53	810	115.81	8442	LOFER	
32	M	A	4555.94 5274.78	850	119.99	8342	OB.MAYRBERG	
32	M	B	4556.04 5274.64	870	120.09	8342	OB.MAYRBERG	
32	M	C	4556.04 5274.64	870	120.09	*8342	OB.MAYRBERG	
31	M	A	4560.39 5270.87	960	124.63	*8442	HIRSCHBICHL	
31	M	B	4560.74 5271.13	950	124.96	8442	HIRSCHBICHL	
31	M	C	4560.96 5271.50	940	125.16	8442	HIRSCHBICHL	
31	M	B		860	128.60	8343	HINTERSEE	
31	M	A	4564.64 5274.23	840	128.75	*8343	HINTERSEE	
31	M	A	4568.65 5275.15	840	132.74	*8343	RAMSAU	
31	M	B		860	132.86	8343	RAMSAU	
31	M	A	4572.67 5275.60	630	136.75	*8343	OB.SCHOENAU	
31	M	B	4572.67 5275.60	630	136.75	8343	OB.SCHOENAU	
31	M	A	4577.25 5274.51	1050	141.35	*8344	KLAUS BICHL	
31	M	B		1040	141.60	8344	KLAUS BICHL	
31	M	C		1060	141.91	8344	KLAUS BICHL	
31	M	A	4581.36 5276.08	1540	145.44	*8344	AHORNBUECHS.	
31	M	B	4581.65 5276.21	1540	145.72	8344	AHORNBUECHS.	
31	M	C		590	149.31	*A094	KUCHL	
31	M	A	G 1308.09 4736.93	580	149.36	A094	KUCHL	
31	M	B	G 1308.09 4736.93	580	149.36	A094	KUCHL	
35	M	A	G 1311.46 4736.13	500	153.68	A094	HINTERKELLAU	
35	M	B	G 1311.70 4736.05	500	153.98	*A094	HINTERKELLAU	
31	M	B		520	155.20	*A094	UNT.SCHEFFAU	
31	M	C	G 1313.47 4735.03	520	156.22	A094	UNT.SCHEFFAU	
31	M	A		520	156.25	A094	UNT.SCHEFFAU	
31	M	A		600	159.49	*A094	LAMMEROEFEN	
31	M	C	G 1316.28 4735.24	580	159.72	A094	LAMMEROEFEN	
31	M	B	G 1316.28 4735.24	580	159.72	A094	LAMMEROEFEN	
35	M	A	G 1318.46 4735.41	700	162.50	A094	GRILLBERG	
35	M	B	G 1318.78 4735.38	650	162.90	*A094	GRILLBERG	
35	M	C	G 1318.78 4735.38	650	162.90	A094	RILLBERG	
31	M	A		680	166.60	A095	RIGAUSBACH	
31	M	B	G 1321.82 4735.63	640	166.63	*A095	RIGAUSBACH	
31	M	B	G 1324.04 4734.59	640	169.50	A095	MOESELBERG	
31	M	A		660	169.57	*A095	MOESELBERG	
31	M	C		690	169.65	A095	MOESELBERG	
31	M	A	G 1328.52 4735.55	860	175.03	A095	RUSSBACH	
31	M	B		860	175.20	A095	RUSSBACH	
31	M	C		860	175.32	*A095	RUSSBACH	
42	M	B	G 1351.56 4738.67	900	203.86	A097	GAISWINKEL	
42	M	C	G 1351.69 4738.64	880	204.02	*A097	GAISWINKEL	
43	M	1	G 1405.57 4738.97	770	221.47	*A099	BAUMSCHLAG.	
43	M	2	G 1405.64 4738.97	730	221.56	A098	BAUMSCHLAG.	
43	M	3	G 1406.12 4739.03	740	222.16	A098	BAUMSCHLAG.	
43	M	1	G 1421.44 4738.50	1050	241.36	*A099	SPITAL	
43	M	2	G 1421.48 4738.63	1050	241.41	A099	SPITAL	
43	M	3	G 1421.65 4738.74	1060	241.62	A099	SPITAL	
43	M	1	G 1438.78 4740.61	580	263.04	A100	ST.GALLEN	
43	M	2	G 1438.86 4740.63	580	263.14	A100	ST.GALLEN	
43	M	3	G 1438.95 4740.64	600	263.25	*A100	ST.GALLEN	
42	KI	1	G 1506.25 4742.90	1290	297.34	A102	FADENKAMP	
42	KI	2	G 1506.38 4742.90	1280	297.51	A102	FADENKAMP	
42	KI	3	G 1506.38 4742.98	1270	297.51	*A102	FADENKAMP	
42	KI	1	G 1511.58 4743.03	1020	304.01	A102	GREITH	
42	KI	2	G 1511.76 4743.05	1020	304.24	A102	GREITH	
42	KI	3	G 1511.92 4743.15	1040	304.44	*A102	GREITH	
42	KI	1	G 1517.53 4742.65	870	311.46	A102	BRUNNBAUER	
42	KI	2	G 1517.73 4742.46	820	311.71	A102	BRUNNBAUER	
42	KI	3	G 1517.92 4742.38	870	311.94	*A102	BRUNNBAUER	
42	KI	1	G 1523.44 4743.59	1130	318.86	A103	SCHOENEBEN	
42	KI	2	G 1523.60 4743.59	1140	319.06	A103	SCHOENEBEN	
42	KI	3	G 1523.80 4743.54	1170	319.31	*A103	SCHOENEBEN	
43	M	1	G 1531.00 4742.78	1230	328.38	*A103	TEUFELSBAD	
43	M	2	G 1531.53 4742.67	1230	329.05	A103	TEUFELSBAD	
43	M	3	G 1531.87 4742.66	1250	329.47	A103	TEUFELSBAD	
43	M	1	G 1536.00 4740.65	1180	334.67	A104	KAPELLEN	
43	M	2	G 1536.52 4740.70	1250	335.32	*A104	KAPELLEN	
43	M	3	G 1536.86 4740.70	1310	335.74	A104	KAPELLEN	
43	M	1	G 1542.90 4741.24	1360	343.29	*A104	PREIN	
43	M	2	G 1542.95 4741.30	1360	343.35	A104	PREIN	
43	M	3	G 1543.04 4741.31	1360	343.47	A104	PREIN	

QUARRY-NO. 01 ESCHENLOHE SMAP 8332
PROFILE 01-260-25 COMPILED BY SCHELIGA.KOSCHYK

B	IN	AC	COORDINATES	ALT	X(KM)	SMAP	LOCALITY	G C
44	M	1	G 1103.93 4736.63	840	6.41	8332	OBERAMMERGAU	
44	M	2	G 1103.89 4736.63	835	6.45	8332	OBERAMMERGAU	
44	M	1	G 1101.33 4736.89	855	9.40	8332	UNTERAMMERG.	
44	M	2	G 1101.31 4736.88	855	9.43	8332	UNTERAMMERG.	
44	M	3	G 1101.25 4736.80	865	9.54	8332	UNTERAMMERG.	
44	M	1	G 1050.89 4736.20	950	22.52	8331	REISELBERGH.	
44	M	2	G 1050.80 4736.27	910	22.61	8331	REISELBERGH.	
44	M	1	G 1045.18 4733.61	830	30.52	8430	NEUSCHWANST.	
44	M	2	G 1045.16 4733.61	830	30.54	8430	NEUSCHWANST.	
44	M	3	G 1044.88 4733.56	815	30.90	8430	NEUSCHWANST.	
44	M	1	G 1044.04 4733.32	860	32.04	8430	HOHENSCHWAN.	
44	M	2	G 1043.87 4733.30	870	32.25	8430	HOHENSCHWAN.	
44	M	1	G 1040.15 4732.81	860	37.00	8430	UNTERPINSWA.	
44	M	2	G 1040.10 4732.79	860	37.07	8430	UNTERPINSWA.	
44	M	3	G 1039.92 4732.79	860	37.28	8430	UNTERPINSWA.	
44	M	1	G 1037.05 4732.78	970	40.78	8429	VILS	
44	M	2	G 1036.96 4732.78	970	40.89	8429	VILS	
44	M	3	G 1036.86 4732.77	970	41.01	8429	VILS	
45	M	1	G 1033.63 4731.28	1260	45.74	8528	GRAEN	
45	M	2	G 1033.50 4731.28	1210	45.89	8528	GRAEN	
45	M	3	G 1033.10 4731.26	1180	46.13	8528	GRAEN	
45	M	1	G 1031.20 4730.60	1200	49.02	8528	BERG	
45	M	2	G 1031.08 4730.43	1130	49.25	8528	BERG	
45	M	3	G 1030.76 4730.40	1110	49.65	8528	BERG	
45	M	1	G 1027.30 4730.16	1360	53.95	8528	STUIBEN-ALM	
45	M	2	G 1027.26 4730.06	1390	54.04	8528	STUIBEN-ALM	
45	M	3	G 1027.21 4729.96	1380	54.16	8528	STUIBEN-ALM	
45	M	1	G 1025.92 4728.22	980	56.66	8528	HINTERSTEIN	
45	M	2	G 1025.68 4728.16	940	56.99	8528	HINTERSTEIN	
45	M	3	G 1025.42 4728.12	900	57.32	8528	HINTERSTEIN	
45	M	1	G 1021.64 4728.14	1190	61.82	8528	MITTERH.ALM	
45	M	2	G 1021.30 4728.10	1080	62.26	8528	MITTERH.ALM	
45	M	3	G 1021.25 4728.14	1070	62.30	8528	MITTERH.ALM	
45	M	3	G 1010.46 4726.19	1400	76.33	8526	OBERMAISEL.	
45	M	2	G 1010.32 4726.20	1420	76.49	8526	OBERMAISEL.	
45	M	1	G 1010.07 4726.22	1410	76.78	8526	OBERMAISEL.	
45	M	1	G 1008.64 4726.49	1180	78.37	8526	BALDERSCHW.	
45	M	2	G 1008.62 4726.48	1170	78.40	8526	BALDERSCHW.	
45	M	3	G 1008.54 4726.46	1140	78.49	8526	BALDERSCHW.	
45	GO	1	G 1005.52 4725.06	1120	82.87	8526	FUGENALM	
45	GO	2	G 1005.20 4725.04	1135	83.27	8526	FUGENALM	
45	GO	3	G 1004.96 4725.04	1110	83.56	8526	FUGENALM	
45	GO	1	G 1002.91 4724.92	940	86.10	8526	SIBRATS.	
45	GO	2	G 1002.56 4724.87	975	86.54	8526	SIBRATS.	
45	GO	3	G 1002.24 4724.81	900	86.95	8526	SIBRATS.	
46	GO	11	G 1000.78 4722.46	1040	89.88	112	SCHOENENBACH-W	
46	GO	12	G 1000.44 4722.48	1050	90.27	112	SCHOENENBACH-W	
46	GO	13	G 1000.17 4722.45	1050	90.61	112	SCHOENENBACH-W	
46	GO	21	G 0956.87 4722.58	890	94.48	112	BIZAU	
46	GO	22	G 0956.55 4722.60	910	94.85	112	BIZAU	
46	GO	23	G 0956.18 4722.59	920	95.30	112	BIZAU	
46	KI	11	G 0919.84 4715.68	1320	142.81	227	SAENTIS	
46	KI	12	G 0919.64 4715.51	1340	143.15	227	SAENTIS	
46	KI	13	G 0919.44 4715.38	1370	143.46	227	SAENTIS	
46	KI	31	G 0917.58 4714.70	1175	146.08	227	DUNKELBODEN	
46	KI	32	G 0917.35 4714.56	1140	146.44	227	DUNKELBODEN	
46	KI	33	G 0917.18 4714.43	1190	146.71	227	DUNKELBODEN	
46	KI	21	G 0913.74 4713.16	1140	151.57	227	SCHLOH/NESSLAU	
46	KI	22	G 0913.50 4713.05	1030	151.92	227	SCHLOH/NESSLAU	
46	KI	23	G 0913.30 4712.95	970	152.22	227	SCHLOH/NESSLAU	
46	HH	11	G 0911.52 4711.83	906	155.01	237	VORDERLAAD	
46	HH	12	G 0911.20 4711.72	960	155.46	237	VORDER LAAD	
46	HH	13	G 0910.82 4711.65	970	155.96	237	VORDER LAAD	
46	HH	21	G 0906.44 4711.32	1185	161.39	237	KALTENBRUNN	
46	HH	22	G 0906.15 4711.83	1100	161.45	237	KALTENBRUNN	
46	HH	23	G 0905.86 4711.75	1105	161.84	237	KALTENBRUNN	
46	F	31	G 0903.81 4710.97	560	164.75	236	RUFI-E	
46	F	32	G 0903.68 4710.98	500	164.90	236	RUFI-E	
46	F	33	G 0903.56 4710.98	465	165.05	236	RUFI-E	
46	F	11	G 0859.50 4709.40	660	170.84	236	REICHENBURG-S	
46	F	12	G 0859.20 4709.45	640	171.17	236	REICHENBURG-S	
46	F	13	G 0859.00 4709.50	640	171.38	236	REICHENBURG-S	
46	F	21	G 0854.90 4709.05	920	176.56	236	SIEBEN-S	
46	F	22	G 0854.68 4709.00	860	176.85	236	SIEBEN-S	
46	F	23	G 0854.42 4708.94	820	177.20	236	SIEBEN-S	

QUARRY-NO. 01 ESCHENLOHE SMAP 8332
PROFILE 01-290 COMPILED BY MUELLER.PETERSCHMITT.FOERTSCH

B	IN	AC	COORDINATES	ALT	X(KM)	SMAP	LOCALITY	G C
08	BO		4435.66 5276.38	690	1.53	8332	SCHWAIGEN	
02	ST		4425.08 5281.72	903	11.82	8332	SAULGRUB	
01	ST		4425.08 5281.72	903	12.38	8332	SAULGRUB	
07	CL		4416.15 5283.32	841	20.65	8331	HIEBLER	
02	ST		4410.97 5283.15	800	25.73	8330	PREM	
01	ST		4410.97 5283.15	800	26.49	8330	PREM	
23	BO		4408.44 5286.16	794	28.87	8230	ECHERSCHWANG	
12	ST		4402.53 5287.49	785	34.87	8230	STEINBACH	
13	ST		4402.53 5287.49	785	35.56	8230	STEINBACH	
12	ST		4397.57 5289.76	792	40.28	8229	SULZSCHNEID	
07	GO		4395.78 5282.95	820	40.52	8329	KIRCHTHAL	
13	ST		4397.57 5289.76	792	40.96	8229	SULZSCHNEID	
32	KR		4391.60 5279.88	900	44.45	8329	RUECKHOLZ	
02	M		G 1028.05 4740.03	905	51.32	8328	UNTERSCHWARZENBERG	
01	M		G 1028.05 4740.03	905	51.99	8328	UNTERSCHWARZENBERG	
07	GO		3602.95 5285.39	790	58.72	8328	OBERHOF	
26	GO		3603.25 5284.95	785	59.17	8328	DURACH-OBERHOF	
26	GO		3595.56 5283.38	785	66.72	8327	HEHLEN	
14	KR		3593.78 5294.52	677	68.65	8227	DEPSRIED	
07	F		3591.70 5287.90	972	70.16	8227	ESCHACHBERG	
02	M		G 1009.88 4744.70	825	74.92	8226	UNTERKUERNACH	
01	M		G 1009.88 4744.70	825	75.57	8226	UNTERKUERNACH	
07	K		3582.06 5289.20	908	80.15	8226	SCHMIEDSFELDEN	
06	KR		3577.37 5294.30	708	85.48	8226	HANSER	
26	H	F	3574.74 5295.85	705	88.95	8225	TAUTENHOFEN	
26	H	E	3574.62 5296.03	695	89.10	8225	TAUTENHOFEN	
26	H	D	3574.41 5296.19	690	89.34	8125	TAUTENHOFEN	
26	H	C	3574.22 5296.42	683	89.57	8125	TAUTENHOFEN	
26	H	B	3574.02 5296.55	684	89.79	8125	TAUTENHOFEN	
26	H	A	3573.77 5296.55	690	90.03	8125	TAUTENHOFEN	
26	H		3570.60 5298.12	670	93.46	8125	ETTERAZHOFEN	
02	M		3565.58 5300.00	685	98.10	8125	IMMENRIED 1	
17	M	B	3566.33 5301.17	685	98.17	8125	IMMENRIED 2	
18	M	B	3566.33 5301.17	685		8125	IMMENRIED 2	
01	M		3565.58 5300.00	685	98.73	8125	IMMENRIED 1	
06	PR		3563.05 5310.16	740	103.03	8025	WURZACH	
06	S		3557.54 5302.20	653	106.35	8124	BERGATREUTE 1	
17	M		3555.30 5300.57	590	108.83	8124	BERGATREUTE 2	
18	M		3555.30 5300.57	590		8124	BERGATREUTE 2	
32	KI		3551.90 5304.35	600	112.35	8124	ENZISREUTE	
32	HH		3548.37 5306.28	518	116.23	8123	MOCHENWANGEN	
06	S		3545.50 5306.65	594	119.00	8123	WOLPERSWENDE	
07	CL		3544.16 5308.25	575	120.54	8023	STUBEN-BLOENRIED	
32	HH		3540.38 5308.10	575	124.42	8023	INGENHARDT	
32	HH		3537.48 5309.80	635	127.64	8023	MAUREN-KREENRIED	
07	CL		3534.84 5308.94	705	130.26	8022	BAUHOF	
14	MZ		3533.00 5315.37	659	133.33	8022	BACHHAUPTEN	

26	M		3525.21	5313.91	635	141.23	8022	MAGENBUCH		
07	MZ		3521.90	5315.00	631	143.99	8021	MOTTSCHIESS		
07	PR		3518.06	5315.79	640	147.62	8021	ETTISWEILER		
26	M		3516.37	5317.03	640	150.57	8021	BITTELSCHIESS		
07	MZ		3513.50	5321.30	660	153.66	7921	GOEGGINGEN		
40	KA	31	3513.38	5326.34	600	155.13	7921	INZIGKOFEN		
40	KA	32	3513.12	5326.54	600	155.44	7921	INZIGKOFEN		
40	KA	33	3512.98	5326.50	630	155.56	7921	INZIGKOFEN		
07	BO		3507.48	5220.83	660	159.39	7920	ROHRDORF		
07	BO		3501.18	5323.28	807	166.11	7920	LEIBERTINGEN		
32	CL		3498.45	5323.15	780	168.77	7919	BEURON		
17	M	A	3498.17	5328.38	810	171.05	7919	IRRENDORF		
18	M	A	3498.17	5328.38	810		7919	IRRENDORF		
27	GO		3493.37	5326.01	835	174.52	7919	KATZENTAL		
17	M		3493.25	5328.79	750	175.88	7919	BAERENTAL		
18	M		3493.25	5328.79	750		7919	BAERENTAL		
07	KR		3491.39	5326.70	877	176.41	7919	RENQUISHAUSEN		
27	GO		3487.33	5328.44	860	180.99	7918	ALLENSPACHER HOF		
17	M	B	3487.16	5331.64	930	182.53	7818	BUBSHEIM		
02	M		3484.55	5324.37	708	182.56	7918	DUERBHEIM		
17	M	A	3487.08	5331.88	900	182.68	7818	BUBSHEIM		
18	M	A	3487.08	5331.88	900		7818	BUBSHEIM		
01	M		3484.55	5324.37	708	183.18	7918	DUERBHEIM		
18	M	B	3487.16	5331.64	930	183.22	7818	BUBSHEIM		
07	S		3483.31	5332.36	990	185.62	7818	GOSHEIM		
14	ST		3481.71	5335.83	920	188.31	7818	SCHOERZINGEN		
17	M	B	3481.86	5335.66	960	188.77	7818	LEMBERG		
18	M	B	3481.86	5335.66	960		7818	LEMBERG		
07	S		3475.31	5333.31	605	193.67	7818	NEUFRA		
17	M	A	3475.74	5335.02	695	194.44	7818	WELLENDINGEN		
18	M	A	3475.74	5335.02	695		7818	WELLENDINGEN		
17	M	B	3475.35	5335.15	695	194.85	7818	WELLENDINGEN		
18	M	B	3475.35	5335.15	695		7818	WELLENDINGEN		
14	S		3464.43	5337.41	708	205.38	7817	STETTEN		
17	M		3459.62	5341.69	705	211.80	7716	HINTERSULGEN		
18	M		3459.62	5341.69	705		7716	HINTERSULGEN		
16	ST	A	3447.22	5344.25	775	223.83	7715	FAHRENBUHL		
16	ST	B	3447.22	5344.25	775	223.83	7715	FAHRENBUHL	X	
16	ST	C	3447.22	5344.25	775	223.83	7715	FAHRENBUHL	Y	
17	M		3447.23	5344.43	765	224.45	7715	FOHRENBUEHL		
18	M		3447.23	5344.43	765		7715	FOHRENBUEHL		
42	KA	62	3439.03	5313.56	1045	224.91	8015	LANGENORDNACH		
42	KA	61	3438.82	5313.93	1045	225.18	8015	LANGENORDNACH		
17	M		3440.12	5345.42	390	231.55	7715	GUTACH		
18	M		3440.12	5345.42	390		7715	GUTACH		
17	MM		3435.12	5346.97	320	236.81	7714	FIDELISHOF		
18	MM		3435.12	5346.97	320		7714	FIDELISHOF		
02	S		3434.03	5347.67	415	237.56	7714	HASLACH		
01	S		3434.03	5347.67	415	238.16	7714	HASLACH		
42	KA	51	3432.35	5345.44	490	238.35	7714	MUEHLENBACH		
11	M		3455.67	5404.85		241.54	7216	LOFFENAU		
17	MM		3428.45	5348.70	350	243.65	7714	WIRTSBERG		
18	MM		3428.45	5348.70	350		7714	WIRTSBERG		

QUARRY-NO. 01 ESCHENLOHE SMAP 8332
PROFILE 01-315-05 COMPILED BY EMTER

B	IN	AC	COORDINATES		ALT	X(KM)	SMAP	LOCALITY	G	C
22	MZ		4422.00	5292.93	695	21.70	8231	SCHNALZ		
10	M		4413.98	5305.00	750	35.35	8131	DIENHAUSEN I		
11	M		4413.98	5305.00	750	35.66	8131	DIENHAUSEN II		
23	GO		4411.80	5309.05	720	39.87	8030	LEEDER		
22	CL		4407.97	5310.89	631	51.00	7930	HAUSEN		
22	CL		4401.38	5325.20	613	59.11	7930	AMBERGER HOELZ		
17	CL		4397.04	5327.62	590	63.78	7929	GUT SCHOENBRUNN I		
18	CL		4397.04	5327.62	590	63.88	7929	GUT SCHOENBRUNN II		
32	GO		4395.57	5331.46	619	67.54	7829	TUSSENHAUSEN		
30	KR		4392.16	5334.12	620	71.43	7829	ZAISERTSHOFEN		
32	GO		3610.14	5340.26	560	79.79	7828	KIRCHHEIM		
45	S	1	3606.58	5342.58	590	83.72	7728	MINDELZELL		
45	S	2	3606.58	5342.58	590	83.72	7728	MINDELZELL		
45	S	3	3606.58	5342.58	590	83.72	7728	MINDELZELL		
23	KI		3603.52	5344.40	580	87.23	7728	NIEDERRAUNAU		
22	KR		3608.17	5350.71	453	89.21	7728	THANNHAUSEN		
45	S	42	3604.08	5349.53	538	90.65	7728	ATTENHAUSEN		
45	S	41	3603.75	5349.30	550	90.67	7728	ATTENHAUSEN		
45	S	43	3603.90	5349.52	530	90.76	7728	ATTENHAUSEN		
44	S	41	3605.66	5352.77	520	92.18	7639	MUENSTERHAUSEN		
44	S	41	3605.66	5352.77	520	92.18	7628	MUENSTERHAUSEN		
44	S	42	3605.66	5352.77	520	92.18	7628	MUENSTERHAUSEN		
44	S	43	3605.66	5352.77	520	92.18	7628	MUENSTERHAUSE		
45	S	11	3604.46	5354.52	542	94.35	7628	KEMNAT		
45	S	31	3604.46	5354.52	542	94.35	7628	KEMNAT		
45	S	12	3604.39	5355.08	540	94.84	7628	KEMNAT		
45	S	32	3604.39	5355.08	540	94.84	7628	KEMNAT		
45	S	13	3604.29	5355.48	542	95.22	7628	KEMNAT		
45	S	33	3604.29	5355.48	540	95.22	7628	KEMNAT		
22	CL		3604.71	5358.29	500	97.37	7628	KLINGENBERG		
20	CL		3603.47	5362.55	525	101.52	7628	HARTBERG I		
21	CL		3603.47	5362.55	525	101.72	7628	HARTBERG II		
21	CL		3601.49	5368.75	473	107.96	7528	REMSHART II		
22	BO		3597.00	5370.30	500	111.52	7527	REISENBURG		
02	M		3592.76	5375.33	440	118.18	7427	RIEDHAUSEN II		
01	M		3592.76	5375.33	440	118.42	7427	RIEDHAUSEN I		
17	S		3591.47	5380.37	497	123.13	7427	NIEDERSTOTZINGEN I		
18	S		3591.47	5380.37	497	123.21	7427	NIEDERSTOTZINGEN II		
22	M		3589.60	5381.97	490	125.43	7427	LONTAL		
20	S		3589.00	5383.59	490	127.06	7427	HUERBEN I		
21	S		3589.00	5383.59	490	127.27	7427	HUERBEN II		
12	M	A	3579.45	5378.97	535	129.08	7426	BALLENDORF I		
13	M	A	3579.45	5378.97	535	129.40	7426	BALLENDORF II		
20	S		3591.95	5388.97	485	129.91	7327	GIENGEN		
20	M		3562.59	5365.96	646	131.53	7525	WIPPINGEN		
15	S		3583.74	5386.95	527	133.12	7326	ANHAUSEN		
17	S		3582.38	5389.79	582	136.07	7326	UGENHOF I		
18	S		3582.38	5389.79	582	136.15	7326	UGENHOF II		
16	MS		3573.01	5383.34	600	136.61	7425	ZAEHRINGEN		
02	M		3582.49	5391.44	597	137.26	7326	MERGELSTETTEN-BUCH		
01	M		3582.49	5391.44	597	137.49	7326	MERGELSTETTEN-BUCH		
15	S		3580.48	5392.96	628	139.90	7326	KUEPFENDORF		
20	M		3570.58	5385.80	655	140.07	7325	BRAEUNISHEIM		
16	MS		3570.53	5386.65	615	140.74	7325	GUSSENSTADT		
02	PR		3579.14	5396.00	530	142.92	7326	STEINHEIM II		
01	PR		3579.14	5396.00	530	143.15	7326	STEINHEIM I		
22	M		3566.54	5387.21	660	143.80	7325	SCHALKSTETTEN		
20	S		3576.40	5397.71	610	145.86	7326	DOSCHENTAL I		
21	S		3576.40	5397.71	610	146.08	7226	DOSCHENTAL II		
16	S	A	3575.32	5400.06	625	148.40	7226	GEMEINTAL		
16	S	S	3575.53	5400.26	625	148.42	7226	GEMEINTAL		
22	M		3553.52	5381.94	720	149.03	7424	BAD DIETZENBACH2		
22	M		3553.51	5381.95	723	149.04	7424	BAD DIETZENBACH1		
16	MS		3564.53	5394.78		150.83	7325	TREFFELHAUSEN		
17	M		3562.42	5392.82	660	150.83	7325	OBERWECKERSTELL I		
18	M		3562.42	5392.82	660	150.94	7325	OBERWECKERSTELL II		
16	S		3573.71	5402.85	635	151.61	7226	ST.BARTHOLOMAE		
22	M		3561.42	5394.42	650	152.62	7325	MESSELBERG		
02	KR		3571.85	5404.87	700	154.37	7225	LAUTERBURG II		
01	KR		3571.85	5404.87	700	154.61	7225	LAUTERBURG I		
22	S		3570.74	5407.05	700	156.79	7225	ROSENSTEIN		
22	M		3557.91	5399.69	620	158.90	7224	ZIRSCHBERG		
02	M		3570.74	5411.17	450	160.17	7125	MOEGGLINGEN II		
01	M		3570.74	5411.17	450	160.39	7125	MOEGGLINGEN I		
20	M		3558.18	5402.55	690	160.86	7224	HOHENRECHENBERG		
22	M		3553.04	5405.11	310	166.18	7224	BEUTENHOF		
22	M		3564.43	5414.38	470	166.42	7125	GOEGGINGEN		
22	M		3565.57	5419.13	512	169.66	7025	UNTERGROENINGEN		
22	M		3563.86	5419.84	400	171.21	7025	WALDMANNSHOFEN		
12	M		3549.45	5409.41	395	171.68	7124	KLOTZENHOF I		
13	M		3549.45	5409.41	395	172.02	7124	KLOTZENHOF II		
22	M	A	3546.96	5408.48	380	172.70	7123	STECHERS HUETTE		
22	M	B	3546.70	5408.63	395	172.99	7123	STECHERS HUETTE		
22	M	C	3546.29	5409.01	370	173.54	7123	STECHERS HUETTE		
20	MS		3565.91	5425.95	495	175.10	7025	HOHENBERG I		
21	MS		3565.91	5425.95	495	175.29	7025	HOHENBERG II		
20	M		3545.86	5411.04	440	175.33	7123	WALKERSBACH I		
20	BO		3552.50	5416.63	525	175.37	7124	MITTELWEILER		
21	M		3545.86	5411.04	440	175.60	7123	WALKERSBACH II		
22	M		3544.56	5411.33	465	176.45	7123	PLUEDERWIESEN HOF		
20	BO		3558.75	5422.86	480	176.58	7024	STEINHOEFLE I		
21	BO		3558.75	5422.86	480	176.80	7024	STEINHOEFLE II		
22	M		3556.89	5424.53	460	179.07	7024	UNTERROT		
22	M		3542.03	5414.32	420	180.37	7123	STEINENBERG		
45	F	21	3542.58	5415.85	525	180.85	7123	EDELMANNSHOF		
45	F	22	3542.37	5415.97	515	181.23	7123	EDELMANNSHOF		
45	F	23	3542.10	5416.23	505	181.60	7123	EDELMANNSHOF		
22	M		3556.99	5432.71	470	185.72	6924	EUTENDORF		
45	F	11	3540.70	5421.52	450	186.53	7023	WALDENWEILE-W		
45	F	12	3540.31	5421.68	450	186.90	7023	WALDENWEILE-W		
45	F	13	3540.04	5421.90	465	187.25	7023	WALDENWEILE-W		
43	S	43	3539.68	5422.10	430	187.25	7023	EBERSBERG		
43	S	41	3539.49	5421.78	430	187.51	7023	EBERSBERG		
43	S	42	3539.60	5421.88	452	187.53	7023	EBERSBERG		
14	M		3537.14	5427.45	460	193.46	7023	ITTENBERG		
43	S	21	3534.27	5429.75	430	196.96	6922	SCHIFFRAIN		
43	S	22	3534.02	5430.04	445	197.34	6922	SCHIFFRAIN		
43	S	23	3533.75	5430.24	460	197.67	6922	SCHIFFRAIN		
14	M		3532.34	5432.14	485	200.14	6922	JUX		
27	M		3530.63	5432.88	470	201.65	6922	NASSACH		
43	S	31	3528.48	5432.77	450	203.02	6922	GRONAU		
43	S	32	3528.17	5433.05	415	203.43	6922	GRONAU		
43	S	33	3527.76	5433.18	350	203.81	6922	GRONAU		
27	M		3524.14	5436.52	370	208.85	6921	HELFENBERG		
15	M		3523.49	5436.96	360	209.91	6921	SCHLOSS WILDECK		
22	MZ		3519.82	5444.32	290	217.58	6821	JAEGERHAUS		
42	S	41	3500.60	5439.08	305	226.70	6920	STOCKHEIM		
42	S	42	3500.32	5439.20	305	226.98	6920	STOCKHEIM		
42	S	43	3500.05	5439.36	315	227.28	6920	STOCKHEIM		
15	M		3505.66	5447.17	251	229.45	6820	MASSENBACH		
42	S	31	3508.46	5458.31	215	235.47	6720	SIEGELSBACH		
42	S	32	3508.47	5458.39	215	235.53	6720	SIEGELSBACH		
42	S	33	3508.18	5458.69	250	235.95	6720	SIEGELSBACH		
36	S	31	3501.93	5460.04	250	241.00	6720	UNTERGIMPERN		
36	S	32	3501.75	5460.46	250	241.53	6720	UNTERGIMPERN		
36	S	33	3501.67	5460.91	242	241.82	6720	UNTERGIMPERN		
14	M		3500.44	5463.28	200	244.55	6620	BARGEN		
15	M		3500.38	5462.98	200	244.71	6620	FLINSBACH		
15	M		4282.25	5467.30	200	244.75	6620	FLINSBACH		
22	F		3500.03	5472.93	295	252.18	6620	NEUNKIRCHEN		
36	S	41	3497.47	5471.36	289	252.48	6619	MICHELBACH		
36	S	42	3497.30	5471.47	275	252.67	6619	MICHELBACH		
14	M		3493.83	5469.96	310	253.93	6619	WALDWIMMERSBACH		
20	MM		3490.74	5476.16	320	260.64	6519	NECKARHAUSEN		
42	S	21	3490.30	5478.07	460	262.29	6519	OESTL.GREIN		
42	S	22	3490.00	5478.39	463	262.73	6519	OESTL. GREIN		
42	S	23	3489.70	5478.72	464	263.03	6519	OESTL.GREIN		
22	SE		3488.86	5478.54	435	263.72	6519	GREIN		
42	S	11	3487.16	5480.60	420	266.26	6518	ALTNEUDORF		
42	S	12	3486.84	5480.81	427	266.63	6518	ALTNEUDORF		
22	SE		3483.66	5483.64	410	270.97	6518	VORDERHEUBACH		
27	S		3482.70	5485.82	350	273.25	6418	BAERSBACH		
35	F	22	3522.38	5515.18	225	273.86	6221	REISTENHAUSEN		
22	SE		3482.60	5488.38	340	275.25	6418	TROESEL		
35	CL	33	3526.20	5522.00	453	277.88	6122	HUNDSRUECKKOPF		
35	CL	32	3526.33	5522.16	465	277.95	6122	HUNDSRUECKKOPF		
35	CL	31	3526.45	5522.30	473	278.02	6122	HUNDSRUECKKOPF		
20	MM		3480.18	5490.04	320	278.08	6418	BUCHKLINGEN I		
21	MM		3480.18	5490.04	320	278.35	6418	BUCHKLINGEN II		
27	S		3477.20	5492.76	320	282.04	6418	NAECHSTENBACH		
22	MM		3477.36	5493.99	240	282.90	6418	NIED.LIEBERSBACH		
35	CL	13	3530.80	5531.16	505	283.77	6022	WEIBERSBRUNN		
35	CL	12	3530.85	5531.33	500	283.90	6022	WEIBERSBRUNN		
35	CL	11	3531.25	5531.63	505	283.98	6022	WEIBERSBRUNN		
43	KA	62	3477.84	5498.02	350	285.83	6318	OBERLAUDERBACH		
22	MM		3475.61	5499.85	200	289.17	6317	HEPPENHEIM		
35	KI	13	3430.32	5494.08	295	315.52	6414	GOELLHEIMER WALD		
35	KI	11	3430.32	5494.10	295	315.53	6414	GOELLHEIMER WALD		
35	F	32	3426.10	5497.37	293	320.85	6313	JAKOBWEILER-E		
35	M	21	3419.88	5507.15	330	332.08	6313	SCHNEEBERGERHOF		
35	M	22	3419.58	5507.01	350	332.20	6313	SCHNEEBERGERHOF		
35	M	24	3419.29	5506.98	370	332.39	6313	SCHNEEBERGERHOF		
35	M	31	3418.02	5510.28	280	335.58	6213	OBERHAUSEN		
35	M	32	3417.67	5510.17	290	335.76	6213	OBERHAUSEN		
35	M	34	3417.54	5510.15	300	335.84	6213	OBERHAUSEN		
37	KR	11	2560.46	5720.82	67	540.50	4307	KIRCHHELLEN		

B	IN	AC	COORDINATES		ALT	X(KM)	SMAP	LOCALITY	G C
37	KR	12	2560.25	5720.93	65	540.72	4307	KIRCHHELLEN	
37	KR	13	2559.83	5720.93	64	540.97	4307	KIRCHHELLEN	

QUARRY-NO. 01 ESCHENLOHE SMAP 8332

PROFILE 01-345-02 COMPILED BY AICHELE.FUCHS

B	IN	AC	COORDINATES		ALT	X(KM)	SMAP	LOCALITY	G C
16	BO	A	4433.90	5279.15	660	2.78	8332	ASCHAU	
16	BO	B	4433.90	5279.15	660	2.78	8332	ASCHAU	X
16	BO	C	4433.90	5279.15	660	2.78	8332	ASCHAU	Y
17	BO	A	4433.90	5279.15	660	3.02	8332	ASCHAU	
17	BO	B	4433.90	5279.15	660	3.02	8332	ASCHAU	X
17	BO	C	4433.90	5279.15	660	3.02	8332	ASCHAU	Y
18	BO	A	4433.90	5279.15	660	3.20	8332	ASCHAU	
18	BO	B	4433.90	5279.15	660	3.20	8332	ASCHAU	X
18	BO	C	4433.90	5279.15	660	3.20	8332	ASCHAU	Y
15	BO		4433.90	5279.15	660	3.31	8332	ASCHAU	
15	CL		4431.94	5286.08	740	9.85	8232	BRAND	
17	CL	A	4433.14	5288.61	650	11.67	8232	BIRNBAUMHOLZ	
16	CL	A	4432.61	5289.47	672	12.64	8232	GRASLEITEN	MA
15	M		4434.13	5291.19	670	13.96	8232	DEIMENRIED	
12	CL		4432.20	5290.95	670	14.15	8232	RECHETSBERG	
13	CL		4432.20	5290.95	670	14.19	8232	RECHETSBERG	
14	CL		4430.88	5299.47	610	22.80	8132	PEISSENBERG	
10	MZ		4428.00	5309.73	678	33.38	8032	RAISTING	
11	MZ		4428.00	5309.73	678	33.40	8032	RAISTING	
13	CL		4426.44	5312.77	690	36.74	8032	DETTENSCHWNG	MA
08	M		4426.40	5312.78	690	36.78	8032	DETTENSCHWNG	MA
09	M		4424.88	5320.92	630	45.00	7931	OBERFINNING	MH
13	MZ		4424.08	5326.90	621	50.99	7931	SCHOEFFELDING	
12	MZ		4424.08	5326.90	621	51.00	7931	SCHOEFFELDING	
08	M		4423.52	5333.90	580	57.95	7831	JEDELSTETTEN	MA
10	CL		4420.17	5337.87	558	62.59	7831	WINKEL	HT
10	GO		4405.00	5343.03	590	72.64	7730	GUGGENBERG	MH
11	GO		4405.00	5343.03	590	72.80	7730	GUGGENBERG	MH
14	CL		4405.20	5343.40	590	72.94	7730	GUGGENBERG	HT
10	GO		4403.90	5348.70	522	78.25	7730	DOEPSHOFEN	MH
11	GO		4403.90	5348.70	522	78.39	7730	DOEPSHOFEN	MH
12	GO		4410.19	5356.82	510	83.58	7630	ANHAUSEN	MH
13	GO		4410.19	5356.82	510	83.64	7630	ANHAUSEN	MH
12	GO		4404.78	5359.06	520	87.49	7630	ROMMELSRIED	MH
13	GO		4404.78	5359.06	520	87.58	7630	ROMMELSRIED	MH
16	CL	C	4410.20	5366.30	488	92.67	7530	RETTENBERGEN	HT
16	CL	D	4410.20	5366.30	488	92.67	7530	RETTENBERG.	HTY
16	CL	E	4410.20	5366.30	488	92.67	7530	RETTENBERG.	HTX
08	GO		4409.63	5371.51	495	97.86	7530	MUTTERSHOFEN	MH
18	CL		4412.40	5373.22	473	98.76	7530	EGGELHOF	
17	CL		4412.40	5373.22	473	98.78	7530	EGGELHOF	
30	MH		4410.41	5374.08	448	100.08	7530	FEIGENHOFEN 3	
15	M		4408.92	5375.23	485	101.64	7430	ALBERTSHOFEN	
22	BO		3600.97	5365.29	480	105.24	7528	HAMMERSTETTEN	
20	CL		3601.49	5368.75	473	107.76	7528	REMSHART I	
29	MH		4408.15	5375.89	480	102.41	7530	FEIGENHOFEN 2	
12	CL		4412.12	5378.37	470	103.84	7427	BIBERACH	
08	GO		4410.00	5379.33	470	105.30	7430	ZEISENRIED	MH
15	M		4409.93	5383.03	460	108.91	7430	AHLINGEN	MA
13	CL		4411.84	5384.90	450	110.25	7430	EHINGEN	
12	CL		4411.84	5384.90	450	110.27	7430	EHINGEN	
18	MZ		4409.66	5389.10	435	114.83	7330	MERTINGEN	
17	MZ		4409.66	5389.10	435	114.85	7330	MERTINGEN	
08	GO		4409.97	5396.45	437	121.95	7230	GREGGENHOF	
13	DO		4407.50	5397.10	445	123.12	7230	STEINBERG	
12	DO		4407.50	5397.10	445	123.13	7230	STEINBERG	
10	CL		4405.29	5400.17	425	126.62	7230	HOERNITZHEIM	
11	CL		4405.29	5400.17	425	126.63	7230	HOERNITZHEIM	
35	MH	13	4406.95	5403.34	425	129.30	7230	EBERMERGEN	
35	MH	12	4406.99	5403.72	430	129.69	7230	EBERMERGEN	
35	MH	11	4406.80	5404.15	427	130.13	7230	EBERMERGEN	
10	CL		4405.49	5406.38	460	132.62	7230	KRATZHOF	
11	CL		4405.49	5406.38	460	132.62	7230	KRATZHOF	
13	MZ		4404.40	5410.13	482	136.51	7130	HUISHEIM	
12	MZ		4404.40	5410.13	482	136.53	7130	HUISHEIM	BS
35	MH	21	4405.50	5413.35	428	139.38	7130	GOSHEIM	
35	MH	22	4405.71	5413.74	430	139.72	7130	GOSHEIM	
35	MH	23	4405.70	5414.18	433	140.15	7130	GOSHEIM	
13	MZ		4406.43	5417.12	520	142.87	7130	AMERBACH	
15	MZ		4406.43	5417.12	520	142.90	7130	AMERBACH	
12	MZ		4406.43	5417.12	520	142.91	7130	AMERBACH	
35	MH	33	4401.75	5420.65	420	147.34	7029	LAUB S	
35	MH	32	4401.70	5421.02	421	147.71	7029	LAUB M	
35	MH	31	4401.62	5421.48	421	148.18	7029	LAUB N	
16	MZ		4402.96	5424.95	482	151.31	7030	STEINHART	
27	MZ		4413.56	5428.38	593	152.74	7030	ROHRACH	
35	MH	41	4397.48	5426.56	465	154.11	7029	OETTINGEN	
35	MH	42	4397.26	5426.91	485	154.50	7029	OETTINGEN	
35	MH	43	4396.99	5427.25	490	154.90	7029	OETTINGEN	
18	KR		4400.05	5429.75	423	156.60	7029	WACHFELD	
17	KR		4400.05	5429.75	423	156.63	7029	WACHFELD	
26	CL		4402.13	5431.34	462	157.71	6929	WESTHEIM	MP
26	CL		4402.13	5431.34	462	157.71	6929	WESTHEIM	
18	F		4399.34	5433.01	433	159.94	6929	WASSERTRUEDGN	
17	F		4399.34	5433.01	433	159.96	6929	WASSERTRUEDGN	
26	CL		4400.66	5435.97	500	162.55	6929	EISELBERG	
18	M		4397.64	5437.11	470	164.32	6929	ALTENTRUEDINGN	
17	M		4397.64	5437.11	470	164.34	6929	ALTENTRUEDINGN	
10	MZ		4393.20	5437.70	662	165.99	6929	HESSELBERG	
11	MZ		4393.20	5437.70	662	166.01	6929	HESSELBERG	
18	M		4396.55	5440.32	470	167.70	6929	DAMBACH	
17	M		4396.55	5440.32	470	167.72	6929	DAMBACH	
15	CL		4399.55	5441.30	475	167.99	6929	DENNENLOHE	HT
16	CL	A	4397.19	5445.58	455	172.71	6829	HEINERSDORF	HT
16	CL	B	4397.19	5445.58	455	172.71	6829	HEINERSDORF	HT
16	CL	C	4397.19	5445.58	455	172.71	6829	HEINERSDORF	
16	CL	D	4397.19	5445.58	455	172.71	6829	HEINERSD.	MAY
16	CL	E	4397.19	5445.58	455	172.71	6829	HEINERSD.	MAX
29	B		4394.11	5447.76	427	175.50	6829	BECHHOFEN	
10	MZ		4394.30	5450.35	443	177.98	6829	LIEBERSDORF	BS
11	MZ		4394.30	5450.35	443	177.99	6829	LIEBERSDORF	BS
15	BO		4390.80	5453.95	430	182.36	6729	LAMMELBACH	
15	BO		4390.80	5453.95	430	182.36	6729	LAMMELBACH	Y
15	BO		4390.80	5453.95	430	182.36	6729	LAMMELBACH	X
16	BO		4390.38	5453.85	435	182.36	6728	LAMMELBACH	
16	BO		4390.38	5453.85	435	182.36	6728	LAMMELBACH	Y
16	BO		4390.38	5453.85	435	182.36	6728	LAMMELBACH	X
13	S		4392.79	5460.11	468	187.82	6729	UNT. DAUTENWND.	
12	S		4392.79	5460.11	468	187.84	6729	UNT. DAUTENWND.	
13	MH		4389.60	5464.69	500	193.03	6628	TIEFENTHAL	
12	MH		4389.60	5464.69	500	193.04	6628	TIEFENTHAL	
16	BO		4392.25	5468.60	482	196.25	6629	LEHRBERG	
16	BO		4392.25	5468.60	482	196.25	6629	LEHRBERG	X
16	BO		4392.25	5468.60	482	196.25	6629	LEHRBERG	Y
15	BO		4392.07	5468.60	476	196.28	6629	LEHRBERG	
15	BO		4392.07	5468.60	476	196.28	6629	LEHRBERG	X
15	BO		4392.07	5468.60	476	196.28	6629	LEHRBERG	Y
12	MH		4388.80	5471.93	420	200.26	6628	GRAEFENBUCH	
12	MH		4388.79	5471.91	420	200.24	6628	GRAEFENBUCH	
14	MH		4387.20	5472.45	455	201.21	6628	UNTERSULZBACH	
14	MH		4387.20	5472.45	455	201.21	6628	UNTERSULZBACH	X
15	F		4387.82	5477.22	488	205.65	6528	URPHERTSHOFEN	
36	MH	31	4390.39	5477.61	448	205.39	6528	EGENHAUSEN	X
36	MH	32	4390.39	5477.61	448	205.39	6528	EGENHAUSEN	
36	MH	33	4390.12	5478.55	422	206.36	6528	EGENHAUSEN	
11	M		4389.49	5482.41	400	210.29	6528	BREITENAU	
10	M		4389.49	5482.41	400	210.30	6528	BREITENAU	
14	MH		4389.49	5482.41	210	210.36	6528	BREITENAU	
14	MH		4389.49	5482.41	210	210.36	6528	BREITENAU	X
36	MH	21	4389.74	5484.40	350	212.16	6528	BEI B.WINDSHEIM	
36	MH	22	4389.46	5485.36	326	213.15	6528	BEI B.WINDSHEIM	
36	MH	23	4389.46	5485.36	326	213.15	6528	WINDSHEIM	X
16	F		4385.72	5490.11	346	218.68	6428	BEROLZHEIM	
34	KA	1	4392.31	5493.55	345	220.59	6429	ALTHEIM	
34	KA	2	4392.31	5493.55	345	220.59	6429	ALTHEIM	
34	KA	3	4392.31	5493.55	345	220.59	6429	ALTHEIM	
10	M		4381.33	5495.55	330	224.97	6428	MARKT NORDHEIM	
11	M		4381.33	5495.55	330	224.98	6428	MARKT NORDHEIM	
40	MH	41	4415.61	5501.47	301	225.09	6330	MECHELWIND	
40	MH	42	4415.62	5501.90	302	225.52	6330	MECHELWIND	
40	MH	31	4416.49	5506.19	273	229.71	6331	GREMSDORF	
40	MH	32	4416.35	5506.60		230.14	6331	GREMSDORF	
40	MH	33	4416.48	5507.09		230.62	6331	GREMSDORF	
10	S		4381.20	5501.50	335	230.78	6328	LIMBURG	
11	S		4381.20	5501.50	335	230.79	6328	LIMBURG	
40	MH	21	4418.13	5510.99	305	234.37	6231	AISCH	
40	MH	22	4418.27	5511.32	314	234.69	6231	AISCH	
40	MH	23	4418.42	5511.72	305	235.08	6231	AISCH	
15	MH		4383.25	5508.40	380	237.04	6228	ZIEGENBACH	
40	MH	11	4418.02	5515.48	272	238.86	6231	WIND	
40	MH	12	4418.19	5515.82	271	239.18	6231	WIND	
40	MH	13	4418.12	5516.17	270	239.54	6231	WIND	
11	S		4381.40	5512.85	326	241.78	6228	CASTELL	
10	S		4381.40	5512.85	326	241.78	6228	CASTELL	
40	GO	11	4417.50	5521.51	258	244.91	6131	VORRA	
40	GO	12	4417.33	5521.91	270	245.32	6131	VORRA	
40	GO	13	4417.20	5522.34	295	245.76	6131	VORRA	
15	MH		4378.87	5518.49	231	247.87	6227	WIESENTHEID	
40	GO	21	4415.03	5527.14	300	250.71	6130	MUEHLENDORF N	
40	GO	22	4414.94	5527.58	321	251.16	6130	MUEHLENDORF N	
40	GO	23	4414.80	5527.96	327	251.55	6130	MUEHLENDORF N	
40	KI	11	4414.92	5534.25	260	257.81	6030	OBERHAID-N	
40	KI	12	4415.00	5534.65	270	258.20	6030	OBERHAID-N	
16	MH		4376.92	5528.72	220	258.28	6127	KRAUTHEIM	
40	KI	13	4415.02	5535.00	292	258.55	6030	OBERHAID-N	
45	B	51	4415.38	5535.15	290	258.65	6030	OBERHAID	
45	B	52	4415.55	5535.44	290	258.93	6030	OBERHAID	
45	B	11	4415.55	5535.44	290	258.93	6030	OBERHAID2	
45	B	12	4415.55	5535.44	290	258.93	6030	OBERHAID2	
45	B	13	4415.55	5535.44	290	258.93	6030	OBERHAID2	
45	B	53	4415.04	5536.54	290	260.06	6030	OBERHAID	
45	B	62	4415.51	5538.70	270	262.18	6030	GODELHOF	
45	B	63	4415.30	5539.14	290	262.63	6030	GODELHOF	
45	B	71	4415.28	5539.40	300	262.89	6030	RUINE STIEFENBERG	
45	B	72	4415.21	5539.83	340	263.33	6030	RUINE STIEFENBERG	
40	KI	21	4412.84	5539.69	340	263.41	6030	PRIEGENDORF-S	
45	B	73	4415.33	5540.09	320	263.58	6030	RUINE STIEFENBERG	
40	KI	22	4412.97	5540.08	355	263.78	6030	PRIEGENDORF-S	
40	KI	23	4413.03	5540.37	340	264.07	6030	PRIEGENDORF-S	
45	KI	41	4413.54	5540.53	342	264.16	6030	PRIEGENDORF	
45	KI	42	4413.48	5540.93	300	264.57	6030	PRIEGENDORF	
45	KI	43	4413.40	5541.26	305	264.90	5930	PRIEGENDORF	
16	MH		4376.62	5536.07	220	265.50	6027	OBERSPIESSHEIM	
45	KI	31	4411.79	5546.20	380	269.96	5930	SCHELMSBRUNNEN	
40	KI	41	4411.68	5546.25	390	270.04	5930	RENTWEINSDORF	
40	KI	42	4411.68	5546.50	413	270.29	5930	RENTWEINSDORF	
45	KI	32	4411.69	5546.54	410	270.31	5930	SCHELMSBRUNNEN	
45	KI	33	4411.48	5546.88	400	270.67	5930	SCHELMSBRUNNEN	
40	KI	43	4411.48	5546.88	403	270.69	5930	RENTWEINSDORF	
40	KI	31	4410.35	5551.06	320	274.95	5930	VORBACH	
40	KI	32	4410.20	5551.29	308	275.20	5930	VORBACH	
40	KI	33	4410.00	5551.56	295	275.49	5930	VORBACH	
45	KI	21	4409.72	5553.29	427	277.21	5830	VORBACH	
45	KI	22	4409.67	5553.48	400	277.41	5830	VORBACH	
45	KI	23	4409.62	5553.70	360	277.63	5830	VORBACH	
40	M	21	4405.60	5558.54	320	282.87	5830	MARBACH	
40	M	22	4405.61	5558.97	345	283.29	5830	MARBACH	
40	M	23	4405.59	5559.37	340	283.69	5830	MARBACH	
45	KI	12	4407.95	5559.69	337	283.75	5830	PFAFFENDORF	
45	KI	13	4407.75	5559.80	345	283.88	5830	PFAFFENDORF	
37	MH	13	4405.76	5565.08	354	289.37	5730	NNW-ALLERTSHAUSEN	
37	MH	12	4405.38	5565.22	357	289.55	5730	NNW-ALLERTSHAUSEN	
37	MH	33	4401.76	5565.89	375	290.62	5729	NNW-ERMERSHAUSEN	
37	MH	32	4401.37	5565.97	383	290.75	5729	NNW-ERMERSHAUSEN	
37	MH	31	4400.96	5566.12	390	290.94	5729	NNW-ERMERSHAUSEN	
37	MH	23	4400.88	5569.13	361	293.94	5729	ZIMMERAU	
37	MH	22	4400.48	5569.41	376	294.27	5729	ZIMMERAU	
37	MH	21	4400.13	5569.56	370	294.46	5729	ZIMMERAU	
40	M	33	4395.20	5576.84	320	302.29	5629	EYERSHAUSEN	
40	M	32	4394.88	5577.05	350	302.54	5629	EYERSHAUSEN	
40	M	31	4394.79	5577.39	360	302.89	5629	EYERSHAUSEN	
40	M	11	3601.50	5581.34	305	307.98	5628	GOLLMUTHHAUSEN	
40	M	12	3601.46	5581.77	350	308.41	5628	GOLLMUTHHAUSEN	
40	M	13	3601.51	5582.14	350	308.76	5628	GOLLMUTHHAUSEN	
15	M		4352.17	5576.41	800	310.67	5625	ERLENBERG	
40	M	41	3595.52	5591.31	340	318.95	5528	MUEHLFELD	
40	M	42	3595.48	5591.76	350	319.40	5528	MUEHLFELD	

B	IN	AC	COORDINATES	ALT	X(KM)	SMAP	LOCALITY	G	C
40	M	43	3595.53 5592.24	370	319.86	5528	MUEHLFELD		
11	MH		4362.70 5591.52	560	322.64	5526	GINOLFS		
10	MH		4362.70 5591.52	560	322.64	5526	GINOLFS		
10	MH		4360.88 5597.88	740	329.24	5426	SANKT BARBARA		
37	MH	43	4364.09 5600.06	770	330.67	5426	RHOEN HUT		
37	MH	42	4363.68 5600.13	778	330.82	5426	RHOEN HUT		
37	MH	41	4363.53 5600.48	783	331.20	5426	RHOEN HUT		
42	M	A	3542.30 5768.93	260	504.00	3923	SALZHEMMENDORF		X
42	M	B	3542.30 5768.93	260	504.00	3923	SALZHEMMENDORF		
42	M	C	3542.30 5768.93	260	504.00	3923	SALZHEMMENDORF		Y

QUARRY-NO. 01 ESCHENLOHE SMAP 8332
ARC 01-130-F COMPILED BY EMTER.GEBRANDE.BEHNKE

B	IN	AC	COORDINATES	ALT	X(KM)	SMAP	LOCALITY	G	C
45	M	41	3530.47 5295.52	665	131.80	8222	FALKENHALDEN		
45	M	42	3530.18 5295.20	600	132.05	8222	FALKENHALDEN		
45	M	43	3529.91 5294.91	540	132.28	8222	FALKENHALDEN		
45	M	11	3530.95 5302.59	690	132.39	8122	WILHELMSDORF		
45	M	12	3531.04 5302.16	705	132.22	8122	WILHELMSDORF		
45	M	13	3531.11 5301.72	710	132.08	8122	WILHELMSDORF		
32	S		3539.15 5328.10	600	131.68	7923	KANZACH		
39	S	21	3541.52 5336.23	765	132.76	7823	BUSSEN		
32	KA		3548.00 5348.38	540	132.83	7723	KIRCHEN		
39	S	11	3552.42 5353.50	525	131.90	7624	ALLMENDINGEN		
32	KA		3557.16 5359.90	565	131.83	7624	SCHELKLINGEN		
20	M		3562.59 5365.96	646	131.53	7525	WIPPINGEN I		
21	M		3562.59 5365.96	646	131.87	7525	WIPPINGEN II		
18	M		3573.18 5372.70	575	128.75	7525	BEIMERSTETTEN I		
19	M		3573.18 5372.70	575	128.87	7525	BEIMERSTETTEN II		
12	M		3579.45 5378.97	535	129.08	7426	BALLENDORF I		
13	M		3579.45 5378.97	535	129.40	7426	BALLENDORF II		
20	S		3589.00 5383.59	490	137.06	7427	HUERBEN I		
21	S		3589.00 5383.59	490	127.27	7427	HUERBEN II		
20	S		3591.95 5388.97	485	129.91	7327	GIENGEN I		
21	S		3591.95 5388.97	485	130.09	7327	GIENGEN II		
20	F		3597.06 5393.41	545	131.09	7327	ZOESCHINGENI		
21	F		3597.06 5393.41	545	131.25	7327	ZOESCHINGEN II		
37	F	11	3605.32 5397.48	570	130.88	7228	DUNSTELKINGEN		
37	F	12	3605.28 5397.98	557	131.37	7228	DUNSTELKINGEN		
37	F	13	3605.06 5398.33	542	131.78	7228	DUNSTELKINGEN		
37	F	31	4391.42 5398.30	510	128.93	7229	ZOLTINGEN		
37	F	32	4391.39 5398.72	500	129.33	7229	ZOLTINGEN		
37	F	33	4391.35 5399.12	500	129.72	7229	ZOLTINGEN		
37	F	21	4398.35 5399.20	457	127.57	7229	BISSINGEN		
37	F	22	4398.25 5399.64	485	128.02	7229	BISSINGEN		
37	F	23	4398.15 5400.07	470	128.46	7229	BISSINGEN		

QUARRY-NO. 01 ESCHENLOHE SMAP 8332
ARC 01-150-F COMPILED BY EMTER

B	IN	AC	COORDINATES	ALT	X(KM)	SMAP	LOCALITY	G	C
40	KA	21	3517.58 5325.67	630	150.78	7921	SIGMARINGEN		
33	KR		3520.59 5331.67	630	150.29	7821	HORNSTEIN		
44	S	31	3520.13 5330.92	650	150.23	7821	BINGEN		
40	KA	11	3522.19 5337.50	735	150.73	7821	EMERFELD		
33	S		3524.31 5341.33	755	150.53	7721	ITTENHAUSEN		
33	S		3527.82 5350.93	700	151.65	7722	AICHELAU		
32	F		3530.71 5355.82	780	151.51	7622	EHESTETTEN		
33	S		3535.48 5365.64	780	152.76	7522	TRAILFINGEN		
40	S	41	3542.53 5374.14	856	151.97	7423	DONNSTETTEN		
40	S	42	3542.48 5374.56	821	152.28	7423	DONNSTETTEN		
40	S	43	3542.14 5374.80	835	152.69	7423	DONNSTETTEN		
32	F		3544.87 5375.14	819	150.96	7423	WESTERNHEIM		
40	S	22	3545.90 5381.30	745	154.01	7423	WIESENSTEIG		
22	M		3453.51 5381.95	720	149.05	7424	BAD DITZENBACH		
17	M		3562.48 5392.82	660	150.83	7325	OBERWECKERSTELL I		
18	M		3562.48 5392.82	660	150.94	7325	OBERWECKERSTELL II		
22	M		3561.42 5394.42	650	152.62	7325	MESSELBERG		
16	MS		3564.53 5394.78	640	150.82	7325	TREFFELHAUSEN		
22	S		3569.49 5400.46	685	152.24	7225	ROETENBACH		
22	S	A	3570.74 5407.05	700	156.77	7225	ROSENSTEIN		
22	S	B	3570.80 5407.05	700	156.74	7225	ROSENSTEIN		
16	S		3573.71 5402.85	635	151.61	7226	ST.BARTHOLOMAE		
16	S		3575.53 5400.26	625	148.42	7226	GEMEINTAL		
22	S		3581.18 5407.35	520	151.24	7226	OBERKOCHEN		
26	M		3588.05 5411.29	605	151.14	7127	WALDHAUSEN		
22	S		3592.00 5414.32	625	151.95	7127	HUELEN		
27	M		3599.73 5414.63	650	149.76	7128	IPF		
44	S	21	3604.53 5414.57	508	146.76	7128	GOLDBURGHAUSEN		
26	S		3603.12 5417.80	490	150.48	7128	DIRGENHEIM		
27	MZ		3606.84 5423.28	485	154.10	7028	MARKOFFINGEN		
26	S		4400.03 5425.02	475	152.05	7029	MAINSFAHRT		
16	MZ		4402.96 5424.95	482	151.31	7030	STEINHART		
27	MZ		4413.56 5428.39	593	152.74	7030	ROHRACH		
30	CL		4523.96 5397.05	470	148.59	7239	GERABACH		
30	CL		4510.45 5404.35	433	147.25	7238	OBERHASELBACH		

QUARRY-NO. 01 ESCHENLOHE SMAP 8332
ARC 01-175-F COMPILED BY SCHICK.EMTER

B	IN	AC	COORDINATES	ALT	X(KM)	SMAP	LOCALITY	G	C
29	S		3495.09 5338.77	890	176.67	7919	HOSSINGEN		
43	S	12	3494.53 5330.71	898	174.51	7819	NUSPLINGEN		
38	S	43	3496.55 5344.41	870	177.21	7719	BURGFELDEN		
38	S	41	3496.94 5344.30	875	176.81	7719	BURGFELDEN		
38	S	42	3497.18 5344.17	870	176.54	7719	BURGFELDEN		
28	S		3503.58 5355.20	645	175.06	7620	JUNGINGEN		
28	S		3509.00 5359.60	730	172.20	7620	TALHEIM		
29	CL		3508.95 5366.04	510	175.46	7520	GOENNINGEN		
29	CL		3516.08 5379.67	310	176.68	7421	OFERDINGEN		
28	GO		3521.80 5388.82	340	177.40	7321	HARDT		
29	GO		3525.39 5395.08	326	178.68	7322	KOENGEN		
29	GO		3533.63 5399.95	360	175.81	7222	BALTMANNSWEILER		
28	GO		3536.31 5401.03	380	174.53	7222	THOMASHARDT		
22	M		3545.56 5410.63	380	175.25	7123	PLUEDERWIESENHOF		
28	M		3550.27 5416.15	540	176.30	7124	BURGHOLZ		
20	BO		3552.50 5416.63	525	175.36	7124	MITTELWEILER		
28	M		3556.77 5418.81	470	174.40	7024	BIRKENLOHE		
20	BO		3558.75 5422.86	480	176.55	7024	STEINHOEFLE		
29	F		3557.80 5422.93	503	177.17	7024	FRICKENHOFEN		
20	MS		3563.66 5418.34		170.06	7025	ESCHACH		
20	MS		3565.91 5425.95	492	175.10	7025	HOHENBERG		
28	M		3569.42 5430.00	480	176.58	6925	BENZENHOF		
28	M		3578.67 5433.76	465	175.26	6926	RIEGELHOF		
29	KI		3587.09 5436.59	465	174.05	6927	MELBERSMUEHLE		
28	M		3588.93 5439.40	500	175.79	6927	ROETLEIN		
29	MZ		3591.70 5439.76	498	175.03	6927	ESBACH		
29	MZ		3597.22 5442.35	475	175.31	6827	DICKERSBRONN		
29	MH		3601.65 5444.79	458	176.04	6828	WEHLMAEUSEL		
29	MH		3605.83 5446.59	457	176.37	6828	OBERMOSBACH		

QUARRY-NO. 01 ESCHENLOHE SMAP 8332
ARC 01-210-F COMPILED BY EMTER

B	IN	AC	COORDINATES	ALT	X(KM)	SMAP	LOCALITY	G	C
34	S		3464.10 5341.69	715	206.93	7717	DUNNINGEN		
30	S		3465.29 5344.41	670	207.62	7717	BOESINGEN		
34	S	1	3468.79 5352.18	555	205.92	7617	AISTAIG		
36	S	11	3467.44 5356.07	590	208.47	7617	MARSCHALKENZIMMERN		
33	S		3469.30 5358.82	560	207.97	7617	NIEDERDOBEL		
30	S		3469.81 5360.58	590	208.17	7617	DUERRENMETTSTETTEN		
34	S		3471.27 5363.39	550	207.91	7517	DETTINGEN		
33	KA		3473.82 5366.67	515	207.00	7517	REXINGEN		
30	S		3476.04 5371.58	570	207.10	7518	UNTERTALHEIM		
36	S	21	3476.04 5371.58	570	206.94	7518	UNTERTALHEIM		
36	S	22	3476.04 5371.58	570	206.94	7518	UNTERTALHEIM		
36	S	23	3476.04 5371.58	570	206.94	7518	UNTERTALHEIM		X
34	S		3476.90 5374.28	525	207.59	7418	SCHIETINGEN		
33	KA		3476.10 5379.55	560	210.75	7418	UNTERSCHWANDORF		
34	S		3482.74 5387.25	550	208.77	7318	WILDBERG		
36	KA	42	3489.99 5400.53	470	209.86	7219	WEIL DER STADT		
43	KA	51	3491.81 5406.61	485	212.01	7219	MALMSHEIM		
43	KA	52	3491.86 5406.79	490	212.11	7219	MALMSHEIM		
43	KA	53	3491.90 5406.88	485	212.16	7219	MALMSHEIM		
43	KA	41	3496.84 5409.56	470	209.83	7119	RUTESHEIM		
43	KA	42	3496.83 5409.64	470	209.95	7119	RUTESHEIM		
43	KA	43	3496.92 5409.90	470	209.98	7119	RUTESHEIM		
36	KA	22	3506.88 5419.24	250	208.41	7020	MARKGROENINGEN		
36	KA	12	3513.09 5429.25	240	210.86	7020	BESIGHEIM		
27	M		3524.14 5436.52	370	208.85	6921	HELFENBERG		
37	S	42	3528.81 5440.12	410	208.48	6922	LOEWENSTEIN		
37	S	23	3538.45 5443.42	450	205.12	6823	GLEICHEN		
37	S	32	3548.54 5449.81	515	204.54	6823	WALDENBURG		
37	S	13	3554.87 5456.40	360	206.75	6724	RUEBLINGEN		
37	KA	32	3563.51 5462.56	435	207.85	6725	BRUECHLINGEN		
37	KA	42	3571.84 5468.69	450	209.59	6625	SCHROZBERG		
37	KA	22	3576.87 5472.91	450	211.35	6626	SPIELBACH		
37	KA	12	3587.04 5474.26	400	208.75	6527	GATTENHOFEN		

QUARRY-NO. 01 ESCHENLOHE SMAP 8332
ARC 01-220-F COMPILED BY AICHELE.FUCHS.KAMINSKI

B	IN	AC	COORDINATES	ALT	X(KM)	SMAP	LOCALITY	G	C
24	M		4565.68 5456.91	895	221.61	6743	BURGSTALL		
24	CL		4553.73 5461.87	447	219.00	6742	RIED		
25	CL		4553.73 5461.87	447	218.39	6742	RIED		
37	KI	43	4549.00 5476.34	640	228.96	6542	UNTERGRAFENRIED		
37	KI	42	4548.99 5476.47	645	229.06	6542	UNTERGRAFENRIED		
37	KI	41	4548.90 5476.39	645	228.95	6542	UNTERGRAFENRIED		
37	KI	23	4549.62 5473.45	530	226.76	6642	HOCHA		
37	KI	22	4549.61 5473.55	520	226.84	6642	HOCHA		
37	KI	21	4549.49 5473.62	520	226.84	6642	HOCHA		
24	CL		4545.60 5467.80	480	219.86	6641	DOEFERING		
25	CL		4545.50 5467.80	480	219.27	6641	DOEFERING		
37	KI	11	4543.89 5477.75	511	227.72	6541	BREITENRIED		
37	KI	12	4543.70 5477.63	521	227.52	6541	BREITENRIED		
37	KI	13	4543.76 5477.38	515	227.33	6541	BREITENRIED		
37	KI	33	4539.82 5482.03	740	229.62	6541	GAISTHAL-SE		
37	KI	32	4539.60 5482.20	719	229.67	6541	GAISTHAL-SE		
37	KI	31	4539.44 5482.34	717	229.73	6541	GAISTHAL-SE		
24	MZ		4533.69 5471.94	565	217.86	6640	THANSTEIN		
25	MZ		4533.69 5471.94	565	217.31	6640	THANSTEIN		
24	S		4526.18 5478.08	462	220.17	6540	HOEFLARN		
25	S		4526.18 5478.08	462	219.65	6540	HOEFLARN		
25	S		4516.60 5482.23	470	219.79	6539	NEUSATH		
24	S		4516.60 5482.23	470	220.28	6539	NEUSATH		
24	F		4496.08 5486.95	475	218.15	6437	KRICKLHOF		
25	F		4496.08 5486.95	475	217.73	6437	KRICKLHOF		
25	MH		4486.95 5491.44	460	219.79	6436	FROHNHOF		
24	MH		4486.95 5491.44	460	220.17	6436	FROHNHOF		
25	MH		4476.85 5493.75	526	219.97	6436	EDELSFELD		
24	MH		4476.85 5493.75	526	220.32	6436	EDELSFELD		
07	M		4470.77 5500.10	430	225.52	6335	KROTTENSEE		
07	M		4468.10 5490.85	550	215.97	6435	RUINE HAUSECK		
34	MH	C	4465.14 5494.24	476	218.89	6435	HARTENSTEIN		
34	MH	B	4464.95 5495.16	493	219.77	6435	HARTENSTEIN		
34	MH	A	4464.95 5495.16	493	219.77	6435	HARTENSTEIN		
33	MH		4458.30 5494.56	627	218.40	6434	ALGERSDORF		
34	MH	12	4456.14 5495.98	545	219.60	6334	GOETZLESBERG		
34	MH	11	4456.14 5495.98	545	219.60	6334	GOETZLESBERG		
34	MH	14	4456.14 5496.80	547	220.42	6334	GOETZLESBERG		
34	MH	B	4448.66 5498.20	500	221.26	6333	OEDHOF		
34	MH	C	4448.61 5497.38	515	220.43	6333	OEDHOF		
34	MH	D	4448.52 5496.46	536	219.51	6333	OEDHOF		
33	F		4448.18 5494.83	415	217.86	6433	KIRCHROETTENBACH		Y
33	F		4448.18 5494.83	415	217.86	6433	KIRCHROETTENBACH		
36	MH	42	4440.98 5497.39	380	220.15	6333	ETLASWIND		
36	MH	41	4440.98 5497.39	380	220.15	6333	ETLASWIND		
36	MH	43	4440.95 5498.18	420	220.94	6333	ETLASWIND		
33	F		4437.75 5490.60	400	213.30	6432	KALCHREUTH		X
33	F		4437.75 5490.60	400	213.30	6432	KALCHREUTH		Y
33	F		4437.75 5490.60	400	213.30	6432	KALCHREUTH		
34	MH	31	4430.57 5497.94	370	220.69	6332	ERLANGEN NO		
34	MH	32	4430.82 5498.61	382	221.36	6332	ERLANGEN NO		
34	MH	34	4430.82 5498.61	382	221.36	6332	ERLANGEN NO		

34	MH	22	4421.20 5497.94	298	221.12	6331	UNTERMEMBACH	
34	MH	24	4421.41 5497.06	318	220.23	6331	UNTERMEMBACH	
34	F		4415.18 5495.05	360	218.72	6430	UNTERREICHENBACH	
34	MH	4	4414.64 5496.82	341	220.54	6330	BUCH	Y
34	MH	4	4414.64 5496.82	341	220.54	6330	BUCH	X
34	MH	4	4414.64 5496.82	341	220.54	6330	BUCH	
34	F		4410.22 5495.58	360	219.78	6430	OBERREICHENBACH	
36	MH	53	4404.97 5495.19	374	220.06	6430	DETTENDORF	
36	MH	52	4404.92 5494.30	381	219.18	6430	DETTENDORF	X
36	MH	51	4404.92 5494.30	381	219.18	6430	DETTENDORF	
36	MH	11	4398.55 5493.26	340	219.15	6429	NEUSTADT.AISCH	
36	MH	12	4398.46 5493.71	325	219.60	6429	NEUSTADT.AISCH	
36	MH	13	4398.30 5494.07	335	219.99	6429	NEUSTADT.AISCH	
34	KA		4392.31 5493.55	345	220.59	6429	ALTHEIM	
34	KA		3604.80 5493.22	370	221.34	6428	RUEDISBRONN	
34	KA		3604.80 5493.22	370	221.34	6428	RUEDISBRONN	
34	KA		3604.80 5493.22	370	221.34	6428	RUEDISBRONN	
16	F		3602.83 5489.89	346	218.68	6428	BEROLZHEIM	
34	KA		3596.73 5490.67	360	221.11	6428	HERBOLZHEIM	
34	KA		3596.73 5490.67	360	221.11	6428	HERBOLZHEIM	
34	KA		3596.73 5490.67	360	221.11	6428	HERBOLZHEIM	
34	KA		3591.00 5487.45	400	219.80	6427	CUSTENLOHR	
34	KA		3591.00 5487.45	400	219.80	6427	CUSTENLOHR	
34	KA		3591.00 5487.45	400	219.80	6427	CUSTENLOHR	

QUARRY-NO. 02 MILDERS SMAP 5426
PROFILE 02-125-09 COMPILED BY VOSS.VETTER.KATZLER

B	IN	AC	COORDINATES	ALT	X(KM)	SMAP	LOCALITY	G C
23	MH	41	3576.23 5599.26	783	3.10	5426	RHOEN HUT	
22	MH	51	3576.93 5599.81	730	3.31	5426	RHOEN HUT	
13	F	A	3576.93 5599.83	715	3.44	5426	RUEDENSCHWINDEN	
23	MH	42	3576.40 5598.92	778	3.45	*5426	RHOEN HUT	
22	MH	52	3577.24 5599.69	692	3.64	5426	RHOEN HUT	
23	MH	43	3576.81 5598.87	770	3.80	5426	RHOEN HUT	
22	MH	53	3577.47 5599.59	695	3.89	5426	RHOEN HUT	
13	F	1	3580.11 5596.25	482	8.00	*5526	STETTEN	
22	MH	21	3582.55 5595.21	375	10.46	5526	NORDHEIM VD RHOEN	
22	MH	22	3582.92 5595.00	361	10.89	5526	NORDHEIM VD RHOEN	
22	MH	23	3583.37 5594.90	347	11.32	*5526	NORDHEIM VD RHOEN	
22	MH	33	3585.99 5593.24	345	14.42	5527	OSTHEIM VD RHOEN	
22	MH	32	3586.33 5593.50	355	14.57	5527	OSTHEIM VD RHOEN	
22	MH	31	3586.48 5593.64	382	14.63	*5527	OSTHEIM VD RHOEN	
21	M	11	3588.29 5590.15	320	18.00	*5527	HAINHOF N	
21	M	12	3588.37 5589.80	340	18.28	5527	HAINHOF N	
21	M	13	3588.62 5589.53	346	18.65	5527	HAINHOF N	
21	M	41	3590.19 5588.52	342	20.50	5527	MELLRICHSTADT W	
21	M	42	3590.62 5588.47	348	20.88	5527	MELLRICHSTADT W	
21	M	43	3591.07 5588.40	340	21.28	*5527	MELLRICHSTADT W	
21	M	21	3593.42 5587.12	280	23.94	5527	MELLRICHSTADT SSE	
21	M	22	3593.63 5587.03	285	24.16	5527	MELLRICHSTADT SSE	
21	M	23	3594.00 5586.95	305	24.51	5527	MELLRICHSTADT SSE	
14	KI	A	3596.17 5585.17	300	27.56	*5628	HENDUNGEN	
21	M	51	3598.50 5582.70	380	30.64	*5628	RAPPERSHAUSEN	
21	M	52	3598.80 5582.46	363	31.03	5628	RAPPERSHAUSEN	
21	M	53	3599.07 5582.20	355	31.40	5628	RAPPERSHAUSEN	
14	KI	A	3601.45 5581.79	350	33.83	*5628	GOLLMUTHHAUSEN	
21	M	31	3604.22 5579.80	339	36.99	5628	OTTELMANNSHAUSERHOF	
21	M	32	3604.30 5579.62	330	37.15	*5628	OTTELMANNSHAUSERHOF	
21	M	33	3604.65 5579.44	315	37.54	5628	OTTELMANNSHAUSERHOF	
22	MH	43	4393.38 5579.66	296	39.34	*5629	HERBSTADT	
22	MH	42	4393.90 5579.56	304	39.83	5629	HERBSTADT	
22	MH	41	4394.26 5579.31	313	40.27	5629	HERBSTADT	
21	S	1	4395.82 5575.72	295	43.58	*5629	EYERSHAUSEN	
21	S	2	4396.25 5575.59	295	44.00	5629	EYERSHAUSEN	
22	H	41	4398.85 5574.81	310	46.62	*5629	ALSLEBEN	
22	H	42	4399.25 5574.65	300	47.03	5629	ALSLEBEN	
23	MH	21	4400.13 5569.56	370	51.01	5729	ZIMMERAU	
23	MH	22	4400.48 5569.41	376	51.38	5729	ZIMMERAU	
23	MH	23	4400.88 5569.13	361	51.86	5729	ZIMMERAU	
23	MH	31	4401.37 5565.97	383	54.31	*5729	ERMERSHAUSEN NNW	
23	MH	11	4405.38 5565.22	357	57.82	5730	ALLERTSHAUSEN	
23	MH	12	4405.76 5565.08	354	58.20	*5730	ALLERTSHAUSEN	
21	MH	1	4410.59 5566.52	344	60.92	*5730	DUERRENRIED	
21	MH	2	4410.93 5566.20	342	61.39	5730	DUERRENRIED	
21	MH	3	4411.31 5566.01	348	61.81	5730	DUERRENRIED	
22	H	11	4412.54 5565.54	350	63.14	5730	MUGGENBACH	
22	H	12	4412.73 5565.22	340	63.48	5730	MUGGENBACH	
22	H	13	4413.08 5564.94	330	63.93	*5730	MUGGENBACH	
27	KI	41	4415.77 5563.95	325	66.87	5730	DIETERSDORF S	
27	KI	42	4416.07 5563.85	312	67.17	5730	DIETERSDORF S	
27	KI	43	4416.43 5563.75	293	67.52	5730	DIETERSDORF S	
21	S	1	4417.83 5562.23	290	69.31	5831	SESSLACH	
21	S	2	4417.77 5561.98	295	69.40	5831	SESSLACH	
21	S	3	4417.79 5561.94	297	69.44	*5831	SESSLACH	
21	F	31	4420.96 5560.60	375	72.81	5831	SW NEUSES AD EICHEN	
21	F	32	4421.33 5560.36	360	73.25	*5831	SW NEUSES AD EICHEN	
21	F	33	4421.64 5560.16	340	73.62	5831	SW NEUSES AD EICHEN	
21	F	11	4424.96 5558.51	315	77.28	*5831	ROSSACH	
21	KI	33	4428.44 5556.00	410	81.57	5832	KLOSTER BANZ	
21	KI	32	4428.67 5555.87	390	81.84	*5832	KLOSTER BANZ	
21	KI	31	4428.73 5555.73	400	81.96	5832	KLOSTER BANZ	
18	KI	1	4434.06 5555.47	330	86.84	*5832	LICHTENFELS	
13	KI	A	4432.92 5553.10	455	87.11	5832	VIERZEHNHEILIGEN	
27	KI	11	4434.21 5552.74	448	88.41	5832	OBERLANGHEIM NW	
27	KI	12	4434.43 5552.39	455	88.79	5832	OBERLANGHEIM NW	
27	KI	13	4434.79 5552.32	443	89.13	5832	OBERLANGHEIM NW	
13	GO	A	4435.85 5550.05	520	91.26	*5932	OBERLANGHEIM	
18	CL	A	4440.80 5549.25	523	95.88	*5933	KOETTEL	
13	KR	A	4443.82 5546.18	480	100.02	*5933	WALLERSBERG	
13	KR	B	4443.91 5546.06	470	100.16	5933	WALLERSBERG	
13	KR	C	4443.91 5546.06	470	100.16	5933	WALLERSBERG	
13	KR	D	4444.07 5545.94	460	100.36	5933	WALLERSBERG	
13	H	A	4446.32 5544.30	485	103.14	*5933	WEIDEN	
13	H	A	4447.78 5540.97	445	106.24	6033	ZEDERSITZ	
13	H	B	4448.00 5540.75	454	106.55	6033	ZEDERSITZ	
13	H	C	4448.27 5540.62	468	106.84	*6033	ZEDERSITZ	
13	H	D	4448.61 5540.46	471	107.21	6033	ZEDERSITZ	
13	H	E	4448.80 5540.30	467	107.46	6033	ZEDERSITZ	
13	H	F	4448.99 5540.12	460	107.72	6033	ZEDERSITZ	
27	H	51	4450.98 5539.34	485	109.83	6033	SANSPAREIL	
27	H	52	4451.36 5539.21	500	110.21	6033	SANSPAREIL	
27	H	53	4451.73 5539.10	510	110.58	6033	SANSPAREIL	
13	H	A	4452.86 5539.70	530	111.70	*6034	GROSSENHUEL	
13	S	A	4455.25 5535.33	472	115.25	*6034	ALLADORF	
18	CL	1	4459.60 5536.12	548	117.98	*6034	LOCHAU	
18	CL	2	4458.77 5535.96	555	118.21	6034	LOCHAU	
18	CL	3	4458.95 5535.67	545	118.52	6034	LOCHAU	
13	S	A	4462.19 5534.25	410	121.92	*6034	OBERWAIZ	
21	KI	11	4465.72 5531.99	385	125.90	6035	DONNDORF SE	
21	KI	12	4465.70 5531.96	385	125.91	*6035	DONNDORF SE	
21	KI	13	4465.84 5531.91	380	126.05	6035	DONNDORF SE	
23	H	42	4467.93 5531.75	415	128.09	*6035	FORKENDORF	
23	H	43	4468.03 5531.39	420	128.37	6035	FORKENDORF	
23	H	41	4468.24 5531.01	420	128.75	6035	FORKENDORF	
13	CL		4467.90 5528.55	480	129.84	6135	GESEES	
27	H	21	4472.13 5530.57	415	132.27	6035	WOLFSBACH	
27	H	22	4472.33 5530.37	390	132.55	6035	WOLFSBACH	
27	H	23	4472.67 5530.22	370	132.92	6035	WOLFSBACH	
22	KI	31	4473.71 5527.52	445	135.10	6135	EMTMANNSBERG	
22	KI	32	4474.04 5527.31	472	135.48	*6135	EMTMANNSBERG	
22	KI	33	4474.42 5527.06	470	135.94	6135	EMTMANNSBERG	
22	KI	23	4475.63 5526.09	445	137.48	6135	TIEFENTHAL	
22	KI	21	4475.56 5525.97	440	137.49	6135	TIEFENTHAL	
22	KI	22	4475.69 5526.03	420	137.57	*6135	TIEFENTHAL	
13	MH	A	4475.67 5522.71	542	139.56	*6135	NEUHOF	
13	CL		4477.98 5525.28	435	140.02	6135	LANKENREUTH	
13	MH	A	4478.86 5522.21	437	142.47	*6136	LOSAU	
16	MH	A	4458.30 5494.56	627	145.18	6434	ALGERSDORF	
13	M	A	4483.02 5521.52	505	146.31	6136	VORBACH	
13	M	B	4483.27 5521.39	503	146.59	*6136	VORBACH	
13	M	C	4483.42 5521.21	495	146.81	6136	VORBACH	
23	H	11	4486.22 5519.07	440	150.34	6136	OBERBIBRACH E	
13	M	A	4486.33 5518.78	450	150.58	6136	OBERBIBRACH	
23	H	12	4486.49 5518.83	450	150.70	6136	OBERBIBRACH E	
13	M	B	4486.54 5518.57	460	150.87	6136	OBERBIBRACH	
23	H	13	4486.78 5518.48	470	151.13	*6136	OBERBIBRACH E	
13	M	C	4486.86 5518.44	470	151.21	6136	OBERBIBRACH	
18	CL	1	4487.57 5519.60	495	151.25	6136	OBERBIBRACH	
13	M	A	4489.93 5517.01	530	154.56	6237	SEITENTHAL	
13	M	B	4490.32 5516.70	518	155.06	*6237	SEITENTHAL	
13	M	C	4490.69 5516.55	505	155.45	6237	SEITENTHAL	
14	M	A	4491.84 5512.31	414	158.84	6237	GOESSENREUTH	
14	M	B	4492.09 5512.20	414	159.11	6237	GOESSENREUTH	
14	M	C	4492.37 5512.08	414	159.41	*6237	GOESSENREUTH	
13	M	A	4497.16 5518.32	540	159.94	6137	HESSENREUTH	
13	M	B	4497.17 5518.12	548	160.06	6137	HESSENREUTH	
13	M	C	4497.26 5517.95	540	160.22	6137	HESSENREUTH	
13	M	A	4499.28 5516.63	535	162.64	6237	RIGGAU	
13	M	B	4499.56 5516.53	545	162.93	6237	RIGGAU	
14	M	A	4495.75 5510.62	433	163.02	6237	PFAFF DICKICHT	
13	M	C	4499.91 5516.41	568	163.29	6237	RIGGAU	
14	M	B	4495.99 5510.44	433	163.32	6237	PFAFF DICKICHT	
14	M	C	4496.18 5510.31	433	163.55	*6237	PFAFF DICKICHT	
14	M	A	4499.58 5507.45	410	167.97	*6237	PARKSTEIN HUETTEN	
14	M	B	4499.87 5507.24	410	168.33	6237	PARKSTEIN HUETTEN	
14	M	C	4500.30 5507.03	410	168.80	6237	PARKSTEIN HUETTEN	
14	M	A	4501.70 5506.12	410	170.47	*6338	WEIDEN	
14	M	B	4502.04 5505.94	410	170.85	6338	WEIDEN	
14	M	C	4502.31 5505.70	410	171.21	6338	WEIDEN	
14	M	A	4504.31 5504.48	417	173.55	6338	WIESENDORF	
14	M	B	4504.48 5504.38	416	173.75	*6338	WIESENDORF	
14	M	C	4504.65 5504.28	418	173.95	6338	WIESENDORF	
14	M	A	4507.41 5502.29	495	177.35	6338	MALLERSRICHT	
14	M	B	4507.66 5502.13	470	177.64	6338	MALLERSRICHT	
14	M	C	4508.10 5501.83	475	178.18	6338	MALLERSRICHT	
22	H	31	4508.96 5502.32	460	178.40	*6338	ERMERSRICHT	
22	H	32	4509.67 5501.81	415	179.28	6338	ERMERSRICHT	
14	M	A	4512.99 5499.19	440	183.70	6339	ENZENRIETH	
14	M	B	4513.41 5498.97	485	184.23	6339	ENZENRIETH	
14	M	C	4513.66 5498.73	470	184.52	*6339	ENZENRIETH	
14	M	A	4516.01 5496.76	420	187.57	*6339	ENGLESHOF	
14	M	B	4516.06 5496.51	425	187.75	6339	ENGLESHOF	
14	M	C	4516.09 5496.32	425	187.87	6339	ENGLESHOF	
18	KI	A	4518.52 5495.51	546	190.36	6439	LEUCHTENBERG	
23	CL	1	4518.93 5495.07	470	190.87	6439	LEUCHTENBERG	
23	CL	2	4519.32 5494.86	470	191.32	*6439	LEUCHTENBERG	
23	CL	3	4519.71 5494.68	510	191.74	6439	LEUCHTENBERG	
22	H	21	4522.83 5493.94	465	194.59	*6439	VOITSBERG	
22	H	22	4523.22 5493.72	475	195.04	6439	VOITSBERG	
22	H	23	4523.63 5493.58	495	195.45	6439	VOITSBERG	
21	CL	1	4525.69 5491.57	566	198.22	6440	BOEHM. BRUCK	
21	CL	2	4526.07 5491.54	587	198.56	*6440	BOEHM. BRUCK	
21	CL	3	4526.34 5491.38	598	198.87	6440	BOEHM. BRUCK	
21	CL	3	4528.13 5490.97	630	200.59	6440	ETZGERSRIETH	
21	CL	2	4528.24 5490.85	645	200.74	6440	ETZGERSRIETH	
21	CL	1	4528.51 5490.62	655	201.10	6440	ETZGERSRIETH	
14	H	A	4525.99 5485.81	595	202.01	*6440	ZEINRIED	
23	CL	1	4530.75 5488.54	726	204.32	6440	PULLENRIED	
23	CL	2	4531.05 5488.32	728	20[illegible].70	*6440	PULLENRIED	
23	CL	3	4531.42 5488.08	704	205.14	6440	PULLENRIED	
14	H	A	4531.24 5484.38	550	207.13	*6540	OBERVIECHTACH	
23	CL	1	4534.94 5484.67	595	209.96	*6540	RACKENTHAL	
23	CL	2	4535.28 5484.42	629	210.39	6540	RACKENTHAL	
23	CL	3	4535.58 5484.19	642	210.77	6540	RACKENTHAL	
23	KI	31	4539.44 5482.34	717	215.00	6541	GAISTHAL SE	
23	KI	32	4539.60 5482.20	719	215.21	*6541	GAISTHAL SE	
23	KI	33	4539.82 5482.03	740	215.48	6541	GAISTHAL SE	
14	H	A	4539.90 5478.01	750	217.98	*6541	ALTENSCHNEEBERG	
23	KI	12	4543.69 5477.63	521	221.16	6541	BREITENRIED	
23	KI	11	4543.89 5477.75	511	221.25	6541	BREITENRIED	
23	KI	13	4543.76 5477.38	515	221.36	*6541	BREITENRIED	
23	KI	41	4548.90 5476.39	645	226.17	6542	UNTERGRAFENRIED	
23	KI	42	4548.98 5476.47	645	226.19	6542	UNTERGRAFENRIED	
23	KI	43	4549.00 5476.34	640	226.28	*6542	UNTERGRAFENRIED	
23	KI	21	4549.49 5473.62	520	228.21	6642	HOCHA	
23	KI	22	4549.61 5473.55	520	228.35	*6642	HOCHA	
23	KI	23	4549.61 5473.44	530	228.41	6642	HOCHA	
15	H	A	4551.90 5469.91	560	232.09	6642	HERZOGAU	
15	H	B	4552.10 5469.94	590	232.31	6642	HERZOGAU	
15	H	C	4552.38 5469.92	600	232.61	6642	HERZOGAU	
15	H	D	4552.56 5469.60	600	232.89	*6642	HERZOGAU	
15	H	E	4552.78 5469.42	610	233.17	6642	HERZOGAU	
15	H	F	4553.12 5469.29	620	233.52	6642	HERZOGAU	
15	H	A	4557.25 5465.34	910	239.17	*6642	VOITHENBERG	

18	M	3	4386.93	5489.54	335	115.73	6429	BAD WINDSHEIM	
14	MM		3605.85	5487.96	310	117.61	6429	OBERNDORF	
04	M		3605.08	5482.86	350	122.21	6528	ICKELHEIM	MA
18	M	1	4389.54	5482.80	424	122.89	6528	BREITENAU	
18	M	2	4389.50	5482.40	410	123.27	6528	BREITENAU	
18	M	3	4389.73	5482.23	425	123.49	6528	BREITENAU	
14	MM		4389.25	5480.04	386	125.56	6528	OBERZENN	
18	M	1	4387.34	5477.75	490	127.33	6528	OBERDACHST.	
18	M	2	4387.42	5477.34	495	127.75	6528	OBERDACHST.	
18	M	3	4387.32	5476.65	495	128.40	6528	OBERDACHST.	
18	M	1	4383.85	5475.60	495	128.77	6528	SPIELBERG	
12	MM		4390.67	5476.47	495	129.33	6528	WIPPENAU	BA
18	M	2	4384.42	5474.45	530	130.00	6628	SPIELBERG	
04	M		3603.98	5474.48	448	130.07	6628	SPIELBERG	MM
18	M	3	4384.37	5473.55	450	130.88	6628	SPIELBERG	
18	M	1	4384.43	5471.56	510	132.84	6628	SCHLOSS COLMBERG	
12	MM		4393.09	5473.05	450	133.23	6629	BOHRSBACH	
18	M	2	4384.60	5471.16	505	133.27	6628	SCHLOSS COLMBERG	
18	M	3	4384.72	5470.89	510	133.55	6628	SCHLOSS COLMBERG	
18	M	1	4385.30	5470.36	500	134.18	6628	COLMBERG	
18	M	2	4385.85	5469.96	525	134.67	6628	COLMBERG	
18	M	3	4386.28	5469.31	500	135.39	6628	COLMBERG	
14	MM		4392.53	5470.05	460	136.04	6629	LEHRBERG	
18	M	1	4386.90	5466.76	510	138.01	6628	MITTELRAMSTADT	
18	M	2	4387.20	5466.15	490	138.67	6628	MITTELRAMSTADT	
18	M	3	4387.67	5465.60	490	139.30	6628	MITTELRAMSTADT	
04	M		3606.89	5465.22	498	139.75	6628	TIEFENTHAL	MM
04	M		3606.89	5465.22	498	139.75	6628	TIEFENTHAL	MA
06	F		4393.64	5461.37	455	144.64	6729	DOMBACH	
06	MZ		4395.48	5456.80	481	149.51	6729	RAUENZELL	
04	CL		3604.70	5453.85	532	150.35	6729	LAMMELBACH	
04	CL		4390.75	5453.85	532	151.38	6729	LAMMELBACH	
06	M	A	4395.15	5451.35	441	154.74	6829	LIEBERSDORF	MA
06	M	B	4395.15	5451.35	441	154.74	6829	LIEBERSDORF	
06	M	C	4395.15	5451.35	435	154.74	6829	LIEBERSDORF	MA
05	CL	B	4393.25	5447.52	433	158.06	6829	ROHRBACH	
05	CL	C	4393.25	5447.52	433	158.06	6829	ROHRBACH	MA
05	CL	D	4393.25	5447.52	433	158.06	6829	ROHRBACH	MA
06	M	C	4397.17	5446.61	425	159.82	6829	ARBERG	MA
06	M	A	4397.34	5446.06	440	160.39	6829	ARBERG	MA
06	M	B	4397.34	5446.06	440	160.39	6829	ARBERG	
06	M	A	4396.93	5444.66	445	161.66	6829	AOETTENBACH	
06	M	B	4397.42	5442.65	463	163.73	6829	ROEM. LIMES	MA
06	M	B	4397.67	5441.34	455	165.06	6829	DENNENLOHE N	
06	M	C	4397.67	5441.34	455	165.06	6829	DENNENLOHE N	MA
06	M	B	4397.85	5440.20	470	166.22	6929	DENNENLOHE S	MA
04	S		4393.09	5437.65	671	167.72	6929	HESSELBERG	
04	S		4393.09	5437.65	671	167.72	6929	HESSELBERG	Y
06	M	A	4397.59	5437.80	475	168.50	6929	ALTENTRUED.	MA
06	M	B	4397.59	5437.80	475	168.50	6929	ALTENTRUEDINGEN	
25	S	41	4400.45	5436.74	472	170.22	6929	OBERMOEGERSHEIM	
25	S	42	4400.46	5436.49	480	170.46	6929	OBERMOEGERSHEIM	
25	S	43	4400.47	5436.07	487	170.87	6929	OBERMOEGERSHEIM	
25	S	31	4401.32	5435.51	495	171.62	6929	GEILSHEIM	
25	S	32	4401.61	5434.76	496	172.41	6929	GEILSHEIM	
06	M	A	4401.75	5431.64	480	175.44	6929	WESTHEIM N	MA
06	M	B	4401.75	5431.64	480	175.44	6929	WESTHEIM N	
06	M	C	4405.87	5429.35	585	178.66	7030	HOHENTRUEDINGEN	
06	M	B	4405.87	5429.35	585	178.66	7030	HOHENTRUED.	MA
20	MM	43	4396.99	5427.25	490	178.78	7029	OETTINGEN	
20	MM	42	4397.26	5426.91	485	179.17	7029	OETTINGEN	
20	MM	41	4397.48	5426.56	465	179.56	7029	OETTINGEN	
05	MZ		4402.96	5424.95	482	182.20	7030	STEINHART	
06	M	A	4406.85	5423.76	565	184.31	7030	URSHEIM	MA
06	M	B	4406.85	5423.76	565	184.31	7030	URSHEIM	
20	MM	31	4401.62	5421.48	421	185.40	7029	LAUB	
11	MM		4404.46	5421.65	485	185.84	7030	TRENDEL	BA
20	MM	32	4401.70	5421.02	421	185.87	7029	LAUB	
20	MM	33	4401.75	5420.65	420	186.24	7029	LAUB	
06	M	A	4405.18	5418.68	465	188.84	7130	AMERBACH	MA
06	M	B	4405.18	5418.68	465	188.84	7130	AMERBACH	
10	M	C	4430.63	5424.03	535	191.43	7032	BIESWANG	
10	M	B	4430.43	5423.61	550	191.75	7032	BIESWANG	
10	M	A	4430.28	5423.19	530	192.08	7032	BIESWANG	
20	MM	23	4405.70	5414.18	433	193.43	7130	GOSHEIM	
11	MM		4404.36	5413.61	435	193.64	7130	WILDBACH	
20	MM	22	4405.71	5413.74	430	193.86	7130	GOSHEIM	
20	MM	21	4405.50	5413.35	428	194.19	7130	GOSHEIM	
05	M		4405.31	5410.57	470	196.73	7130	HUISHEIM	
05	S		4406.75	5405.95	450	201.56	7230	HARTHOF	
07	M	A	4408.19	5403.89	470	203.89	7230	EBERMERGEN	MA
07	M	B	4408.19	5403.89	470	203.89	7230	EBERMERGEN	
20	MM	11	4406.80	5404.15	427	203.45	7230	EBERMERGEN	
20	MM	12	4406.89	5403.72	430	203.89	7230	EBERMERGEN	
07	M	C	4408.25	5403.55	500	204.24	7230	EBERMERGEN	MA
20	MM	13	4408.95	5403.34	425	204.27	7230	EBERMERGEN	
23	F	13	3605.06	5398.33	543	205.14	7228	DUNSTELKINGEN	
23	F	33	4391.35	5399.12	500	205.48	7229	ZOLTANGEN	
23	F	12	3605.28	5397.98	558	205.52	7228	DUNSTELKINGEN	
23	F	23	4398.15	5400.07	470	205.66	7229	NOERDL.BISSINGEN	
23	F	32	4391.39	5398.72	500	205.89	7229	ZOLTANGEN	
23	F	11	3605.32	5397.46	573	206.04	7228	DUNSTELKINGEN	
23	F	22	4398.25	5399.64	485	206.11	7229	NOERDL.BISSINGEN	
23	F	31	4391.42	5398.30	510	206.31	7229	ZOLTANGEN	
23	F	21	4398.35	5399.20	457	206.56	7229	NOERDL.BISSINGEN	
05	S		4404.55	5400.10	440	206.79	7230	WOERNITZSTEIN	
05	S		4404.55	5400.10	440	206.79	7230	WOERNITZSTEIN	X
05	S		4404.55	5400.10	440	206.79	7230	WOERNITZSTEIN	Y
07	M	B	4406.97	5398.26	455	209.10	7230	RAMBERG	
07	M	A	4407.04	5398.25	455	209.13	7230	RAMBERG	MA
07	M	C	4407.05	5397.89	435	209.48	7230	RAMBERG	MA
07	M		4407.50	5397.10	445	210.35	7230	STEINBERG	MA
07	M	A	4409.97	5387.53	435	220.23	7330	ALLMANNSHOF.	MA
07	M	B	4410.16	5387.18	470	220.62	7330	ALLMANNSHOFEN	
07	M	C	4410.16	5387.18	470	220.62	7330	ALLMANNSHOF.	MA
05	M		4410.26	5385.55	455	222.23	7330	EHINGEN	
07	M		4409.93	5383.03	460	224.61	7430	AHLINGEN	
07	M	A	4409.42	5380.02		227.44	7430	LANGENREICHEN	MA
07	M	B	4409.42	5380.02		227.44	7430	LANGENREICHEN	
07	M	C	4409.18	5379.66		227.75	7430	LANGENREICHEN	MA
07	M	A	4409.43	5378.14	480	229.28	7430	RIEBLINGEN	MA
07	M	B	4409.43	5378.14	480	229.28	7430	RIEBLINGEN	
07	M	C	4408.63	5375.57	485	231.63	7430	ALBERTSHOF.	MA
07	M	A	4409.39	5374.88	485	232.46	7430	ALBERTSHOF.	MA
07	M	B	4409.44	5370.35	520	236.91	7530	LUETZELBURG	MA
07	M	C	4409.54	5369.88	520	237.39	7530	LUETZELBURG	MA
07	M	A	4409.54	5369.54	520	237.72	7530	LUETZELBURG	
07	M	D	4409.54	5369.54	520	237.72	7530	LUETZELBURG	MA
07	M	E	4409.45	5369.26	520	237.98	7530	LUETZELBURG	MA
07	M	F	4409.23	5368.76	520	238.42	7530	LUETZELBURG	MA
10	M	B	4411.57	5356.44	550	251.05	7630	ANHAUSEN	
10	M	A	4411.55	5356.16	550	251.32	7630	ANHAUSEN	
10	M	A	4410.17	5352.72	555	254.42	7630	BANNACKER	
10	M	B	4410.23	5352.49	550	254.66	7630	BANNACKER	
10	M	A	4408.53	5348.93	560	257.83	7730	STRASSBERG	
10	M	B	4408.40	5348.64	560	258.09	7730	STRASSBERG	
10	M	A	4405.90	5344.19	590	262.03	7730	REINHARTSHOFEN	
10	M	B	4405.70	5344.02	590	262.16	7730	REINHARTSHOFEN	
10	M	C	4405.48	5343.57	580	262.57	7730	REINHARTSHOFEN	
10	M	A	4401.63	5339.05	600	266.40	7830	SCHWABEGG	
10	M	B	4401.69	5338.59	610	266.87	7830	SCHWABEGG	
05	M		4421.07	5337.67	568	271.32	7831	WINKE	
04	M		4446.15	5336.53	570	279.08	7833	FUERSTENFELD.	MA
10	M	A	4423.96	5328.88	600	280.62	7931	RAMSACH	
10	M	B	4424.35	5328.42	608	281.16	7931	RAMSACH	
05	FF		4425.41	5323.48	650	286.13	7931	SCHOEFFELDING	
10	M	A	4426.86	5318.62	640	291.28	7932	ENTRACHING	
10	M	B	4426.88	5318.14	645	291.75	8032	ENTRACHING	
10	M	C	4427.03	5317.86	650	292.06	8032	ENTRACHING	
05	M		4429.66	5307.98	615	302.19	8032	RAISTING	
10	M	A	4428.40	5306.57	700	303.37	8132	HAID	
10	M	B	4428.36	5306.28	700	303.64	8132	HAID	
10	M		4429.05	5298.03	685	311.84	8132	SCHLAG	
10	M	A	4429.17	5297.52	650	312.37	8132	SCHLAG	
10	M	A	4432.48	5289.47	680	320.95	8232	GAISLEITEN	
10	M	B	4432.39	5289.41	680	320.99	8232	GAISLEITEN	
10	M	A	4430.50	5282.09	795	327.72	8332	BAD KOHLGRUB	
10	M	C	4430.50	5282.09	795	327.72	8332	BAD KOHLGRUB	
10	M	B	4430.65	5281.96	795	327.88	8332	BAD KOHLGRUB	X
05	M	A	4435.34	5272.07	800	338.47	8432	SCHARFECK	
05	M	B	4435.34	5272.07	800	338.47	8432	SCHARFECK	
05	M	A	4436.23	5268.02	800	342.62	8432	OBERAU	MA
05	M	B	4436.23	5268.02	800	342.62	8432	OBERAU	
05	FF	A	4433.48	5263.61	850	346.33	8432	DAX KAPELLE	MA
05	FF	B	4433.48	5263.61	850	346.33	8432	DAX KAPELLE	MA
10	M	A	4435.32	5264.06	740	346.37	8432	WANK	MA
10	M	B	4435.32	5264.06	740	346.37	8432	WANK	
05	M		4433.25	5259.64	800	350.17	8532	PARTNACHKLAMM	
20	M	11	3598.85	5246.43	1040	355.55	8627	SPIELMANNSAU	
20	M	12	3598.85	5246.43	040	355.55	8627	SPIELMANNSAU	

QUARRY-NO. 02 HILDERS SMAP 5426
PROFILE 02-215 COMPILED BY BAIER.BERCKHEMER.MEISSNER

B	IN	AC	COORDINATES		ALT	X(KM)	SMAP	LOCALITY	G C
19	F	11	3549.48	5561.66	400	46.23	5824	HEILIGKREUZ W	
19	F	12	3549.25	5561.29	434	46.67	5824	HEILIGKREUZ W	
19	F	13	3549.08	5560.90	439	47.09	5824	HEILIGKREUZ W	
19	F	22	3548.90	5559.16	300	48.67	5824	BURGSINN	
19	F	21	3548.80	5558.73	300	49.09	5834	BURGSINN	
19	F	23	3548.63	5558.32	300	49.53	5824	BURGSINN	
18	F	21	3544.32	5553.06	230	56.16	5823	TROCKENBACHHOF	
18	F	22	3544.30	5553.04	230	56.19	5823	TROCKENBACHHOF	
18	F	23	3544.26	5553.00	230	56.24	5823	TROCKENBACHHOF	
12	MZ		3538.08	5545.80	320	65.69	5923	PARTENSTEIN	
19	F	31	3536.65	5540.05	480	71.36	6023	RECHTENBACH	
19	F	32	3536.58	5539.13	410	72.19	6023	RECHTENBACH	
17	F	22	3533.98	5535.56	337	76.69	6022	LICHTENAU	
17	F	23	3533.98	5535.56	337	76.69	6022	LICHTENAU	
12	M	A	3532.11	5532.01	506	80.55	6022	WEIBERSBRUNN	
12	M	B	3531.95	5531.83	480	80.79	6022	WEIBERSBRUNN	
12	M	C	3531.76	5531.57	518	81.11	6022	WEIBERSBRUNN	
20	CL	11	3531.25	5531.63	505	81.42	6022	WEIBERSBRUNN	
20	CL	12	3530.85	5531.33	500	81.88	6022	WEIBERSBRUNN	
20	CL	13	3530.80	5531.16	505	82.06	6022	WEIBERSBRUNN	
12	MZ		3641.20	5547.96	420	85.93	5923	RUPPERTSHUETTEN	
12	M	B	3527.70	5527.62	375	86.60	6122	ROHRBRUNN	
12	M	A	3527.70	5527.62	375	86.60	6122	ROHRBRUNN	
12	M	C	3527.51	5527.25	445	87.01	6122	ROHRBRUNN	
12	M	A	3527.28	5523.57	390	90.27	6122	ALTENBUCH	
12	M	C	3527.05	5523.47	365	90.48	6122	ALTENBUCH	
12	M	B	3527.03	5523.48	365	90.48	6122	ALTENBUCH	
14	CL		3527.04	5523.45	350	90.55	6122	ALTENBUCH	
17	F	12	3534.60	5519.22	410	90.78	6122	WILDENSEE	
20	CL	21	3526.45	5522.30	473	91.89	6122	HUNDSRUECKKOPF	
20	CL	22	3526.33	5522.16	465	92.07	6122	HUNDSRUECKKOPF	
20	CL	23	3526.20	5522.00	453	92.27	6122	HUNDSRUECKKOPF	
12	M	A	3525.03	5519.85	380	94.62	6122	WILDENSEE	
12	M	B	3524.89	5519.46	370	95.03	6122	WILDENSEE	
12	M	C	3524.60	5519.22	410	95.38	6122	WILDENSEE	
17	F	13	3524.48	5519.04	410	95.74	6122	WILDENSEE	
20	F	21	3522.56	5515.36	180	99.84	6221	REISTENHAUSEN	
20	F	22	3522.38	5515.18	225	100.09	6221	REISTENHAUSEN	
20	F	23	3522.13	5514.93	280	100.43	6221	REISTENHAUSEN	
12	M	A	3517.39	5511.96	224	105.34	6221	GROSSHEUBACH	
12	M	B	3517.55	5511.82	230	105.38	6221	GROSSHEUBACH	
12	M	C	3517.61	5511.54	222	105.58	6221	GROSSHEUBACH	
12	F		3516.40	5505.90	150	111.02	6321	BREITENDIEL	
12	F		3514.32	5501.50	215	115.87	6321	AMORBACH	
12	M	A	3513.82	5495.95	355	120.91	6421	PREUNSCHEN	
12	M	B	3513.81	5495.59	360	121.23	6421	PREUNSCHEN	
12	M	C	3513.77	5495.44	385	121.38	6421	PREUNSCHEN	
12	M	A	3510.91	5490.37	500	127.20	6420	SCHLOSSAU	
12	M	B	3510.89	5490.24	485	127.33	6420	SCHLOSSAU	
12	M	C	3510.83	5490.08	440	127.49	6420	SCHLOSSAU	
12	M	C	3507.85	5487.32	450	131.37	6420	KAILBACH	
12	M	B	3507.81	5487.13	410	131.55	6420	KAILBACH	
12	M	A	3507.68	5487.00	480	131.73	6420	KAILBACH	
12	S		3507.10	5483.18	535	135.34	6520	STRUEMPFELBRUNN	
12	S		3504.51	5476.64	422	142.31	6520	SCHOLLBRUNN	
11	M		3504.75	5473.32	250	145.18	6620	GUTTENBACH	
23	S	13	3554.87	5456.40	360	145.88	6724	RUEBLINGEN	
23	S	12	3555.10	5456.25	400	146.00	6724	RUEBLINGEN	
23	S	11	3555.24	5455.96	400	146.27	6724	RUEBLINGEN	

B	IN	AC	COORDINATES		ALT	X(KM)	SMAP	LOCALITY	G	C
25	M	11	4563.32	5463.45	480	245.25	6643	FURTH I. W.		
25	M	12	4563.78	5463.03	480	245.87	6643	FURTH I. W.		
25	M	31	4567.74	5459.67	480	251.04	6743	OBER FAUSTERN		
25	M	32	4567.98	5459.62	480	251.27	6743	OBER FAUSTERN		
25	M	33	4568.08	5459.23	480	251.57	6743	OBER FAUSTERN		
27	KI	33	4573.72	5460.48	575	255.67	6744	HOFBERG		
27	KI	31	4573.84	5460.43	590	255.79	6744	HOFBERG		
27	KI	32	4573.79	5460.32	575	255.81	6744	HOFBERG		
25	M	41	4572.61	5452.81	600	258.96	6743	OTTENZELL		
25	M	42	4572.65	5452.82	600	258.99	6743	OTTENZELL		
27	KI	21	4576.85	5456.02	810	260.75	6744	RITTSTEIG		
27	KI	22	4576.89	5455.94	825	260.83	6744	RITTSTEIG		
27	KI	23	4577.15	5455.69	890	261.19	6744	RITTSTEIG		
28	M	A	4578.66	5451.78	650	264.76	*6844	LAM		
25	KI	11	4582.42	5449.27	780	269.03	6844	ALTLOHBERGERHUETTE		
25	KI	12	4582.68	5449.40	835	269.17	6844	ALTLOHBERGERHUETTE		
25	KI	13	4582.74	5449.41	845	269.21	6844	ALTLOHBERGERHUETTE		
28	KI	41	4584.50	5446.21	062	272.56	6844	HINDENBURG KANZEL		
25	KI	21	4585.61	5443.78	910	274.79	6845	BAY. EISENSTEIN		
25	KI	22	4585.82	5443.58	890	275.08	6845	BAY. EISENSTEIN		
25	KI	23	4586.00	5443.54	875	275.25	6845	BAY. EISENSTEIN		
25	KI	41	4591.05	5440.77	710	280.97	6945	ZWIESLER WALDHAUS		
25	KI	42	4591.08	5440.75	713	281.01	6945	ZWIESLER WALDHAUS		
25	KI	43	4591.29	5440.60	725	281.26	6945	ZWIESLER WALDHAUS		
25	KI	31	4595.97	5433.85	785	288.99	6945	JUNGMAIER BERG		
25	KI	32	4596.07	5433.78	775	289.11	6945	JUNGMAIER BERG		
25	KI	33	4596.21	5433.78	778	289.22	6945	JUNGMAIER BERG		
28	KI	32	4604.65	5425.02	874	301.27	7046	RACHEL DIENSTHUETTE		
28	KI	31	4604.67	5425.02	876	301.29	7046	RACHEL DIENSTHUETTE		
28	KI	33	4604.65	5424.72	846	301.45	7046	RACHEL DIENSTHUETTE		
28	KI	22	G 1516.41	4819.29	400	452.82	36	MANNERSDORF		

QUARRY-NO. 02 HILDERS SMAP 5426
PROFILE 02-140 COMPILED BY AICHELE

B	IN	AC	COORDINATES		ALT	X(KM)	SMAP	LOCALITY	G	C
31	M	71	3603.75	5568.32	450	44.48	5728	SEMBACHSHOF		
31	M	72	3603.92	5568.32	450	44.73	5728	SEMBACHSHOF		
31	M	73	3604.13	5567.92	450	45.03	5728	SEMBACHSHOF		
31	F	11	4396.61	5560.67	333	54.84	5829	WALCHENECK		
31	F	12	4396.69	5560.32	350	55.16	5829	WALCHENECK		
31	F	13	4396.88	5559.95	328	55.56	5829	WALCHENECK		
29	F	31	4398.72	5557.42	360	58.66	5829	GROSSMANNDORF		
29	F	32	4398.82	5557.17	360	58.92	5829	GROSSMANNDORF		
29	F	33	4398.92	5556.88	360	59.21	5829	GROSSMANNDORF		
31	F	21	4403.19	5553.43	395	64.62	5829	BRAMBERG		
31	F	22	4403.58	5553.25	410	65.01	5829	BRAMBERG		
31	F	23	4403.96	5553.08	416	65.39	5829	BRAMBERG		
31	F	31	4407.59	5549.58	355	70.42	5930	WEISSENBRUNN		
31	F	32	4407.47	5548.97	390	70.79	5930	WEISSENBRUNN		
31	F	33	4407.89	5548.47	380	71.45	5930	WEISSENBRUNN		
28	KA	51	4409.87	5546.10	350	74.48	5930	GOGGELGEREUTH		
28	KA	52	4410.17	5545.88	330	74.83	5930	GOGGELGEREUTH		
28	KA	53	4410.40	5545.61	310	75.19	5930	GOGGELGEREUTH		
28	KA	61	4413.52	5542.33	360	79.71	5930	PRIEGENDORF		
28	KA	62	4413.88	5542.08	335	80.13	5930	PRIEGENDORF		
28	KA	63	4414.23	5541.87	330	80.52	5930	PRIEGENDORF		
28	KA	42	4415.69	5537.74	280	84.60	6030	GODELSDORF		
28	KA	41	4415.32	5537.99	270	84.71	6030	GODELSDORF		
28	KA	43	4416.00	5537.48	330	85.00	6030	GODELSDORF		
31	B	32	4416.59	5536.85	377	85.93	6031	DOERFLEINS		
31	B	33	4416.93	5536.61	376	86.33	6031	DOERFLEINS		
28	F	31	4421.74	5535.70	290	90.12	6031	LAUBEND		
28	F	32	4421.98	5535.36	270	90.54	6031	LAUBEND		
28	F	33	4422.23	5535.00	255	90.97	6031	LAUBEND		
31	B	41	4423.10	5535.06	285	91.59	6031	GUNDELSHEIM		
31	B	42	4423.19	5534.72	305	91.90	6031	GUNDELSHEIM		
31	B	43	4423.52	5534.63	303	92.19	6031	GUNDELSHEIM		
29	S	11	4425.61	5531.68	286	95.75	6031	POEDELDORF		
29	S	12	4425.80	5531.39	305	96.10	6031	POEDELDORF		
29	S	13	4425.94	5531.00	300	96.48	6031	POEDELDORF		
28	F	21	4426.22	5528.46	261	98.51	6131	ROSSDORF		
28	F	22	4426.53	5528.30	300	98.83	6131	ROSSDORF		
28	F	23	4426.83	5528.12	310	99.17	6131	ROSSDORF		
29	S	41	4427.11	5524.20	298	102.36	6131	AMLINGSTADT		
29	S	42	4427.40	5523.97	305	102.72	6131	AMLINGSTADT		
29	S	43	4427.71	5523.71	300	103.12	6131	AMLINGSTADT		
26	F	31	4429.08	5521.00	280	106.03	6132	SEIGENDORF		
26	F	32	4429.39	5520.68	280	106.48	6132	SEIGENDORF		
27	B	11	4430.98	5520.73	340	107.50	6132	DREUSCHENDORF		
27	B	12	4431.28	5520.39	305	107.96	6132	DREUSCHENDORF		
27	B	13	4431.56	5520.06	315	108.39	6132	DREUSCHENDORF		
28	S	21	4434.02	5518.00	310	111.52	6232	WEIGELSHOFEN		
28	S	22	4434.37	5517.72	310	111.96	6232	WEIGELSHOFEN		
28	S	23	4434.67	5517.50	300	112.33	6232	WEIGELSHOFEN		
26	F	21	4435.58	5512.86	375	116.45	6232	KIRCHEHRENBACH		
26	F	22	4435.89	5512.58	359	116.86	6232	KIRCHEHRENBACH		
26	F	23	4436.20	5512.32	356	117.26	6232	KIRCHEHRENBACH		
26	F	11	4440.89	5510.59	450	121.61	6233	LEUTENBACH		
26	F	12	4441.28	5510.32	490	122.07	6233	LEUTENBACH		
26	F	13	4441.58	5510.03	492	122.49	6233	LEUTENBACH		
26	KA	11	4442.60	5507.43	555	125.13	6333	HAIDHOF		
26	KA	12	4442.77	5507.09	550	125.49	6333	HAIDHOF		
26	KA	13	4443.03	5506.70	540	125.96	6333	HAIDHOF		
29	B	31	4445.06	5505.41	478	128.31	6333	THUISBRUNN		
29	B	32	4445.15	5505.14	461	128.57	6333	THUISBRUNN		
29	B	33	4445.33	5504.96	480	128.83	6333	THUISBRUNN		
26	KA	21	4447.53	5503.04	500	131.67	6333	KEMNATHEN		
26	KA	22	4447.60	5502.62	480	132.03	6333	KEMNATHEN		
26	KA	23	4447.82	5502.32	487	132.40	6333	KEMNATHEN		
26	KA	31	4451.16	5499.37	490	136.82	6333	WINTERSTEIN		
26	KA	32	4451.39	5499.11	500	137.17	6333	WINTERSTEIN		
26	KA	33	4451.58	5498.90	495	137.45	6333	WINTERSTEIN		
29	B	41	4453.02	5498.60	503	138.68	6434	DIEPOLTSDORF		
29	B	42	4453.29	5498.33	490	139.06	6434	DIEPOLTSDORF		
29	B	43	4453.48	5498.19	470	139.29	6434	DIEPOLTSDORF		
28	S	31	4455.05	5494.77	510	142.89	6434	OSTERNOHE		
26	KA	41	4453.88	5493.58	470	142.98	6434	HEDERSDORF		
28	S	32	4455.39	5494.61	540	143.19	6434	OSTERNOHE		
26	KA	42	4454.04	5493.17	430	143.40	6434	HEDERSDORF		
28	S	33	4455.59	5494.39	560	143.49	6434	OSTERNOHE		
26	KA	43	4454.34	5492.98	430	143.74	6434	HEDERSDORF		
16	MM		4458.30	5494.57	627	145.19	6434	ALGERDORF		
26	KA	51	4456.83	5491.22	517	146.69	6434	KIRCHENSITTENBACH		
26	KA	52	4457.07	5491.18	517	146.88	6434	KIRCHENSITTENBACH		
26	KA	53	4457.31	5490.92	510	147.23	6434	KIRCHENSITTENBACH		
26	KA	61	4460.50	5487.95	490	151.57	6434	GROSSVIEBERG		
26	KA	62	4460.62	5487.84	485	151.73	6434	GROSSVIEBERG		
26	KA	63	4461.04	5487.76	460	152.07	6434	GROSSVIEBERG		
26	M	11	4461.75	5487.10	490	153.03	6434	HOHENSTADT		
26	M	12	4462.00	5487.05	490	153.23	6434	HOHENSTADT		
26	M	13	4462.46	5486.62	490	153.86	6434	HOHENSTADT		
26	M	31	4465.87	5483.21	600	158.67	6535	MITTELBURG		
26	M	32	4466.15	5482.92	600	159.07	6535	MITTELBURG		
26	M	33	4466.58	5482.78	600	159.46	6535	MITTELBURG		
26	M	21	4468.62	5479.83	490	163.02	6535	KIRCHTHALMUEHLE		
26	M	22	4468.75	5479.60	490	163.27	6535	KIRCHTHALMUEHLE		
26	M	23	4468.79	5479.23	490	163.59	6535	KIRCHTHALMUEHLE		
26	M	51	4472.10	5475.20	580	168.80	6535	BAUMGARTEN		
26	M	53	4472.54	5474.68	580	169.48	6535	BAUMGARTEN		
26	M	61	4476.26	5472.90	500	173.28	6636	KASTL		
26	M	62	4476.53	5472.57	500	173.70	6636	KASTL		
26	M	63	4476.87	5472.45	500	174.02	6636	KASTL		
26	M	41	4480.25	5468.96	530	178.87	6636	AICHA		
26	M	42	4480.55	5468.93	530	179.10	6636	AICHA		
26	M	43	4480.77	5468.89	530	179.28	6636	AICHA		
27	M	41	4482.20	5465.12	520	183.08	6636	RAUSBACH		
27	M	42	4482.49	5464.98	520	183.37	6636	RAUSBACH		
27	M	43	4482.80	5464.90	520	183.64	6636	RAUSBACH		
27	M	51	4487.08	5462.91	465	187.98	6636	ALLERTSHOFEN		
27	M	52	4487.20	5462.74	465	188.18	6636	ALLERTSHOFEN		
27	M	53	4487.54	5462.67	465	188.46	6636	ALLERTSHOFEN		
27	M	11	4491.09	5461.46	440	191.72	6737	WINDBUCH		
27	M	12	4491.07	5461.24	440	191.90	6737	WINDBUCH		
27	M	13	4491.04	5460.74	440	192.37	6737	WINDBUCH		
27	M	31	4495.44	5458.09	450	197.18	6737	SCHMIDMUEHLEN		
27	M	32	4495.47	5457.81	450	197.41	6737	SCHMIDMUEHLEN		
27	M	33	4495.45	5457.43	450	197.67	6737	SCHMIDMUEHLEN		
27	M	21	4498.48	5453.12	370	202.89	6737	MACHTLWIES		
27	M	22	4498.94	5452.77	370	203.46	6737	MACHTLWIES		
27	M	23	4499.15	5452.35	370	203.91	6737	MACHTLWIES		
28	M	31	4501.59	5450.76	370	206.70	6838	GREINHOF		
28	M	32	4501.64	5450.48	360	206.94	6838	GREINHOF		
28	M	33	4501.82	5450.18	355	207.28	6838	GREINHOF		
27	M	71	4502.96	5448.48	270	209.34	6838	BURGLENGEF.FORST		
27	M	72	4503.32	5448.25	270	209.75	6838	BURGLENGEF.FORST		
27	M	73	4504.11	5447.98	270	210.49	6838	BURGLENGEF.FORST		
28	M	11	4502.82	5443.70	420	212.75	6838	BUCHENLOHE		
28	M	12	4503.08	5443.44	415	213.12	6838	BUCHENLOHE		
28	M	13	4503.13	5443.16	420	213.36	6838	BUCHENLOHE		
28	M	21	4503.15	5441.51	410	214.56	6838	HOLZ/STEINSBERG		
28	M	22	4503.42	5441.12	400	215.08	6838	HOLZ/STEINSBERG		
28	M	23	4503.79	5440.66	360	215.67	6838	HOLZ/STEINSBERG		
28	M	42	4508.31	5436.97	395	221.41	6938	ZEITLARN		
28	M	43	4508.42	5436.72	390	221.69	6938	ZEITLARN		
28	M	61	4514.75	5434.90	480	227.28	6939	PROBSTBERG		
28	M	62	4515.05	5434.58	510	227.72	6939	PROBSTBERG		
28	M	63	4515.13	5434.44	505	227.87	6939	PROBSTBERG		
28	M	51	4519.78	5433.29	370	231.86	6939	BACH A.D.DONAU		
28	M	52	4520.83	5432.78	405	232.95	6939	BACH A.D.DONAU		
28	M	53	4521.12	5432.58	395	233.29	6939	BACH A.D.DONAU		

QUARRY-NO. 02 HILDERS SMAP 5426
PROFILE 02-165-01 COMPILED BY AICHELE.FUCHS

B	IN	AC	COORDINATES		ALT	X(KM)	SMAP	LOCALITY	G	C
04	K		3576.71	5595.15	771	6.57	5526	ROTHER KUPPE		
04	MZ		3577.20	5589.40	590	12.10	5526	GINOLFS	HA	
04	MZ		3579.40	5580.20	345	21.54	5626	BURG WALBACH	B4	
05	MZ		3579.45	5578.42	406	23.25	5626	STRUTHOF	B4	
06	MZ	A	3579.45	5578.42	406	23.28	5626	STRUTHOF	B4	
06	MZ	B	3579.45	5578.42	406	23.28	5626	STRUTHOF	B4	
06	MZ	C	3579.45	5578.42	406	23.28	5626	STRUTHOF	B4	
04	CL		3580.38	5574.20	228	27.60	5726	UNTEREBERSBACH		
25	M	51	3581.43	5573.40	300	28.69	5726	OBER EBERSBACH		
25	M	52	3581.43	5572.97	300	29.11	5726	OBER EBERSBACH		
25	M	53	3581.50	5572.83	300	29.26	5726	OBER EBERSBACH		
25	M	61	3583.39	5569.03	380	33.43	5727	MUENNERSTADT		
25	M	62	3583.67	5568.86	380	33.68	5727	MUENNERSTADT		
25	M	63	3583.71	5568.56	380	33.98	5727	MUENNERSTADT		
04	CL		3582.75	5567.85	320	34.34	5726	BURGHAUSEN		
25	M	21	3583.40	5561.78	330	40.44	5827	RANNUNGEN		
25	M	22	3583.39	5561.62	330	40.59	5827	RANNUNGEN		
25	M	23	3583.49	5560.94	330	41.27	5827	RANNUNGEN		
04	M		3586.00	5561.00	340	41.83	5827	RANNUNGEN		
04	MM		3586.16	5553.96	310	48.64	5827	HAMBACH		
10	GO		3588.17	5549.18	291	53.85	5927	DITTELBRUNN	BA	
04	M		3588.21	5541.75	206	60.98	5927	GRAFENRHEINF		
06	CL		3587.63	5532.18	254	70.20	6027	KOLITZHEIM		
25	KA	11	3591.34	5529.33	245	73.84	6127	KRAUTHEIM		
25	KA	12	3591.47	5528.87	253	74.32	6127	KRAUTHEIM		
25	KA	13	3591.69	5528.49	240	74.74	6127	KRAUTHEIM		
05	CL		3593.68	5523.69	248	79.81	6127	EICHFELD		
25	KA	21	3594.24	5518.84	232	84.72	6227	FEUERBACH		
25	KA	22	3594.26	5518.39	230	85.16	6227	FEUERBACH		
25	KA	23	3594.26	5517.94	220	85.60	6227	FEUERBACH		
04	M		3594.72	5513.45	260	90.01	6227	WIESENBRONN		
25	KA	33	3598.20	5506.95	330	97.22	6328	ENZLAR		
25	KA	32	3598.30	5506.58	325	97.61	6328	ENZLAR		
25	KA	31	3598.47	5506.30	320	97.92	6328	ENZLAR		
04	M		3596.90	5501.55	330	102.10	6328	DORNHEIM		
04	M		3600.62	5494.90	362	109.44	6429	KRAUTOSTHEIM	MA	
18	M	1	4385.11	5492.87	330	112.08	6428	BEROLZHEIM		
18	M	2	4385.19	5492.45	325	112.51	6428	BEROLZHEIM		
18	M	3	4385.29	5492.04	340	112.93	6428	BEROLZHEIM		
05	F		3602.83	5489.89	346	114.83	6428	BEROLZHEIM		
05	F		3602.83	5489.89	346	114.83	6428	BEROLZHEIM		X
05	F		3602.83	5489.89	346	114.83	6428	BEROLZHEIM		Y
18	M	1	4386.83	5490.35	328	114.92	6428	BAD WINDSHEIM		
18	M	2	4386.93	5489.95	322	115.33	6428	BAD WINDSHEIM		

B	IN	AC	COORDINATES	ALT	X(KM)	SMAP	LOCALITY	G C
11	F		3501.58 5466.36	275	152.82	6620	ASBACH	
23	S	33	3548.24 5449.94	500	153.25	6823	WALDENBURG	
23	S	32	3548.54 5449.81	515	153.33	6823	WALDENBURG	
23	S	31	3548.35 5449.51	510	153.66	6823	WALDENBURG	
11	F		3499.25 5462.65	220	157.19	6619	FLINSBACH	
14	F		3497.09 5459.50	280	160.96	6719	ADERSBACH	
23	S	23	3538.45 5443.42	430	161.53	6823	GLEICHEN	
23	S	22	3538.57 5443.08	465	161.84	6823	GLEICHEN	
23	S	21	3538.80 5442.89	440	161.97	6823	GLEICHEN	
11	S		3495.88 5455.15	185	165.39	6719	STEINSFURT	
23	S	43	3528.72 5440.37	382	166.87	6922	LOEWENSTEIN	
23	S	42	3528.81 5440.12	405	167.08	6922	LOEWENSTEIN	
23	S	41	3529.10 5439.91	410	167.21	6922	LOEWENSTEIN	
11	S		3491.76 5449.13	210	172.64	6819	ADELSHOFEN	
11	M	B	3490.34 5445.09	205	176.87	6819	ROHRBACH	
11	M	A	3490.38 5444.90	210	177.02	6819	ROHRBACH	
11	M	B	3492.13 5439.03	320	181.43	6919	OCHSENBURG	
11	M	A	3491.93 5438.60	310	181.90	6919	OCHSENBURG	
11	M	A	3487.22 5435.65	265	186.68	6918	DERDINGEN	
11	M	B	3487.10 5435.50	260	186.86	6918	DERDINGEN	X
11	M	C	3487.10 5435.50	260	186.86	6918	DERDINGEN	
11	M	A	3485.70 5431.25	310	191.28	6918	HOHENKLINGEN	
11	M	B	3485.64 5431.13	315	191.41	6918	HOHENKLINGEN	
11	M	C	3485.56 5430.90	320	191.66	6918	HOHENKLINGEN	
11	M	A	3482.83 5426.98	375	196.39	7018	OELBRONN	
11	M	B	3482.81 5426.76	375	196.60	7018	OELBRONN	
11	M	C	3482.78 5426.59	375	196.76	7018	OELBRONN	
11	M	A	3482.90 5421.45	325	201.28	7018	KIESELBRONN	
11	M	B	3482.90 5421.45	325	201.28	7018	KIESELBRONN	
11	M	C	3483.00 5421.00	350	201.64	7018	KIESELBRONN	
11	M	C	3482.02 5415.80	405	206.73	7118	PFORZHEIM	
11	M	A	3481.94 5415.62	420	206.92	7118	PFORZHEIM	
11	M	B	3481.84 5415.42	410	207.15	7118	PFORZHEIM	
11	M	A	3478.84 5412.13	470	211.43	7118	HUCHENFELD	
11	M	B	3478.78 5411.85	480	211.71	7118	HUCHENFELD	
11	M	C	3478.70 5411.65	460	211.92	7118	HUCHENFELD	
11	M	A	3478.29 5409.55	400	213.98	7118	KAPFENHARDT	
11	M	B	3478.04 5409.12	470	214.48	7118	KAPFENHARDT	
11	M	C	3478.04 5408.85	500	214.72	7118	KAPFENHARDT	
11	M	D	3477.99 5408.47	500	215.08	7118	KAPFENHARDT	
11	M	E	3477.74 5407.78	500	215.81	7118	KAPFENHARDT	
11	M	A	3473.15 5405.07	650	220.30	7217	SCHOENBERG	
11	M	B	3473.09 5404.94	650	220.44	7217	SCHOENBERG	
11	M	A	3473.76 5400.18	680	224.39	7217	OBERREICHENBACH	
11	M	B	3473.65 5399.89	680	224.70	7217	OBERREICHENBACH	
11	M	C	3473.37 5399.54	680	225.14	7217	OBERREICHENBACH	
16	KA		3476.10 5379.55	560	242.09	7418	UNTERSCHWANDORF	
16	KA		3473.82 5366.67	515	254.82	7517	REXINGEN	

QUARRY-NO. 02 HILDERS SMAP 5426
PROFILE 02-220-05 COMPILED BY STROBACH

B	IN	AC	COORDINATES	ALT	X(KM)	SMAP	LOCALITY	G C
03	M	F	3569.84 5595.72	690	6.54	5525	WUESTENSACHSEN	
03	M	E	3569.68 5595.58	739	6.75	5525	WUESTENSACHSEN	
03	M	D	3569.58 5595.38	765	6.97	5525	WUESTENSACHSEN	
03	M	C	3569.37 5595.10	825	7.32	5525	WUESTENSACHSEN	
03	M	A	3558.38 5586.75	700	20.92	5524	DALHERDA	MH
03	M	B	3558.38 5586.75	700	20.92	5524	DALHERDA	HA
03	CL		3552.57 5575.50	350	33.12	5624	ZUENTERSBACH	
03	M	A	3537.69 5552.25	506	60.60	5823	BAYER. SCHANZ	MH
03	M	B	3537.69 5552.25	506	60.60	5823	BAYER. SCHANZ	HA
03	KR		3526.17 5544.16	320	74.09	5922	HEIGENBRUECKEN	
06	SE		3523.65 5541.90	220	77.44	5921	HAIN	
06	SE		3517.79 5538.00	245	84.23	6021	KEILBERG	
06	SE		3518.10 5533.50	255	87.45	6021	OBERHESSENBACH	
03	CL	A	3515.00 5526.10	212	95.16	6121	HAUSEN	
03	CL	B	3515.00 5526.10	212	95.16	6121	HAUSEN	BAX
03	CL	C	3515.00 5526.10	212	95.16	6121	HAUSEN	BAY
03	K		3503.35 5516.84	230	109.69	6220	RIMHORN	
05	HH		3496.68 5512.65	240	117.25	6318	BIRKERT	
03	MZ	A	3496.75 5503.10	331	124.52	6319	ROSSBACH	B4
03	MZ	B	3496.75 5503.10	331	124.52	6319	ROSSBACH	B4
03	MZ	C	3496.75 5503.10	331	124.52	6319	ROSSBACH	B4X
05	HH		3493.13 5506.12	435	124.52	6318	OBER MOSSAU	
05	HH		3488.31 5501.04	410	131.51	6318	HAMMELBACH	
03	HH		3485.05 5493.72	390	139.17	6418	STALLENKANDEL	
14	GO		3470.11 5506.24	93	140.42	6317	SCHWANHEIM	
14	GO		3475.66 5499.81	210	140.92	6317	HEPPENHEIM	
05	CL		3485.69 5487.10	525	143.98	6418	UNTERABTSTEINACH	
03	HH		3477.79 5485.35	270	150.24	6418	HEILIGKREUZ	
14	S		3451.43 5422.17	117	216.68	7016	BRUCHHAUSEN W	

QUARRY-NO. 02 HILDERS SMAP 5426
PROFILE 02-240-19 COMPILED BY MEISSNER.BERCKHEMER

B	IN	AC	COORDINATES	ALT	X(KM)	SMAP	LOCALITY	G C
17	S	11	3510.75 5565.48	265	72.42	5720	LIEBLOS	
17	S	12	3510.75 5565.48	265	72.42	5720	LIEBLOS	
17	S	13	3510.75 5565.48	265	72.42	5720	LIEBLOS	X
17	H	22	3506.58 5562.23	170	77.65	5820	ROTHENBERGEN	
17	S	21	3505.37 5560.87	170	79.38	5820	LANGENSELBOLD	
17	S	22	3505.25 5560.82	167	79.51	5820	LANGENSELBOLD	
17	S	23	3505.12 5560.80	162	79.63	5820	LANGENSELBOLD	
17	S	11	3504.07 5556.85	220	82.59	5820	NIEDERRODENBACH	
17	S	12	3504.07 5556.85	220	82.59	5820	NIEDERRODENBACH	
17	S	13	3504.07 5556.85	220	82.59	5820	NIEDERRODENBCH	X
17	H		3502.16 5554.56	130	85.43	5820	NIEDERRODENBACH	
18	F	13	3492.33 5550.61	112	95.60	5919	HAUSEN/OFFENB.	
18	F	12	3492.05 5550.33	109	95.98	5919	HAUSEN/OFFENB.	
17	HH	23	3487.80 5549.03	120	100.53	5918	HEUSENSTAMM NE	
17	HH	22	3487.47 5548.83	119	100.92	5918	HEUSENSTAMM NE	
17	HH	21	3487.12 5548.58	118	101.35	5918	HEUSENSTAMM NE	
17	HH	C	3484.32 5545.32	133	105.44	5918	HEUSENSTAMM W	
17	HH	B	3483.92 5545.13	138	105.88	5918	HEUSENSTAMM W	
17	HH	A	3483.60 5544.91	159	106.27	5918	HEUSENSTAMM W	
17	HH	13	3481.56 5544.52	136	108.21	5918	NEU ISENBURG	
17	HH	12	3481.20 5544.24	139	108.66	5918	NEU ISENBURG	
17	HH	11	3480.78 5543.96	141	109.17	5918	NEU ISENBURG	
04	F		3465.20 5565.30	815	114.15	5716	KL.FELDBERG	
18	S	21	3477.80 5538.70	170	114.25	6018	LANGEN	
18	S	22	3477.80 5538.70	170	114.25	6018	LANGEN	
18	S	23	3477.80 5538.70	170	114.25	6018	LANGEN	X
12	H		3475.57 5535.36	120	118.02	6017	EGELSBACH	
17	CL		3473.48 5537.93	110	118.55	6017	LANGEN	
17	CL		3460.48 5528.34	85	134.67	6116	GEINSHEIM	
17	H		3451.61 5520.22	165	146.53	6115	UELVERSHEIM	
18	S	31	3445.86 5520.07	142	151.19	6115	WEINOLSHEIM	
18	S	32	3445.86 5520.07	142	151.19	6115	WEINOLSHEIM	
18	S	33	3445.86 5520.07	142	151.19	6115	WEINOLSHEIM	X
17	M	A	3440.89 5518.68	175	156.39	6115	BECHTOLSHEIM	
17	M	B	3440.77 5518.62	175	156.52	6115	BECHTOLSHEIM	
17	M	C	3440.77 5518.62	175	156.52	6115	BECHTOLSHEIM	
18	S	11	3439.24 5516.40	190	158.75	6214	ALZEY	
18	S	12	3439.24 5516.40	190	158.75	6214	ALZEY	X
18	S	13	3439.24 5516.40	190	158.75	6214	ALZEY	
17	KA	11	3437.66 5514.49	250	161.35	6214	SCHAFHAUSEN	
17	KA	12	3437.66 5514.49	250	161.35	6214	SCHAFHAUSEN	
17	KA	21	3434.67 5514.06	235	164.11	6214	ALZEY	
17	KA	22	3434.67 5514.06	235	164.11	6214	ALZEY	
17	M	A	3432.77 5510.77	220	167.48	6214	WEINHEIM	
17	M	B	3432.72 5510.76	220	167.53	6214	WEINHEIM	
17	M	C	3432.70 5510.72	220	167.56	6214	WEINHEIM	
17	M	A	3428.76 5508.70	330	171.97	6214	BECHENHEIM	
17	M	B	3428.48 5508.42	330	172.36	6214	BECHENHEIM	
17	M	C	3428.15 5508.17	330	172.77	6214	BECHENHEIM	
17	M	A	3426.62 5504.10	360	176.27	6313	LEITHOF	
17	M	B	3426.43 5504.18	345	176.38	6313	LEITHOF	
17	M	C	3426.02 5503.92	360	176.87	6313	LEITHOF	
22	F	31	3414.68 5517.18	275	179.92	6212	FEIL	
22	F	32	3414.70 5516.77	268	180.09	6212	FEIL	
22	F	33	3414.88 5516.25	272	180.18	6212	FEIL	
17	M	A	3423.14 5502.01	420	180.33	6313	DANNENFELS	
17	M	B	3422.99 5501.94	430	180.49	6313	DANNENFELS	
17	M	A	3420.28 5499.97	460	183.84	6313	MARIENTHAL	
17	M	B	3420.18 5499.66	460	184.09	6313	MARIENTHAL	
17	M	C	3420.18 5499.33	470	184.27	6313	MARIENTHAL	
17	M	1	3418.05 5498.38	475	186.57	6313	FALKENSTEIN	
17	M	2	3417.76 5498.09	465	186.97	6313	FALKENSTEIN	
17	M	3	3417.46 5497.76	420	187.41	6313	FALKENSTEIN	
17	M	B	3415.21 5492.86	330	192.01	6412	WINGERTSWEILERHOF	
17	M	A	3414.72 5493.13	310	192.26	6412	WINGERTSWEILERHOF	
17	M	1	3412.43 5492.20	315	194.68	6412	HEILIGENMOSCHEL	
17	M	2	3411.76 5491.80	345	195.46	6412	HEILIGENMOSCHEL	
17	M	3	3411.04 5491.24	355	196.37	6412	HEILIGENMOSCHEL	
17	M	A	3407.96 5489.24	350	200.04	6412	MEHLBACH	
17	M	B	3407.59 5489.11	355	200.42	6412	MEHLBACH	
17	M	C	3407.39 5488.77	355	200.77	6412	MEHLBACH	
17	M	A	3405.56 5488.81	290	202.27	6412	HIRSCHHORN	
17	M	B	3405.40 5488.66	290	202.49	6412	HIRSCHHORN	
17	M	C	3405.19 5488.35	290	202.83	6412	HIRSCHHORN	
17	M	1	3402.40 5489.18	340	204.70	6411	FRANKELBACH	
17	M	2	3402.07 5489.07	360	205.04	6411	FRANKELBACH	
17	M	3	3402.04 5489.06	360	205.07	6411	FRANKELBACH	
17	M	A	3400.54 5487.76	430	207.04	6411	EULENBIS	
17	M	B	3400.52 5487.66	430	207.11	6411	EULENBIS	
17	M	C	3400.30 5487.57	445	207.34	6411	EULENBIS	
17	CL		3395.65 5483.60	310	213.41	6511	KOLTWEILER	
18	HH	21	2598.76 5475.96	280	229.33	6510	KUEBELBERG	
18	HH	22	2598.76 5475.96	280	229.33	6510	KUEBELBERG	
18	HH	23	2598.43 5475.76	280	229.71	6510	KUEBELBERG	
18	HH	13	2588.26 5465.33	262	244.05	6609	NEUNKIRCHEN	
18	HH	12	2588.26 5465.33	262	244.05	6609	NEUNKIRCHEN	
18	HH	11	2588.18 5465.30	260	244.14	6609	NEUNKIRCHEN	
18	HH	31	2585.15 5465.22	299	246.64	6608	SPIESEN	
18	HH	33	2585.15 5465.22	299	246.64	6608	SPIESEN	
18	HH	32	2585.08 5465.14	295	246.75	6608	SPIESEN	
18	HH	42	2579.33 5462.70	350	252.85	6708	ST. INGBERT	

QUARRY-NO. 02 HILDERS SMAP 5426
PROFILE 02-265 COMPILED BY HAENEL.STEIN.BEHNKE

B	IN	AC	COORDINATES	ALT	X(KM)	SMAP	LOCALITY	G C
08	GO		3564.60 5600.60	735	9.05	5425	DANZWIESEN	
08	GO		3559.00 5599.97	495	14.68	5425	FINKENHAIN	
09	MZ		3554.05 5598.05	385	19.85	5424	LOHELAND	
09	GO		3542.65 5599.50	340	31.07	5423	MITTELRODE	
09	GO		3537.68 5600.82	340	36.00	5423	KLEINLUEDER	
02	H		3508.68 5598.10	398	65.13	5420	BETZENROD	
02	H		3501.65 5597.98	196	72.16	5420	GONTERSKIRCHEN	
02	H		3497.40 5598.23	181	76.40	5419	RUPPERTSBURG	
02	H		3493.50 5598.33	191	80.29	5419	NONNENROTH	
02	GO		3487.50 5596.35	210	86.37	5418	LICH	
02	GO		3482.45 5596.39	200	91.41	5418	DORF GUELL	
08	CL		3482.26 5594.11	220	91.64	5518	EBERSTADT	
02	GO		3478.60 5596.94	250	95.23	5418	LUDWIGSHOEHE	
08	CL		3477.19 5600.19	170	96.45	5418	LEIHGESTERN	
02	PR		3473.25 5596.68	190	100.59	5417	HOCHELHEIM	
08	CL		3466.47 5599.43	380	107.18	5417	STOPPEL BERG	
10	H		3440.73 5596.07	260	133.04	5514	LAHR	
10	H		3425.61 5599.34	355	148.08	5413	GUCKHEIM	
10	H		3418.36 5598.97	420	155.33	5413	SAINERHOLZ	
10	H		3418.17 5598.97	435	155.52	5413	SAINERHOLZ	
10	H		3417.98 5598.98	440	155.71	5413	SAINERHOLZ	
10	H		3417.70 5599.01	440	155.99	5413	SAINERHOLZ	
09	KR		3417.71 5597.40	390	156.01	5413	SAINERHOLZ	
10	H		3417.50 5599.03	430	156.19	5413	SAINERHOLZ	
10	H		3417.25 5599.12	430	156.44	5413	SAINERHOLZ	
09	MZ		3410.36 5598.26	295	163.34	5412	ELLENHAUSEN	
09	M		3404.36 5595.70	190	169.40	5511	KAUSEN	
09	M		3404.30 5595.52	210	169.47	5511	KAUSEN	
09	M		3399.32 5595.32	170	174.45	5511	ISENBURG	
09	M		3399.27 5595.29	170	174.50	5511	ISENBURG	
09	M		3398.56 5595.04	240	175.22	5511	ISENBURG	
09	M		3394.35 5595.71	150	179.41	5511	OBERBIEBER	
09	M		3394.34 5595.37	180	179.43	5511	OBERBIEBER	

B	IN	AC	COORDINATES		ALT	X(KM)	SMAP	LOCALITY
09	M		3387.90	5599.11	180	195.79	5410	DATZEROTH
09	M		3387.81	5599.13	210	185.88	5410	DATZEROTH
09	M		3387.56	5599.28	240	186.13	5410	DATZEROTH
09	M		3378.78	5597.16	185	194.94	5509	OBERBREISIG
09	M		3378.33	5597.16	185	195.39	5509	OBERBREISIG
09	M		3370.20	5596.51	310	203.53	5509	KOENIGSFELD
09	M		3370.10	5596.51	310	203.63	5509	KOENIGSFELD
09	M		2576.77	5595.78	445	209.79	5508	RAMERSBACH
10	M		2576.77	5595.79	440	209.79	5508	RAMERSBACH
10	M		2566.39	5596.78	400	220.10	5407	KREUZBERG

QUARRY-NO. 02 HILDERS SMAP 5426
PROFILE 02-300 COMPILED BY HEEP.GUTHEIL

B	IN	AC	COORDINATES		ALT	X(KM)	SMAP	LOCALITY	G	C
22	GO	21	3542.73	5616.75	288	34.87	5323	HUTZDORF		
22	GO	22	3542.73	5616.75	288	34.87	5323	HUTZDORF		X
22	GO	23	3542.73	5616.75	288	34.87	5323	HUTZDORF		Y
21	H	1	3534.91	5620.51	405	43.60	5222	GREBENAU		
21	H	2	3534.67	5620.82	415	43.96	5222	GREBENAU		
21	H	3	3534.31	5621.11	390	44.41	5222	GREBENAU		
22	GO	11	3523.13	5626.44	318	56.74	5221	ELBENROD		
22	GO	12	3522.73	5626.58	317	57.16	5221	ELBENROD		
22	GO	13	3522.38	5626.77	313	57.56	5221	ELBENROD		
21	H	1	3516.60	5628.22	330	63.44	5221	VOCKENROD		
21	H	2	3516.41	5628.38	345	63.68	5221	VOCKENROD		
21	H	3	3516.37	5628.56	355	63.79	5221	VOCKENROD		
21	GO	11	3506.31	5635.01	295	75.67	5120	NEUSTADT		
21	GO	12	3505.89	5635.18	309	76.12	5120	NEUSTADT		
21	GO	13	3505.47	5635.32	320	76.65	5120	NEUSTADT		
21	GO	23	3498.75	5638.20	342	83.86	5119	EMSDORF		Y
21	GO	21	3498.75	5638.20	342	83.86	5119	EMSDORF		
21	GO	22	3498.75	5638.20	342	83.86	5119	EMSDORF		X
21	MS		3490.03	5644.17	280	94.33	5019	BRACHT	MA	
21	MS		3489.58	5644.36	259	94.82	5019	BRACHT	MA	
23	MS		3484.65	5646.53	329	99.98	5018	BURGWALD	MA	
22	MS		3478.43	5647.39	260	106.09	5018	LANGENGRUND	MA	
22	MS		3477.97	5647.41	265	106.52	5018	LANGENGRUND	MA	
23	MS		3471.63	5655.42	375	115.64	4917	DODENAU	MA	
22	MS		3471.57	5655.45	370	115.86	4917	DODENAU	MA	
23	MS		3471.06	5655.68	375	116.27	4917	DODENAU	MA	
21	MS	11	3465.57	5658.34	530	122.58	4917	ALERTSHAUSEN	BS	
21	MS	12	3465.44	5658.41	537	122.73	4917	ALERTSHAUSEN	BS	
21	MS	13	3465.30	5658.46	545	122.87	4917	ALERTSHAUSEN	BS	
21	MS		3460.90	5660.55	475	127.74	4916	HEMLIGHAUSEN	MA	
21	MS	23	3453.54	5663.90	510	135.81	4816	LATROP	BS	
21	MS	22	3453.39	5664.04	500	136.01	4816	LATROP	BS	
21	MS	21	3453.12	5664.26	520	136.35	4815	LATROP	BS	
25	BO	21	3451.07	5664.81	430	138.34	4815	FLECKENBG SE	BS	
25	BO	22	3450.81	5665.01	425	138.65	4815	FLECKENBG SE	BS	
25	BO	23	3450.57	5665.20	405	138.96	4815	FLECKENBG SE	BS	
25	BO	11	3446.70	5666.13	380	142.82	4815	LENNE	BS	
25	BO	12	3446.43	5666.45	385	143.21	4815	LENNE	BS	
25	BO	13	3446.09	5666.67	390	143.61	4815	LENNE	BS	
21	MS		3443.53	5668.50	550	146.81	4815	AUERGANG	MA	
21	MS		3443.44	5668.54	555	146.92	4815	AUERGANG		X
21	MS		3443.34	5668.59	560	147.02	4815	AUERGANG	MA	
21	MS		3443.21	5668.68	560	147.18	4815	AUERGANG	MA	
21	MS		3443.12	5668.74	570	147.29	4815	AUERGANG		X
21	MS		3443.03	5668.81	580	147.40	4815	AUERGANG		
23	MS	21	3439.86	5669.47	500	150.31	4814	BRENSCHEDE	BS	
23	MS	22	3439.46	5669.59	495	150.72	4814	BRENSCHEDE		X
23	MS	23	3439.16	5669.62	465	151.00	4814	BRENSCHEDE	BS	
23	MS		3436.61	5671.41	400	154.09	4814	OEDINGEN	MA	
23	MS		3436.08	5671.66	440	154.67	4814	OEDINGEN	MA	
21	BO	1	3432.68	5673.15	325	158.59	4814	OSTENTROP	BS	
21	BO	2	3432.45	5673.25	327	158.84	4814	OSTENTROP		X
21	BO	3	3432.30	5673.32	325	159.00	4814	OSTENTROP		Y
23	MS		3428.24	5678.13	380	164.62	4713	ROENKHAUSEN	BS	
23	MS		3428.12	5678.16	380	164.74	4713	ROENKHAUSEN		X
23	MS		3428.03	5678.18	380	164.82	4713	R[illegible]NKHAUSEN	BS	
21	BO	3	3422.66	5680.60	465	170.91	4713	PLETTENBERG	BS	
21	BO	2	3422.39	5680.70	470	171.20	4713	PLETTENBERG	BS	
21	BO	1	3422.23	5680.75	175	171.36	4713	PLETTENBERG	BS	
22	MS	11	3416.70	5682.47	375	177.00	4712	NEUENRADE 1	BS	
22	MS	12	3416.52	5682.67	359	177.25	4712	NEUENRADE 1		X
22	MS	13	3416.34	5682.79	370	177.47	4712	NEUENRADE 1	BS	
23	MS	11	3413.09	5683.80	440	180.67	4712	NEUENRADE 2	BS	
23	MS	12	3412.73	5683.96	425	181.06	4712	NEUENRADE 2		X
23	MS	13	3412.50	5684.08	405	181.32	4712	NEUENRADE 2	BS	
22	MS	21	3412.21	5684.44	410	181.90	4712	S. DAHLE	BS	
22	MS	22	3412.02	5684.53	410	182.11	4712	S. DAHLE		X
22	MS	23	3411.88	5684.76	400	182.34	4712	S. DAHLE	BS	
22	BO	11	3408.54	5687.93	450	186.76	4612	N. ALTENA	BS	
22	BO	12	3408.19	5688.04	450	187.12	4612	N. ALTENA		X
22	BO	13	3407.81	5688.15	410	187.51	4612	N. ALTENA	BS	
22	BO	22	3403.45	5686.88	470	190.81	4611	WIBLINGWERDE		X
22	BO	23	3403.45	5686.88	470	190.81	4611	WIBLINGWERDE	BS	
23	H	31	3398.45	5689.07	280	196.11	4611	HOHENLIMBURG		
23	H	32	3398.12	5689.36	330	196.53	4611	HOHENLIMBURG		
23	H	33	3397.98	5689.64	330	196.78	4611	HOHENLIMBURG		
23	H	21	2600.80	5689.72	340	202.49	4610	HAGEN		
23	H	22	2600.41	5689.95	350	202.94	4610	HAGEN		
23	H	23	2599.99	5690.06	280	203.37	4610	HAGEN		
22	KR	11	3388.58	5697.66	230	208.96	4510	ARDEY GEB.		
22	KR	12	3388.26	5697.84	273	209.33	4510	ARDEY GEB.		
22	KR	13	3387.90	5697.95	205	209.70	4510	ARDEY GEB.		
23	BO	11	3379.94	5701.32	115	218.16	4509	BOCHUM QBG	BS	
23	BO	12	3379.74	5701.36	118	218.36	4509	BOCHUM QBG	BS	
23	BO	13	3379.38	5701.39	130	218.69	4509	BOCHUM QBG	BS	
23	BO	21	3372.92	5714.96	48	230.84	4408	EMSCHERBRUCH	BS	
23	BO	22	3372.76	5715.13	48	231.06	4408	EMSCHERBRUCH	BS	
23	BO	23	3372.57	5715.34	49	231.33	4408	EMSCHERBRUCH	BS	
23	KR	13	3353.08	5722.70	64	251.93	4307	KIRCHHELLEN		
23	KR	11	3352.71	5722.60	67	252.21	4307	KIRCHHELLEN		
23	KR	12	3352.51	5722.73	65	252.45	4307	KIRCHHELLEN		

QUARRY-NO. 02 HILDERS SMAP 5426
PROFILE 02-315 COMPILED BY DEGUTSCH.THYSSEN.QUERFURTH

B	IN	AC	COORDINATES		ALT	X(KM)	SMAP	LOCALITY	G C
29	CL	41	3499.20	5666.40	557	99.05	4819	FREBERSHAUSEN	
29	CL	42	3498.96	5666.62	539	99.38	4819	FREBERSHAUSEN	
29	CL	43	3498.78	5666.80	570	99.63	4819	FREBERSHAUSEN	
29	CL	31	3497.08	5668.09	320	101.76	4819	ALTENLOTHEIM	
29	CL	32	3496.74	5668.38	446	102.21	4819	ALTENLOTHEIM	
29	CL	33	3496.44	5668.64	380	102.61	4819	ALTENLOTHEIM	
29	CL	21	3494.38	5670.52	406	105.39	4819	HARBSHAUSEN	
29	CL	22	3494.06	5670.72	391	105.77	4819	HARBSHAUSEN	
29	CL	23	3493.81	5671.06	386	106.18	4819	HARBSHAUSEN	
29	CL	11	3491.04	5673.63	362	109.96	4819	IMMIGHAUSEN	
29	CL	12	3490.70	5673.86	366	110.36	4719	IMMIGHAUSEN	
29	CL	13	3490.40	5674.18	395	110.80	4719	IMMIGHAUSEN	
29	GO	21	3488.70	5675.72	423	113.09	4719	IMMIGHAUSEN NW	
29	GO	22	3488.40	5675.98	419	113.49	4719	IMMIGHAUSEN NW	
29	GO	23	3488.13	5676.21	435	113.85	4719	IMMIGHAUSEN NW	
29	GO	11	3486.67	5677.87	470	116.04	4718	GOLDHAUSEN SW	
29	GO	12	3486.35	5678.11	465	116.44	4718	GOLDHAUSEN SW	
29	GO	13	3486.09	5678.31	475	116.76	4718	GOLDHAUSEN SW	
29	H	41	3478.30	5685.70	545	127.50	4618	EIMELROD	
29	H	42	3477.93	5685.93	545	127.92	4618	EIMELROD	
29	H	43	3477.52	5685.86	530	128.19	4618	EIMELROD	
29	MS	12	3473.51	5681.53	610	128.46	4617	BRILON S	
29	H	31	3475.56	5688.02	625	131.08	4617	RATTLAR	
29	H	32	3475.50	5688.10	625	131.18	4617	RATTLAR	
29	H	33	3475.43	5688.18	600	131.29	4617	RATTLAR	
29	MS	11	3473.88	5689.40	615	133.26	4617	BRILON S	
29	MS	13	3473.27	5689.89	600	134.04	4617	BRILON S	
29	MS	21	3471.59	5690.65	520	135.80	4617	BRILON	
29	MS	22	3471.59	5690.65	520	135.80	4617	BRILON	
29	MS	23	3471.53	5691.10	530	136.14	4617	BRILON	
29	BO	21	3469.55	5691.68	625	138.02	4617	PETERSBORN	
29	BO	22	3469.43	5692.00	600	138.32	4617	PETERSBORN	
29	BO	23	3469.29	5692.42	553	138.70	4617	PETERSBORN	
29	BO	11	3467.33	5692.97	575	140.54	4617	ALTENBUEREN	
29	BO	12	3466.97	5693.17	570	140.94	4617	ALTENBUEREN	
29	BO	13	3466.60	5693.45	555	141.40	4617	ALTENBUEREN	
29	KR	11	3465.90	5696.32	584	143.82	4517	SCHARFENBERG	
29	KR	12	3465.72	5696.46	550	144.05	4517	SCHARFENBERG	
29	KR	13	3465.60	5696.60	520	144.23	4517	SCHARFENBERG	
29	KI	41	3464.67	5697.81	430	145.73	4516	ALTBUER.WALD	
29	KI	42	3464.43	5698.02	428	146.05	4516	ALTBUER.WALD	
29	KI	43	3464.14	5698.09	457	146.31	4516	ALTBUER.WALD	
29	KI	31	3462.78	5698.80	468	147.80	4516	GR.STEINBERG	
29	KI	32	3462.52	5699.10	425	148.19	4516	GR.STEINBERG	
29	KI	33	3462.30	5699.48	435	148.61	4516	GR.STEINBERG	
29	KI	21	3461.14	5700.37	365	150.07	4516	KALLENHARDT	
29	KI	22	3460.97	5700.50	403	150.28	4516	KALLENHARDT	
29	KI	23	3460.65	5700.67	402	150.63	4516	KALLENHARDT	
29	KI	11	3458.77	5702.89	330	153.51	4516	KOERTLINGHAUSEN	
29	KI	12	3458.48	5703.15	320	153.90	4516	KOERTLING	
29	KI	13	3458.13	5703.35	312	154.30	4516	KOERTLING	
28	GO	11	3456.03	5704.15	320	156.47	4516	BELECKE	
28	GO	12	3455.74	5704.52	332	156.93	4516	BELECKE	
28	GO	13	3455.50	5704.83	341	157.31	4516	BELECKE	
28	GO	21	3453.24	5708.00	305	161.10	4415	UELDE	
28	GO	22	3452.90	5708.30	334	161.56	4415	UELDE	
28	GO	23	3452.61	5708.58	318	161.96	4415	UELDE	
28	CL	41	3450.77	5710.07	220	164.33	4415	ALTENMELLERICH	
28	CL	42	3450.52	5710.36	210	164.71	4415	ALTENMELLERICH	
28	CL	43	3450.25	5710.64	210	165.10	4415	ALTENMELLERICH	
28	CL	31	3447.00	5710.71	240	167.59	4415	HERRINGSEN	
28	CL	32	3446.79	5711.00	225	167.93	4415	HERRINGSEN	
28	CL	33	3446.56	5711.28	205	168.29	4415	HERRINGSEN	
29	CL	21	3446.73	5714.21	161	170.10	4415	ALTENGESEKE	
28	CL	22	3446.28	5714.42	160	170.58	4415	ALTENGESEKE	
28	CL	23	3445.96	5714.63	15[illegible]	170.95	4415	ALTENGESEKE	
28	CL	11	3443.42	5715.77	120	173.61	4415	LOHNE	
28	H	11	3440.62	5719.24	92	178.00	4314	HEPPEN	
28	H	12	3440.33	5719.58	85	178.45	4314	HEPPEN	
28	H	13	3439.94	5719.87	83	178.93	4314	HEPPEN	
28	H	41	3439.11	5721.13	80	180.39	4314	OESTINGHAUSEN	
28	H	42	3438.74	5721.39	76	180.84	4314	OESTINGHAUSEN	
28	H	43	3438.35	5721.59	76	181.26	4314	OESTINGHAUSEN	
28	H	31	3436.83	5723.13	75	183.42	4314	WILTROP	
28	H	32	3436.69	5723.38	75	183.69	4314	WILTROP	
28	H	33	3436.56	5723.79	75	184.06	4314	WILTROP	
28	H	21	3433.88	5723.69	75	186.00	4314	HULTROP	
29	H	22	3433.72	5724.05	75	186.36	4314	HULTROP	
28	H	23	3433.48	5724.36	75	186.75	4314	HULTROP	
28	HH	43	3431.66	5724.92	73	188.48	4314	BUENINGHAUSEN	
28	HH	42	3431.36	5725.20	74	188.89	4314	BUENINGHAUSEN	
28	HH	41	3431.18	5725.59	75	189.29	4314	BUENINGHAUSEN	
28	HH	31	3430.80	5729.25	75	192.00	4313	NAEHE PLATTENBER	
28	HH	32	3430.46	5729.53	78	192.44	4313	NAEHE PLATTENBER	
28	HH	33	3430.18	5729.74	90	192.78	4313	NAEHE PLATTENBER	
28	HH	11	3427.62	5731.07	69	195.58	4213	E. DOLBERG	
28	HH	12	3427.31	5731.38	72	196.02	4213	E. DOLBERG	
28	HH	13	3426.92	5731.59	83	196.45	4213	E. DOLBERG	
28	HH	21	3425.86	5732.47	97	197.83	4213	DOLBERG	
2	HH	22	3425.44	5732.65	96	198.26	4213	DOLBERG	
28	HH	23	3425.07	5732.83	93	198.66	4213	DOLBERG	
28	KR	11	3420.62	5734.75	80	203.27	4213	AHLEN	
28	KR	12	3420.19	5734.92	80	203.70	4213	AHLEN	
29	KR	13	3419.76	5735.00	80	204.08	4213	AHLEN	
28	BO	11	3419.15	5738.40	75	206.78	4212	NE. HALSTEDDE	
28	BO	12	3418.93	5738.55	74	207.05	4212	NE. HALSTEDDE	
28	BO	13	3418.65	5738.85	80	207.45	4212	NE. HALSTEDDE	
27	KR	12	3418.56	5738.98	72	207.56	4212	HALSTEDDE	
28	BO	21	3416.03	5741.70	66	211.31	4112	E. DRENSTEINFURT	
28	BO	22	3419.67	5741.99	65	211.77	4112	E. DRENSTEINFURT	
28	BO	23	3415.35	5742.26	64	212.19	4112	E. DRENSTEINFURT	
27	BO	11	3414.28	5743.50	62	213.77	4112	N. DRENSTEINFURT	
27	BO	12	3414.15	5743.67	62	213.98	4112	N. DRENSTEINFURT	
27	BO	13	3413.91	5743.90	61	214.30	4112	N. DRENSTEINFURT	
27	MS	11	3412.61	5745.37	58	216.25	4112	RINKERODE 1	
27	MS	12	3412.50	5745.44	58	216.38	4112	RINKERODE 1	
27	MS	13	3412.32	5745.63	58	216.64	4112	RINKERODE 1	
27	MS	21	3410.40	5747.14	54	219.08	4112	RINKERODE 2	
27	MS	22	3410.06	5747.40	56	219.51	4112	RINKERODE 2	

B	IN	AC	COORDINATES	ALT	X(KM)	SMAP	LOCALITY	G C
27	MS	23	3409.84 5747.60	58	219.80	4112	RINKERODE 2	
28	MS	21	3391.54 5758.00	170	240.47	4010	SCHAPDETTEN	
28	MS	22	3391.30 5758.25	175	240.81	4010	SCHAPDETTEN	
28	MS	23	3391.04 5758.63	175	241.26	4010	SCHAPDETTEN	BS
28	MS	11	3390.30 5758.90	180	241.99	4010	LASBECK	
28	MS	12	3390.00 5759.25	156	242.45	4010	LASBECK	
28	MS	13	3389.57 5759.45	167	242.91	4010	LASBECK	

QUARRY-NO. 02 HILDERS SMAP 5426
PROFILE 02-325 COMPILED BY DEGUTSCH.WIEHLE

B	IN	AC	COORDINATES	ALT	X(KM)	SMAP	LOCALITY	G C
19	MS		3557.28 5624.50	375	28.80	5224	LICHTENBERG	BS
12	CL		3554.28 5631.13	290	35.77	5124	FISCHBACH	
19	MS	22	3549.58 5631.70	380	39.17	5124	S. BAD HERSFELD	X
19	MS		3549.40 5631.75	355	39.32	5124	B.HERSFELD S	HA
19	MS		3549.13 5631.99	380	39.68	5124	B.HERSFELD S	HA
18	MS		3546.62 5635.94	365	44.13	5123	E. HERSFELD	BS
12	CL		3543.50 5637.50	481	47.29	5123	SSW. UNTERGEISS	
19	MS		3545.81 5640.74	340	48.68	5023	N. GITTERDORF	X
15	KR	B	3543.06 5644.32	410	53.11	5023	OBERTHALHAUSEN	
15	KR	C	3542.98 5644.37	422	53.20	5023	OBERTHALHAUSEN	
15	KR	A	3542.84 5644.42	432	53.33	5023	OBERTHALHAUSEN	
19	MS		3541.00 5647.50	480	56.97	5023	HAINRODE	BS
12	H	A	3538.32 5648.79	460	59.38	5023	NENTERODE	
12	H	B	3538.17 5649.07	440	59.70	5023	NENTERODE	
12	H	C	3538.02 5649.30	400	59.97	5023	NENTERODE	
12	H	D	3537.84 5649.55	390	60.28	5023	NENTERODE	
12	H	E	3537.81 5649.85	390	60.54	5023	NENTERODE	
12	H	F	3537.75 5650.12	390	60.79	5023	NENTERODE	
18	MS		3539.35 5652.79	465	62.07	4923	RENGSHAUSEN	BS
18	MS		3536.41 5654.12	410	64.83	4923	LICHTENHAGEN	BS
19	MS	12	3533.04 5656.80	380	69.17	4922	BERNDSHAUSEN	BS
16	MS		3533.46 5658.19	320	69.86	4922	S. SIPPERHAUS	BS
12	KR	D	3532.20 5660.00	320	72.05	4922	HOSHEIM	
16	MS		3528.92 5661.83	325	75.47	4922	HESSERODE	BS
15	MS		3524.90 5667.13	217	82.24	4822	CAPPEL	HA
20	MS		3526.84 5673.21	290	86.02	4822	ODENBERG	HA
20	MS		3526.68 5673.52	290	86.38	4822	ODENBERG	HA
15	MS		3523.11 5675.35	245	89.98	4721	METZE	HA
15	MS		3522.54 5675.93	265	90.78	4721	METZE	HA
16	MS		3519.72 5678.88	325	94.69	4721	SAND	HA
16	MS		3519.48 5679.19	325	95.08	4721	SAND	HA
18	KR	A	3520.63 5680.19	390	95.26	4721	ELMSHAGEN	
18	KR	B	3520.50 5680.22	380	95.35	4721	ELMSHAGEN	
18	KR	F	3520.39 5680.42	375	95.58	4721	ELMSHAGEN	
12	MS		3517.55 5681.67	355	98.21	4721	BALHORN	HA
16	MS		3514.56 5687.16	341	104.43	4621	W. WOLFHAGEN	HA
14	MS	A	3512.75 5693.22	275	110.45	4621	NIEDERELSUNGN	HA
14	MS	E	3512.68 5694.02	260	111.16	4621	NIEDERELSUNGN	HA
15	KI		3508.27 5695.30	238	114.83	4620	EHRINGEN	BS
12	MS		3507.60 5700.54	210	119.41	4520	N. VOLKMARSEN	BS
11	GO		3503.25 5700.00	274	121.49	4520	NW. HERBSEN	
14	MS		3502.53 5706.11	370	126.85	4520	RHODEN	BS
11	GO		3503.03 5706.97	310	127.36	4520	NE. RHODEN/WA	BS
11	MS		3497.59 5710.64	285	133.45	4419	WREXEN	HA
24	MS	21	3495.50 5713.09	355	136.62	4419	MARSCHALLSHGN	HA
24	MS	23	3494.83 5713.80	395	137.58	4419	MARSCHALLSHGN	HA
11	MS		3495.20 5714.80	375	138.23	4419	SE. HOLTHEIM	
18	BO	3	3493.86 5716.32	359	140.19	4419	HOLTHEIM	BS
18	BO	1	3491.43 5718.72	343	143.54	4319	SW. LICHTENAU	BS
24	MS	11	3490.55 5721.92	305	146.69	4319	LICHTENAU	HA
12	MS		3489.89 5722.45	260	147.49	4319	GRUNDSTEINHM SW	
24	MS	13	3490.06 5722.74	300	147.65	4319	LICHTENAU	HA
11	H	F	3487.98 5722.95	308	·149.04	4318	EBBINGHAUSEN	
11	H	E	3487.80 5723.21	308	149.36	4318	EBBINGHAUSEN	
11	H	D	3487.62 5723.46	308	149.67	4318	EBBINGHAUSEN	
13	MS		3491.92 5726.53	343	149.77	4319	IGGENHAUSEN	HA
11	H	C	3487.48 5723.71	300	149.95	4318	EBBINGHAUSEN	
11	H	B	3487.33 5723.98	300	150.26	4318	EBBINGHAUSEN	
11	H	A	3487.22 5724.25	300	150.54	4318	EBBINGHAUSEN	
13	MS		3491.48 5727.29	322	150.65	4319	IGGENHAUSEN	HA
20	MS		3490.00 5728.08	281	152.11	4319	DAHL	HA
15	HH		3496.92 5744.89	425	163.13	4119	S. HORN	Y
12	KI		3478.29 5738.52	103	167.30	4218	SENNELAGER	BS
12	MS		3476.14 5741.02	105	170.58	4117	HOEVELHOF	BS
15	HH		3479.44 5756.27	280	181.69	4018	RABENSBERG	
11	H		3460.17 5761.22	100	196.35	4016	STEINHAGE	
14	MS		3461.05 5766.78	306	200.34	3916	BUSSBERG	BS
11	H		3458.59 5770.17	260	204.60	3916	W. HALLE	
25	MS	11	3451.06 5775.10	291	212.98	3815	BORGHOLZHAUSEN	
25	MS	12	3451.03 5775.32	280	213.18	3815	BORGHOLZHAUSEN	
25	MS	13	3451.03 5775.44	275	213.28	3815	BORGHOLZHAUSEN	
11	MS		3446.83 5778.00	210	217.75	3815	DISSEN	BS
15	MS		3444.55 5780.26	235	220.99	3815	EPPENDORF	HA
25	MS	21	3445.17 5786.39	180	225.61	3715	SE. BISSENDORF	
11	MS		3445.35 5786.66	185	225.69	3715	EBBENDORF	BS
25	MS	22	3445.29 5786.71	182	225.80	3715	SE. BISSENDORF	
25	MS	23	3445.49 5787.11	180	226.02	3715	SE. BISSENDORF	
13	MS		3437.77 5799.40	77	240.45	3614	WITTEKINDSBG.	HA
15	MS		3432.71 5805.32	35	248.30	3614	SCHLEPTRUPER	BS
11	MS		3426.67 5809.16	61	254.83	3513	ACHMER	BS
13	MS		3424.45 5811.90	90	258.32	3513	GEHN	HA
11	MS		3419.00 5822.02	95	269.76	3412	SCHWAGSTORF	BS
13	MS		3415.19 5825.02	125	274.38	3412	S. BIPPEN	BS
13	MS		3412.26 5831.71	78	281.54	3312	GRAFELD	
13	MS		3405.66 5837.23	30	289.85	3311	FELSEN	BS

QUARRY-NO. 02 HILDERS SMAP 5426
PROFILE 02-350-06 COMPILED BY HINZ.KAMINSKI

B	IN	AC	COORDINATES	ALT	X(KM)	SMAP	LOCALITY	G C
05	H	AB	3574.31 5602.18	778	1.31	5426	FRANKENHEIM	
05	H	C	3574.31 5602.18	778	1.31	5426	FRANKENHEIM	
26	HH	21	3571.45 5606.11	600	5.54	*5426	SIMMERSHAUSEN	
26	HH	22	3571.42 5606.54	660	5.94	5426	SIMMERSHAUSEN	
26	HH	23	3571.30 5606.86	735	6.28	5426	SIMMERSHAUSEN	
06	H		3573.30 5608.06	480	7.08	*5326	DIPPACH	
06	H	B	3573.27 5609.06	440	8.08	5326	DIPPACH	
26	HH	31	3568.82 5613.99	373	13.83	5325	NEUWARTS	
26	HH	32	3568.67 5614.36	354	14.22	5325	NEUWARTS	
26	HH	33	3568.45 5614.70	345	14.62	5325	NEUWARTS	
25	HH	42	3563.83 5619.43	397	20.91	5225	RASDORF S	
25	HH	43	3563.79 5619.83	380	21.29	5225	RASDORF S	
26	HH	11	3566.57 5627.28	308	27.17	5225	MANSBACH	
26	HH	12	3566.72 5627.67	330	27.51	5225	MANSBACH	
26	HH	13	3566.77 5628.00	310	27.82	5225	MANSBACH	
06	CL		3565.26 5630.65	350	30.84	*5125	OBERBREITZBACH	
06	CL		3565.26 5630.65	350	30.84	5125	OBERBREITZBACH	
25	HH	31	3565.22 5636.94	390	36.89	5125	UNTERNEURODE	
25	HH	33	3565.34 5637.77	382	37.67	5125	UNTERNEURODE	
26	HH	41	3565.74 5639.76	325	39.51	5125	HERFA	
26	HH	42	3565.81 5640.14	345	39.87	5125	HERFA	
26	HH	43	3565.98 5640.49	375	40.18	5125	HERFA	
05	GO		3563.90 5648.10	320	48.15	*5025	MACHTLOS	
25	GO	11	3563.85 5652.86	395	52.74	4925	DENS	
25	GO	12	3563.68 5653.29	422	53.20	4925	DENS	
25	GO	13	3563.68 5653.69	375	53.59	4925	DENS	
25	GO	21	3562.50 5660.82	330	60.82	4925	METZLAR	
25	GO	22	3562.50 5661.25	357	61.24	4925	METZLAR	
25	GO	23	3562.42 5661.62	370	61.62	4925	METZLAR	
05	GO		3561.08 5663.00	305	63.31	4925	THURNHOSBACH	MH
07	CL		3561.46 5670.50	288	70.61	*4825	HARTMUTSHAUSEN	
07	CL		3560.98 5674.85	560	74.97	*4725	SCHWALBENTHAL	
07	CL		3560.98 5674.85	560	74.97	4725	SCHWALBENTHAL	MT
07	CL		3560.98 5674.85	560	74.97	4725	SCHWALBENTHAL	MT
05	GO		3557.96 5684.89	260	85.39	*4724	FAHRENBACH	MH
07	F		3557.09 5693.57	175	94.09	*4624	GERTENBACH	
25	CL	11	3554.34 5699.47	310	100.31	4524	MEENSEN	
25	CL	13	3554.25 5700.12	310	100.97	4524	MEENSEN	
05	SE		3551.20 5699.90	400	101.47	4524	BRACKENBERG	
19	GO	11	3570.28 5702.49	305	101.60	4526	GOETTINGEN	
19	GO	12	3570.28 5702.49	305	101.60	4526	GOETTINGEN	X
19	GO	13	3570.28 5702.49	305	101.60	4526	GOETTINGEN	X
19	GO	21	3570.28 5702.49	305	101.60	4526	GOETTINGEN	
19	GO	22	3570.28 5702.49	305	101.60	4526	GOETTINGEN	
19	GO	23	3570.28 5702.49	305	101.60	4526	GOETTINGEN	
25	CL	21	3553.91 5704.02	402	104.86	4524	GAUSSTURM	
25	CL	22	3553.88 5704.47	427	105.31	4524	GAUSSTURM	
25	CL	23	3553.85 5704.90	430	105.74	4524	GAUSSTURM	
05	H		3553.20 5704.70	500	105.74	4524	H.HAGEN/DRANSF.	
25	CL	31	3552.47 5709.35	285	110.37	4424	DRANSFELD	
25	CL	32	3552.50 5709.72	269	110.73	4424	DRANSFELD	
25	CL	33	3552.49 5710.15	272	111.15	4424	DRANSFELD	
05	H		3552.65 5713.55	330	114.54	4424	GREFENBURG	
25	CL	41	3552.44 5713.78	315	114.72	4424	ADELEBSEN S	
25	CL	42	3552.30 5714.14	310	115.11	4424	ADELEBSEN S	
25	CL	43	3552.41 5714.45	278	115.38	4424	ADELEBSEN S	
07	H		3551.26 5719.71	370	120.85	*4324	ADELEBSEN	
26	GO	11	3551.09 5724.13	330	125.13	4324	SCHLARPE	
26	GO	12	3551.05 5724.47	305	125.47	4324	SCHLARPE	
26	GO	13	3551.00 5724.86	285	125.86	4324	SCHLARPE	
06	S		3549.66 5727.90	400	129.17	4324	DELLIEHAUSEN	
06	S		3549.66 5727.90	400	129.17	*4324	DELLIEHAUSEN	
06	S		3549.66 5727.90	400	129.17	4324	DELLIEHAUSEN	
10	HH		3549.05 5732.29	437	133.54	4224	GRASBORN	
25	HH	12	3548.46 5732.84	423	134.19	4224	STAATSF.SULZERTH.	
25	HH	13	3548.65 5733.23	415	134.54	4224	STAATSF.SULZERTH.	
26	GO	21	3548.82 5736.93	256	138.12	4224	DELLIEHAUSEN	
06	S	A	3547.56 5736.82	190	138.32	4224	RELLIEHAUSEN	
06	S	B	3547.56 5736.82	190	138.32	*4224	RELLIEHAUSEN	X
06	S	C	3547.56 5736.82	190	138.32	4224	RELLIEHAUSEN	
26	GO	22	3548.72 5737.36	225	138.57	4224	DELLIEHAUSEN	
26	GO	23	3548.65 5737.77	200	138.99	4224	DELLIEHAUSEN	
06	H		3545.24 5744.52	280	146.33	4123	MACKENSEN	
25	HH	21	3547.64 5746.96	288	148.22	4124	LINNENKAMP	
25	HH	23	3547.42 5747.69	260	148.98	4124	LINNENKAMP	
07	CL		3543.71 5752.58	280	154.56	*4023	STADTOLDENDORF	
06	H		3544.91 5758.17	300	159.80	*4023	SCHARFOLDENDORF	
10	HH		3544.21 5761.98	275	163.62	*4023	FOELZIEHAUSEN	
26	H	21	3544.37 5764.50	170	166.05	3923	WALLENSEN	
26	H	22	3544.29 5764.90	180	166.46	3923	WALLENSEN	
26	H	23	3544.20 5765.32	180	166.89	3923	WALLENSEN	
09	CL		3544.15 5769.13	390	170.67	*3923	SALZHEMMENDORF 2	
06	H	B	3542.65 5768.88	300	170.74	3923	SALZHEMMENDORF 1	
06	H	C	3542.65 5768.88	300	170.74	3923	SALZHEMMENDORF 1	
26	H	41	3542.11 5772.89	120	174.71	3923	SALZHEMMENDORF	
26	H	43	3542.08 5773.76	130	175.57	3923	SALZHEMMENDORF	
07	H		3541.48 5775.90	250	177.88	*3823	OSTERWALD	
06	H	A	3539.98 5779.65	250	181.82	*3823	DOERPE	
06	H	C	3539.98 5779.65	250	181.82	3823	DOERPE	
09	CL		3541.37 5780.50	180	182.35	*3823	OSTERWALD N	MP
26	H	41	3538.89 5783.33	280	185.58	3823	SAUPARK/SPR.	
25	H	42	3538.60 5783.57	270	185.87	3823	SAUPARK/SPR.	
25	H	43	3538.41 5783.76	265	186.09	3823	SAUPARK/SPR.	
06	SE		3539.31 5789.70	300	191.82	3723	SPRINGE	
07	H	B	3535.97 5789.33	320	192.11	*3723	SPRINGE	
19	H	12	3535.96 5789.33	315	192.15	3723	SPRINGE	
19	H	14	3535.13 5789.19	315	192.18	3723	SPRINGE	
19	H	11	3536.78 5789.56	340	192.22	3723	SPRINGE	
25	H	11	3535.70 5792.20	190	194.90	3723	WENNIGSER MARK	
25	H	12	3535.64 5792.40	175	195.11	3723	WENNIGSER MARK	
25	H	13	3535.58 5792.59	165	195.30	3723	WENNIGSER MARK	
25	H	21	3538.66 5797.66	75	199.71	3623	LEVESTE	
25	H	22	3538.36 5797.94	70	200.04	3623	LEVESTE	
25	H	23	3538.28 5798.38	70	200.49	3623	LEVESTE	
25	H	31	3538.52 5801.80	59	203.81	3623	LATHWEHREN	
25	H	32	3538.29 5802.21	58	204.26	3623	LATHWEHREN	
06	H		3536.16 5802.00	80	204.49	*3623	STEMMEN	
25	H	33	3538.09 5802.60	56	204.68	3623	LATHWEHREN	

QUARRY-NO. 02 HILDERS SMAP 5426
PROFILE 02-178-F COMPILED BY BAIER

B	IN	AC	COORDINATES	ALT	X(KM)	SMAP	LOCALITY	G C
22	F	31	3414.68 5517.18	275	179.92	6212	FEIL	
22	F	32	3414.70 5516.77	268	180.09	6212	FEIL	
22	F	33	3414.88 5516.25	272	180.18	6212	FEIL	
20	M	32	3417.67 5510.17	290	180.61	6213	OBERHAUSEN	
20	M	31	3418.02 5510.28	280	180.26	6213	OBERHAUSEN	
22	F	11	3418.72 5508.69	310	180.55	6213	GANGREHWEILER E	
22	F	12	3418.94 5508.47	330	180.47	6213	GANGREHWEILER E	
22	F	13	3418.97 5508.20	325	180.58	6213	GANGREHWEILER E	
22	F	21	3416.79 5511.35	285	180.87	6213	KALKOFEN SE	
22	F	22	3416.94 5511.12	322	180.86	6213	KALKOFEN SE	
22	F	23	3417.22 5510.80	327	180.77	6213	KALKOFEN SE	
20	M	21	3419.88 5507.15	330	180.26	6313	SCHNEEBERGERHOF	
20	M	22	3419.58 5507.01	350	180.59	6313	SCHNEEBERGERHOF	
20	M	24	3419.29 5506.98	370	180.85	6313	SCHNEEBERGERHOF	
22	CL	11	3420.71 5504.94	405	180.80	6313	GERBACH	
22	CL	12	3420.71 5504.56	395	181.01	6313	GERBACH	
22	CL	13	3420.55 5504.91	455	180.96	6313	GERBACH	
20	F	32	3426.10 5497.37	293	180.42	6313	JAKOBWEILER E	
22	CL	21	3427.88 5494.53	255	180.71	6414	GOELLHEIM	
22	CL	22	3428.13 5494.30	260	180.64	6414	GOELLHEIM	
22	CL	23	3428.37 5493.86	260	180.71	6414	GOELLHEIM	
20	KI	11	3430.32 5494.10	295	178.92	6414	GOELLHM.WALD	
20	KI	13	3430.32 5494.08	295	178.93	6414	GOELLHM.WALD	
22	S	41	3434.22 5490.41	218	178.15	6414	TIEFENTHAL NW	
22	S	42	3434.56 5490.15	221	178.05	6414	TIEFENTHAL NW	
22	S	43	3434.82 5490.93	235	177.36	6414	TIEFENTHAL NW	

QUARRY-NO. 03 GERSFELD(SCHWARZER ACKER) SMAP 5525
PROFILE 03-250 COMPILED BY HAENEL.STEIN

B	IN	AC	COORDINATES	ALT	X(KM)	SMAP	LOCALITY	G C
02	BO		3567.21 5591.88	610	1.19	5525	SANDBERG	
02	M		3564.71 5590.60	528	3.99	5525	GERSFELD	
02	M		3560.38 5589.00	470	8.60	5525	GICHENBACHSH.	
02	BO		3552.33 5585.81	429	17.26	5524	ANBERG	
02	M		3549.21 5584.20	510	20.76	5624	OBERKALBACH	
03	MZ		3546.90 5587.33	420	22.02	5624	MITTELKALBACH	
02	MM		3537.81 5580.56	235	32.69	5623	HOF REITH	
02	CL		3532.20 5577.15	305	39.17	5622	HUNDSR.BERG	
02	KR		3527.10 5577.20	220	43.90	5622	ROMSTHAL	
02	SE		3514.60 5573.35	340	56.97	5721	BUEDINGEN	
02	SE		3512.12 5571.84	326	59.81	5721	BUEDINGEN	
02	SE		3510.04 5571.00	156	62.05	5720	BUEDINGEN	
03	KR		3507.28 5569.99	205	65.02	5720	VONHAUSEN	
02	SE		3506.10 5570.32	170	65.99	5720	BUEDINGEN	
02	M		3502.74 5568.48	227	69.77	5720	ECKHARTSHAUSEN	
02	M		3497.42 5569.42	160	74.50	5719	ROMMELSHAUSEN	
02	M		3492.23 5568.60	120	79.70	5719	EICHEN	
02	M		3488.07 5566.44	158	84.32	5718	BUEDESHEIM	
02	F		3485.33 5564.78	153	87.44	5718	RENDEL	
03	CL		3482.65 5561.91	110	90.93	5818	DOTTENF.HOF	
02	M		3463.60 5558.88	250	109.93	5816	MAMMOLSHAIN	
03	CL		3459.36 5557.50	257	114.41	5816	FISCHBACH	
02	M		3451.16 5554.62	230	123.07	5815	NAUROD	
02	M		3438.21 5550.25	339	136.74	5914	SCHLANGENBAD	
03	M		3430.05 5550.64	360	144.44	5914	OBERGLADBACH	
03	M		3430.01 5550.56	360	144.50	5914	OBERGLADBACH	
02	MZ		3427.55 5549.46	459	147.15	5913	KIEDRICHER H.	
03	M		3424.42 5548.76	230	150.37	5913	N STEPHANSK.	
03	M		3424.17 5548.49	230	150.69	5913	N STEPHANSK.	
03	M		3423.77 5548.73	215	151.00	5913	N STEPHANSK.	
03	M		3423.31 5548.79	220	151.43	5913	N STEPHANSK.	
03	M		3423.06 5549.02	200	151.60	5913	N STEPHANSK.	
02	MZ		3419.65 5547.33	405	155.33	5913	PRESBERG	
03	M		3418.40 5546.88	180	156.68	5913	W PRESBERG	
03	M		3412.66 5545.52	230	162.57	5912	OBERDIEBACH	
03	M		3412.53 5545.40	245	162.73	5912	OBERDIEBACH	
03	M		3405.18 5543.57	410	170.29	5912	RHEINBOELLEN	
03	M		3405.14 5543.45	410	170.36	5912	RHEINBOELLEN	
03	M		3403.61 5541.01	405	172.54	6011	KL.WEIDELBACH	
02	M		3400.15 5544.30	435	174.89	5911	BENZWEILER	
03	M		3397.00 5540.05	408	179.14	6011	MUTTERSCHIED	
03	M		3396.84 5540.14	395	179.27	6011	MUTTERSCHIED	
02	S		3391.61 5540.08	368	184.27	6010	FRONHOFEN K.	
03	M		3390.60 5540.06	370	185.27	6010	FRONHOFEN	
03	M		2598.25 5535.31	360	193.86	6010	NIEDER KOSTENZ	
02	S		2595.17 5536.67	407	196.30	6009	SCHWARZEN	
03	M		2595.18 5536.67	410	196.32	6009	SCHWARZEN	
02	S		2595.13 5536.71	411	196.33	6009	SCHWARZEN	

QUARRY-NO. 03 GERSFELD(SCHWARZER ACKER) SMAP 5525
ARC 03-118-F COMPILED BY HAENEL.STEIN

B	IN	AC	COORDINATES	ALT	X(KM)	SMAP	LOCALITY	G C
03	S		3485.45 5509.73	440	117.01	6218	LAUDENAU	
03	S		3476.13 5516.62	430	119.30	6218	NIEDER BEERBACH	
03	MZ		3456.85 5549.58	220	119.40	5916	HOFHEIM	
03	M		3453.25 5566.20	370	118.02	5716	NIEDERROD	
03	M		3449.78 5575.13	260	119.81	5615	SCHWICKERSHAUSEN	

QUARRY-NO. 04 GROSSENRITTE SMAP 4722
PROFILE 04-045 COMPILED BY BEHNKE.STEIN.SCHROEDER

B	IN	AC	COORDINATES	ALT	X(KM)	SMAP	LOCALITY	G C
01	M	--	3528.55 5682.63	290	4.72	4722	SCHWAELMERHAUS	
01	M	--	3541.04 5690.62	285	19.38	4623	LANDWEHRHAGEN	
01	M	--	3550.09 5699.29	210	31.86	4524	WIERSHAUSEN	
01	GO	--	3566.90 5712.71	272	53.37	4425	GOETTINGEN	
01	M	--	3568.48 5719.07	245	58.74	3425	REYERSHAUSEN	
01	SE	-F	3579.24 5724.21	150	70.21	4326	LINDAU/HARZ	
01	MM	--	3582.97 5732.62	250	78.58	4227	HELLEN BERG	
01	CL	--	3592.58 5742.20	560	92.14	4128	CLAUSTHAL	
01	CL	--	3592.23 5750.48	475	97.77	4128	GLOCKENBERG	
01	M	--	3604.42 5761.83	165	114.39	4029	BEUCHTE	
01	M	--	3613.36 5767.27	115	124.52	3929	HORNBURG	
01	PR	--	3628.52 5773.75	185	139.95	3931	SCHOEPPENSTEDT	

QUARRY-NO. 05 BIRKENAU SMAP 6418
PROFILE 05-040-02 COMPILED BY STROBACH

B	IN	AC	COORDINATES	ALT	X(KM)	SMAP	LOCALITY	G C
01	M		3483.98 5498.62	210	9.06	6318	LUETZEL RIMBACH	
01	M		3492.22 5504.70	350	18.85	6319	ROHRBACH	
01	M	AB	3494.81 5510.19	355	24.73	6219	HEMBACH	MM
01	M	C	3494.81 5510.19	355	24.73	6219	HEMBACH	MA
01	M		3503.00 5516.84	310	35.08	6220	RIMHORN	MM
01	M		3508.35 5523.22	175	43.41	6120	LAUTERHOF	
01	M		3519.20 5540.08	265	63.30	6021	STEIGER	
01	M		3525.92 5547.80	450	73.53	5922	HEINRICHSTHAL	
01	MZ	A	3529.37 5552.56	465	79.40	5822	MOSBORN	B4
01	MZ	B	3529.37 5552.56	465	79.40	5822	MOSBORN	B4X
01	MZ	C	3529.37 5552.56	465	79.40	5822	MOSBORN	B4Y
01	M		3535.58 5559.95	470	89.05	5822	PFAFFENHAUSEN	
01	MM		3548.20 5570.66	380	105.36	5724	ZEITLOFS	
01	KR		3554.42 5577.80	510	114.83	5624	VOLKERS	
01	M		3559.96 5588.95	460	126.93	5525	GICHENBACH	
01	M		3564.18 5593.76	640	133.30	5525	SOMMERBERG	

QUARRY-NO. 06 ADELEBSEN SMAP 4324
PROFILE 06-080 COMPILED BY HAENEL.STEIN

B	IN	AC	COORDINATES	ALT	X(KM)	SMAP	LOCALITY	G C
01	GO		3554.80 5720.90	220	3.51	4324	HETTENSEN	
02	CL		3555.38 5720.04	245	3.87	4324	ASCHE	
02	CL		3558.66 5719.37	180	7.11	4325	MARSTE	
01	GO		3562.95 5721.85	140	11.56	4325	LUETGENRODE	
03	GO		3566.95 5712.70	270	16.72	4425	INST.GOETTINGEN	
01	GO		3568.90 5727.95	175	19.27	4325	ST.MARGARETHEN	
02	GO		3572.93 5722.23	281	21.55	4326	GILLERSHEIM	
01	MZ		3577.48 5724.74	141	26.37	4326	LINDAU	
02	GO		3584.15 5727.95	207	33.68	4327	SCHWIEGERSHAUSEN	
01	SE		3590.53 5724.57	249	39.21	4327	ELBINGERODE	
02	CL		3596.90 5726.91	505	45.95	4328	HERZBERG	
02	CL		3602.20 5730.57	395	51.85	4228	OESTL.SIEBER	
01	MM		4404.59 5735.04	700	62.22	4229	BRAUNLAGE	

QUARRY-NO. 06 ADELEBSEN SMAP 4324
PROFILE 06-170-02 COMPILED BY HINZ.GRUBBE

B	IN	AC	COORDINATES	ALT	X(KM)	SMAP	LOCALITY	G C
03	M		3552.55 5717.30	200	2.30	4424	ADELEBSEN	
03	BO		3552.47 5714.00	310	5.47	4424	BARTERODE	
03	BO		3552.47 5714.00	310	5.47	4424	BARTERODE	Y
03	BO		3552.47 5714.00	310	5.47	4424	BARTERODE	X
03	BO		3553.47 5706.50	350	13.04	4524	DRANSFELD SE	
03	BO		3553.47 5706.50	350	13.04	4524	DRANSFELD SE	X
03	BO		3553.47 5706.50	350	13.04	4524	DRANSFELD SE	Y
03	GO		3566.95 5712.70	270	16.72	4425	GEO.INST.GOETT.	
02	M		3553.62 5702.35	345	17.27	4524	JUEHNDE	
03	MZ		3553.96 5698.93	342	20.61	4524	MEENSEN	

B	IN	AC	COORDINATES	ALT	X(KM)	SMAP	LOCALITY	G	C
03	GO	A	3555.00 5691.50	260	28.11	4624	ERMSCHWERD W		
03	GO	B	3555.00 5691.50	260	28.11	4624	ERMSCHWERD W		
02	M		3557.96 5684.89	300	35.20	4724	FAHRENBACH		
03	KR		3559.70 5675.70	740	44.45	4725	BRANSRODE		
05	GO		3560.52 5674.11	550	46.30	4825	SCHWALBENTHAL		
05	GO		3560.52 5674.11	550	46.30	4825	SCHWALBENTHAL		Y
05	GO		3560.52 5674.11	550	46.30	4825	SCHWALBENTHAL		X
05	GO		3561.33 5670.26	308	50.24	4825	HARMUTHSACHSEN	BS	
05	GO		3561.33 5670.26	308	50.24	4825	HARMUTHSACHSEN	BS	Y
05	GO		3561.33 5670.26	308	50.24	4825	HARMUTHSACHSEN	BS	X
03	CL		3563.43 5665.08	235	55.59	4825	KIRCHHOSBACH		
03	CL		3562.04 5656.52	321	63.74	4925	MOENCHHOSBACH		
02	F	A	3564.90 5648.90	355	71.85	5025	MACHTLOS		
02	F	B	3564.90 5648.90	355	71.85	5025	MACHTLOS		X
02	F	C	3564.90 5648.90	355	71.85	5025	MACHTLOS		Y
04	HH	A	3565.74 5643.22	463	77.61	5025	HERFA		
03	H		3563.92 5640.11	320	80.24	5125	HERFA		
03	MZ		3565.38 5630.72	347	89.75	5125	OBERBREITZBACH		
03	F		3562.56 5621.39	455	98.63	5225	GEHILFENBERG		
03	F		3562.56 5621.39	455	98.63	5225	GEHILFENBERG	B4	
02	M		3572.83 5614.90	545	106.74	5326	THEOBALDSHOF		
02	M		3572.83 5614.90	545	106.74	5326	THEOBALDSHOF		X
02	M		3571.89 5606.97	670	114.35	5426	SIMMERSBACH	MA	
02	M		3571.89 5606.97	670	114.35	5426	SIMMERSBACH	MA	
02	M		3571.89 5606.97	670	114.35	5426	SIMMERSBACH	MA	
03	CL		3573.75 5602.33	725	119.15	5426	BATTEN		
03	CL		3573.75 5602.33	725	119.15	5426	BATTEN	MP	
02	M	B	3573.65 5596.80	730	124.67	5426	ST.BARBARA	MA	
03	S	A	3577.24 5592.38	640	129.58	5526	OBER ELSBACH		
02	FF	A	3575.88 5588.12	565	133.61	5526	GINOLFS		X
03	S		3581.45 5582.19	370	140.42	5626	SCHOENAU		
02	MZ		3579.45 5578.42	406	143.81	5626	WINDSHAUSEN		
03	M	A	3580.32 5575.10	280	147.13	5626	UNTEREBERSBACH	MA	
03	M	B	3580.32 5575.10	280	147.13	5626	UNTEREBERSBACH		
03	M	A	3580.95 5572.32	350	149.98	5726	REICHENBACH	MA	
03	M	B	3581.21 5571.62	370	150.72	5726	REICHENBACH		
03	M	C	3581.21 5571.62	370	150.72	5726	REICHENBACH	MA	
03	M	A	3583.00 5569.35	400	153.30	5726	BURGHAUSEN	MA	
03	M	B	3583.00 5569.35	400	153.30	5726	BURGHAUSEN		
03	M	C	3583.05 5569.22	390	153.44	5726	BURGHAUSEN	MA	
02	S		3583.09 5569.19	381	153.58	5726	BURGHAUSEN		
02	S		3583.09 5569.19	381	153.58	5726	BURGHAUSEN		
02	S		3583.09 5569.19	381	153.58	5726	BURGHAUSEN		X
03	M	A	3583.57 5568.17	345	154.57	5727	MUENNERSTADT	MA	
03	M	B	3583.57 5568.17	345	154.57	5727	MUENNERSTADT		
03	M	C	3583.78 5567.79	360	154.99	5727	MUENNERSTADT	MA	
03	M	D	3584.73 5564.90	290	158.01	5727	POPPENLAUER		
03	M	A	3585.80 5563.08	280	160.02	5827	DIANENLUST	MA	
03	M	B	3585.96 5562.84	280	160.29	5827	DIANENLUST		
03	M	C	3584.22 5559.27	320	163.42	5827	RANNUNGEN	MA	
03	M	A	3584.30 5558.92	320	163.78	5827	RANNUNGEN	MA	
03	M	B	3584.30 5558.92	320	163.78	5827	RANNUNGEN		
03	M	A	3587.56 5556.97	395	166.36	5827	PFAENDHAUSEN	MA	
03	M	B	3588.14 5556.97	395	166.49	5827	PFAENDHAUSEN		
03	M	C	3588.14 5556.97	395	166.49	5827	PFAENDHAUSEN	MA	
03	M	A	3589.01 5555.07	325	168.53	5827	BROENNHOF	MA	
03	M	B	3589.00 5554.94	325	168.66	5827	BROENNHOF		
03	M	C	3589.07 5554.73	320	168.88	5827	BROENNHOF	MA	
03	M	A	3589.15 5553.46	300	170.14	5827	ZELL N	MA	
03	M	B	3589.19 5553.04	290	170.55	5827	ZELL N		
03	M	C	3589.29 5552.94	290	170.67	5827	ZELL N	MA	
03	M	B	3589.35 5552.71	285	170.91	5827	ZELL S	MA	
03	M	A	3589.51 5552.58	280	171.07	5827	ZELL S	MA	
02	S	A	3589.69 5552.58	288	171.22	5827	WEIPOLTSHAUSEN		
02	S	B	3589.69 5552.58	288	171.22	5827	WEIPOLTSHAUSEN		
02	S	C	3589.69 5552.58	288	171.22	5827	WEIPOLTSHAUSEN		
02	S	D	3589.69 5552.58	288	171.22	5827	WEIPOLTSHAUSEN		X
02	S	E	3589.69 5552.58	288	171.22	5827	WEIPOLTSHAUSEN		X
05	F		3594.17 5547.96	298	176.79	5926	HAUSEN		
05	F		3595.97 5546.32	355	178.82	5928	NAEHE FORST		
05	KA	A	3596.56 5541.45	240	183.69	5928	OBER EUERHEIM		
05	KA	B	3596.56 5541.45	240	183.69	5928	OBER EUERHEIM		
05	KA	C	3596.56 5541.45	240	183.69	5928	OBER EUERHEIM		
05	S		3598.90 5535.79	291	189.75	6028	EICHELBERG		
05	S	A	3602.90 5531.97	350	194.47	6028	MICHELAU		
05	S	B	3602.90 5531.97	350	194.47	6028	MICHELAU		
05	S	C	3602.90 5531.97	350	194.47	6028	MICHELAU		
05	S	D	3602.90 5531.97	350	194.47	6028	MICHELAU		
05	M	A	3591.07 5528.23	250	195.35	6127	OBERVOLKACH		
05	M	B	3590.92 5527.77	260	195.77	6127	OBERVOLKACH		
05	M	C	3590.92 5527.77	260	195.77	6127	OBERVOLKACH		
05	M	A	3593.88 5524.69	250	199.39	6127	EICHFELD		
05	M	B	3594.05 5524.39	250	199.72	6127	EICHFELD	MA	
05	M	C	3594.22 5524.18	255	199.96	6127	EICHFELD		
05	M	C	3594.88 5519.99	247	204.20	6127	REUPELSDORF		
05	M	A	3594.76 5519.22	247	204.92	6127	REUPELSDORF	MA	
05	M	B	3594.76 5519.22	247	204.92	6127	REUPELSDORF		
05	M	A	3600.07 5518.31	300	206.99	6228	UNTERSAMBACH		
05	M	B	3600.14 5517.73	300	207.57	6228	UNTERSAMBACH		
05	M	A	3597.94 5511.10	400	213.54	6228	WUESTENFELD.	MA	
05	M	B	3597.93 5510.85	400	213.78	6228	WUESTENFELDEN		
05	M	C	3598.01 5510.67	400	213.97	6228	WUESTENFELDEN		
05	M	C	3600.19 5506.65	350	218.37	6328	OBERAMBACH		
05	M	B	3600.25 5506.46	350	218.57	6328	OBERAMBACH		
05	M	A	3600.20 5506.28	350	218.74	6328	OBERAMBACH		
06	KA	1	3602.45 5500.61	320	224.59	6328	BIBART		
06	KA	2	3602.45 5500.61	320	224.59	6328	BIBART		
06	KA	3	3602.45 5500.61	320	224.59	6328	BIBART		
06	KA	1	3602.96 5495.47	350	229.72	6428	DEUTENHEIM		
06	KA	2	3602.96 5495.47	350	229.72	6428	DEUTENHEIM		
06	KA	3	3602.96 5495.47	350	229.72	6428	DEUTENHEIM		
06	KA	1	3601.62 5489.37	360	235.38	6428	ERKENBRECHTSHOFEN		
06	KA	2	3601.62 5489.37	300	235.38	6428	ERKENBRECHTSHOFEN		
06	KA	3	3601.62 5489.37	360	235.38	6428	ERKENBRECHTSHOFEN		
06	KA	1	4392.11 5488.18	415	238.28	6429	SCHL.HOHENECK EICH.		
06	KA	2	4392.11 5488.18	415	238.28	6429	SCHL.HOHENECK EICH.		
06	KA	3	4392.11 5488.18	415	238.28	6429	SCHL.HOHENECK EICH.		
06	S	1	3604.42 5480.34	412	244.80	6528	OBERNZENN		
06	S	2	3604.42 5480.34	412	244.80	6528	OBERNZENN		
06	S	3	3604.42 5480.34	412	244.80	6528	OBERNZENN		
06	S	1	3608.21 5475.00	505	250.84	6528	MITTELDACHSTETTEN		
06	S	2	3608.21 5475.00	505	250.84	6528	MITTELDACHSTETTEN		
06	S	3	3608.21 5475.00	505	250.84	6528	MITTELDACHSTETTEN		
06	S	1	3607.20 5469.87	470	255.62	6628	ZAILACH		
06	S	2	3607.00 5469.69	490	255.76	6628	ZAILACH		
06	S	3	3607.00 5469.69	490	255.76	6628	ZAILACH		
06	M	A	3612.18 5468.40	475	258.18	6629	STRUEHT		
06	M	B	3612.40 5468.04	485	258.58	6629	STRUEHT		
06	M	C	3612.46 5467.94	480	258.70	6629	STRUEHT		
06	M	1	3609.58 5463.12	485	262.73	6729	GEISENGRUND		
06	M	2	3609.64 5462.87	485	262.99	6729	GEISENGRUND		
06	M	3	3609.81 5462.43	485	263.46	6729	GEISENGRUND		
06	M	1	3614.03 5458.26	480	268.47	6729	ROES		
06	M	2	3614.00 5457.76	489	268.95	6729	ROES		
06	M	3	3613.91 5457.26	477	269.42	6729	ROES		
06	M	A	3613.08 5452.36	435	274.00	6829	LIEBERSDORF		
06	M	B	3613.09 5451.70	447	274.64	6829	LIEBERSDORF		
06	M	C	3613.00 5451.40	451	274.92	6829	LIEBERSDORF		
06	M	A	3616.10 5447.57	423	279.35	6829	WIESETHBRUCK		
06	M	B	3616.11 5447.26	425	279.65	6829	WIESETHBRUCK		
06	M	C	3616.13 5446.86	430	280.05	6829	WIESETHBRUCK		
06	F	2	3619.71 5437.67	450	289.85	6929	OBERMOEGERSHEIM		
06	F	2	3620.64 5430.58	435	296.96	7029	NAEHE OETTINGEN		

QUARRY-NO. 06 ADELEBSEN SMAP 5426
PROFILE 06-220 COMPILED BY STEIN, KAMINSKI

B	IN	AC	COORDINATES	ALT	X(KM)	SMAP	LOCALITY	G	C
11	H	1A	3532.37 5666.41	345	56.45	4822	HEILIGENBERG		
11	H	1A	3532.37 5666.41	345	56.45	4822	HEILIGENBERG		
09	KR	11	3414.56 5551.66	390	216.59	5912	KAUB		
09	KR	12	3414.62 5551.44	400	216.72	5912	KAUB		
09	KR	13	3414.50 5550.99	385	217.15	5912	KAUB		
09	HH	51	3397.66 5539.49	433	236.77	6011	MUTTERSCHIED S		
09	HH	52	3397.24 5539.48	410	237.05	6011	MUTTERSCHIED S		
09	HH	53	3397.03 5539.29	400	237.33	6011	MUTTERSCHIED S		
09	HH	61	3396.68 5538.83	420	237.91	6011	MUTTERSCHIED		
09	HH	62	3396.53 5538.67	424	238.13	6011	MUTTERSCHIED		
09	HH	63	3396.18 5538.38	413	238.57	6011	MUTTERSCHIED		
09	HH	21	3396.11 5538.24	412	238.73	6011	DREI FICHTEN		
09	HH	22	3395.84 5537.94	437	239.13	6011	DREI FICHTEN		
09	HH	23	3395.56 5537.73	425	239.47	6011	DREI FICHTEN		
09	HH	31	3394.88 5537.77	390	239.88	6011	EICHHOLZ		
09	HH	32	3394.62 5537.41	410	240.33	6011	EICHHOLZ		
09	HH	33	3394.29 5537.37	411	240.57	6011	EICHHOLZ		
09	HH	41	3394.07 5536.57	378	241.32	6011	HOLZBACH		
09	HH	42	3393.77 5536.31	392	241.71	6011	HOLZBACH		
09	HH	43	3393.33 5536.19	367	242.09	6011	HOLZBACH		
09	HH	11	3392.70 5536.09	352	242.58	6011	WIMMERSPACHER		
09	HH	12	3392.42 5535.94	370	242.88	6011	WIMMERSPACHER		
09	HH	13	3392.15 5535.71	371	243.23	6010	WIMMERSPACHER		
09	MS	11	3391.88 5535.35	365	243.68	6010	BELGWEILER		
09	MS	12	3391.54 5535.10	340	244.09	6010	BELGWEILER		
09	MS	13	3391.48 5534.84	360	244.32	6010	BELGWEILER		
09	MS	21	3391.16 5534.23	330	244.99	6010	RAVENGIERSBURG		
09	MS	22	3390.73 5534.40	340	245.15	6010	RAVENGIERSBURG		
09	MS	23	3390.51 5534.10	320	245.52	6010	RAVENGIERSBURG		
09	KI	41	3390.26 5533.56	305	246.09	6010	MAITZBORN E		
09	KI	42	2605.38 5533.22	341	246.51	6010	MAITZBORN E		
09	KI	43	2605.05 5532.95	365	246.92	6010	MAITZBORN E		
09	KI	21	2604.26 5533.15	363	247.26	6010	WOMRATH N		
09	KI	22	2604.03 5532.86	375	247.63	6010	WOMRATH N		
09	KI	23	2603.73 5532.66	382	247.98	6010	WOMRATH N		
09	KI	11	2603.61 5532.34	380	248.30	6010	WOMRATH W		
09	KI	12	2603.47 5531.98	370	248.67	6010	WOMRATH W		
09	KI	13	2603.30 5531.66	375	249.02	6010	WOMRATH W		
09	KI	31	2602.52 5531.68	406	249.50	6010	DICKENSCHIED		
09	KI	32	2602.25 5531.45	398	249.85	6010	DICKENSCHIED		
09	KI	33	2601.97 5531.20	395	250.22	6010	DICKENSCHIED		
09	H	31	2601.19 5531.07	430	250.81	6010	DICKENSCHIED2		
09	H	32	2600.96 5530.92	420	251.07	6010	DICKENSCHIED2		
09	H	33	2600.62 5530.70	400	251.46	6010	DICKENSCHIED2		
09	H	21	2600.78 5529.97	410	251.92	6110	OBERKIRN		
09	H	22	2600.45 5529.72	410	252.32	6110	OBERKIRN		
09	H	23	2600.06 5529.49	410	252.75	6110	OBERKIRN		
09	KA	22	2599.22 5529.15	390	253.54	6110	OBERKIRN2		
09	KA	21	2599.40 5529.33	390	253.29	6110	OBERKIRN2		
09	KA	23	2599.00 5528.87	330	253.90	6110	OBERKIRN2		
09	KA	11	2598.42 5528.84	370	254.29	6110	HUEHNERBERG		
09	KA	12	2598.14 5528.53	380	254.70	6110	HUEHNERBERG		
09	KA	13	2597.78 5528.35	365	255.07	6110	HUEHNERBERG		
09	S	41	2597.63 5527.84	370	255.56	6110	OBERKIRN3		
09	S	42	2597.36 5527.56	385	255.95	6110	OBERKIRN3		
09	S	43	2597.03 5527.30	390	256.36	6110	OBERKIRN3		
09	S	31	2596.35 5526.85	320	257.14	6110	RHAUNEN		
09	S	32	2596.09 5526.64	345	257.46	6110	RHAUNEN		
09	S	33	2595.87 5526.48	365	257.73	6110	RHAUNEN		
09	F	11	2580.44 5514.51	585	276.80	6208	HOXEL		
09	F	12	2580.44 5514.51	585	276.80	6208	HOXEL		X
09	F	13	2580.44 5514.51	585	276.80	6208	HOXEL		Y
09	F	31	2549.24 5489.79	450	315.92	6406	BRITTEN N		
09	F	32	2549.24 5489.79	450	315.92	6406	BRITTEN N		X
09	F	33	2549.24 5489.79	450	315.92	6406	BRITTEN N		Y
09	F	21	2564.80 5402.25	560	377.50	6307	HERMESKEIL W		
09	F	22	2564.80 5402.25	560	377.50	6307	HERMESKEIL W		X
09	F	23	2564.80 5402.25	560	377.50	6307	HERMESKEIL W		Y

QUARRY-NO. 06 ADELEBSEN SMAP 4324
TEST 06-260-V COMPILED BY WIEHLE,HEEP

B	IN	AC	COORDINATES	ALT	X(KM)	SMAP	LOCALITY	G	C
07	MS		3471.68 5710.39	270	80.42	4417	BUEREN	MA	
07	MS		3459.07 5708.68	355	93.14	4416	MENZEL	BS	
07	MS		3459.00 5708.70	360	93.21	4416	MENZEL		X
07	MS		3458.95 5708.72	362	93.25	4416	MENZEL	BS	
07	MS		3451.73 5708.95	290	100.41	4415	WALDHAUSEN	MA	
07	MS		3451.28 5708.94	295	100.85	4415	WALDHAUSEN	MA	
07	MS		3438.08 5707.92	255	114.09	4414	WIPPRINGSEN	MA	
07	MS		3437.85 5707.89	255	114.32	4414	WIPPRINGSEN	MA	

QUARRY-NO. 06 ADELEBSEN SMAP 4324
PROFILE 06-260 COMPILED BY HAENEL.PLAUMANN.STEIN

B	IN	AC	COORDINATES	ALT	X(KM)	SMAP	LOCALITY	G C
01	H		3544.35 5718.76	240	7.34	4323	ARENBORN	
01	H		3535.06 5716.86	270	16.79	4423	GOTTSBUEREN	
08	GO	21	3527.97 5715.05	142	24.03	4422	TRENDELBURG	
08	GO	22	3527.52 5715.04	173	24.48	4422	TRENDELBURG	
08	GO	23	3527.12 5715.00	190	24.88	4422	TRENDELBURG	
08	GO	11	3524.55 5714.55	275	27.49	4422	SIETEN	
08	GO	12	3524.13 5714.50	290	27.91	4422	SIETEN	
08	GO	13	3523.58 5714.49	300	28.45	4422	SIETEN	
01	H		3520.42 5715.44	242	31.49	4421	BUEHNE	
01	H		3508.20 5713.61	215	43.84	4420	IKENHAUSEN	
02	H		3506.51 5713.10	250	45.49	4420	IKENHAUSEN	
08	KI	31	3500.69 5710.89	303	51.62	4420	SCHERFEDE	
08	KI	32	3500.57 5711.00	298	51.72	4420	SCHERFEDE	
08	KI	33	3500.41 5711.40	298	51.81	4420	SCHERFEDE	
04	KR		3496.60 5710.06	364	55.75	4419	SCHERFELDER H.	
01	H		3486.60 5712.81	375	65.39	4418	ELISENHOF	
02	H		3475.22 5708.82	240	77.07	4417	LEIBERG	
08	KI	41	3469.00 5705.15	390	83.82	4517	RINGELSTEIN.HALD	
08	KI	42	3468.76 5705.'03	405	84.08	4517	RINGELSTEIN.HALD	
08	KI	43	3468.64 5704.97	412	84.21	4517	RINGELSTEIN.HALD	
01	H		3465.11 5706.19	365	87.55	4516	JHS.MIEBACH	
08	KI	11	3461.38 5702.46	315	91.80	4516	KALLENHARDT	
08	KI	12	3461.06 5702.56	310	92.10	4516	KALLENHARDT	
08	KI	13	3460.81 5702.82	320	92.30	4516	KALLENHARDT	
02	H		3451.06 5700.39	390	102.29	4515	HIRSCHBERG	
01	H		3439.84 5704.01	240	112.87	4514	HILHELMSRUH	
04	BO		3436.36 5707.73	270	115.78	4414	HESTRICH	
08	KI	21	3435.51 5699.26	300	117.83	4514	NIEDEREIMER	
08	KI	22	3435.05 5699.30	260	118.28	4514	NIEDEREIMER	
08	KI	23	3434.85 5699.42	250	118.45	4514	NIEDEREIMER	
04	BO		3429.32 5705.88	258	122.98	4513	HOEINGEN	
01	S		3425.51 5702.24	250	127.31	4513	DREIHAUSEN	
02	K		3425.45 5701.94	255	127.32	4513	DREIHAUSEN	
04	BO		3417.20 5702.52	187	135.41	4512	MENDEN E	
01	BO		3412.33 5699.23	227	140.78	4512	GAXBERG	
08	H	51	3409.83 5690.99	380	144.60	4612	KESBERN	
08	H	52	3409.59 5691.06	380	144.82	4612	KESBERN	
08	H	53	3409.14 5691.12	365	145.25	4612	KESBERN	
08	H	31	3403.31 5696.89	215	149.99	4611	LETMATHE	
08	H	32	3402.90 5696.68	230	150.43	4611	LETMATHE	
02	BO		2605.00 5701.15	200	156.23	4511	HOLZEN	
08	H	21	3392.20 5688.75	350	162.33	4610	HAGEN	
08	H	22	3391.79 5688.62	350	162.76	4610	HAGEN	
08	H	23	3391.45 5688.38	350	163.14	4610	HAGEN	
02	HH		2594.13 5695.06	185	167.98	4610	HENGERN	
08	MS	1	2585.13 5688.66	250	178.10	4609	HERZKAMP	
08	MS	2	2584.86 5688.56	255	178.39	4609	HERZKAMP	
08	MS	3	2584.43 5688.48	250	178.83	4609	HERZKAMP	
08	MS	1	2579.30 5688.53	168	183.83	4608	LANGENBERG	
08	MS	2	2579.26 5688.43	145	183.89	4608	LANGENBERG	
08	MS	3	2579.13 5688.50	148	184.01	4608	LANGENBERG	
02	HH		2573.39 5692.23	130	188.89	4608	VELBERT	
04	H		2517.58 5679.78	50	246.03	4703	DILKRATH	
04	H		2506.42 5677.63	45	257.39	4702	SHALMEN	

QUARRY-NO. 06 ADELEBSEN SMAP 4324
TEST 06-280-V COMPILED BY HEEP.DEGUTSCH

B	IN	AC	COORDINATES	ALT	X(KM)	SMAP	LOCALITY	G C
07	MS		3473.09 5733.02	90	79.65	4217	SO. DELBRUECK	MA
07	MS		3472.43 5733.20	90	80.33	4217	SO. DELBRUECK	MA
07	MS	11	3460.89 5735.85	80	92.15	4216	O. MASTHOLTE	BS
07	MS	12	3460.71 5735.80	80	92.32	4216	O. MASTHOLTE	X
07	MS	13	3460.53 5735.78	80	92.50	4216	O. MASTHOLTE	BS
07	MS	21	3456.12 5735.34	75	96.77	4216	H. MASTHOLTE	MA
07	MS	22	3455.88 5735.37	75	97.01	4216	H. MASTHOLTE	X
07	MS	23	3455.64 5735.44	75	97.26	4216	H. MASTHOLTE	MA
07	BO	21	3450.78 5736.92	78	102.30	4215	ALLERBECK	BS
07	BO	22	3450.41 5737.06	80	102.68	4215	ALLERBECK	X
07	BO	23	3450.00 5737.21	85	103.11	4215	ALLERBECK	BS
07	BO	11	3439.43 5739.47	120	113.91	4214	VELLERN E	BS
07	BO	12	3439.06 5739.60	115	114.30	4214	VELLERN E	X
07	BO	13	3438.65 5739.75	129	114.73	4214	VELLERN E	BS

QUARRY-NO. 06 ADELEBSEN SMAP 4324
PROFILE 06-300 COMPILED BY DEGUTSCH.THYSSEN.HEEP

B	IN	AC	COORDINATES	ALT	X(KM)	SMAP	LOCALITY	G C
05	CL		3546.99 5722.45	185	5.39	4324	SCHONINGEN	
05	CL		3542.35 5725.65	245	11.02	4323	USLAR	
04	M	3	3539.23 5727.33	315	14.58	4323	SCHOENHAGEN	
04	M	2	3538.75 5727.53	315	15.09	4323	SCHOENHAGEN	
04	M	1	3538.35 5727.40	315	15.36	4323	SCHOENHAGEN	
05	HH	4	3524.23 5736.61	250	32.20	4222	SH.HOEXTER	
05	HH	2	3524.22 5736.61	250	32.20	4222	SH.HOEXTER	
04	GO	3	3518.30 5739.20	300	38.63	4221	ALTENBERGEN	
04	GO	1	3514.30 5744.00	220	44.57	4121	NE.BREDENBORN	BS
06	MS		3515.15 5745.84	235	45.13	4121	BORN	MA
06	MS		3514.90 5746.00	235	45.42	4121	BORN	MA
06	MS		3512.52 5747.59	173	48.28	4121	RUENSIEK	BS
06	MS		3511.24 5748.80	175	50.03	4120	LOTHE	X
05	KI	2	3509.35 5750.90	140	52.56	4120	STEINHEIM	BS
05	MS		3509.99 5752.36	160	52.94	4020	S.SCHIEDER	MA
05	MS		3509.72 5752.47	160	53.22	4020	S.SCHIEDER	MA
04	HH		3501.43 5752.94	200	60.23	4020	REELKIRCHEN S	
05	KR	1	3503.06 5755.49	172	60.34	4020	HOENTRUP	
05	KR	2	3502.89 5755.55	172	60.52	4020	HOENTRUP	
05	KR	3	3502.80 5755.60	172	60.62	4020	HOENTRUP	
04	MS		3498.39 5754.78	220	63.78	4019	E.DETMOLD	MA
05	KI	2	3496.68 5759.67	132	67.96	4019	BIESEN	BS
06	MS		3494.52 5757.70	245	68.84	4019	BROCKHAUSEN	BS
06	MS		3494.44 5757.72	240	68.92	4019	BROCKHAUSEN	MA
06	MS		3494.39 5757.76	240	68.98	4019	BROCKHAUSEN	BS
05	MS		3491.40 5760.93	206	73.00	4019	BENTRUP	MA
04	KI	3	3485.75 5766.30	80	80.73	3918	S.HOELSEN	BS
06	MS		3485.06 5765.98	99	81.32	3918	HOELSERHEIDE	BS
05	H	1	3482.72 5767.07	98	83.63	3918	HOLZHAUSEN	
04	MS		3478.90 5768.43	110	87.57	3918	LOCKHAUSEN	MA
06	BO	13	3479.96 5770.25	88	87.95	3918	KRIEGERHEIDE	BS
04	MS		3478.35 5768.99	110	88.34	3918	LOCKHAUSEN	
06	BO	13	3475.89 5773.10	88	92.92	3917	HILLEHALSEN	BS
04	CL	1'	3474.50 5775.10	80	95.00	3817	ELIM DIEBROCK	
05	H	2	3471.94 5777.33	100	98.36	3817	ENGER	
04	S	2	3468.55 5779.40	115	102.34	3817	ENGER	
05	HH	1	3464.64 5783.13	82	107.67	3816	HOYEL	
04	CL	1	3461.88 5784.27	96	110.60	3816	NH.BENNIEN	
04	S	1	3457.62 5787.96	124	116.21	3716	BARKHAUSEN	
04	H	3	3453.23 5793.24	215	122.88	3715	GRAMBERGEN	
05	MS		3445.56 5797.66	135	131.65	3615	KLEIN HELLERN	MA
05	MS		3444.68 5798.16	135	132.66	3615	KLEIN HELLERN	MA
05	HH	3	3437.89 5800.74	93	139.67	3614	ICKER RULLE	
04	F	3	3436.38 5805.53	85	143.73	3614	ENGTER	
05	MS		3432.71 5805.32	110	146.55	3614	N.HALLENHORST	BS
05	MS		3427.19 5811.49	90	154.64	3513	HESTERHAUSEN	MA
04	F	2	3423.69 5814.22	75	159.11	3513	UEFFELN	
04	MS		3422.20 5816.23	78	161.50	3513	HESTERHOLTE	BS
04	MS		3419.12 5818.66	94	165.42	3512	SCHHAGSTORF	MA
05	MS		3405.40 5823.98	37	179.61	3411	H.LENGERICH	BS

QUARRY-NO. 06 ADELEBSEN SMAP 4324
PROFILE 06-350 COMPILED BY HINZ.KAMINSKI

B	IN	AC	COORDINATES	ALT	X(KM)	SMAP	LOCALITY	G C
08	CL	31	3550.65 5726.16	240	6.62	4324	VOLPRIEHAUSEN	
08	CL	32	3550.56 5726.48	240	6.95	*4324	VOLPRIEHAUSEN	
08	CL	33	3550.51 5726.84	260	7.32	4324	VOLPRIEHAUSEN	
10	GO	11	3550.51 5729.18	290	9.70	4324	HAJE N	
10	GO	12	3550.57 5729.63	325	10.14	4224	HAJE N	
10	GO	13	3550.49 5730.04	325	10.55	4224	HAJE N	
07	GO	11	3549.11 5734.12	362	14.83	4224	GRASBORN E	
07	GO	12	3549.06 5734.51	357	15.22	4224	GRASBORN E	
07	GO	13	3549.02 5734.93	355	15.64	4224	GRASBORN E	
08	CL	21	3548.60 5736.13	300	16.79	*4224	HILHARTSHAUSEN	
08	CL	22	3548.64 5736.51	273	17.16	4224	HILHARTSHAUSEN	
07	CL	1	3548.14 5736.68	254	17.52	4224	RELLIEHAUSEN	
08	CL	23	3548.42 5736.88	248	17.56	4224	HILHARTSHAUSEN	
07	CL	2	3548.03 5737.09	229	17.95	4224	RELLIEHAUSEN	
07	CL	3	3547.88 5737.48	207	18.36	*4224	RELLIEHAUSEN	
07	CL	1	3546.89 5743.53	315	24.48	4124	MACKENSEN	
07	CL	2	3546.60 5743.85	325	24.85	*4124	MACKENSEN	
07	CL	3	3546.70 5744.27	335	25.25	4124	MACKENSEN	
07	CL	11	3546.56 5748.11	308	29.05	4124	HANGELNSTEDT H	
07	CL	12	3546.48 5748.58	283	29.52	4124	HANGELNSTEDT H	
07	CL	13	3546.41 5749.02	275	29.97	4124	HANGELNSTEDT H	
08	H	41	3546.35 5750.50	270	31.34	*4124	LENNE	
08	H	42	3546.33 5750.91	270	31.75	4124	LENNE	
08	H	43	3546.18 5751.30	280	32.16	4124	LENNE	
10	GO	21	3543.58 5753.65	320	35.03	4023	ESCHERSHAUSEN	
10	GO	22	3543.38 5754.02	280	35.44	4023	ESCHERSHAUSEN	
10	GO	23	3543.28 5754.38	250	35.81	4023	ESCHERSHAUSEN	
06	CL	1	3544.90 5758.18	310	39.38	4023	LUEERDISSEN	X
06	CL	2	3544.90 5758.18	310	39.38	4023	LUEERDISSEN	X
06	CL	3	3544.90 5758.18	310	39.38	4023	LUEERDISSEN	Y
06	CL	1	3544.29 5761.75	260	43.00	*4023	FOELZIEHAUSEN	
06	CL	2	3544.22 5761.92	275	43.18	4023	FOELZIEHAUSEN	
06	CL	3	3544.22 5761.92	275	43.18	4023	FOELZIEHAUSEN	X
07	F	11	3541.31 5764.56	185	46.22	*3923	HALLENSEN H	
07	F	12	3541.22 5764.98	180	46.64	3923	HAL:.ENSEN H	
07	F	13	3541.02 5765.13	160	46.84	3923	HALLENSEN H	
06	H	1	3542.67 5768.89	320	50.31	3923	SALZHEMMENDORF	
08	H	11	3542.29 5770.81	150	52.04	3923	HEMMENDORF	
08	H	12	3542.17 5771.21	130	52.46	*3923	HEMMENDORF	
08	H	13	3542.06 5771.62	120	52.88	3923	HEMMENDORF	
07	F	21	3542.38 5771.64	117	52.95	3923	HEMMENDORF E	
07	F	22	3542.33 5772.08	107	53.39	3923	HEMMENDORF E	
07	F	23	3542.29 5772.35	103	53.66	3923	HEMMENDORF E	
06	GO	1	3541.48 5775.89	245	57.41	3823	OSTERHALD	
06	GO	2	3541.48 5775.89	245	57.41	3823	OSTERHALD	X
06	GO	3	3541.48 5775.89	245	57.41	3823	OSTERHALD	Y
10	CL	11	3541.40 5777.48	355	58.82	3823	COPPENBRUEGGE	
10	CL	12	3541.31 5777.83	343	59.18	3823	COPPENBRUEGGE	
10	CL	13	3541.22 5778.20	304	59.56	3823	COPPENBRUEGGE	
06	GO	1	3540.37 5779.42	245	61.09	3823	HOHEN BERG	
06	GO	2	3540.37 5779.42	245	61.09	3823	HOHEN BERG	X
06	GO	3	3540.37 5779.42	245	61.09	3823	HOHEN BERG	Y
06	KI	1	3538.60 5783.60	270	65.53	3823	SAUPARK	BSY
06	KI	2	3538.60 5783.60	270	65.53	3823	SAUPARK	BS
06	KI	3	3538.60 5783.60	270	65.53	3823	SAUPARK	BSX
08	HH	41	3539.97 5787.58	130	68.96	3723	SPRINGE NE	
08	HH	42	3540.02 5788.05	140	69.42	*3723	SPRINGE NE	
08	HH	43	3539.76 5788.30	150	69.71	3723	SPRINGE NE	
05	H		3538.45 5790.18	260	71.85	3723	HENNIGSEN	X
05	H		3538.45 5790.18	260	71.85	3723	HENNIGSEN	
05	H		3538.45 5790.18	260	71.85	3723	HENNIGSEN	Y
10	CL	41	3538.49 5790.22	257	71.88	3723	HENNIGSEN	
10	CL	42	3538.47 5790.63	198	72.29	3723	HENNIGSEN	
10	CL	43	3538.45 5791.05	175	72.70	3723	HENNIGSEN	
07	F	33	3537.71 5794.53	80	76.30	3723	REDDERSE	
07	F	32	3537.64 5795.08	87	76.86	*3723	REDDERSE	
07	F	31	3537.76 5795.60	96	77.34	3723	REDDERSE	
07	H	41	3539.57 5796.68	90	78.11	3623	GEHRDEN	
07	H	43	3539.58 5797.59	95	79.01	*3623	GEHRDEN	
07	H	42	3539.58 5797.14	90	78.56	3623	GEHRDEN	
08	HH	32	3536.95 5800.12	58	81.84	3623	GOEXE	
08	HH	31	3537.07 5799.78	58	81.48	*3623	GOEXE	
08	HH	33	3536.94 5800.52	63	82.23	3623	GOEXE	
07	H	11	3536.93 5803.92	56	85.68	3623	ALMHORST	
07	H	12	3536.93 5804.36	55	86.12	3623	ALMHORST	
07	H	13	3536.78 5804.84	56	86.61	3623	ALMHORST	
10	H	31	3536.00 5806.58	50	88.42	3623	DEDENSEN	

B	IN	AC	COORDINATES	ALT	X(KM)	SMAP	LOCALITY	G C
10	M	32	3536.01 5807.02	50	88.85	3623	DEDENSEN	
10	M	33	3536.08 5807.48	50	89.29	3523	DEDENSEN	
07	M	51	3534.93 5811.50	48	93.49	3523	RICKLINGEN	
07	M	52	3534.99 5811.94	48	93.92	3523	RICKLINGEN	
07	M	53	3534.96 5812.38	48	94.36	3523	RICKLINGEN	
07	M	31	3534.33 5814.61	45	96.66	3523	FRIELINGEN	
07	M	32	3534.39 5815.05	45	97.08	*3523	FRIELINGEN	
07	M	33	3534.52 5815.43	45	97.43	3523	FRIELINGEN	
07	M	21	3534.64 5817.53	40	99.48	3523	OTTERNHAGEN	
07	M	22	3534.45 5817.93	40	99.91	3523	OTTERNHAGEN	
07	M	23	3534.29 5818.32	40	100.32	3523	OTTERNHAGEN	
06	CL	1	3533.60 5820.15	40	102.37	3422	NEUSTADT A RBG.	
06	CL	2	3533.48 5820.46	40	102.69	*3422	NEUSTADT A RBG.	
06	CL	3	3533.36 5820.80	40	103.05	3422	NEUSTADT A RBG.	
06	M	1	3533.73 5823.22	34	105.37	*3422	MARIENSEE	
06	M	2	3533.69 5824.11	34	106.25	3422	MARIENSEE	
06	M	4	3533.60 5824.71	34	106.86	*3422	MARIENSEE	
10	M	41	3532.47 5826.73	40	108.88	3422	WULFELADE	
10	M	42	3532.63 5827.18	40	109.29	3422	WULFELADE	
10	M	43	3532.72 5827.50	40	109.59	3422	WULFELADE	
05	M	A	3530.50 5827.34	49	109.84	3422	DUDENSEN	
05	M	B	3530.47 5827.62	54	110.12	3422	DUDENSEN	
05	M	C	3530.46 5827.84	51	110.34	3422	DUDENSEN	
05	M	D	3530.46 5828.13	52	110.62	3422	DUDENSEN	
05	M	E	3530.50 5828.35	53	110.83	3422	DUDENSEN	
05	M	F	3530.60 5828.63	52	111.09	3422	DUDENSEN	
06	M		3532.18 5828.92	60	111.25	3422	BUEREN	X
06	M		3532.18 5828.92	60	111.25	3422	BUEREN	
06	M		3532.18 5828.92	60	111.25	3422	BUEREN	Y
06	KI	21	3533.61 5830.10	55	112.17	*3322	BEVENSEN	
06	KI	22	3533.48 5831.69	40	113.76	*3322	BEVENSEN	
06	MM	31	3532.94 5833.68	25	115.81	3322	RODEWALD OB.B.	
06	MM	32	3532.16 5834.14	25	116.39	3322	RODEWALD OB.B.	
06	MM	33	3531.64 5834.86	25	117.19	3322	RODEWALD OB.B.	
06	MM	23	3530.73 5836.50	25	118.97	3322	RODEWALD MITTL.B.	
06	MM	22	3530.22 5837.29	25	119.83	3322	RODEWALD MITTL.B.	
06	MM	21	3530.32 5838.23	25	120.74	3322	RODEWALD MITTL.B.	
06	MM	41	3531.41 5839.83	26	122.13	3322	RODEWALD UNT.B.	
06	MM	42	3531.38 5840.67	26	122.96	*3322	RODEWALD UNT.B.	
06	MM	43	3531.28 5841.45	25	123.75	3222	RODEWALD UNT.B.	
06	MM	12	3531.07 5842.88	23	125.20	3222	LICHTENHORST	
06	MM	13	3531.07 5842.88	23	125.20	3222	LICHTENHORST	
06	MM	11	3530.78 5843.68	23	126.03	3222	LICHTENHORST	
08	MM	11	3529.82 5843.87	23	126.16	3222	LICHTENHORST	
08	MM	12	3529.71 5844.26	23	126.56	3222	LICHTENHORST	
08	MM	13	3529.77 5844.71	23	127.00	3222	LICHTENHORST	
08	MM	21	3529.52 5846.90	20	129.20	*3222	FRANKENFELD	
08	MM	22	3529.65 5847.36	20	129.62	3222	FRANKENFELD	
08	MM	23	3529.74 5847.64	20	129.89	3222	FRANKENFELD	
07	MM	51	3529.40 5851.82	21	134.17	3222	ALTENWAHLINGEN	
07	MM	52	3529.48 5852.26	25	134.59	3122	ALTENWAHLINGEN	
07	MM	53	3529.38 5852.67	25	135.01	3122	ALTENWAHLINGEN	
07	MM	31	3529.58 5853.33	21	135.63	*3122	GROSSEILSTORF SW	
07	MM	32	3529.59 5853.73	22	136.02	3122	GROSSEILSTORF SW	
07	MM	33	3529.45 5854.11	25	136.42	3122	GROSSEILSTORF SW	
07	MM	21	3529.17 5854.83	27	137.17	3122	GROSSEILSTORF NW	
07	MM	22	3529.13 5855.17	28	137.51	3122	GROSSEILSTORF NW	
07	MM	23	3529.05 5855.62	26	137.97	3122	GROSSEILSTORF NW	
07	MM	11	3529.00 5856.16	27	138.52	3122	SUEDKAMPEN S	
07	MM	12	3528.96 5856.58	30	138.93	3122	SUEDKAMPEN S	
07	MM	13	3528.71 5856.93	27	139.32	3122	SUEDKAMPEN S	
07	MM	41	3528.71 5857.47	32	139.85	3122	SUEDKAMPEN	
07	MM	42	3528.78 5857.92	32	140.29	3122	SUEDKAMPEN	
07	MM	43	3528.81 5858.35	34	140.71	3122	SUEDKAMPEN	
07	MM	61	3528.53 5858.85	36	141.24	3122	NORDKAMPEN	
07	MM	62	3528.50 5859.34	38	141.73	3122	NORDKAMPEN	
07	MM	63	3528.52 5859.74	37	142.12	3122	NORDKAMPEN	
07	KI	41	3528.22 5860.52	35	142.94	*3122	KRONSNEST	
07	KI	42	3528.12 5860.92	25	143.35	3122	KRONSNEST	
07	KI	43	3528.16 5861.32	25	143.74	3122	KRONSNEST	
07	KI	31	3527.83 5861.65	30	144.12	3122	IHLDEN	
07	KI	32	3527.62 5861.98	35	144.48	3122	IHLDEN	
07	KI	33	3527.50 5862.37	33	144.89	3122	IHLDEN	
07	KI	21	3527.32 5862.95	33	145.49	3122	VERDENER MOOR S	
07	KI	22	3527.11 5863.29	34	145.86	3022	VERDENER MOOR S	
07	KI	23	3526.98 5863.69	36	146.27	3022	VERDENER MOOR S	
07	KI	11	3526.78 5864.24	33	146.85	3022	VERDENER MOOR	
07	KI	12	3526.52 5864.47	35	147.12	*3022	VERDENER MOOR	
07	KI	13	3526.25 5864.93	38	147.62	3022	VERDENER MOOR	

QUARRY-NO. 07 GERSFLED/WALLENBERG SMAP 5525
PROFILE 07-010 COMPILED BY VETTER

B	IN	AC	COORDINATES	ALT	X(KM)	SMAP	LOCALITY	G C
01	M		3568.62 5614.00	370	24.56	5325	NEUSWARTS	BA
01	M		3568.62 5614.00	370	24.56	5325	NEUSWARTS	BAX
01	M		3568.62 5614.00	370	24.56	5325	NEUSWARTS	BAY
01	MZ		3565.22 5622.65	300	32.60	5225	GRUESSELBACH	
01	M		3572.48 5640.69	230	51.47	5126	LEIMBACH	MM
01	M		3575.38 5654.91	270	65.98	4926	UNHAUSEN	
01	M		3572.62 5661.43	410	71.96	4926	GRANDENBORN	MM
01	M		3572.56 5669.78	260	80.24	4826	BLAUE KUPPE	
01	M		3575.61 5676.96	360	87.76	4726	KELLA	
01	GO		3575.05 5700.15	350	110.69	4526	VOGLESANG	MM
01	M		3573.41 5703.42	345	113.78	4526	SEMIKERODE	MM
01	GO		3578.75 5709.70	260	120.62	4426	RIEKENRODE	MM
01	S		3580.03 5719.48	218	130.48	4326	BODENSEE/HARZ	
01	S		3578.56 5723.74	189	134.53	4326	LINDAU/HARZ	
01	MM		3582.97 5732.62	250	143.90	4227	WELLENBERG	
01	M		3580.93 5743.10	230	154.03	4127	STAUFFENBERG	
01	M		3590.40 5753.98	280	166.15	4027	WOLFSHAGEN	
01	PR		3586.55 5777.37		188.73	3827	OSTERLINDE	

QUARRY-NO. 08 KIRCHHEIMBOLANDEN SMAP 6313
PROFILE 08-000 COMPILED BY STEIN.HAENEL.PLAUMANN

B	IN	AC	COORDINATES	ALT	X(KM)	SMAP	LOCALITY	G C
01	CL		3425.40 5514.10	220	9.37	6213	WENDELSHEIM	
01	M		3426.17 5519.61	145	14.92	6113	GUMBSHEIM	
01	S		3428.69 5527.88	247	23.45	6114	SPRENDLINGEN	
01	MZ		3427.34 5534.14	195	29.50	6013	OCKENHEIM	
01	M		3424.56 5541.08	230	36.33	5913	GEISENHEIM	
01	M		3425.12 5544.70	360	39.95	5913	STEPHANSHAUSEN	
01	M		3424.50 5549.69	410	44.94	5913	SEIMESDELL	
01	KI		3424.61 5554.58	438	49.83	5813	DICKSCHIED	
01	CL		3423.15 5563.53	397	58.80	5713	MARTENROTH	
01	CL		3424.68 5576.90	257	72.15	5613	STEINSBERG	
01	SE	B	3424.36 5584.06	350	79.31	5613	HIRSCHBERG	
01	MZ		3423.40 5597.10	315	92.36	5413	WERSCHBACH	
01	KR		3424.32 5604.42	480	99.67	5413	HINTERMUEHLEN	
01	M		3424.19 5613.83	510	109.08	5313	UNNAU	
01	M		3418.65 5622.54	370	117.95	5213	BIESENSTUECK	
01	M		3423.25 5629.40	440	124.66	5213	DRUIDENSTEIN	
01	M		3417.50 5637.61	305	133.06	5112	HINHAUSEN	
01	MM		3418.19 5645.74	380	141.15	5013	ROEMERSHAGEN	
01	MM		3419.70 5658.46	460	153.79	4913	RHODE	
01	BO		3419.02 5668.28	490	163.63	4813	WELTRINGHAUSEN	

QUARRY-NO. 08 KIRCHHEIMBOLANDEN SMAP 6313
PROFILE 08-195 COMPILED BY MUELLER. PETERSCHMITT

B	IN	AC	COORDINATES	ALT	X(KM)	SMAP	LOCALITY	G C
02	F	31	3423.74 5498.27	425	6.51	6313	JAKOBSWEILER	
02	F	32	3423.73 5497.85	415	6.92	6313	JAKOBSWEILER	
02	F	33	3423.55 5497.54	460	7.26	6313	JAKOBSWEILER	
02	F	11	3422.12 5493.57	290	11.45	6413	BOERRSTADT	
02	F	12	3421.96 5493.29	310	11.76	6413	BOERRSTADT	
02	F	13	3421.76 5492.96	330	12.13	6413	BOERRSTADT	
02	KA	31	3416.76 5482.65	280	23.49	6513	NEUKIRCHEN	
02	KA	32	3416.84 5482.62	280	23.49	6513	NEUKIRCHEN	
02	KA	33	3416.88 5482.33	280	23.75	6513	NEUKIRCHEN	
02	KA	52	3412.39 5472.57	410	34.46	6612	MOELSCHBACH	
02	KA	61	3409.88 5467.69	425	39.93	6612	TRIPPSTADT	
02	KA	62	3409.63 5467.28	420	40.40	6612	TRIPPSTADT	
02	KA	63	3409.27 5466.87	435	40.92	6612	TRIPPSTADT	
02	KA	71	3407.77 5463.97	355	44.17	6612	WELTERSBERG	
02	KA	72	3407.64 5463.68	350	44.49	6612	WELTERSBERG	
02	KA	73	3407.55 5463.41	340	44.77	6612	WELTERSBERG	
02	KA	21	3406.45 5459.38	370	48.92	6712	CLAUSEN	
02	KA	22	3406.05 5459.00	340	49.42	6712	CLAUSEN	
02	KA	23	3405.68 5458.77	330	49.77	6712	CLAUSEN	
02	KA	11	3403.48 5455.52	280	53.63	6712	IMSBACHER MUEHLE	
02	KA	12	3403.36 5455.14	300	54.02	6712	IMSBACHER MUEHLE	
02	KA	13	3403.14 5454.82	330	54.40	6712	IMSBACHER MUEHLE	
02	S	21	3401.68 5451.22	330	58.29	6811	RUHBANK	
02	S	22	3401.39 5450.92	350	58.68	6811	RUHBANK	
02	S	23	3401.14 5450.87	350	58.83	6811	RUHBANK	
02	S	31	3398.49 5446.63	420	63.78	6811	GLASHUETTE	
02	S	32	3398.08 5446.55	450	64.02	6811	GLASHUETTE	
02	S	33	3397.93 5446.30	430	64.31	6811	GLASHUETTE	
02	S	41	3396.70 5441.89	330	68.84	6811	EPPENBRUNN	
02	S	43	3396.84 5442.11	340	68.58	6811	EPPENBRUNN	

QUARRY-NO. 09 BOEHMISCHBRUCK SMAP 6440
PROFILE 09-135 COMPILED BY GIESE

B	IN	AC	COORDINATES	ALT	X(KM)	SMAP	LOCALITY	G C
08	M	A	4573.83 5446.89	740	66.19	6843	EISMANNSBERG	
08	M	B	4573.93 5446.67	720	66.41	6843	EISMANNSBERG	
08	M	A	4573.60 5446.22	740	66.49	6844	ARNBRUCK/NE	
08	M	C	4574.21 5446.53	720	66.71	6843	EISMANNSBERG	
08	M	B	4573.76 5445.94	725	66.80	6844	ARNBRUCK/NE	
08	M	C	4574.02 5445.78	735	67.09	6844	ARNBRUCK/NE	
08	M	A	4574.75 5443.90	630	68.93	6844	GASHUETTE	
08	M	B	4574.99 5443.64	690	69.28	6844	GASHUETTE	
08	M	1	4576.44 5441.90	680	71.54	6844	OBERRIED	
08	M	2	4576.61 5441.65	680	71.84	6844	OBERRIED	
08	M	3	4576.86 5441.49	650	72.13	6844	OBERRIED	
08	M	1	4578.05 5440.81	690	73.45	6944	RIEDLBURG	
08	M	2	4578.30 5440.64	780	73.75	6944	RIEDLBURG	
08	M	3	4578.62 5440.48	790	74.09	6944	RIEDLBURG	
08	M	A	4580.03 5439.10	860	76.06	6944	BODENMAIS/N	
08	M	B	4580.01 5439.00	840	76.12	6944	BODENMAIS/N	
08	M	C	4580.13 5438.65	840	76.45	6944	BODENMAIS/N	
08	M	1	4582.33 5437.74	920	78.66	6944	BODENMAIS/E	
08	M	2	4582.66 5437.48	950	79.07	6944	BODENMAIS/E	
08	M	3	4583.19 5436.97	910	79.81	6944	BODENMAIS/E	
08	M	A	4583.92 5436.82	890	80.44	6944	SCHUELLERMN.	
08	M	B	4584.14 5436.58	910	80.77	6944	SCHUELLERMN.	
08	M	C	4584.48 5436.63	910	80.98	6944	SCHUELLERMN.	
08	M	A	4586.31 5435.42	875	83.14	6945	RABENSTEIN	
08	M	B	4586.79 5435.37	840	83.52	6945	RABENSTEIN	
08	M	C	4587.28 5435.38	800	83.87	6945	RABENSTEIN	
09	M	C	G 1432.99 4818.58	600	213.12	35	DIMBACH	
09	M	A	G 1452.64 4818.57	600	231.76	35	DIMBACH	
09	M	B	G 1452.83 4818.58	600	231.93	35	DIMBACH	
09	M	A	G 1459.50 4816.39	500	240.93	35	ISPER MUEHLE	
09	M	B	G 1459.54 4816.25	500	241.13	35	ISPER MUEHLE	
09	M	C	G 1459.64 4816.24	500	241.24	35	ISPER MUEHLE	
09	M	1	G 1504.96 4814.79	850	248.10	54	ANRATSBERG	
09	M	2	G 1505.14 4814.66	850	248.42	54	ANRATSBERG	
09	M	3	G 1505.18 4814.62	850	248.51	54	ANRATSBERG	
09	M	A	G 1505.73 4813.99	350	249.76	54	ARTSTETTEN	
09	M	B	G 1505.94 4813.98	350	249.98	54	ARTSTETTEN	
09	M	C	G 1506.02 4813.87	350	250.18	54	ARTSTETTEN	
09	M	A	G 1517.96 4810.99	450	265.20	54	GR.WEICHSELB.	
09	M	B	G 1518.08 4810.93	450	265.39	54	GR.WEICHSELB.	
09	M	C	G 1518.28 4810.88	450	265.64	54	GR.WEICHSELB.	
09	M	A	G 1521.44 4809.18	340	270.67	55	SIMONSBERG	
09	M	B	G 1521.60 4809.08	300	270.94	55	SIMONSBERG	

B	IN	AC		COORDINATES		ALT	X(KM)	SMAP	LOCALITY	G C
09	M	1	G	1527.16	4806.00	400	279.88	55	CHRISTENBG.	
09	M	2	G	1527.24	4805.87	380	280.11	55	CHRISTENBG.	
09	M	3	G	1527.48	4805.76	360	280.47	55	CHRISTENBG.	
09	M	1	G	1533.08	4804.26	480	287.71	55	STEUBACH	
09	M	2	G	1533.20	4804.15	500	287.95	55	STEUBACH	
09	M	3	G	1533.40	4804.07	520	288.23	55	STEUBACH	
09	M	A	G	1535.24	4802.49	480	291.82	56	TRAISEN	
09	M	B	G	1535.48	4802.44	460	292.11	56	TRAISEN	
09	M	C	G	1535.87	4802.33	480	292.62	56	TRAISEN	
09	M	B	G	1541.47	4759.02	1000	301.88	74	EBENWALD	
09	M	C	G	1541.66	4758.88	1025	302.22	74	EBENWALD	
09	M	A	G	1542.39	4759.05	1000	302.76	74	EBENWALD	
09	M	A	G	1546.26	4757.24	580	308.63	74	SPECKMANN	
09	M	B	G	1546.39	4757.18	590	308.82	74	SPECKMANN	
09	M	C	G	1546.56	4757.16	600	309.01	74	SPECKMANN	
09	M	1	G	1549.64	4754.49	600	315.06	74	HASELRAST	
09	M	2	G	1549.85	4754.30	600	315.48	74	HASELRAST	
09	M	3	G	1550.00	4754.18	600	315.76	74	HASELRAST	
09	M	1	G	1554.73	4752.19	550	322.69	75	GUTENSTEIN	
09	M	2	G	1554.75	4752.06	550	322.86	75	GUTENSTEIN	
09	M	3	G	1554.77	4751.97	600	322.98	75	GUTENSTEIN	
09	M	1	G	1601.15	4749.89	1040	331.65	75	HOHE WAND	
09	M	2	G	1601.32	4749.75	1000	331.98	75	HOHE WAND	
09	M	3	G	1601.48	4749.63	980	332.27	75	HOHE WAND	

QUARRY-NO. 09 BOEHMISCHBRUCK SMAP 6440
PROFILE 09-200-01 COMPILED BY GIESE, PRODEHL

B	IN	AC		COORDINATES		ALT	X(KM)	SMAP	LOCALITY	G C
03	BO			4525.66	5493.19	485	.81	6440	BOEHM BRUCK	
02	BO			4525.66	5493.19	485	.86	6440	BOEHM BRUCK	
03	BO			4522.63	5492.93	555	3.17	6439	VOITSBERG	
01	M	B		4524.05	5488.89	640	3.83	6439	TAENNESBERG	
02	CL			4526.54	5487.25	668	5.15	6440	TAENNESB.FORST	
01	M	A		4523.02	5487.40	560	5.63	6439	TAENNESB.BNH	
01	M			4521.96	5484.22	510	8.94	6539	STOECKELHOF	
01	M			4520.53	5481.36	490	12.14	6539	WEIDENTHAL	
01	M			4519.63	5478.37	410	15.23	6539	GUTENECK	
01	M			4518.52	5475.20	400	18.59	6539	WILLHOF	
01	M			4517.70	5472.00	460	21.86	6639	UNTER AUERBACH	
01	M			4516.51	5470.41	393	23.78	6639	STANGELHOF	
02	GO			4515.33	5468.61	377	25.89	6639	WEIHERHAUS	
02	GO			4514.85	5467.59	372	27.01	6639	RAEUBERWEIHERH	
02	GO			4515.10	5463.40	451	30.81	6639	GRAFENRICHT	
01	M			4517.64	5462.48	412	30.93	6639	ALTENSCHWAND	
01	M			4514.45	5460.13	400	34.12	6739	STEINBERG	
02	KR			4511.00	5457.06	383	38.21	6738	LOISNITZ	
01	KR			4509.85	5454.38	398	41.14	6738	MESSNERSKREITH	
02	CL			4511.50	5448.42	350	46.15	6838	RAMSPAU	
01	MZ			4507.80	5447.85	370	47.96	6838	PONHOLZ	
02	CL			4505.44	5445.19	367	51.31	6838	OBERHUB	
01	MZ			4505.40	5442.30	365	54.00	6838	EITLBRUNN	
01	CL			4502.75	5436.55	397	60.33	6938	SCHWETZENDORF	
02	MZ			4499.75	5433.88	338	63.95	6937	ETTERZHAUSEN	
01	MM			4499.00	5427.45	350	70.17	7037	ALLING	
01	MM			4495.96	5421.93	440	76.44	7037	LINDACH	
08	M	A		4482.53	5427.60	350	77.87	7036	BAIERSDORF	
08	M	B		4482.57	5426.95	350	78.39	7036	BAIERSDORF	
08	M	B		4487.68	5423.71	480	78.52	7036	SAUSTHAL	
08	M	A		4487.23	5423.77	470	78.68	7036	SAUSTHAL	
01	S			4498.00	5418.15	380	79.19	7037	SAAL	
02	MZ			4493.37	5413.92	414	84.81	7137	TEUERTING	
01	S			4492.02	5408.96	360	89.93	7137	ABENSBERG	
01	F			4489.90	5403.31	403	95.96	7237	SIEGENBURG	
02	MZ			4488.65	5399.83	430	99.64	7237	SIEGENBURG	
05	M	B		4483.43	5397.75	405	103.66	7236	ELSENDORF	
02	M			4483.46	5393.10	440	107.85	7336	LINDKIRCHEN	
05	M	B		4483.26	5393.23	445	107.87	7376	LINDKIRCHEN	
06	M	B		4482.91	5390.89	445	110.16	7336	SCHLEISSBACH	
01	M			4483.23	5387.73	455	112.90	7336	MAINBURG	
05	M	B		4482.35	5387.90	460	113.14	7336	UNTEREMPFENB.	
06	M	A		4481.54	5385.69	480	115.49	7336	STEINBACH	
06	M	B		4479.73	5383.67	495	118.05	7436	ENZELHAUSEN	
05	M	C		4478.02	5381.87	505	120.38	7436	RUDERTSHAUSN	
05	M	A		4478.99	5379.09	480	122.56	7436	GUENZENHAUSN	
01	M			4478.43	5377.44	480	124.25	7436	SILLERTSHAUSEN	
05	M	A		4478.79	5375.84	515	125.65	7436	SILLERTSHSN	
03	S			4476.00	5372.16	465	130.11	7536	GEIERLAMBACH	
06	M	A		4474.33	5372.63	500	130.32	7535	AMPERTSHAUSN	
05	M	A		4471.09	5370.61	480	133.48	7535	AUFHAM	
01	M			4473.88	5366.00	480	136.56	7535	SCHOENBICHL	
05	M	B		4467.60	5364.93	485	140.09	7535	EGLHAUSEN	
05	M	A		4464.79	5361.93	505	143.99	7635	LAUTERBACH	
03	S			4472.41	5358.37	508	144.24	7635	SCHAIDENHAUSEN	
03	M	A		4460.39	5356.79	493	150.52	7634	SCHOENBRUNN	
03	M	B		4454.52	5347.32	505	161.60	7734	GUENDING	
03	M	B		4447.57	5336.82	545	174.10	7833	EICHENAU	
01	M			4446.24	5336.56	570	174.88	7833	FUERSTENFELDBR.	
03	M	B		4444.83	5330.81	565	180.70	7833	THALHOF	
03	M	A		4442.93	5321.87	630	189.56	7933	HACKEN	
03	M	B		4442.93	5321.87	630	189.56	7933	HACKEN	
03	M	A		4442.19	5313.38	710	197.54	8033	MACHTLFING	
03	M	A		4439.79	5301.52	640	209.32	8133	STEINBERG	
03	M	B		4442.10	5292.23	683	216.93	8233	EGENRIED	
02	M	A		4438.75	5282.47	690	227.16	8333	MURNAU	
02	M	A		4438.13	5277.87	650	231.65	8333	WEGHAUSKGL	
02	M	B		4438.13	5277.87	650	231.65	8333	WEGHAUSKGL	X
02	M			4435.34	5272.07	760	238.07	8432	ETTALER MDL.Z	
02	M	A		4435.32	5264.06	1740	245.51	8432	WANK	
02	M	A		4431.85	5257.61	1250	252.79	8532	REINTHALDORF	
02	M	A		4432.30	5254.36	1010	255.64	8532	BOCKHUETTE	
02	M	A	G	1106.95	4723.28	1850	259.17	FB34	WANG ALPE	
02	M	B	G	1106.32	4720.22	1260	264.76	117	BUCHEN	
02	M	B	G	1106.82	4716.89	954	270.36	117	FLAURLING/N	
02	M	B	G	1106.63	4715.89	1160	272.19	117	FLAURLING/S	
03	M	A	G	1104.45	4713.34	1810	277.61	FB25	ZIRMBACH ALM	
03	M	B	G	1104.45	4713.34	1810	277.61	FB25	ZIRMBACH ALM	
11	GO	11		3579.03	5252.93	990	294.42	8526	SIBRATSGFAELL	
11	GO	12		3578.83	5252.63	960	294.78	8526	SIBRATSGFAELL	
03	M	A	G	1101.67	4704.11	1650	294.91	FB25	GRIES	
03	M	B	G	1101.67	4704.11	1650	294.91	FB25	GRIES	
11	GO	13		3578.60	5252.30	945	295.18	8526	SIBRATSGFAELL	
03	M	A	G	1101.27	4657.87	1620	306.03	FB25	SOELDERN	
05	M	A	G	1055.90	4643.90	1650	332.73	FB25	PFOSSENTHAL	
05	M	A	G	1051.89	4638.26	1800	344.25	FB46	ST.MARTIN	
05	M	A	G	1044.61	4631.48	1650	359.13	FB46	MAHDER	

QUARRY-NO. 09 BOEHMISCHBRUCK SMAP 6440
PROFILE 09-240 COMPILED BY EMTER

B	IN	AC		COORDINATES		ALT	X(KM)	SMAP	LOCALITY	G C
02	BO			4524.33	5491.55	582	1.58	6440	KLEINSCHWAND	
04	MZ			4507.50	5482.75	565	20.59	6538	WINDPAISSING	
04	GO			4490.16	5471.00	436	41.52	6637	THEUERN WOLFSBACH	
04	KI			4482.67	5466.26	440	50.91	6636	MALSBACH	
07	M	A		4474.91	5461.90	540	59.28	6735	ZIEGELHUETTE	
07	M	B		4474.54	5461.83	570	59.63	6735	ZIEGELHUETTE	
07	M	C		4474.32	5461.61	570	59.93	6735	ZIEGELHUETTE	
04	CL			4474.05	5461.40	575	60.27	6735	SCHWAND	
07	M	A		4470.44	5460.02	555	64.08	6735	FRICKENHOFEN	
07	M	B		4470.06	5460.02	543	64.41	6735	FRICKENHOFEN	
07	M	C		4469.92	5459.79	543	64.65	6735	FRICKENHOFEN	
07	M	A		4467.51	5458.32	562	67.47	6735	OBERBUCHFELD	
07	M	B		4467.13	5458.23	566	67.84	6735	OBERBUCHFELD	
07	M	C		4466.91	5458.19	530	68.05	6735	OBERBUCHFELD	
04	CL			4466.41	5457.40	525	68.88	6735	ARZTHOFEN	
06	M	B		4467.53	5454.88	555	69.25	6735	DEINING	
06	M	A		4467.41	5454.77	560	69.71	6735	DEINING	
07	M	A		4463.40	5456.36	535	72.01	6734	WINNBERG	
07	M	B		4463.13	5456.08	560	72.38	6734	WINNBERG	
07	M	C		4462.88	5455.81	590	72.73	6734	WINNBERG	
07	M	A		4461.98	5450.77	535	76.15	6834	WAPPERSDORF	
07	M	B		4461.79	5450.83	510	76.27	6834	WAPPERSDORF	
07	M	C		4461.62	5450.80	480	76.43	6834	WAPPERSDORF	
07	M	A		4458.60	5452.97	417	77.86	6734	FORST	
07	M	B		4458.34	5452.64	419	78.25	6734	FORST	
07	M	C		4458.17	5452.59	425	78.42	6734	FORST	
08	M	C		4462.47	5445.98	520	78.47	6834	POLLANTEN	
08	M	B		4462.21	5446.08	520	78.62	6834	POLLANTEN	
08	M	A		4461.99	5446.18	520	78.74	6834	POLLANTEN	
08	M	A		4470.34	5436.34	460	78.81	6935	PREMERZHOFEN	
08	M	C		4470.76	5435.66	400	79.00	6935	PREMERZHOFEN	
08	M	B		4470.47	5435.88	400	79.05	6935	PREMERZHOFEN	
08	M	C		4473.85	5431.52	570	79.98	6935	SCHWEINKOFEN	
08	M	B		4473.55	5431.54	560	80.16	6935	SCHWEINKOFEN	
08	M	A		4473.21	5431.65	580	80.30	6935	SCHWEINKOFEN	
06	M	C		4456.72	5449.27	500	81.39	6834	SULZBURG	
06	M	B		4456.25	5448.91	500	81.98	6834	SULZBURG	
04	CL			4455.81	5449.10	515	82.25	6834	GALGENBERG	
06	M	A		4455.71	5444.79	535	84.68	6834	BURGGRIESBACH	
06	M	B		4455.36	5444.78	535	84.97	6834	BURGGRIESBACH	
07	M	A		4453.09	5445.77	455	86.32	6834	OBERNRICHT	
07	M	B		4453.06	5445.66	450	86.41	6834	OBERNRICHT	
07	M	C		4453.00	5445.41	435	86.60	6834	OBERNRICHT	
07	M	A		4448.70	5444.21	450	90.87	6833	HAGENBUCH	
07	M	B		4448.56	5444.18	460	91.00	6833	HAGENBUCH	
07	M	C		4448.39	5444.02	495	91.23	6833	HAGENBUCH	
04	F			4448.11	5443.43	552	91.78	6833	OBERMAESSING	
04	M			4445.77	5443.56	545	93.70	6833	OFFENBAU	
07	M	A		4445.73	5439.84	530	95.73	6933	AUE	
07	M	B		4445.54	5439.88	530	95.86	6933	AUE	
07	M	A		4442.73	5437.50	568	99.21	6933	THALMAESSING	
07	M	B		4442.68	5437.50	568	99.56	6933	THALMAESSING	
04	M	C		4439.29	5438.19	590	102.04	6932	DANNHAUSEN	
04	M	A		4438.91	5438.14	590	102.39	6932	DANNHAUSEN	
07	M	A		4435.85	5436.53	553	105.84	6932	BERGEN	
07	M	B		4435.74	5436.60	563	105.89	6932	BERGEN	
04	M	A		4432.44	5437.39	595	108.31	6932	GEYERN	
04	M	B		4432.36	5437.34	595	108.40	6932	GEYERN	
04	M			4427.27	5434.44	580	114.26	6932	WEISSENBURG	
04	M			4424.39	5430.83	530	118.59	6931	WEISSENBURGER WAL	
04	M			4424.17	5430.57	530	118.91	6931	WEISSENBURGER WAL	
04	M	A		4419.58	5426.91	570	124.73	7031	TREUCHTLINGEN	
04	M	B		4418.97	5426.69	575	125.37	7031	TREUCHTLINGEN	
04	M	A		4413.01	5426.79	570	130.43	7030	AUERHEIM	
04	M	C		4412.47	5426.38	610	131.10	7030	AUERHEIM	
04	M	E		4412.23	5426.42	605	131.29	7030	AUERHEIM	
04	M	B		4407.28	5423.34	530	137.12	7030	URSHEIM	
04	M	A		4407.05	5423.05	500	137.46	7030	URSHEIM	
04	M	A		4407.24	5415.98	475	141.00	7130	WEMDING	
04	M	B		4406.75	5415.91	470	141.45	7130	WEMDING	
04	M	A		4401.25	5409.47	460	149.58	7129	HEROLDINGEN	
04	M	B		4401.24	5409.37	460	149.64	7129	HEROLDINGEN	
04	M	A		4398.59	5406.84	450	153.25	7229	KLEINSORHEIM	
04	M	B		4398.36	5406.81	473	153.46	7229	KLEINSORHEIM	
04	MZ			4390.08	5410.23	560	158.61	7129	REIMLINGEN	
06	M	B		3606.61	5408.41	550	162.79	7128	THALMUEHLE	
04	M			3606.10	5409.00	540	162.90	7128	EDERHEIM	
06	M	A		3606.40	5408.48	500	162.93	7128	THALMUEHLE	
06	M	A		3603.63	5407.54	610	165.71	7228	MOERTINGEN	
06	M	B		3602.83	5407.02	590	166.66	7228	MOERTINGEN	
06	M	A		3602.16	5405.48	605	168.12	7228	HOHLENSTEIN	
06	M	B		3601.76	5405.48	590	168.45	7228	HOHLENSTEIN	
06	M			3599.56	5403.33	605	171.47	7228	SCHL. NERESHEIM	
04	M			3598.61	5401.00	520	173.55	7228	NERESHEIM	
08	S	1		3564.23	5481.40	620	179.16	7425	REUTTI	
08	S	2		3563.99	5481.20	630	179.42	7425	REUTTI	
08	S	4		3563.60	5481.25	630	179.81	7425	REUTTI	
04	MM			3588.15	5395.44	590	185.34	7327	NATTHEIM	
04	MM			3581.94	5391.78	517	192.54	7326	HEIDENHEIM	
04	S			3577.34	5388.81	625	198.02	7326	ERPFENHAUSER HOF	
04	S			3573.66	5386.98	520	202.10	7326	SONTBERGEN	
08	S			3570.70	5383.92	665	206.25	7425	ZAEHRINGEN	
04	S			3568.58	5382.55	650	208.78	7425	HOFST. EMERBUCH	
06	MZ			3563.24	5378.28	652	215.59	7425	RADELSTETTEN	
08	S			3559.18	5374.44	655	221.10	7424	MERKLINGEN	
08	KA			3555.66	5371.76	685	225.46	7524	MACHTOLSHEIM	
06	MZ			3551.34	5371.42	730	229.30	7524	LAICHINGEN	
08	KA			3551.26	5369.94	735	230.18	7524	SUPPINGEN	
08	KA			3549.88	5366.74	745	233.11	7524	SONTHEIM	
09	F	21		3542.59	5360.92	708	242.40	7623	MEHRSTETTEN	

B	IN	AC	COORDINATES		ALT	X(KM)	SMAP	LOCALITY	G C
09	F	2?	3542.30	5360.68	710	242.77	7623	MEHRSTETTEN	
09	F	23	3541.96	5360.32	748	243.26	7623	MEHRSTETTEN	
09	F	31	3536.85	5359.49	791	247.95	7622	APFELSTETTEN	
09	F	32	3536.51	5359.27	770	248.36	7622	APFELSTETTEN	
09	F	33	3536.20	5358.99	748	248.77	7622	APFELSTETTEN	
06	F		3533.03	5356.23	790	252.96	7622	EGLINGEN	
09	F	11	3526.19	5351.98	777	260.99	7622	OBERSTETTEN	
09	F	12	3525.94	5351.63	750	261.39	7622	OBERSTETTEN	
09	F	13	3525.65	5351.42	737	261.76	7622	OBERSTETTEN	
09	KA	11	3522.56	5348.39	760	266.01	7721	WILSINGEN	
09	KA	12	3522.51	5348.24	750	266.13	7721	WILSINGEN	
09	KA	22	3515.60	5343.88	722	274.33	7721	GAMMERTINGEN	
09	KA	23	3515.32	5343.71	702	274.67	7721	GAMMERTINGEN	
09	KA	31	3509.92	5341.57	765	280.30	7720	HARTHAUSEN	
09	KA	32	3509.92	5341.53	740	280.32	7720	HARTHAUSEN	
09	KA	33	3509.68	5341.20	808	280.78	7720	HARTHAUSEN	
09	KA	42	3502.00	5340.20	910	287.66	7720	EBINGEN	
06	S		3498.34	5327.96	810	297.50	7919	FINSTERTAL	
09	S	31	3494.83	5331.14	900	298.64	7819	DIETSTAIG H.	
09	S	32	3494.53	5330.71	898	299.13	7819	DIETSTAIG H.	
09	S	33	3494.36	5330.17	921	299.58	7819	DIETSTAIG H.	
06	S		3490.60	5325.72	810	305.14	7919	MAHLSTETTEN	
09	S	43	3481.74	5324.06	910	313.41	7918	SPAICHINGEN S	
09	S	42	3481.43	5324.08	900	313.66	7918	SPAICHINGEN S	
09	S	41	3481.38	5324.10	900	313.69	7918	SPAICHINGEN S	
11	M	11	3473.10	5314.97	795	325.65	8017	UNTER BALDINGEN	
11	M	12	3472.65	5314.87	800	326.08	8017	UNTER BALDINGEN	
11	M	13	3472.30	5314.61	770	326.51	8017	UNTER BALDINGEN	
07	S		3457.26	5309.21	780	342.03	8016	BRAEUNLINGEN2	
07	S		3457.22	5309.13	780	342.11	8016	BRAEUNLINGEN1	
11	M	41	3458.46	5305.17	735	343.27	8116	DOEGGINGEN	
11	M	42	3458.20	5304.83	725	343.67	8116	DOEGGINGEN	
11	M	43	3457.88	5304.54	680	344.10	8116	DOEGGINGEN	
09	S	11	3443.54	5298.35	960	359.45	8115	SOMMERAU	
09	S	12	3443.16	5298.26	920	359.82	8115	SOMMERAU	
09	S	13	3442.76	5298.04	950	360.28	8115	SOMMERAU	
11	S	11	3439.51	5294.95	1020	364.69	8215	OB.SCHWARZWALDN	
11	S	12	3439.50	5294.91	1010	364.72	8215	OB.SCHWARZWALDN	
11	S	13	3439.49	5294.89	1005	364.74	8215	OB.SCHWARZWALDN	
09	S	22	3432.91	5289.55	998	373.18	8214	WITTENSCHWAND	
09	S	23	3432.64	5289.08	1000	373.67	8214	WITTENSCHWAND	
11	S	41	3430.72	5290.75	1020	374.34	8214	OBER IBACH	
11	S	42	3430.64	5290.77	1040	374.39	8214	OBER IBACH	
11	S	43	3430.57	5290.73	1060	374.47	8214	OBER IBACH	
11	S	1	3427.23	5287.73	950	378.92	8214	LINDAU	
11	S	2	3427.23	5287.73	950	378.92	8214	LINDAU	
11	S	3	3427.05	5287.80	970	379.03	8214	LINDAU	

QUARRY-NO. 09 BOEHMISCHBRUCK SMAP 6440
PROFILE 09-275-08 COMPILED BY AICHELE

B	IN	AC	COORDINATES		ALT	X(KM)	SMAP	LOCALITY	G C
11	CL	13	4520.26	5492.39	492	5.42	6439	KLESSBERG	
11	CL	21	4515.99	5492.43	550	9.70	6439	DEINDORF	
11	CL	22	4515.59	5492.34	525	10.10	6439	DEINDORF	
11	CL	23	4515.18	5492.34	498	10.51	6439	DEINDORF	
11	CL	32	4509.96	5493.15	445	15.74	6438	LUHE	
11	CL	33	4509.48	5493.17	440	16.22	6438	LUHE	
11	CL	41	4505.55	5493.40	528	20.15	6438	WEISSENBRUNN	
11	CL	42	4505.09	5493.38	538	20.61	6438	WEISSENBRUNN	
11	CL	43	4504.62	5493.42	549	21.08	6438	WEISSENBRUNN	
11	KI	41	4500.34	5493.89	482	25.39	6437	KOHLBERG	
11	KI	42	4500.00	5493.91	487	25.69	6437	KOHLBERG	
11	KI	43	4499.70	5493.81	510	26.02	6437	KOHLBERG	
11	KI	32	4495.33	5493.99	503	30.40	6437	EHENFELD	
11	KI	33	4494.94	5494.11	523	30.79	6437	EHENFELD	
11	KI	21	4490.72	5494.10	478	35.00	6437	ADLHOLZ	
11	KI	22	4490.34	5494.11	486	35.38	6437	ADLHOLZ	
11	KI	23	4490.00	5494.30	480	35.73	6437	ADLHOLZ	
10	M	41	4486.37	5493.91	510	39.41	6436	HOHENZAUNT	
10	M	42	4486.00	5493.72	510	39.77	6436	HOHENZAUNT	
10	M	43	4485.59	5493.81	510	40.08	6436	HOHENZAUNT	
10	M	51	4481.43	5494.16	510	44.36	6436	WEISSENBERG	
10	M	52	4481.14	5494.35	510	44.65	6436	WEISSENBERG	
10	M	53	4480.97	5494.53	510	44.83	6436	WEISSENBERG	
10	M	61	4476.12	5494.62	510	49.68	6435	VOEGELAS	
10	M	62	4475.89	5494.64	510	49.90	6435	VOEGELAS	
10	M	62	4475.61	5494.56	510	50.18	6435	VOEGELAS	
10	M	21	4472.05	5494.69	510	53.75	6435	RATZENHOFEN	
10	M	22	4471.80	5494.80	510	54.01	6435	RATZENHOFEN	
10	M	23	4471.56	5494.91	510	54.25	6435	RATZENHOFEN	
11	KI	11	4470.23	5495.41	520	55.73	6435	RATZENHOF	
11	KI	12	4470.00	5495.41	510	55.76	6435	RATZENHOF	
11	KI	13	4469.74	5495.41	502	56.02	6435	RATZENHOF	
10	M	11	4466.54	5495.79	520	59.29	6435	HARTENSTEIN	
10	M	12	4466.18	5495.78	520	59.67	6435	HARTENSTEIN	
10	M	13	4465.94	5495.76	520	59.92	6335	HARTENSTEIN	
10	M	31	4461.42	5496.15	500	64.44	6334	RAITENBERG	
10	M	32	4461.08	5496.00	500	64.77	6334	RAITENBERG	
10	M	33	4460.80	5495.90	500	65.05	6334	RAITENBERG	
10	B	11	4456.07	5496.28	530	69.79	6334	GOETZLESBERG	
10	B	12	4455.96	5496.31	525	69.88	6334	GOETZLESBERG	
10	B	13	4455.70	5496.34	510	70.16	6334	GOETZLESBERG	
10	B	51	4450.10	5496.68	520	75.78	6333	HUETTENBACH	
10	B	53	4449.48	5496.70	539	76.39	6333	HUETTENBACH	
10	B	61	4446.68	5498.48	520	79.31	6333	RUESSELBACH	
10	B	62	4446.32	5498.38	495	79.62	6333	RUESSELBACH	
10	B	63	4445.91	5498.37	470	80.06	6333	RUESSELBACH	
11	B	32	4443.70	5496.70	330	82.10	6333	AFFALTERBACH	
10	B	71	4441.40	5498.50	440	84.55	6333	ETLASWIND	
10	B	72	4441.04	5498.57	455	84.92	6333	ETLASWIND	
10	B	73	4440.55	5498.53	390	85.42	6333	ETLASWIND	
11	B	42	4438.10	5497.27	328	87.72	6332	NEUNKIRCHEN A BRAND	
10	B	81	4435.60	5497.52	335	90.31	6332	ROSENBACH	
10	B	82	4435.15	5497.59	358	90.75	6332	ROSENBACH	
10	M	21	4431.35	5497.81	335	94.56	6332	ERLANGEN	
10	M	22	4431.01	5497.78	365	94.90	6332	ERLANGEN	
10	M	23	4430.63	5497.93	370	95.28	6332	ERLANGEN	
10	M	42	4425.38	5496.85	285	100.47	6331	ALT ERLANGEN	
10	M	43	4424.96	5496.88	290	100.89	6331	ALT ERLANGEN	
10	M	11	4422.25	5497.58	330	103.63	6331	UNTER MEMBACH	
10	M	12	4421.86	5497.54	330	104.02	6331	UNTER MEMBACH	
10	M	13	4421.46	5497.41	315	104.41	6331	UNTER MEMBACH	
10	M	51	4416.69	5497.88	355	109.20	6331	NAUKENDORF	
10	M	52	4416.20	5498.00	345	109.70	6331	NAUKENDORF	
10	M	53	4415.80	5498.09	350	110.10	6331	NAUKENDORF	
10	CL	11	4411.53	5499.38	347	114.44	6330	REZELSDORF	
10	CL	12	4411.13	5499.38	357	114.84	6330	REZELSDORF	
10	CL	13	4410.78	5499.37	338	115.18	6330	REZELSDORF	
10	CL	21	4406.76	5499.61	317	119.21	6330	GERHARDSHOFEN	
10	CL	22	4406.35	5499.55	307	119.62	6330	GERHARDSHOFEN	
10	CL	23	4405.97	5499.58	288	120.00	6330	GERHARDSHOFEN	
10	CL	31	4400.80	5499.97	335	125.19	6329	GUTENSTETTEN	
10	CL	32	4400.42	5500.00	330	125.56	6329	GUTENSTETTEN	
10	CL	33	4400.42	5500.00	330	125.56	6329	GUTENSTETTEN	X
10	F	31	4396.10	5499.60	315	129.83	6329	BAUDENBACH	
10	F	32	4395.68	5499.65	312	130.29	6329	BAUDENBACH	
10	F	33	4395.09	5499.61	309	130.86	6329	BAUDENBACH	
10	F	21	4390.04	5500.33	323	135.95	6328	FRANKENFELD	
10	F	22	4389.60	5500.41	338	136.39	6328	FRANKENFELD	
10	F	23	4389.17	5500.41	355	136.83	6328	FRANKENFELD	
10	F	11	3602.97	5500.77	368	139.72	6328	SUEDLICH BIBART	
10	F	12	3602.55	5500.67	380	140.14	6328	SUEDLICH BIBART	
10	F	13	3602.09	5500.71	388	140.60	6328	SUEDLICH BIBART	
11	B	51	3595.50	5500.63	350	147.22	6327	DORNHEIM	
11	B	52	3595.06	5500.78	350	147.56	6327	DORNHEIM	
11	B	53	3594.62	5500.85	325	148.00	6327	DORNHEIM	
11	B	62	3590.62	5501.30	280	152.01	6327	HUETTENHEIM	
11	B	63	3590.25	5501.30	280	152.38	6327	HUETTENHEIM	
11	B	73	3585.93	5501.10	250	156.70	6327	WAESSERNDORF	
11	B	12	3580.14	5501.36	300	162.49	6326	GNODSTADT	
11	B	13	3579.68	5501.34	300	162.95	6326	GNODSTADT	
11	F	11	3575.72	5500.52	295	166.89	6326	HOHESTADT	
11	F	12	3575.34	5500.48	275	167.27	6326	HOHESTADT	
11	F	13	3574.97	5500.53	260	167.64	6326	HOHESTADT	
11	F	21	3571.13	5500.48	281	171.48	6325	ACHOLSHAUSEN	
11	F	22	3570.70	5500.48	282	171.91	6325	ACHOLSHAUSEN	
11	F	23	3570.29	5500.44	288	172.32	6325	ACHOLSHAUSEN	
11	F	31	3565.29	5500.80	303	177.33	6325	SULZDORF	
11	F	32	3564.91	5500.72	298	177.71	6325	SULZDORF	
11	F	33	3564.50	5500.70	298	178.12	6325	SULZDORF	

QUARRY-NO. 09 BOEHMISCHBRUCK SMAP 6440
PROFILE 09-300-13 COMPILED BY BEHNKE.STEIN

B	IN	AC	COORDINATES		ALT	X(KM)	SMAP	LOCALITY	G C
02	BO		4525.66	5493.19	485	.81	6440	BOEHMISCHBRUCK1	
03	BO		4525.66	5493.19	485	.81	6440	BOEHMISCHBRUCK2	
02	BO		4524.33	5491.55	580	1.58	6440	KLEINSCHWAND	
03	BO		4522.62	5492.92	555	3.18	6439	VOITSBERG	
03	M		4521.47	5495.62	555	5.37	6439	STEINACH	
06	MM	A	4520.14	5498.89	530	8.59	6339	UNTERLIND	MA
03	M		4517.77	5497.96	495	9.74	6339	MICHLDORF	
03	M		4512.45	5500.46	420	15.56	6339	SCHIRMITZ	
08	M		4509.77	5502.10	450	18.70	6338	ERMERSRICHT	
03	MZ		4507.38	5502.25	500	20.85	6338	NEUNKIRCHEN1	
06	KR		4507.39	5502.41	485	20.92	6338	NEUNKIRCHEN 2	
08	M	A	4504.42	5504.45	415	24.51	6338	WIESENDORF	
08	M	B	4504.21	5504.59	410	24.76	6338	WIESENDORF	
08	M	C	4503.55	5505.01	410	25.54	6338	WIESENDORF	
08	M	D	4503.12	5505.28	410	26.05	6338	WIESENDORF	
08	M		4500.53	5506.91	410	29.11	6338	PARKSTEIN	
06	GO		4498.45	5507.54	410	31.23	6237	PECHHOF	
08	M		4496.52	5509.96	425	34.11	6237	DIESSFURTH	
08	M		4492.22	5512.43	410	39.07	6237	BAERNWINKEL	
06	GO		4490.82	5515.27	445	41.76	6237	PICHLBERG	
08	M		4488.19	5515.15	420	43.92	6237	ESCHENBACH	
06	KI		4484.50	5519.20	450	49.20	6136	OBERBIBRACH 2	
08	GO		4482.31	5518.08	465	50.47	6136	VORBACH	
03	KR		4482.83	5519.13	462	50.57	6136	OBERBIBRACH 1	
08	GO		4479.79	5522.34	435	54.86	6136	LOSAU	
06	MS		4475.30	5523.50	510	59.28	6135	NEUHOF	
06	MS		4474.84	5525.38	495	60.67	6135	CREUSSEN	
08	F		4471.10	5526.52	440	64.44	6135	BOCKSRUECK	
03	MZ		4470.10	5528.68	440	66.44	6135	KRODELSBERG	
08	F		4468.10	5529.10	435	68.35	6135	ROEDENSDORF	
08	MM	1	4466.85	5530.28	395	70.04	6035	MISTELBACH	
08	MM	2	4466.29	5530.38	400	70.57	6035	MISTELBACH	
08	MM	3	4465.85	5530.57	400	71.04	6035	MISTELBACH	
09	KR	1	4462.64	5532.01	445	74.53	6034	MISTELGAU	
09	KR	2	4462.20	5532.49	445	74.73	6034	MISTELGAU	
09	KR	3	4462.20	5532.49	440	75.17	6034	MISTELGAU	
06	MS	A	4463.01	5538.02	420	77.58	6034	NEUSTAEDTLEIN 1	
08	MM	2	4459.91	5533.44	500	77.59	6034	SCHANZ	
06	MS	B	4462.65	5538.28	414	78.03	6034	NEUSTAEDTLEIN1	
06	MS	C	4462.63	5538.29	413	78.05	6034	NEUSTAEDTLEIN1	
06	MS		4462.63	5538.29	415	78.05	6034	NEUSTAEDTLEIN 2	
08	MM	3	4459.47	5533.62	455	78.06	6034	SCHANZ	
06	MS	D	4462.62	5538.31	412	78.07	6034	NEUSTAEDTLEIN1	
06	MS	E	4462.27	5538.56	372	78.50	6034	NEUSTAEDTLEIN1	
08	MM	3	4456.83	5535.97	500	81.55	6034	ALLADORF 2	
08	MM	2	4456.33	5535.95	475	81.96	6034	ALLADORF 2	
08	MM		4456.33	5535.95	475	81.96	6034	ALLADORF 3	
08	MM	1	4455.94	5536.20	465	82.42	6034	ALLADORF 2	
03	F		4455.24	5537.64	475	83.79	6034	ALLADORF 1	
08	MM	2	4453.47	5537.29	510	85.10	6034	KLEINHUEL	
08	MM	4	4453.16	5537.64	510	85.54	6034	KLEINHUEL	
08	MM	3	4452.78	5537.95	510	86.03	6034	KLEINHUEL	
03	M		4450.71	5541.20	450	89.52	5933	SCHIRRADORF	
08	KI		4447.08	5544.04	475	94.12	5933	MODSCHIEDEL	
03	M		4443.89	5545.59	460	97.63	5933	WALLERSBERG	
09	GO	1	4441.53	5547.72	490	100.79	5933	KOETTEL	
09	GO	2	4441.20	5547.98	500	101.20	5933	KOETTEL	
09	GO	3	4441.05	5548.33	445	101.52	5933	KOETTEL	
06	M	A	4437.89	5550.41	495	105.29	5932	LAHM	
06	M	B	4437.61	5550.40	496	105.52	5932	LAHM	
06	M	C	4437.32	5550.48	500	105.81	5932	LAHM	
06	M	D	4437.02	5550.54	505	106.09	5932	LAHM	
06	M	E	4436.74	5550.65	502	106.39	5932	LAHM	
06	M	F	4436.44	5550.74	490	106.69	5932	LAHM	

02	M			4432.92 5553.10	450	110.91	5832	WOLFSDORF	
02	M			4428.29 5556.31	420	116.54	5831	ALTENBANZ	
09	B	1		4424.57 5556.90	320	120.01	5831	PUECHITZ 2	
06	M			4424.54 5556.93	320	120.04	5831	PUECHITZ 1	
09	B	2		4424.37 5557.04	320	120.25	5831	PUECHITZ 2	
09	B	3		4424.21 5557.17	310	120.46	5831	PUECHITZ 2	
06	M			4421.35 5559.86	330	124.31	5831	WELSBERG 1	
09	B	1		4421.35 5559.85	330	124.31	5831	WELSBERG 2	
09	B	2		4421.35 5559.85	330	124.31	5831	WELSBERG 2	
09	B	3		4421.35 5559.85	330	124.31	5831	WELSBERG 2	
02	M			4417.84 5562.01	300	128.41	5831	SESSLACH	
09	KI	3		4415.90 5564.68	280	131.52	5730	DIETERSDORF	
09	KI	2		4415.74 5564.64	305	131.63	5730	DIETERSDORF	
09	KI	1		4415.61 5564.66	310	131.75	5730	DIETERSDORF	
06	M			4413.18 5565.71	375	134.35	5730	MUGGENBACH	
02	M			4409.84 5567.58	305	138.15	5730	GLEISMUTHMAUSEN	
03	BO			4405.55 5565.00	355	140.43	5730	ALLERTSHAUSEN	
09	KI	1		4401.20 5565.05	365	144.21	5729	ERMERSDORF	
09	KI	2		4400.93 5565.14	360	144.49	5729	ERMERSDORF	
09	KI	3		4400.63 5565.14	355	144.75	5729	ERMERSDORF	
03	BO			4399.50 5570.16	367	148.29	5729	ZIMMERAU	
06	CL			4399.49 5572.78	365	149.69	5729	ST.URSULA	
02	MM			4398.58 5577.55	320	153.04	5629	TRAPPSTADT	
02	MM			4394.15 5580.11	365	158.15	5629	MERBSTADT	
07	M	A		4391.36 5579.28	310	160.04	5628	OTTELMANNSHAUS	
07	M	B		4391.12 5579.45	315	160.33	5628	OTTELMANNSHAUS	
09	KI	1		4391.60 5580.69	315	160.62	5628	IRMELSHAUSEN	
07	M	C		4390.86 5579.60	325	160.63	5628	OTTELMANNSHAUS	
09	KI	2		4391.21 5580.20	330	160.68	5628	IRMELSHAUSEN	
07	M	D		4390.67 5579.82	335	160.91	5628	OTTELMANNSHAUS	
07	M	E		4390.46 5580.00	330	161.19	5628	OTTELMANNSHAUS	
09	KI	3		4390.91 5580.91	345	161.32	5628	IRMELSHAUSEN	
07	M	F		4390.21 5580.08	325	161.44	5628	OTTELMANNSHAUSEN	
07	M			4388.04 5581.98	350	164.29	5628	GOLLMUTHMAUSEN 2	
06	MM			4388.15 5582.13	350	164.28	5628	GOLLMUTHMAUSEN 1	
07	M			4383.83 5585.83	330	169.92	5628	MENDUNGEN 2	
09	KI	1		4386.70 5584.92	385	167.04	5628	RAPPERSHAUS.	
09	KI	2		4386.29 5585.12	395	167.49	5628	RAPPERSHAUS.	
09	KI	3		4385.97 5585.20	395	167.80	5628	PAPPERSHAUS.	
06	MM			4383.00 5584.94	310	170.13	5628	MENDUNGEN 1	
09	M	1		4377.61 5588.97	345	176.85	5527	MAINHOF	
09	M	2		4377.21 5589.00	330	177.20	5527	MAINHOF	
09	M	3		4376.83 5589.14	335	177.59	5527	MAINHOF	
09	M	1		4373.47 5594.29	340	183.25	5527	OSTHEIM	
09	M	2		4373.18 5594.49	390	183.60	5527	OSTHEIM	
09	M	3		4372.80 5594.78	400	184.07	5527	OSTHEIM	
09	MM	31		3582.16 5595.46	420	187.92	5526	STETTEN	
09	MM	32		3581.81 5595.46	427	188.22	5526	STETTEN	
09	MM	33		3581.41 5595.53	418	188.60	5526	STETTEN	
09	MM	13		3579.93 5597.04	455	190.65	5426	MAUSEN	
09	MM	12		3579.65 5597.16	490	190.95	5426	MAUSEN	
09	MM	11		3579.36 5597.36	500	191.30	5426	MAUSEN	
09	MM	41		3576.77 5598.64	785	194.18	5426	LEUBACH	
09	MM	42		3576.62 5598.95	770	194.47	5426	LEUBACH	
09	MM	43		3576.24 5599.00	783	194.82	5426	LEUBACH	
09	MM	23		3573.97 5600.81	780	197.70	5426	STBR.HILDERS	
09	MM	22		3573.68 5601.07	788	198.08	5426	STBR.HILDERS	
09	MM	21		3573.53 5601.28	780	198.32	5426	STBR.HILDERS	

QUARRY-NO. 09 BOEHMISCHBRUCK SMAP 6440
PROFILE 09-325 COMPILED BY PETERS

B	IN	AC		COORDINATES	ALT	X(KM)	SMAP	LOCALITY	G C
11	M	92		4503.97 5526.90	640	40.74	6138	WAELDERN	
11	M	93		4503.97 5526.90	640	40.74	6138	WAELDERN	
11	M	81		4500.12 5527.94	625	43.76	6138	GRAETSCHENREUTH	
11	M	82		4500.18 5528.37	630	44.08	6138	GRAETSCHENREUTH	
11	M	83		4500.40 5528.82	640	44.32	6138	GRAETSCHENREUTH	
11	M	42		4493.38 5538.97	620	56.66	6037	NAGEL	
11	M	62		4489.69 5544.71	895	63.48	5937	SCHNEEBERG S	
11	M	51		4483.84 5553.70	553	74.20	5836	ZETTLITZ	
11	M	52		4483.66 5554.13	590	74.66	5836	ZETTLITZ	
11	M	53		4483.63 5554.53	585	75.00	5836	ZETTLITZ	
11	M	72		4481.09 5558.34	587	79.58	5836	OELSCHNITZ	
11	M	71		4480.69 5558.60	590	80.02	5836	OELSCHNITZ	
11	M	11		4477.02 5562.18	620	85.05	5736	EISENBUEHL	
11	M	12		4477.04 5562.51	640	85.31	5736	EISENBUEHL	
11	M	13		4476.63 5562.77	641	85.76	5736	EISENBUEHL	
11	M	22		4474.90 5565.51	630	89.00	5735	WALBENGRUEN	
11	M	31		4472.41 5567.31	683	91.90	5735	BRUMBERG	
11	M	32		4472.17 5569.40	690	93.75	5735	BRUMBERG	
11	M	33		4472.09 5569.55	680	93.92	5735	BRUMBERG	

QUARRY-NO. 10 VOGGENDORF SMAP 6540
PROFILE 10-135 COMPILED BY GIESE.WOLBER

B	IN	AC		COORDINATES	ALT	X(KM)	SMAP	LOCALITY	G C
01	BO			4527.66 5481.25	460	.59	6540	BRUECKLINGHOF	
01	BO			4527.66 5481.25	460	.59	6540	BRUECKLINGHOF	X
01	BO			4527.66 5481.25	460	.59	6540	BRUECKLINGHOF	Y
03	M	B		4530.17 5479.12	570	3.69	6540	EIGELSBERG	
03	M	B		4533.63 5475.97	555	8.31	6540	KULZ	
03	M	B		4536.17 5473.51	488	11.83	6641	PILLMERSRIED	
03	M	B		4539.72 5471.08	485	16.12	6641	DIEPOLTSRIED	
03	M	B		4542.69 5468.87	470	19.82	6641	SCHOENTHAL	
03	M	B		4546.66 5465.25	480	25.18	6641	NW OBERNRIED	
03	M	B		4551.36 5462.24	550	30.74	6742	GSCHIESS	
04	M	A		4553.36 5459.40	460	34.07	6742	DALKING	
03	M	B		4553.36 5459.40	460	34.07	6742	DALKING	
04	M	C		4553.36 5459.40	460	34.07	6742	DALKING	
03	M	B		4558.57 5455.66	510	40.47	6742	ZENCHING	
03	M	B		4562.69 5452.89	520	45.42	6743	THENRIED	
03	M	C		4565.22 5449.28	540	49.66	6843	ARNDORF	
03	M	F		4565.22 5449.28	540	49.66	6843	ARNDORF	
03	M	B		4569.80 5446.36	560	55.05	6843	KL.RIEDELSTN	
03	CL	D		4574.64 5443.86	620	60.40	6844	POSCHINGER M	
05	M	B		4574.68 5443.86	620	60.43	6844	ARNBRUCK	
05	M	A		4577.44 5442.65	865	63.35	6844	SCHOENBACH	
05	M	B		4577.67 5442.25	865	63.78	6844	SCHOENBACH	
05	M	C		4577.86 5442.32	890	63.89	6844	SCHOENBACH	
03	CL			4578.95 5440.90	880	65.62	6844	BODENMAIS	
04	M	A		4582.00 5440.36	1080	68.38	6944	BODENMAIS	
03	CL			4581.90 5438.01	863	69.72	6944	BODENMAIS	
03	CL			4582.56 5437.56	945	70.52	6944	BODENMAIS	
04	CL			4583.25 5436.94	905	71.44	6944	BODENMAIS	
05	M	A		4585.61 5435.62	910	74.12	6945	RABENSTEIN	
05	M	B		4585.61 5435.62	910	74.12	6945	RABENSTEIN	
04	M	A		4587.70 5433.51	740	77.06	6945	RABENSTEIN	
04	M	B		4587.70 5433.51	740	77.06	6945	RABENSTEIN	
01	MZ			4591.21 5432.24	640	80.39	6945	ZWIESEL	
04	M	A		4593.17 5432.87	635	81.82	6945	UT.ZWIESELAU	
04	M	B		4593.17 5432.87	635	81.82	6945	UT.ZWIESELAU	
04	M	C		4593.17 5432.87	635	81.82	6945	UT.ZWIESELAU	
05	M	B		4596.09 5427.61	700	87.31	7045	FRAUENAU	
04	M	A		4596.15 5427.51	700	87.41	7045	FRAUENAU	
03	M			4598.69 5425.90	820	90.41	7046	KLINGENBRUNN 2	
05	M	B		4600.91 5425.00	930	92.72	7046	LAERCHENBERG	
04	M	A		4604.72 5422.82	840	97.08	7046	GUGLOED	
04	M	B		4604.99 5422.62	870	97.41	7046	GUGLOED	
04	M	C		4605.14 5422.57	850	97.56	7046	GUGLOED	
03	M			4606.69 5419.64	720	100.57	7046	ALTSCHOENAU 1	
04	M	A		4610.65 5419.02	790	104.10	7147	SAEGWASSER M	
04	M	B		4612.49 5412.93	770	109.27	7147	BIERHUETTE	
04	M	A		4612.67 5413.00	780	109.37	7147	BIERHUETTE	
05	M	A		4620.73 5410.96	973	117.00	7147	HERZOGSREUT	
04	M	A		4622.70 5408.48	890	120.07	7148	OBERGRAINET	
04	M	B		4622.70 5408.48	890	120.07	7148	OBERGRAINET	
04	M	C		4623.12 5408.61	970	120.32	7148	OBERGRAINET	
05	M	A		4625.45 5404.36	803	124.75	7248	ALTREICHENAU	
05	M	B		4634.57 5397.16	605	136.37	7348	BREITENBERG	
05	M	B	G	1350.19 4839.35	730	139.85	13	KAISERHAEUSL	
05	M	A	G	1355.32 4836.73	590	147.80	14	PEILSTEIN	
05	M	A	G	1357.10 4834.49	630	152.17	14	RUMERSDORF	
05	M	A	G	1402.16 4833.12	500	158.54	14	HASLACH	
05	M	B	G	1408.04 4829.77	790	168.08	32	ST.JOHANN	
06	M	B	G	1412.05 4829.46	830	172.22	32	WAXENBERG	
06	M	C	G	1414.81 4828.46	730	176.03	32	OBERNENKIR.	
05	M	B		4690.60 5403.00	716	181.25	7248	SCHIMMELBACH	
06	M	B	G	1419.71 4827.93	790	181.34	32	HELLMONSOEDT	
06	M	B	G	1422.52 4828.08	830	183.89	33	OTTENSCHLAG	
06	M	B	G	1426.90 4826.70	530	189.72	33	SCHALL	
06	M	B	G	1429.80 4824.86	640	194.64	33	REMPLDORF	
06	M	C	G	1429.80 4824.86	640	194.64	33	REMPLDORF	
06	M	A	G	1434.11 4824.19	540	199.62	33	NEUSTADT/KFM	
06	M	A	G	1438.54 4823.04	500	205.27	34	ERDLEITEN	
06	M	A	G	1441.93 4822.66	600	209.05	34	ZELL	
06	M	A	G	1445.73 4820.57	550	215.13	34	PIERBACH	
06	M	B	G	1452.89 4818.58	630	224.45	35	DIMBACH	
06	M	B	G	1459.71 4816.27	500	233.80	35	ST.OSWALD	
07	M	42	G	1504.96 4814.79	850	240.69	36	NEUE WALDHAEUSER	
07	M	61	G	1527.16 4806.00	400	272.57	55	CHRISTENBERG	
07	M	62	G	1527.24 4805.87	380	272.79	55	CHRISTENBERG	
07	M	63	G	1527.48 4805.78	360	273.13	55	CHRISTENBERG	
07	M	51	G	1533.08 4804.26	480	280.45	55	STEUBACH	
07	M	52	G	1533.20 4804.15	500	280.69	55	STEUBACH	
07	M	53	G	1533.40 4804.07	520	280.98	55	STEUBACH	
07	M	11	G	1549.64 4754.49	600	307.73	74	HASELRAST 1	
07	M	12	G	1549.85 4754.30	600	308.14	74	HASELRAST 1	
07	M	33	G	1549.85 4754.30	600	308.14	74	HASELRAST 2	
07	M	13	G	1550.00 4754.18	600	308.43	74	HASELRAST 1	
07	M	21	G	1554.36 4752.30	500	314.88	75	GUTENSTEIN	
07	M	22	G	1554.61 4752.24	570	315.19	75	GUTENSTEIN	
07	M	23	G	1554.77 4751.97	600	315.65	75	GUTENSTEIN	

QUARRY-NO. 10 VOGGENDORF SMAP 6540
PROFILE 10-305 COMPILED BY GEBRANDE

B	IN	AC		COORDINATES	ALT	X(KM)	SMAP	LOCALITY	G C
01	BO			4526.70 5481.49	490	.79	6540	VOGGENDORF	
01	BO			4526.70 5481.49	490	.79	6540	VOGGENDORF	X
01	BO			4526.70 5481.49	490	.79	6540	VOGGENDORF	Y
07	KR	11	G	1128.77 4955.46	446	82.15	6034	MISTELGAU	
07	KR	12	G	1128.67 4955.56	445	82.36	6034	MISTELGAU	
07	KR	13	G	1128.40 4955.73	440	82.81	6034	MISTELGAU	
07	GO	11		4441.38 5547.72	490	108.56	5933	KOESTEL	
07	GO	12		4441.15 5548.05	475	108.94	5933	KOESTEL	
07	GO	13		4441.05 5548.33	447	109.19	5933	KOESTEL	
07	B	21		4424.75 5557.09	300	127.46	5831	PUECHITZ	
07	B	22		4424.61 5557.16	300	127.62	5831	PUECHITZ	
07	B	23		4424.21 5557.26	300	128.00	5831	PUECHITZ	
07	B	11		4421.35 5559.85	330	131.84	5831	WELSBERG	
07	B	12		4421.35 5559.85	330	131.84	5831	WELSBERG	
07	B	13		4421.35 5559.85	330	131.84	5831	WELSBERG	
07	KI	21		4415.61 5564.66	312	139.32	5730	DIETERSDORF	
07	KI	22		4415.74 5564.64	305	139.20	5730	DIETERSDORF	
07	KI	23		4415.90 5564.68	280	139.10	5730	DIETERSDORF	
07	KI	41		3600.01 5584.63	385	174.61	5628	RAPPERSHAUSEN	
07	KI	42		3599.58 5584.82	393	175.07	5628	RAPPERSHAUSEN	
07	KI	43		3599.27 5584.89	395	175.36	5628	RAPPERSHAUSEN	
07	MM	32		3581.92 5595.21	407	195.52	5526	STETTEN	
07	MM	12		3579.22 5596.93	545	198.72	5426	MAUSEN	
07	MM	42		3576.92 5599.82	730	202.24	5426	LEUBACH	
07	MM	22		3573.69 5602.05	740	206.17	5426	STBR.HILDERS	

QUARRY-NO. 11 BISCHOFSHEIM SMAP 5526
PROFILE 11-120-09 COMPILED BY VOSS.KATZLER

B	IN	AC		COORDINATES	ALT	X(KM)	SMAP	LOCALITY	G C
01	M			3571.44 5588.78	720	.25	5526	BISCHOFSHEIM	
01	CL	A		3572.11 5588.03	580	1.17	5526	BAUERSBERG	
01	CL	B		3572.73 5587.56	610	1.95	5526	BAUERSBERG	
01	K	A		4409.86 5567.58	310	56.52	*5730	GLEISMUTHMAUSEN	
01	CL	A		4417.70 5561.60	300	66.14	*5831	SESSLACH	
01	CL	A		4432.92 5553.10	450	83.55	*5832	WOLFSDORF	
01	M	A		4446.17 5543.86	475	99.55	*5933	WEIDEN	

B	IN	AC	COORDINATES	ALT	X(KM)	SMAP	LOCALITY	G C
01	M	A	4461.95 5535.72	400	117.31	*6034	OBERHAIZ	
01	M	A	4474.28 5567.75	715	118.37	*5735	ENCHENREUTH	
01	M	A	4482.06 5567.04	560	126.14	*5736	HELMBRECHTS	
01	M	A	4485.95 5564.95	555	130.36	*5736	MACKERSREUTH	
01	MZ	A	4475.41 5524.01	510	134.73	*6135	NEUHOF	
01	FF	A	4493.58 5562.86	520	138.26	*5737	BURGSTALL	
01	MM	A	4507.61 5502.48	486	173.38	*6338	NEUNKIRCHEN	
01	MM	A	4525.68 5492.34	510	194.10	*6440	BOEHM. BRUCK	
01	S	A	4545.25 5478.68	571	217.89	*6541	BREITENRIED	
01	S	B	4545.37 5478.83	575	217.92	6541	BREITENRIED	
01	S	A	4557.40 5465.29	800	235.24	6642	DACHSRIEGEL	
01	S	B	4557.44 5465.22	800	235.32	*6642	DACHSRIEGEL	

QUARRY-NO. 12 ROMSTHAL SMAP 5622
PROFILE 12-260 COMPILED BY MAENEL.STEIN

B	IN	AC	COORDINATES	ALT	X(KM)	SMAP	LOCALITY	G C
01	CL		3526.53 5576.94	324	.66	5622	KATHOLISCHWILL.	
01	CL		3525.73 5576.91	324	1.45	5622	KATHOLISCHWILL.	
01	M	--	3523.50 5575.36	300	4.08	5621	UDENHAIN	
01	GO	--	3517.25 5573.55	362	10.54	5721	WITTGENBORN	
01	GO	--	3512.45 5569.52	170	16.58	5721	BREITENBORN	
01	GO	--	3506.01 5570.14	160	22.28	5720	VONHAUSEN	
01	M	--	3502.81 5568.48	230	25.85	5720	ECKHARTSHAUSEN	
01	M	--	3497.70 5568.54	168	30.69	5719	ROMMELSHAUSEN	
01	M	--	3492.34 5566.65	171	36.37	5719	OSTHEIM	
01	M	-A	3488.16 5566.42	155	40.45	5719	BUEDESHEIM	
01	CL	--	3484.77 5564.22	141	44.32	5718	RENDEL	
01	CL	--	3469.50 5560.05	194	60.14	5817	STEINBACH	
01	M		3463.60 5558.88	250	66.13	5816	MAMMOLSHEIM	
01	M	-B	3456.53 5556.79	260	73.50	5816	VOCKENHAUSEN	
01	M	-A	3451.16 5554.62	230	79.27	5815	NAUROD	
01	PR	--	3444.40 5553.30	250	86.13	5815	RABENGRUND	
01	M	--	3438.21 5550.25	339	92.93	5914	SCHLANGENBAD	
01	MZ	--	3427.55 5549.46	459	103.39	5913	KIEDRICHER WALD	
01	MZ	--	3419.70 5547.33	405	111.52	5913	PRESBERG	
01		--	3410.36 5544.76	280	121.21	5912	MANUBACH	
01	M	--	3406.23 5539.78	430	126.57	6012	RHEINBOELLEN	
01	M	--	3396.69 5542.18	360	135.07	5911	PLEIZENHAUSEN	
01	M	--	3390.69 5539.98	380	141.44	6010	FRONHOFEN	
01	S	--	2601.52 5538.99	392	145.81	6010	HEINZENBACH	
01	S	-A	2595.17 5536.67	407	152.57	6009	SCHWARZEN	
01	S	-B	2595.15 5536.73	411	152.57	6009	SCHWARZEN	
06	F	11	2582.26 5531.42	390	166.40	6008	KAUTENBACH	
06	F	12	2581.86 5531.28	402	166.82	6008	KAUTENBACH	
06	F	13	2581.64 5531.06	400	167.10	6008	KAUTENBACH	
03	S	--	2578.70 5530.18	280	170.23	6008	BERNKASTEL	
06	F	33	2575.05 5529.39	342	173.88	6108	ANDEL	
06	F	32	2574.80 5529.40	367	174.12	6108	ANDEL	
06	F	31	2574.59 5529.40	367	174.32	6108	ANDEL	
03	S	--	2570.46 5529.11	250	178.41	6107	FILZEN	
06	F	21	2565.60 5527.52	137	183.45	6107	NIEDEREMMEL	
06	F	22	2565.24 5527.27	140	183.87	6107	NIEDEREMMEL	
06	F	23	2564.93 5526.96	132	184.26	6107	NIEDEREMMEL	
06	S	31	2562.71 5525.70	315	186.76	6107	KLUESSERATH	
06	S	32	2562.41 5525.40	295	187.14	6107	KLUESSERATH	
06	S	33	2562.20 5525.36	307	187.35	6107	KLUESSERATH	
06	S	11	2558.79 5522.87	220	191.36	6106	BEKOND	
06	S	12	2558.46 5522.68	240	191.73	6106	BEKOND	
06	S	13	2558.12 5522.79	290	192.02	6106	BEKOND	
03	M	--	2548.59 5522.90	365	201.13	6106	KORDEL	
06	M	11	2543.89 5520.20	395	206.36	6105	BUTZWEILER	
06	M	12	2543.48 5520.10	360	206.78	6105	BUTZWEILER	
06	M	13	2543.06 5520.07	350	207.19	6105	BUTZWEILER	
03	M	--	2538.18 5520.62	260	211.75	6105	MAHLENDORF	

QUARRY-NO. 12 ROMSTHAL SMAP 5622
ARC 12-115-F COMPILED BY MAENEL.STEIN

B	IN	AC	COORDINATES	ALT	X(KM)	SMAP	LOCALITY	G C
03	M	--	3478.23 5483.29	270	105.87	6518	SCHRIESHEIM	
03	GO	--	3445.23 5498.62	137	113.49	6315	HOHEN SULZEN	
03	GO	--	3435.83 5503.28	285	117.47	6314	EINSELTHUM	
03	CL	--	3431.85 5512.75	285	115.03	6214	WEINHEIM	
03	CL	--	3422.75 5519.13	165	119.44	6113	VOLXSHEIM	
03	MM	--	3419.32 5534.76	230	115.86	6013	MUENSTER SARMSH.	
03	MM	--	3415.46 5550.30	320	114.86	5912	LORCH	
03	MM	--	3417.27 5555.94	282	111.90	5812	BORNICH	
03	MM	--	3411.94 5562.19	310	116.16	5812	FEUERBACH	
03	M		3412.17 5570.93	310	115.13	5712	DORNHOLZHAUSEN	
03	M	--	3412.71 5581.86	440	114.52	5612	WELSCHNEUDORF	

QUARRY-NO. 13 DORHEIM SMAP 5021
PROFILE 13-060-17 COMPILED BY VEES

B	IN	AC	COORDINATES	ALT	X(KM)	SMAP	LOCALITY	G C
05	KI	41	3523.75 5654.09	224	10.59	4922	BATZENBERG	
05	KI	42	3524.12 5654.14	229	10.91	4922	BATZENBERG	
05	KI	43	3524.43 5654.29	226	11.25	4922	BATZENBERG	
05	KI	31	3528.53 5656.96	215	16.14	4922	MARDORF E	
05	KI	32	3528.86 5657.15	230	16.52	4922	MARDORF E	
05	KI	33	3529.24 5657.34	255	16.94	4922	MARDORF E	
05	KI	11	3530.28 5658.35	360	18.37	4922	EILERTHOF	
05	KI	12	3530.48 5658.66	381	18.72	4922	EILERTHOF	
05	KI	13	3530.64 5659.00	361	19.05	4922	EILERTHOF	
05	KI	21	3532.14 5659.72	350	20.68	4922	MOSHEIM	
05	KI	22	3532.45 5659.94	310	21.06	4922	MOSHEIM	
05	KI	23	3532.80 5660.10	281	21.44	4922	MOSHEIM	
06	KI	41	3534.27 5660.71	300	23.02	4922	OSTHEIM E	
06	KI	42	3534.53 5660.93	293	23.36	4922	OSTHEIM E	
06	KI	43	3534.83 5661.25	273	23.79	4922	OSTHEIM E	
06	KI	31	3535.80 5661.64	260	24.81	4923	ELFERSHAUSEN S	
06	KI	32	3535.91 5661.82	267	25.00	4923	ELFERSHAUSEN S	
06	KI	33	3536.08 5662.04	250	25.27	4923	ELFERSHAUSEN S	
06	KI	21	3537.61 5662.94	168	27.04	4823	FAHRE	
06	KI	22	3537.83 5663.29	179	27.42	4823	FAHRE	
06	KI	23	3538.19 5663.50	173	27.84	4823	FAHRE	
05	GO	11	3547.17 5669.75	505	38.74	4824	SCHNELLRODE	
05	GO	12	3547.55 5669.90	510	39.14	4824	SCHNELLRODE	
05	GO	13	3547.88 5670.11	535	39.53	4824	SCHNELLRODE	
05	GO	21	3548.91 5670.90	495	40.83	4824	RETTERODE	
05	GO	22	3549.15 5671.16	475	41.18	4824	RETTERODE	
05	GO	23	3549.42 5671.41	480	41.54	4824	RETTERODE	
06	GO	11	3551.52 5671.74	382	43.49	4824	RETTERODE NE	
06	GO	12	3551.88 5671.79	392	43.82	4824	RETTERODE NE	
06	GO	13	3552.31 5671.81	390	44.19	4824	RETTERODE NE	
06	GO	21	3556.12 5675.17	425	49.22	4724	VELMEDEN	
06	GO	22	3556.45 5675.53	435	49.70	4724	VELMEDEN	
06	GO	23	3556.53 5675.80	440	49.92	4724	VELMEDEN	
06	B	51	3559.75 5678.13	620	53.89	4725	BRANSRODE	
06	B	52	3559.81 5678.34	593	54.06	4725	BRANSRODE	
06	B	53	3560.22 5678.59	560	54.54	4725	BRANSRODE	
06	B	62	3561.65 5679.14	470	56.03	4725	FRANKENHAIN W	
06	B	63	3562.04 5679.20	430	56.38	4725	FRANKENHAIN W	
06	B	11	3563.32 5679.33	310	57.52	4725	FRANKENHAIN E	
06	B	13	3563.48 5679.47	317	57.73	4725	FRANKENHAIN E	
06	B	71	3564.52 5680.46	305	59.14	4725	ORFERODE	
06	B	72	3564.65 5680.48	300	59.26	4725	ORFERODE	
06	B	73	3564.99 5680.59	308	59.61	4725	ORFERODE	

QUARRY-NO. 13 DORHEIM SMAP 5021
PROFILE 13-120-09 COMPILED BY STEIN

B	IN	AC	COORDINATES	ALT	X(KM)	SMAP	LOCALITY	G C
01	BO	-	3514.29 5648.18	210	1.18	5021	SCHLIERBACH	
01	BO	-	3516.63 5647.43	215	1.31	5021	MICHELSBERG	
01	M		3519.22 5643.90	240	5.40	5021	SCHOENBORN	
01	M		3519.77 5643.30	255	6.21	5021	SCHOENBORN	
01	MS	-	3527.74 5638.79	490	15.24	5122	CHRISTERODE	
01	MZ	-	3529.97 5637.05	510	18.07	5122	OLBERODE	
01	BO	-	3538.83 5630.78	355	28.92	5123	HATTENBACH	
01	BO	-	3545.90 5623.68	285	38.84	5223	WEHRDA	
04	GO	11	3547.43 5620.20	392	42.20	5224	FORST BURGHAUN	
04	GO	12	3547.71 5619.90	375	42.61	5224	FORST BURGHAUN	
04	GO	13	3547.90 5619.63	360	42.91	5224	FORST BURGHAUN	
01	SE	-	3556.40 5614.10	300	53.02	5324	MACKENZELL	
04	KI	11	3559.66 5612.88	450	56.30	5325	HOFASCHENBACH	
04	KI	12	3559.82 5612.66	462	56.56	5325	HOFASCHENBACH	
04	KI	13	3560.11 5612.33	413	56.99	5325	HOFASCHENBACH	
01	PR	-	3563.65 5608.90	430	61.93	5325	GOTTHARDS	
04	KI	21	3567.72 5606.25	530	66.73	5425	ECKWEISSB.KUPPE	
04	KI	22	3567.69 5605.93	552	66.91	5425	ECKWEISSB.KUPPE	
04	KI	23	3567.76 5605.60	571	67.17	5425	ECKWEISSB.KUPPE	
01	CL	1	3572.92 5604.57	710	71.91	5426	BUCHSCHIRMBERG	
01	CL	24	3573.53 5604.01	741	72.74	5426	BUCHSCHIRMBERG	
01	CL	-	3578.41 5600.54	540	78.72	5426	LEUBACH	
03	F	-	3578.82 5597.22	530	81.06	5426	HAUSEN/RH	
03	S	-	3583.93 5595.36	365	86.24	5527	NORDHEIM V.D.RH.	
03	F	-	3591.45 5598.85	275	90.40	5527	MELLRICHSTADT	
03	S	-	3588.90 5591.33	380	92.63	5527	OSTHEIM V.D.RH.	
01	F	24	3597.35 5588.90	380	100.88	5528	ROSSRIETH	
04	KI	41	3598.40 5585.47	366	103.72	5628	RAPPERSHAUSEN	
04	KI	42	3598.79 5585.26	377	104.16	5628	RAPPERSHAUSEN	
04	KI	43	3599.11 5584.98	394	104.59	5628	RAPPERSHAUSEN	
01	MZ	-	3602.55 5584.80	320	107.50	5728	ROTHAUSEN	
03	M	A	3600.73 5582.01	335	107.69	5628	GOLLMUTHHAUSEN	
03	C	C	3601.08 5581.80	320	108.10	5628	GOLLMUTHHAUSEN	
03	M	B	4390.69 5581.38	310	110.84	5628	IRMELSHAUSEN	
03	M	A	4390.78 5581.31	305	110.96	5628	IRMELSHAUSEN	
03	M	A	4393.49 5580.02	310	113.88	5629	HERBSTADT	
03	M	B	4393.90 5579.80	325	114.33	5629	HERBSTADT	
03	M	C	4394.01 5579.86	340	114.38	5629	HERBSTADT	
01	MM	-	4393.13 5578.70	310	114.46	5628	HERBSTADT	
03	M	A	4395.45 5577.47	350	117.00	5629	EYERSHAUSEN	
03	M	B	4395.71 5577.32	340	117.30	5629	EYERSHAUSEN	
03	M	C	4395.92 5577.21	345	117.53	5629	EYERSHAUSEN	
03	M	A	4397.99 5578.24	312	118.51	5629	TRAPPSTADT	
03	M		4398.32 5578.01	315	118.91	5629	TRAPPSTADT	
01	MM	-	4399.66 5572.80	390	123.25	5729	ST.URSULA	
03	M	A	4398.92 5571.23	390	123.66	5729	STERNBERG	
03	M	B	4399.07 5571.16	425	123.82	5729	STERNBERG	
03	M	C	4399.34 5571.11	405	124.06	5729	STERNBERG	
01	CL	-	4402.54 5565.52	340	130.16	5729	ERMERSHAUSEN	
01	M	-	4409.84 5567.58	305	134.45	5730	GLEISMUTSHAUSEN	
02	M	-	3625.62 5565.81	340	137.31	5730	LECHENROTH	
01	M	-	4417.84 5562.01	300	144.18	5831	SESSLACH	
02	M	-	3635.64 5561.00	330	148.24	5831	WELSBERG	
01	M	-	4428.29 5556.31	420	155.92	5831	KLOSTER BANZ	
01	M	-	4432.92 5553.10	450	161.55	5832	VIERZEHNHEILIGEN	
02	M	A	4436.76 5554.52	420	163.75	5832	KLOSTERLANGHEIM	
02	M	B	4436.98 5554.21	415	164.11	5832	KLOSTERLANGHEIM	
02	M	C	4437.16 5553.86	395	164.46	5832	KLOSTERLANGHEIM	
01	CL	-	4436.01 5547.56	521	167.39	5932	HOHER STEIN	
02	M	A	4439.03 5547.57	515	169.74	5932	EICHING	
02	M	B	4439.12 5547.55	515	169.82	5932	EICHING	
02	M	A	4442.76 5545.84	460	173.75	5933	ARNSTEIN	
02	M	C	4442.69 5545.69	445	173.79	5933	ARNSTEIN	
02	M	B	4442.85 5545.74	450	173.88	5933	ARNSTEIN	
01	S	-	4445.42 5544.78	450	176.54	5933	WEIDEN/FR.ALB	
02	M	A	4446.52 5543.72	485	178.03	5933	WEIDEN	
02	M	B	4446.64 5543.67	485	178.15	5933	WEIDEN	
02	M	A	4449.66 5542.45	467	181.30	5933	SCHIRRADORF	
02	M	B	4449.80 5542.31	468	181.49	5933	SCHIRRADORF	
02	M	A	4454.37 5539.18	530	187.03	6034	GROSSENHUEL	
02	M	B	4454.58 5539.10	520	187.24	6034	GROSSENHUEL	
02	M	C	4454.89 5539.03	500	187.53	6034	GROSSENHUEL	
02	M	A	4458.08 5536.39	445	191.67	6034	LOCHAU	
02	M	B	4458.29 5536.21	445	191.95	6034	LOCHAU	
02	M		4458.73 5536.03	450	192.42	6034	LOCHAU	
02	M	D	4458.93 5535.94	550	192.62	6034	LOCHAU	
02	M	E	4459.37 5535.69	540	193.12	6034	LOCHAU	

B	IN	AC	COORDINATES	ALT	X(KM)	SMAP	LOCALITY	G C
01	S		4461.95 5535.73	415	195.20	6034	OBERHAIZ	
04	M	21	4463.49 5535.20	370	196.71	6034	ECKERSDORF	
04	M	22	4463.58 5534.77	405	197.04	6034	ECKERSDORF	
04	M	23	4463.91 5534.54	380	197.44	6034	ECKERSDORF	
04	M	31	4465.85 5530.56	400	201.37	6035	FORKENDORF	
04	M	32	4466.31 5530.41	400	201.83	6035	FORKENDORF	
04	M	33	4466.63 5530.28	400	202.16	6035	FORKENDORF	
04	M	41	4469.50 5528.88	450	205.30	6135	ROEDENSDORF	
04	M	42	4469.79 5528.82	440	205.57	6135	ROEDENSDORF	
04	M	43	4470.12 5528.68	440	205.92	6135	ROEDENSDORF	
04	M	11	4471.04 5525.96	485	208.28	6135	LANKENREUTH	
04	M	12	4471.42 5525.68	450	208.76	6135	LANKENREUTH	
04	M	13	4471.81 5525.41	450	209.23	6135	LANKENREUTH	
01	M		4475.04 5524.30	495	212.52	6135	NEUHOF	
01	M		4486.75 5520.09	480	224.45	6136	OBERBIBRACH	
01	M	B	4499.96 5519.50	715	235.63	6137	MESSERBERG	
01	M	B	4507.37 5502.22	495	251.64	6338	NEUNKIRCHEN	
01	M	B	4515.88 5504.22	595	257.40	6339	TROEGLERSRICHT	
04	M	22	4515.04 5499.48	440	259.42	6339	IRCHENRIETH	
04	M	12	4515.33 5499.33	430	259.74	6339	IRCHENRIETH	
04	M	13	4515.73 5499.08	425	260.21	6339	IRCHENRIETH	
01	M		4518.68 5496.82	490	263.96	6339	MICHLDORF	
04	M	21	4522.92 5493.66	520	269.21	6440	BOEHMISCHBRUCK	
04	M	22	4523.83 5493.54	500	270.02	6440	BOEHMISCHBRUCK	
04	M	23	4524.22 5493.35	480	270.45	6440	BOEHMISCHBRUCK	
01	M	B	4526.03 5491.36	565	273.12	6440	BOEHMISCHBRUCK	
04	M	31	4525.78 5486.10	610	275.99	6440	ZEINRIED	
04	M	32	4525.99 5485.82	610	276.33	6440	ZEINRIED	
04	M	33	4526.04 5485.59	600	276.51	6440	ZEINRIED	
01	M	A	4528.46 5487.67	685	277.25	6440	DREIFELSEN	
04	M	51	4530.84 5484.63	570	280.93	6540	GARTENRIED	
04	M	52	4531.04 5484.44	540	281.21	6540	GARTENRIED	
04	M	53	4531.22 5484.36	549	281.40	6540	GARTENRIED	
01	M		4536.64 5484.98	720	285.46	6441	SCHWAND	
04	M	41	4538.90 5484.64	635	287.45	6541	MUGGENTHAL	
04	M	42	4539.14 5484.25	655	287.88	6541	MUGGENTHAL	
04	M	43	4539.52 5484.06	549	288.30	6541	MUGGENTHAL	
01	M		4540.78 5481.78	765	290.69	6541	WEIDING	
04	M	61	4541.31 5481.61	685	291.18	6541	KAGERN	
04	M	62	4541.59 5481.45	655	291.50	6541	KAGERN	
04	M	63	4541.90 5481.19	650	291.90	6541	KAGERN	
01	M		4545.89 5479.32	600	296.27	6541	SILBERBERG	
05	M		4548.94 5471.43	520	303.34	6642	WALDMUENCHEN	

QUARRY-NO. 13 DORMEIM SMAP 5021
PROFILE 13-240-20 COMPILED BY BAIER.MEISSNER.GLOCKE

B	IN	AC	COORDINATES	ALT	X(KM)	SMAP	LOCALITY	G C
06	M	31	3507.40 5643.92	310	8.76	5020	SACHSENHAUSEN	
06	M	32	3507.19 5643.82	330	8.99	5020	SACHSENHAUSEN	
06	M	33	3506.86 5643.65	370	9.36	5020	SACHSENHAUSEN	
06	M	21	3505.64 5642.22	330	11.10	5020	APPENHAIN	
06	M	22	3505.39 5642.11	300	11.37	5020	APPENHAIN	
06	M	23	3505.21 5642.00	320	11.58	5020	APPENHAIN	
06	M	41	3503.35 5640.46	340	13.96	5120	SPECKSWINKEL	
06	M	42	3502.95 5640.25	330	14.41	5120	SPECKSWINKEL	
06	M	43	3502.61 5639.98	330	14.84	5120	SPECKSWINKEL	
06	M	12	3500.79 5639.55	260	16.63	3500	MATZBACH	
06	M	13	3500.36 5639.29	260	17.13	3500	MATZBACH	
06	MM	41	3499.24 5639.26	232	18.14	5119	WOLFERODE	
06	MM	42	3498.84 5639.31	235	18.47	5119	WOLFERODE	
06	MM	43	3498.62 5638.98	258	18.82	5119	WOLFERODE	
06	MM	13	3495.90 5637.71	235	21.82	5119	BURGHOLZ	
06	MM	21	3494.31 5636.86	228	23.62	5119	HIMMELSBERG	
06	MM	22	3493.99 5636.58	254	24.03	5119	HIMMELSBERG	
06	MM	23	3493.58 5636.48	282	24.44	5119	HIMMELSBERG	
06	MM	31	3491.80 5635.50	278	26.47	5119	ANZEFAHR NE	
06	MM	32	3491.48 5635.41	260	26.80	5119	ANZEFAHR NE	
06	MM	33	3491.12 5635.20	218	27.21	5119	ANZEFAHR NE	
05	M	31	3488.85 5632.52	275	30.51	5119	BAUERBACH	
05	M	32	3488.41 5632.10	275	31.10	5119	BAUERBACH	
05	M	33	3488.33 5631.71	275	31.37	5119	BAUERBACH	
05	M	11	3487.56 5630.42	300	32.70	5118	MARBURG E	
05	M	12	3486.98 5629.86	300	33.48	5118	MARBURG E	
05	M	13	3486.76 5629.56	300	33.83	5118	MARBURG E	
05	M	21	3486.21 5628.64	330	34.79	5218	SCHROECK	
05	M	22	3485.87 5628.00	330	35.43	5218	SCHROECK	
05	M	23	3486.02 5627.57	330	35.54	5218	SCHROECK	
05	M	42	3481.91 5627.38	320	39.10	5218	MARBURG SW	
05	M	41	3481.78 5627.59	320	39.11	5218	MARBURG SW	
05	M	43	3481.72 5626.68	320	39.63	5218	MARBURG SW	
05	M	61	3479.68 5625.93	260	41.76	5218	OBERWEIMAR	
05	M	62	3479.25 5625.82	260	42.19	5218	OBERWEIMAR	
05	M	63	3479.00 5625.87	260	42.37	5218	OBERWEIMAR	
05	M	51	3477.78 5625.99	240	43.37	5218	ALLNA	
05	M	52	3477.57 5626.01	240	43.54	5218	ALLNA	
05	M	53	3477.20 5625.80	240	43.96	5218	ALLNA	
06	M	63	3475.92 5625.35	285	45.27	5217	NANZ WILLERSHAUSEN	
06	M	62	3475.50 5625.14	305	45.74	5217	NANZ WILLERSHAUSEN	
06	M	41	3473.96 5623.92	280	47.68	5217	LOHRA 3	
06	M	42	3473.65 5623.75	280	48.03	5217	LOHRA 3	
06	M	51	3473.60 5623.25	248	48.33	5217	LOHRA 2	Y
06	M	52	3473.60 5623.25	248	48.33	5217	LOHRA 2	X
06	M	53	3473.60 5623.25	248	48.33	5217	LOHRA 2	
06	M	43	3473.39 5623.46	245	48.40	5217	LOHRA 3	
06	CL	11	3472.55 5622.76	255	49.48	5217	LOHRA 1	
06	CL	12	3472.23 5622.57	248	49.85	5217	LOHRA 1	
06	CL	13	3471.83 5622.30	297	50.33	5217	LOHRA 1	
06	CL	21	3470.50 5621.48	315	51.90	5217	RODENHAUSEN 2	
06	CL	22	3470.10 5621.23	325	52.37	5217	RODENHAUSEN 2	
06	CL	23	3469.73 5621.10	310	52.75	5217	RODENHAUSEN 2	
06	M	11	3467.78 5620.78	400	54.60	5217	RODENHAUSEN 1	
06	M	12	3467.54 5620.62	410	54.89	5217	RODENHAUSEN 1	
06	CL	31	3467.66 5619.74	347	55.23	5217	ROSSBACH 2	
06	M	13	3467.22 5620.36	405	55.30	5217	RODENHAUSEN 1	
06	M	21	3467.00 5620.28	410	55.53	5217	ROSSBACH NE	
06	CL	32	3467.30 5619.53	342	55.64	5217	ROSSBACH 2	
06	M	22	3466.68 5619.97	400	55.96	5217	ROSSBACH NE	
06	CL	33	3466.98 5619.24	340	56.07	5217	ROSSBACH 2	
06	M	23	3466.26 5619.82	370	56.40	5217	ROSSBACH NE	
06	M	31	3466.14 5619.89	365	56.47	5217	ROSSBACH 1	
06	M	32	3466.02 5619.49	345	56.77	5217	ROSSBACH 1	
06	M	33	3465.86 5619.35	345	56.98	5217	ROSSBACH 1	
06	F	13	3463.26 5617.17	280	60.32	5316	MUDERSBACH	
06	F	12	3462.90 5616.98	290	60.73	5316	MUDERSBACH	
06	F	11	3462.62 5616.75	295	61.09	5316	MUDERSBACH	
06	F	23	3459.10 5614.72	350	65.15	5316	BELLERSDORF	
06	F	22	3458.84 5614.48	360	65.50	5316	BELLERSDORF	
06	F	21	3458.34 5614.36	360	65.99	5316	BELLERSDORF	
06	F	33	3456.14 5612.22	325	68.97	5316	SINN	
06	F	32	3455.86 5611.98	320	69.33	5316	SINN	
06	F	31	3455.57 5611.80	315	69.67	5316	SINN	
05	MM	31	3450.94 5607.29	355	76.01	5315	HOLZHAUSEN	
05	MM	32	3450.61 5607.01	300	76.44	5315	HOLZHAUSEN	
05	MM	33	3450.44 5606.75	280	76.72	5315	HOLZHAUSEN	
05	MM	43	3449.36 5605.49	410	78.31	5415	ULM W	
05	MM	42	3448.97 5605.62	383	78.57	5415	ULM W	
05	MM	41	3448.62 5605.77	350	78.78	5415	ULM W	
05	MM	11	3447.63 5604.66	425	80.21	5415	OBERSHAUSEN	
05	MM	12	3447.38 5604.30	385	80.61	5415	OBERSHAUSEN	
05	MM	13	3446.97 5604.28	335	80.97	5415	OBERSHAUSEN	
05	MM	21	3445.16 5603.12	340	83.12	5415	EPSTEINS KOPF	
05	MM	22	3444.90 5602.88	346	83.47	5415	EPSTEINS KOPF	
04	MM	13	3442.94 5601.18	265	86.06	5415	PROBACH	
04	MM	12	3442.48 5600.42	348	86.86	5415	PROBACH	
04	MM	11	3441.97 5599.89	363	87.58	5415	PROBACH	
04	MM	43	3440.24 5599.03	327	89.50	5414	LAHR	
04	MM	42	3439.86 5598.57	340	90.07	5414	LAHR	
04	MM	41	3439.17 5598.36	333	90.76	5414	LAHR	
04	MM	33	3437.75 5597.69	265	92.32	5414	ELLAR	
04	MM	32	3436.95 5597.28	282	93.21	5414	ELLAR	
04	MM	31	3436.19 5596.76	282	94.13	5414	ELLAR	
04	F	21	3430.71 5594.11	175	100.18	5514	NIEDERZEUZHEIM	
04	F	22	3430.32 5593.83	205	100.66	5514	NIEDERZEUZHEIM	
04	F	23	3429.80 5593.68	230	101.18	5514	NIEDERZEUZHEIM	
04	F	11	3427.03 5593.75	350	103.49	5513	MOLSBERG	
04	F	12	3426.75 5593.39	325	103.92	5513	MOLSBERG	
04	F	13	3426.40 5593.20	325	104.32	5513	MOLSBERG	
05	KA	11	3413.65 5590.49	300	116.66	5512	ELGENDORF	
05	KA	12	3413.39 5590.40	305	116.93	5512	ELGENDORF	
05	KA	13	3412.94 5590.33	340	117.36	5512	ELGENDORF	
05	KA	23	3411.48 5589.89	420	118.85	5512	HOEHR GRENZHS.3	
05	KA	22	3411.07 5589.83	455	119.24	5512	HOEHR GRENZHS.3	
05	KA	21	3410.67 5589.70	460	119.65	5512	HOEHR GRENZHS.3	
05	KA	33	3408.97 5588.80	430	121.57	5512	HOEHR GRENZHS.2	
05	KA	32	3408.55 5588.78	400	121.95	5512	HOEHR GRENZHS.2	
05	KA	31	3408.10 5588.82	380	122.32	5512	HOEHR GRENZHS.2	
05	KA	41	3406.52 5587.96	240	124.12	5512	HOEHR GRENZHS.1	
05	KA	42	3406.28 5588.02	230	124.31	5512	HOEHR GRENZHS.1	
05	KA	43	3405.86 5587.86	225	124.75	5512	HOEHR GRENZHS.1	
06	BO	11	2597.70 5579.50	204	147.63	5610	OCHTENDUNG S	
06	BO	12	2597.38 5579.26	190	148.03	5610	OCHTENDUNG S	
06	BO	13	2597.05 5579.15	193	148.37	5610	OCHTENDUNG S	
06	BO	21	2596.08 5578.95	223	149.31	5610	OCHTENDUNG W	
06	BO	22	2595.67 5578.94	218	149.67	5610	OCHTENDUNG W	
06	KA	41	2594.29 5581.26	160	149.76	5609	REGINARISBRUNNEN	
06	KA	42	2594.06 5581.05	162	150.06	5609	REGINARISBRUNNEN	
06	BO	3	2595.25 5578.82	221	150.10	5610	OCHTENDUNG W	
06	KA	43	2593.70 5580.86	165	150.47	5609	REGINARISBRUNNEN	
06	KA	51	2590.71 5579.41	195	153.79	5609	KOTTENHEIM	
06	KA	52	2590.27 5579.30	195	154.23	5609	KOTTENHEIM	
06	KA	53	2590.19 5579.10	200	154.39	5609	KOTTENHEIM	
05	CL	41	2581.04 5575.19	365	164.32	5608	WEILER	
05	CL	42	2580.73 5575.01	403	164.68	5608	WEILER	
05	CL	43	2580.29 5574.89	424	165.13	5608	WEILER	
05	CL	11	2577.35 5573.88	395	168.20	5708	ANSCHAU	
05	CL	12	2576.96 5573.73	360	168.61	5708	ANSCHAU	
05	CL	13	2576.63 5573.60	380	168.96	5708	ANSCHAU	
05	CL	21	2575.59 5573.24	420	170.05	5708	ARBACH 2	
05	CL	22	2575.26 5573.13	380	170.40	5708	ARBACH 2	
05	CL	23	2574.86 5572.92	423	170.85	5708	ARBACH 2	
04	CL	1	2573.57 5572.30	410	172.30	5708	ARBACH 1	
04	CL	2	2573.25 5572.09	465	172.69	5708	ARBACH 1	
04	CL	3	2572.94 5571.92	470	173.04	5708	ARBACH 1	
04	CL	1	2565.50 5568.46	495	181.23	5707	KATZWINKEL	
04	CL	2	2565.07 5568.33	511	181.68	5707	KATZWINKEL	
04	CL	3	2564.75 5568.22	527	182.01	5707	KATZWINKEL	
04	CL	71	2558.92 5565.63	480	188.38	5706	STEINBORN	
04	CL	72	2558.60 5565.44	495	188.75	5706	STEINBORN	
04	CL	73	2558.24 5565.25	505	189.16	5706	STEINBORN	
05	S	41	2554.07 5564.96	625	192.97	5706	KIRCHWEILER 1	
05	S	42	2553.77 5564.80	650	193.31	5706	KIRCHWEILER 1	
05	S	43	2553.45 5564.65	645	193.66	5706	KIRCHWEILER 1	
05	S	21	2552.30 5564.70	590	194.67	5706	GEROLSTEIN	
05	S	22	2551.83 5564.69	595	195.09	5706	GEROLSTEIN	
05	MS	11	2531.08 5555.57	515	217.73	5804	WEISDORF	
05	MS	12	2530.74 5555.64	510	218.00	5804	WEISDORF	
05	MS	13	2530.38 5555.68	520	218.30	5804	WEISDORF	
05	MS	21	2527.18 5554.69	505	221.61	5804	EILSCHEID	
05	MS	22	2526.79 5554.72	495	221.95	5804	EILSCHEID	
05	MS	23	2526.38 5554.86	450	222.26	5804	EILSCHEED	
06	MS	21	2517.71 5551.24	520	231.62	5903	ARZFELDERHOEHE 2	
06	MS	12	2517.54 5551.12	513	231.82	5903	ARZFELDER HOEHE	
06	MS	23	2517.54 5551.12	513	231.82	5903	ARZFELDERHOEHE 2	
06	MS	22	2517.33 5551.03	500	232.05	5903	ARZFELDERHOEHE 2	
06	MS	13	2517.16 5550.95	487	232.24	5903	ARZFELDERHOEHE 1	

QUARRY-NO. 14 MEHRBERG SMAP 5309
PROFILE 14-010 COMPILED BY DEGUTSCH.WIEHLE.WEEP

B	IN	AC	COORDINATES	ALT	X(KM)	SMAP	LOCALITY	G C
07	F	31	2593.12 5615.60	210	6.91	5309	E.BRUNGSBERG	
07	F	32	2593.18 5615.89	200	7.21	5309	E.BRUNGSBERG	
07	F	33	2593.27 5616.15	200	7.48	5309	E.BRUNGSBERG	
05	F	31	2591.30 5623.42	190	14.63	5209	HOMMERICH	
05	F	32	2591.37 5623.91	175	15.12	5209	HOMMERICH	
05	F	33	2591.42 5624.38	145	15.59	5209	HOMMERICH	
05	F	11	2592.75 5633.32	202	24.54	5109	W.WOLPERATH	
05	F	12	2592.83 5633.68	182	24.91	5109	W.WOLPERATH	
07	F	11	2595.53 5637.48	210	28.90	5110	SOENTGERATH	

07	F	12	2595.46 5637.92	222	29.32	5110	SOENTGERATH	
07	F	13	2595.55 5638.35	215	29.77	5110	SOENTGERATH	
03	MS	11	2596.20 5640.98	220	32.45	5110	HETZENHOLZ	MA
03	MS		2596.29 5641.34	230	32.82	5110	HETZENHOLZ	MA
07	F	21	2596.00 5643.87	250	35.30	5010	BIRKEN	
07	F	22	2596.08 5644.25	250	35.69	5010	BIRKEN	
07	F	23	2596.18 5644.64	260	36.09	5010	BIRKEN	
02	MS		2598.86 5647.50	275	39.21	5010	ENGELSKIRCHEN	MA
02	MS		2598.40 5653.43	320	44.99	4910	SE.LINDLAR	MA
02	MS		2598.45 5653.88	320	45.44	4910	SE.LINDLAR	MA
07	M		2598.37 5656.25	305	47.88	4910	FRIELINGSDORF	
07	M		2598.55 5656.57	280	48.23	4910	FRIELINGSDORF	
07	M		2598.73 5656.80	240	48.48	4910	FRIELINGSDORF	
03	MS		2599.16 5661.71	315	53.39	4910	FAEHNRICHSSTU	BS
07	M	11	2601.58 5666.33	330	58.33	4810	WIPPERFUERTH	
07	M	12	2601.70 5666.70	350	58.72	4810	WIPPERFUERTH	
07	M	13	2601.73 5667.13	340	59.15	4810	WIPPERFUERTH	
03	MS	21	2599.93 5670.94	340	62.63	4810	GARDEWEGERMUE	BS
07	M	21	2602.55 5672.89	350	64.97	4810	HALVERFUERTH	
07	M	22	2602.62 5673.25	380	65.33	4810	HALVERFUERTH	
07	M	23	2602.67 5673.58	395	65.67	4810	HALVERFUERTH	
06	M	21	2602.23 5675.72	365	67.73	4710	KREISCH	
06	M	22	2602.37 5676.08	390	68.10	4710	KREISCH	
06	M	23	2602.37 5676.46	390	68.47	4710	KREISCH	
05	KR	11	2600.97 5678.80	340	70.60	4710	ALTENBRECKERFELD	
05	KR	13	2601.30 5679.57	390	71.40	4710	ALTENBRECKERFELD	
07	M	41	2604.33 5682.25	330	74.50	4710	EPSCHEID	
07	M	42	2604.42 5682.58	350	74.83	4710	EPSCHEID	
07	M	43	2604.42 5682.98	315	75.23	4710	EPSCHEID	
02	MS		2603.05 5684.09	350	76.00	4710	N.BRECKERFELD	MA
02	MS		2603.04 5684.41	350	76.32	4710	N.BRECKERFELD	MA
02	MS		2603.03 5684.99	380	76.89	4710	N.BRECKERFELD	MA
03	KR	14	2605.40 5686.83	312	79.17	4611	DAHL	
06	M	41	2608.58 5688.88	395	81.82	4611	BRECHTEFELD	
06	M	42	2608.63 5689.29	370	82.22	4611	BRECHTEFELD	
06	M	43	2608.70 5689.68	350	82.62	4611	BRECHTEFELD	
02	MS		2607.95 5697.43	185	89.96	4511	S.WESTHOFEN	MA
06	BO	11	2609.30 5705.70	172	98.46	4511	SW.DORTMUND S	BS
06	BO	12	2609.35 5706.06	155	98.82	4511	SW.DORTMUND S	BS
06	BO	13	2609.16 5706.41	148	99.13	4511	SW.DORTMUND S	BS
07	BO	21	2609.82 5709.81	128	102.59	4411	NEU ASSELN	
07	BO	22	2609.74 5710.20	120	102.96	4411	NEU ASSELN	
07	BO	23	2609.81 5710.60	107	103.36	4411	NEU ASSELN	
06	BO	22	2609.62 5716.04	73	108.70	4411	DORTMUND KURL	BS
06	BO	23	2609.69 5716.30	74	108.97	4411	DORTMUND KURL	BS
07	BO	11	2611.47 5722.18	65	115.06	4311	W.BERGKAMEN	
07	BO	12	2611.53 5722.48	65	115.36	4311	W.BERGKAMEN	
07	BO	13	2611.55 5722.60	60	115.48	4311	W.BERGKAMEN	
03	MS		2608.01 5727.39	102	119.66	4311	NE.CAPPENBERG	BS
02	MS		2608.85 5731.71	75	123.97	4211	E.SUEDKIRCHEN	BS
05	BO	21	2612.45 5731.78	78	124.69	4211	SW.HERBERN	BS
05	BO	22	2612.38 5732.06	83	124.96	4211	SW.HERBERN	BS
05	BO	23	2612.28 5732.29	90	125.17	4211	SW.HERBERN	BS
05	BO	11	2612.51 5735.90	78	128.77	4211	NW.HERBERN	BS
05	BO	12	2612.58 5736.25	79	129.13	4211	NW.HERBERN	BS
05	BO	13	2612.67 5736.68	80	129.57	4211	NW.HERBERN	BS
03	MS	11	2611.50 5739.71	68	132.34	4211	ASCHEBERG	BS
03	MS	12	2611.50 5739.71	68	132.35	4211	ASCHEBERG	X
03	MS	13	2611.50 5739.80	68	132.44	4211	ASCHEBERG	BS
07	MM	41	2613.89 5744.75	60	137.71	4111	W.ALTENDORF	
07	MM	42	2613.97 5745.09	60	138.06	4111	W.ALTENDORF	
07	MM	43	2613.94 5745.50	60	138.46	4111	W.ALTENDORF	
02	MS		2612.70 5747.47	60	140.12	4111	NE.DAVENSBERG	BS
05	MS	21	2617.08 5747.70	60	141.18	4112	N.RINKERODE	
05	MS	22	2617.18 5748.11	57	141.60	4112	N.RINKERODE	
05	MS	23	2617.18 5748.47	58	141.95	4112	N.RINKERODE	
07	MM	31	2614.85 5752.38	57	145.40	4112	WENTRUP	
07	MM	32	2614.89 5752.72	55	145.75	4112	WENTRUP	
07	MM	33	2614.90 5753.13	52	146.14	4012	WENTRUP	
07	MM	21	2616.28 5756.28	50	149.47	4012	NE.GREMMENDORF	
07	MM	22	2616.40 5756.64	55	149.85	4012	NE.GREMMENDORF	
07	MM	23	2616.48 5757.10	54	150.32	4012	NE.GREMMENDORF	
07	MM	12	2618.00 5758.81	53	152.26	4012	W.MS.MOELLENBECK	
07	MM	13	2618.05 5759.15	52	152.61	4012	W.MS.MOELLENBECK	
07	MS		2617.44 5763.02	51	156.32	4012	E.MANDORF	MA
07	MS		2617.46 5763.10	51	156.41	4012	E.MANDORF	MA
07	MS		2617.47 5763.24	51	156.53	4012	E.MANDORF	MA
07	KI	11	2620.34 5769.88	53	163.58	3912	SOMMER POHLKOETT	
07	KI	12	2620.41 5770.28	53	163.98	3912	SOMMER POHLKOETT	
07	KI	13	2620.43 5770.68	53	164.37	3912	SOMMER POHLKOETT	
07	KI	21	2620.68 5772.24	51	165.95	3912	BROCK	
07	KI	22	2620.60 5772.63	51	166.34	3912	BROCK	
07	KI	23	2620.67 5773.03	51	166.73	3912	BROCK	
07	KI	31	2618.92 5775.81	48	169.18	3812	S.LADBERGEN	
07	KI	32	2619.02 5776.20	49	169.57	3812	S.LADBERGEN	
07	KI	33	2619.10 5776.55	50	169.94	3812	S.LADBERGEN	
07	KI	41	2619.64 5783.92	55	177.31	3812	STILLE	
07	KI	42	2619.79 5784.33	55	177.73	3812	STILLE	
07	KI	43	2619.89 5784.76	55	178.17	3812	STILLE	
07	MS	11	2620.55 5788.59	75	182.06	3712	BROCHTERBECK	
06	MS	11	2620.40 5788.78	79	182.23	3712	BROCHTERBECK	
06	MS	12	2620.26 5788.96	95	182.39	3712	BROCHTERBECK	
07	MS	12	2620.61 5788.97	120	182.45	3712	BROCHTERBECK	
06	MS	13	2620.51 5789.21	90	182.67	3712	BROCHTERBECK	
07	MS	13	2620.82 5789.26	80	182.76	3712	BROCHTERBECK	
05	MS	13	2618.93 5790.33	130	183.54	3712	BROCHTERBECK	
05	MS	12	2618.52 5790.50	150	183.65	3712	BROCHTERBECK	
05	MS	11	2618.29 5790.59	145	183.71	3712	BROCHTERBECK	
07	MS	21	2623.20 5792.55	110	186.40	3712	E.LAGGENBECK	
07	MS	22	2623.26 5792.83	114	186.68	3712	E.LAGGENBECK	
07	MS	23	2623.32 5793.15	87	187.01	3712	E.LAGGENBECK	
06	MS	21	2621.65 5794.45	100	188.03	3712	LAGGENBECK	
06	MS	22	2621.71 5794.77	110	188.35	3712	LAGGENBECK	
06	MS	23	2621.75 5795.22	120	188.81	3712	LAGGENBECK	

QUARRY-NO. 14 MEHRBERG SMAP 5309
PROFILE 14-090-02 COMPILED BY HAENEL.STEIN

B	IN	AC	COORDINATES	ALT	X(KM)	SMAP	LOCALITY	G C
02	M	F	2592.74 5610.03	345	1.37	5309	KALENBORN	
02	M	E	2592.79 5610.33	340	1.65	5309	KALENBORN	
02	M	D	2592.83 5610.62	325	1.92	5309	KALENBORN	
02	M	C	2592.89 5610.90	325	2.20	5309	KALENBORN	
02	M	B	2592.93 5611.16	320	2.46	5309	KALENBORN	
02	M	A	2592.97 5611.46	305	2.75	5309	KALENBORN	
02	KR		3401.82 5608.72	270	22.19	5311	PUDERBACH	
01	KR		3411.52 5606.04		32.19	5412	HERSCHBACH	
02	GO		3418.80 5605.45	446	39.37	5413	DREIFELDEN	
02	GO		3424.19 5606.60	490	44.64	5413	AILERTCHEN	
01	GO		3433.05 5604.40	430	53.77	5414	SECK	
02	F		3443.82 5605.26	315	64.32	5415	NENDERROTH	
01	GO		3448.06 5603.96	340	68.77	5415	ALLENDORF W	
02	F		3453.68 5603.88	250	74.25	5416	LEUNER BERG	
02	CL		3461.42 5607.76	215	81.80	5316	ASSLAR	
02	CL		3466.75 5606.97	232	87.15	5417	NAUNHEIM	
01	M		3471.72 5606.10	230	92.26	5417	KINZENBACH	
01	M		3477.18 5602.97	160	97.88	5418	GIESSEN	
01	M		3484.31 5606.50	270	104.83	5418	ANNEROD	
01	CL		3487.23 5606.28	238	107.76	5418	REISKIRCHEN	
02	S		3489.00 5605.58	245	109.43	5419	LINDENSTRUTH	
02	S		3494.93 5603.79	250	115.43	5419	QUECKBORN	
01	S		3498.26 5604.35	247	118.85	5419	LAUTER	
01	S		3503.39 5604.46	301	123.97	5420	LARDENBACH	
02	M		3503.84 5603.49	300	124.34	5420	FREIENSEN	
02	M		3509.62 5603.44	450	130.12	5420	BOBENHAUSEN	
01	MZ		3510.56 5605.28	455	131.11	5420	BOBENHAUSEN	
02	M	-A	3512.27 5605.72	480	132.68	5421	ULRICHSTEIN	
02	M	-B	3512.68 5605.66	505	133.09	5421	ULRICHSTEIN	
02	M	-C	3513.08 5605.59	525	133.50	5421	ULRICHSTEIN	
02	M	-A	3513.56 5605.79	570	133.97	5421	ECKMANNSHAIN	
02	M	-B	3513.98 5605.73	590	134.39	5421	ECKMANNSHAIN	
02	M	-C	3514.47 5605.71	585	134.88	5421	ECKMANNSHAIN	
02	M	-A	3515.31 5605.89	595	135.72	5421	LANGERAIN	
02	M	-B	3515.86 5605.80	575	136.27	5421	LANGERAIN	
02	M	-C	3516.28 5605.90	570	136.69	5421	LANGERAIN	
01	MZ		3516.55 5605.95	560	137.08	5421	ENGELROD	
02	M	-B	3518.20 5606.12	530	138.60	5421	ENGELROD	
02	M	-A	3518.29 5606.30	535	138.69	5421	ENGELROD	
02	M	-A	3519.40 5606.64	505	139.79	5421	HOERGENAU	
02	M	-B	3519.55 5606.71	505	139.94	5421	HOERGENAU	
02	M	-C	3519.95 5606.88	495	140.33	5421	HOERGENAU	
02	M	-A	3521.24 5606.22	490	141.64	5421	HOPFMANNSFELD	
02	M	-B	3521.67 5606.41	500	142.06	5421	HOPFMANNSFELD	
02	M	-C	3521.91 5606.40	495	142.30	5421	HOPFMANNSFELD	
02	M	-A	3523.12 5606.28	480	143.52	5421	WEHRBERG	
01	M	B	3522.98 5605.28	460	143.52	5421	HOPFMANNSFELD	
01	M	A	3522.98 5605.16	470	143.53	5421	HOPFMANNSFELD	
01	M	C	3523.04 5605.30	460	143.58	5421	HOPFMANNSFELD	
02	M	-B	3523.56 5606.34	460	143.95	5421	WEHRBERG	
02	M	-C	3523.90 5606.56	475	144.29	5421	WEHRBERG	
02	M	-A	3526.12 5606.39	415	146.51	5422	RIXFELD 1	
02	M	-B	3526.67 5606.52	425	147.06	5422	RIXFELD 1	
02	M	-C	3526.99 5606.71	410	147.38	5422	RIXFELD 1	
01	M	A	3527.25 5605.34	460	147.79	5422	RIXFELD	
02	M	-A	3527.72 5606.19	440	148.12	5422	RIXFELD 2	
01	M	B	3527.79 5605.41	470	148.33	5422	RIXFELD	
02	M	-B	3528.01 5606.26	450	148.40	5422	RIXFELD 2	
01	M	C	3528.04 5605.45	465	148.58	5422	RIXFELD	
02	M	-C	3528.43 5606.42	460	148.82	5422	RIXFELD 2	
01	M	C	3534.18 5605.93	345	154.70	5422	STOCKHAUSEN	
01	M	B	3534.43 5605.96	350	154.95	5422	STOCKHAUSEN	
01	M	A	3534.68 5606.02	340	155.20	5422	STOCKHAUSEN	
01	M	A	3546.20 5606.26	450	166.71	5423	GLAESERZELL	
01	M	C	3546.21 5606.28	260	166.72	5423	GLAESERZELL	
01	M	B	3546.31 5606.26	455	166.82	5423	GLAESERZELL	
01	M	A	3549.00 5609.53	370	169.48	5324	MARBACH	
01	M	B	3549.30 5609.61	360	169.78	5324	MARBACH	
01	M	C	3549.63 5609.65	340	170.11	5324	MARBACH	
01	M	C	3554.88 5603.97	330	175.45	5424	MARGRETENHAUN	
01	M	B	3555.15 5604.01	345	175.72	5424	MARGRETENHAUN	
01	M	A	3555.58 5604.06	370	176.14	5424	MARGRETENHAUN	
01	M	A	3559.15 5603.28	415	179.74	5425	LANGENBIEBER	
01	M	B	3559.25 5603.25	415	179.84	5425	LANGENBIEBER	
01	M	A	3565.61 5606.03	600	186.12	5425	OBERGRUBEN	
01	M	B	3565.90 5606.26	585	186.41	5425	OBERGRUBEN	
01	M	A	3571.69 5605.94	600	192.20	5426	HILDERS	
01	M	B	3571.95 5605.88	620	192.46	5426	HILDERS	

QUARRY-NO. 14 MEHRBERG SMAP 5309
TEST 14-195-V COMPILED BY STEIN

B	IN	AC	COORDINATES	ALT	X(KM)	SMAP	LOCALITY	G C
04	F	11	3397.89 5539.60	440	72.17	6011	SIMMERN E	
04	F	12	3397.89 5539.60	440	72.17	6011	SIMMERN E	
04	F	13	3397.89 5539.60	440	72.17	6011	SIMMERN E	
04	KR	11	3446.64 5573.00	240	76.37	5715	CAMBERG	
04	KR	13	3446.55 5572.69	243	76.44	5715	CAMBERG	
04	KR	12	3446.78 5572.58	235	76.69	5715	CAMBERG	
04	F	21	2596.39 5527.10	345	81.84	6110	RHAUNEN N	
04	F	22	2596.39 5527.10	345	81.84	6110	RHAUNEN N	
04	F	23	2596.39 5527.10	345	81.84	6110	RHAUNEN N	
04	MM	11	2580.77 5514.68	641	94.78	6208	MOXEL	
04	MM	12	2580.46 5514.52	625	94.97	6208	MOXEL	
04	MM	13	2580.41 5514.32	585	95.18	6208	MOXEL	
04	MM	43	2580.40 5514.32	600	95.18	6208	MORBACH	
04	MM	42	2580.03 5514.26	600	95.28	6208	MORBACH	
04	MM	41	2579.74 5514.16	600	95.42	6208	MORBACH	
04	MM	51	2579.40 5513.52	655	96.10	6208	MORSCHEID	
04	MM	52	2579.04 5513.29	631	96.37	6208	MORSCHEID	
04	MM	53	2578.71 5513.12	607	96.59	6208	MORSCHEID	
04	MM	61	2578.28 5512.92	571	96.84	6208	DEUSELBACH	
04	MM	62	2578.14 5512.73	548	97.05	6208	DEUSELBACH	
04	MM	63	2577.91 5512.47	603	97.34	6208	DEUSELBACH	
04	MM	21	2577.29 5512.14	636	97.76	6208	SCHWARZER STEIN	

B	IN	AC	COORDINATES	ALT	X(KM)	SMAP	LOCALITY	G C
04	MM	22	2577.06 5512.02	624	97.91	6208	SCHWARZER STEIN	
04	MM	23	2576.74 5511.82	604	98.16	6208	SCHWARZER STEIN	
04	MM	31	2576.32 5511.40	570	98.64	6208	SCHW.STEIN SW	
04	MM	32	2576.05 5511.20	557	98.88	6208	SCHW.STEIN SW	
04	MM	33	2575.77 5511.05	540	99.07	6208	SCHW.STEIN SW	
04	MS	11	2575.39 5510.60	535	99.58	6208	FORST DHRONECK	
04	MS	21	2574.61 5510.30	500	100.00	6208	HILSCHEID	
04	MS	13	2574.82 5510.21	525	100.06	6208	FORST DHRONECK	
04	MS	22	2574.23 5509.84	485	100.52	6208	HILSCHEID	
04	MS	23	2573.79 5509.57	526	100.87	6208	HILSCHEID	
04	KI	31	2573.83 5509.26	545	101.16	6208	MARBORN	
04	KI	32	2573.51 5509.00	610	101.48	6208	MARBORN	
04	KI	33	2573.22 5508.74	665	101.79	6208	MARBORN	
04	KI	41	2572.68 5508.40	670	102.22	6208	MALBORNER WALD	
04	KI	42	2572.37 5508.20	660	102.47	6208	MALBORNER WALD	
04	KI	43	2572.08 5507.94	640	102.79	6208	MALBORNER WALD	
04	KI	11	2571.66 5507.83	620	102.97	6207	TIEFENTALERHOF SE	
04	KI	12	2571.55 5507.53	615	103.29	6207	TIEFENTALERHOF SE	
04	KI	13	2571.37 5507.35	610	103.50	6207	TIEFENTALERHOF SE	
04	KI	21	2571.29 5507.24	666	103.63	6307	THIERGARTEN	
04	KI	22	2571.11 5507.13	607	103.77	6307	THIERGARTEN	
04	KI	23	2570.75 5506.90	577	104.07	6307	THIERGARTEN	
04	H	31	2569.77 5506.27	540	104.89	6307	ABTEI	
04	H	32	2569.49 5506.14	540	105.07	6307	ABTEI	
04	H	33	2569.13 5506.04	540	105.25	6307	ABTEI	
04	H	21	2568.56 5505.70	520	105.70	6307	ABTEI	
04	H	22	2568.47 5505.63	520	105.79	6307	ABTEI	
04	H	23	2568.09 5505.41	520	106.09	6307	ABTEI	
04	KA	21	2567.96 5504.71	558	106.80	6307	HERMESKEIL 3	
04	KA	22	2567.56 5504.47	537	107.13	6307	HERMESKEIL 3	
04	KA	23	2567.16 5504.30	505	107.38	6307	HERMESKEIL 3	
04	KA	11	2566.88 5503.94	495	107.80	6307	MUEHLENBERG	
04	KA	12	2566.67 5503.69	525	108.09	6307	MUEHLENBERG	
04	KA	13	2566.55 5503.52	530	108.28	6307	MUEHLENBERG	
04	S	31	2566.05 5503.10	540	108.81	6307	HERMESKEIL 1	
04	S	32	2565.80 5503.00	540	108.96	6307	HERMESKEIL 1	
04	S	33	2565.61 5502.87	550	109.14	6307	HERMESKEIL 1	
04	S	41	2565.16 5502.47	550	109.63	6307	HERMESKEIL 2	
04	S	42	2564.94 5502.27	560	109.88	6307	HERMESKEIL 2	
04	S	43	2564.55 5502.06	530	110.18	6307	HERMESKEIL 2	
04	F	31	2549.24 5489.79	450	126.41	6406	BRITTEN N	
04	F	32	2549.24 5489.79	450	126.41	6406	BRITTEN N	
04	F	33	2549.24 5489.79	450	126.41	6406	BRITTEN N	

QUARRY-NO. 15 BUEDINGEN SMAP 5621
PROFILE 15-240 COMPILED BY BERCKHEMER, MEISSNER

B	IN	AC	COORDINATES	ALT	X(KM)	SMAP	LOCALITY	G C
02	B		3504.80 5567.00	232	11.01	5720	RONNEBERG	
02	B		3500.74 5563.42	195	16.42	5720	HUETTENGESAESS	
02	GO		3497.28 5561.39	147	20.39	5819	OBERINIGHEIM	
02	GO		3491.35 5553.32	112	30.15	5819	HANAU	
02	F		3483.42 5544.84	150	41.74	5918	NEU ISENBURG	
02	F		3483.42 5544.84	150	41.74	5918	NEU ISENBURG	
01	F		3483.00 5538.90	205	46.42	6018	OFFENTHAL	
01	F		3478.90 5502.40	375	79.43	6318	KIRSCHHAUSEN	
01	H		3451.61 5520.23	160	81.92	6115	UELVERSHEIM	
02	CL		3438.22 5509.49	237	98.99	6214	DINTESHEIM	
02	CL		3434.82 5507.82	280	102.66	6214	FREIMERSHEIM	
02	KI		3424.40 5502.40	420	114.15	6313	SCHWARZFELSEN	
01	H		3423.61 5498.35	480	117.42	6313	DANNENFELS	
02	KI		3421.67 5499.29	685	118.23	6313	DONNERSBERG	
02	HH	A	3409.48 5487.31	328	135.25	6412	OTTERBERG	
02	HH	B	3409.05 5486.99	342	135.79	6412	OTTERBERG	
02	HH	C	3408.71 5486.64	335	136.27	6412	OTTERBERG	
02	HH	A	3405.99 5484.00	250	140.05	6512	OTTERBACH	
02	HH	B	3405.82 5483.91	278	140.24	6512	OTTERBACH	
02	HH	C	3405.52 5483.63	256	140.65	6512	OTTERBACH	
02	HH	A	3401.30 5481.86	226	145.03	6511	WEILERBACH	
02	HH	B	3400.81 5481.68	227	145.52	6511	WEILERBACH	
02	HH	C	3400.35 5481.59	230	145.54	6511	WEILERBACH	

QUARRY-NO. 16 TABEN-RODT SMAP 6405
PROFILE 16-080 COMPILED BY MEISSNER, BAIER

B	IN	AC	COORDINATES	ALT	X(KM)	SMAP	LOCALITY	G C
03	H	31	2590.96 5503.90	362	48.70	6309	KRONWEILER	
03	H	33	2591.47 5504.16	395	49.27	6309	KRONWEILER	
03	H	34	2591.47 5504.16	395	49.27	6309	KRONWEILER	
03	H	24	2595.93 5505.15	505	53.83	6309	AUSWEILER	
03	H	21	2596.22 5505.26	520	54.14	6309	AUSWEILER	
04	MS	1	2602.58 5507.56	440	60.84	6310	KIRCHENBOTTENBACH	
04	MS	2	2602.91 5507.77	476	61.22	6310	KIRCHENBOTTENBACH	
04	MS	3	2603.23 5507.40	485	61.42	6310	KIRCHENBOTTENBACH	
03	KI	11	2605.49 5508.39	337	63.91	6210	OBERREIDENBACH	
03	KI	13	2605.50 5508.39	335	63.92	6210	OBERREIDENBACH	
07	GO	11	3396.30 5510.21	375	71.31	6211	HUNDSBACH	
07	GO	12	3396.70 5510.35	365	71.73	6211	HUNDSBACH	
07	GO	13	3397.15 5510.48	340	72.20	6211	HUNDSBACH	
07	GO	21	3400.47 5511.30	377	75.62	6211	LAUSCHIED	
07	GO	22	3400.84 5511.44	375	76.02	6211	LAUSCHIED	
07	GO	23	3401.29 5511.66	378	76.51	6211	LAUSCHIED	
02	CL		3402.40 5511.36	273	77.45	6211	RAUMBACH	
02	HH		3404.01 5511.73	220	79.10	6212	REHBORN	
07	GO	31	3406.10 5511.45	235	81.13	6212	REHBORN	
07	GO	32	3406.42 5511.51	250	81.45	6212	REHBORN	
02	HH		3410.01 5512.43	370	85.10	6212	NEUDORFERHOF	
02	HH		3411.67 5513.15	355	86.88	6212	DREIWEIHERHOF	
07	F	11	3414.05 5513.78	260	89.40	6212	HOLLGARTEN	
07	F	12	3414.54 5513.88	290	89.90	6212	HOLLGARTEN	
07	F	13	3414.92 5514.02	293	90.30	6212	HOLLGARTEN	
02	M	21	3416.90 5513.81	295	92.12	6213	WINTERBORN	
02	M	22	3417.10 5513.30	595	92.20	6213	WINTERBORN	
02	M	23	3417.54 5513.31	595	92.63	6213	WINTERBORN	
02	M	A	3420.53 5516.67	250	96.32	6213	FUERFELD	
02	M	B	3420.55 5516.64	250	96.34	6213	FUERFELD	
02	M	C	3420.84 5516.67	310	96.62	6213	FUERFELD	
07	HH	21	3422.05 5516.73	210	97.88	6213	FUERFELD E	
07	HH	22	3422.49 5516.85	213	98.33	6213	FUERFELD E	
02	M	A	3424.86 5517.37	200	100.69	6213	WONSHEIM	
02	M	B	3425.29 5517.42	220	101.12	6213	WONSHEIM	
02	M	C	3425.54 5517.43	210	101.37	6213	WONSHEIM	
07	HH	31	3427.13 5518.19	159	103.16	6213	ECKELSHEIM	
07	HH	32	3427.56 5518.36	158	103.62	6213	ECKELSHEIM	
02	M	A	3429.84 5518.83	178	105.88	6114	ARMSHEIM	
02	M	B	3430.04 5518.84	175	106.08	6114	ARMSHEIM	
02	M	C	3430.28 5518.97	170	106.34	6114	ARMSHEIM	
03	CL	21	3432.93 5518.65	138	108.89	6114	ARMSHEIM S	
03	CL	22	3433.13 5518.65	138	109.08	6114	ARMSHEIM S	
03	CL	23	3433.29 5518.66	145	109.24	6114	ARMSHEIM S	
02	M	31	3434.44 5519.92	175	110.61	6114	ARMSHEIM E	
02	M	32	3434.57 5519.82	175	110.71	6114	ARMSHEIM E	
02	M	33	3434.97 5519.88	200	111.11	6114	ARMSHEIM E	
07	GO	33	3436.78 5511.50	260	111.23	6212	REHBORN	
03	CL	13	3435.33 5519.75	257	111.48	6114	ENSHEIM	
03	CL	12	3435.60 5520.00	252	111.80	6114	ENSHEIM	
03	CL	11	3435.77 5520.15	246	112.00	6114	ENSHEIM	
02	M	11	3437.79 5520.92	225	114.10	6114	SPIESHEIM	
02	M	12	3438.08 5520.97	208	114.39	6114	SPIESHEIM	
02	M	13	3438.44 5521.02	202	114.75	6114	SPIESHEIM	
07	H	21	3442.02 5522.68	150	118.70	6115	UNDENHEIM	
07	H	22	3442.43 5522.82	150	119.13	6115	UNDENHEIM	
07	H	23	3442.88 5522.93	145	119.59	6115	UNDENHEIM	
07	H	31	3443.88 5524.17	150	120.87	6115	UNDENHEIM 2	
07	H	32	3444.31 5524.34	146	121.33	6115	UNDENHEIM 2	
07	H	33	3444.77 5524.46	140	121.81	6115	UNDENHEIM 2	
07	H	11	3446.99 5525.62	132	124.26	6115	SELZEN	
07	H	12	3447.42 5525.77	150	124.71	6115	SELZEN	
07	H	13	3447.86 5525.79	155	125.14	6115	SELZEN	
07	H	41	3450.14 5526.18	155	127.44	6115	SCHWABSBURG	
07	H	42	3450.56 5526.38	170	127.90	6115	SCHWABSBURG	
07	H	43	3450.97 5526.58	180	128.35	6115	SCHWABSBURG	
02	F	12	3477.80 5538.68	175	157.42	6018	LANGEN	
02	F	22	3483.04 5538.89		162.50	6018	OFFENTHAL	
02	F	21	3483.25 5538.96		162.72	6018	OFFENTHAL	
06	F	11	3491.09 5551.80	124	174.13	5819	STEINHEIM S	
06	F	12	3491.39 5552.00	126	174.48	5819	STEINHEIM S	
06	F	13	3491.65 5552.22	128	174.80	5819	STEINHEIM S	
02	H		3491.88 5551.91	120	174.95	5819	STEINHEIM	
02	H		3502.16 5554.56	130	185.53	5820	NIEDERRODENBACH	
06	F	33	3506.22 5562.80	160	192.17	5820	LANGENSELBOLD NE	
02	H		3506.58 5562.23	170	192.33	5820	ROTHENBERG	
06	F	32	3506.56 5562.66	170	192.43	5820	LANGENSELBOLD NE	
06	F	31	3506.81 5562.28	165	192.53	5820	LANGENSELBOLD NE	
06	F	21	3511.45 5564.49	260	197.65	5720	ROTH	
06	F	22	3511.65 5564.74	290	197.93	5720	ROTH	
06	F	23	3511.84 5565.04	260	198.22	5720	ROTH	
02	CL		3516.03 5562.45	250	201.25	5821	ALTENHASSLAU	
02	CL		3516.03 5562.45	250	201.25	5821	ALTENHASSLAU	X
03	F	13	3517.12 5568.55	409	204.50	5721	WAECHTERSBACH W	
03	F	22	3522.92 5572.27	250	211.27	5721	WEILERS	
03	F	32	3528.19 5574.97	254	217.17	5622	ROMSTAL	
01	KI		3575.47 5470.83	40	249.20	6626	ILLINGNEN	

QUARRY-NO. 16 TABEN-RODT SMAP 6405
PROFILE 16-130 COMPILED BY MUELLER, PETERSCHMITT

B	IN	AC	COORDINATES	ALT	X(KM)	SMAP	LOCALITY	G C
04	HH	21	2549.61 5487.29	351	6.18	6406	BRITTEN	
04	HH	22	2549.99 5487.18	323	6.56	6406	BRITTEN	
04	HH	23	2550.40 5486.95	284	7.03	6406	BRITTEN	
04	HH	41	2554.91 5482.52	375	13.23	6506	WAHLEN	
04	HH	42	2555.31 5482.32	368	13.67	6506	WAHLEN	
04	HH	43	2555.63 5482.10	370	14.05	6506	WAHLEN	
04	HH	32	2561.86 5479.64	288	20.61	6507	SCHMELZ	
04	HH	12	2569.63 5476.35	305	28.98	6507	BUBACH	
04	HH	13	2570.00 5476.19	328	29.39	6507	BUBACH	
04	KA	11	2574.29 5472.47	310	34.96	6608	UCHTELFANGEN	
04	KA	12	2574.27 5472.42	310	34.97	6608	UCHTELFANGEN	
04	KA	13	2574.32 5472.40	310	35.02	6608	UCHTELFANGEN	
04	KA	21	2580.56 5470.14	320	41.56	6608	HEILIGENWALD	
04	KA	23	2580.56 5470.14	320	41.56	6608	HEILIGENWALD	
04	KA	31	2586.27 5466.27	285	48.43	6609	NEUNKIRCHEN	
04	KA	32	2586.27 5466.27	285	48.43	6609	NEUNKIRCHEN	
04	KA	33	2586.27 5466.27	285	48.43	6609	NEUNKIRCHEN	
01	MS		2585.70 5464.21	300	48.98	6609	KLEBERBACH	
01	MS		2589.12 5464.34		51.81	6609	KOHLKOPF	
03	HH	1	2590.82 5461.62	375	54.75	6709	KIRKEL NEUHAEUSEL 2	
03	HH	2	2591.16 5461.40	335	55.15	6709	KIRKEL NEUHAEUSEL 2	
03	HH	3	2591.50 5461.16	360	55.57	6709	KIRKEL NEUHAEUSEL 2	
01	MS	A	2592.30 5461.40	337	56.07	6709	KIRKEL NEUHAEUSEL 1	
01	MS	G	2593.22 5461.23	360	56.93	6709	KIRKEL NEUHAEUSEL 1	
01	MS	F	2593.28 5461.23	360	57.44	6709	KIRKEL NEUHAEUSEL I	
01	MS		2594.90 5457.86		60.12	6709	BIERBACH	BSX
01	MS		2594.90 5457.86		60.12	6709	BIERBACH	BS
03	HH	3	2596.50 5460.56	285	60.18	6709	EINOED	
03	HH	2	2596.72 5460.23	300	60.53	6709	EINOED	
03	HH	1	2597.08 5460.08	320	60.92	6709	EINOED	
04	KA	43	2600.92 5459.46	300	64.55	6710	NIEDERAUERBACH	
04	KA	41	2600.94 5459.40	290	64.59	6710	NIEDERAUERBACH	
04	KA	42	2600.94 5459.40	290	64.59	6710	NIEDERAUERBACH	
03	HH	1	2602.14 5457.40	340	66.64	6710	ZWEIBRUECKEN	
03	HH	2	2602.40 5457.23	320	66.95	6710	ZWEIBRUECKEN	
01	F		2604.16 5454.32	320	69.88	6710	WALSHAUSEN 1	
02	HH	A	2606.84 5453.81	305	72.44	6710	WALSHAUSEN 2	
02	HH	B	2607.05 5453.73	292	72.67	6710	WALSHAUSEN 2	
02	HH	C	2607.40 5453.74	271	72.96	6710	WALSHAUSEN 2	
04	KI	21	2608.32 5452.76	300	74.29	6710	HUBERHOF	
04	KI	22	2608.43 5452.76	280	74.38	6710	HUBERHOF	
04	KI	23	2608.63 5452.63	260	74.62	6710	HUBERHOF	
04	KI	31	3392.37 5451.79	290	77.06	6811	EICHELSBACHERMUEHLE	
04	KI	32	3392.48 5451.64	318	77.23	6811	EICHELSBACHERMUEHLE	
02	HH	C	3395.21 5450.20	380	80.28	6811	WINZELN 2	
01	F		3395.72 5449.58	397	80.40	6811	WINZELN 1	
02	HH	B	3395.53 5450.03	389	80.64	6811	WINZELN 2	
02	HH	A	3395.94 5449.82	392	81.09	6811	WINZELN 2	
04	KI	11	3396.96 5447.68	360	83.14	6811	SIMTEN	
04	KI	12	3397.10 5447.67	320	83.26	6811	SIMTEN	
04	KI	13	3397.15 5447.59	360	83.35	6811	SIMTEN	
03	HH	1	3400.11 5447.96	325	84.98	6811	ERLENBRUNN	

03	MM 2	3399.47 5448.27	350	85.69	6811	ERLENBRUNN
03	MM 3	3399.47 5448.27	350	85.69	6811	ERLENBRUNN
01	S	3398.54 5442.18	360	87.50	6811	EPPENBRUNN
04	KI 41	3402.96 5444.79	330	89.75	6812	GLASHUETTE
04	KI 42	3403.14 5444.72	280	89.94	6812	GLASHUETTE
04	KI 43	3403.14 5444.64	280	89.98	6812	GLASHUETTE
02	S	3403.41 5441.21	252	92.06	6812	LUDWIGSWINKEL
01	S	3407.56 5440.17	285	96.08	6912	FISCHBACH
01	S	3411.14 5438.07	400	100.23	6912	NOTHWEILER
02	ST	3411.68 5434.54	260	102.70	6912	GIMBELHOF
03	S	3416.55 5438.08	235	104.87	6913	BOBENTHAL 2
01	M	3417.47 5435.15	415	107.12	6913	BOBENTHAL 1
03	S	3419.56 5438.27	210	107.35	6913	REISDORF
01	M	3423.04 5436.74	245	111.03	6913	RECHTENBACH 1
02	S	3423.53 5436.36	240	111.67	6913	RECHTENBACH 2
07	KA 61	3438.71 5432.47	137	126.90	6914	BUECHELBERG
07	KA 62	3438.71 5432.47	137	126.90	6914	BUECHELBERG
07	KA 63	3438.71 5432.47	137	126.90	6914	BUECHELBERG
01	MM	3449.64 5411.21	198	147.14	7115	ROTENFELS
02	KA	3453.00 5414.43	350	148.24	7116	MALSCH
01	MM	3448.17 5406.80	175	148.44	7215	WALDSEE GAGGENAU
01	CL	3457.50 5421.16	230	148.69	7016	ETTLINGEN
01	CL	3458.31 5416.80	285	151.54	7116	SCHOELLBRUNN
02	KA	3443.17 5393.32	750	152.88	7315	BUEHLERHOEHE
02	KA	3459.25 5413.57	310	154.45	7116	MARXZELL
01	MM	3455.60 5404.86	430	155.63	7216	LOFFENAU
02	KA	3458.40 5404.45	470	158.20	7216	HERRENALB
01	MM	3460.30 5404.46	521	159.75	7216	ASCHENHUETTE
02	KA	3462.58 5403.05	540	162.45	7216	EYACHMUEHLE
06	KA 51	3465.46 5402.54	670	165.13	7217	WILDBAD
06	KA 52	3465.95 5402.52	640	165.55	7217	WILDBAD
06	KA 53	3466.42 5402.44	600	165.99	7217	WILDBAD
03	KA 3	3468.87 5400.76	560	169.04	7217	KLEINENZHOF
02	M A	3469.36 5399.23	722	170.09	7217	BECHERKOPF
02	M B	3469.42 5399.17	725	170.30	7217	BECHERKOPF
02	M C	3469.44 5399.13	726	170.34	7217	BECHERKOPF
02	M A	3473.38 5396.70	690	174.96	7217	WUERZBACH
02	M B	3473.49 5396.80	690	175.00	7217	WUERZBACH
02	M C	3473.49 5396.80	690	175.00	7217	WUERZBACH
03	KA 4	3476.61 5399.58	580	176.20	7218	OBERREICHENBACH
04	S 31	3479.25 5401.38	575	177.48	7218	HIRSAU
04	S 32	3479.51 5401.22	540	177.78	7218	HIRSAU
04	S 33	3479.93 5401.16	500	178.17	7218	HIRSAU
06	KA 61	3482.36 5400.15	545	180.73	7218	NEUHENGSTETT
06	KA 62	3482.70 5399.91	530	181.15	7218	NEUHENGSTETT
06	KA 63	3483.06 5399.71	530	181.56	7218	NEUHENGSTETT
03	KA 6	3484.08 5394.90	550	185.02	7318	STAMMHEIM
01	M A	3483.13 5392.78	570	185.25	7318	GUELTINGEN
01	M B	3483.46 5392.62	540	185.62	7318	GUELTINGEN
04	S 21	3486.66 5493.82	530	187.74	7318	GECHINGEN
04	S 22	3487.00 5493.76	515	188.06	7318	GECHINGEN
04	S 23	3487.26 5493.47	544	188.44	7318	GECHINGEN
03	KA 2	3489.96 5390.98	510	192.08	7319	DACHTEL
04	S 11	3492.18 5392.26	513	193.26	7319	AIDLINGEN
04	S 12	3492.56 5392.22	510	193.61	7319	AIDLINGEN
04	S 13	3492.90 5392.00	510	194.01	7319	AIDLINGEN
01	M B	3492.85 5385.89	490	197.16	7319	NUFRINGEN
01	M C	3493.05 5385.85	510	197.35	7319	NUFRINGEN
03	KA 1	3496.76 5390.66	450	198.04	7319	EHNINGEN
06	KA 31	3500.55 5386.38	510	203.43	7320	SCHAICHHOF
06	KA 32	3500.82 5386.19	520	203.76	7320	SCHAICHHOF
06	KA 33	3501.08 5385.81	505	204.18	7320	SCHAICHHOF
01	M A	3501.79 5382.35	410	206.59	7420	BEBENHAUSEN
01	M B	3502.41 5382.29	400	207.15	7420	BEBENHAUSEN
06	KA 21	3505.52 5385.76	430	208.00	7320	WEIL IM SCHOENBUCH
06	KA 22	3505.77 5385.72	433	208.23	7320	WEIL IM SCHOENBUCH
06	KA 23	3506.17 5385.58	430	208.65	7320	WEIL IM SCHOENBUCH
03	S	3510.27 5382.58	410	213.79	7420	WALDORF
01	M A	3511.06 5380.18	415	215.61	7420	KIRCHENTELLINSFURT
01	M B	3511.24 5380.06	415	215.83	7420	KIRCHENTELLINSFURT
07	S 21	3515.62 5382.26	400	218.57	7421	ALTENRIET
07	S 22	3516.00 5382.34	375	218.86	7421	ALTENRIET
07	S 23	3516.29 5382.41	355	219.07	7421	ALTENRIET
06	KA 12	3517.42 5383.09	330	219.60	7421	NECKARTENZLGN
06	KA 11	3517.42 5383.04	333	219.62	7421	NECKARTENZLGN
06	KA 13	3517.47 5383.12	330	219.63	7421	NECKARTENZLGN
07	S 11	3518.46 5379.87	384	222.23	7421	RIEDERICH
07	S 13	3518.68 5379.62	380	222.55	7421	RIEDERICH
04	CL 21	3522.60 5379.82	415	225.76	7421	KAEPPISHAEUSEN
03	M 11	3522.65 5379.80	420	225.85	7421	METZINGEN
06	KA 41	3522.87 5379.57	522	226.08	7421	FLORIAN
04	CL 22	3522.90 5379.54	520	226.16	7421	KAEPPISHAEUSEN
07	S 31	3522.89 5379.54	510	226.22	7421	KAEPPISHAEUSERN SE
06	KA 42	3523.03 5379.50	400	226.25	7421	FLORIAN
03	M 12	3523.03 5379.62	480	226.27	7421	METZINGEN
06	KA 43	3523.05 5379.48	470	226.28	7421	FLORIAN
07	S 32	3523.07 5379.55	480	226.37	7421	KAEPPISHAEUSERN SE
04	CL 23	3523.25 5379.44	450	226.52	7421	KAEPPISHAEUSEN
07	S 33	3523.10 5379.25	422	226.55	7421	KAEPPISHAEUSERN SE
03	M 13	3523.50 5379.44	460	226.77	7421	METZINGEN
04	CL 71	3528.04 5377.51	708	231.64	7422	DETTINGEN
04	CL 72	3528.45 5377.44	704	232.03	7422	DETTINGEN
04	CL 73	3528.73 5377.20	700	232.39	7422	DETTINGEN
03	M 2	3530.95 5376.28	660	234.80	7422	HUELBEN
03	M 1	3530.68 5376.40	680	234.81	7422	HUELBEN
03	M 3	3531.04 5376.37	640	234.83	7422	HUELBEN
03	M 1	3532.59 5374.45	640	237.14	7422	GRABENSTETTEN
03	M 2	3532.81 5374.27	685	237.42	7422	GRABENSTETTEN
03	M 3	3533.13 5374.23	695	237.71	7422	GRABENSTETTEN
06	KA 71	3542.13 5374.87	820	245.17	7423	DONNSTETTEN
06	KA 72	3542.25 5374.72	830	245.35	7423	DONNSTETTEN
06	KA 73	3542.48 5374.35	835	245.73	7423	DONNSTETTEN
06	S 21	3545.24 5374.08	807	248.28	7423	WESTERHEIM
06	S 22	3545.42 5373.96	805	248.49	7423	WESTERHEIM
06	S 23	3545.73 5373.73	755	248.88	7423	WESTERHEIM
06	S 41	3549.85 5366.74	748	255.87	7524	SONTHEIM
06	S 42	3549.90 5366.74	747	255.92	7524	SONTHEIM
06	S 43	3550.03 5366.66	730	256.29	7524	SONTHEIM
06	S 31	3554.04 5363.74	690	261.00	7524	SEISSEN
06	S 32	3554.22 5363.53	680	261.26	7524	SEISSEN
06	S 33	3554.20 5363.45	675	261.28	7524	SEISSEN
06	S 11	3562.62 5368.77	625	266.10	7525	BERMARINGEN
06	S 12	3562.80 5368.61	620	266.34	7525	BERMARINGEN
06	S 13	3563.16 5368.49	615	266.71	7525	BERMARINGEN

QUARRY-NO. 16 TABENRODT SMAP 6405
PROFILE 16-195 COMPILED BY PRODEHL.PETERSCHMITT

B	IN AC	COORDINATES	ALT	X(KM)	SMAP	LOCALITY	G C
08	F 21	2542.56 5480.24	195	10.24	6505	SCHWEMLINGEN	
08	F 22	2542.64 5479.88	205	10.58	6505	SCHWEMLINGEN	
08	F 23	2542.82 5479.53	205	10.90	6505	SCHWEMLINGEN	
08	F 31	2537.30 5474.79	346	17.03	6505	MONDORF	
08	F 32	2537.27 5474.36	338	17.43	6505	MONDORF	
08	F 33	2537.15 5473.96	323	17.85	6505	MONDORF	
08	F 11	2541.86 5468.96	225	21.51	6605	NIEDALDORF	
08	F 12	2541.82 5468.54	265	21.93	6605	NIEDALDORF	
08	F 13	2541.88 5468.12	290	22.34	6605	NIEDALDORF	
08	M 71	G 0635.22 4919.01	325	25.92	35-12	HEINING L.B.VILLE	
08	M 72	G 0635.21 4918.89	330	26.14	35-12	HEINING L.B.VILLE	
08	M 73	G 0635.23 4918.73	325	26.43	35-12	HEINING L.B.VILLE	
08	M 81	G 0633.81 4916.46	290	30.78	35-12	ALZING	
08	M 82	G 0633.82 4916.21	300	31.24	35-12	ALZING	
08	M 83	G 0633.82 4916.00	290	31.63	35-12	ALZING	
08	M 91	G 0633.41 4913.73	235	35.86	35-12	TETERCHEN	
08	M 92	G 0633.46 4913.58	240	36.13	35-12	TETERCHEN	
08	M 93	G 0633.36 4913.39	240	36.49	35-12	TETERCHEN	
08	M101	G 0631.88 4911.14	300	40.86	35-12	BOULAY MOSELLE	
08	M102	G 0631.86 4911.07	310	40.99	35-12	BOULAY MOSELLE	
08	M103	G 0631.77 4910.87	320	41.37	35-12	BOULAY MOSELLE	
10	MM 11	G 0632.10 4905.83	304	50.48	35-13	BASSE VIGNEUL	
10	MM 12	G 0631.98 4905.59	304	50.94	35-13	BASSE VIGNEUL	
10	MM 13	G 0632.08 4905.37	304	51.33	35-13	BASSE VIGNEUL	
08	KI 11	G 0631.37 4900.73	290	60.09	35-13	MANY	
08	KI 12	G 0631.33 4900.51	280	60.50	35-13	MANY	
08	KI 13	G 0631.29 4900.32	260	60.86	35-13	MANY	
08	KI 21	G 0630.74 4857.60	265	65.94	35-14	HOLACOURT	
08	KI 22	G 0630.74 4857.37	290	66.36	35-14	HOLACOURT	
08	KI 23	G 0630.74 4857.17	320	66.73	35-14	HOLACOURT	
07	KI 13	G 0630.38 4855.76	258	69.37	35-14	CHICOURT	
07	KI 12	G 0630.43 4855.49	290	69.86	35-14	CHICOURT	
07	KI 11	G 0630.42 4855.23	320	70.34	35-14	CHICOURT	
07	KI 21	G 0629.41 4852.25	280	75.97	35-14	CHATEAU SALIN	
07	KI 22	G 0629.27 4852.06	335	76.34	35-14	CHATEAU SALIN	
07	KI 23	G 0629.11 4851.89	310	76.67	35-14	CHATEAU SALIN	
07	KI 31	G 0628.59 4849.56	270	81.04	34-14	COUTURES	
07	KI 32	G 0628.59 4849.41	260	81.31	34-14	COUTURES	
07	KI 33	G 0628.59 4849.20	280	81.70	34-14	COUTURES	
07	KI 42	G 0628.64 4847.49	228	84.84	34-14	SALONNES	
07	KI 43	G 0628.64 4847.28	208	85.23	34-14	SALONNES	
08	KA 41	G 0627.58 4844.43	250	90.62	34-15	BEZANGE LA GRANDE	
08	KA 42	G 0627.34 4844.23	274	91.03	34-15	BEZANGE LA GRANDE	
08	KA 43	G 0627.11 4844.11	265	91.28	34-15	BEZANGE LA GRANDE	
07	KA 41	G 0627.60 4842.36	280	94.42	34-15	SERRES	
07	KA 42	G 0627.75 4842.11	288	94.86	34-15	SERRES	
07	KA 43	G 0627.64 4841.86	280	95.34	34-15	SERRES	
07	KA 21	G 0627.16 4839.23	250	100.25	34-15	MAIXE	
07	KA 22	G 0627.05 4839.08	260	100.54	34-15	MAIXE	
07	KA 23	G 0627.08 4838.92	250	100.83	34-15	MAIXE	
08	KA 13	G 0626.81 4836.70	260	104.96	34-15	VITRIMONT	
08	KA 12	G 0626.71 4836.45	285	105.43	34-15	VITRIMONT	
08	KA 11	G 0626.92 4836.38	290	105.53	34-15	VITRIMONT	
07	M 61	G 0626.47 4833.83	250	110.28	34-16	MONT SUR MEURTHE	
07	M 62	G 0626.50 4833.72	255	110.48	34-16	MONT SUR MEURTHE	
07	M 51	G 0626.02 4830.96	240	115.63	34-16	LAMATH	
07	M 52	G 0625.81 4830.84	245	115.88	34-16	LAMATH	
07	M 53	G 0625.65 4830.67	255	116.22	34-16	LAMATH	
07	M 42	G 0625.34 4828.14	280	120.92	34-16	CLAYEURES	
07	M 43	G 0625.25 4828.01	270	121.17	34-16	CLAYEURES	
07	M 31	G 0625.09 4826.14	290	124.64	34-16	ST.BOINGT E	
07	M 32	G 0625.20 4825.99	295	124.90	34-16	ST.BOINGT E	
07	M 33	G 0625.20 4825.88	300	125.10	34-16	ST.BOINGT E	
07	M 21	G 0624.00 4822.97	330	130.63	34-17	DAMAS AUX BAIS	
07	M 22	G 0624.04 4822.76	345	131.01	34-17	DAMAS AUX BAIS	
07	M 23	G 0623.87 4822.59	360	131.34	34-17	DAMAS AUX BAIS	
07	M 11	G 0623.78 4820.38	350	135.42	34-17	CHATEL N	
07	M 12	G 0623.84 4820.19	360	135.77	34-17	CHATEL N	
07	M 13	G 0623.90 4819.98	355	136.14	34-17	CHATEL N	
07	MM 11	G 0623.10 4817.73	350	140.42	34-17	FRIZON	
07	MM 12	G 0623.10 4817.51	360	140.82	34-17	FRIZON	
07	MM 13	G 0623.20 4817.27	355	141.25	34-17	FRIZON	
07	MM 43	G 0623.22 4815.32	372	144.84	34-17	ONCOURT	
07	MM 42	G 0623.26 4815.11	372	145.22	34-17	ONCOURT	
07	MM 41	G 0623.32 4814.86	372	145.67	34-17	ONCOURT	
09	M 11	G 0623.90 4812.31	375	150.29	34-18	UXEGNEY	
09	M 12	G 0623.89 4812.06	375	150.75	34-18	UXEGNEY	
09	M 13	G 0623.90 4811.83	370	151.17	34-18	UXEGNEY	
09	M 41	G 0623.44 4809.68	420	155.20	34-18	BOUZEY	
09	M 42	G 0623.26 4809.48	430	155.59	34-18	BOUZEY	
09	M 43	G 0623.06 4809.28	450	155.98	34-18	BOUZEY	
10	M102	G 0620.29 4801.25	535	171.05	34-19	CHAPELLE BOIS	
10	M103	G 0620.23 4801.11	545	171.32	34-19	CHAPELLE BOIS	
10	M 81	G 0619.12 4758.77	495	175.79	34-19	LA ROCHEREE	
10	M 82	G 0619.13 4758.54	485	176.21	34-19	LA ROCHERE	
10	M 83	G 0619.06 4758.39	485	176.49	34-19	LA ROCHERE	
10	M 71	G 0619.74 4756.37	305	180.11	34-19	AILLESVILLERS	
10	M 72	G 0619.75 4756.32	300	180.20	34-19	AILLESVILLERS	
10	M 73	G 0619.76 4756.19	305	180.44	34-19	AILLESVILLERS	
10	M 91	G 0618.70 4753.82	258	184.96	34-19	GORBENAY	
10	M 92	G 0618.70 4753.82	258	184.96	34-19	GORBENAY	
10	M 93	G 0618.59 4753.71	259	185.17	34-19	GORBENAY	
09	MS 11	G 0617.09 4751.23	300	190.07	34-20	HAUTEVELLE	
09	MS 12	G 0617.08 4751.07	285	190.36	34-20	HAUTEVELLE	
09	MS 13	G 0617.07 4750.87	275	190.73	34-20	HAUTEVELLE	
09	MS 21	G 0616.44 4745.43	290	200.83	34-20	VISONCOURT	
09	MS 22	G 0616.32 4745.25	288	201.18	34-20	VISONCOURT	
09	MS 23	G 0616.28 4745.02	280	201.61	34-20	VISONCOURT	
09	CL 11	G 0616.62 4740.35	280	210.15	34-21	COLOMBOTTE	
09	CL 12	G 0616.56 4740.16	295	210.51	34-21	COLOMBOTTE	
09	CL 13	G 0616.48 4739.94	320	210.93	34-21	COLOMBOTTE	
09	CL 21	G 0617.04 4734.83	325	220.26	34-21	BELL.BARAQUES	
09	CL 22	G 0617.11 4734.56	295	220.75	34-21	BELL.BARAQUES	
09	CL 23	G 0617.00 4734.35	300	221.15	34-21	BELL.BARAQUES	

QUARRY-NO. 17 BRANSRODE SMAP 4725
PROFILE 17-170-01 COMPILED BY AICHELE.KAMINSKI

B	IN	AC	COORDINATES		ALT	X(KM)	SMAP	LOCALITY	G	C
05	CL	41	3563.27	5634.66	450	43.15	5125	AUSBACH		
05	CL	42	3563.27	5634.66	450	43.15	5125	AUSBACH		X
05	CL	43	3563.27	5634.66	450	43.15	5125	AUSBACH		Y
05	CL	21	3561.45	5608.51	450	69.20	5325	MORLES/RHOEN		
05	CL	22	3561.45	5608.51	450	69.20	5325	MORLES/RHOEN		X
05	CL	23	3561.45	5608.51	450	69.20	5325	MORLES/RHOEN		Y
01	F	1	3594.03	5548.00	280	134.13	5927	SCHONUNGEN		
01	F	2	3594.03	5547.90	280	134.22	5927	SCHONUNGEN		
01	F	3	3594.03	5547.90	280	134.22	5927	SCHONUNGEN		
01	F		3595.83	5546.30	355	136.23	5928	FORST		
02	KA		3596.56	5541.46	240	141.08	5928	OBEREUERHEIM 67		
01	KA		3596.56	5541.45	240	141.11	5928	OBEREUERHEIM 66		
01	KA		3596.56	5541.45	240	141.11	5928	OBEREUERHEIM 66		
01	KA		3596.56	5541.45	240	141.11	5928	OBEREUERHEIM 66		
01	S		3602.90	5531.97	350	151.95	6028	MICHELAU		
02	F	1	3601.35	5527.32	305	155.98	6128	OBERSCHWARZACH		
02	F	3	3601.35	5526.47	305	156.80	6128	OBERSCHWARZACH		
02	S	1	3600.36	5514.94	464	167.71	6228	ABSTWIND		
02	S	2	3600.36	5514.94	464	167.71	6228	ABSTWIND		
02	S	3	3600.36	5514.94	464	167.71	6228	ABSTWIND		
02	S	1	3600.06	5512.56	400	169.95	6228	WUESTENFELDEN		
02	S	2	3600.06	5512.56	400	169.95	6228	WUESTENFELDEN		
02	S	1	3600.22	5506.21	400	176.16	6328	ENZLAR		
02	S	2	3600.22	5506.21	400	176.16	6328	ENZLAR		
02	S	3	3600.22	5506.21	400	176.16	6328	ENZLAR		
02	KA	1	3602.45	5500.61	320	182.13	6328	BIBART		
02	KA	2	3602.45	5500.61	320	182.13	6328	BIBART		
02	KA	3	3602.45	5500.61	320	182.13	6328	BIBART		
02	KA	2	3602.96	5495.47	350	187.25	6428	DEUTENHEIM		
02	KA	3	3602.96	5495.47	350	187.25	6428	DEUTENHEIM		
02	KA	1	3601.62	5489.37	360	192.90	6428	ERKENBRECHTSHOFEN		
02	KA	2	3601.62	5489.37	360	192.90	6428	ERKENBRECHTSHOFEN		
03	F	12	3607.05	5481.31	375	201.91	6528	OBERZENN		
03	F	13	3607.15	5480.86	385	202.37	6528	OBERZENN		
03	F	31	3607.05	5476.30	450	206.78	6528	MITTELDACHSTETTEN		
03	F	32	3607.29	5475.86	436	207.27	6528	MITTELDACHSTETTEN		
03	F	22	3608.03	5470.07	425	213.07	6628	WESTL.HESSBACH		
03	F	23	3608.13	5469.70	437	213.46	6628	WESTL.HESSBACH		
03	B	12	4391.71	5466.36	440	217.05	6629	SCHMALENBACH		
03	B	11	4394.00	5461.29	480	222.47	6729	ANSBACH		
03	B	12	4394.24	5461.05	490	222.75	6729	ANSBACH		
03	M	51	4395.26	5458.98	480	224.98	6729	RAUENZELL		
03	M	53	4395.47	5458.09	490	225.90	6729	RAUENZELL		
03	M	61	4394.03	5452.40	430	231.20	6729	WEIDENDORF		
03	M	62	4394.10	5452.09	440	231.52	6829	WEIDENDORF		
03	M	63	4394.18	5451.72	450	231.90	6829	WEIDENDORF		
03	M	A	4396.41	5447.51	425	236.47	6829	HEINERSDORF		
03	M	B	4396.47	5447.43	425	236.56	6829	HEINERSDORF		
03	M	C	4396.68	5447.24	425	236.78	6829	HEINERSDORF		
03	M	11	4397.53	5442.43	460	241.66	6829	HAMMERSCHMIEDE		
03	M	12	4397.66	5442.02	465	242.08	6829	HAMMERSCHMIEDE		
03	M	13	4397.72	5441.76	465	242.35	6829	HAMMERSCHMIEDE		
03	M	21	4397.65	5436.79	455	247.21	6929	ALTENTRUEDINGEN		
03	M	22	4397.50	5436.49	460	247.48	6929	ALTENTRUEDINGEN		
03	M	41	4400.67	5431.69	450	252.81	6929	AUHAUSEN		
03	M	42	4400.90	5431.43	450	253.11	6929	AUHAUSEN		
03	M	43	4401.04	5431.22	450	253.34	6929	AUHAUSEN		
03	M	A	4400.90	5427.04	470	257.42	7029	HAINSFARTH		
03	M	B	4400.84	5426.82	495	257.62	7029	HAINSFARTH		
03	M	C	4400.72	5426.53	500	257.88	7029	HAINSFARTH		
03	M	31	4402.91	5422.86	440	261.91	7030	LERCHENBUEHL		
03	M	32	4403.16	5422.59	440	262.22	7030	LERCHENBUEHL		
03	M	33	4403.47	5422.38	460	262.49	7030	LERCHENBUEHL		
03	KA	23	4405.45	5419.03	495	266.18	7030	AMERBACH		
03	KA	22	4405.34	5418.64	460	266.54	7130	AMERBACH		
03	KA	21	4405.34	5418.64	460	266.54	7130	AMERBACH		
03	KA	33	4409.90	5414.32	510	271.73	7130	HEIDMERSBRUNN		
03	KA	32	4409.84	5414.26	510	271.77	7130	HEIDMERSBRUNN		
03	KA	31	4409.84	5413.94	505	272.09	7130	HEIDMERSBRUNN		
03	KA	43	4405.11	5407.80	510	277.10	7130	SALCHHOF		
03	KA	42	4405.11	5407.75	510	277.15	7130	SALCHHOF		
03	KA	41	4405.09	5407.72	510	277.18	7130	SALCHHOF		
03	S	41	4406.37	5403.34	450	281.72	7230	EBERMERGEN		
03	S	42	4406.33	5403.00	440	282.05	7230	EBERMERGEN		
03	S	43	4406.33	5402.60	425	282.44	7230	EBERMERGEN		
03	S	23	4407.48	5397.58	420	287.59	7230	RIEDLINGEN		
03	S	22	4407.48	5397.12	445	288.04	7230	RIEDLINGEN		
03	S	21	4407.48	5396.70	430	288.45	7230	RIEDLINGEN		
03	S	31	4409.67	5388.57	420	296.86	7330	LAUTERBACH		
03	S	32	4409.71	5388.26	420	297.17	7330	LAUTERBACH		
03	S	33	4409.84	5387.81	430	297.64	7330	LAUTERBACH		

QUARRY-NO. 17 BRANSRODE SMAP 4725
PROFILE 17-230 COMPILED BY STEIN

B	IN	AC	COORDINATES		ALT	X(KM)	SMAP	LOCALITY	G	C
04	HH	43	3479.39	5597.56	281	113.49	5418	NEUHOF		
04	HH	42	3479.10	5597.36	270	113.84	5418	NEUHOF		
04	HH	41	3478.96	5597.03	264	114.17	5418	NEUHOF		
04	HH	13	3478.91	5597.02	262	114.21	5418	NEUHOF		
04	HH	12	3478.59	5596.82	247	114.58	5418	NEUHOF		
04	HH	11	3478.26	5596.58	250	114.98	5418	NEUHOF		
04	HH	33	3477.76	5596.14	238	115.65	5418	LANG GOENS E		
04	HH	32	3477.54	5595.96	225	115.93	5418	LANG GOENS E		
04	HH	31	3477.26	5595.78	215	116.26	5418	LANG GOENS E		
04	HH	23	3476.83	5595.46	222	116.79	5518	LANG GOENS S		
04	HH	22	3476.68	5595.20	225	117.08	5518	LANG-GOENS S		
04	HH	21	3476.39	5595.06	198	117.38	5518	LANG GOENS S		
04	HH	63	3475.82	5594.80	216	117.97	5517	LANG GOENS		
04	HH	62	3475.35	5594.69	205	118.38	5517	LANG GOENS		
04	HH	61	3474.96	5594.49	222	118.80	5517	LANG GOENS		
04	MS	13	3473.82	5593.38	233	120.39	5517	NIEDERKLEEN 1		
04	MS	11	3473.21	5593.70	228	120.60	5517	NIEDERKLEEN 1		
04	MS	12	3473.44	5593.42	240	120.63	5517	NIEDERKLEEN 1		
04	MS	22	3472.47	5592.43	225	122.02	5517	NIEDERKLEEN 2		
04	MS	21	3472.16	5592.27	230	122.35	5517	NIEDERKLEEN 2		
04	KI	23	3471.98	5592.10	225	122.60	5517	OBERKLEEN		
04	KI	22	3471.64	5591.92	228	122.97	5517	OBERKLEEN		
04	KI	21	3471.35	5591.84	230	123.23	5517	OBERKLEEN		
04	KI	13	3470.74	5591.19	295	124.12	5517	OBERKLEEN SW		
04	KI	12	3470.47	5591.13	265	124.36	5517	OBERKLEEN SW		
04	KI	11	3470.34	5590.90	310	124.61	5517	OBERKLEEN SW		
04	KI	33	3469.90	5590.74	335	125.04	5517	CLEEBERG S		
04	KI	32	3469.58	5590.50	330	125.44	5517	CLEEBERG S		
04	KI	31	3469.23	5590.32	330	125.81	5517	CLEEBERG S		
04	KI	43	3468.73	5590.04	320	126.37	5517	CLEEBERG SW		
04	KI	42	3468.48	5589.87	360	126.67	5517	CLEEBERG SW		
04	KI	41	3468.20	5589.68	385	127.00	5517	CLEEBERG SW		
04	H	33	3467.21	5589.20	330	128.05	5517	CLEEBERG		
04	H	31	3466.94	5589.06	320	128.34	5517	CLEEBERG		
04	H	32	3466.94	5589.06	320	128.34	5517	CLEEBERG		
04	H	23	3466.65	5589.06	310	128.55	5517	GRIEDELBACH		
04	H	21	3466.01	5588.53	330	129.38	5517	GRIEDELBACH		
04	KA	13	3465.73	5587.87	285	130.04	5517	AM STOCKER		
04	KA	12	3465.42	5587.56	355	130.47	5517	AM STOCKER		
04	KA	11	3465.17	5587.25	370	130.87	5517	AM STOCKER		
04	KA	22	3464.91	5587.30	355	131.02	5517	HOCHWARDT		
04	KA	23	3464.91	5587.30	355	131.02	5517	HOCHWARDT		
04	KA	21	3464.63	5587.25	300	131.26	5517	HOCHWARDT		
04	S	33	3463.74	5586.54	380	132.40	5516	HASSELBORN		
04	S	32	3463.54	5586.35	390	132.67	5516	HASSELBORN		
04	S	31	3463.32	5586.25	380	132.90	5516	HASSELBORN		
04	S	43	3463.00	5585.87	385	133.39	5516	WELLERSTR		
04	S	42	3462.68	5585.72	440	133.73	5516	WELLERSTR		
04	S	41	3462.38	5585.52	430	134.08	5516	WELLERSTR		
04	F	31	3446.34	5573.96	255	153.70	5615	CAMBERG W		
04	F	32	3446.34	5573.96	255	153.70	5615	CAMBERG W		
04	F	33	3446.34	5573.96	255	153.70	5615	CAMBERG W		
04	F	21	3429.75	5562.37	452	173.79	5814	HUPPERT		
04	F	22	3429.75	5562.37	452	173.79	5814	HUPPERT		
04	F	23	3429.75	5562.37	452	173.79	5814	HUPPERT		
04	F	11	3413.40	5550.68	320	193.79	5912	KAUB		
04	F	12	3413.40	5550.68	320	193.79	5912	KAUB		
04	F	13	3413.40	5550.68	320	193.79	5912	KAUB		

QUARRY-NO. 17 BRANSRODE SMAP 4725
PROFILE 17-240-20 COMPILED BY PRODEHL.FUCHS.KAMINSKI

B	IN	AC	COORDINATES		ALT	X(KM)	SMAP	LOCALITY	G	C
05	GO	11	3557.18	5676.33	442	3.18	4724	VELMEDEN		
05	GO	12	3556.82	5676.18	462	3.57	4724	VELMEDEN		
05	GO	13	3556.53	5675.81	440	3.99	4724	VELMEDEN		
05	GO	21	3554.66	5675.58	459	5.79	4724	WALBURG		
05	GO	22	3554.29	5675.30	450	6.24	4724	WALBURG		
05	GO	23	3553.97	5675.06	380	6.62	4724	WALBURG		
06	GO	11	3549.58	5671.51	447	12.05	4824	RETTERODE W		
07	GO	11	3549.60	5671.50	440	12.24	4824	RETTERODE W		
06	GO	12	3549.31	5671.22	480	12.43	4824	RETTERODE W		
06	GO	13	3548.85	5671.11	500	12.88	4824	RETTERODE W		
07	GO	13	3548.87	5671.09	500	13.07	4824	RETTERODE W		
07	GO	21	3547.87	5670.13	540	14.43	4824	GUENSTERODE		
07	GO	22	3547.53	5669.91	509	14.83	4824	GUENSTERODE		
07	GO	23	3547.23	5669.55	525	15.28	4824	GUENSTERODE		
06	KI	41	3546.19	5668.93	450	16.29	4823	GUESTERODE SW		
06	KI	42	3545.81	5668.78	410	16.69	4823	GUESTERODE SW		
06	KI	43	3545.50	5668.58	410	17.06	4823	GUESTERODE SW		
06	KI	31	3544.35	5667.80	330	18.45	4823	ALTES GEHEGE		
06	KI	32	3544.02	5667.58	350	18.84	4823	ALTES GEHEGE		
06	KI	33	3543.76	5667.24	320	19.25	4823	ALTES GEHEGEE		
07	KI	21	3538.19	5663.51	173	26.15	4823	FAHRE		
07	KI	22	3537.92	5663.27	180	26.51	4823	FAHRE		
07	KI	23	3537.62	5663.05	170	26.88	4823	FAHRE		
07	KI	11	3534.80	5661.18	265	30.26	4962	OSTHEIM		
07	KI	12	3534.55	5660.90	285	30.63	4962	OSTHEIM		
07	KI	13	3534.25	5660.72	290	30.97	4962	OSTHEIM		
06	H	41	3515.81	5649.50	250	52.35	5021	DORHEIM		
06	H	42	3515.38	5649.30	245	52.82	5021	DORHEIM		
06	H	43	3515.04	5649.07	230	53.23	5021	DORHEIM		
06	H	51	3513.53	5648.55	220	54.78	5021	SCHLIERBACH		
06	H	52	3513.05	5648.33	250	55.31	5021	SCHLIERBACH		
06	H	53	3512.65	5648.15	315	55.74	5021	SCHLIERBACH		
07	H	41	3511.84	5647.07	290	57.20	5020	ELNRODE S		
07	H	11	3509.22	5645.09	380	60.48	5020	SACHSENHAUSEN		
07	H	12	3508.77	5645.07	380	60.87	5020	SACHSENHAUSEN		
07	H	13	3508.39	5644.88	370	61.29	5020	SACHSENHAUSEN		
06	H	31	3507.70	5643.63	310	62.35	5020	SACHSENHAUSEN		
06	H	32	3507.28	5643.53	320	62.75	5020	SACHSENHAUSEN		
06	H	33	3506.86	5643.38	350	63.19	5020	SACHSENHAUSEN		
06	H	11	3505.76	5642.28	340	64.71	5020	APPENHAIN		
06	H	12	3505.39	5642.11	310	65.11	5020	APPENHAIN		
06	H	13	3504.96	5641.98	320	65.54	5020	APPENHAIN		
07	M	51	3481.93	5627.31	280	93.05	5218	MARBURGER STADTWALD		
07	M	52	3481.68	5627.06	240	93.40	5218	MARBURGER STADTWALD		
07	M	53	3481.68	5626.68	270	93.60	5218	MARBURGER STADTWALD		
07	M	11	3479.60	5625.92	210	95.76	5218	GERMERSHAUSEN		
07	M	12	3479.30	5625.83	230	96.06	5218	GERMERSHAUSEN		
07	M	13	3479.00	5625.73	235	96.37	5218	GERMERSHAUSEN		
07	M	21	3477.94	5626.08	280	97.08	5218	ALLNA		
07	M	22	3477.56	5626.00	280	97.44	5218	ALLNA		
07	M	23	3477.32	5626.13	285	97.57	5218	ALLNA		
06	B	61	3475.92	5625.35	285	98.97	5217	NANZ WILLERSHAUSEN		
06	B	62	3475.49	5625.14	305	99.45	5217	NANZ WILLERSHAUSEN		
06	B	63	3475.08	5624.94	295	99.90	5217	NANZ WILLERSHAUSEN		
06	B	51	3474.02	5623.96	273	101.32	5217	LOHRA		
06	B	52	3473.65	5623.68	270	101.78	5217	LOHRA		
06	B	53	3473.41	5623.45	250	102.11	5217	LOHRA		
06	B	71	3471.19	5621.84	302	104.85	5217	SEELBACH		
06	B	72	3470.98	5621.65	300	105.12	5217	SEELBACH		
06	B	73	3470.65	5621.50	300	105.48	5217	SEELBACH		
06	B	11	3466.89	5620.35	400	109.28	5217	ROSSBACH		
06	B	12	3466.61	5620.12	375	109.64	5217	ROSSBACH		
06	B	13	3466.21	5619.89	355	110.10	5217	ROSSBACH		
06	HH	31	3458.30	5612.83	295	120.56	5316	NIEDERLEMP		
06	HH	32	3457.97	5612.56	227	120.98	5316	NIEDERLEMP		
06	HH	33	3457.66	5612.42	236	121.32	5316	NIEDERLEMP		

B	IN	AC	COORDINATES		ALT	X(KM)	SMAP	LOCALITY	G C
06	HH	21	3456.40	5611.31	247	122.98	5316	KOELSCHHAUSEN	
06	HH	23	3455.66	5610.92	258	123.81	5316	KOELSCHHAUSEN	
06	HH	41	3455.25	5610.77	283	124.24	5316	KATZENFURT	
06	HH	42	3454.93	5610.57	264	124.61	5316	KATZENFURT	
06	HH	43	3454.63	5610.23	240	125.05	5316	KATZENFURT	
06	HH	11	3453.32	5608.58	250	127.04	5316	GREIFENTHAL	
06	HH	12	3452.91	5608.41	277	127.48	5316	GREIFENTHAL	
06	HH	13	3452.54	5608.24	283	127.88	5315	GREIFENTHAL	
05	HH	31	3450.91	5607.26	355	129.89	5415	HOLZHAUSEN E	
05	HH	33	3450.42	5606.72	255	130.60	5415	HOLZHAUSEN E	
05	HH	21	3449.31	5605.69	360	132.09	5415	ULM W	
05	HH	22	3448.95	5605.61	383	132.43	5415	ULM W	
05	HH	23	3448.61	5605.51	410	132.77	5415	ULM W	
05	HH	11	3447.63	5604.66	425	134.06	5415	GRUBE WOHLFEIL	
05	HH	12	3447.38	5604.30	385	134.47	5415	GRUBE WOHLFEIL	
05	HH	13	3446.97	5604.28	335	134.82	5415	GRUBE WOHLFEIL	
05	HH	41	3445.26	5603.09	320	136.90	5415	EPPSTEINS KOPF	
05	HH	42	3444.88	5602.88	347	137.33	5415	EPPSTEINS KOPF	
05	HH	43	3444.65	5602.60	328	137.68	5415	EPPSTEINS KOPF	
05	F	11	3440.32	5600.17	325	142.63	5414	WALDERNBACH	
05	F	12	3440.01	5599.87	325	143.06	5414	WALDERNBACH	
05	F	13	3439.79	5599.66	345	143.36	5414	WALDERNBACH	
05	F	21	3434.44	5598.83	303	148.31	5414	MUEHLBACH	
05	F	22	3434.10	5598.70	300	148.67	5414	MUEHLBACH	
05	F	23	3433.86	5598.52	270	148.97	5414	MUEHLBACH	
05	F	31	3432.67	5597.80	180	150.36	5414	FRICKHOFEN E	
05	F	32	3432.39	5597.38	200	150.82	5414	FRICKHOFEN E	
05	F	33	3431.96	5597.27	205	151.24	5414	FRICKHOFEN E	
07	F	31	3424.84	5594.02	302	159.10	5513	DAHLEN S	
07	F	11	3422.50	5593.15	301	161.54	5513	DAHLEN S	
07	F	12	3422.25	5592.86	305	161.91	5513	DAHLEN S	
07	F	13	3421.89	5592.76	317	162.27	5513	DAHLEN S	
06	KA	31	3416.78	5591.66	280	167.01	5512	MONTABAUR	
06	KA	32	3416.39	5591.60	280	167.37	5512	MONTABAUR	
06	KA	33	3416.03	5591.56	290	167.70	5512	MONTABAUR	
05	KA	21	3413.66	5590.50	300	170.39	5512	ELGENDORF	
05	KA	22	3413.40	5590.40	303	170.66	5512	ELGENDORF	
05	KA	23	3412.94	5590.33	340	171.09	5512	ELGENDORF	
05	KA	13	3411.35	5589.95	430	172.66	5512	HOEHR GRENZHS.3	
05	KA	12	3411.07	5589.84	460	172.95	5512	HOEHR GRENZHS.3	
05	KA	11	3410.67	5589.70	460	173.37	5512	HOEHR GRENZHS.3	
05	KA	41	3408.96	5588.70	437	175.35	5512	HOEHR GRENZHS.2	
05	KA	42	3408.59	5588.82	400	175.61	5512	HOEHR GRENZHS.2	
05	KA	43	3408.17	5588.79	380	175.99	5512	HOEHR GRENZHS.2	
05	KA	32	3407.72	5588.54	355	176.50	5512	HOEHR GRENZHS.1	
05	KA	31	3407.38	5588.33	340	176.90	5512	HOEHR GRENZHS.1	
05	KA	33	3407.26	5588.21	320	177.06	5512	HOEHR GRENZHS.1	
06	KA	11	3404.79	5587.77	265	179.31	5511	WANDHOF	
06	KA	12	3404.30	5587.82	270	179.71	5511	WANDHOF	
06	KA	13	3403.85	5587.85	270	180.09	5511	WANDHOF	
07	KA	51	3398.78	5576.33	245	190.57	5611	KOBLENZ	
07	KA	52	3398.47	5576.23	300	190.89	5611	KOBLENZ	
07	KA	53	3398.00	5576.11	330	191.35	5611	KOBLENZ	
07	KA	21	3396.38	5576.06	316	192.75	5611	LAY	
07	KA	22	3396.01	5575.76	305	193.22	5611	LAY	
07	BO	21	3392.91	5580.80	211	193.28	5610	HENGSHOF	
07	KA	23	3395.63	5575.44	290	193.71	5611	LAY	
07	BO	22	3392.54	5580.50	220	193.75	5610	HENGSTHOF	
07	BO	23	3392.23	5580.32	228	194.11	5610	HENGSTHOF	
06	BO	11	3390.80	5579.98	272	195.32	5610	WOLKEN NW	
06	BO	12	3390.48	5579.73	272	195.72	5610	WOLKEN NW	
06	BO	13	3390.18	5579.63	265	196.03	5610	WOLKEN NW	
06	BO	21	2601.40	5579.28	333	198.12	5610	OCHTENDUNG SW	
06	BO	22	2601.10	5579.03	330	198.51	5610	OCHTENDUNG SW	
06	BO	23	2600.63	5578.99	323	198.93	5610	OCHTENDUNG SW	
05	BO	11	2597.70	5579.50	204	201.24	5610	OCHTENDUNG S	
05	BO	12	2597.37	5579.29	190	201.63	5610	OCHTENDUNG S	
05	BO	13	2597.05	5579.15	192	201.98	5610	OCHTENDUNG S	
05	BO	21	2596.08	5578.95	223	202.92	5610	EMMINGERHOF	
05	BO	22	2595.66	5578.93	218	203.27	5610	EMMINGERHOF	
05	BO	23	2595.25	5578.81	221	203.69	5610	EMMINGERHOF	
07	CL	41	2593.48	5580.28	186	204.56	5609	WELLING N	
07	CL	42	2593.16	5579.99	190	204.98	5609	WELLING N	
07	CL	43	2592.78	5579.82	210	205.39	5609	WELLING N	
07	CL	31	2591.18	5579.15	212	207.11	5609	HAUSEN	
07	CL	32	2591.18	5579.15	212	207.11	5609	HAUSEN	X
07	CL	33	2591.18	5579.15	212	207.11	5609	HAUSEN	Y
07	CL	11	2585.34	5577.43	355	213.02	5609	MAYEN W	
07	CL	13	2585.34	5577.43	355	213.02	5609	MAYEN W	Y
07	CL	21	2583.89	5576.65	435	214.67	5609	GEISBUESCHOF	
07	CL	22	2583.66	5576.54	432	214.92	5609	GEISBUESCHOF	
07	CL	23	2583.25	5576.40	405	215.34	5609	GEISBUESCHHOF	
06	CL	51	2581.04	5575.19	340	217.67	5608	WEILER E	
06	CL	52	2581.04	5575.19	340	217.67	5608	WEILER E	X
06	CL	53	2581.04	5575.19	340	217.67	5608	WEILER E	Y
07	CL	51	2581.04	5575.19	340	217.86	5608	WEILER E	
07	CL	52	2581.04	5575.19	340	217.86	5608	WEILER E	X
07	CL	53	2581.04	5575.19	340	217.86	5608	WEILER E	Y
06	CL	41	2579.08	5574.69	376	219.62	5608	ANSCHAU E	
07	CL	41	2579.08	5574.69	376	219.81	5608	ANSCHAU E	
06	CL	42	2578.78	5574.57	397	219.94	5608	ANSCHAU E	
07	CL	42	2578.78	5574.57	397	220.13	5608	ANSCHAU E	
06	CL	43	2578.59	5574.50	427	220.14	5608	ANSCHAU E	
07	CL	43	2578.59	5574.50	427	220.33	5608	ANSCHAU E	
07	CL	31	2577.26	5573.76	380	221.85	5708	ANSCHAU W	
06	CL	32	2576.91	5573.62	375	222.04	5708	ANSCHAU W	
07	CL	32	2576.91	5573.62	375	222.22	5708	ANSCHAU W	
06	CL	33	2576.68	5573.56	390	222.26	5708	ANSCHAU W	
07	CL	33	2576.68	5573.56	390	222.45	5708	ANSCHAU W	
07	CL	21	2575.60	5573.24	420	223.55	5708	DITSCHEID	
07	CL	22	2575.36	5573.02	380	223.87	5708	DITSCHEID	
07	CL	23	2574.88	5572.98	411	224.30	5708	DITSCHEID	
06	CL	11	2573.55	5572.25	420	225.63	5708	ARBACH W	
06	CL	12	2573.24	5572.09	460	225.98	5708	ARBACH W	
06	KR	11	2562.26	5567.82	515	237.64	5707	NERDLEN N	
06	KR	12	2561.88	5567.62	515	238.07	5707	NERDLEN N	
06	KR	13	2561.50	5567.39	480	238.51	5707	NERDLEN N	
07	S	31	2557.94	5566.70	540	242.14	5706	WALDKOENIGEN	
07	S	32	2558.00	5566.27	535	242.30	5706	WALDKOENIGEN	
07	S	33	2557.75	5566.20	560	242.55	5706	WALDKOENIGEN	
06	S	21	2555.22	5566.47	640	244.45	5706	ERNSTBERG	
06	S	22	2555.07	5566.15	645	244.73	5706	ERNSTBERG	
06	S	23	2554.84	5566.11	645	244.95	5706	ERNSTBERG	
06	S	11	2554.13	5564.88	640	246.17	5706	KIRCHWEILER	
06	S	12	2553.77	5564.80	650	246.52	5706	KIRCHWEILER	
06	S	13	2553.51	5564.60	640	246.85	5706	KIRCHWEILER	
06	S	31	2552.30	5564.70	590	247.86	5706	FORST GEROLSTEIN	
06	S	32	2551.83	5564.69	595	248.27	5706	FORST GEROLSTEIN	
06	S	33	2551.44	5564.60	550	248.66	5706	FORST GEROLSTEIN	
07	S	21	2550.47	5563.92	550	250.02	5706	PELMERWALD	
07	S	22	2549.98	5563.78	550	250.52	5706	PELMERWALD	
07	S	23	2549.65	5563.65	540	250.87	5706	PELMERWALD	
07	S	41	2548.17	5562.96	530	252.50	5706	BUESCHEICH	
07	S	42	2548.17	5562.96	530	252.50	5706	BUESCHEICH	
07	S	43	2548.08	5563.10	535	252.51	5706	BUESCHEICH	
07	MS	12	2530.70	5555.67	510	271.33	5804	HEISDORF	
07	MS	13	2530.34	5555.58	520	271.69	5804	HEISDORF	
07	MS	21	2529.02	5555.05	560	273.10	5804	EILSCHEID E	
07	MS	22	2528.64	5554.90	540	273.51	5804	EILSCHEID E	
07	MS	23	2528.32	5554.79	550	273.84	5804	EILSCHEID E	
05	MS	21	2524.45	5553.65	410	277.72	5804	KINZENBERG	
05	MS	22	2524.29	5553.63	405	277.87	5804	KINZENBURG	
05	MS	23	2524.00	5553.45	450	278.21	5804	KINZENBURG	
05	MS	11	2522.56	5553.24	490	279.58	5803	LICHTENBORN	
05	MS	12	2522.25	5553.06	505	279.94	5803	LICHTENBORN	
05	MS	13	2521.82	5552.93	520	280.37	5803	LICHTENBORN	

QUARRY-NO. 17 BRANSRODE SMAP 4725
TEST 17-320-V COMPILED BY DEGUTSCH

B	IN	AC	COORDINATES		ALT	X(KM)	SMAP	LOCALITY	G C
02	MS		3494.48	5743.25	415	92.66	4119	VELDROM	HA
02	MS		3494.30	5743.35	420	92.86	4119	VELDROM	HA
02	MS		3494.21	5743.38	415	92.94	4119	VELDROM	HA
02	BO	11	3475.69	5756.86	170	115.63	4017	OERLINGHSN SW	BS
02	BO	21	3471.47	5761.49	200	121.87	4017	HILLEGOSSN SW	BS
02	MS		3463.56	5764.73	260	129.88	3916	HUENENBURG	BS
02	MS		3456.30	5773.17	200	140.93	3916	WICHLINGHSN	BS
02	MS		3448.06	5778.61	140	150.69	3815	DISSEN	BS
02	MS	13	3432.46	5805.50	110	180.53	3614	PENTERKNAPP	BS
02	MS	11	3432.38	5805.53	105	180.61	3614	PENTERKNAPP	BS
02	MS	23	3427.97	5811.00	80	187.60	3513	GEHN	HA
02	MS	22	3427.93	5811.05	80	187.66	3513	GEHN	BS
02	MS	21	3427.88	5811.13	80	187.75	3513	GEHN	HA

QUARRY-NO. 17 BRANSRODE SMAP 4725
PROFILE 17-350-06 COMPILED BY HINZ,KAMINSKI,STEIN

B	IN	AC	COORDINATES		ALT	X(KM)	SMAP	LOCALITY	G C
03	MS	11	3558.23	5685.35	260	7.89	*4625	DOHRENBACH	BS
03	MS	12	3558.25	5685.68	235	8.21	4625	DOHRENBACH	BS
03	MS	13	3558.25	5686.00	221	8.53	4625	DOHRENBACH	BS
03	MS	21	3556.30	5689.23	206	12.14	4624	KL.ALMERODE E	BS
03	MS	22	3556.15	5689.56	263	12.50	*4624	KL.ALMERODE E	BS
03	MS	23	3556.05	5690.00	280	12.95	4624	KL.ALMERODE E	BS
05	CL	11	3549.52	5689.26	445	15.64	4624	ZIEGENHAGEN	
05	CL	12	3549.52	5689.26	445	15.64	4624	ZIEGENHAGEN	X
05	CL	13	3549.52	5689.26	445	15.64	4624	ZIEGENHAGEN	Y
03	H	51	3557.97	5694.35	235	16.83	4624	ALBSHAUSEN	
03	H	52	3557.80	5694.73	280	17.23	4624	ALBSHAUSEN	
03	H	53	3557.87	5695.12	290	17.60	*4624	ALBSHAUSEN	
03	H	41	3556.18	5697.73	370	20.44	*4524	ATZENHAUSEN	
03	H	42	3556.15	5698.18	345	20.88	4524	ATZENHAUSEN	
03	H	43	3556.04	5698.60	315	21.32	4524	ATZENHAUSEN	
03	H	11	3553.68	5702.20	335	25.34	4524	MEENSEN	
03	H	12	3553.56	5702.62	365	25.77	*4524	MEENSEN	
03	H	13	3553.52	5703.05	360	26.20	4524	MEENSEN	
03	H	22	3555.07	5707.08	400	29.84	4524	BOERDEL	
03	H	23	3554.87	5707.40	425	30.19	4524	BOERDEL	
05	CL	31	3546.62	5706.87	327	32.12	4524	BUEHREN	
05	CL	32	3546.62	5706.87	327	32.12	4524	BUEHREN	X
05	CL	33	3546.62	5706.87	327	32.12	4524	BUEHREN	Y
03	H	31	3553.78	5711.06	310	33.98	4424	BARTERODE	
03	H	32	3553.68	5711.53	320	34.46	4424	BARTERODE	
03	H	33	3553.65	5711.97	320	34.90	*4424	BARTERODE	
03	GO	11	3553.63	5714.52	277	37.42	4424	ADELEBSEN SE	
03	GO	12	3553.50	5715.00	307	37.91	*4424	ADELEBSEN SE	
03	GO	13	3553.42	5715.41	260	38.33	4424	ADELEBSEN SE	
03	KR	12	3552.76	5721.58	347	44.52	4324	HETTENSEN	
03	KR	13	3552.72	5721.96	350	44.90	*4324	HETTENSEN	
03	CL	12	3550.32	5723.80	420	47.15	*4324	VOLPRIEHAUSEN SW	
03	CL	13	3550.28	5724.24	370	47.59	4324	VOLPRIEHAUSEN SW	
03	CL	21	3549.71	5727.50	367	50.89	4324	DELLIEHAUSEN W	
03	CL	22	3549.66	5727.89	400	51.28	*4324	DELLIEHAUSEN W	
03	CL	23	3549.79	5728.52	420	51.88	4324	DELLIEHAUSEN W	
03	CL	31	3549.85	5729.85	413	53.17	4224	ESPOL W	
03	CL	32	3549.77	5730.31	425	53.64	*4224	ESPOL W	
03	CL	33	3549.73	5730.70	450	54.03	4224	ESPOL W	
05	KR	11	3548.65	5733.26	415	56.73	4224	HILWARTSHSN S	Y
05	KR	12	3548.65	5733.26	415	56.73	4224	HILWARTSHSN S	
05	KR	13	3548.65	5733.26	415	56.73	4224	HILWARTSHSN S	X
03	KI	41	3549.17	5733.57	385	56.95	4224	HILWARTSHSN.S	
03	KI	12	3548.01	5737.26	220	60.80	4224	HILWARTSHSN WSW	
03	KI	13	3547.92	5737.67	205	61.22	4224	HILWARTSHSN WSW	
03	KI	21	3549.31	5739.64	305	62.90	*4224	DASSEL	
03	KI	22	3549.31	5740.05	292	63.31	4224	DASSEL	
03	KI	23	3549.28	5740.52	235	63.77	4224	DASSEL	
03	KI	31	3548.88	5744.10	235	67.38	*4124	HUNNESRUCK	
03	KI	32	3549.02	5744.52	235	67.76	4124	DELLIEHAUSEN	
02	KI		3546.84	5748.59	250	72.08	*4124	WANGELNSTEDT	BS
02	KI		3546.84	5748.59	250	72.08	4124	WANGELNSTEDT	BSX
02	GO	1	3546.00	5751.60	290	75.19	*4124	LENNE	X
02	GO	2	3546.00	5751.60	290	75.19	4124	LENNE	
02	GO	1	3545.18	5756.88	220	80.53	4023	SCHARFOLIEND.	BSX
02	GO	3	3545.18	5756.88	220	80.53	4023	SCHARFOLIEND.	BSX
02	CL	1	3544.90	5758.18	300	81.86	4023	LUEERDISSEN	X
02	CL	2	3544.90	5758.18	300	81.86	*4023	LUEERDISSEN	
02	CL	3	3544.90	5758.18	300	81.86	4023	LUEERDISSEN	Y
02	CL	1	3544.31	5761.81	270	85.54	*4023	FOELZIEHAUSEN	
02	CL	2	3544.22	5761.93	260	85.67	4023	FOELZIEHAUSEN	

B	IN	AC	COORDINATES		ALT	X(KM)	SMAP	LOCALITY	G C
02	CL	3	3544.22	5761.93	260	85.67	*4023	FOELZIEHAUSEN	X
01	H		3541.05	5765.35	158	89.64	*3923	OCKENSEN	
01	H		3542.54	5769.29	360	93.20	*3923	SALZHEMMENDORF	
01	KR		3542.01	5773.05	124	96.99	*3923	HEMMENDORF	
01	GO		3541.51	5775.90	243	99.89	*3823	OSTERWALD	
01	GO		3540.37	5779.42	242	103.56	*3823	HOHENBERG	
01	H		3538.61	5783.34	285	107.75	*3823	SAUPARK	
02	HH	11	3539.98	5787.56	128	111.65	3723	SPRINGE	
02	HH	12	3540.00	5788.02	140	112.10	*3723	SPRINGE	
02	HH	14	3539.73	5788.35	149	112.47	3723	SPRINGE	
01	H		3538.45	5790.18	260	114.49	*3723	WENNIGSEN	
01	H		3538.03	5795.45	80	119.74	*3723	REDDERSE	
02	HH	31	3538.98	5796.82	77	120.94	*3623	GEHRDEN	
02	HH	32	3538.99	5797.08	76	121.19	3623	GEHRDEN	
02	HH	34	3539.01	5797.46	79	121.56	3623	GEHRDEN	
02	HH	44	3539.28	5798.51	63	122.55	*3623	LEVESTE	
02	HH	42	3539.38	5798.91	70	122.93	3623	LEVESTE	
02	HH	41	3539.50	5799.33	80	123.32	3623	LEVESTE	
02	KR		3537.24	5803.23	56	127.56	*3623	KIRCHWEHREN	
01	CL		3536.50	5805.54	53	129.94	*3623	ALMHORST	
02	HH	1	3534.87	5811.30	48	135.92	3523	RICKLINGEN	
02	HH	2	3534.96	5811.75	48	136.35	*3523	RICKLINGEN	
02	HH	4	3535.05	5812.16	48	136.73	3523	RICKLINGEN	
01	H		3534.76	5814.74	45	139.31	*3523	FRIELINGEN	
02	H	12	3534.43	5818.05	41	142.64	*3523	OTTERNHAGEN	
02	H	14	3534.05	5818.84	41	143.48	3523	OTTERNHAGEN	
02	CL	1	3533.60	5820.16	40	144.86	*3422	NEUSTADT/RBG	
02	CL	2	3533.36	5820.36	40	145.10	3422	NEUSTADT/RBG	
02	CL	3	3533.48	5820.46	40	145.18	3422	NEUSTADT/RBG	
01	HH		3533.68	5822.51	37	147.14	*3422	SUTTORF	
02	H	31	3533.73	5823.22	34	147.85	3422	MARIENSEE	
02	H	32	3533.64	5824.01	34	148.64	3422	MARIENSEE	
02	H	34	3533.61	5824.68	34	149.31	3422	MARIENSEE	
02	H	21	3533.31	5828.40	54	153.02	3422	BUEREN	
02	H	22	3533.31	5828.40	54	153.02	3422	BUEREN	
02	H	24	3533.32	5828.81	55	153.43	*3422	BUEREN	
02	KI	21	3533.62	5830.10	55	154.64	3322	BEVENSEN	
02	KI	22	3533.62	5830.86	48	155.39	3322	BEVENSEN	
03	HH	51	3529.44	5851.88	20	176.89	*3122	ALTENWAHLINGEN	
03	HH	52	3529.50	5852.26	23	177.25	3122	ALTENWAHLINGEN	
03	HH	53	3529.36	5852.64	25	177.65	3122	ALTENWAHLINGEN	
03	HH	43	3529.62	5853.57	22	178.52	3122	GR.EILSTORF SW	
03	HH	42	3529.65	5853.84	23	178.78	3122	GR.EILSTORF SW	
03	HH	41	3529.40	5854.20	20	179.18	3122	GR.EILSTORF SW	
03	HH	61	3529.19	5854.70	25	179.71	3122	GR.EILSTORF NW	
03	HH	62	3529.11	5855.18	27	180.19	*3122	GR.EILSTORF NW	
03	HH	63	3529.05	5855.61	24	180.63	3122	GR.EILSTORF NW	
03	HH	11	3529.00	5856.16	27	181.18	3122	SUEDKAMPEN S	
03	HH	12	3528.95	5856.60	30	181.62	3122	SUEDKAMPEN S	
03	HH	13	3528.67	5856.95	25	182.01	3122	SUEDKAMPEN S	
03	HH	21	3528.71	5857.52	30	182.57	3122	SUEDKAMPEN	
03	HH	22	3528.78	5857.94	30	182.97	3122	SUEDKAMPEN	
03	HH	23	3528.81	5858.36	30	183.38	3122	SUEDKAMPEN	
03	HH	31	3528.54	5858.93	30	183.99	*3122	NORDKAMPEN	
03	HH	32	3528.48	5859.33	32	184.39	3122	NORDKAMPEN	
03	HH	33	3528.55	5859.70	35	184.74	3122	NORDKAMPEN	

QUARRY-NO. 19 MERLENBACH
PROFILE 19-055 COMPILED BY NN

B	IN	AC	COORDINATES		ALT	X(KM)	SMAP	LOCALITY	G C
07	KI		2582.49	5463.90	270	29.78	6608	SPIESER	
06	KI		2582.49	5463.90	270	29.85	6608	SPIESER	
05	KI		2582.49	5463.90	270	29.91	6608	SPIESER	
06	KI		2588.10	5467.60	265	36.57	6609	NEUNKIRCHEN	
07	MS		2603.43	5480.63	384	56.52	6510	NANZWEILER	
06	MS		2603.43	5480.63	384	56.59	6510	NANZWEILER	
05	MS		2603.43	5480.63	384	56.65	6510	NANZWEILER	
05	MS		3391.10	5482.26	280	61.76	6510	REUSCHBACH	
07	HH		3399.04	5488.26	425	71.55	6411	GALGENBERG	
06	HH		3399.04	5488.26	425	71.62	6411	GALGENBERG	
06	HH		3399.04	5488.26	425	71.62	6411	GALGENBERG	
05	HH		3399.04	5488.26	425	71.68	6411	GALGENBERG	
06	HH		3403.67	5491.55	336	77.29	6412	HORSBACH	
05	HH		3406.11	5493.93	305	80.69	6412	HEIMKIRCHEN	
06	HH		3411.00	5498.84	395	87.43	6312	DOERRMOSCHEL	
02	F		3417.42	5496.11	345	90.85	6413	HAMBACHER HOF	
03	F		3417.42	5496.11	345	90.87	6413	HAMBACHER HOF	
03	F		3417.42	5496.11	345	90.87	6413	HAMBACHER HOF	
02	F		3424.30	5502.50	400	100.06	6313	DANNENFELS	
03	F		3424.30	5502.50	400	100.08	6313	DANNENFELS	
04	F		3424.30	5502.50	400	100.09	6313	DANNENFELS	
05	CL		3426.58	5508.16	325	105.62	6213	OBERWIESEN	
05	CL		3429.34	5510.60	270	109.27	6214	BECHENHEIM	
07	CL		3434.78	5518.40	200	118.06	6214	ALZEY	
07	H		3504.38	5566.91	240	202.89	5720	RONNEBERG	
06	H		3504.38	5566.91	240	202.96	5720	RONNEBERG	
06	H		3507.14	5569.56	200	206.75	5720	VONHAUSEN	

QUARRY-NO. 19 MERLENBACH
PROFILE 19-110 COMPILED BY MUELLER.PETERSCHMITT

B	IN	AC		COORDINATES		ALT	X(KM)	SMAP	LOCALITY	G C
02	ST		G	651.45	4907.57	250	5.80	36-13	FAREBERSVILLER	
01	ST		G	651.45	4907.57	250	5.85	36-13	FAREBERSVILLER	
04	ST		G	659.38	4908.15	345	14.02	36-13	ROULING	
03	ST		G	659.38	4908.15	345	14.06	36-13	ROULING	
04	KA		G	713.05	4912.58	225	24.49	36-13	FOLPERSVILLER	
03	KA		G	713.05	4912.58	225	24.53	36-13	FOLPERSVILLER	
04	KA		G	718.75	4910.64	262	28.96	36-13	BLIESBRUCK	
03	KA		G	718.75	4910.64	262	29.00	36-13	BLIESBRUCK	
02	KA		G	726.36	4908.97	340	34.86	37-13	RIMLING	
01	KA		G	726.36	4908.97	340	34.89	37-13	RIMLING	
02	S		G	732.75	4907.37	315	39.83	37-13	PETIT REDERDIN	
01	S		G	732.75	4907.37	315	39.86	37-13	PETIT REDERDIN	
04	S		G	737.21	4902.67	370	43.02	37-13	HOTTWILLER	
03	S		G	737.21	4902.67	370	43.06	37-13	HOTTWILLER	
07	ST		G	720.23	4857.42	300	46.60	36-14	SOUCHT	
06	ST		G	720.23	4857.42	300	46.70	36-14	SOUCHT	
05	ST		G	725.68	4905.00	300	48.00	37-13	BITCHE	
04	KA		G	751.47	4902.67	300	54.32	37-13	EGUELSHARDT	
03	KA		G	751.47	4902.67	300	54.37	37-13	EGUELSHARDT	
02	S		G	755.06	4903.72		56.61	37-13	STURZELBRONN	
01	S		G	755.06	4903.72		56.65	37-13	STURZELBRONN	
04	S		G	761.10	4902.71	265	61.10	38-13	NEUNHOFFEN	
03	S		G	761.10	4902.71	265	61.14	38-13	NEUNHOFFEN	
02	KA		G	765.74	4901.56	290	64.78	38-13	WINECKETHAL	
01	KA		G	765.74	4901.56	290	64.81	38-13	WINECKETHAL	
05	ST		G	740.87	4859.17	240	68.62	37-13	WINDSTEIN	
04	KA		G	771.94	4898.95	265	69.86	38-13	LANGENSOULTZBACH	
03	KA		G	771.94	4898.95	265	69.89	38-13	LANGENSOULTZBACH	
02	ST		G	747.53	4859.63	240	74.84	37-13	FORT DE TOUR A CH.	
03	ST		G	749.67	4858.68	450	77.82	37-13	SOULTZERKOPF	
06	ST		G	753.40	4900.87	320	82.48	37-13	CLEEBOURG	
07	ST		G	753.40	4900.87	320	82.48	37-13	CLEEBOURG	
06	KA			3446.21	5409.68	210	115.31	7115	KUPPENHEIM	
05	KA			3446.21	5409.68	210	115.34	7115	KUPPENHEIM	
02	M			3447.71	5408.64	250	115.64	7115	ROTENFELS	
01	M			3447.71	5408.64	250	115.67	7115	ROTENFELS	
04	M	A		3449.61	5405.32	230	118.51	7215	SELBACH	
04	M	B		3449.61	5405.32	230	118.51	7215	SELBACH	
03	M	A		3449.61	5405.32	230	118.54	7215	SELBACH	
03	M	B		3449.61	5405.32	230	118.54	7215	SELBACH	
01	M			3455.13	5408.89	440	122.56	7116	MICHELBACH	
07	KA			3453.39	5405.32	263	122.90	7216	MOERDEN	
00	S			3456.37	5408.77	698	123.67	7116	HERRENALB(11.2.66)	
05	KA			3455.59	5404.83	450	125.20	7216	LOFFENAU	
02	M	A		3458.66	5404.41	540	127.37	7216	GAISTAL	
01	M	A		3458.66	5404.41	540	127.41	7216	GAISTAL	
02	M	C		3459.07	5404.20	600	127.82	7216	GAISTAL	
01	M	C		3459.07	5404.20	600	127.86	7216	GAISTAL	
04	M	A		3461.32	5403.06	690	130.27	7216	ROSSKOPF	
03	M	A		3461.32	5403.06	690	130.30	7216	ROSSKOPF	
04	M	B		3461.56	5403.04	700	130.50	7216	ROSSKOPF	
03	M	B		3461.56	5403.04	700	130.52	7216	ROSSKOPF	
04	M	C		3461.73	5402.96	720	130.68	7216	ROSSKOPF	
03	M	C		3461.73	5402.96	720	130.71	7216	ROSSKOPF	
05	KA			3462.62	5403.07	550	132.40	7216	EYACHMUEHLE	
07	KA			3468.87	5400.76	560	139.05	7217	WILDBAD	
06	KA			3468.87	5400.76	560	139.09	7217	WILDBAD	
02	M	A		3472.55	5399.19	675	142.21	7217	WUERZBACH	
02	M	C		3472.59	5399.19	675	142.24	7217	WUERZBACH	
02	M	B		3472.61	5399.23	675	142.25	7217	WUERZBACH	
01	M	A		3472.55	5399.19	675	142.25	7217	WUERZBACH	
01	M	C		3472.59	5399.19	675	142.28	7217	WUERZBACH	
01	M	B		3472.61	5399.23	675	142.29	7217	WUERZBACH	
05	KA			3476.61	5399.58	580	146.73	7218	OBERREICHENBACH	
02	M	B		3478.16	5397.06	580	148.21	7218	SPESSHARDT	
01	M	B		3478.16	5397.06	580	148.25	7218	SPESSHARDT	
07	KA			3480.96	5396.36	440	151.84	7218	CALW	
02	M	A		3482.56	5392.70	580	153.88	7318	STAMMHEIM 2	
01	M	A		3482.56	5392.70	580	153.92	7318	STAMMHEIM 2	
03	M	B		3482.81	5392.55	590	154.12	7318	STAMMHEIM 2	
02	M	B		3482.81	5392.55	590	154.16	7318	STAMMHEIM 2	
03	M	C		3482.79	5392.37	575	154.18	7318	STAMMHEIM 2	
01	M	B		3482.81	5392.55	590	154.20	7318	STAMMHEIM 2	
02	M	C		3482.79	5392.37	575	154.22	7318	STAMMHEIM 2	
01	M	C		3482.79	5392.37	575	154.26	7318	STAMMHEIM 2	
05	S			3484.07	5395.90	555	154.99	7318	STAMMHEIM 1	
03	M	A		3485.94	5392.92	550	156.91	7318	DECKENPFRONN	
02	M	A		3485.94	5392.92	550	156.95	7318	DECKENPFRONN	
01	M	A		3485.94	5392.92	550	156.99	7318	DECKENPFRONN	
03	M	B		3486.32	5392.84	540	157.29	7318	DECKENPFRONN	
02	M	B		3486.32	5392.84	540	157.33	7318	DECKENPFRONN	
01	M	B		3486.32	5392.84	540	157.37	7318	DECKENPFRONN	
02	M	C		3486.84	5392.59	520	157.92	7318	DECKENPFRONN	
01	M	C		3486.84	5392.59	520	157.96	7318	DECKENPFRONN	
03	M	C		3487.09	5392.48	505	158.14	7318	DECKENPFRONN	
02	M	A		3489.96	5390.98	515	161.40	7319	AIDLINGEN	
01	M	A		3489.96	5390.98	515	161.44	7319	AIDLINGEN	
02	M	B		3490.46	5390.88	505	161.90	7319	AIDLINGEN	
01	M	B		3490.46	5390.88	505	161.94	7319	AIDLINGEN	
06	S			3496.66	5390.66	460	168.57	7319	EHNINGEN	
04	M	A		3497.53	5389.10	470	169.08	7319	HILDRIZHAUSEN	
03	M	A		3497.53	5389.10	470	169.11	7319	HILDRIZHAUSEN	
04	M	B		3497.77	5388.97	480	169.34	7319	HILDRIZHAUSEN	
03	M	B		3497.77	5388.97	480	169.37	7319	HILDRIZHAUSEN	
07	S			3503.82	5389.03	435	175.81	7320	BREITENSTEIN	
06	S			3503.82	5389.03	435	175.85	7320	BREITENSTEIN	
05	S			3510.26	5382.58	410	184.19	7420	WALDDORF	
07	M	A		3522.67	5379.78	430	196.70	7421	KAPPISHAEUSERN	
06	M	A		3522.67	5379.78	430	196.74	7421	KAPPISHAEUSERN	
05	M	A		3522.67	5379.78	430	196.78	7421	KAPPISHAEUSERN	
07	M	B		3522.99	5379.78	485	197.00	7421	KAPPISHAEUSERN	
05	M	B		3522.99	5379.69	485	197.11	7421	KAPPISHAEUSERN	
07	M	C		3523.31	5379.60	470	197.36	7421	KAPPISHAEUSERN	
06	M	C		3523.31	5379.60	470	197.41	7421	KAPPISHAEUSERN	
05	M	C		3523.31	5379.60	470	197.44	7421	KAPPISHAEUSERN	
07	M	A		3525.36	5375.52	620	200.75	7422	DETTINGEN	
07	M	B		3525.64	5375.26	680	201.11	7422	DETTINGEN	
07	M	C		3530.02	5374.28	680	205.54	7422	HUELBEN	
07	M	A		3530.18	5374.60	665	205.57	7422	HUELBEN	
06	M	C		3530.02	5374.28	680	205.58	7422	HUELBEN	
06	M	A		3530.18	5374.60	665	205.61	7422	HUELBEN	
05	M	C		3530.02	5374.28	680	205.62	7422	HUELBEN	
05	M	A		3530.18	5374.60	665	205.65	7422	HUELBEN	
07	M	B		3530.23	5374.44	685	205.68	7422	HUELBEN	
06	M	B		3530.23	5374.44	685	205.72	7422	HUELBEN	
05	M	B		3530.23	5374.44	685	205.75	7422	HUELBEN	
07	M	A		3533.60	5373.94	580	209.00	7422	GRABENSTETTEN	
06	M	A		3533.60	5373.94	580	209.04	7422	GRABENSTETTEN	
05	M	A		3533.60	5373.94	580	209.08	7422	GRABENSTETTEN	
07	M	B		3533.77	5373.68	600	209.25	7422	GRABENSTETTEN	
06	M	B		3533.77	5373.68	600	209.29	7422	GRABENSTETTEN	
05	M	B		3533.77	5373.68	600	209.33	7422	GRABENSTETTEN	
07	M	A		3537.57	5371.03	750	213.76	7523	BOEHRINGEN	
06	M	A		3537.57	5371.03	750	213.80	7523	BOEHRINGEN	
07	M	B		3537.75	5371.07	740	213.91	7523	BOEHRINGEN	
06	M	B		3537.75	5371.07	740	213.95	7523	BOEHRINGEN	
07	M	C		3537.96	5371.10	735	214.09	7523	BOEHRINGEN	
06	M	C		3537.96	5371.10	735	214.14	7523	BOEHRINGEN	
07	M	A		3542.66	5370.07	830	218.85	7523	MUEHLHALDE	

B	IN	AC	COORDINATES		ALT	X(KM)	SMAP	LOCALITY
06	M	A	3542.66	5370.07	830	218.89	7523	MUEHLHALDE
07	M	B	3542.74	5370.13	830	218.90	7523	MUEHLHALDE
05	M	A	3542.66	5370.07	830	218.92	7523	MUEHLHALDE
06	M	B	3542.74	5370.13	830	218.94	7523	MUEHLHALDE
05	M	B	3542.74	5370.13	830	218.98	7523	MUEHLHALDE
07	M	C	3543.04	5370.14	810	219.17	7523	MUEHLHALDE
06	M	C	3543.04	5370.14	810	219.22	7523	MUEHLHALDE
05	M	C	3543.04	5370.14	810	219.25	7523	MUEHLHALDE
07	M	A	3545.82	5368.91	815	222.21	7523	ENNABEUREN
06	M	A	3545.82	5368.91	815	222.25	7523	ENNABEUREN
05	M	A	3545.82	5368.91	815	222.29	7523	ENNABEUREN
07	M	B	3546.07	5368.89	800	222.45	7523	ENNABEUREN
06	M	B	3546.07	5368.89	800	222.49	7523	ENNABEUREN
05	M	B	3546.07	5368.89	800	222.53	7523	ENNABEUREN
07	M	C	3546.45	5368.88	780	222.81	7523	ENNABEUREN
06	M	C	3546.45	5368.88	780	222.85	7523	ENNABEUREN
05	M	C	3546.45	5368.88	780	222.89	7523	ENNABEUREN
07	M	B	3551.21	5366.34	745	228.17	7524	SUPPINGEN
06	M	B	3551.21	5366.34	745	228.21	7524	SUPPINGEN
05	M	B	3551.21	5366.34	745	228.24	7524	SUPPINGEN
07	M	A	3551.48	5366.36	750	228.41	7524	SUPPINGEN
06	M	A	3551.48	5366.36	750	228.45	7524	SUPPINGEN
05	M	A	3551.48	5366.36	750	228.49	7524	SUPPINGEN
07	M	C	3551.96	5366.40	725	228.84	7524	SUPPINGEN
06	M	C	3551.96	5366.40	745	228.88	7524	SUPPINGEN
05	M	C	3551.96	5366.40	725	228.92	7524	SUPPINGEN
07	M	A	3555.40	5363.59	695	233.07	7524	SEISSEN
06	M	A	3555.40	5363.59	695	233.12	7524	SEISSEN
05	M	A	3555.40	5363.59	695	233.15	7524	SEISSEN
07	M	B	3555.70	5363.77	705	233.29	7524	SEISSEN
06	M	B	3555.70	5363.77	705	233.33	7524	SEISSEN
05	M	B	3555.70	5363.77	705	233.36	7524	SEISSEN
07	M	C	3556.36	5363.44	685	234.02	7524	SEISSEN
06	M	C	3556.36	5363.44	685	234.06	7524	SEISSEN
05	M	C	3556.36	5363.44	685	234.10	7524	SEISSEN
07	M	A	3564.48	5360.31	615	242.72	7625	ERMINGEN
06	M	A	3564.48	5360.31	615	242.76	7625	ERMINGEN
05	M	A	3564.48	5360.31	615	242.80	7625	ERMINGEN
07	M	B	3564.72	5360.28	605	242.96	7625	ERMINGEN
06	M	B	3564.72	5360.28	605	243.00	7625	ERMINGEN
05	M	B	3564.72	5360.28	605	243.03	7625	ERMINGEN
07	M	C	3564.96	5360.24	780	243.20	7625	ERMINGEN
06	M	C	3564.96	5360.24	780	243.24	7625	ERMINGEN
05	M	C	3564.96	5360.24	780	243.27	7625	ERMINGEN

QUARRY-NO. 20 BIRRESBORN SMAP 5805
PROFILE 20-070 COMPILED BY VEES

B	IN	AC	COORDINATES		ALT	X(KM)	SMAP	LOCALITY	G C
04	S	31	2549.79	5563.75	550	5.06	5706	PELMERHALD	
04	S	32	2549.97	5563.76	552	5.23	5706	PELMERHALD	
04	S	33	2550.35	5563.76	558	5.59	5706	PELMERHALD	
04	S	41	2551.83	5564.69	595	7.29	5706	GEROLSTEIN	
04	S	42	2552.21	5564.57	610	7.61	5706	GEROLSTEIN	
04	S	43	2552.47	5564.89	570	7.96	5706	GEROLSTEIN	
01	S	31	2554.84	5566.20	655	10.58	5706	ERNSTBERG	
01	S	32	2555.15	5566.20	635	10.87	5706	ERNSTBERG	
01	S	33	2555.16	5566.49	660	10.99	5706	ERNSTBERG	
04	S	21	2557.80	5566.21	555	13.43	5706	HALDKOENIGEN	
04	S	22	2558.08	5566.40	535	13.76	5706	HALDKOENIGEN	
04	S	23	2558.40	5566.65	515	14.14	5706	HALDKOENIGEN	
03	KR	11	2559.68	5567.63	499	15.46	5707	NERDLEN	
03	KR	12	2560.10	5567.56	488	15.82	5707	NERDLEN	
03	KR	13	2560.53	5567.66	475	16.26	5707	NERDLEN	
04	KR	11	2561.53	5567.41	480	17.35	5707	NERDLEN	
04	KR	12	2561.88	5567.62	515	17.75	5707	NERDLEN	
04	KR	13	2562.26	5567.82	515	18.17	5707	NERDLEN	
01	CL	11	2572.94	5571.92	470	29.54	5708	ARBACH 1	
01	CL	12	2573.25	5572.09	465	29.89	5708	ARBACH 1	
01	CL	13	2573.57	5572.30	410	30.26	5708	ARBACH 1	
01	CL	21	2574.88	5572.99	410	31.73	5708	ARBACH 2	
01	CL	22	2575.23	5573.12	410	32.10	5708	ARBACH 2	
01	CL	23	2575.57	5573.23	415	32.46	5708	ARBACH 2	
01	CL	31	2576.62	5573.60	380	33.57	5708	ANSCHAU 1	
01	CL	32	2576.96	5573.72	360	33.93	5708	ANSCHAU 1	
01	CL	33	2577.35	5573.87	395	34.35	5708	ANSCHAU 1	
03	CL	31	2578.41	5574.36	427	35.35	5608	ANSCHAU E	
03	CL	32	2578.77	5574.51	392	35.74	5608	ANSCHAU E	
03	CL	33	2579.06	5574.61	365	36.05	5608	ANSCHAU E	
03	CL	41	2580.73	5575.01	403	37.75	5608	HEILER	
03	CL	42	2581.08	5575.21	360	38.15	5608	HEILER	
03	CL	43	2581.28	5575.27	395	38.38	5608	HEILER	
03	CL	21	2583.26	5576.41	405	40.61	5609	GEISBUESCHHOF	
03	CL	22	2583.66	5576.54	431	41.03	5609	GEISBUESCHHOF	
03	CL	23	2583.89	5576.65	435	41.29	5609	GEISBUESCHHOF	
03	H	11	2592.68	5579.79	220	50.62	5609	THUER	
03	H	12	2593.07	5579.98	190	51.05	5609	THUER	
03	H	13	2593.36	5580.28	190	51.43	5609	THUER	
04	BO	21	2595.25	5578.83	221	52.95	5610	EMMINGERHOF	
04	BO	22	2595.65	5578.94	218	53.37	5610	EMMINGERHOF	
04	BO	23	2596.08	5578.95	223	53.78	5610	EMMINGERHOF	
04	BO	11	2597.05	5579.15	193	54.76	5610	OCHTENDUNG SH	
04	BO	12	2597.38	5579.31	190	55.13	5610	OCHTENDUNG SH	
04	BO	13	2597.70	5579.55	203	55.50	5610	OCHTENDUNG SH	
05	BO	21	2599.57	5579.30	255	56.94	5610	OCHTENDUNG SE	
05	BO	22	2599.88	5579.37	275	57.25	5610	OCHTENDUNG SE	
05	BO	23	2600.13	5579.55	275	57.55	5610	OCHTENDUNG SE	
03	BO	11	2600.03	5582.63	175	58.50	5610	SAFFIG	
03	BO	12	2600.35	5582.80	178	58.86	5610	SAFFIG	
03	BO	13	2600.63	5582.95	170	59.17	5610	SAFFIG	
05	BO	11	2601.95	5580.00	280	59.42	5610	BASSENHEIM SH	
05	BO	12	2602.15	5580.10	270	59.64	5610	BASSENHEIM SH	
03	BO	21	2602.68	5583.99	190	61.46	5610	KETTIG	
03	BO	22	2602.93	5584.13	175	61.75	5610	KETTIG	
03	BO	23	2603.23	5584.26	170	62.07	5610	KETTIG	
01	KA	23	3407.25	5588.21	320	79.92	5512	GRENZHAUSEN1	
01	KA	21	3407.70	5588.54	355	80.45	5512	GRENZHAUSEN1	
01	KA	13	3408.27	5588.76	385	81.06	5512	GRENZHAUSEN2	
01	KA	12	3408.60	5588.87	400	81.41	5512	GRENZHAUSEN2	
01	KA	11	3408.89	5588.96	410	81.71	5512	GRENZHAUSEN2	
05	F	11	3420.88	5593.78	315	94.42	5513	DAHLEN H	
05	F	12	3421.24	5593.92	326	94.81	5513	DAHLEN H	
05	F	13	3421.52	5594.06	330	95.12	5513	DAHLEN H	
02	F	13	3422.11	5593.54	307	95.49	5513	DAHLEN S 2	
01	F	13	3422.11	5593.54	307	95.69	5513	DAHLEN S 1	
02	F	12	3422.60	5593.48	315	95.94	5513	DAHLEN S 2	
01	F	12	3422.60	5593.48	315	96.14	5513	DAHLEN S 1	
02	F	11	3423.05	5593.40	315	96.35	5513	DAHLEN S 2	
01	F	11	3423.05	5593.40	315	96.55	5513	DAHLEN S 1	
02	F	21	3424.78	5594.11	313	98.21	5513	HALLMEROD SH 2	
01	F	31	3424.74	5593.98	305	98.33	5513	HALLMEROD SH 1	
02	F	22	3425.05	5594.18	324	98.49	5513	HALLMEROD SH 2	
01	F	32	3425.08	5594.22	323	98.73	5513	HALLMEROD SH 1	
01	F	33	3425.22	5594.32	320	98.89	5513	HALLMEROD SH 1	
02	F	23	3425.40	5594.42	313	98.89	5513	HALLMEROD SH 2	
02	F	31	3427.81	5596.00	205	101.66	5513	MOLSBERG NE 2	
01	F	21	3427.81	5596.00	205	101.87	5513	MOLSBERG NE 1	
02	F	32	3428.18	5596.12	230	102.05	5513	MOLSBERG NE 2	
01	F	22	3428.18	5596.12	230	102.26	5513	MOLSBERG NE 1	
02	F	33	3428.48	5596.38	240	102.42	5513	MOLSBERG NE 2	
01	F	23	3428.48	5596.38	240	102.63	5513	MOLSBERG NE 1	
04	F	11	3429.44	5596.00	210	103.49	5514	THALHEIM	
04	F	12	3429.74	5596.16	210	103.83	5514	THALHEIM	
04	F	13	3430.20	5596.27	195	104.30	5514	THALHEIM	
05	F	22	3431.54	5595.89	195	105.21	5514	THALHEIM	
05	F	21	3431.44	5596.28	200	105.24	5514	THALHEIM	
05	F	23	3431.77	5595.68	205	105.37	5514	THALHEIM	
03	F	31	3431.96	5597.27	205	106.05	5414	FRICKHOFEN	
03	F	33	3432.67	5597.80	180	106.89	5414	FRICKHOFEN	
04	F	21	3434.02	5597.76	240	108.39	5414	ELBGRUND	
04	F	22	3434.34	5597.86	260	108.73	5414	ELBGRUND	
04	F	23	3434.62	5597.98	275	109.03	5414	ELBGRUND	
04	F	31	3437.11	5599.46	388	111.85	5414	FUSSINGEN N	
04	F	32	3437.42	5599.68	385	112.22	5414	FUSSINGEN N	
04	F	33	3437.73	5599.90	383	112.58	5414	FUSSINGEN N	
05	F	31	3439.71	5599.77	338	114.18	5414	HALDENBACH	
03	F	21	3439.75	5599.75	340	114.22	5414	HALDERNBACH	
03	F	22	3440.01	5599.87	325	114.51	5414	HALDERNBACH	
05	F	32	3440.10	5599.84	323	114.57	5414	HALDERNBACH	
05	F	33	3440.20	5599.90	318	114.69	5414	HALDENBACH	
03	F	23	3440.32	5600.17	325	114.90	5414	HALDERNBACH	
03	F	32	3432.39	5607.38	200	115.93	5414	FRICKHOFEN	
04	HH	11	3452.52	5608.25	283	129.28	5315	GREIFENTHAL	
04	HH	12	3452.91	5608.41	277	129.71	5316	GREIFENTHAL	
04	HH	13	3453.32	5608.53	245	130.13	5316	GREIFENTHAL	
04	HH	43	3454.63	5610.24	241	131.95	5316	KATZENFURT	
04	HH	42	3454.93	5610.57	260	132.34	5316	KATZENFURT	
04	HH	41	3455.25	5610.77	275	132.71	5316	KATZENFURT	
04	HH	31	3457.66	5612.42	236	135.55	5316	NIEDERLEMP	
04	HH	32	3457.97	5612.56	227	135.89	5316	NIEDERLEMP	
04	HH	33	3458.26	5612.82	296	136.25	5316	NIEDERLEMP	
04	CL	11	3466.80	5617.58	350	145.93	5317	HILSBACH	
04	CL	12	3466.80	5617.58	350	145.93	5317	HILSBACH	X
04	CL	13	3466.80	5617.58	350	145.93	5317	HILSBACH	Y
04	CL	21	3469.39	5619.31	325	148.97	5217	SEELBACH	
04	CL	22	3469.69	5619.50	280	149.32	5217	SEELBACH	
04	CL	23	3470.17	5619.73	290	149.85	5217	SEELBACH	
04	CL	31	3471.88	5620.82	260	151.84	5217	ROLLSHAUSEN	
04	CL	32	3471.88	5620.82	260	151.84	5217	ROLLSHAUSEN	X
04	CL	33	3471.88	5620.82	260	151.84	5217	ROLLSHAUSEN	Y
04	CL	41	3475.14	5623.11	285	155.72	5217	LOHRA	
04	CL	42	3475.62	5623.24	260	156.22	5217	LOHRA	
04	CL	43	3476.02	5623.27	245	156.60	5217	LOHRA	
05	M	81	3477.22	5624.24	240	157.88	5218	KEHNA N	
05	M	82	3477.38	5624.33	255	158.07	5218	KEHNA N	
05	M	83	3477.42	5624.35	255	158.11	5218	KEHNA N	
05	M	101	3479.18	5625.82	240	160.30	5218	GERMERSHAUSEN	
05	M	102	3479.30	5625.85	230	160.42	5218	GERMERSHAUSEN	
05	M	103	3479.54	5625.92	210	160.67	5218	GERMERSHAUSEN	
05	M	71	3481.95	5627.33	280	163.44	5218	MARBG.STADTH.	
05	M	72	3482.24	5627.50	210	163.77	5218	MARBG.STADTH.	
05	M	73	3482.36	5627.57	230	163.91	5218	MARBG.STADTH.	
04	H	11	3504.96	5641.98	320	190.55	5020	APPENHAIN	
04	H	12	3505.39	5642.11	310	191.00	5020	APPENHAIN	
04	H	13	3505.76	5642.28	340	191.40	5020	APPENHAIN	
04	H	41	3506.86	5643.38	350	192.86	5020	SACHSENHAUSEN	
04	H	42	3507.28	5643.53	320	193.30	5020	SACHSENHAUSEN	
04	H	43	3507.70	5643.63	310	193.72	5020	SACHSENHAUSEN	
05	H	11	3508.39	5644.88	370	194.70	5020	SACHSENHAUSEN	
05	H	12	3508.77	5645.07	382	195.13	5020	SACHSENHAUSEN	
05	H	13	3509.22	5645.09	380	195.54	5020	SACHSENHAUSEN	
05	H	42	3511.54	5646.76	317	198.35	5020	ELNRODE	
05	H	43	3511.84	5647.07	287	198.75	5020	ELNRODE	
01	H	21	3512.65	5648.15	315	200.02	5021	SCHLIERBACH	
01	H	22	3513.05	5648.33	250	200.46	5021	SCHLIERBACH	
01	H	23	3513.53	5648.55	220	200.98	5021	SCHLIERBACH	
01	H	11	3515.04	5649.07	230	202.57	5021	DORHEIM	
01	H	12	3515.38	5649.30	245	202.98	5021	DORHEIM	
01	H	13	3515.81	5649.50	250	203.45	5021	DORHEIM	
05	KI	31	3540.43	5665.30	340	232.39	4823	MELSUNGEN E	
05	KI	32	3540.74	5665.35	315	232.69	4823	MELSUNGEN E	
05	KI	33	3541.08	5665.54	340	233.08	4823	MELSUNGEN E	
05	KI	41	3542.28	5665.87	400	234.30	4823	ENTENPFUHL	
05	KI	42	3542.57	5666.09	400	234.66	4823	ENTENPFUHL	
05	KI	43	3542.82	5666.36	405	235.00	4823	ENTENPFUHL	
04	KI	31	3543.85	5667.52	310	236.58	4823	ALTES GEHEGE	
04	KI	32	3544.02	5667.58	350	236.76	4823	ALTES GEHEGE	
04	KI	33	3544.35	5667.80	330	237.15	4823	ALTES GEHEGE	
04	KI	43	3545.50	5668.58	410	238.53	4823	GUESTERODE	
04	KI	42	3545.81	5668.78	410	238.89	4823	GUESTERODE S	
04	KI	41	3546.19	5668.93	450	239.30	4823	GUESTERODE SH	
04	GO	21	3547.20	5669.55	515	240.48	4824	GUENSTERODE	
04	GO	22	3547.52	5669.90	509	240.92	4824	GUENSTERODE	
04	GO	23	3547.86	5670.13	540	241.33	4824	GUENSTERODE	
04	GO	11	3548.86	5670.89	510	242.56	4824	RETTERODE H	
04	GO	12	3548.99	5671.19	500	242.81	4824	RETTERODE H	
04	GO	13	3549.30	5671.20	470	243.09	4824	RETTERODE H	
05	CL	41	3592.64	5717.40	245	303.04	4428	RHUMSPRINGE	
05	CL	42	3592.87	5717.75	281	303.42	4428	RHUMSPRINGE	
05	CL	43	3593.14	5718.12	263	303.84	4428	RHUMSPRINGE	
05	CL	31	3594.80	5718.58	280	305.50	4428	RHUMSPRINGE E	
05	CL	32	3594.80	5718.58	280	305.50	4428	RHUMSPRINGE E	X
05	CL	22	3596.26	5719.41	296	307.18	4328	BARBIS 2	
05	CL	21	3596.51	5719.50	275	307.44	4328	BARBIS 2	
05	CL	11	3596.54	5719.50	300	307.47	4328	BARBIS	
05	CL	12	3596.54	5719.50	300	307.47	4328	BARBIS	X

B	IN	AC	COORDINATES	ALT	X(KM)	SMAP	LOCALITY	G C
05	CL	13	3596.54 5719.50	300	307.47	4328	BARBIS	Y
05	CL	51	3594.52 5746.16	670	320.11	4128	KAHLER BERG	
05	CL	52	3594.52 5746.16	670	320.11	4128	KAHLER BERG	X
05	CL	53	3594.52 5746.16	670	320.11	4128	KAHLER BERG	Y
04	CL	51	3594.52 5746.16	670	320.15	4128	KAHLER BERG	
04	CL	52	3594.52 5746.16	670	320.15	4128	KAHLER BERG	X
04	CL	53	3594.52 5746.16	670	320.15	4128	KAHLER BERG	Y

QUARRY-NO. 20 BIRRESBORN SMAP 5805
PROFILE 20-250 COMPILED BY VEES

B	IN	AC	COORDINATES	ALT	X(KM)	SMAP	LOCALITY	G C
01	S	41	2538.80 5556.51	515	8.47	5805	MUERLENBACH	
01	S	42	2538.45 5556.50	510	8.74	5805	MUERLENBACH	
01	S	43	2538.22 5556.20	510	9.11	5805	MUERLENBACH	
03	H	41	2533.77 5555.91	540	12.98	5804	NIMSREULAND	
03	H	42	2533.40 5555.66	515	13.43	5804	NIMSREULAND	
03	H	43	2532.94 5555.62	480	13.86	5804	NIMSREULAND	
01	MS	1	2529.14 5555.05	590	17.44	5804	EILSCHEID	
01	MS	2	2528.82 5554.95	570	17.77	5804	EILSCHEID	
01	MS	3	2528.68 5554.88	550	17.93	5804	EILSCHEID	
04	MS	22	2526.83 5554.73	495	19.62	5804	EILSCHEID W	
04	MS	23	2526.34 5554.60	420	20.12	5804	EILSCHEID W	
04	MS	11	2524.60 5553.73	400	22.06	5804	KINZENBURG	
04	MS	12	2524.29 5553.63	410	22.39	5804	KINZENBURG	
04	MS	13	2523.90 5553.55	475	22.78	5804	KINZENBURG	
05	MS	11	2522.68 5553.33	460	24.23	5803	LICHTENBORN	
05	MS	22	2522.30 5553.10	505	24.66	5803	LICHTENBORN	
05	MS	21	2518.33 5551.50	487	28.94	5803	HICKESHAUSEN	
03	H	21	2515.39 5551.13	500	31.82	5903	REIFF	
03	H	22	2515.00 5551.02	485	32.23	5903	REIFF	
03	H	23	2514.61 5550.90	440	32.64	5903	REIFF	
01	MS	2	2512.69 5549.86	517	34.63	5903	REIPELDINGEN	
01	MS	3	2512.36 5549.72	360	34.98	5903	REIPELDINGEN	

QUARRY-NO. 22 THALHEIM SMAP 5414
PROFILE 22-060-13 COMPILED BY STEIN, THUEMMEL

B	IN	AC	COORDINATES	ALT	X(KM)	SMAP	LOCALITY	G C
02	F	21	3434.52 5597.95	270	6.35	5414	DORCHHEIM	
02	F	22	3434.75 5598.10	285	6.62	5414	DORCHHEIM	
02	F	23	3434.95 5598.24	310	6.86	5414	DORCHHEIM	
02	F	31	3439.60 5599.50	355	11.41	5414	WALDERNBACH S	
02	F	32	3439.85 5599.65	335	11.70	5414	WALDERNBACH S	
02	F	33	3440.30 5599.75	320	12.14	5414	WALDERNBACH S	
01	MM	31	3452.54 5608.24	283	26.99	5315	GREIFENTAL	
01	MM	32	3452.91 5608.41	277	27.39	5315	GREIFENTAL	
01	MM	33	3453.25 5608.58	258	27.77	5315	GREIFENTAL	
01	MM	41	3454.58 5610.17	236	29.75	5316	KATZENFURTH	
01	MM	42	3454.87 5610.48	255	30.16	5316	KATZENFURTH	
01	MM	43	3455.18 5610.78	282	30.58	5316	KATZENFURTH	
01	MM	22	3456.02 5611.14	274	31.48	5316	KOELSCHHAUSEN	
01	MM	11	3457.68 5612.40	233	33.56	5316	NIEDERLEMP	
01	MM	12	3457.94 5612.59	235	33.88	5316	NIEDERLEMP	
01	MM	13	3458.23 5612.74	286	34.21	5316	NIEDERLEMP	
02	M	81	3477.22 5625.78	255	57.27	5218	ALLNA	
02	M	82	3477.56 5626.00	280	57.67	5218	ALLNA	
02	M	83	3477.94 5626.09	285	58.04	5218	ALLNA	
02	M	73	3479.10 5625.79	245	58.84	5218	OBERWEIMAR	
02	M	72	3479.32 5625.86	230	59.06	5218	OBERWEIMAR	
02	M	71	3479.56 5625.94	210	59.31	5218	OBERWEIMAR	
02	M	103	3481.69 5626.78	280	61.55	5218	MARBURGER STADTWALD	
02	M	102	3481.68 5627.06	240	61.70	5218	MARBURGER STADTWALD	
02	M	101	3481.93 5627.31	280	62.04	5218	MARBURGER STADTWALD	
01	KI	21	3508.42 5644.92	366	93.83	5020	SACHSENHAUSEN	
01	KI	22	3508.79 5645.10	383	94.23	5020	SACHSENHAUSEN	
01	KI	23	3509.18 5645.11	381	94.57	5020	SACHSENHAUSEN	
01	GO	11	3512.65 5648.15	315	99.13	5021	SCHLIERBACH	
01	GO	12	3513.05 5648.33	250	99.57	5021	SCHLIERBACH	
01	GO	13	3513.53 5648.55	220	100.09	5021	SCHLIERBACH	
01	H	41	3515.04 5649.07	230	101.64	5021	DORHEIM	
01	H	42	3515.38 5649.30	245	102.05	5021	DORHEIM	
01	H	43	3515.81 5649.50	250	102.52	5021	DORHEIM	
01	KI	31	3523.84 5653.70	260	111.55	4922	BATZENBERG	
01	KI	32	3524.19 5653.94	238	111.97	4922	BATZENBERG	
01	KI	33	3524.45 5654.18	230	112.32	4922	BATZENBERG	
01	KI	41	3525.39 5654.97	237	113.54	4992	HYNBURG	
01	KI	42	3525.64 5655.24	230	113.89	4922	HYNBURG	
01	KI	43	3525.85 5655.44	200	114.18	4922	HYNBURG	
02	KI	11	3530.18 5658.44	353	119.46	4922	EILERTHOF	
02	KI	12	3530.50 5658.68	385	119.86	4922	EILERTHOF	
02	KI	13	3530.86 5658.84	377	120.25	4922	EILERTHOF	
02	KI	21	3532.14 5659.78	348	121.84	4922	MOSHEIM	
02	KI	22	3532.33 5659.88	320	122.05	4922	MOSHEIM	
02	KI	23	3532.66 5659.98	293	122.38	4922	MOSHEIM	

QUARRY-NO. 22 THALHEIM SMAP 5414
PROFILE 22-250-20 COMPILED BY PRODEHL

B	IN	AC	COORDINATES	ALT	X(KM)	SMAP	LOCALITY	G C
03	F	21	3425.40 5594.28	315	4.29	5513	WALLMERODE	
03	F	22	3425.03 5594.18	323	4.65	5513	WALLMERODE	
03	F	23	3424.85 5594.03	310	4.82	5513	WALLMERODE	
03	F	11	3422.50 5593.15	301	7.18	5513	DAHLEN	
03	F	12	3422.25 5592.86	305	7.46	5513	DAHLEN	
03	BO	11	3392.95 5580.70	213	38.96	5610	BASSENHEIM 1	
03	BO	12	3392.58 5580.53	219	39.36	5610	BASSENHEIM 1	
03	BO	13	3392.23 5580.30	230	39.77	5610	BASSENHEIM 1	
03	BO	21	3390.73 5579.90	290	41.32	5610	FASANERIEWALD	
03	BO	22	3390.20 5579.89	250	41.82	5610	FASANERIEWALD	
03	BO	23	3390.00 5580.18	236	41.91	5610	FASANERIEWALD	
01	BO	11	2602.55 5580.28	249	42.80	5610	BASSENHEIM SW	
01	BO	12	2602.15 5580.10	270	43.23	5610	BASSENHEIM SW	
01	BO	13	2601.95 5580.00	280	43.46	5610	BASSENHEIM SW	
01	BO	21	2600.13 5579.55	275	45.32	5610	OCHTENDUNG SE	
01	BO	22	2599.88 5579.37	275	45.62	5610	OCHTENDUNG SE	
01	BO	23	2599.57 5579.30	255	45.93	5610	OCHTENDUNG SE	
02	BO	11	2597.70 5579.55	203	47.60	5610	OCHTENDUNG SW	
02	BO	12	2597.38 5579.31	190	47.98	5610	OCHTENDUNG SW	
02	BO	13	2597.05 5579.15	193	48.35	5610	OCHTENDUNG SW	
02	BO	21	2596.08 5578.95	223	49.33	5610	EMMINGERHOF	
02	BO	22	2595.67 5578.94	218	49.72	5610	EMMINGERHOF	
02	BO	23	2595.25 5578.82	221	50.15	5610	EMMINGERHOF	
01	CL	21	2575.60 5573.31	420	70.54	5708	DITSCHEID	
01	CL	22	2575.36 5573.07	400	70.84	5708	DITSCHEID	
01	CL	23	2574.93 5572.98	410	71.28	5708	DITSCHEID	
01	CL	31	2573.60 5572.32	410	72.75	5708	ARBACH W	
01	CL	32	2573.28 5572.12	460	73.12	5708	ARBACH W	
01	CL	33	2572.95 5571.92	470	73.49	5708	ARBACH W	

QUARRY-NO. 24 STEINACH SMAP 7714
PROFILE 24-170 COMPILED BY MUELLER, PETERSCHMITT

B	IN	AC	COORDINATES	ALT	X(KM)	SMAP	LOCALITY	G C
03	KA	51	3432.35 5345.44	490	6.00	7714	MUEHLENBACH	
03	KA	52	3432.35 5345.44	490	6.00	7714	MUEHLENBACH	
03	KA	53	3432.35 5345.44	490	6.00	7714	MUEHLENBACH	
02	KA	21	3434.78 5339.19	690	12.57	7814	UNTERPRECHTAL	
02	KA	22	3434.83 5339.05	710	12.72	7814	UNTERPRECHTAL	
02	KA	23	3434.71 5338.89	720	12.83	7814	UNTERPRECHTAL	
02	KA	31	3434.88 5336.03	710	15.60	7814	YACH	
02	KA	32	3434.79 5335.91	680	15.69	7814	YACH	
02	KA	33	3434.65 5335.89	670	15.67	7814	YACH	
03	KA	71	3435.17 5329.66	760	21.97	7814	HINTERGRIESBACH	
03	KA	72	3435.17 5329.66	760	21.97	7814	HINTERGRIESBACH	
03	KA	73	3435.17 5329.66	760	21.97	7814	HINTERGRIESBACH	
02	KA	41	3434.75 5325.46	690	25.88	7914	OBERSIMONSWALD	
02	KA	42	3434.63 5325.31	690	26.01	7914	OBERSIMONSWALD	
02	KA	43	3434.34 5325.10	670	26.17	7914	OBERSIMONSWALD	
01	S	41	3437.17 5319.69	770	31.98	7914	MUEHLELOCH	
01	S	42	3437.01 5319.33	840	32.30	7914	MUEHLELOCH	
01	S	43	3437.08 5319.16	860	32.48	7914	MUEHLELOCH	
02	KA	11	3436.02 5317.82	980	33.63	8014	REDECKHOF WALDAU	
02	KA	12	3436.02 5317.82	980	33.63	8014	REDECKHOF WALDAU	
02	KA	13	3435.96 5317.71	950	33.73	8014	REDECKHOF WALDAU	
03	KA	61	3438.82 5313.93	1045	38.12	8015	LANGENORDACH	
03	KA	62	3439.03 5313.56	1045	38.52	8015	LANGENORDACH	
03	KA	63	3438.98 5313.12	1050	38.94	8015	LANGENORDACH	
01	S	33	3438.80 5310.06	890	41.74	8015	FEUERBERGER HOF	
01	S	31	3438.58 5309.98	920	41.78	8015	FEUERBERGER HOF	
01	S	32	3438.58 5309.98	920	41.78	8015	FEUERBERGER HOF	
01	S	21	3438.57 5302.14	920	49.48	8115	SCHWENDE	
01	S	22	3438.57 5302.14	920	49.48	8115	SCHWENDE	
01	S	23	3438.57 5302.14	920	49.48	8115	SCHWENDE	
02	KA	61	3443.17 5298.43	990	54.08	8115	SOMMERAU	
02	KA	63	3443.22 5298.38	990	54.13	8115	SOMMERAU	
01	KA	41	3440.69 5293.18	1000	58.68	8215	SCHOENENBACH	
01	KA	42	3440.69 5293.18	1000	58.68	8215	SCHOENENBACH	
01	KA	43	3440.69 5293.18	1000	58.68	8215	SCHOENENBACH	
01	KA	31	3441.73 5287.05	795	64.89	8215	BRENDEN	
01	KA	32	3441.91 5286.83	770	65.15	8215	BRENDEN	
01	KA	33	3442.07 5286.81	770	65.19	8215	BRENDEN	
02	KA	73	3444.33 5283.57	600	68.83	8315	WITZNAU	
02	KA	71	3444.28 5283.51	600	68.88	8315	WITZNAU	
02	KA	72	3444.28 5283.51	600	68.88	8315	WITZNAU	
02	KA	52	3442.77 5279.04	530	72.99	8315	BUERGLEN	
02	KA	51	3442.77 5279.04	530	72.99	8315	BUERGLEN	
02	KA	53	3442.77 5279.04	530	72.99	8315	BUERGLEN	

3.4 Presentation of Record Sections

P. GIESE, C. PRODEHL, H. SCHRÖDER, and A. STEIN

ABSTRACT

This chapter deals with questions concerning the preparation of the record sections presented in the appendix. All sections are plotted in a reduced time scale using the reduction velocity 6 km/s.

ZUSAMMENFASSUNG

Dieses Kapitel behandelt die Fragen, welche die Herstellung der Seismogramm-Montagen (Anhang) betreffen. Alle Montagen sind in reduziertem Zeitstab dargestellt. Es wurde eine Reduktionsgeschwindigkeit von 6 km/s benutzt.

One of the main purposes of this publication is the presentation of the basic data obtained from seismic-refraction measurements in western Germany since 1958 in a form which also allows an evaluation of the material by other authors. One part of the basic data, e.g., shotpoint data, coordinates, distances etc. is presented in 3.1 and 3.3. The recordings must also be shown. The simplest way is to display all original seismograms. Such a presentation would be extremely expensive and would go far beyond the frame work of any one book. Therefore it seems reasonable to choose the form of record sections, a presentation which has become more and more common.

A record section means the display of seismic traces in a time-distance graph. For the presentation of the travel time two possibilities exist. In the simplest way full travel time is plotted and the shot time is used as zero point for each distance. This type of presentation is widely used in continuous profiling (e.g., SOLLOGUB et al., 1972).

The other form of display uses reduced travel time instead of full travel time which was first introduced by FÖRTSCH (1951). The reduced travel time $\bar{t}$ is defined by the difference between the full travel time t and the quantity: distance Δ divided by a reduction velocity V: $\bar{t} = t - \Delta/V$. This kind of display offers the great advantage of better time resolution over smaller space.

The correlation of seismic events from trace to trace is facilitated if the reduction velocity is close to the apparent velocities of the waves of interest thus producing a display showing the phases arranged more or less horizontally. In this way the correlation procedure is comparable with that in reflection-seismic surveys.

The main interest of the investigations presented here concerns the crustal structure by means of the evaluation of P waves. The average velocity of compressional waves in the crust is around 6 km/s. Therefore a reduction velocity of 6 km/s best fulfills the requirements mentioned above and is used for all record sections presented in this monograph.

As outlined, the display of data should allow a detailed reinterpretation without being forced to go back to the original material. Due to the frequency range mainly generated by quarry explosions, a time scale 1 cm ≙ 2 s is adequate for this purpose and can be realized within a reasonable page size. The time axis comprises between 10 and 20 s depending on the length of the profiles presented.

The correlation is also facilitated by the right choice of a proper distance scale. Proper means that the spacing of traces must be so that a clear readability is assured and on the other hand the interval between the traces is as small as possible. Thus a uniform distance scale is not favorable since the density of observations varies.

No elevation corrections have been applied for two reasons. Such a correction requires the knowledge of the near-surface velocity which is usually unknown in detail. On the other hand the influence of the correction would be small because in the area under investigation the elevation difference from station to station is in general smaller than 100 m.

All record sections are shown in the appendix.

All record sections show only the trace of the vertical component. As described in 3.3 many stations were equipped with more than one vertical component spaced some hundred m apart. The average spacing of stations is in general, however, five to ten times higher. A record section containing all traces in a readable manner would show groups of two or three traces separated by large intervals without records. As outlined above, this would embarrass the correlation procedure. Therefore, only one trace of each group was selected in order to obtain a proper distance scale.

The fans (Fächer) comprise all observations which were obtained from one shotpoint at nearly the same distance as indicated in the code. Two types are shown: one uses full travel time whereas the other contains reduced times. The distance on the horizontal axis is measured along the arc, the zero point being arbitrary. In the case that a station is not located exactly in the fan distance, a time correction t_c is applied on the zero time of the corresponding trace which is given by the difference $\delta\Delta$ between station distance Δ_s and the arc distance Δ_a divided by 6 km/s: $t_c = \delta\Delta/6$.

Due to the fact that the observation material has been obtained over a period of about 15 years, it is obvious that the quality and kind of records differ very much. Most of the profiles were observed with heterogeneous equipment having different film speeds. For the preparation of record sections seismic traces with uniform time scale are necessary. Three methods were applied. By the first the records were normalized photographically and redrawn by hand. Another method of normalization was offered by an affinograph developed by Mr. Ball at the Institut de Physique du Globe, Strasbourg (FUCHS et al., 1963). Here the seismic traces are followed by hand. This procedure allows an independent treatment of time and amplitude axis. Thirdly it is possible to digitize the seismic traces by hand, after which they can be processed by computer techniques. For profiles which were recorded with magnetic-tape equipments record sections can be produced without plotting traces by hand. From tapes records can be displayed with uniform speed in a reproduceable form or a reproduction applying computer techniques is feasible.

In complete record sections amplitude information should be included. There are several reasons why this has not been done here. The main difficulty is the fact that in quarry blasts the quantitative determination of amplitude-charge relations is very difficult and was attempted only in a few cases (BURCKHARDT and VEES, 3.2 of this volume). The second difficulty arises by the different amplitude-frequency characteristics of the older instruments used. In addition, some of these instruments were not calibrated. By aid of the original records, the registration cards, and the calibration curves stored in the DFG-Zentralarchiv, Stuttgart, the true amplification can be determined.

The record sections shown here are furnished by numerous authors. The principles of correlation applied by them differ more or less. In order to give objective information and to avoid prejudgment of any given correlation the record sections are presented without any travel-time curves.

3.5 Standard Equipment for Deep-Seismic Sounding

H. BERCKHEMER

ABSTRACT

Portable equipment for deep-seismic sounding is described in this article. Use is made of frequency-multiplex-modulation recording technique on 1/4" magnetic tape. Each recording unit consists of three 2 cps-universal field seismometers, the corresponding amplifiers, modulators, the tape recorder and the time signal receiver. The signal frequency range is 0.3-100 cps.

The high dynamic range of 60 db is reached by a very effective flutter and wow compensation. A playback center has been established. The geophysical institutes in the Federal Republic of Germany are equipped with 60 calibrated instruments of this type.

ZUSAMMENFASSUNG

Es werden Aufbau und Wirkungsweise einer tragbaren Apparatur für seismische Tiefensondierung beschrieben. Die Aufnahme erfolgt nach dem Frequenz-Multiplex-Modulations-Verfahren auf 1/4" Magnetband. Jede Aufnahmeeinheit besteht aus drei 2-Hz-Universalseismometern, den zugehörigen Verstärkern, Modulatoren, dem Magnetbandgerät und dem Zeitsignalempfänger. Signalfrequenzbereich 0,3-100 Hz. Der große Dynamikumfang von 60 db wird durch wirkungsvolle Kompensation der Bandgleichlaufschwankungen erreicht. Die Abspielung der Magnetbänder erfolgt in einer ortsfesten Zentrale. Die geophysikalischen Institute in der Bundesrepublik sind mit ca. 60 einheitlich kalibrierten Geräten dieses Typs ausgestattet.

3.5.1 Introduction

In the pioneer days of seismic crustal investigations with explosions E. WIECHERT, L. MINTROP, B. BROCKAMP, and others used portable pendulum seismographs of high sensitivity with direct photographic recording. An improved version of this category designed by G. SCHULZE and H. FÖRTSCH was succesfully applied for the big Helgoland explosion in 1947. In the course of the systematic crustal survey of central Europe and the Alpine region, initiated by a priority program of the German Research Society in 1958, several types of field seismometers with electromechanical transducers and electronic amplifiers, developed e.g. by H. BAULE, W. BEUERMANN, H. BERCKHEMER, A. STEIN came into use. These allowed the separation of seismic sensor and photorecording galvanometer and facilitated the amplitude- and frequency control of the system.

In the middle of the sixties it became more and more evident that full use of the information contained in the seismic wave field could be made only if the stations of the eleven institutes participating in this research were equipped with a uniform and improved type of instrument. The interpretation of seismograms was no longer restricted to the travel time of the first onset but extended also to later arrivals in particular to the usually well developed wide-angle reflection from the crust/mantle boundary. Intensity and frequency content of the respective arrivals were also found useful for the interpretation. Unambiguous correlation of arrivals is only possible by phase correlation of adjacent seismograms. This requires a dense coverage of the observation profiles with seismic receivers. To meet all these requirements in the most economic way and to make full use of the modern methods of electronic data treatment, a working group of the Forschungskollegium Physik des Erdkörpers, in close cooperation with the manufacturer of the electronic equipment Ing. A. LENNARTZ, developed in 1966 the magnetic tape recording seismograph system MARS 66. Thanks to the generous financial support by the Stiftung Volkswagenwerk more than 40 field recording sets and the necessary playback facilities have been made available to the German institutes and have been successfully used under all environmental conditions. A playback center and a tape library have been established at Frankfurt and automatic digitization on IBM compatible tape is possible at the Institutes for Geophysics at Hamburg, Karlsruhe, and Kiel.

Today approximately 250 recording sets of this type are in use in geophysical institutions of Germany, France, Italy, Switzerland, Austria, England, Norway, Sweden, Denmark and Finland. This instrumental development marks an important progress in deep-seismic sounding. A brief description will, therefore, be given below.

3.5.2 The Recording Unit

After numerous preliminary experiments the principle of multiplex-frequency modulation has been chosen for magnetic-tape recording of the data. The signal from each of the three seismic receivers modulates, after passing a calibrated wide-band amplifier, its individual carrier frequency. A fourth carrier serves for the time signal. All carriers together with a crystal-controlled fixed frequency are added and stored on a single track of the 1/4" battery-supplied UHER 4000-Report-K audio tape recorder. During or after reception the intensity of ground noise and seismic signal can easily be estimated without demodulation in a qualitative manner by listening to the frequency variations of the carrier.

Three 2 Hz electrodynamic seismometers of type FS 60 are connected to the recording set by unshielded field cable. With the 40 stations available a profile of 50 km length can be covered during one explosion with a seismograph separation of 400 m, which is adequate for phase correlation.

In central and southern Europe the Swiss time signal HBG, broadcast continuously on 75 kHz, and the German time signal DCF (77.5 kHz) are successfully recorded with special receivers. Short-wave receivers are available for other regions.

A block diagram of the recording set is shown in Fig. 1 and technical data given in Table 1.

3.5.3 The Playback Center

Although each of the eleven cooperating institutes has a playback facility of its own a playback center has been established at the Institute of Geophysics of the Frankfurt University. Here all recorded events are copied on larger reels of 1/4" tape and stored for further use in a tape library. The block diagram of the playback unit including the copying machine is shown in Fig. 2. The frequency mixture of the output of the playback machine is separated by phase-compensated band-pass filters into the

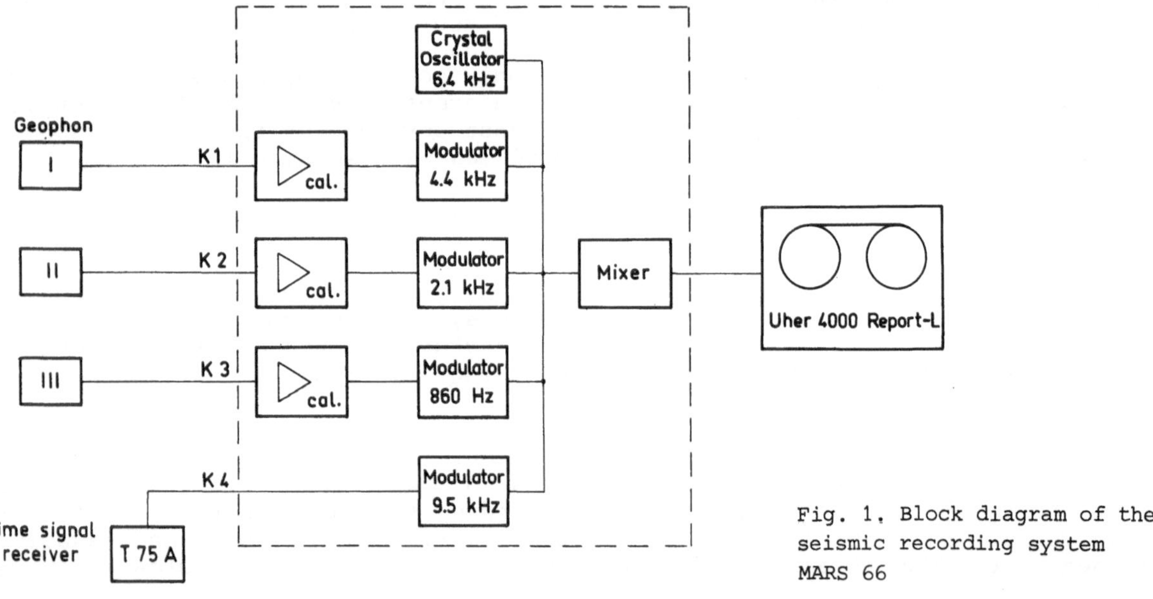

Fig. 1. Block diagram of the seismic recording system MARS 66

Fig. 2. Block diagram of the seismic playback system

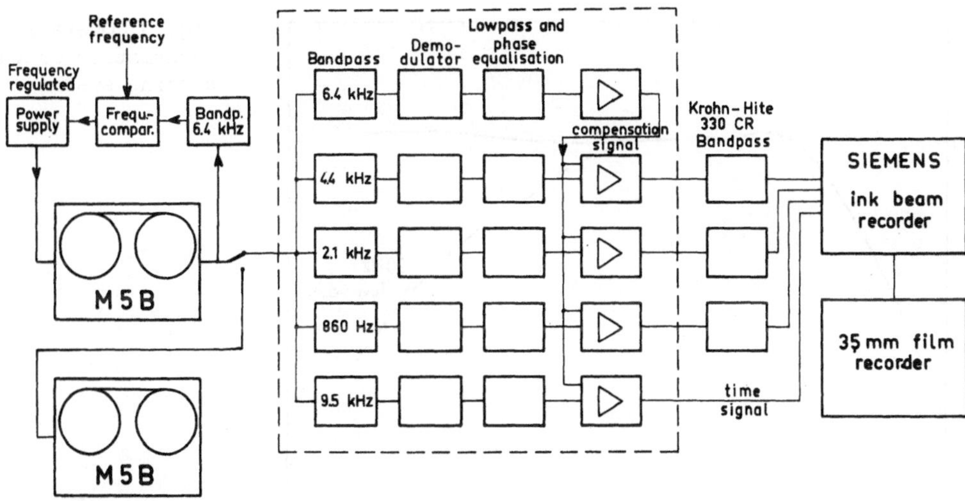

Table 1. Technical data of the recording unit

Seismometer:
STROPPE FS 60 (after Berckhemer), moving coil seismometer, convertibel for vertical and horizontal operation.

Natural frequency:	2 Hz
Sensitivity:	2.0 $Vscm^{-1}$
Coil impedance:	1 kOhm
Weight:	7 kg

Amplifier - modulator - unit:
LENNARTZ FM3/1K-M, 3 seismic signal channels with calibrated amplifiers, 1 time signal channel.
Amplifier:
Maximum gain: 2048
Gain setting: by step attenuator 2^{11}, 2^{10}, ...2^{-5}
Input resistance: 0.0-10 kOhm, adjustable for optimum seismometer damping
Frequency range: 0.2-200 Hz (-3 db)
Noise: equivalent input < 1 μV_{pp}
calibration signal, high-frequency noise filter.
Modulator:
Carrier frequencies: 0.86 kHz, 2.1 kHz, 4.4 kHz, 9.5 kHz
Reference frequency: 6.4 kHz (crystal controlled)
Signal frequency range: 0-100 Hz
Input sensitivity: ± 200 mV for ± 15% frequency deviation
Output voltage: > 0.1 V for each carrier
Operating temperature: -10°C - +50°C
Built-in output- and frequency-meter, power supply by batteries.

Tape recorder:
UHER 4000 Report L, battery supplied
Tape: 1/4" wide, 13 cm reels
Tape speed: 9.5 or 19 cm/s
Operating temperature: -15°C - +50°C

original carriers. The respective demodulator generates an output signal proportional to the frequency variation of the carrier. It is a superposition of the original signal and some electric noise due to flutter and wow of the recording and the playback machine. By subtracing the demodulated pilot tune from the disturbed signal an almost perfect flutter and wow compensation is achieved with the result of a dynamic recording range of approximately 60 db, almost independent of the quality of the tape recorder. In addition the pilot frequency is used to adjust the tape speed of the playback machine automatically according to the recording speed. Technical data of the playback unit are given in Table 2.

Table 2. Technical data of the playback unit

Demodulator unit:
LENNARTZ FM4/1K-D, 4 signal channels, compensated for flutter and wow
Singal frequency range: 0-100 Hz
Output voltage: ± 2V

Playback tapedeck:
TELEFUNKEN M5B with speed control unit LENNARTZ R91T or UHER 4000 Report L with automatic speed control

The overall frequency characteristics of the recording and playback system shown in Fig. 3 is flat with respect to ground velocity from 3 to 50 Hz (solid line). In the presence of strong high-frequency infiltration at the recording site the 20 Hz lowpass filter of the recording amplifier can be switched on. This changes the

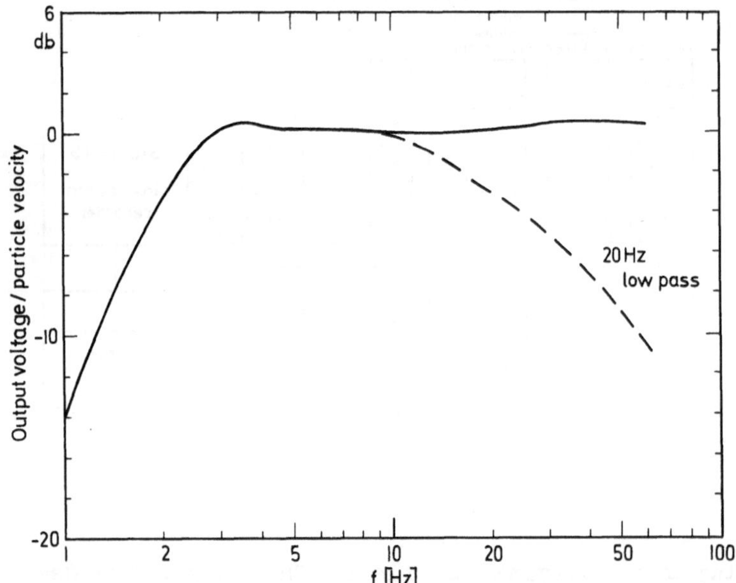

Fig. 3. Overall frequency characteristics of the MARS 66 recording system

characteristics according to the dashed line. The output signals can be fed through variable band-pass filters into a multichannal ink-beam recorder for visible display or into an analog computer or an analog-to-digital converter for further analysis.

For more details of the instruments and some examples of data treatment the reader is referred to BERCKHEMER (1970).

3.6 Seismic Long-Distance Observations of Quarry Blasts - Test Measurements in Western Germany

H. BURKHARDT, B. BUTTKUS, F. KELLER, and R. VEES

ABSTRACT

The results of seismic test measurements at distances from 300 to 670 km indicate that satisfactory data can be expected with quarry blasts as seismic sources using suitable seismometer arrays and applying various filter methods.

ZUSAMMENFASSUNG

Die Ergebnisse seismischer Versuchsmessungen in Entfernungen von 300 bis 670 km zeigen, daß brauchbare Daten von Steinbruchsprengungen als Energiequelle zu erwarten sind, wenn geeignete Seismometerarrays benutzt werden und verschiedene Filterverfahren angewendet werden.

3.6.1 Introduction

Longrange explosion seismic measurements can provide valuable information about the deeper part of the earth's crust and the upper mantle, yielding data for detailed investigations complementary to earth-quake seismology.

During the last few years several investigations, such as the Lake superior experiments in the Canadian Shield (MEREU and HUNTER, 1969) and recordings up to distances of 800 km in central Europe during the Lac Nègre experiments in 1966 (ANSORGE et al., 1970) have shown that underwater explosions are suitable seismic signal sources for long-distance recordings.

In this paper we investigate if industrial quarry blasts can also be used for such experiments.

For this purpose test recordings were made in western Germany at distances $\Delta > 300$ km utilizing shot points which were used up to now for seismic profiles with distances up to 250 km (GIESE and STEIN, 1971).

3.6.2 Compilation of Present Data

The position of shotpoints and recording sites is shown in Fig. 1.

The recording sites were arranged as follows:

1. along the extension of the profile Eschenlohe-Hilders (01-345-02) in the distance interval from 340 to 670 km,
2. on a fan around the shotpoint Eschenlohe (01), $\Delta \approx 400$ km. $\alpha = 353^{\circ}-313^{\circ}$,
3. single test points for different shotpoints.

All measurements were made with the portable magnetic-tape recording equipment MARS 66 (velocity transducers) (BERCKHEMER, 1970).

In most cases a three-component seismometer arrangement was chosen, facilitating the search for locations with small ambient seismic noise.

All seismograms were filtered by means of a band-pass-filter 0.5-20 c/s during analog playback of magnetic tapes and were subsequently digitized with a sample interval of 13.3 ms ($f_{Nyquist} = 37.5$ cps) for further digital data processing.

Fig. 2 shows a record section of the vertical component of ground motion (Z) for the profile Eschenlohe-Hilders with recording distances between 340 and 670 km, using four explosions with charge weights from 7 to 22 t. The time scale is reduced with a velocity of 6 km/s.

For recording distances up to 430 km the first arrival can be determined exactly and as a rule is perceptible up to a distance of 530 km. In the North German Basin ($\Delta > 530$ km) the ambient seismic noise increases considerably and the signal amplitude decreases rapidly as expected because of the prevalent sediment thickness. Therefore no signals can be detected in this distance interval.

The onset times of first arrivals scatter up to ± 200 ms with respect to a straight line as approximate time-distance curve, its slope cor-

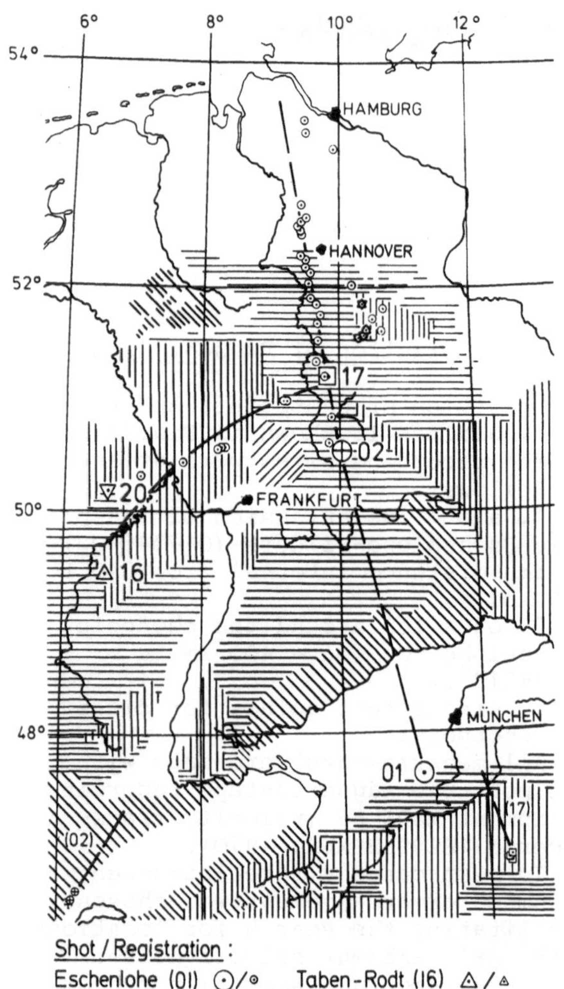

Fig. 1. Shotpoints and recording sites

Quaternary + Pleistocene (Molasse, North German Plain, Rhine Graben)

Neozoic volcanics (Vogelsberg)

Cretaceous + Jurassic (Munsterland, Franconian-Swabian Alb, Swiss Jura)

Triassic (Middle and Southern Germany)

Paleozoic + Pre Cambrian (Rhenish Massif, Harz, Thuringian Forest, Bohemian Massif, Central Alps)

responding to an apparent velocity of 8.3 km/s.

In addition to the wave group of the first arrivals with dominant spectral components in the frequency range between 6 and 14 cps, later arrivals are detectable in many cases with frequencies in the range 3 to 8 cps (t_{red} = +4 s, approximately) which, however, cannot be correlated satisfactorily across several traces.

The evaluation of time-distance curves for later arrivals is difficult, because of relatively large station spacings and of the missing connection to less distant recordings.

The results of a fan-recording to the west of the profile for the shotpoint Eschenlohe (01), charge weight 12.8 t, mean radius 400 km, is given in Table 1. For technical reasons a seismometer array on the profile 17-240 (shotpoint Bransrode) had to be used. Despite the relatively large ambient seismic noise first-signal arrivals are detectable [signal-to-noise ratio (SNR) ≈ 2]. The arrival times were reduced tentatively to a distance of 400 km with velocities between 7.8 and 8.7 km/s.

The two records to the west of 325° show longer reduced arrival times (1-2 s) than the other seismograms. The relatively unfavorable station positions, however, preclude a detailed interpretation of these time differences.

Data from 28 single station recordings for different shotpoints are compiled in Table 2, covering a distance interval from 300 to 530 km and a charge weight range from 3.3 to 22 t.

Most of these seismograms were recorded by a reference station in the Harz Mountains (Devonian anticlinic). The mean SNR is 2-3, with comparatively small ambient noise.

The comparison of these measurements with seismograms from the Harz foreland, the Alps, and the Swiss Jura definitely indicates a stronger dependence of the SNR on the choice

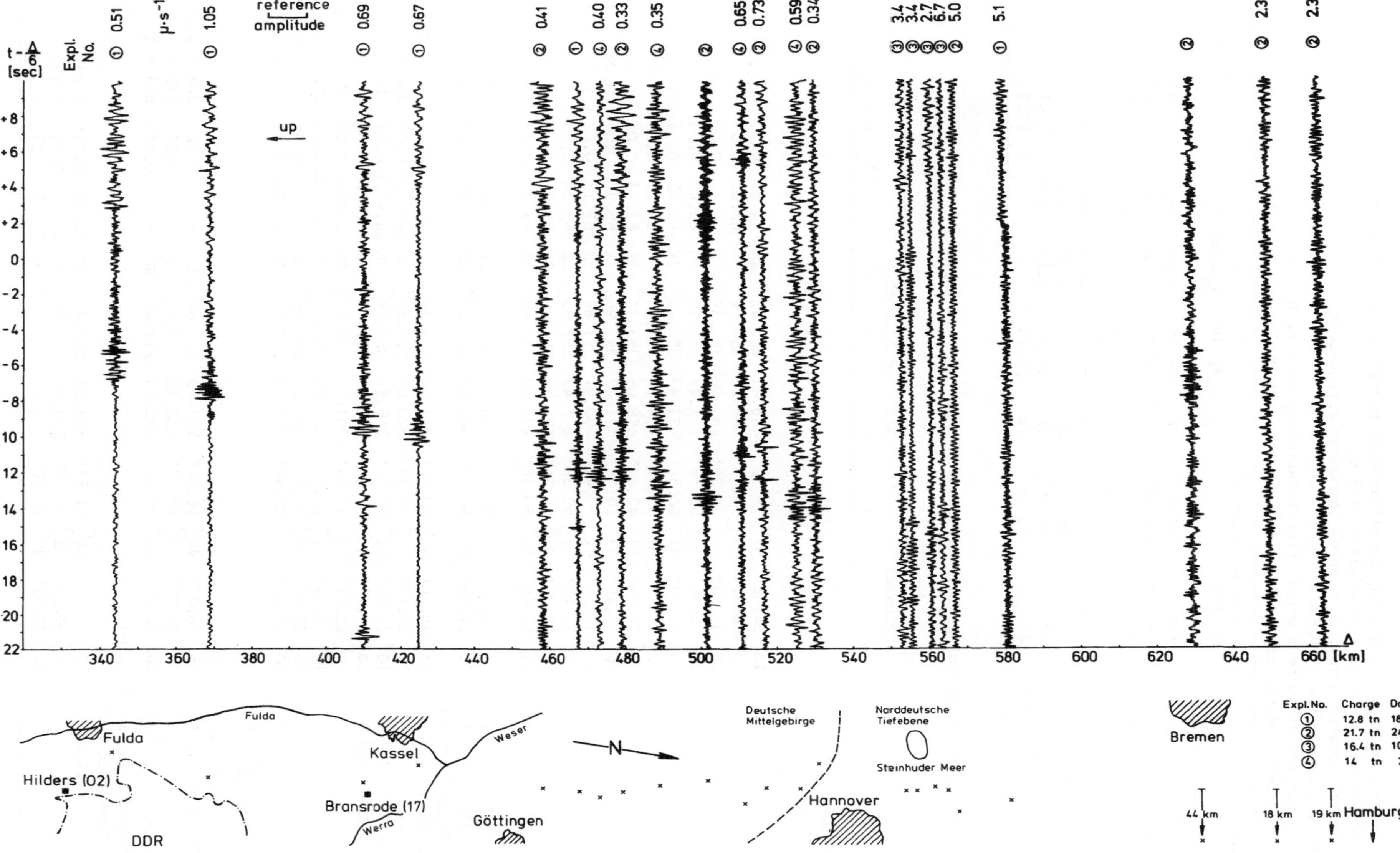

Fig. 2. Record section profile Eschenlohe (O1) - Hilders (vertical component Z). Values of ground velocity (μ/s) refer to indicated reference amplitude

Table 1. Fan-recordings $\Delta \approx 400$ km, coordinates of recording sites and some seismogram data shot point: Eschenlohe (01) - 18.07.69 - 12.8 t.
L = linear seismometer array, 3K = three component arrangement
t_1 = travel time of first arrival
A_n = RMS amplitude of ambient noise, time interval t_1-20 s to t_1
A_s = Zero-to-peak maximum amplitude of signal, time interval t_1 to t_1 + 3 s

Coordinates λ	φ	Distance Δ (km)	Azimuth α		t_1 (s)	A_n (μ/s)	A_s (μ/s)	$t_1 - \frac{\Delta-400}{8.1}$ (s)
07° 06.69'	50° 18.25'	418.47	313.63°	L	60.30	0.03	0.07	58.02
07° 42.81'	50° 25.76'	399.11	319.87°	3K	58.76	0.08	0.16	58.87
08° 13.32'	50° 33.61'	388.94	325.70°	3K	55.82	0.06	0.11	57.19
08° 15.44'	50° 34.39'	388.73	326.15°	3K	55.57	0.05	0.16	56.96
08° 16.75'	50° 35.05'	388.90	326.46°	3K	-	0.13	0.18	-
09° 11.15'	50° 58.27'	397.51	338.30°	3K	-	0.04	0.22	-
09° 13.15'	50° 58.78'	397.53	338.69°	3K	56.80	0.11	0.40	57.11
09° 46.63'	51° 12.67'	410.06	345.50°	3K	58.50			57.23

Table 2. Single test recordings for different shot points (see legend of Table 1 for explanation)

	Coordinates λ	φ	Distance Δ (km)	Azimuth α		t_1 (s)	A_n (μ/s)	A_s (μ/s)
1. Recording sites in the Harz region								
a) S.P. Eschenlohe (01)								
06.10.67 - 7.0 t	10° 24.28'	51° 38.19'	448.41	353.43°	2K	ca.62	0.17	SNR < 1
26.07.68 - 9.1 t	10° 36.33'	51° 37.83'	446.28	354.78°	L	62.32	0.03	0.02
	10° 35.50'	51° 49.26'	467.46	354.89°	L	-	0.02	SNR < 1
	10° 22.45'	51° 50.56'	471.44	352.94°	3K	65.45	0.04	0.13
06.12.68 - 7.1 t	10° 22.45'	51° 50.56'	471.43	352.93°	3K	65.42	0.04	0.12
14.03.69 - 7.7 t	10° 22.11'	51° 35.74'	444.22	352.47°	3K	62.21	0.03	0.04
	10° 28.80'	51° 42.67'	456.05	353.37°	3K	63.33	0.02	0.02
	10° 22.45'	51° 50.56'	471.42	352.96°	3K	65.37	0.04	0.06
	10° 13.22'	52° 00.41'	490.91	351.88°	3K	67.13	0.06	0.03
24.10.69 - 21.7 t	10° 22.45'	51° 50.56'	471.42	352.96°	3K	65.38	0.03	0.11
10.07.70 - 16.4 t	10° 22.45'	51° 50.56'	471.42	352.96°	3K	65.31	0.05	0.13
02.07.71 - 14.0 t	10° 22.45'	51° 50.56'	471.42	352.96°	3K	65.38	0.04	0.16
b) S.P. Taben-Rodt (16)								
04.10.67 - 3.3 t	10° 24,28'	51° 38.19'	354.27	047.63°	2K	ca.49.5	0.04	0.1
11.06.69 - 4.5 t	10° 22.45'	51° 50.56'	367.67	044.67°	3K	51.77	0.03	0.07
c) S.P. Birresborn (20)								
13.07.68 - 4.3 t	10° 22.45'	51° 50.56'	318.99	053.71°	3K	46.23	0.02	0.08
	10° 23.17'	51° 50.93'	320.05	053.69°	L	ca.46.5	0.04	SNR < 1
19.06.70 - 4.2 t	10° 22.45'	51° 50.56'	319.51	053.58°	3K	ca.47.5	0.03	SNR < 1
30.06.70 - 7.3 t	10° 20.37'	51° 35.28'	302.62	057.67°	L	ca.44.6	0.04	0.18
	10° 22.11'	51° 35.74'	304.77	057.72°	3K	ca.44.8	0.02	0.05
	10° 23.39'	51° 36.13'	306.67	057.74°	L	-	ca.0.05	SNR < 1
	10° 23.62'	51° 36.17'	306.69	057.75°	3K	-	0.05	SNR < 1
	10° 22.45'	51° 50.56'	319.48	053.45°	3K	ca.47.5	0.03	SNR < 1
2. Recording sites in Swiss Jurassic								
S.P. Hilders (02)								
25.08.67 - 4.1 t	06° 18.09'	46° 36.64'	516.19	213.62°	L	-	0.02	SNR < 1
	06° 13.21'	46° 30.92'	528.48	213.61°	L	-	0.02	SNR < 1
	06° 12.35'	46° 30.04'	530.43	213.62°	3K	-	0.03	SNR < 1
3. Recording sites in the Alps								
S.P. Bransrode (17)								
02.10.68 - 14.0 t	12° 20.70'	46° 51.83'	518.32	160.46°	3K	72.55	0.01	0.03
	12° 20.27'	46° 50.37'	520.69	160.61°	L	ca.72.5	0.05	0.05
	12° 20.51'	46° 49.86'	521.68	160.61°	3K	72.95	0.03	0.02

of a suitable recording location than on charge weight and recording distance. For large SNR it is essential not only to have a small ambient noise at the seismometer location but above all a sufficient coupling of surface layers to the substratum.

In general the SNR of seismic traces, recorded in distances with $\Delta >$ 300 km is of the order of 1. For a more exact determination of arrival times it is therefore necessary to apply filter methods for SNR-improvement.

3.6.3 Application of Filter Methods to Improve the Signal-to-Noise Ratio

To improve the signal-to-noise ratio of seismograms, differences in the following properties of seismic and noise signal can be used: frequency spectra, apparent velocities, wave polarization, and spatial correlation.

The most important filter methods with their criteria and conditions for seismic applications are given in Table 3.

Frequency filtering of long-range recordings proved to be suitable in some cases by application of narrow band-pass filters with upper and lower cutoff frequencies ranging from 6 to 14 cps to separate signal and noise. As an example, Fig. 3 shows a seismic record trace (Δ = 446 km) in original form as well as after application of an analog band-pass filter 10 to 14 cps. After filtering, the first arrival is clearly detectable, the low-frequency parts of the noise being preferably suppressed.

As a rule, a sufficient signal-to-noise separation by frequency filtering is, however, not achievable, be-

Table 3. Some filter methods for seismic SNR-improvement

Method	Conditions for application		Determination of filter operator	
	Recording arrangement of seismometers	Properties of signal and noise	Criteria	Known/estimated defining quantities
1. Frequency filter	arbitrary	different amplitude spectra of signal and noise	maximum filter-transmissivity in the frequency range of the signal	estimate of the amplitude spectrum of the signal
2. Velocity filter	array in profile direction	different apparent velocities	maximum filter transmissivity in the velocity range of the signal	estimate of signal velocity
3. Correlation methods	separated in space	correlating signal, poorly correlating noise		
4. Optimum filter (single and multi-channel filters)	see 1-3	see 1-3	mathematical-statistical criteria. Maximum of transmission characteristic for parts of seismogram with signal properties, minimum characteristic for noise propterties	see 1 and 2 plus determination of statistical noise properties
5. Polarization method	Three-component arrangement	linearly polarized signal, elleptically polarized noise		

cause the two spectra overlap in a wide frequency band.

Polarization filter methods are applicable for SNR-improvement with three-component recordings. In this case the filter is designed to suppress the elliptically polarized noise-component and to transmit only the linearly polarized seismic signal.

For numerical applications the method developed by SHIMSHONI and SMITH (1964) has proved to be convenient (Fig. 4).

The weight function M depends on the amplitude squared so that the maximum of the linearly polarized wave group is primarily emphasized. For SNR $\geq$ 1, moreover, the onset time of the signal can be determined more exactly after filtering (Fig. 4a), whereas for seismograms with SNR < 1 the signal maximum alone is extracted from the noise (Fig. 4b).

The application of polarization filters is limited above all by signal-generated noise for complicated layered station substratum. Additionally, with the recordings in question the SNR of horizontal components ($H_{\parallel}$ and $H_{\perp}$) is in many cases considerably smaller than that of the vertical components.

In case of seismograms recorded in a linear array (standard configuration:

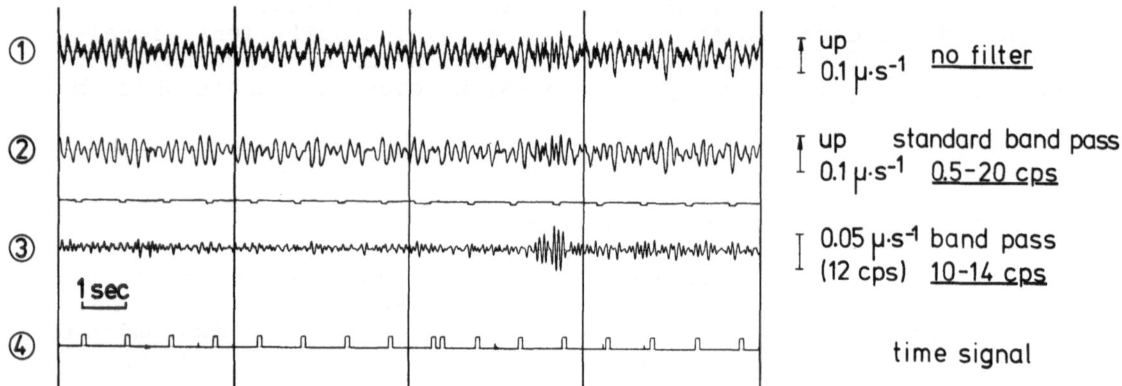

Fig. 3. Application of frequency filter (vertical component Z); S.P. Eschenlohe (01) (26.07.68) 9.1 t, Δ = 446.28 km, α = 354.78°

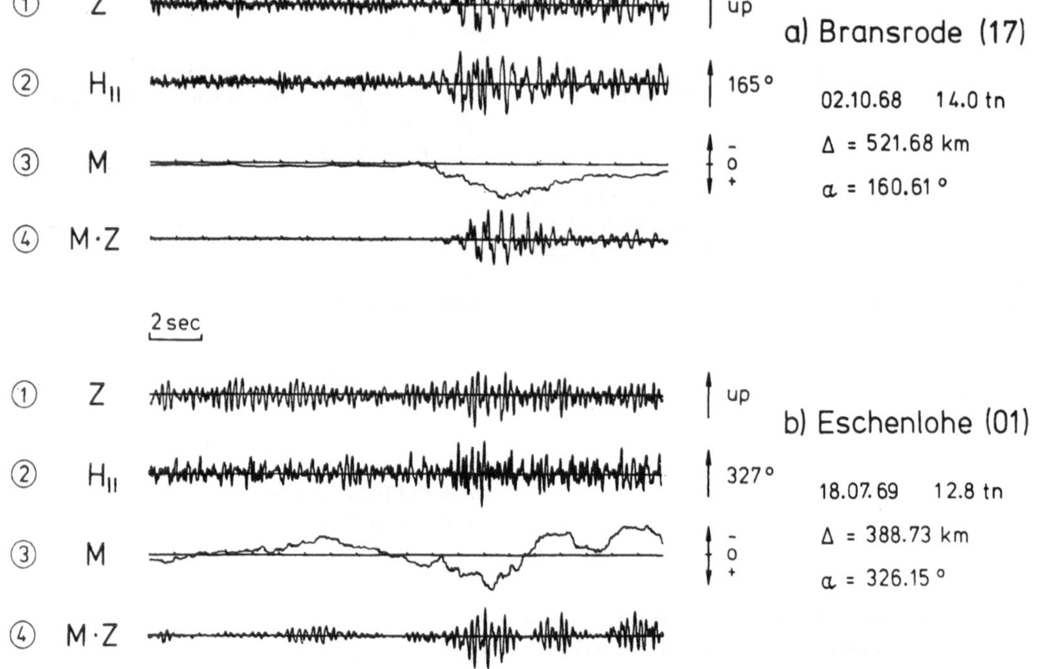

Fig. 4. Application of polarization filter, (Method of SHIMSHONI and SMITH, 1964). $M = \overline{Z \cdot H_{\parallel}}$, integration interval 2 s

3 vertical-component seismometers with 400 m spacing in the direction of the profile) multi-channel filters can be used. With only three seismometers arranged within a total distance of 800 m and using the velocity filters (frequency-wave number filter) by BURG (1964), however, the required transfer function can only be approximated in such a rough manner that seismic signal and noise cannot be sufficiently well separated.

For seismometer spacings of several hundred m the coherence of the ambient seismic noise is very small in the interesting seismic frequency interval 0.5 to 20 cps. On the other hand the signal coherence is expected to be close to 1 with similar substratum of seismometer locations.

There are many filter methods for signal-to-noise ratio improvement based on linear and nonlinear processes, (e.g. RYALL, 1964), which make use of differences in the spatial correlation of seismic signals and ambient noise.

With regard to the test measurements in question the method of multiple correlation proved to be applicable in particular (Fig. 5). For records with SNR ≥ 1 the method yields considerable SNR-improvements, and signal onset times can be determined with greater accuracy. For records with SNR < 1 the number of seismometers used (i.e. three) is too small to achieve accurate arrival times. In addition our measurements show that in many cases the signal coherence was smaller than one (different substratum of seismometer locations within one array).

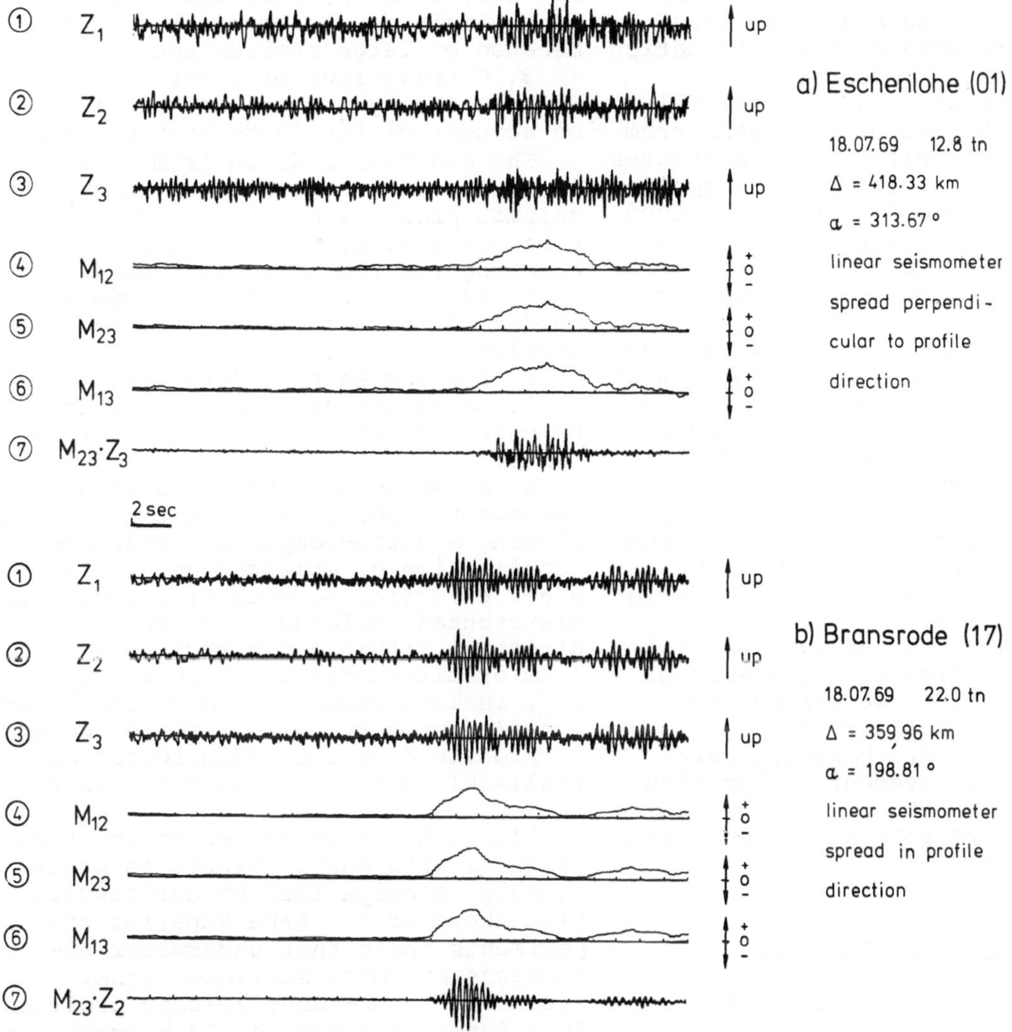

Fig. 5. Application of multiple correlation (vertical components Z); Δ and α for reference signal Z_2; time scales for Z_1 and Z_3 shifted (velocity 8.0 km/s) $M_{ij} = \overline{Z_i \cdot Z_j}$, integration interval 2 s

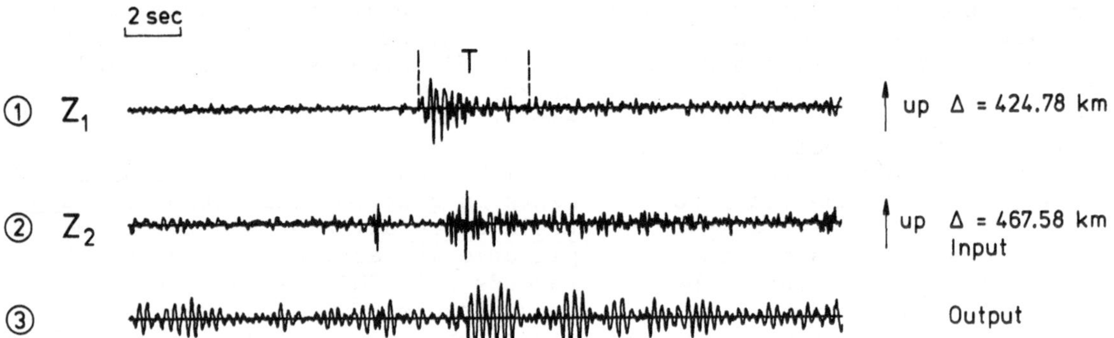

Fig. 6. Application of maximum output energy filter S.P. Eschenlohe (01) (18.07.69) 12.8 t. Filter operator derived from interval T of signal Z_1

A necessary condition for the application of optimum filters is the knowledge of the amplitude spectra of seismic signals.

Thus these filters can only be applied for seismic purposes in cases where the SNR is much greater than 1, in order to obtain this information directly from the records.

For records with SNR $\leq$ 1 the filter operator has to be estimated from adjacent records with SNR much greater than 1. For our field data the application of filter operators determined in such a way yields no improvement in general, the signal spectra obviously changing too fast along the profile.

An example is given in Fig. 6: the operator for the maximum-output-energy filter (ROBINSON, 1967) was computed using the signal from trace 1 and was subsequently applied to trace 2. Although both seismograms have been recorded for the same shot event with a range difference of 43 km the result in trace 3 shows no improvement but on the contrary a distinctive deterioration of the SNR.

As an example for an optimum filter, which only requires the knowledge of the noise spectrum, we tested the prediction-error filter. The application of this filter, however, gave no essential improvement either with SNR < 1 or with SNR > 1, because the power spectra of both signal and noise overlap too much.

3.6.4 Results and Conclusions

The result of our investigations can be summarized as follows:

1. There is a stronger dependence of the SNR on the choice of station location than on recording distance and charge weight. In favorable cases the SNR is of the order of 2 to 3.
2. A substantial SNR improvement is achievable by filtering methods. The particular convenient correlation methods, however, were applicable only within one seismometer spread because of large station spacing.
3. Correlations of later arrivals in the seismograms are not possible on account of the large station spacing.

The conclusion drawn from these investigations is that long-range seismic profiles with Δ > 300 km are possible also with quarry blasts, allowing for the following points:

1. Sufficient coupling of overburden to the bedrock (cf. record section in Fig. 2).
2. Station spacing less than 5 km.
3. Selection of station locations by means of preceding noise measurements.
4. Extension of the prevailing seismometer spread to larger arrays including a three-component arrangement (application of multichannel filters), e.g., 8 vertical-component seismometers distributed preferably in profile direction within a distance range of 3 km or alternatively in an areal array, in both cases including one horizontal-component seismometer oriented in profile direction (configuration realizable with three MARS-66 instruments).

Although the extension of existing profiles with quarry blasts is practicable, a comparison of our results with those of the Lake Superior experiments shows that underwater explosions are more suitable signal sources of long-range seismic profiles. Such investigations should be realized in the scope of large-scale experiments taking into account the abovementioned recommendations.

3.7 Cooperation in Deep-Seismic Sounding between the Geophysical Institutes, the Geological Surveys, and the Oil Industry

H. CLOSS, G. DOHR, and H. MENZEL

The main tasks of geophysical institutes of the universities are research and teaching. The geological surveys deal with the geological and geophysical mapping of the Federal Republic of Germany. The companies of the oil industry are commercially interested in exploring the sedimentary cover in order to detect structures where oil and gas can be expected. Fortunately, it has been possible to coordinate the different interests with respect to the investigation of the deeper structure beneath the basement. This common interest helped to avoid duplicating work and has led to a fruitful exchange of information and experience.

The geophysical departments of the regional geological surveys have jointly cooperated in field and interpretation work with the geophysical institutes of the universities. In addition, they have taken over a number of tasks in administration and organization of the joint research programs.

The oil companies announced commercial investigations to the research institutions when results interesting for crustal studies were to be expected, and they were also willing, in addition, to cooperate in various ways. Owing to this cooperation it was possible to extend the recording time within the routine work of the commercial exploration in order to obtain deep-seismic reflection data. In this way, a very great number of reflection seismograms could be used for crustal studies.

Furtheron the observation of larger commercial explosions by the geophysical institutions could be arranged, which served their studies. In the beginning of crustal investigation, various funds were contributed to these programs.

The geophysical exploration companies PRAKLA and SEISMOS have always shown interest in these scientific programs and supported the measurements in a very generous manner. A number of scientific experiments were used by them to test new equipment.

On the application of the seismic reflection method for investigations of crust and upper mantle, several summarizing reports have been published (SCHULZ, 1959; DOHR, 1957a, b, 1959, 1967, 1968, 1970; LIEBSCHER, 1962, 1964; HEHN, 1964; DEMNATI and DOHR, 1965; HADJEBI, 1966; DOHR and FUCHS, 1967; DOHR et al., 1967; KERTZ et al., 1972; DOHR and MEISSNER, 1975).

This monograph, therefore, does not contain a special contribution on the results of deep-seismic reflection studies.

4. Problems of Evaluation of Seismic-Refraction Data for Crustal and Upper-Mantle Studies

P. GIESE and C. PRODEHL

This chapter deals with methods of evaluation of deep-seismic sounding data. The procedure of data interpretation starts with the presentation of the data observed and ends with the display of a cross section which can reversely generate the primary observations. This procedure passes through several stages and it is quite evident that the final stage can be reached only in a more or less satisfactory approximation. This chapter contains contributions to some problems of evaluation of deep-seismic sounding data.

The first basic task should be the presentation of the data observed. This topic has already been discussed in Chapter 3. The next step of evaluation concerns the correlation of wave groups in the record sections. Just this task is the most critical step of the whole procedure, and some basic problems are discussed in 4.1.

As an intermediate step in the interpretation process, methods may be useful that allow the presentation of traveltimes in a form which already reflects qualitatively the main features of crustal structure including their lateral variations. 4.2 shows some possibilities for this task.

A quantitative interpretation of the records must take into consideration kinematic as well as dynamic aspects. Methods to invert travel times into depth values are well developed, as described in 4.3. Difficulties arise if the profile under discussion shows strong lateral inhomogeneities. Here, only the indirect method of interpretation is possible up to the present. Sections 4.4 and 4.5 deal with the calculation of traveltimes in lateral inhomogeneous media. GEBRANDE (4.4) combines different arctg-functions in order to describe the inhomogeneous velocity field. WILL (4.5) divides the inhomogeneous profile into polygons, each of which is characterized by constant velocity gradients being arbitrarily directed.

The dynamic aspects of interpretation are treated in the last part of this chapter. MÜLLER and FUCHS (4.6) describe briefly the basic ideas of the methods for the computation of synthetic seismograms, the ray-theoretical method and the reflectivity method, and present examples for the inversion of observed data. They show that the consideration of the amplitude-frequency characteristics of the consideration of the phases in addition to the travel-time interpretation may lead to considerable improvements of the resulting velocity-depth relation, assuming, however, that lateral inhomogeneities along the main part of the profile can be neglected. VETTER and MEISSNER (4.7) discuss the principal character of theoretical traveltime and amplitude curves, calculated for some typical models in Europe.

4.1 General Remarks on Travel Time Data and Principles of Correlation

P. GIESE

ABSTRACT

Beginning with the velocity gradient, the various types of velocity models will be discussed. The cases of constant velocity and first-order discontinuity must be regarded as extreme cases. Low-velocity layers as well as too-short observation lines cause an incompleteness of travel-time data, preventing a unique solution of the problem to determine the velocity distribution v (x,y,z).

In the second part, the term correlation will be treated, with its meaning which must be differentiated in reflection and refraction seismics. Finally the importance of phase and group correlation will be outlined.

ZUSAMMENFASSUNG

Ausgehend von Geschwindigkeitsgradienten werden die verschiedenen Modelltypen diskutiert. Die Fälle konstanter Geschwindigkeit und einer Diskontinuität erster Ordnung sind als Grenzfälle anzusehen. Zonen geringer Geschwindigkeit und auch zu kurze Beobachtungslinien bedingen eine Unvollständigkeit des Datenmaterials, und somit ist eine eindeutige Lösung des Problems der Ermittlung des Geschwindigkeitsfeldes v(x,y,z) nicht möglich.

4.1.1 General Remarks on Travel Time Data

Crustal studies of the last decade revealed that a two- or three-layered crust with constant velocities within each layer and with first-order boundaries between the layers can be regarded only as a rough first approximation. From the geological and petrological points of view, low-velocity layers and transition zones are of great importance, and they do exist, as demonstrated by several authors. Data generalization meets with difficulties when finding out the main features of crustal-velocity distribution including velocity inversions as well as transition zones, using, however, only a few typical travel-time curves. In order to state clearly the contents of information and statements derived from the data observed and selected for data generalization, some brief basic considerations will be covered first.

All considerations are based on the assumption that the medium under study is isotropic and inhomogeneous only in a vertical direction. An extension to the three-dimensional case is possible in principle. For the purpose of a worldwide generalization of seismic-refraction data, one has to confine oneself mainly to a kinematic interpretation of the records. This statement need not exclude the possibility and necessity of evaluating the data under dynamic aspects.

The source of the rays as well as the receivers are located at the surface. In order to avoid misinterpretations, it should be mentioned again that the expression "refracted" means that a corresponding ray is turned at a certain depth-level z characterized by the velocity v(z). In this case, the velocity of the vertex dx/dt is equal to the velocity v(z). These waves can also be described by the expressions "diving" or "penetrating" waves. Reflected waves are caused by sudden changes of wave impedance; the quantity dx/dt is, in general, an apparent velocity. A head wave (Mintrop-wave) is a refracted wave, guided at the top of a constant-velocity layer. It is evident that the overburden must show a lower velocity in this case.

In the interpretation procedure, the curvature of the travel-time curve, which is strongly influenced by the velocity gradient dv/dz in the depth of the vertex, is of great importance. In general, the velocity gradient dv/dz can be situated within:

$$-\infty \leq dv/dz \leq +\infty \ .$$

Two cases can be distinguished:

1. $0 \leq dv/dz \leq +\infty$ (positive gradient)

2. $-\infty \leq dv/dz \leq 0$ (negative gradient)

1. Positive Velocity Gradient

Possible cases are:

a) $dv/dz = 0$ (constant-velocity layer)
There is no vertex within this layer. Head waves are possible at the top of the layer and reflections are generated at the upper and lower boundary.

b) $0 < dv/dz < \infty$ (layer with an arbitrary positive velocity gradient)
Rays are bottoming within this layer except in those cases where the same velocity interval is already present in overlying layers. To each depth z there belongs one ray. Except from special cases each ray differs in distance and travel time.

c) $dv/dz = \infty$ (discontinuity)
A distinct velocity interval is associated with the same depth value z. Thus more than one ray is bottoming at this depth z and more than one point represents this depth value in the travel-time diagram.

From these three cases, two main types of travel-time curves, normal or reversed, can be derived.

In (a) a constant velocity layer is characterized by a linear travel-time curve which can be regarded as the extreme case of the normal type.

In (b) the general case of an arbitrary positive velocity gradient for refracted waves, curved travel-time branches of concave as well as of convex curvature with respect to the x-axis are possible. A "weak" velocity gradient generates a travel-time curve of convex curvature (normal type), while a "strong" velocity gradient causes a travel-time curve of concave curvature (reversed or retrograde type).
In (c) the travel times of a reflected wave are always arranged on a travel-time curve of concave curvature (reversed type).

From these considerations it is evident that the generally used assumptions $dv/dz = 0$ and $dv/dz = \infty$ are only the extreme cases of the more general case $0 \leq dv/dz \leq \infty$. Starting from this general assumption, it has to be deduced from the recorded data whether the extreme cases are realistic or not. It is evident that this question is related to the scatter of data available. Whereas in the beginning of crustal studies in the fifties, a clear decision on this question was not possible, today quantity and quality of modern observations do allow in principle a clear statement concerning this problem.

2. Negative Velocity Gradient

Today, no interpretation can exclude the possibility of a negative velocity gradient existing within the crust. In the depth interval between z_A and z_B which is characterized by the condition:

$$v(z) < v(z_A) = v(z_B) \text{ for } z_A < z < z_B$$

no refracted ray can bottom and, therefore, no information from this velocity interval is available from the travel-time diagram. Thus, in a certain area of the x,t-field, there are no arrivals possible (shadow or interruption zone).

If such velocity inversion exists in the velocity-depth function, its unique determination from the travel-time curves is, on principle, impossible for its underlying zones; only the range of possible solutions can be given. Introducing additional assumptions, this range can be narrowed.

Incomplete travel-time data. The existence of a velocity inversion is not the only reason for a lack of travel time data. In order to obtain the complete interval for the quantity dx/dt, e.g. between 5.0 and 8.2 km/s, the refraction profile must have a length of about 200 km for a crustal thickness of about 30 km. This length has to be extended to 300 km if the crust is 40 km thick. Shorter profiles may show a lack of true-velocity data in the interval of about 6.5 to 7.0 km/s.

In practice, there is still another reason for incomplete velocity data. The energy of the waves returning to the surface is strongly influenced by geometrical spreading. Waves bottoming strong gradient zones or discontinuities near the critical point are characterized by large amplitudes whereas weak gradient zones generate only small amplitudes mostly undetectable when occurring as later arrivals. Thus a lack of velocity data can exist, too.

A further complication is caused near intersections of travel-time curves. A clear and unique identification and separation of the several wave groups are mostly doubtful here, due to the interference. Finally, converted, multiple and diffracted waves can destroy the regular wave pattern. The cases of incomplete traveltime data are graphically shown in Fig. 1.

4.1.2 Correlation in Reflection and Refraction Work

When evaluating seismic data, correlation plays an important role, perhaps the most important one. Before dealing with the specific problems of data generalization, the term "correlation" and its meaning and use must be clearly defined.

The expression "correlation" means that a distinct feature is associated with a distinct term. In order to avoid difficulties, a clear definition by aid of the properties of the object under discussion must be given. Consequently, features equally termed can be joined or, less precisely expressed, "correlated".

The use of the word "correlation" in the seismic language means mainly the identification of certain events in the records and their joining or correlation in the time-distance domain. From reflection work it is very common to identify the correlation in the travel-time diagram with that in the cross section. It will be shown, however, that in refraction seismics, the correlation in the cross sections which is closely connected with the problem of nomenclature of distinct features of the crust, such as discontinuities, has to be considered separately.

Reflected waves can be characterized by their property that the derivative dx/dt means in general an apparent velocity. Here, no information on the velocity at the turning point, the reflecting element, can be obtained. The reflection work aims at tracing sudden changes of wave impedance in the medium under study. The sudden change of wave impedance is the feature that must be traced or correlated from point to point in the cross section. The tracing of the arrival times of the reflected wave in the time-distance domain, applying the principle of phase correlation, corresponds directly to the tracing of the interface, characterized by change of wave impedance in the cross section. Thus, the correlation in the time sections coincides with that in the depth sections.

Refracted waves can be described by the feature that the quantity dx/dt enables the determination of the true velocity at the depth where the ray bottoms. In general, each refracted ray - each point on the corresponding travel-time curve - reaches its point of maximum penetration, i.e. the ray's turning point, at different velocity levels. Consequently, one has to distinguish between

- the correlation in the record sections (time sections) and
- the correlation in the cross sections (depth sections).

The correlation in the record sections means the tracing of the arrival times of a wave group from receiver to receiver, just as in reflection work.

When, however, comparing and compiling the results of two or more corresponding travel-time curves, a common and joining parameter is necessary. The only possible parameter is the velocity at the depth of the ray-turning point. In cross sections,

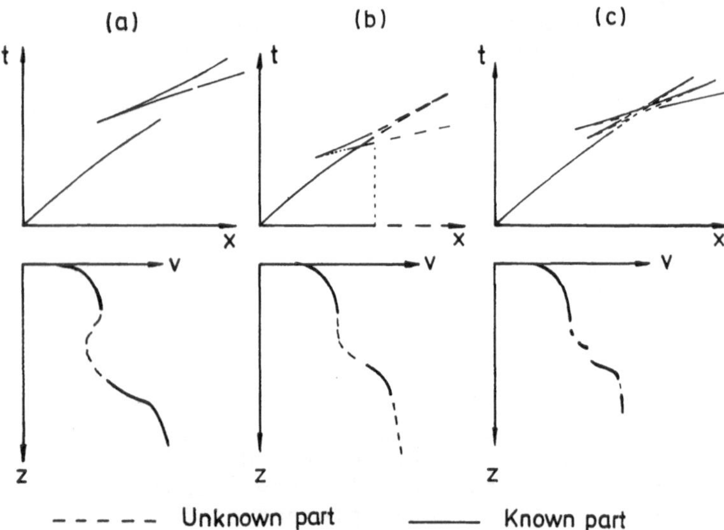

Fig. 1a-c. Cases of incomplete travel-time data for a continuously varying velocity-depth function. (a) Low-velocity layer. (b) Incomplete system of observations, e.g., a profile being too short. (c) A unique correlation is not possible due to a complex interaction of travel-time curves. t travel-time, x distance, z depth, v velocity

the correlation of points showing the same velocity leads automatically to a display with lines of equal velocity. It is quite evident that a special case is involved here when the velocity gradient becomes infinite, meaning a first-order discontinuity. If the inversion of the travel-time data into the depth-velocity domain reveals that more than one refracted ray was turned in the same depth, then an interface exists. Converging isolines indicate the development of an interface. In principle, each sudden velocity increase can be displayed if the chosen velocity interval of the isolines is sufficiently small. With minor modifications and extensions, this general statement is also valid for inclined interfaces.

Cross sections are one kind of seismic-data display, contour maps are another. The feature which can be correlated here is the depth of the velocity. Thus a contour map can show isolines of depth for a distinct velocity value, or reversely, the velocity isolines are contoured for a constant depth value.

4.1.3 Principles of Phase and Group Correlation

The first step of interpretation aims at identifying the arrival times of a wave group in the record sections. A wave group can be briefly defined by the conditions that the travel-time curve is not interrupted; the travel time increases with distance, and the quantity dx/dt is a continuous function. The basic principle of correlation is the phase correlation. In practice, the phase correlation sometimes does not use the beginning of the event but the next pronounced amplitude minimum or maximum. Evidently a sure phase correlation requires a spacing of detectors smaller than one wave length.

The crust passed by a wave field generates many secondary events returning to the surface. A first simplification has to be introduced because many arrivals are only caused by local inhomogeneities. In order to separate this "noise" from the regular signal, some criteria must be applied when identifying wave groups in record sections.

1. Amplitudes of the events being correlated must exceed those of the noise.

2. The apparent velocity must show values within a possible and reasonable range.

3. The travel-time branches must be of some length.

Due to the complex and heterogeneous structure of the crust, numerous shorter and longer travel-time branches can be detected in a time-distance diagram. A simplification, first step of data generalization, was already introduced by the criterion "certain length of travel-time branch". There arises the question of how long a segment must be. Here one of the main difficulties in the evaluation of crustal data is touched.

The principle of phase correlation is the base for data evaluation in exploration-reflection work. Due to narrow-spaced detectors and the large apparent wave length of the reflected pulses, a phase correlation is easily feasible. Using continuous profiling in seismic-refraction studies with receiver distances of some 100 m,

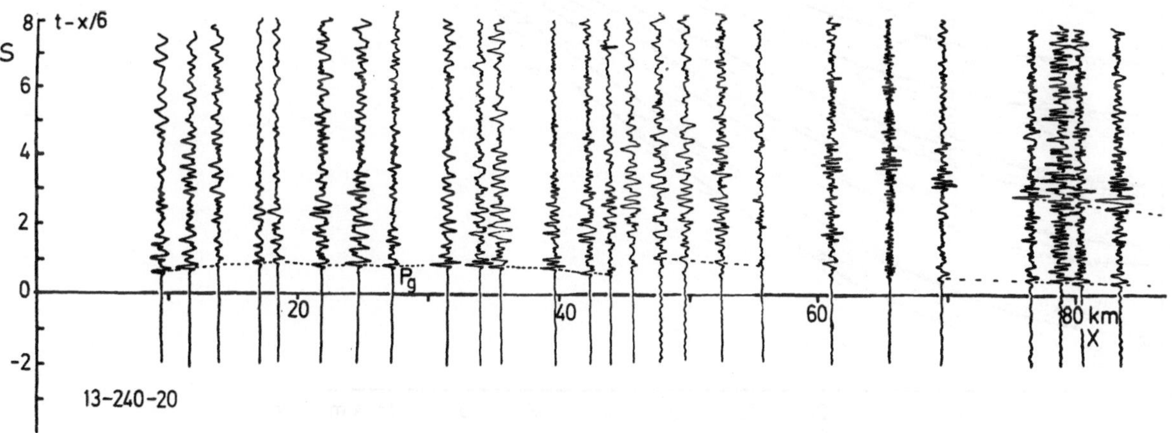

Fig. 2. Example for a phase-correlated P_g-curve (profile 13-240-20)

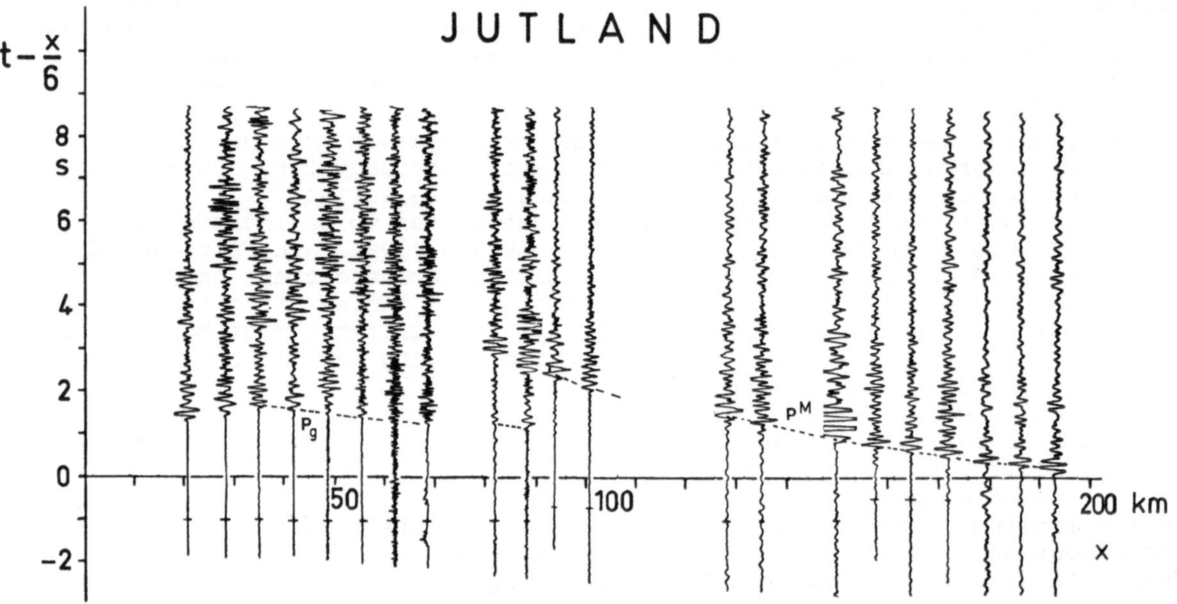

Fig. 3. Example for a phase-correlated P^M-curve (profile Jutland, HIRSCHLEBER et al., 1966). The records are related to a unique depth of water under the shotpoint

phase correlation is in general still possible. It has to be mentioned, however, that a phase correlation can be applied in many cases, too, if the receivers are spaced more widely. Figs. 2 and 3 show two examples.

The problem of the "short elements", established by phase correlation, leads to the principle of group correlation. The crust, as outlined in the next section, shows a sandwich-like structure with components larger

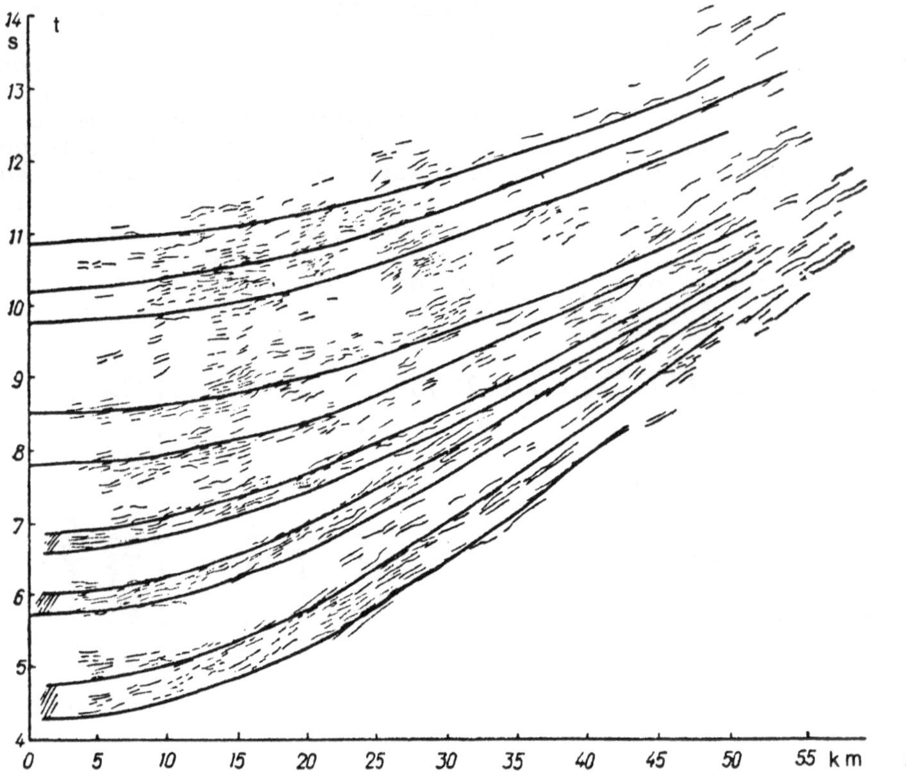

Fig. 4. The group-correlated branch runs parallel to the individual phase-correlated elements

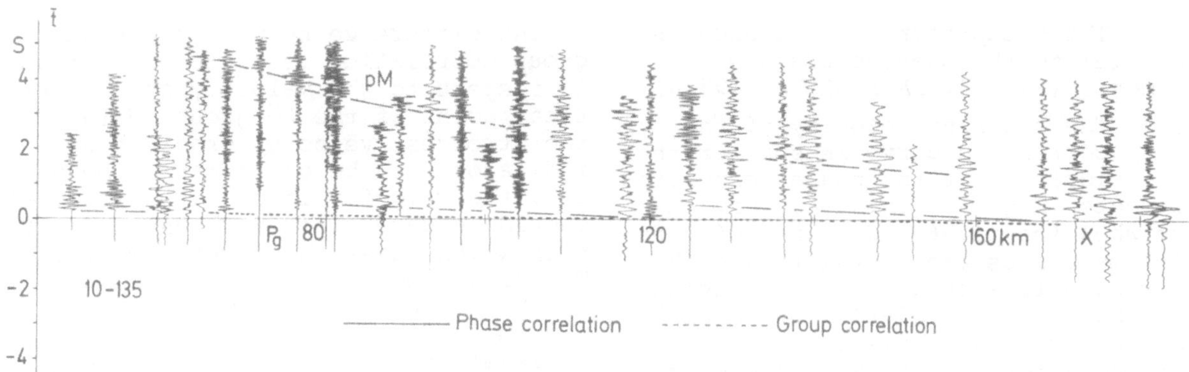

Fig. 5. The apparent velocity of the group-correlated curve is smaller than that of the phase-correlated segments (profile 10-135)

and smaller than the wavelength of the seismic pulse. The classical discontinuities are primarily composed of lamellae causing many wave groups and a corresponding number of travel-time branches. These segments are situated in a more or less wide stripe. When representing these separated segments by a mean travel-time curve, the principle of group correlation is applied, whereby the points of support are individual segments, indicated by phase correlation, or even by groups of large amplitudes. The application of the principles of group correlation results in a more or less simplified presentation of the real crustal structure. It is evident that the simplification increases even if only a few travel-time branches are introduced.

As in phase correlation, some criteria are necessary for the application of group correlation:

1. Joining of separated phase-correlated segments and/or joining of arrivals characterized by clear amplitudes.

2. The resulting apparent velocity must be within a possible and reasonable range.

3. The travel-time branch must show "some length".

With the third criterion, the problem of data generalization is again directly touched.

Some remarks are necessary concerning the apparent velocity of the group-correlated travel-time curve. Three cases are possible:

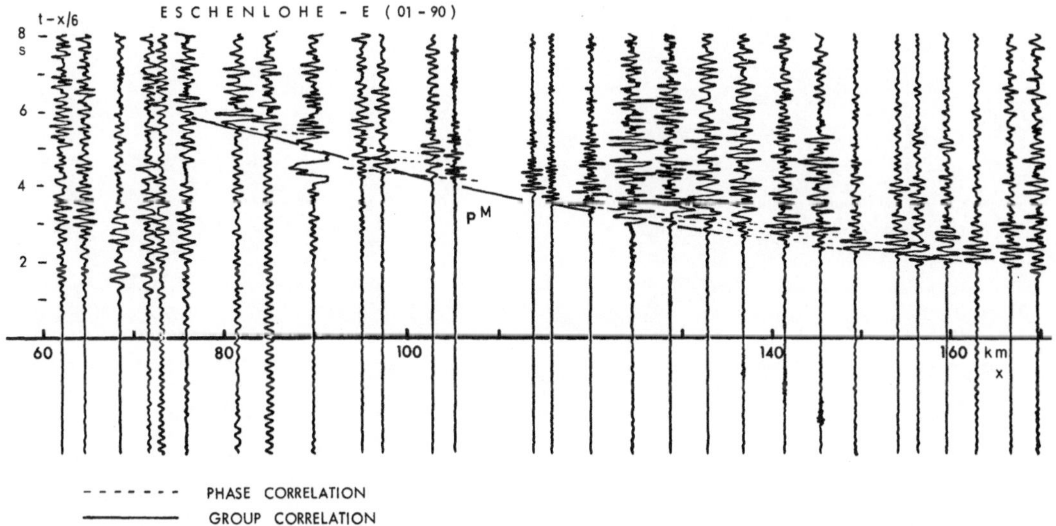

Fig. 6. The apparent velocity of the group-correlated curve is greater than that of the phase-correlated branches. The record section is a portion of the profile 01-090. The sections of the profiles Eschenlohe-SE, Eschenlohe-S, and Eschenlohe-SW show the same feature as the split P^M-curve (see Chapter 7)

1. The group-correlated branch is parallel to the phase-correlated elements (Fig. 4, PAVLENKOVA, 1973a).

2. The apparent velocity of the group-correlated curve is smaller than the velocities of the phase-correlated elements (Fig. 5). This case may occur when the right-side parts of travel-time curves are correlated, each showing a time delay with increasing distance. The travel-time branches of the intercrustal waves between the P_g- and P^M-group can be regarded as a typical example (KOSMINSKAYA and RIZNICHENKO, 1964).

3. The apparent velocity of the group-correlated curve is greater than the velocities of the phase-correlated elements (Fig. 6). This case is relevant when the left-side parts of reversed travel-time curves, which are delayed with decreasing distance are correlated. The P^M-group can show such an arrangement of travel-time branches (GIESE, 1972).

The difference between phase and group correlation is of some importance in respect to the velocity at the critical point of the P^M-group. Using the strongest velocity gradient as an indicator for the crust/mantle boundary, the velocity itself at this level should be determined by phase correlation which gives the layer velocity. The difference between both velocities may be in the order of 0.5 km/s.

The expressions "phase" and "group" correlation are not of absolute but only of relative meaning. Large wave lengths veil the detailed structure by integration, yielding an uninterrupted travel-time curve, thereby allowing a phase correlation. Using shorter wave lengths, the resolving power is higher and the previous travel-time branch may dissolve into numerous separated shorter segments. In the terminology of filter processes, group correlation is identical with a higher cutoff frequency (low pass filter).

4.2 Problems and Tasks of Data Generalization

P. GIESE

ABSTRACT

The evaluation of seismic refraction data is a process of different stages. The first stage comprises the presentation of the records; in the second, some characteristic parameters or even complete travel-time curves are presented in a form giving qualitative information on crustal structure. In this contribution some corresponding methods are described, principally the method of reduced travel-time curves. The presentation of intercept-time sections with velocity isolines gives a detailed picture of crustal structure. Such a display corresponds to the t_o-section in reflection seismics. Finally the problem of how to define the crust/mantle boundary is discussed.

ZUSAMMENFASSUNG

Die Bearbeitung und Auswertung refraktionsseismischer Meßdaten ist ein mehrstufiger Prozeß. Die erste Stufe beinhaltet die Darstellung der Meßergebnisse. Im zweiten Schritt sollten einige charakteristische Parameter oder aber auch gesamte Laufzeitkurven so dargestellt werden, daß ein qualitatives Bild von der Struktur der Erdkruste erhalten wird. Es werden hier eine Reihe von entsprechenden Methoden beschrieben. Insbesondere werden die Möglichkeiten erläutert, die die Methode der reduzierten Laufzeitkurven bietet. Die Darstellung der Laufzeiten in intercept-time Profilen mit Linien gleicher Geschwindigkeit gibt bereits ein recht detailliertes Bild von der Krustenstruktur. Diese Darstellung entspricht den t_o-Profilen in der Reflexionsseismik. Am Ende dieses Abschnitts wird die Frage diskutiert, wie die Grenze Kruste/Mantel definiert werden kann.

4.2.1 Introduction

During the meeting of the Study Group of Experts on Explosion Seismology at Leningrad in 1968, the problem of how to find methods for a generalizing interpretation and presentation of deep-seismic sounding data was discussed.

The term "data generalization" summarizes different aspects of seismic-data evaluation. The starting position is as follows: there are inhomogeneous data which are interpreted under different conceptions, and the results obtained are heterogeneous. Comparisons and compilations become difficult and may not be objective in all cases. Therefore, data generalization should comprise the following aspects:

1. Presentation of data observed in such a form as to allow a possible later reinterpretation, i.e., any subjective selection of data recorded should be avoided as much as possible.
2. Presentation of data observed applying simple transformation procedures thus giving qualitative, easily understood information on crustal structure.
3. Presentation of velocity cross sections containing the main and typical features of crustal-velocity distribution.

These principles should be applicable to all kinds of deep-seismic sounding techniques, for continuous profiling as well as for pointwise technique.

4.2.2 Presentation of Records

The first step toward data generalization concerns the display of recorded data. In Western Europe, the pointwise field technique is widely used. Here, the presentation of data observed as record section with a reduced travel-time scale (reducing velocity 6.0 km/s) is now very common, and any publications dealing with seismic crustal studies should contain this basic and objective information.

Most of the records and necessary basic data, obtained within the framework of the research program for seismic crustal studies in the Federal Republic of Germany are presented in Chapter 3. Later (Chap. 7) many of

Chapter 3. In Chap. 7, the main features of crustal structure in the Alps are described and here, many record sections are presented, some for the first time. Thus the requirement for data generalization is fully met by this presentation.

In continuous profiling with reversed and overtaking travel-time curves, it is difficult and laborious to display all traces. Therefore, phase-correlated travel-time curves (unreduced) should be presented in a scale and manner allowing for a reinterpretation.

4.2.3 Presentation of Some Basic Parameters

4.2.3.1 Some General Remarks on the Problem of Correlation

The aim of correlation is to trace the front of a wave group thus giving the several travel-time branches. In reflection seismics, the principle of phase correlation is quite common and is the basis of all further interpretations.

In refraction seismics, especially in crustal studies, the situation is more complex and difficult due to receiver distances larger than in reflection work. So the principle of group correlation must be widely used. Group correlation means that arrivals, characterized by large amplitudes, are fitted by straight or curved travel-time curves. Such a group-correlated travel-time curve can be composed of events having no similarity to those in adjacent traces or of groups of shorter segments which are internally established by phase correlation.

It is quite evident that group correlation is more subjective than phase correlation. For purposes of data generalization, data should be read from travel-time segments which are phase-correlated. The methods described in the following do not require long branches, they can be applied to segments having a length of only some 10 km. The importance of distinguishing phase and group correlation is discussed in detail in 4.1.

4.2.3.2 Presentation of Some Typical Basic Parameters

The following methods are based on the widely evidenced experience that in crustal studies three main wave groups can be observed (Fig. 1):

1. The P_g-group, penetrating into the upper part of the basement.

2. The P^M-group, bottoming the crust/mantle transition or boundary.

3. The P_n-group penetrating the uppermost part of the mantle.

It is evident that position and shape of these main curves in the travel-time diagram depend on the velocity distribution in crust and upper mantle. Parameters which characterize position and shape of these travel-time curves can already give some qualitative information on crustal and upper mantle structures. There are several possibilities how to select and define some of these typical parameters.

a) Critical Point of the P^M-wave

The critical point x_c, t_c of the P^M-wave is ray-geometrically defined as the point at which the reflection time is equal to the refraction time. This definition is also valid for continuously refracted waves. Considering only ray-geometrics, the critical point is characterized by a maximum of amplitudes, but taking into account finite wave length, the point of largest amplitudes is shifted to larger distances (CERVENY, 1966). The amount of shift depends on the wave-length used, crustal velocity and thickness. To demonstrate the order of shift, two models, 30 and 50 km thick, will be considered. The ray-geometrically defined critical distance is 68 or 113 km respectively. Taking into account a finite wave length of λ = 1 km,

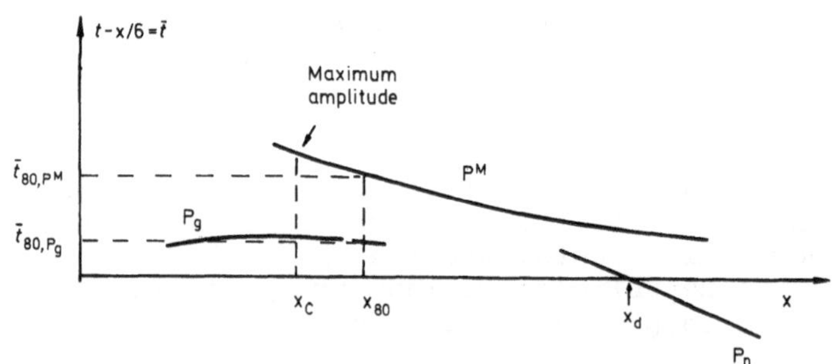

Fig. 1. Schematic travel-time diagram with reduced time axis showing the three main wave groups P_g, P^M, and P_n and the basic parameters x_c, x_d, and $\bar{t}$ (x = 80)

this results in 81 or 129 km respectively for the distance of largest amplitudes. The differences between the amplitude maxima for $\lambda = 0$ and $\lambda = 1$ km are 13 and 16 km. So the relative shift is only 3 km, a value which can be neglected. Thus, in principle, the amplitude maximum may serve as a parameter which characterizes variations of crustal thickness. The practical difficulties are caused by the broad amplitude maximum and the scatter of the amplitudes themselves which impedes the precise determination of the largest-amplitude distance.

The values of critical distance can be plotted in contour maps or profiles whereby each value is associated with its half-distance (PRODEHL, 1970a, b; CHOUDHURY et al., 1971; GIESE and STEIN, 1971).

In a homogeneous crust and with a wavelength $\lambda = 0$, the quantity x_c is linearly related to the crustal thickness z by the equation

$$z = \frac{x_c}{2} \operatorname{ctg} i \qquad \text{with } \sin i = v_o/v_1$$

v_o: crustal velocity
v_1: upper-mantle velocity.

Assuming 6 and 8 km/s respectively for the velocities, the results is:

$$z = 0.44 \cdot x_c.$$

Thus the quantity x_c reflects, in first approximation, variations of crustal thickness.

Although this quantity x_c has a clear meaning, difficulties arise in practice. The determination x_c in the records includes an uncertainty of ± 10 to 15%, that means about ± 10 km. Thus only variations of crustal thickness greater than at least 5 km are clearly detectable.

Fig. 2 shows the frequency distribution of the quantity x_c for profiles observed in western Germany and the Alps. In the Alpine area, the range of observed data clearly exceeds the width of uncertainty of x_c, whereas for the data of western Germany, their interval is only slightly larger than the range of uncertainty.

b) Intersection of the P_n-curve with the P_g-branch or with the Curve v = 6 km/s

The cross-over distance between two curves has the advantage of precise reading of values. Owing to the experience that the P_g-wave mostly dies out within 100 km distance and thus a direct intersection between the P_g- and the P_n-branch does not occur, it would be better to use the average crustal velocity v = 6 km/s as a mean P_g-curve.

Assuming again a homogeneous crust with a velocity v_o and a constant upper-mantle velocity v_1, the relation between crustal thickness z and cross-over distance x_d is linear

$$z = \frac{x_d}{2} \sqrt{\frac{v_1 - v_o}{v_1 + v_o}} .$$

Assuming v_o = 6 km/s and v_1 = 8 km/s, the result is

$$z = 0.17 \cdot x_d.$$

So the quantity x_d reflects like x_c variations of crustal thickness. This parameter plotted at half its distance can be displayed in sections or contour maps (PRODEHL, 1970a, b; CHOUDHURY et al., 1971; GIESE and STEIN, 1971).

Although this quantity x_d is clearly defined, there is the disadvantage of its low resolving power due to the

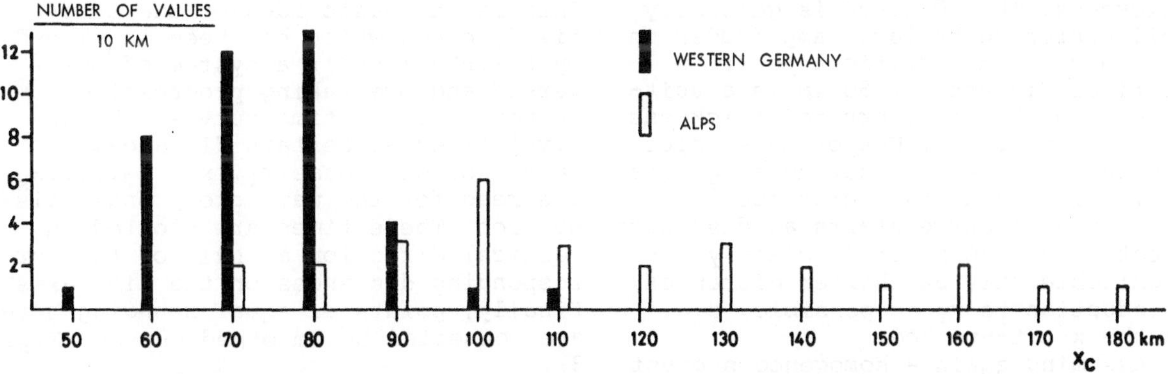

Fig. 2. Histogram of the critical distance x_c from profiles obtained in western Germany and the Alps (GIESE and STEIN, 1971)

great values of the cross-over distance which are in the range beyond about 100-150 km distance. Inhomogeneities smaller than 100-150 km remain undetected. Therefore, the presentation of this quantity is only reasonable in large-scale structures with distinct variations of crustal thickness.

c) Traveltimes at Fixed Distances

Whereas in the previous section the position of the parameters is defined by the properties of the curves themselves, in this section a principle is described which starts from a fixed coordinate. As the seismic-refraction method was still widely used for oil exploration, the principle of fan shooting was quite common, i.e. the determination of travel times at fixed distances x_f.

This simple method may serve for the purposes of data generalization. After having correlated the main wave groups, the travel times for the several branches can be read and displayed for one or more fixed distances. The complete or reduced travel time of a certain branch at a fixed distance is plotted at half of the corresponding distance in sections or contour maps. Due to the general use of the reduced time axis in record sections, it is reasonable to present reduced times. An increase of travel times may be caused by an increase of overburden thickness and/or a decrease of velocity in the overburden.

The P_g-wave is generally observable up to a distance of 60-100 km. Reading the travel times at 60 or 80 km distance, a corresponding display reflects the variations of thickness and average velocity of the sedimentary cover and/or the weathered basement.

The use of P^M data may serve to characterize the variations of crustal thickness. The P^M curve is generally well expressed by large amplitudes in the range of the critical point. A critical distance of 80 km is a suitable value having a crustal thickness up to 30 to 35 km. Use of this value has the advantage of having a P_g wave still visible at this distance.

If the P^M-curve starts at distances greater than 80 km or is clearly recognizable only at shorter distances, a reduction to x_f is possible, e.g. by the x^2, t^2-method.

Assuming again a homogeneous crust with the velocity v_o, the relation between the crustal thickness z and the travel time t(x) at a fixed distance x_f is given by the simple relation

$$z = \frac{x_f}{2}\sqrt{\left(\frac{v_o}{x_f/t(x_f)}\right)^2 - 1}$$

or with $dx_f/dt = v_p$
and $\sin i = v_o/v_p$

$$z = \frac{v_o \cdot t(x_f) \cdot \cos i}{2} \quad .$$

Here, the depth z depends on the travel-time $t(x_f)$ and the corresponding apparent velocity dx/dt. Because both quantities increase with depth, the travel time $t(x_f)$ can serve as an indicator for variations of crustal thickness.

A sedimentary cover can cause a large delay for the travel time of waves penetrating into the crust and upper mantle. In order to eliminate this effect, the difference of travel times between the P^M- and P_g-curve must be read at the same distance. The only overlapping distance interval for both wave groups is near 60 to 80 km, for crustal thicknesses up to about 35 km. Therefore, the distance of 80 km can be regarded as an optimum for this method of data generalization.

4.2.4 Presentation of Transformed Travel Time Curves

4.2.4.1 Display of Times at Fixed Distances

In the previously described methods only one parameter is used from each curve. An extension of the last method to several fixed distances x_1, x_2, x_3 ....x_i offers the possibility to transform a complete travel-time branch into another domain of display. This is the basic idea of the time-field method which has been developed by PUZIREV (1968): a system of reversed and overtaking progressive (normal) travel-time curves, the arrival times at certain distances, named sounding base x_1, x_2, x_3.....x_k are read for the wave group under discussion. These times are plotted in vertical direction at half of the corresponding distances on the time axis. Finally, points of equal sounding base are connected by smoothed curves (Fig. 3).

These isolines show qualitatively lateral variations of crustal velo-

Fig. 3. Principle of the time-field method after PUZIREV (1968). The upper part shows some schematic travel-time curves; in the lower part, the "time field" is constructed. *SP* shotpoint, *x* distance, *t* traveltime

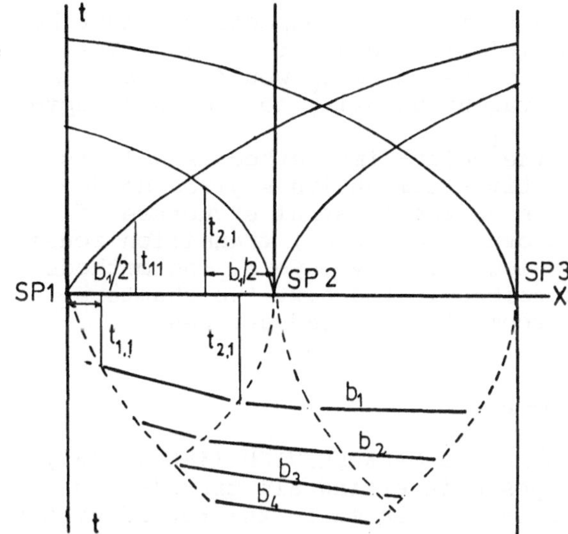

city distribution. For progressive (normal) branches, the quotient $(x_{i+1} - x_i)/(t_{i+1} - t_i)$ has the meaning of apparent velocity. For reflected waves, the square of the time must be plotted. Here, the quotient $(x_{i+1}{}^2 - x_i{}^2)/(t_{i+1}{}^2 - t_i{}^2)$ gives the average velocity of the overburden, as it is proved in the x^2, t^2-method.

The principle of the time-field method is also applicable for travel-time differences between two curves, so indicating variations of the internal structure of the crust. The term "time-field" is also in use for another method, described e.g. by THORNBURGH (1930), RIZNICHENKO (1945), and HAGEDOORN (1959). Here, wave fronts are constructed by using the principles of Huygens. In order to prevent confusion, the method of PUZIREV should be called fixed-distance-time method.

4.2.4.2 Presentation of Time Residuals

The principle of presentation of time residuals has been well known for some decades and several modifications have been developed for different purposes.

In exploration-refraction work, the technique of delay times is widely used. It is based on the simple relation which subdivides the traveltime of a refracted (head) wave into three components:

$$t = x/v_p + \bar{t}_{sh} + \bar{t}_{rec}$$

- x : distance shotpoint - recording point
- v_p : refractor velocity
- $\bar{t}_{sh}$: delay-time at the shotpoint
- $\bar{t}_{rec}$: delay-time at the recording point.

In a system of reversed and overtaking travel-time curves $\bar{t}_{sh}$ and $\bar{t}_{rec}$ can be determined point for point and displayed in profiles or contour maps showing qualitative variations of the thickness of the layer under discussion.

a) Delay-Time Method and Time-Term Analysis

The simple delay-time method has been extended by SCHEIDEGGER and WILLMORE (1957) and WILLMORE and BANCROFT (1960) to the time-term least-square method with the goal to evaluate seismic crustal data obtained from an observation system with arbitrarily distributed shot and recording points. The basic idea is the following: there is a system of m stations and n shots, thus m + n time terms can be obtained. The maximum are m · n records. Because the system is overestimated, the least-square principle can be applied. The determination of these time terms can be carried out by solving a set of normal equations. BERRY and WEST (1966) and SMITH et al. (1966) have applied this method in their interpretation of the first-arrival data of the Lake Superior experiments in 1963. MEREU (1966) has proposed an iterative method for solving the time-term equations for which no computers with large storage capacity are needed.

BAMFORD (1973; Chap. 5.3 of this volume) has developed a further extension and modification now allowing also the determination of anisotropy of the refractor.

Hitherto the delay-time method as well as the time-term analysis are applicable only for first arrivals. The evaluation of reflected arrivals by the delay-time method or by the time-term method respectively must be possible in the x^2, t^2 domain. The zero-offset time t_o would be the equivalent quantity to the intercept-time t_i.

The delay-time method as well as the time-term analysis gives no details on the internal structure of the crust, as, e.g., transition zones and low-velocity layers. The methods described in the next two chapters overcome these disadvantages.

b) τ-method

GERVER and MARKUSHEVICH (1966, 1967) proposed inverting all travel-time curves correlated in the record by aid of the equation

$$\tau = t - x \cdot dt/dx.$$

Whereas, in the previous methods, the intercept time is split into the two terms $\bar{t}_{sh}$ and $\bar{t}_{rec}$, this separation is neglected in the following.

GERVER and MARKUSHEVICH (1966, 1967) as well as KEILIS-BOROL (1971) and BESSONOVA et al. (1974) treated the τ-method mainly from a mathematical point of view in connection with the Herglotz-Wiechert equation. Especially BESSONOVA et al. discuss the problems concerning the nonuniqueness of solutions, due to the existence of a low-velocity layer and the inaccuracy of data. All considerations are based on horizontal layering. As the basic idea of the τ-method is included in the method of reduced travel-time curves, which is the more general one, all further details are described in the next section.

c) Method of Reduced Travel-Time Curves

The procedure proposed by PAVLENKOVA (1973a, b) has other main objects and is applicable to lateral inhomogeneous media as well. The first aim is to determine the nature of the waves and to correlate them to velocity levels. The procedure is as follows: all travel-time curves are transformed into a $x/2$, $t - x/v_r$ display, using different reduction velocities v_r, as e.g. v_1, v_2, v_3, etc. (Fig. 4). The time axis, having the meaning of intercept time, is directed downward again, where the distances between the shotpoints remain unchanged. From each travel-time branch results a family of curves, each having a maximum or minimum if the reduction velocity is within the interval of slope values of the curve under discussion. Progressive curves show a minimum, whereas retrograde branches give a maximum. Linear travel-time curves do not change their form, only the slope varies. It is now possible to find out from the pattern of the reduced travel-time curves those branches belonging to the same velocity level v_i (Fig. 4).

Progressive (normal) travel-time curves are always associated with layer velocities. The case is different for reversed segments. Because reversed branches can be generated by sharp discontinuities and/or strong gradient zones, the nature of the quantity dx/dt is open. In combination with progressive branches, the critical point part and the far-distant end can be used for the construction of velocity isolines, meaning layer velocities.

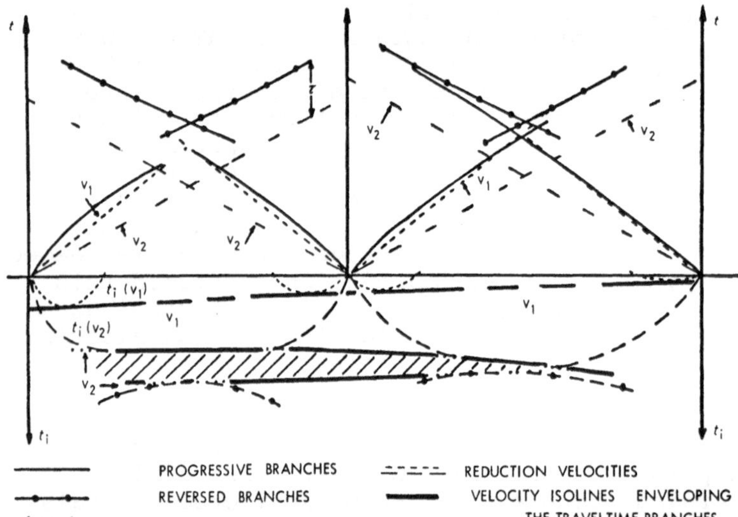

Fig. 4. Principle of construction of iso-velocity lines in the distance-intercept-time section. The reduced travel-time results from the relation $\tau = t - x/v_r$ with v_r as reduction velocity. Progressive branches show a curvature directed to the distance axis, whereas the reversed branches are bent in opposite direction

It is also possible to transform reversed branches completely and to construct the envelopes. But such envelopes associated with reversed travel-time curves should be plotted with different signatures in order to demonstrate their possible apparent-velocity nature (Fig. 5).

The second aim is to obtain a qualitative picture of the main features of vertical as well as of horizontal velocity distribution by means of the

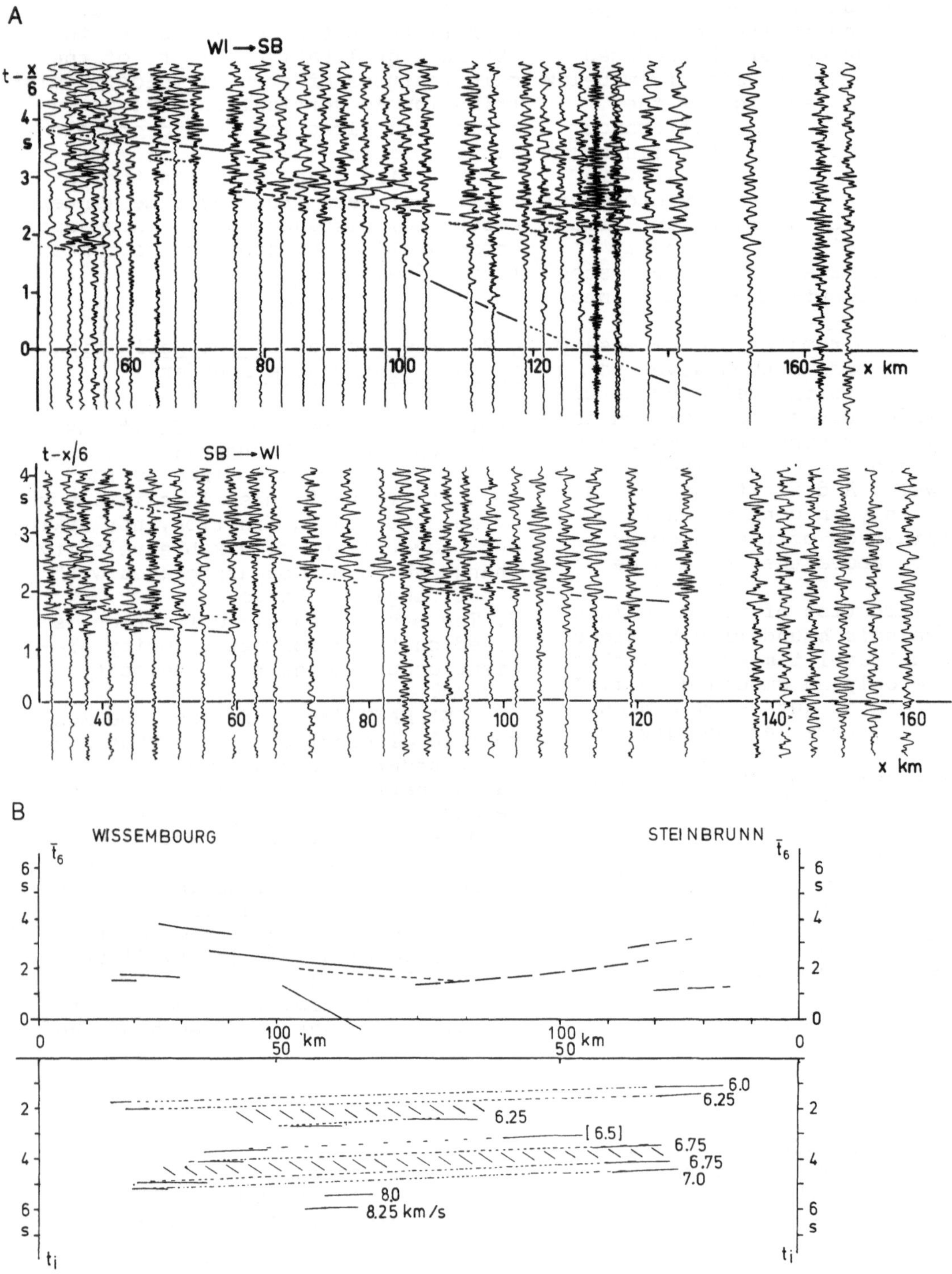

Fig. 5A and B. Intercept-time section of the profile Wissembourg-Steinbrunn. The upper part (A) shows the record sections of this reversed line (EDEL et al., 1975). In the lower part (B) the traveltime curves used and the resulting time section are presented. The velocity value in brackets indicates that the nature of this velocity level is open (a layer velocity from a penetrating wave or an apparent velocity from a reflected wave)

velocity isolines in the distance-intercept-time section.

Reduced travel-time curves also enable the detection of low-velocity zones and the determination of their lateral variations. The line connecting the points of termination of the first arrivals determines the shape of the top of the inversion zone whereas the envelope of the reduced travel-time curves of waves from below the low-velocity layer gives the shape of its bottom, the difference in time Δt between these two lines is the variation of velocity inversion (thickness and velocity decrease) along the profile. An intercept-time section with isolines of velocity can be used as base for the next step to invert intercept-time values into depth values.

Finally, two remarks are necessary. The intercept-time method gives no information as to the detailed nature of retrograde curves; no decision is possible whether they are composed of pure reflected and/or of penetrating waves as well. A definite answer is obtainable only in connection with the determination of the velocity distribution above the level under discussion.

The construction of velocity isolines or seismic boundaries by aid of the reduced travel-time-curve method is possible only in smoothed form. If the dimensions of quasi-homogeneous blocks or structures at the given level are less than the length of the travel-time curves of the waves penetrating to this depth, the lateral inhomogeneities are smoothed in respect to their relief.

In conclusion it can be said that the intercept-time method is a powerful tool for evaluating deep seismic-sounding data obtained by quite different field techniques.

d) Summary of Conclusions and Recommendations

<u>1st stage</u>: 1) Presentation of record sections or phase-correlated travel-time curves (continuous profiling).
<u>2nd stage</u>: 1a) Presentation of the reduced travel time (v = 6 km/s) of the P^M-group at 80 km distance. If the P^M-group appears only at greater distance, this travel time should be reduced to 80 km distance. In case of no detectable P^M-group, the travel time can be read by substitution of the prolongated P_n-wave.
1b) Presentation of the reduced travel time of the P_g-wave at 80 km distance.
1) Presentation of the difference of reduced travel times of the P^M- and P_g-wave at 80 km distance.
1d) Analogous displays for other distances (sounding bases) may be presented if necessary.
2) Presentation of intercept-time data as cross sections, aiming to clarify the nature of waves and to obtain a qualitative picture of the main features of crustal velocity distribution in vertical as well as in horizontal direction by aid of velocity isolines.

The third stage presents the inversion of data from the time (intercept-time) distance domain into depth cross sections or depth contour maps respectively.

<u>4.2.5 The Problem of the Definition of the Crust/Mantle Boundary</u>

There are two possible definitions of the crust/mantle boundary:
1. The boundary is defined by the velocity value 8 km/s. This is a purely mathematical definition, not taking into account any remarkable velocity change which may be associated with a petrological change.
2. The other possible definition is based on the strongest velocity gradient occurring near the velocity 8 km/s.

STEINHART (1967) defines the Mohorovičić discontinuity, the crust/mantle boundary, as that level in the earth where the compressional wave velocity increases rapidly or discontinuously to a value between 7.6 and 8.6 km/s. In the absence of an identifiable rapid increase in velocity, the Mohorovičić discontinuity is taken to be the level at which the compressional wave velocity first exceeds 7.6 km/s.

This definition is based upon the velocity gradient. The depth of the strongest velocity gradient is regarded as the level of the crust/mantle boundary. Some difficulties arise if the velocity range mentioned above is exceeded, especially at the lower limit. Therefore, a modification of the definition given by STEINHART seems necessary. Having a P^M-group, the highest velocity indicated on the left side should be used for the crust/mantle boundary definition. The difference between phase and group correlation must also be taken into account. In the past, in most of the record sections, the P^M group has

been correlated under the concept of group correlation, thus yielding velocity values near 8.0 km/s. A close inspection of the left-side end of the P^M group, however, reveals in many cases its split nature requiring a phase correlation. From the phase correlation, distinctly smaller velocity values result for the level of the strongest velocity gradient. There are, at any rate, numerous cases giving velocity values between 7.0 and 7.6 km/s. Therefore, the crust/mantle boundary should be defined by aid of that part of the P^M group that gives the highest velocity and, consequently, the greatest penetration depth.

This gradient may have a finite value but it can become infinite in the case of a sharp discontinuity. This depth is associated with the critical point of the P^M group and can therefore be determined with relatively high accuracy. Using this definition, it is necessary to characterize the crust/mantle boundary by a second quantity - the velocity value at the depth of strongest gradient. A rough classification in steps of 0.5 km/s would be sufficient (< 7.5 km/s, 7.5-8.0 km/s, > 8.0 km/s, Figs. 8 (5.2) and 9 (5.2)).

In cases with normal crustal structure, both definitions will give practically the same result. Differences may exist in anomalous regions, e.g., rift zones. In order to demonstrate clearly this anomalous behavior, both possibilities should be used and displayed.

4.3 Depth Calculation

P. GIESE

ABSTRACT

This contribution deals with the task of inverting travel-time data into depth-velocity values. It gives a short review of the main methods used in seismic crustal studies. The first requirement is the determination of the apparent velocity treated in the first section. The next section deals with the question of determining limiting values for single points as well as for separated segments. The nature of reversed travel-time curves is treated in 4.3.4. The possibility of application of the x^2, t^2-method is discussed in 4.3.5. The well-known Herglotz-Wiechert method and its extension when low-velocity layers are present is outlined in 4.3.6. The final section deals with approximation methods.

ZUSAMMENFASSUNG

Dieser Beitrag befaßt sich mit der Aufgabe, die Laufzeitkurven in Tiefenwerte umzuwandeln. Ein kurzer Überblick über die in der Krustenseismik üblichen Methoden wird gegeben. Als erstes ist die Bestimmung der scheinbaren Geschwindigkeit erforderlich. Der nächste Abschnitt behandelt die Frage, auf welche Weise Extremwerte für einzelne Punkte und auch für Laufzeitkurvenabschnitte zu bestimmen sind. Im Anschluß daran wird die Natur von retrograden Laufzeitkurven untersucht. Die Möglichkeit der Anwendung des x^2, t^2-Verfahrens wird diskutiert. Die bekannte Herglotz-Wiechert-Methode einschließlich ihrer Erweiterung bei Vorhandensein von Geschwindigkeitsinversion ist das Thema des Abschnitts 4.3.6. Im letzten Abschnitt werden Näherungsmethoden diskutiert.

4.3.1 Introduction

When the work of crustal seismic studies was started on a large scale 20 years ago, two different groups of geophysicists evolved: one group was educated in pure seismology and trained in continuous refraction and the Wiechert-Herglotz method. The other group was experienced in exploration seismics and familiar with the interpretation of reflections and head waves. Correspondingly, the methods for data evaluation used were different, and the types of models as well. During the last five to ten years, however, this difference became less and less important, and the methods of both groups are now in use in deep-seismic sounding. This report deals with some of the main methods suitable for the evaluation of deep-seismic-sounding data. Although, in principle, the methods are applicable to all types of field techniques, the author's experience is based mainly on pointwise observations. All the following considerations are based on the assumption of horizontel layering and isotropy.

4.3.2 Determination of Apparent Velocity

The determination of the apparent velocity $v_a = dx/dt$ is of fundamental importance and appears quite simple in principle, but in practice difficulties arise due to the scatter of data caused by lateral and vertical inhomogeneities of local nature.

Near-surface and locally limited inhomogeneities can be recognized in the travel-time data by corresponding bulges of subsequent travel-time curves. Due to their local nature, they are generally only of special interest. For the purpose of deep-seismic sounding, the investigation of large-scale lateral variations is interesting. Thus, the local fluctuations of slope along the curve under discussion must be eliminated in order to get a monotonously increasing or decreasing apparent velocity function with distance for the whole or, at least, for a longer part of the curve.

There are two methods to determine the apparent velocity: interpolation of functions or graphical construction.

4.3.2.1 Interpolation Method

Having three adjacent data points, the derivative dx/dt or dt/dx respectively is determined by a second degree polynomial. The derivative can be derived from the Lagrangian form of a polynomial (STEWART, 1966)

$$f'(x) = \frac{2x - (x_2 + x_3)}{(x_1-x_2)(x_1-x_3)} f(x_1) + \frac{2x - (x_1 + x_3)}{(x_2-x_1)(x_2-x_3)} f(x_2) + \frac{2x - (x_1 + x_2)}{(x_3-x_1)(x_3-x_2)} f(x_3) \quad (1)$$

x is the distance where the derivative is required.

In practice, the data points x_1, x_2, x_3 are, in most cases, equally spaced; thus this equation can be reduced to the following form

$$f'(x) = \frac{(x-x_2)[2f(x_1)+4f(x_2)+2f(x_3)]}{2\Delta x^2} + \frac{f(x_3) - f(x_1)}{2\Delta x} \quad (2)$$

with $x_1 = x_2 - \Delta x$ and
$x_3 = x_2 + \Delta x$.

If point x coincides with midpoint x_2 this equation has the very simple form

$$f'(x_2) = \frac{f(x_3) - f(x_1)}{2\Delta x} \quad (3)$$

or

$$f'(x_2) = \frac{t_3 - t_1}{x_3 - x_1} = \frac{1}{v_a(x_2)} . \quad (4)$$

This method does not contain any aspects of the least-square principle. BARTELSEN (1970) has proposed a polynom of second degree, the parameters of which are derived from five data points, i.e. the system is overestimated and the least-square method can be introduced. The expression for the apparent velocity becomes very simple if the distances x_i are equally spaced (Δx). The apparent velocity is related to the center of this 5-point segment

$$v_a = \frac{10\ \Delta x}{-2t_2 - t_{-1} + t_{+1} + 2t_{+2}} . \quad (5)$$

Due to the application of the least-square method, local fluctuations are suppressed. There is only one disadvantage: the first two and the last

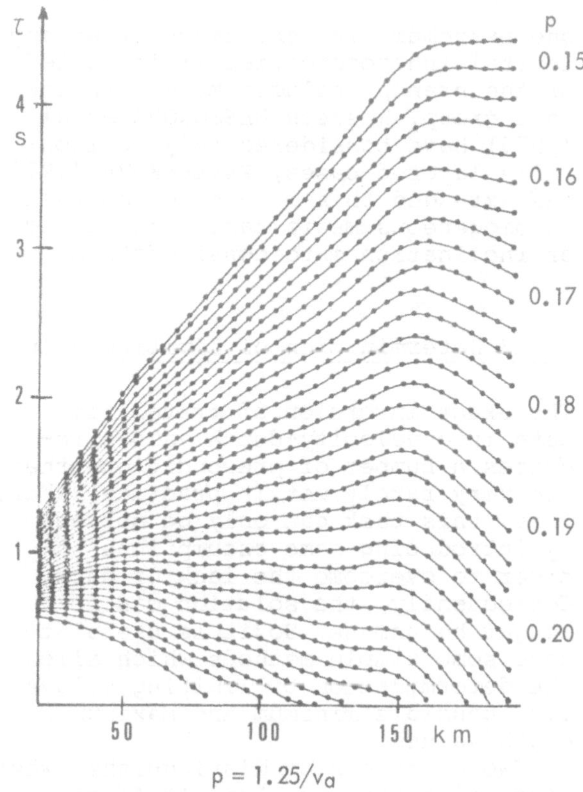

Fig. 1. Example showing the method of determining graphically the apparent velocity (BESSONOVA et al., 1974). First arrivals from the P_g- and P_n-wave are plotted. Each curve is characterized by its reduction velocity

two points are lost. For these parts, the Lagrangian equations must be applied.

4.3.2.2 Graphical Method

PAVLENKOVA (1973a, b) and BESSONOVA et al. (1974) propose a graphical method in order to determine the apparent velocity. Each branch of the travel-time diagram is transformed by aid of the equation

$$\tau = t - x/v_r \quad (6)$$

where v_r is the reduction velocity. A sequence of stepwise increasing (or decreasing) velocity values being within the range of the slope of the curve under discussion is used for plotting this set of transformed travel-time curves (Fig. 1). For lateral homogeneous media or media with only weak lateral inhomogeneities, each curve will have an extremum. It becomes evident from Eq. (6) that the position of the extremum gives the distances with the apparent velocity $v_a = v_r$. When a curve shows more than one extremum, the existence of strong lateral inhomogeneities is indicated or the branch includes more than one wave group. Whereas BESSONOVA et al. (1974) have considered only horizontally layered cases, PAVLENKOVA (1975) has extended this method to laterally inhomogeneous media taking the effect of inclination into consideration.

4.3.3 Determination of Limiting Values

The exact inversion of travel-time data into velocity-depth values requires a number of preconditions that are very rarely met in practice. Therefore, this task can only be solved by introducing some assumptions in order to overcome the lack of data. Consequently, the solution can no longer be unique. So it is useful to have some simple methods which allow the determination of limiting values, i.e. possible minimum and maximum depth values.

Two cases can be distinguished when discussing this problem. At first, we can ask for limiting values if only one individual point characterized by distance x, travel time t, and apparent velocity v_a is regarded. Secondly, the analogous question stands for travel-time branches separated from each other.

4.3.3.1 Individual Points

Maximum Depth. From each point of a travel-time curve, three quantities can be taken:

1. distance x,
2. corresponding traveltime t, and
3. apparent velocity $dx/dt = v_a$

Methods how to determine the apparent velocity are described in 4.3.2.

Assuming a homogeneous overburden of constant thickness z, the three quantities x, t, v_a, taken from the reflection branch, allow the determination of the quantity z:

$$z = \frac{x}{2}\sqrt{\frac{v_a \cdot t}{x} - 1} \leq z_{max}. \qquad (7a)$$

If the overburden is homogeneous, the depth is correct. In vertically inhomogeneous media which, in practice, are the general rule, the real depth z must be smaller than the value resulting from Eq. (7a). Consequently, Eq. (7a) gives the possible maximum depth z_{max} (GIESE, 1968).

A corresponding consideration can be made for average velocity $\bar{c}$ of the overburden. For a homogeneous medium results the relation:

$$\bar{c} = \sqrt{v_a \cdot \frac{x}{t}} \leq \bar{c}_{max}. \qquad (7b)$$

Here again, in inhomogeneous media, the real average velocity is smaller, thus Eq. (7b) gives the possible maximum value.

It can, therefore be stated that z_{max} and $\bar{c}_{max}$ give the maximum values independently of the knowledge of velocity structure in the overburden.

Proof of this statement can be given in the following way (GEBRANDE, pers. comm.):

The basic equations are the well-known relations between the velocity function v = v(z) and distance x and travel time t with the parameter $p = 1/v_p$

$$x(p) = 2\int_0^z \frac{p \cdot v}{\sqrt{1 - p^2 \cdot v^2}}\,dz \qquad (8a)$$

and

$$t(p) = 2\int_0^z \frac{1}{v \cdot \sqrt{1-p^2 \cdot v^2}}\,dz. \qquad (8b)$$

Eq. (8b) can be transformed to the following form with t(p) = t:

$$t = 2\int_0^z \left(\frac{1}{v}\sqrt{1-p^2v^2} + p\frac{p \cdot v}{\sqrt{1-p^2v^2}}\right)dz \qquad (9)$$

and

$$t = p \cdot x + 2\int_0^z \frac{1}{v}\sqrt{1 - p^2v^2}\,dz. \qquad (10)$$

The second term is the intercept-time part. In order to get the expression $t/p \cdot x - 1$, Eq- (10) is written as

$$\frac{t}{p \cdot x} - 1 = \frac{2}{p \cdot x}\int_0^z \frac{\sqrt{1-p^2v^2}}{v}\,dz \qquad (11)$$

and

$$z_{max} = \frac{x}{2}\sqrt{\frac{2}{p \cdot x}\int_0^z \frac{\sqrt{1-p^2v^2}}{v}\,dz} = \sqrt{\frac{x}{2p}\int_0^z \frac{\sqrt{1-p^2v^2}}{v}\,dz}. \qquad (12)$$

Substituting the relation 8a in Eq. (12), we obtain

$$z_{max} = \sqrt{\int_0^z \frac{v \cdot dz}{\sqrt{1-p^2v^2}} \int_0^z \frac{\sqrt{1-p^2v^2}}{v}\,dz}. \qquad (13)$$

Applying the relation of Schwarz, it follows

$$\int_o^z \frac{v \cdot dz}{\sqrt{1-p^2v^2}} \cdot \int_o^z \frac{\sqrt{1-p^2v^2}}{v} dz \tag{14}$$

$$\geq \left[\int_o^z \sqrt{\frac{v}{\sqrt{1-p^2v^2}} \cdot \frac{\sqrt{1-p^2v^2}}{v}} dz \right]^2 = z^2 . \tag{15}$$

So the statement is proved

$$z_{max} = \frac{x}{2} \sqrt{\frac{t}{p \cdot x} - 1} \geq z. \tag{16}$$

This equation is graphically shown in Figs. 2, 11, 12, and 13.

These three quantities x, t, dx/dt allow also the determination of the total lengths s of the incident and reflected ray:

$$s = \sqrt{x \cdot v_p \cdot t} \; . \tag{17}$$

Having this total length of the down- und upgoing ray, all possible reflection points can be constructed by aid of the ellipse method, so obtaining in a cross section the domain x, z of possible positions of the reflecting elements.

4.3.3.2 Upper Bound for Data Points of Progressive Branches

Test calculations show that points $v_p \cdot t/x$, z/x, coming from retrograde branches, are situated near the curve MAX in Fig. 2 whereas points associated with progressive branches have their position far below the curve MAX (see 4.3.7).

Progressive branches including the case of pure head waves are connected with weak gradient (or constant velocity) zones which are situated under a discontinuity or a strong gradient zone. A progressive curve starts at the left-side critical point of retrograde branch. Consequently, the start of a progressive curve is situated between the curves MAX and GRAD. From numerical examples, however, it can be stated that this point moves rapidly below the curve GRAD. Before reaching the cross-over distance with a previous branch (e.g. P_n with P_g), it is situated within the lower domain. Thus the curve GRAD can be practically regarded as upper bound for progressive branch values.

SLICHTER (1932) and BERRY (1971) give another relation for getting the maximum depth for data points of progressive branches. They regard the point under discussion as focus point with regard to both time and distance, so that the traveltime curve degenerates into a single point. For this condition, the following equation results:

$$\frac{z}{x} = \frac{1}{\pi} \cosh^{-1} \frac{v_p \cdot t}{x}. \tag{18a}$$

This relation is graphically shown in Fig. 4 as curve SP (Single Point). It is situated between the curves MAX and GRAD and represents the maximum depth for a first arrival, associated with the velocity function:

$$v = v_o \cosh \frac{\pi \cdot z}{x} \tag{18b}$$

with $v_o = v(z = o)$.

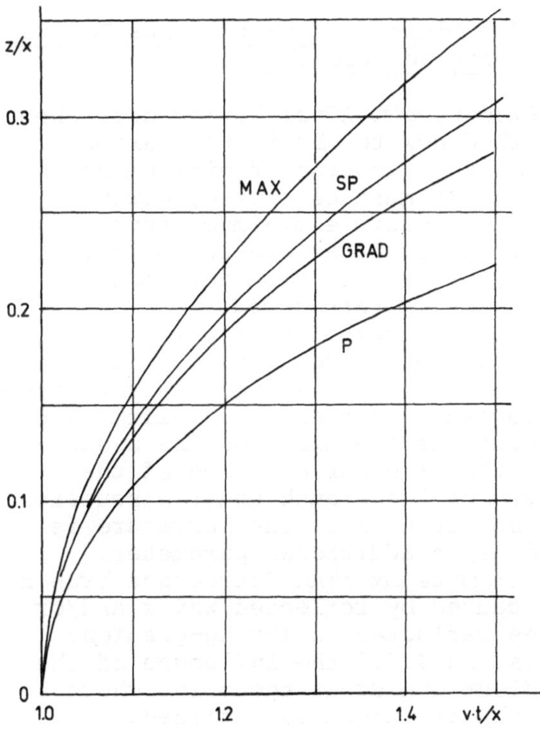

Fig. 2. Diagram z/x versus v · t/x showing

curve MAX $\frac{z}{x} = \frac{1}{2} \sqrt{\frac{dx}{dt} \frac{t}{x} - 1}$ with $\frac{dx}{dt} = v_p$ resp. v_a

curve SP $\frac{z}{x} = \frac{1}{\pi} \cosh^{-1} \frac{v_p \cdot t}{x}$

curve GRAD $\frac{z}{x} = \frac{1 - \sin\phi}{2 \cdot \cos\phi}$ with $\frac{v_p \cdot t}{x} = \frac{\cosh^{-1} (1/\sin\phi)}{\cos\phi}$ and $\sin\phi = v_o/v_p$

curve P $\frac{z}{x} = \frac{1}{2} \sqrt{\frac{\frac{dx}{dt} \cdot \frac{t}{x} - 1}{\frac{dx}{dt} \cdot \frac{t}{x} + 1}}$

4.3.3.3 Minimum Depth for Data Points of Retrograde Branches

Minimum Depth. After having described a method how to obtain the maximum depth, the question of minimum depth arises. It must be stated, however, that the possible minimum depth can be chosen as small as wanted. A restriction is possible if an additional parameter is introduced.

The z_{max}-method is applicable to all types of travel-time curves, progressive (direct) as well as retrograde (reversed) branches. The following considerations are based on the assumption that a clear decision of the nature of the branch under discussion can be given, i.e. the curvature is used as an additional parameter.

Retrograde Branches. Retrograde branches are caused by reflected waves and/or waves reflacted in strong-gradient zones. In 4.3.7 the influence of the gradient dv_p/dz at the point where the ray is turned is outlined.

For a constant-gradient model $v(o) = v_o$, $v(z_p) = v_p$, the following two equations allow the determination of the quantities z/x and vt/v_p with v_o/v_p as parameter

$$\frac{v_p \cdot t}{x} = \frac{\cosh^{-1}(1/\sin\phi)}{\cos\phi} \tag{19a}$$

$$\frac{z}{x} = \frac{1 - \sin\phi}{2 \cdot \cos\phi.} \tag{19b}$$

with $\sin\phi = v_o/v_p$.

For our purposes, the fact is of importance that these equations do not contain the gradient as parameter, i.e. in a diagram z/x versus $v_p \cdot t/x$, there is only one curve for all constant-gradient models (Fig. 2, curve GRAD).

A retrograde segment is generated if the gradient changes and becomes greater than the upper one; thus the point $v_p \cdot t/x$, z/x moves upward towards the curve z_{max}. Therefore, in a model composed of two constant-gradient layers and where the upper gradient is smaller than the second one, the curve GRAD in Fig. 2 gives the minimum depth for retrograde branches. Test calculations using realistic crustal velocity distributions show that the values z/x versus $v_o \cdot t/x$ for reversed segments are generally situated in the domain between the two curves MAX and GRAD in Fig. 2. Thus the curve GRAD gives, in practice, the minimum depth for arbitrary crustal velocity distributions. Having a concrete value or limit on a probable minimal average velocity of the overburden, the known intercept-time depth relation allows the determination of minimal depth.

4.3.4 Nature of Retrograde Branches

A progressive curve is associated with a weak gradient medium. If the smoothed curve degenerates into a straight line, a medium with constant velocity is present. The case is more complicated having retrograde branches. A retrograde branch can be generated by a strong-gradient zone and/or by a sharp velocity change. So the question arises how to distinguish these cases from the travel-time curve without a detailed depth determination.

It is obvious that, for the examination of this question, the full length of the travel-time curve under discussion is required. The display of the quantity z_{max} versus the apparent velocity v_a offers some qualitative information on the nature of the retrograde curve. In a similar way, the average velocity $\bar{c}$ can be used.

Depending on the velocity function $v(z)$, the curve z_{max} versus v_a can show different forms. In the following discussion, the gradient dv_a/dz_{max} plays an important rôle. Three cases are distinguishable (Fig. 3):

1. dv_a/dz_{max} becomes infinite, i.e. z_{max} = const and $\bar{c}$ = const = v_o. The depth z_{max} as well as the average velocity of the overburden is constant and thus independent of the angle of incidence. This is only possible for a homogeneous overburden.

In inhomogeneous media, the near-vertical reflections show a similar behavior, z_{max} and $\bar{c}$ are practically constant. Therefore, for near-vertical rays, an inhomogeneous medium can be regarded as a quasi-homogeneous one. In consequence, the x^2, t^2-method is applicable.

2. $dv_a/dz_{max} < 0$, i.e. z_{max} decreases with increasing v_a. The overburden is inhomogeneous, whereas the boundary is sharp or, at least, a strong velocity gradient is present. With increasing apparent velocity, the curve approaches the real depth z.

3. $dv_a/dz_{max} > 0$. The overburden is inhomogeneous, too, but the real gradient dv_a/dz is also positive, a transition zone exists. With increas-

ing apparent velocity v_a, the curve z_{max} approaches the depth of strongest velocity gradient. This assertion can be proved as follows: if going along a retrograde travel-time curve in direction of the origin, the angle of incidence decreases, and the vertical component of the ray path increases in respect to the horizontal one. Consequently, the difference between the maximum depth and the real depth also decreases. So it can be stated that, if the gradient dv_a/dz_{max} is positive, the gradient dv_p/dz must be positive as well.

It must be noted here that the non-existence of a positive gradient dv_a/dz_{max} does not exclude the possible existence of a gradual zone. Qualitatively, however, is can be said that this zone must have a relatively strong velocity gradient.

There are also models possible that show all three cases described above. Starting from the critical point of a retrograde branch, the curve z_{max} can show a negative gradient. With decreasing values v_a, the function z_{max} passes a point of inflection and reaches the largest depth of z_{max}. The following part is characterized by a positive gradient. Such a shape indicates that near the critical point the corresponding velocity interval is within a relatively thin depth range, whereas the velocity interval of the far-distance part must be associated with a wide depth range.

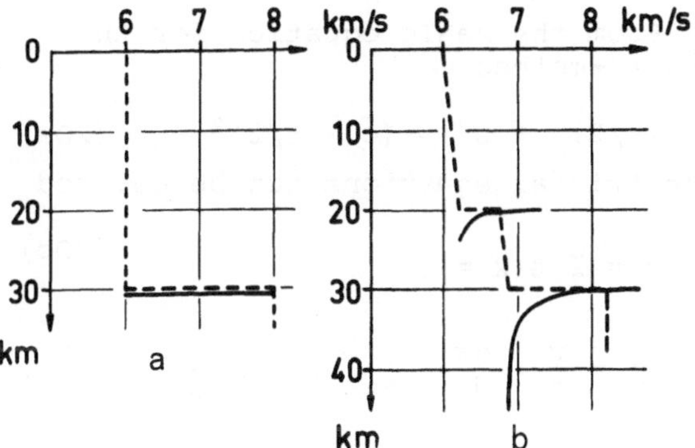

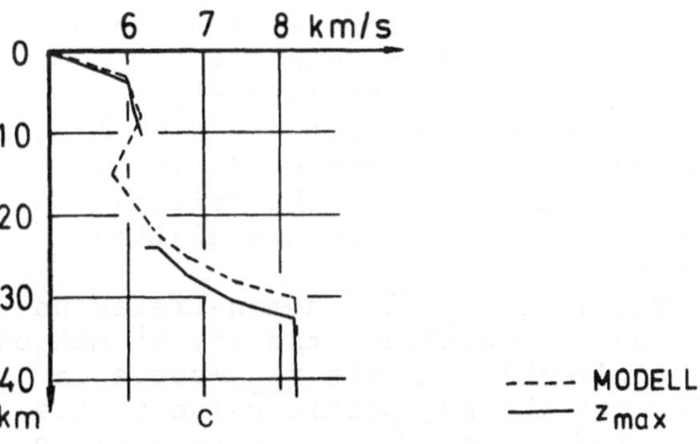

Fig. 3. Diagram showing the three main cases for the behavior of the curve z_{max} versus v_a or v_p respectively. *Dashed curve:* model, *solid curve:* curve for z_{max}. (a) Constant velocity layer over a halfspace. The curve z_{max}, associated with the reflection, coincides with the model. (b) Inhomogeneous model with first-order discontinuities. The resulting gradient for the curve z_{max} is negative for both discontinuities. (c) Model with continuious velocity-depth function characterized by a wide transition between crust and mantle. The velocity gradient of the curve z_{max} is positive

4.3.5 The x^2, t^2-Method

In the previous sections (4.3.2 and 4.3.3), methods for determination of extreme depth for separated data points have been discussed. This section deals with a method aiming at approximate depth values for reversed branches with a negative gradient dv/dz_{max}.

The well-known x^2, t^2-method is widely used in reflections seismics. It is based on the assumption of straight ray paths in the overburden and the existence of a first-order discontinuity as reflecting horizon. If exploring sedimentary suites by near-vertical reflections, both conditions are met in good approximation. If necessary, dip corrections can be made as proposed for instance by DIX (1955), BORTFELD (1957), and SATTLEGGER (1965). In seismic crustal studies, the situation is more differentiated and complex. Generally, only the overcritical range of a retrograde travel-time curve is observed, i.e. the P^M wave, whereas the subcritical portion is completely suppressed or only short travel-time segments, separated from each other, can be recognized.

Nevertheless a careful use of this method in crustal studies may produce useful results, if attention is paid to some conditions.

It is evident that the x^2, t^2-method is closely related to the z_{max}-method because both have been derived from the same model: a homogeneous layer over a halfspace.

The plot t^2 versus x^2 gives a straight line with the intercept $(2z/\bar{c})^2$, whereas the slope is equal to $1/\bar{c}^2$.

From the basic equation for the x^2, t^2-method

$$(x/2)^2 + z^2 = (\bar{c} \cdot t/2)^2 \qquad (20a)$$

the trivial equations can be derived

$$z = \frac{\bar{c}}{2}\, t(x = 0) \qquad (20b)$$

$$\bar{c}^2 = \frac{x}{t} \cdot \frac{dx}{dt} \quad [\text{see Eq. (7b)}]. \qquad (20c)$$

Thus the z_{max} and x^2, t^2-methods are principally identical.

When applying the x^2, t^2-method in crustal studies, one has to confine oneself to rays refracted as little as possible but clearly to be observed in the records. So it is quite obvious to use the critical range of a retrograde travel-time curve.

The z_{max}, v_a-plot demonstrates under which conditions the x^2, t^2-method is applicable. If the z_{max}-curve is close to the asymptotic value z, i.e. the gradient dz_{max}/dv_p is very large or already practically infinite, the x^2, t^2-method produces a depth value very near to the real one. With a positive gradient dv_a/dz_{max}, however, indicating a wide transition zone, the calculated depth is distinctly greater than the depth of the greatest velocity gradient.

In practical use, two requirements are contradictory, the necessity to have a large distance interval for a precise determination of slope and intercept time of the linear regression curve and, on the other hand, to make the distance interval as short as possible in order to avoid refraction effect and to approach the depth of strongest gradient as near as possible. From experience it can be stated that a distance range of 30-40 km beyond the critical point gives reliable results.

The depth values obtained are, however, in any case too large. STEWART (1966) who used the full length (more than 100 km) of retrograde branches recommends a reduction of 10% for the calculated depth, whereas for the average velocity, a correction of 3% may be applied.

Based on experience with shorter branches (30-40 km length), a reducing factor of 3% for the depth and 1.5% for average velocity may be applied in order to get a better approximation to the real value. If evaluating subcritical reflections by the x^2, t^2-method, no corrections are necessary.

4.3.6 Exact Solutions

4.3.6.1 The Wiechert-Herglotz Method

In the previous sections, some methods have been described which allow the estimation of some crustal parameters but they cannot give the complete velocity-depth function.

HERGLOTZ (1907) and BATEMAN (1910) have independently given the solution how to invert travel times into velocity-depth values. WIECHERT (1910) simplified the formulation; thus, the application becomes easier.

Independently from the final form of the theorem, the following conditions must be met:

1. The function t(x) must be continuous between x = 0 and $x = x_p$, i.e. an interruption is forbidden.
2. The apparent velocity along the curve, starting from x = 0 must increase monotonously.
3. The derivative dx/dt must be continuous along the travel-time curve.

There are different forms to present the solution of the integral equation:

$$z(v_p) = \frac{1}{\pi} \int_0^{x_p} \cosh^{-1} \frac{v_p}{v(x)} \cdot dx \qquad (21a)$$

or with $v_p = 1/p$ and $v(x) = 1/q(x)$

$$z(p) = \frac{1}{\pi} \int_0^{x_p} \cosh^{-1} \frac{q(x)}{p}\, dx. \qquad (21b)$$

The following form uses as variable q instead of v:

$$z(p) = \frac{1}{\pi} \int_p^{p(o)} \frac{x(q) \cdot dq}{\sqrt{q^2(x) - p^2}} \qquad (22a)$$

or

$$z(p) = \frac{1}{\pi} \int_p^{p(o)} x(q)\, d\left(\cosh^{-1} \frac{q(x)}{p}\right). \qquad (22b)$$

GERVER and MARKUSHEVICH (1966, 1967) have extended the integral equation by a term taking into account low-velocity zones:

$$z(p) = \frac{1}{\pi} \int_p^{1} \frac{x(q)}{\sqrt{q^2(x)-p^2}}\, dq + \sum_i \Phi_i(p). \qquad (23a)$$

Φ_i is the contribution from the i-th low-velocity layer and is given by the expression

$$\Phi_i(p) = \frac{2}{\pi} \int_{z_i^u}^{z_i^l} \text{arctg} \sqrt{\frac{r^2 - p_i^2}{p_i^2 - p^2}}\, dz \tag{23b}$$

where

z_i^u upper bound of the i-th low velocity zone

z_i^l lower bound of the i-th low-velocity zone

p_i ray parameter at which the i-th low-velocity zone produces an interruption of the travel-time curve (shadow-zone)

$r = 1/v(z)$ function giving the velocity distribution within the low-velocity zone.

The additional term is a formal completion of the basic Wiechert-Herglotz equation. It cannot, of course, overcome the principal indetermination of the solution if a low-velocity layer is present. This special problem is discussed in detail in 4.3.6.2.

There are several ways to approach the integration of the Wiechert-Herglotz equation for practical purposes. The simplest method uses the application of numerical quadrature formulae, e.g. the Simpson rule. STEWART (1966) proposes a method which allows the expression of the integral in a closed form. The advantage is that this method may be considered "exact" to the extent that the chosen approximating function is a good representation of the true functional relation.

An assumption convenient for this aim is that the relation between the velocity v_p and distance x consists of a sequence of connected linear segments. That is

$$v_p(x) = A_i + B_i x \quad \text{for } x_i \leq x \leq x_{i+1} \tag{24a}$$

Setting this relation into the Wiechert-Herglotz equation it results in

$$z = \frac{1}{\pi} \sum_{i=1}^{p-1} \frac{v_p}{B_i} \left[\frac{A_i + B_i x}{v_p} \cdot \cosh^{-1} \frac{v_p}{A_i + B_i x} + \sin^{-1} \frac{A_i + B_i x}{v_p} \right]_{x_i}^{x_{i+1}} \tag{24b}$$

When the terms are written out, some simplifications can be introduced:

$$z = \frac{v_p}{\pi} \left[-\frac{1}{B_i} F(R_i) + \frac{\pi}{2B_{p-1}} + \sum_{i=2}^{p-1} \left(\frac{1}{B_{i-1}} - \frac{1}{B_i}\right) F(R_i) \right] \tag{24c}$$

with $R_i = \frac{v_p}{v_i}$ and

$$F(R_i) = \frac{1}{R_i} \cosh^{-1} (R_i) + \sin^{-1} \left(\frac{1}{R_i}\right).$$

4.3.6.2 Detection and Determination of Low-Velocity Layers

GUTENBERG (1955) suggested the existence of low-velocity layers in the crust. This idea was picked up again in the beginning of the sixties when detailed data from explosion seismology became available. There are two possibilities of detecting the existence of velocity inversions within the profile under study.

Having a series of subcritical reflections from each travel-time curve, the average velocity of the overburden can be calculated by the x^2, t^2-method. Thus, the interval velocity between two reflecting horizons is determined. Depending on the number of reflections, the resulting interval velocity function is more or less detailed and low-velocity zones may be revealed by this method (PAVLENKOVA, 1969). It is evident that only a stepwise interval velocity function can be obtained.

The other possibility is based on the behavior of refracted waves in presence of low-velocity. According to Snell's law, a low-velocity zone causes an interruption of the travel-time curve. A jump in time and distance is generated between the two branches which is spent in passing the low-velocity layer (Fig. 4). For the detection of an inversion zone, the fact that both ends of the interrupted curves must be parallel is important.

It depends on the type of velocity distribution under the inversion zone how the travel-time curve beyond the interruption continues. If there is a strong increase of velocity under this zone, a retrograde branch is generated which can even overlap the branch before the shadow zone. Its critical point is nearer to the origin in the case of an inversion zone than without. From this criterion, MÜLLER and LANDISMAN (1966) derive the existence of an inversion zone between upper and middle crust. An interruption is also possible between two retrograde segments (GIESE, 1971). Here, the classical shadow zone is absent, only the two parallel travel-time segments, showing a delay of intercept times, indicate the existence of an inversion zone.

Owing to the existence of a low-velocity zone, the task of inverting the data t(x) into the domain v(z) leads to a nonunique solution. The ambiguity can be restricted if additional assumptions are introduced.

This problem was first discussed by SLICHTER (1932) who also offered a solution. The existence of an inversion zone is indicated by the presence of a jump in time and distance between two curves. This interruption is characterized and defined by the break-off point A and the starting point B of the continuation (Fig. 4). It is evident that at A and B, the slope dx/dt must be equal.

These data give the time and distance which the ray, appearing at point B, spent in the inversion zone:

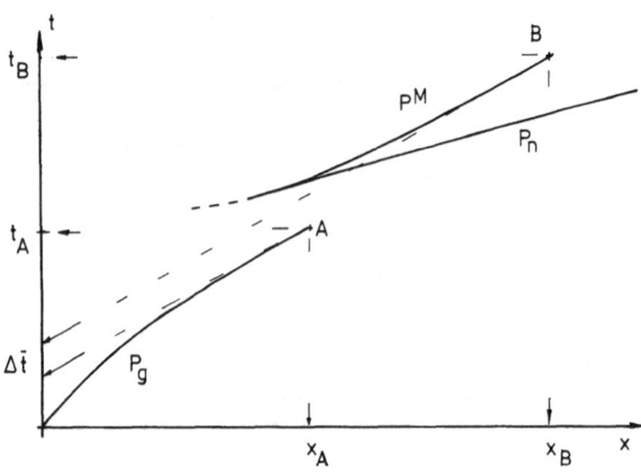

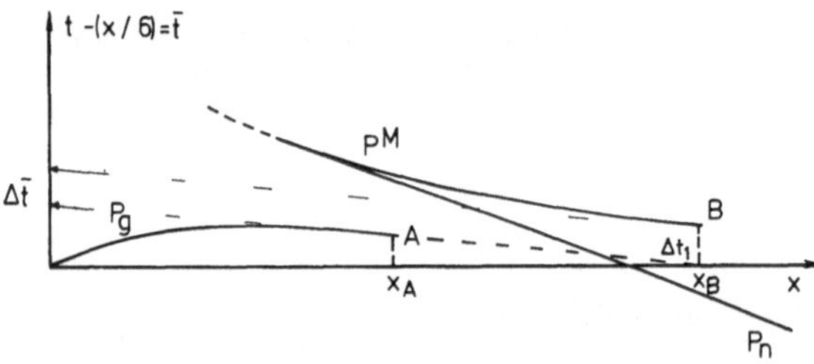

Fig. 4. Fundamental arrangement of travel-time curves if a low-velocity gradient is present. In the travel-time diagram, the inversion zone is characterized by an interruption of travel-time curves. The interruption is delimited by the points *A* and *B*. At both points, the apparent velocity is equal: $v_A = v_B$

$$\Delta x = x_B - x_A \tag{25a}$$

$$\Delta t = t_B - t_A \quad \text{and} \tag{25b}$$

$$\frac{dx(A)}{dt} = \frac{dx(B)}{dt} = v_l^u = v_t \tag{25c}$$

with $v_l^u = v_t$ velocity at the upper and lower bound of the inversion zone.

Based on the considerations in 4.3.3.1, the possible maximum thickness Δz_{max} of the inversion zone is given by Eq. (7a) and Eq. (7b) allows the determination of the average velocity $\bar{c}$. Any other velocity distribution within the inversion zone, being nonrectangular, causes a value smaller than Δz_{max}. According to this statement, there is only one rectangular distribution possible, associated with the maximum thickness Δz_{max}.

In principle, any velocity distribution, arbitrarily shaped, is possible and can be assumed within the channel. A constant velocity function is the simplest possible distribution, the next step is the assumption of linearly increasing velocity, i.e. a constant gradient model.

When introducing a constant-gradient distribution within the inversion zone, the thickness Δz and the minimum velocity v_m can be determined by aid of Eqs. (19a) and (19b) or by curve GRAD in Fig. 2. The fact that, in a constant-gradient model, the quantities $v_p \cdot t/x$ and z/x do not depend on the gradient itself leads to a surprising consequence in respect to other possible velocity distributions within the inversion zone. Fig. 5 shows a series of equivalent velocity distributions with regard to the quantities $v_p \cdot t/x$ and z/x. The case a triangular-shaped velocity distribution within the inversion zone, discussed by KOSCHYK (1973), is, of course, included in this general statement.

It may be argued that each peak reaching the top velocity v_t is bound to generate an event at the surface. Speaking in ray geometrics, only singular points can be produced, arranged on a line between A and B (Fig. 6). Taking subcritical reflections and diffractions into consideration, travel-time branches of some length may be produced, being arranged practically parallel between both interrupted curves. Just this picture is observed in many record sections. Therefore, such a comb-like velocity distribution seems possible. The interpretation of near-vertical reflections yields, in principle, the same results regarding the fine structure of the crust (MEISSNER, 1967a; FUCHS, 1968, 1969a).

Using functions of higher order, thickness and minimum velocity can be further reduced. From the mathematical point of view, both quantities may approach zero without becoming zero.

Independently of the chosen velocity distribution within the inversion zone, it can be stated that the domain of possible solutions is within the area A'B'C' (Fig. 7). This is so because point A is fixed whereas point B approaches A with decreasing minimal velocity within the inversion zone.

SLICHTER (1932) assumes a quasirectangular shape of the function within the inversion in order to determine the minimal thickness if the minimum velocity is given. He argues that, in any actual problem, a limiting value v_m may always be fixed below which the velocity will not fall. This minimum velocity will evidently yield a minimum thickness for the inversion zone. However, in order to satisfy the observed discontinuity, an infinitesimal thickness must be added to the lower or upper part of the low-velocity layer in which the velocity increases very rapidly from the value v_m to the top value v_t

$$\frac{\Delta z}{\Delta x} = \frac{\text{tg } i}{2}\left(\frac{\Delta t \cdot v_t}{\Delta x} - 1\right) \qquad (26)$$

with $\sin i = \dfrac{v_m}{v_t}$ (notations see Eq. 25).

4.3.6.3 Exact Solutions with Low-Velocity Layers and the Domain of Possible Solutions

Having assumed a velocity distribution within the velocity inversion and thus calculated its parameters, it is possible to obtain a unique solution for those travel-time curves following the interruption zone.

By aid of the ray parameter $p = 1/v_p$ and the parameters of the velocity zones, the time- and distance-portions which are spent in passing the low-velocity zone can be calculated. In the case of a rectangular-shaped function, the relations are quite simple

$$\Delta x(v_p) = 2\Delta z \cdot \text{tg } i \qquad (27a)$$

and

$$\Delta t(v_p) = \frac{2\ \Delta z}{\bar{c} \cdot \cos i} \qquad (27b)$$

with $\sin i = \dfrac{\bar{c}}{v_p}$.

These portions $\Delta x(v_p)$, $\Delta t(v_p)$ are subtracted from the values $x(v_p)$, $t(v_p)$ of the later travel-time curve. This procedure, carried out point by point,

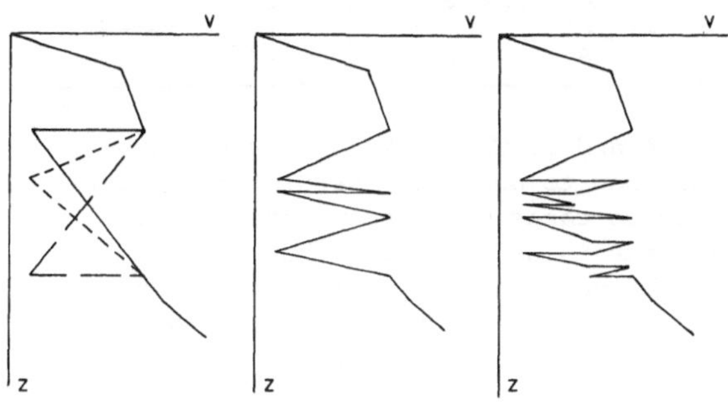

Fig. 5. Equivalent velocity models derived from the basic model of a constant velocity gradient within the inversion zone

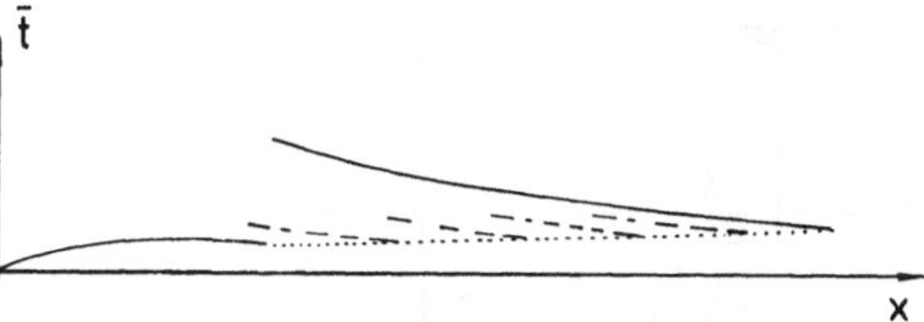

Fig. 6. Generation of subcritical reflections in the interruption zone by a comb-like velocity distribution within the low-velocity layer. *Dashed line:* subcritical reflections

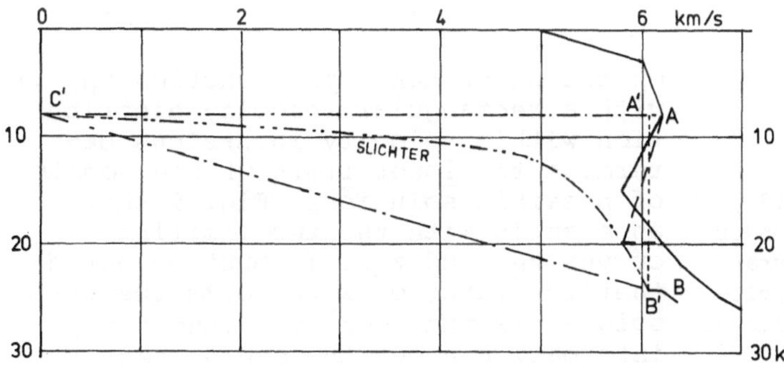

Fig. 7. The area *A'B'C'* shows the domain of possible velocity distribution within the velocity inversion. The curve SLICHTER derived from Eq. (26) gives the lower limit of possible solutions for the presented model (solid line)

gives a new travel-time curve which is now joined to the previous one. Thus, the Wiechert-Herglotz equation is applicable because the condition of continuity is complied with. It is plausible that this procedure can also be repeated for a second and third etc. inversion zone, provided that the necessary parameters can be determined.

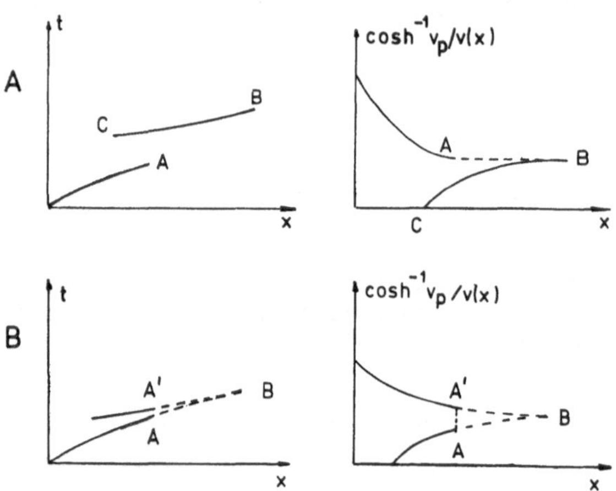

Fig. 8A and B. The figures show the procedure how to apply the Herglotz-Wiechert method to cases with incomplete travel-time data in order to get the upper limit of possible depth values. (A) low-velocity layer; (B) too-short observation lines

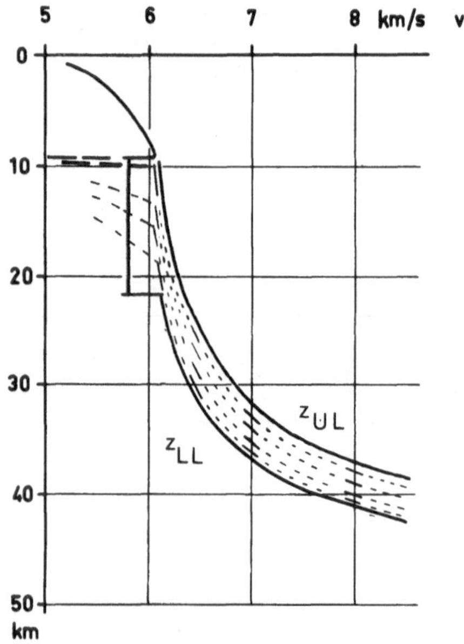

Fig. 9. Example showing the upper and lower depth limit for a reversed segment (solid curves). The dashed curves are derived under the assumption that at v_p the velocity gradient varies linearly between the two limiting curves. The dashed curves are constructed by the method demonstrated in Fig. 10

The method just described and first proposed by SLICHTER (1932) is very concrete, whereas the method of GERVER and MARKUSHEVICH (1966) is more formal Eqs. [(23a) and (23b)].

Assuming a constant velocity $\bar{c}$ within the inversion, for term Φ_i the equation is obtained:

$$\Phi_i = \frac{\frac{2(z_i^{l} - z_i^{u})}{\pi}}{\operatorname{arctg} \sqrt{\frac{(v_i^{l,u}/\bar{c})^2 - 1}{1 - (v_i^{l,u}/v_p)^2}}} \tag{28}$$

with $v_i^{l,u}$ as velocity at the top and bottom of the inversion zone.

The use of the Gerver-Markushevich equation does not need the intermediate step concerning the connection of the separated travel-time curve in order to get a continuous curve. This method may be faster than the Slichter-method if the solution of Eq. (23b) is available in a closed form. In the case Eq. (23b) cannot be solved in a closed form, the method of SLICHTER is easier to handle.

The extension of the Herglotz-Wiechert equation as given by GERVER and MARKUSHEVICH allows the determination of a minimal depth if a velocity inversion is present. The quantity Φ_i is always positive, its minimal value is zero. Having a system of interrupted and/or incomplete traveltime curves, the following equation gives the upper limit of possible depths (GIESE and STEIN, 1971) (Fig. 8)

$$z > z_{UL} = \frac{1}{\pi}\left[\int_0^{x_1} \cosh^{-1} v_p/v(x)\,dx + \int_{x_2}^{x_3} \cosh^{-1} v_p/v(x)\,dx + \ldots\ldots + \int_{x_{p-1}}^{x_p} \cosh^{-1} v_p/v(x)\,dx.\right] \tag{29}$$

On the other hand, the function $z_{UL}(v)$ with a rectangular velocity distribution within velocity inversions determine the lower limit of the domain of possible solutions. Fig. 9 shows an example with the two limiting curves z_{UL} and z_{LL}. It must be noted that the function z_{max} forms the absolute maximum, derived without any information about the overburden.

All other velocity distributions must be situated within this domain. In order to obtain the shape of possible curves, the assumption is introduced that, in first approximation, the quantity dv/dz changes linearly between the corresponding values at z_{UL} and z_{LL} (Fig. 10). These tangential elements determine the shape of solution curves (Fig. 9).

4.3.6.4 Depth Determination from Intercept-Time Presentations

GERVER and MARKUSHEVICH (1966) introduced the function

$$\tau(p) = t(p) - p \cdot x(p) \qquad (30a)$$

$$\tau(p) = \int_0^{z(p)} \sqrt{1/v^2(z) - p^2} \qquad (30b)$$

with $p = 1/v_p$.

The function $\tau(p)$ has a simple geometrical interpretation: the tangent of the travel-time curve at the point with the coordinates $x(p)$, $t(p)$ is characterized by the slope p and the intercept-time $\tau(p)$. GERVER and MARKUSHEVICH (1966), KEILIS-BOROK (1971), and BESSONOVA et al. (1974) regard this function mainly from a mathematical point of view, whereas PAVLENKOVA (1973a, b) uses the intercept-time method for the generalization of seismic data, i.e. identification of travel-time branches in lateral inhomogeneous profiles and the construction of cross-sections with isolines of velocity.

In the following, the inversion problem under the aspect of the τ-method is discussed. The normal presentation t(x) can show a triplication of curves, i.e. cusps. The same data plotted in a $\tau,(v_p)$-diagram show monotonously increasing τ-values with increasing values of $v_p = 1/p$. If no low-velocity layer is present, the $\tau,(v_p)$ diagram contains the same information as the t(x) plot.

The case is different if a low-velocity layer is indicated in the t(x)-diagram by an interruption zone with a jump in travel-time Δt and distance Δx between two travel-time curves with the same endslope. When transforming the interruption zone into a $\tau,(v_p)$ diagram, the travel-time jump Δt is inverted into the intercept-time jump $\Delta\tau$, whereas the distance jump is not displayed; thus one information, present in the primary data, is lost. BESSONOVA et al. (1974) describes the transformation of the Wiechert-Herglotz equation into a form containing the quantity $\tau(p)$. These considerations are very general.

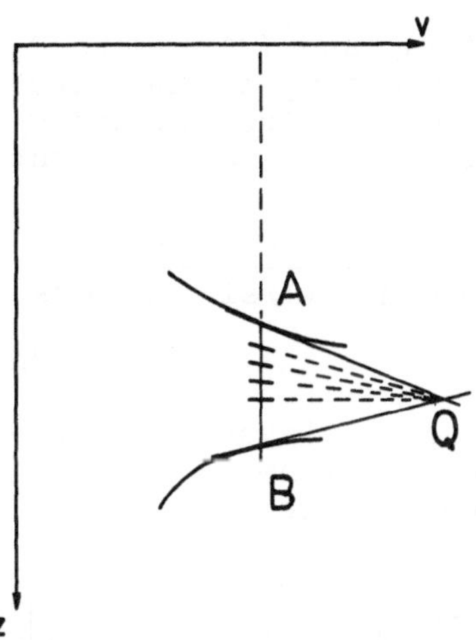

Fig. 10. Sketch showing a simple graphical method how to determine the slope of tangential element of solution curves within the two bounds. It is assumed that the velocity varies linearly between the two limits

The following method is mainly orientated to the task of inverting intercept-time cross sections into velocity-depth cross sections. Having a system of parallel isolines $\tau(v_p)$, the assumption is introduced that between two isolines $\tau(v_{p,j})$ and $\tau(v_{p,j+1})$ the velocity increases linearly in the corresponding depth interval.

Within two iso-velocity lines, the velocity is given by the equation (STEINHART and MEYER, 1961):

$$v = v_j + b_j z \qquad (31a)$$

$$\text{with} \quad z_j \le z \le z_{j+1} \quad j = 0,1,2,...$$

The horizontal distance x_j traversed by a ray with the parameter v_p, in penetrating from z_j to z_{j+1} is determined by the equation:

$$x_j = \frac{2v_p}{b_j} (\cos \phi_j - \cos \phi_{j+1}). \qquad (31b)$$

The corresponding traveltime t_j follows from the equation:

$$t_j = \frac{2}{b_j} \left[\cosh^{-1} \left(\frac{1}{\sin \phi_j}\right) - \cosh^{-1} \left(\frac{1}{\sin \phi_{j+1}}\right) \right] \qquad (31c)$$

with $\sin \phi_j = v_j/v_p$
$\sin \phi_{j+1} = v_{j+1}/v_p$.

Setting Eqs. (31b) and (31c) into Eq. (30a), the thickness m_1 of the first layer between the isolines v_o and v_1 can be calculated by the expression:

$$m_1 = \frac{\tau_1}{2} - \frac{v_1 - v_o}{[\cosh^{-1}(1/\sin\phi_1) - \cos\phi_1]}. \tag{32a}$$

The thickness of the layer m_k, situated between the isolines v_{k-1} and $v_k = v_p$ is given by the sum:

$$m_k = \left[\frac{\tau_k}{2} - \sum_1^{k-1} \frac{m_n}{v_n - v_{n-1}} \cdot F(v_{n,n-1})\right] \cdot \frac{v_k - v_{k-1}}{F(v_k, v_{k-1})} \tag{32b}$$

with $F(v_n, v_{n-1}) = \cos\phi_n - \cos\phi_{n-1} + \cosh^{-1}(1/\sin\phi_{n-1}) - \cosh^{-1}(1/\sin\phi_n)$.

$$\tag{32c}$$

The depth $z(v_p)$ is obtained by summing up the thicknesses m_k

$$z(v_p) = \sum_1^k m_k. \tag{32d}$$

This procedure is applicable to isolines associated with progressive as well as retrograde branches. When evaluating retrograde travel-time curves by this method, it is possible that for a certain velocity range $v_{p,i}$, $v_{p,i+1}$ a thickness results that is very small or even zero. Thus the existence of a velocity jump in the corresponding velocity interval is indicated. Due to the scatter of data, small negative values cannot be excluded, they have to be set zero.

If a low-velocity zone is present, the characteristic data must be determined from the data of the shadow zone, as described in 4.3.6.2. As Eq. (32) is based on the assumption of a linear increase of velocity between two velocity-isolines, it is reasonable to use the constant-gradient distribution in the inversion zone as well.

KEILIS-BOROK (1971), McMECHAN and WIGGINS (1972), WIGGINS et al. (1973), and BESSONOVA et al. (1974) have extended this τ-method under the aspect of obtaining depth limits in connection with the presence of low-velocity layers and the scatter of data.

4.3.7 Approximation Methods

In the previous sections, the possibilities of obtaining exact solutions for the inversion problem have been described. Difficulties arise if the function x, v_p is incomplete owing to the existence of low-velocity channels, too-short profiles or uncertainties in correlation. It has also been outlined how to determine extreme values for single points or travel-time branches, so obtaining the domain of possible solutions. In practice this domain is even smaller, e.g. in inversion zones the minimum velocity cannot be smaller than that in fluid media for physical reasons. Therefore approximation methods suitable for crustal models may be a useful tool for obtaining the main data of crustal structure.

The x^2, t^2-method can be regarded as such method. Another method, proposed by GIESE (1968), takes the velocity gradient at the depth of ray turning into consideration. In 4.3.3 the meaning of the quantity z_{max} has been demonstrated. It can be proved by test calculations on possible crustal models with and without low-velocity layers as well as transition zones of various thicknesses that the correction applied to z_{max}, in order to get a better approximation to the real depth z_{real}, is strongly depending on the gradient dv_p/dz. Fig. 11 shows a diagram with z/x versus $v_p \cdot t/x$, derived from quite different models, containing curves characterized by the parameter dv_p/dz. Points of equal values dv_p/dz can be approximated by a slightly bent curve. This diagram demonstrates that high-gradient curves are situated closer to the limit curve z_{max} than low-gradient curves. Furtheron the low-gradient curves, derived from progressive curves, are more widely spaced than high-gradient ones. It can, therefore, be stated that the correction for depth values with a high-velocity gradient is smaller. The diagram shows that the points, derived from assumed velocity functions, are less scattered for strong gradients than for weak ones.

The importance of the velocity gradient at the bottoming depth in respect to depth calculations for reversed segments by approximation methods can be proved by a model consisting of two layers, each having a constant velocity gradient. A reversed segment is generated if the second

Fig. 11. This diagram shows curves z/x versus $v_p \cdot t/x$ with the parameter v_p derived from assumed but possible crustal models of different thickness and velocity distribution. For the abbreviations see Fig. 2

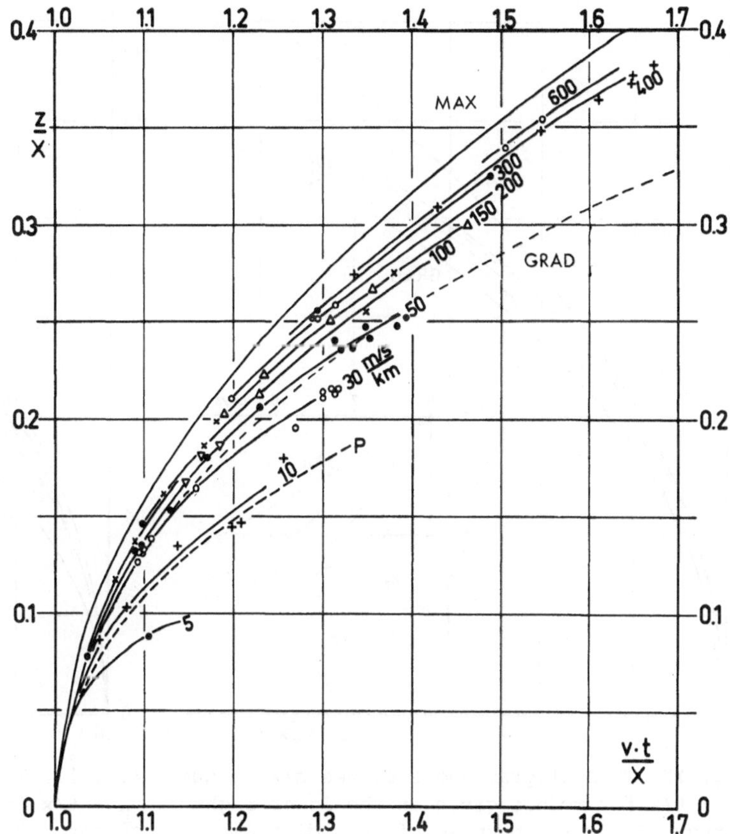

gradient is steeper than the upper one. Such models are defined by the following parameters:

1. the ratio of the both velocity gradient b_1/b_2
2. the ratio between the surface velocity v_o and the velocity v_p
3. the ratio between the velocity v_1 at the boundary between the two layers and the velocity v_p.

All models start with the velocity v_o = 6 km/s. Fig. 12 demonstrates the influence of varying velocity gradient in the upper layer while the gradient is constant in the lower one. The depth of the boundary is fixed. For crustal studies, average gradients of up to 0.02 km/s/km are relevant. The relative width $\Delta z/z$ of this stripe is only between 1% and 3% in the interval of importance ($v \cdot t/x$ between 1.1 and 1.5).

In Fig. 13 the depth of the boundary between the two layers and the gradient in the lower layer vary, whereas the gradient in the upper layer remains constant (0.01 km/s/km). For intermediate gradients b_2 (0.1 km/s) the influence of depth is relatively large.

With increasing gradient b_2 the curves for different depth approach each other as well as become nearer to the limiting curve MAX. In other direction they move to the curve GRAD.

Comparing the empirical curve of Fig. 11 with those of Fig. 13 it can be seen that the empirical curves are very close to that derived from the two-gradient models. Therefore these curves enable the approximative determination of depths for reversed segments of case 3 in 4.3.4 (positive gradient dv/dz_{max}).

The procedure of obtaining a solution starts with the determination of the curve for z_{max}. This curve gives values for depth and velocity gradient which enable an improvement of the solution by aid of the curves in Fig. 13. The more simple set of curves in Fig. 1 needs only the velocity gradient. This process has to be repeated twice or three times until the curve v(z) is not shifted any more. It has to be noted quite clearly that the principal ambiguity is not removed by this method, but it can be stated that, outside the curve determined, the probability that reasonable crustal solutions exist decreases. It is evident that the solution is relatively stable when the velocity gradient is strong but it becomes unstable when the gradient is weak. Examples for the application of this

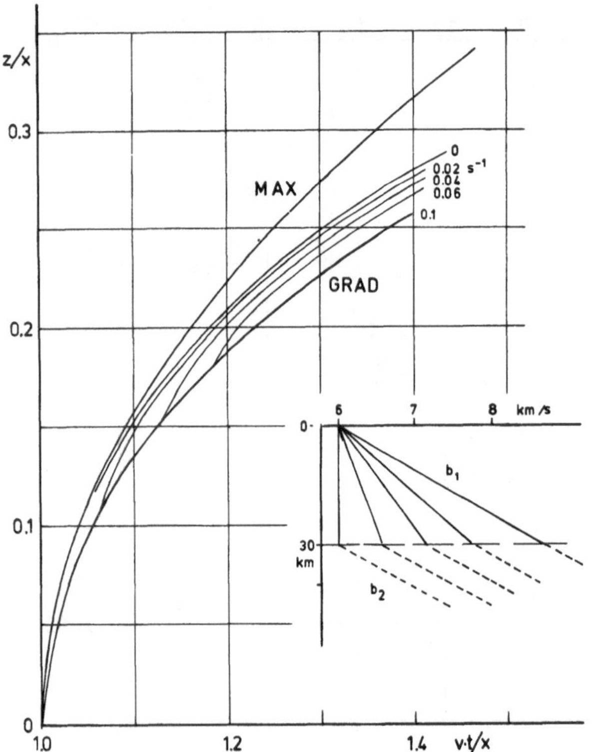

Fig. 12. This diagram shows curves z/x versus $v_p \cdot t/x$ for models given in the lower right corner. It is the purpose of this diagram to demonstrate the relatively low influence of average crustal velocity gradients (0.01-0.02 km/s/km) on depth determination for reversed segments

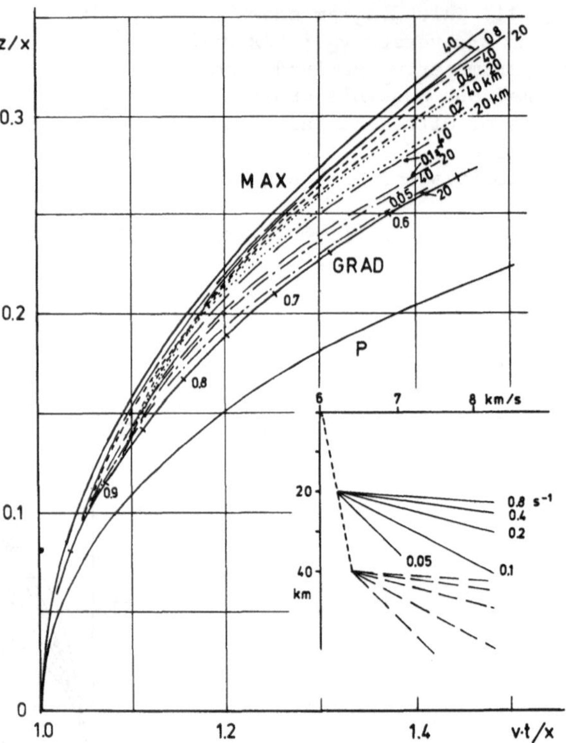

Fig. 13. This diagram shows curves z/x versus $v_p \cdot t/x$ for models shown at the lower right corner. It demonstrates the influences of depth and velocity gradients on depth determination of reversed segments. The abbreviations are outlined in Fig. 2. The values 0.6, 0.7 etc. at the curve GRAD can be used for the determination of the lowest velocity in a gradient model, e.g. in the zone of velocity inversion. This parameter is defined by the relation lowest velocity/velocity v_{AB} at the top (bottom) of the velocity inversion

method are given by GIESE (1968) and PRODEHL (1970a, b).

Whereas the domain between the curves MAX and GRAD is occupied by points associated with retrograde curves, most of the points belonging to progressive branches are situated below the curve GRAD. In a first approximation Eq. (33) allows the depth determination for progressive branches (PAVLENKOVA, 1968).

$$\frac{z}{x} = \frac{1}{2} \sqrt{\frac{\frac{dx}{dt} \cdot \frac{t}{x} - 1}{\frac{dx}{dt} \cdot \frac{t}{x} + 1}} \,. \tag{33}$$

This relation is shown graphically as curve P in Figs. 2, 11 and 13.

It is derived from the simplest refraction model, the one-layer model with constant velocity. The quantities x,t are the coordinates of the cross-over-point. Test calculations, based on a discontinuous two-gradient model with values for the gradient of the lower layer between 0.02 and 0.005 s show that the resulting values for $v_p \cdot t/x$ and z/x are close to the curve P.

Finally a summary is given how to obtain a depth velocity function in first appromation:

1. Progressive branches can be evaluated by Eq. (33).

2. Reversed branches are inverted to the z_{max} form in order to check their nature

2a) application of the x^2, t^2-method for negative gradients v/z_{max}

2b) application of the diagram 11 or 13 for positive gradients dv/dz_{max}.

3. Determination of the velocity inversion by aid the characteristic parameters for an interruption zone using the curves MAX or GRAD in Fig.2.

It is quite clear that the final step should involve the recalculation of the travel-time curves from the derived velocity depth function in order to demonstrate the agreement between the observed data and the model. If the model shows a strong lateral inhomogeneity this should be performed by a method as proposed by GEBRANDE (4.4 of this volume) or WILL (4.5 of this volume). After the kinematic interpretation the dynamic evaluation of the data must follow, e.g. the calculation of synthetic seismograms as developed by MÜLLER (1968), FUCHS (1969b), and MUELLER and FUCHS (4.6 of this volume).

4.4 A Seismic-Ray Tracing Method for Two-Dimensional Inhomogeneous Media

H. GEBRANDE

ABSTRACT

The method of model calculations for two-dimensional inhomogeneous media used in Chapter 6.11 is outlined. The basic formulas are given to trace rays and compute travel times for arbitrary two-dimensional velocity functions, which may contain transition zones as well as discontinuities with strong relief. A rather simple method is described to construct realistic two-dimentional models of the subsoil, which allows ray-tracing and travel-time calculations even on small desk-top calculators. A simple example is presented, which may facilitate the understanding of the more complex applications given in 6.11.

ZUSAMMENFASSUNG

Die im Kapitel 6.11 angewendete Methode der Modell-Rechnung für zweidimensional inhomogene Medien wird beschrieben. Das Formelwerk, das für die Berechnung von Strahlen und Laufzeitkurven bei beliebig vorgegebener zweidimensionaler Geschwindigkeitsverteilung benötigt wird, wird angegeben. Es wird eine besonders einfache und vielseitig anwendbare Methode zur Konstruktion realistischer zweidimensionaler Modelle des Untergrunds beschrieben, die die Berechnung der zugehörigen Strahl-Wege und Laufzeitkurven selbst auf einigen elektronischen Tischrechnern ermöglicht. Ein einfaches Beispiel wird vorgestellt, welches das Verständnis für die komplizierteren Modelle im Kapitel 6.11 erleichtern dürfte.

4.4.1 Introduction

Up to the present deep-seismic sounding data are usually evaluated by direct inversion techniques based on travel times. Standard inversion methods however are available only for some relatively simple classes of models of the earth's interior. For instance the well-known Herglotz-Wiechert method is restricted to the case of horizontally stratified media with seismic wave velocities increasing monotonically with depth. The case of few layers of constant velocities with dipping or curved interfaces can be successfully treated by some wave-front or related methods.

Generally, however, the properties of the earth's interior may deviate more or less from the basic assumptions inherent to the different inversion methods. Therefore these methods sometimes break down entirely or may give only crude results, when applied to real travel-time curves. To evaluate the accuracy of the results it is therefore necessary to compute exactly the travel times for the model obtained and to compare them with the travel times observed.

If appropriate methods are available for the computation of exact travel times for complex and realistic models, it is furthermore possible to improve first-order models iteratively by comparison of computed and observed travel times. In the present article a method suited for ray tracing and travel-time calculation for two-dimensional inhomogeneous media is described. The method may also be applied approximatively to three-dimensional structures, provided that the different rays emerging along a given profile are confined essentially to a common plane through the profile.

4.4.2 Principles of Ray Tracing

The physical basis of ray tracing and traveltime calculation for arbitrary twodimensional models is the same as in the case of a horizontally stratified medium: only Snell's law is needed. If a mathematical model is defined, which unequivocally attributes a velocity value to any point of the subsoil (by this assumption we restrict ourselves to isotropy), any ray path can be computed stepwise, starting at the source of the seismic wave with the respective angle of incidence and successively applying Snell's law. The computation is repeated and the travel-time increments are summed up, until the ray reaches the earth's surface or any other boundary of the model.

Since the velocities of the seismic waves in the earth's interior may vary continuously as well as discontinuously, it is appropriate to allow for both kinds of variations when constructing models. (In principle it would be possible to approximate transition zones by series of small discontinuities or discontinuities by transition zones with steep velocity gradients, but both approximations prove to be inconvenient and uneconomical with respect to computer time.)

In the case of discontinuities the variation of the direction of the ray is described by Snell's law in the usual form: If i_1 is the angle between the ray and the normal to the interface in the first medium with the velocity v_1, the corresponding angle i_2 in the second medium with the velocity v_2 follows from the condition

$$\frac{\sin i_1}{v_1} = \frac{\sin i_2}{v_2} \tag{1}$$

provided that $\sin i_1 < v_1/v_2$. Otherwise the ray does not penetrate into the second medium, but is reflected at the discontinuity with the angle $i_2 = -i_1$ (Fig. 1a). In the case of continuously changing velocities the rays are not refracted discontinuously, but curved continuously. The radius of curvature (Fig. 1b) may be described by the equation

$$\rho = \frac{v}{|\mathrm{grad} v| \sin i}\,, \tag{2}$$

which can be obtained from Snell's law Eq. (1) by a limiting process. In Eq. (2) i is the angle between the unit-vector $\vec{s}$ in the direction of the ray and the velocity gradient grad v. It should be noted that the radius of curvature is a signed quantity; it is positive (negative), if the center of curvature is located on the left (right) side relative to the direction of the ray.

Generally the direction and the absolute value of the velocity gradient may vary from point to point. For the practical calculation of ray paths and travel times the gradient is assumed as piece-wise constant. As long as the velocity gradient is constant, the radius of curvature is also constant. One therefore can proceed along a ray from a point A to a neighboring point B by tracing the circle of curvature through a small angular increment $\Delta\phi$, or a small arc segment $\rho\Delta\phi$ (see Fig. 1b). In most cases it turned out to be of sufficient accuracy, if one proceeds with angular increments of about 3°.

For any incremental step in tracing a ray the following calculations are necessary:

1. At a given point A grad v has to be computed. Since we assume that the velocity may be known as an unique function of the coordinates, a numerical determination of grad v is always possible by comparing the velocity at the point A with the velocities at at least two neighboring points. (How two-dimensional models of the subsoil can be formulated mathematically is described later.)
2. The angle i between the ray-vector $\vec{s}$ and grad v must be determined. Subsequently the radius of curvature is calculated using Eq. (2).
3. The new point B is reached by proceeding in the direction of the ray-vector by

$$\xi = \rho \sin\Delta\phi \tag{3}$$

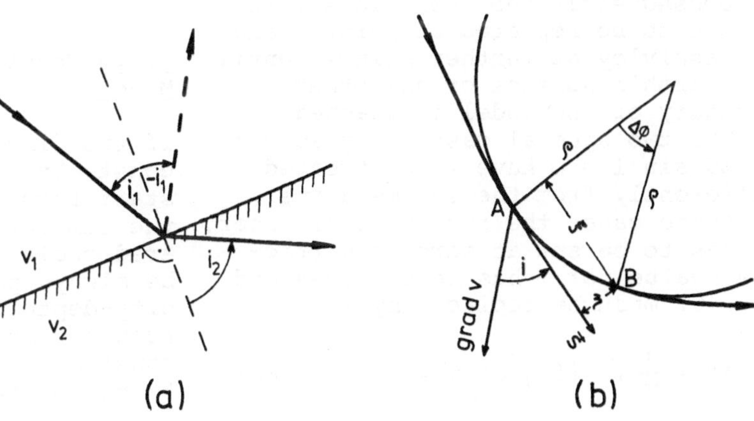

Fig. 1a and b. Refraction (a) and curvature (b) of a seismic ray under the influence of a discontinuous (a), and continuous (b) velocity variation. Symbols are explained in the text

and in the direction perpendicular to the ray-vector by

$$\eta = \rho(1 - \cos\Delta\phi). \qquad (4)$$

A positive sign of η indicates a counter-clockwise, a negative sign of η a clockwise curvature of the ray. The sign of $\Delta\phi$ must be identical with the sign of ρ.

ξ and η may be regarded as coordinates of the point B in a local cartesian system centered at the point A with axis in the direction and perpendicular to the ray. The conversion into the absolute coordinates of the model - these may be cartesian coordinates in refraction seismic studies, but geocentric polar coordinates in large-scale seismological studies - can be done by simple coordinate transformation. For example if x(A), z(A) are cartesian coordinates of the point A and if ϕ is the angle between the x-axis and the ray-vector at A, the corresponding coordinates of B are:

$$\begin{aligned} x(B) &= x(A) + \xi\cos\phi - \eta\sin\phi \\ z(B) &= z(A) + \xi\sin\phi + \eta\cos\phi \end{aligned} \qquad (5)$$

4. The travel time from A to B is given by the equation

$$\Delta t = \frac{1}{|\text{grad } v|}\{\text{Arcth} \cos(i) - \text{Arcth} \cos(i+\Delta\phi)\} \qquad (6)$$

If the angles i and $\Delta\phi$ are used as introduced above, this quantity is always positive. In contrast to other equations Eq. (6) is also valid, if the ray somewhere between A and B is perpendicular to the velocity gradient.

5. By rotating the ray vector by the angle $\Delta\phi$ the new ray vector at the point B is obtained.

Subsequently the operations 1 to 5 have to be repeated at point B and successivley at further points, until the earth's surface or any other boundary of the model is reached.

The two special cases of grad v = 0 and sin i = 0 have to be treated differently from the scheme above. In these cases the ray is rectilinear, $\Delta\phi$ has to be set at zero, a particular value for ξ has to be fixed and Eq. (6) must be replaced by

$$\Delta t = \frac{\xi}{v}, \quad \text{if grad } v = 0 \qquad (6a)$$

or alternatively

$$\Delta t = \frac{1}{|\text{grad } v|} \ln\{\frac{v(B)}{v(A)}\}, \quad \text{if } \sin i = 0. \qquad (6b)$$

Moreover, right after step 3 of the scheme above it must be checked, whether a discontinuity has been crossed between A and B. In such a case the point B must be abandoned and be replaced by the piercing point of the ray through the discontinuity, before proceeding with steps 4 and 6. Subsequently Snell's law has to be applied prior to reentering in the scheme above at step 1.

Because of the frequently recurring operations the described method of ray tracing and travel-time calculation is well suited to be programed for electronic computers or even for some desk-top calculators. The examples presented in MILLER and GEBRANDE (6.11 of this volume), for instance, have been computed and plotted on a HP 9810A desk-top calculator. For the practical interpretation of travel-time curves by model calculations the direct access to desk-top calculators proves to be more valuable than the high speed of big computers, which generally is not felt by the user.

Since the source of the seismic wave can be treated like any other point, it can be located everywhere within the model. By appropriate programing, not only single rays and travel times but arbitrary wisps of rays and complete travel-time curves can be either calculated or plotted. In principle the method described is applicable to any two-dimensional velocity function. The following section deals with the problem of defining velocity functions, which may represent realistic two-dimensional models of the subsoil.

4.4.3 Construction of Two-Dimensional Models

If two-dimensional models have to be constructed, it seems suitable to start from experiences gained with the simpler case of horizontally layered media. In this case it proved to be always possible to represent velocity-depth function with sufficient accuracy by series of layers with constant velocity gradients, that is to say by functions of the type:

$$v = v_i + \frac{v_{i+1} - v_i}{z_{i+1} - z_i}(z - z_i) \qquad (7)$$

for $z_i < z \leq z_{i+1}$

A velocity-depth function of this type represents a polygon in the v,z-plane through the corner-points v_i, z_i. If for any i z_i and z_{i+1} are identical, a discontinuity is modeled.

In the case of a horizontally stratified medium the levels of constant velocities are horizontal planes. The transition to two-dimensional models can easily be done by replacing the depth-values z_i in Eq.(7) by functions of the horizontal coordinate. Depending on the scale of the problem one may introduce cartesian or geocentric polar coordinates; for simplicity cartesian coordinates x,z will be used in the following discussion.

The functions $z_i = z(x,v_i)$ represent lines of constant velocity (isolines) in the x,z-plane. With Eq. (7) these lines then define a two-dimensional velocity function. Between neighboring isolines the velocity follows by linear interpolation in the vertical direction. It may be noteworthy that the velocity gradient grad v is no longer a purely vertical one, but now has in general also a horizontal component. Both components may be functions of x; moreover the horizontal component, even between neighboring isolines, may depend also from the vertical coordinate.

It may be noticed that different approaches have been used by STEIN (1969) and more recently by WILL (1975). In contrast to the present method, their models are composed of layers with constant velocity or constant velocity gradient. This case is more favorable with respect to computer time, since the rays within any particular layer are now straight lines or circular arc segments, and the calculations can be simplified. The method is well suited to model certain sedimentary structures. It is however less appropriate to model continuous velocity functions. At the boundaries between layers of constant velocity gradient, unwelcome discontinuities may arise, which can only be suppressed by extensive refinements of the model. With the present method of modeling two-dimensional velocity functions, these difficulties are avoided.

It remains to specify how the lines of equal velocity should be constructed. In principle this could be done by different types of functions. However, in the interest of uniqueness, care must be taken to avoid intersections of isolines. The use of normalized inverse tangent functions of the type

$$\Delta z_k \left(\frac{1}{2} + \frac{1}{\pi}\operatorname{arctg}\frac{x-x_k}{b_k}\right) \qquad (8)$$

proved to be especially well suited to build up isolines. Each term of this type represents a flexure with its center at the x-coordinate x_k and the throw Δz_k (see Fig. 2). The slope of the flexure is determined by the parameter b_k. Half of the total throw is confined solely to the interval $2b_k$ around x_k; the quantity $2b_k$ therefore may be termed the half-width of the flexure. By combining flexures with different parameters, syn- and anticlines, horsts, throughs and other realistic geologic structures can be modeled. By adding different terms of the type Eq. (8) to a normal depth z_{io} according to

$$z_i = z(x,v_i) = z_{io} + \sum_{k=1}^{n_i} \Delta z_k \left(\frac{1}{2} + \frac{1}{\pi}\operatorname{arctg}\frac{x-x_{ik}}{b_k}\right) \qquad (9)$$

nearly any desired morphology of isolines can be produced. The subscript i refers to a particular isoline, the subscript k indicates parameters of the k-th flexure of the i-th isoline; for instance Δz_{ik} is the throw of the k-th flexure of the i-th isoline. Unlike coefficients of polynomials or

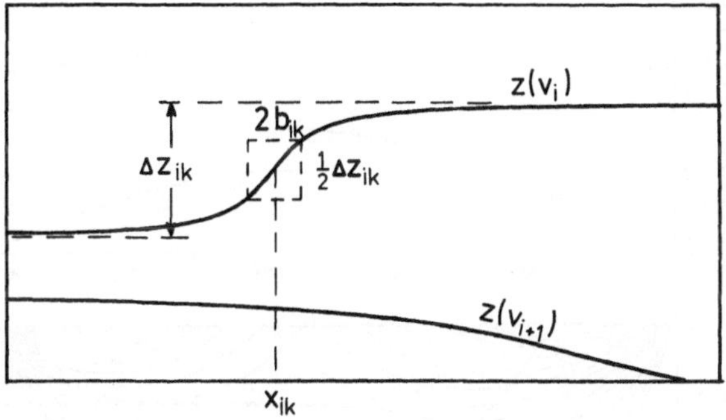

Fig. 2. Simple example illustrating the modeling of isolines by inverse tangent terms. The geometrical significance of the different parameters is indicated. Δz_{ik} and $2b_{ik}$ determine the throw, respectively half-width of a "flexure" centered at the x-coordinate x_{ik}

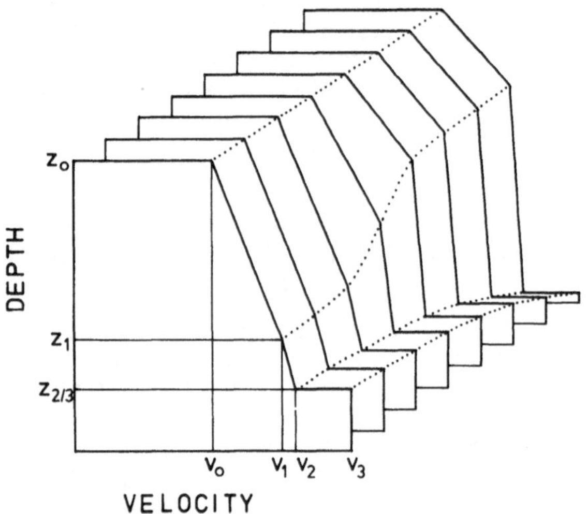

Fig. 3. Block diagram of a two-dimensional velocity distribution obtained by assigning the velocities v_o to v_3 to the isolines shown in Fig. 2. A discontinuity has been modeled by attributing two velocities to the lower isoline

Fourier series, which are frequently used for the representation of arbitrary functions, the different parameters entering Eq. (9) have a clear and simple geometric meaning. This greatly facilitates the construction and variation of models. A further advantage of the present method consists in the relatively small memory capacity required. This allows calculations of rather complex models even with cheap desk-top calculators.

Fig. 2 represents a simple example with only two curved isolines. The upper boundary, which in this case is a straight line, may generally be treated like an isoline and may therefore have any kind of topography. To specify the model fully, velocity values have to be attributed to the isolines. Fig. 3 gives a perspective view of a model, which is obtained by assigning the velocities v_o and v_1 to the upper boundary and the upper isoline and the velocities v_2 and v_3 to the lower isoline of Fig. 2. By attributing two velocities to the lower isoline, a discontinuity has been defined. Underneath the lowest isoline the velocity is automatically regarded as constant by the present program.

By the method described so far the velocity-depth functions at constant x-coordinate are some kind of polygon (Fig. 3). Sometimes more smoothly varying velocity functions are required. This can be achieved by letting the velocities depend upon z in the same manner as the isolines depend upon x, that is by putting:

$$v = v_o + \sum_i \Delta v_i \left(\frac{1}{2} + \frac{1}{\pi}\mathrm{arctg}\frac{z-z_i}{a_i}\right) \quad (10)$$

If all Δv_i, z_i and a_i are constants Eq. (10) represents a one-dimensional velocity-depth function. By indroducing these quantities as functions of the horizontal coordinate two-dimensional smoothly varying velocity distributions are obtained. As before, the dependence of z_i upon the x-coordinate may be described by Eq. (9) and analogous representations may be used for Δv_i and a_i. This method has been applied by MILLER and GEBRANDE (6.11 of this volume) to construct their model A presented in Fig. 9a (6.11).

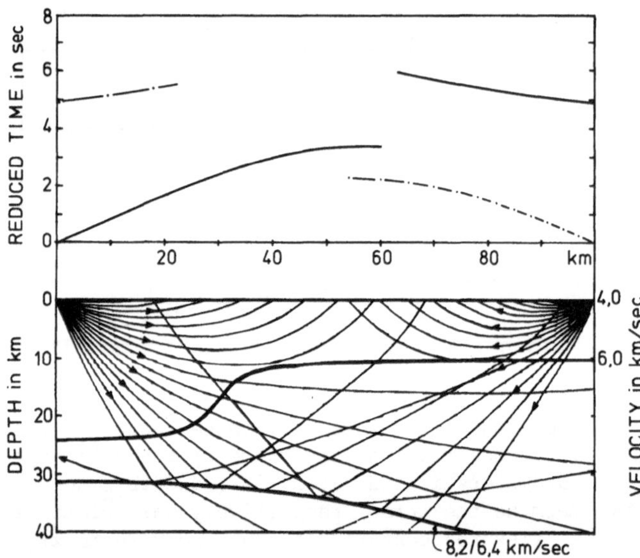

Fig. 4. Ray paths and reduced travel-time curves for the model of Fig. 3. Numerical values of the velocities v_o to v_3 are 4.0, 6.0, 6.4 and 8.2 km/s respectively

Quite similar two-dimensional models may be constructed within the system of geocentric polar coordinates by replacing z by the radius r and x by $R_0\theta$ in Eqs. (7) to (9) or (10), respectively. R_0 is the mean outer radius of the earth and θ is the polar distance. (In general of course this will not be the polar distance within the system of geographic coordinates, but within a system of plane polar coordinates defined in a vertical section through the profile.) In this case one obtains spherical layered earth models, if no flexures are introduced (GEBRANDE, 1975).

4.4.4 Application

Detailed examples of complex applications of the method outlined are given elsewhere (MILLER and GEBRANDE, 6.11 of this volume; GEBRANDE, 1975). It is therefore deemed sufficient to present ray paths and travel time curves for the simple model of Fig. 3. The results are plotted in Fig. 4. The numbers attached on the isolines indicate the seismic velocities along these lines used for the calculation. Rays and travel-time curves have been calculated for two shotpoints 100 km away from each other. Since Fig. 4 by itself is rather expressive, no detailed comment seems necessary. Rays have been plotted with angular distances of 4°, but for the determination of the travel-time curves of the waves reflected at the discontinuity in the lower part of the model, additional travel times have been calculated. The missing symmetry of the reversed travel-time curves clearly reflects the lateral heterogeneity of the model. Yet the reversed times at the shotpoints are identical, as is to be expected for a correctly working program.

Acknowledgments. The author wishes to thank Dr. H. Miller and Dr. M. Will for fruitful discussions. Prof. Dr. G. Angenheister's furtherance of the present work is gratefully acknowledged.

4.5 Calculation of Travel Times and Ray Paths for Lateral Inhomogeneous Media

M. WILL

ABSTRACT

This article describes a ray-tracing method for two-dimensional inhomogeneous media. This method was first developed for near-surface sedimentary structures with discontinuous distributions of velocity. The subsoil models consist of layers with constant velocity or constant velocity gradient of arbitrary direction. Their boundaries are polygons with any kind of topography; even outcrops and pillow-formed structures are possible. Thus, diverse models of the subsoil can easily be constructed. As applications of the method show, satisfactorily exact approximations of continuous velocity distributions are possible. Furthermore, the simple basic structure of the models allows a very quick calculation of ray-paths and travel times by the ALGOL computer program. The method is demonstrated by the model of a salt dome in northern Germany. As examples of application models derived from refraction-seismic measurements at the northern margin of the Eastern Alps are shown. The most important structure of these models is a layer of about 7 km maximum thickness underlying the Northern Calcareous Alps up to at least 15 km southward of their northern margin, probably containing younger sediments.

ZUSAMMENFASSUNG

Ein Verfahren der Modellrechnung für zweidimensional inhomogene Medien wird vorgestellt. Dieses Verfahren wurde zunächst für oberflächennahe sedimentäre Strukturen mit diskontinuierlicher Geschwindigkeits-Verteilung entwickelt. Die Modelle des Untergrundes bestehen aus Schichten mit konstanter Geschwindigkeit oder mit beliebig gerichtetem konstantem Gradienten der Geschwindigkeit. Die Schichtgrenzen sind Polygonzüge mit beliebiger Topographie. Auch Auskeilen von Schichten und linsenförmige Strukturen sind möglich. So können vielfältige Modelle des Untergrundes auf anschauliche und einfache Weise dargestellt werden. Anwendungen des Verfahrens zeigen, daß auch kontinuierliche Geschwindigkeits-Verteilungen mit befriedigender Genauigkeit angenähert werden können. Der einfache Aufbau der Modelle bewirkt weiterhin, daß mit dem in ALGOL beschriebenen Rechenprogramm die Strahlen und Laufzeiten in sehr kurzer Zeit errechnet werden. Das Verfahren wird am Modell eines Salzstocks aus Norddeutschland erläutert. Als Beispiele einer Anwendung werden Modelle gezeigt, die aus refraktionsseismischen Messungen am Nordrand der Ostalpen abgeleitet wurden. Die wichtigste Struktur dieser Modelle ist eine Unterlagerung der Nördlichen Kalkalpen bis mindestens 15 km von ihrem Nordrand nach Süden durch eine maximal etwa 7 km mächtige Schicht, die mit größter Wahrscheinlichkeit jüngere Sedimente enthält.

4.5.1 Introduction

The interpretation of refraction seismic data is often difficult and sometimes even impossible because of lateral velocity variations which affect the observed travel-time curves. The existing inversion methods generally require laterally homogeneous media or allow only restricted lateral variations. Yet the deviations of a horizontal stratification are of particular interest. The computation of ray paths and travel times for two-dimensional inhomogeneous media is a possibility to interpret seismic measurements in such areas: a starting model is varied until a satisfactory fit between computed and observed travel times is found. And though the results of model calculations are not unique, the consideration of known parameters in the area under investigation will decrease the number of possible models.

A ray-tracing method programed for a desk-top calculator and mainly suitable for continuous two-dimensional models with the possibility of certain discontinuities is described by GEBRANDE (4.4 of this volume). The ray-tracing method presented here was first developed for discontinuous velocity distributions with the possibility of approximating continuous models and the computer program is written in ALGOL. The main features of this method, the simple basic model structures facilitating various model constructions and the method of calculation on principle will be demonstrated in this paper.

Several seismic measurements have already been interpreted with the aid of this method, e.g. by GRUBBE (6.4 of this volume), KOSCHYK (1973), and PETERS (1974). Results are shown which were derived from refraction seismic measurements in a northern part of the Eastern Alps.

4.5.2 Construction of Models

The seismic models consist of several layers with constant velocity or constant velocity gradient. This construction is similar to the one used by STEIN (1969). The main differences are: here not only vertical but arbitrary gradient directions are possible and the layer boundaries are polygons with any kind of topography.

An example of the possibilities of model construction is given in Fig. 1. The geological cross section shown above was derived from gravimetric and reflection seismic measurements and from drillings (REICH, 1960c). Part of this section, between 1 and 13.5 km, was transferred into the seismic model shown below. In this model smaller layers with only minor velocity differences were lumped together and some boundaries were slightly simplified. The values of P-wave velocities and gradients were inserted after MÜHLEN and TUCHEL (1953), CLASEN (1958), REICH (1960a), KREY and NODOP (1970) and BORTFELD (1971).

As mentioned above, the layer boundaries in a model are polygons. These polygons have a common field of x-coordinates so that for every boundary only the z-coordinates must be given. This diminishes the number of input data. In Fig. 1 the seismic model has a surface with topography which is not shown in the geological cross section. This is to demonstrate that topographic effects are calculated automatically and therefore no topographic reduction in the seismogram section is necessary.

The velocity in a layer may be defined as a constant value. In case of a velocity gradient two points of a straight line, the velocity value on this line and the gradient value are presented. The gradient direction is perpendicular to the given line and the velocity will increase or decrease with depth according to the gradient sign.

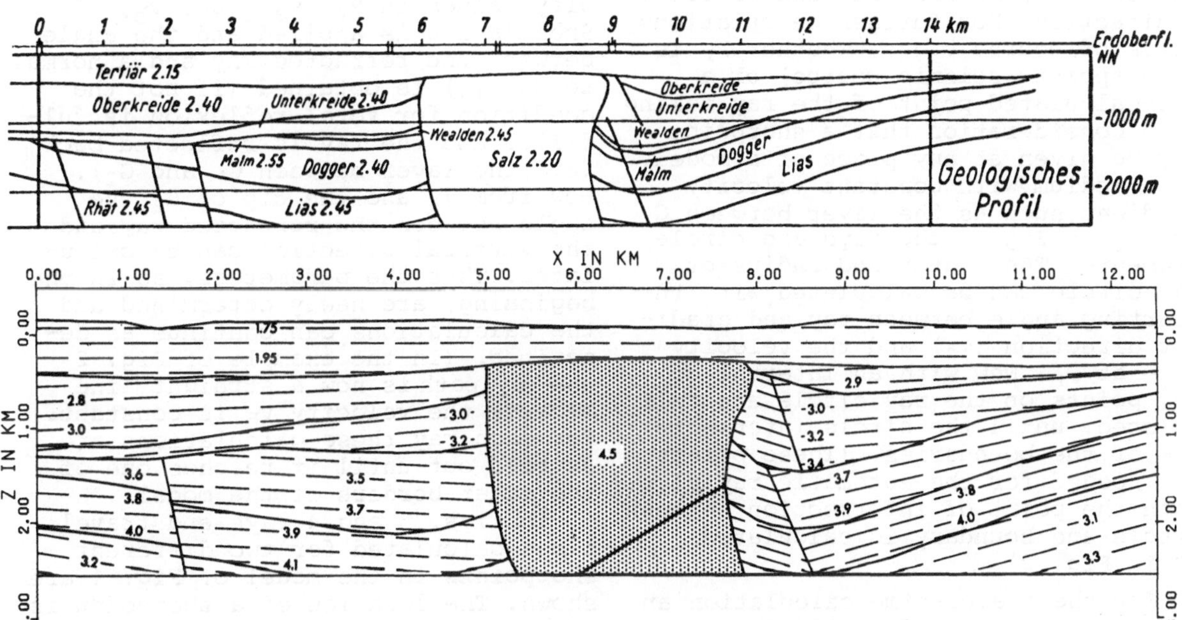

Fig. 1. *Upper part:* Geological cross section with salt dome in northern Germany. Numbers = density in g/cm^3. (From REICH, 1960c). *Lower part:* Model with two-dimensional velocity distribution tallying with the part between 1 and 13.5 km of the above geological section. *Dotted:* salt dome. *Dashed lines:* lines of equal P-wave velocity with value in km/s

In the seismic model of Fig. 1 the velocity is constant only in the salt dome itself. The salt dome is divided into two parts because the boundaries must be unique functions of the horizontal distance x. So the right boundary cannot be represented by only one polygone. Yet as the velocity is the same in both of the two parts, this division does not affect the calculations.

4.5.3 Method of Calculation

For the kind of model described above, the calculations of ray paths and travel times follow the same rules as in geometrical optics so that the equations can be derived from Snell's law. Since these and the other necessary equations (e.g. for coordinate transformation etc.) are well known, in the following brief description of the principle way of calculation they are not included (with one exception).

A part of a model is shown in Fig. 2: between the dashed lines marked with x_1 and x_2 (two values of the common x-field) the boundaries G_1 and G_2 are straight lines. Between G_1 and G_2 the velocity v_a increases with depth, its constant gradient grad v_a is inclined. Below G_2 the velocity v_b is constant.

A ray starts at the point P_1 with the angle γ_1 between ray and vertical z-direction. For further calculations it is of course eager, whether P_1 is a shotpoint (seismic source) or a pre-calculated point of the ray taking into consideration that a shotpoint may be given at any place of a model. In a medium with constant velocity gradient such as the layer between G_1 and G_2 in Fig. 2 the rays are circle segments. The center and radius of the circle can be calculated with the starting angle between ray and gradient direction (γ_2) and the velocity $v_{a,1}$ at P_1. Now with an interval of 3°, points on the ray circle are calculated, until the circle intersects one of the boundaries. (These points are needed for the automatic ray plot.) Then the point of intersection between circle and boundary is calculated (P_2 in Fig. 2).

For the travel-time calculation an equation was derived, valid for any two points P_1, P_2 on a ray circle segment, even if the ray between the two points is perpendicular to the velocity gradient:

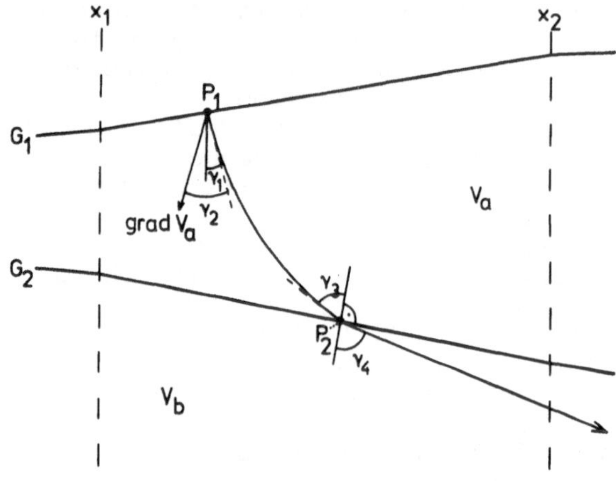

Fig. 2. Sketch to illustrate the method of calculation

$$\Delta t_{1,2} = \frac{1}{2 \cdot |\text{grad } v|} \cdot \ln \frac{R^2 - x_M^2 + |x_2 \cdot R| + x_2 \cdot x_M}{R^2 - x_M^2 - |x_2 \cdot R| + x_2 \cdot x_M}$$

In this equation R is the radius of the circle and the coordinates x_M of the center of the circle and x_2 of P_2 must be given in a cartesian system, centered at P_1 with axis parallel and perpendicular to grad v.

After adding the travel time between P_1 and P_2 to the travel time needed until P_1, the angle between the ray and a normal to G_2 in P_2 (γ_3) is calculated. With γ_3 and the velocity values in P_2, $v_{a,2}$ and v_b, Snell's law is applied and the angle between the refracted ray and a normal to G_2 (γ_4) is obtained (if not the condition for total reflexion is fulfilled and the ray is reflected back into the layer between G_1 and G_2). Now from γ_4 and the dip of G_2 the angle between the refracted ray and the vertical direction can be calculated. Thus the parameters, as in the beginning, are newly determined and the calculations can continue as described. (In the example of Fig. 2 the ray path is now a straight line because the velocity v_b is constant.) For each ray these calculations are carried out until it reaches one of the outer borders of the model.

In Figs. 3 and 4 rays and travel times calculated for two different shotpoints in the model of Fig. 1 are shown. The location of a shotpoint is marked by a star; rays are not plotted near the shotpoints for technical reasons. In Fig. 3 the shotpoint is situated at the surface; the travel-time curve is plotted in a reduced

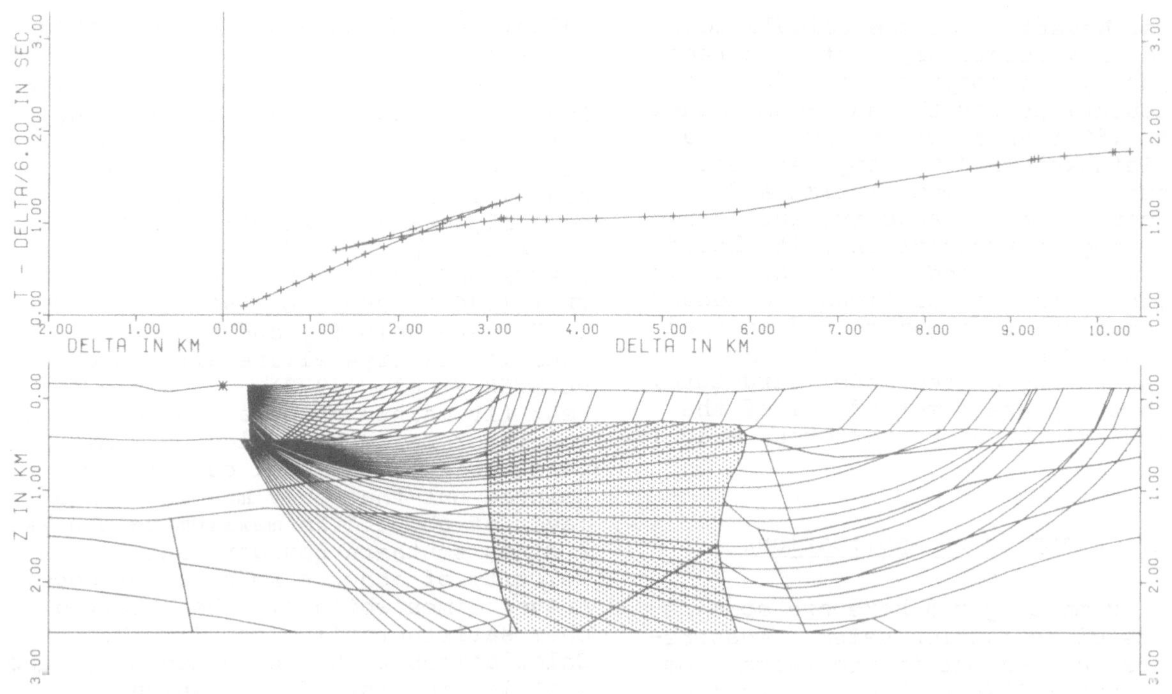

Fig. 3. Ray paths and travel times calculated with the model of Fig. 1 and a seismic source at the surface. *Dotted:* salt dome

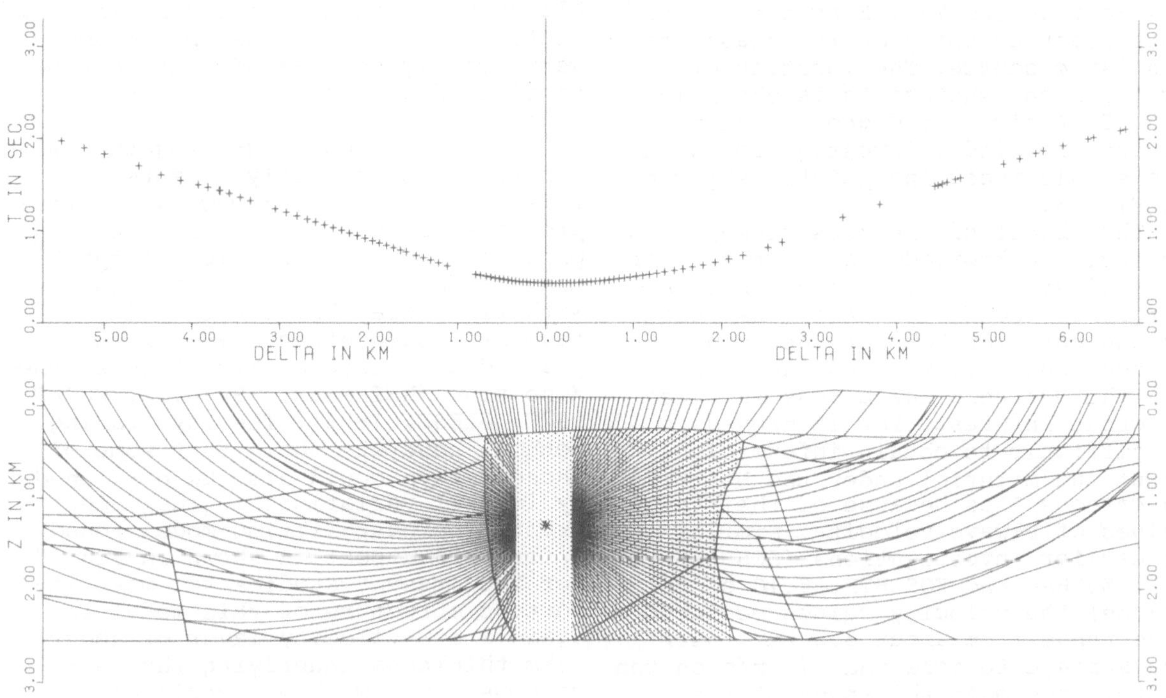

Fig. 4. Ray paths and travel times calculated with the model of Fig. 1 and a seismic source within the salt dome. *Dotted:* salt dome

time scale. In Fig. 4 the results for a seismic source in the salt dome are shown; the travel-time curve here is plotted in an unreduced time scale.

According to these short explanations, the calculations are simple. Yet the consideration of all special cases, e.g. ray parallel velocity gradient etc., complicate the program structure. The program parts, which make input and output of data as comfortable as possible, contribute to this. Nevertheless the calculations need only relatively little storage capacity and computer time. (e.g. the calculations for the salt-dome model with 180 rays shown in Fig. 4 on a "Telefunken TR 440" computer need about 35 s of computer time and 30 k of core-storage capacity, including the program translation.) It should still be mentioned that no approximations in the calculations are needed, in contrast to the method described by GEBRANDE (4.4 of this volume). Owing to the chosen model structure, with the present method all of the calculations can be performed exactly.

4.5.4 Examples of Application

The examples given here are some results of refraction seismic measurements carried out in the years from 1970 until 1974 within the program "Geodynamics of the Mediterranean Region" in a northern part of the Eastern Alps, mainly between the rivers Salzach and Inn. Eleven profiles were measured using quarry blasts as a seismic source. The location of profiles and shotpoints is shown in Fig. 5. A simplified geological map of the area under investigation with the single measuring points is given in Fig. 6.

The object of the measurements was to increase knowledge about the upper crustal structure at the northern margin of the Eastern Alps. Especially the question how far the Molasse, Flysch, and Helvetic sediments underlie the Northern Calcareous Alps was of great interest. Yet it proved to be difficult to obtain information about the material under the Northern Calcareous Alps because of their high values of P-wave velocity, similar to earlier investigations (SCHELIGA, 6.9; ZSCHAU and KOSCHYK, 6.10 of this volume) the velocity values measured here range from about 5.5 km/s near the surface to more than 7 km/s on the bottom (the mean thickness of the Calcareous Alps is about 3.1 km). Hence it was clear that only special profiles can give the desired information: either long profiles, where P^M-phases can be observed in the Calcareous Alps (such as Golling-W) or profiles crossing the Calcareous Alps with shotpoint and part of the measuring points outside them (such as Rohrdorf-S). For these two profiles the models with the best fit between calculated and observed travel times are shown here.

The starting model for the profile Rohrdorf-S was constructed with the results of the other profiles and with data obtained from literature: e.g. velocity and thickness of Molasse-, Flysch- and Helvetic zone in the Alpine Foreland were given after BREYER and DOHR (1959), REICH (1960a), VEIT (1963), PRODEHL (1964) and LEMCKE (1973) and velocity and approximate location of crystalline after ANGENHEISTER et al. (1972). To obtain the best fit mainly shape and velocity of the layer below the Calcareous Alps were varied. The model calculations pointed out that only models with a layer of about 7 km maximum thickness, a steep southern boundary and a mean velocity of about 4 km/s can produce travel times which fit the observed ones satisfactorily (see Fig. 7). Calculations with such a model for the profile Kiefersfelden-N, which is a reverse profile to Rohrdorf-S (see Figs. 5 and 6) also show a fit between observed and calculated travel times. Although the onsets in the seismogram section of Kiefersfelden-N are not very clearly correlated, this can be considered as a confirmation of the model.

The profile Golling-W together with a part of the formerly measured profile Eschenlohe-E (01-090) was interpreted as a reversed profile. Both profiles show a separation of the P^M-travel-time curves into segments with time delay between the single segments. Such effects are often explained by a special structure of the Moho zone (MEISSNER, 1967a, b; FUCHS, 1970; GIESE, 1972). The ray-tracing method, however, shows that some of these delays could also be explained by lateral variations in the upper crust. The best-fitting model found under this aspect is shown in Fig. 8. For the present studies the most important structure of this model is again a low-velocity layer of about 6 km thickness underlying the Calcareous Alps, with a transition to Helvetic and Flysch zone in the west, and outcropping in the east. Below this layer respectively below the Calcareous Alps in the east the model in effect corresponds to a velocity distribution derivable from two cross sections given by ANGENHEISTER et al. (1972) after former measurements.

As the profile Eschenlohe-E is discussed in detail by ZSCHAU and KOSCHYK (6.10 of this volume) we renounce to

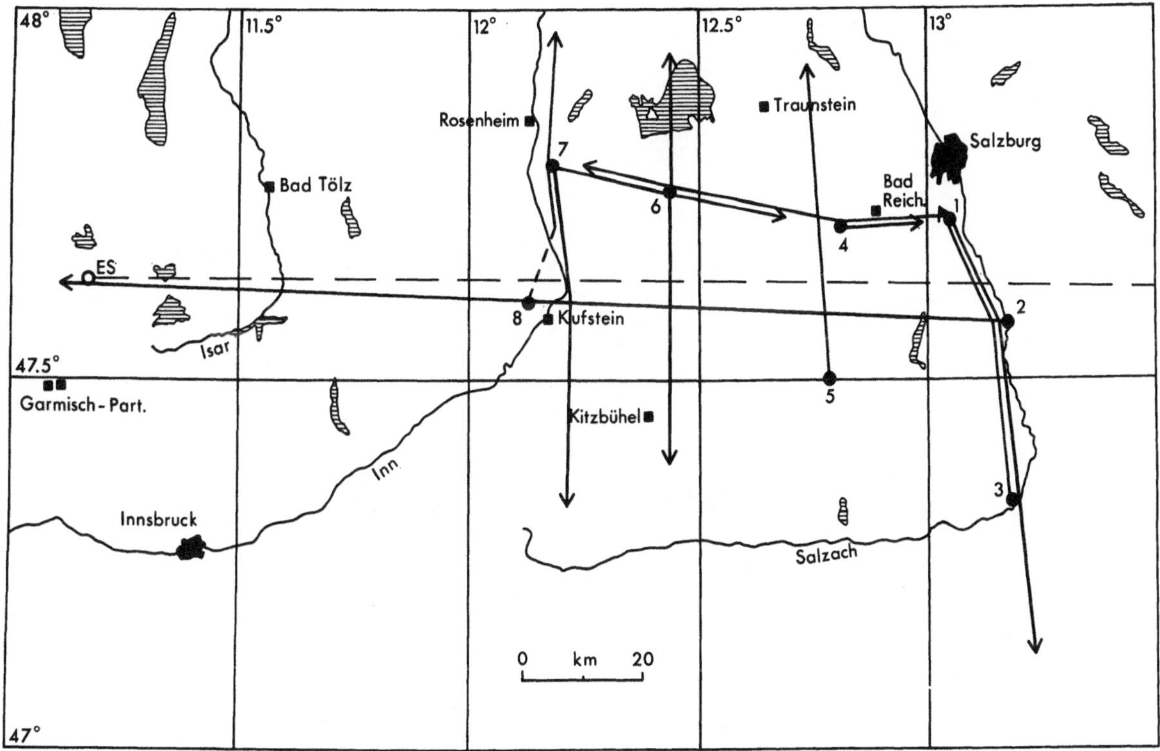

Fig. 5. Location of the refraction seismic profiles measured within the program "Geodynamics of the Mediterranian Region" (*solid lines*). *Dashed line:* profile Eschenlohe-E (01-090). *Round dots:* location of quarries: *1* Gartenau, *2* Golling, *3* St. Johann i. P., *4* Antoniberg, *5* Lofer, *6* Maquartstein, *7* Rohrdorf, *8* Kiefersfelden

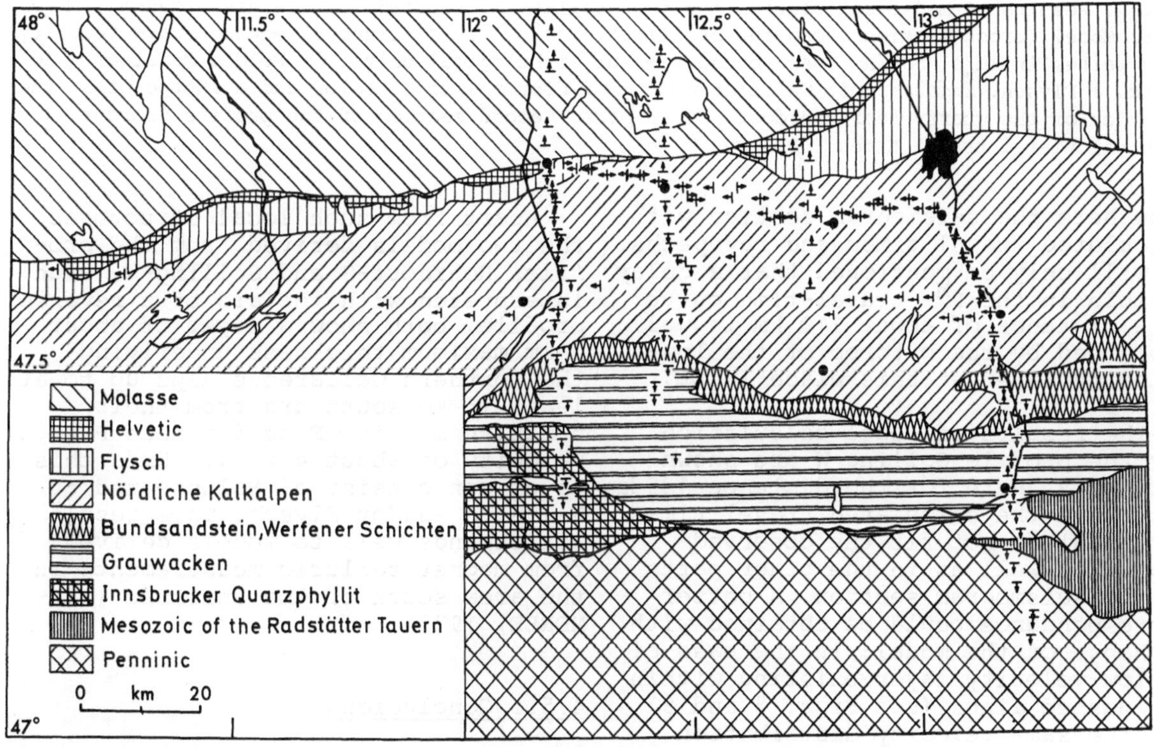

Fig. 6. Simplified map of the surface geology (same area as in Fig. 5) with single measuring points. *Arrows:* the profile direction. *Round dots:* quarries (see Fig. 5)

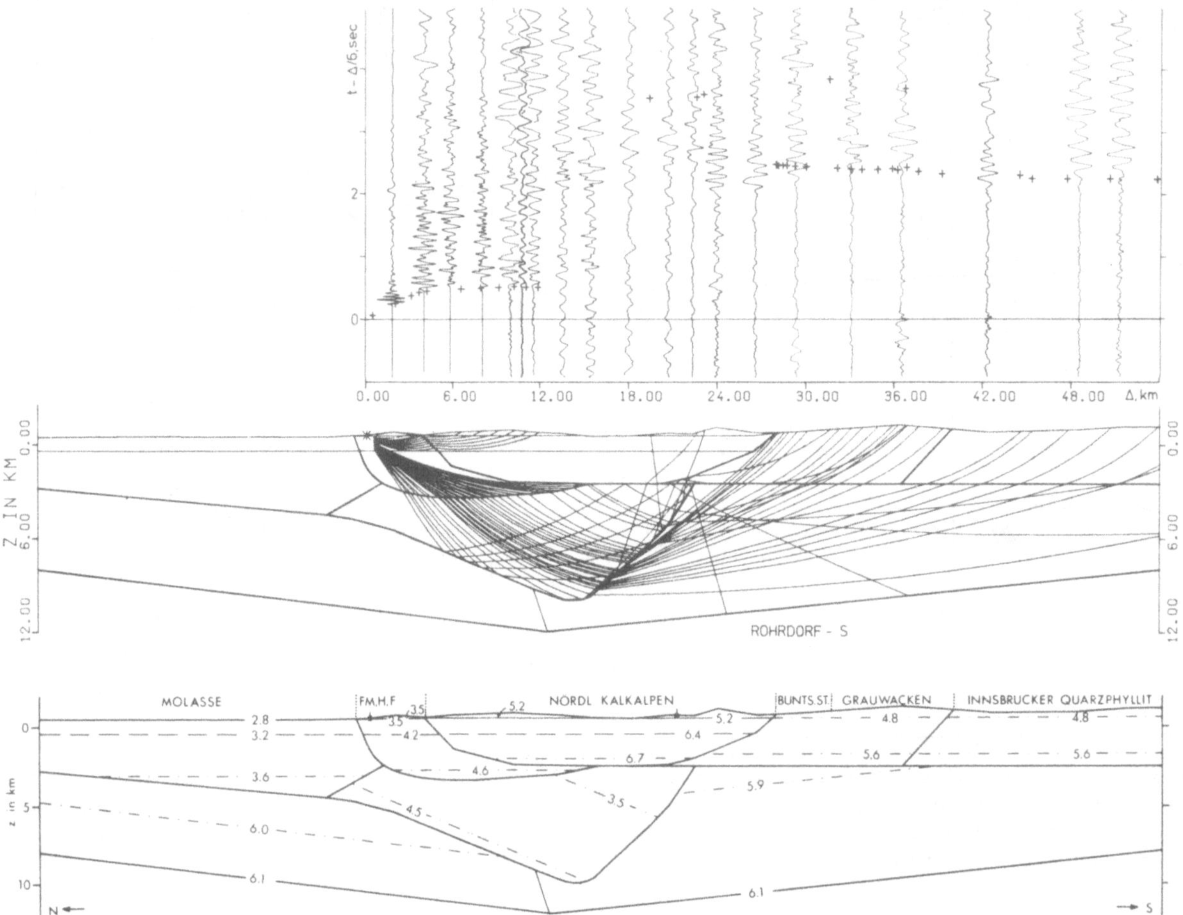

Fig. 7. Results of model calculation for the profiles Rohrdorf-S and Kiefersfelden-N. *Lower part:* Model (cross section along the profiles). *Dashed lines:* lines of equal P-wave velocity with values in km/s. *Solid lines:* boundaries of lithological units. *FM* Folded Molasse, *H* Helvetic, *F* Flysch. *Middle part:* Calculated ray paths for the model below and a shotpoint at the location of the quarry near Rohrdorf. *Upper part:* Record section of the profile Rohrdorf-S with calculated travel times (*crosses*)

show here the seismogram section with calculated travel times and ray paths. It will only be mentioned that a delay visible in the P^M- travel-time curve at a distance of about 110 km is well explained by the outcrop of the low-velocity layer at this place (WILL, 1975).

The results of model calculation for the profile Golling-W are shown in Fig. 9. Here for the distance 125 km to 140 km the calculated travel times (crosses) do not fit the observed onsets as well as in the other parts of the seismogram section. A better correspondance could be obtained with a change of the model: in the case of no low-velocity layer existing within the dashed lines (see Figs. 8 and 9) but the same velocity distribution as east of the layer, the travel times shown as dots in Fig. 9 are calculated.

Apart from such local structures the two models derived from the profiles Rohrdorf-S and Golling-W, each interpreted with the aid of a reverse profile, both show a layer of about 7 km maximum thickness underlying the Northern Calcareous Alps up to at least 15 km southward from their northern margin. From its mean P-wave velocity of about 4 to 4.8 km/s this layer can consist of Molasse and/or Helvetic and/or Flysch. This result corresponds well to models derived from magnetotelluric measurements in the area south of the Chiemsee (KEMMERLE, 1973).

4.5.5 Conclusions

The examples of application of the ray-tracing method presented in this paper show especially its usefulness

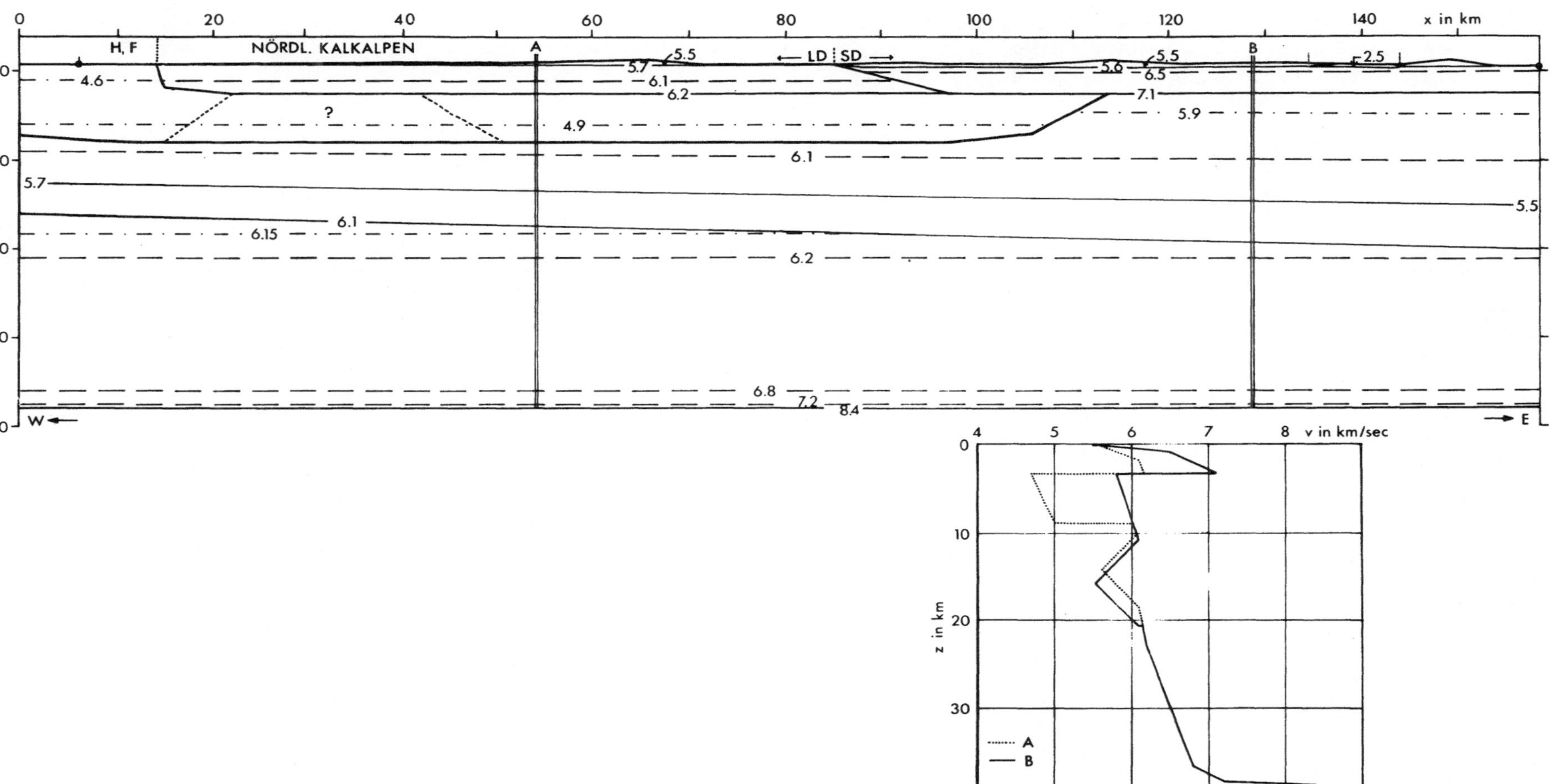

Fig. 8. *Above:* Model derived from the profiles Golling-W and Eschenlohe-E. *Thick solid lines:* boundaries of lithological units. *Other lines:* lines of equal P-wave velocity with value in km/s. (*LD* and *SD* indicate two different nappes of the Northern Calcareous Alps). *Lower right:* Velocity-depth functions at the spots *A* and *B* of the above model

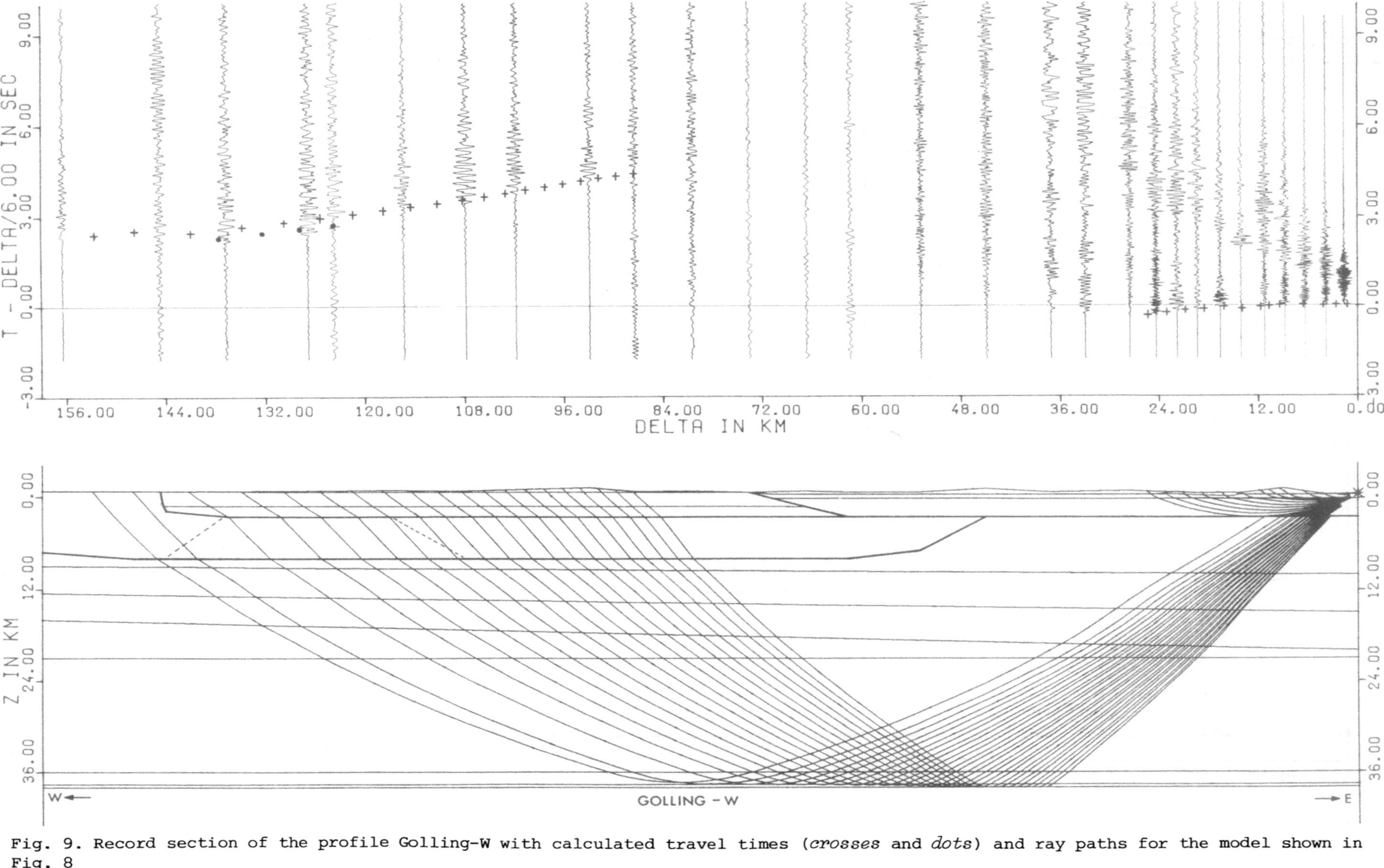

Fig. 9. Record section of the profile Golling-W with calculated travel times (*crosses* and *dots*) and ray paths for the model shown in Fig. 8

for models with discontinuous velocity distributions. Yet in Fig. 8 it can be seen that even a continuous velocity distribution can be approximated satisfactorily. For example, in this model below the line with a velocity of 6.1 km/s and a mean depth of about 18 km, the velocity changes discontinuously. As expected from the small velocity differences, however, model calculations showed the rather small influence of more detailed approximations in the ray paths and travel times, especially when compared to the influence of model parameters such as the total thickness of the crust. It is obvious that only zones with considerable changes of the velocity gradient must be approximated with more accuracy. As such zones are relatively rare in realistic models, the method mostly without complicated extensions of models is also available for continuous velocity distributions.

Acknowledgments. The author wishes to thank Dr. H. Gebrande and Dipl.-Geophys. K. Grubbe for useful discussions and many helpful ideas.

4.6 Inversion of Seismic Records with the Aid of Synthetic Seismograms

G. MÜLLER and K. FUCHS

ABSTRACT

Two methods for the computation of synthetic seismograms, the ray-theoretical method and the reflectivity method, are applied to the inversion of observed records. Both the travel times and the amplitude-frequency characteristics of seismic phases are interpreted. This procedure leads to considerable improvements of earlier inversions which are based on travel times alone. New depth distributions of P velocities are derived for the crust/mantle transition zone beneath the profile Hilders S and for the upper crystalline basement beneath the profile Voggendorf SE.

ZUSAMMENFASSUNG

Zwei Methoden zur Berechnung synthetischer Seismogramme, die strahlentheoretische und die Reflektivitäts-Methode, werden zur Inversion von beobachteten Seismogrammen benutzt. Die Interpretation erstreckt sich sowohl auf die Laufzeiten als auch auf die Amplituden- und Frequenzcharakteristiken seismischer Phasen. Dieses Vorgehen führt zu erheblichen Verbesserungen früherer Inversionen, die auf Laufzeiten allein beruhen. Neue Geschwindigkeits-Tiefenverteilungen werden abgeleitet für die Übergangzone von der Erdkruste zum Erdmantel unterhalb des Profils Hilders S und für das obere Kristallin unter dem Profil Voggendorf SE.

4.6.1 Introduction

Studies of the earth's crust and mantle with the aid of body waves are of first-order importance for the investigation of the physical properties and chemical constitution of the earth. This is especially true for explosion seismological studies, since in this case the location and time of the excitation are accurately known, and since explosions normally have simple radiation patterns and time functions. It is an urgent need that the inversion methods should be able to extract all relevant informations from the observations. Up to now, most interpretations of body-wave observations have been restricted to travel times or apparent velocities. When amplitudes of arrivals were taken into account, they were compared either with exact theoretical amplitudes for oversimplified models or with approximate theoretical amplitudes, calculated for realistic models with geometric ray theory. In both cases, rather large deviations between theory and observations must be accepted. Consequently, such simplified amplitude studies do not effectively reduce the number of solutions to the inversion problem. Recently, computational methods, based on exact solutions of the equation of motion of an elastic medium, have been so far developed that synthetic body wave seismograms can be calculated for models which are as complicated as those estimated from modern travel-time inversions (FUCHS, 1968, 1970; HELMBERGER, 1968; MÜLLER, 1968, 1970, 1971; HELMBERGER and MORRIS, 1969, 1970; FUCHS and MÜLLER, 1971). These methods allow an effective use of the amplitude-frequency information of observed body wave phases.

In this paper, we want to present inversions of observed data which were performed with the aid of synthetic body-wave seismograms. We have used two computational schemes, the reflectivity method (FUCHS, 1968; FUCHS and MÜLLER, 1971) and the ray-theoretical method (MÜLLER, 1970, 1971), as we have called them for brevity. Both methods assume an elastic half-space consisting of homogeneous, plane and isotropic layers (Fig. 1). The i-th layer is characterized by the P velocitiy α_i, the S velocity β_i, the density ρ_i, and the thickness h_i. The source is an explosive point source at the free surface z = 0. The interaction with this surface is omitted.

For the use of the reflectivity method, it is appropriate to subdivide the layered medium into a reflecting zone and a stack of layers on top of this zone which produces only elastic transmission losses and time shifts (see Fig. 1). The Fourier transformation of the vertical displacement for P waves is an integral over the angle of incidence γ at the top of the reflecting zone:

$$\bar{w}(r,o,\omega)=\bar{F}(\omega)\frac{\omega^2}{\alpha_m^2}\int_{\gamma_1}^{\gamma_2}\sin\gamma\cos\gamma\, J_o\left(\frac{\omega}{\alpha_m} r\sin\gamma\right)R_{pp}(\omega,\gamma)\, H(\gamma)\cdot$$
$$\cdot\exp\left[-2j\frac{\omega}{\alpha_m}\sum_{i=1}^{m} h_i\left(\frac{\alpha_m^2}{\alpha_i^2}-\sin^2\gamma\right)^{1/2}\right]d\gamma \quad (1)$$

$\bar{F}(\omega)$ is the spectrum of the spherical wave from the point source, J_o the Bessel function of the first kind and order zero, $R_{pp}(\omega,\gamma)$ the complex reflectivity (or plane-wave reflection coefficient) of the reflecting zone for an incident and a reflected P wave, $H(\gamma)$ is the product of the transmission coefficients of the refracting interfaces, and j the imaginary unit. Details about the derivation and numerical calculation of Eq. (1) are given in FUCHS and MÜLLER (1971).

The ray-theoretical method is based on exact or generalized ray theory and sums up the elementary seismograms of all possible rays from the source to the receiver. This is an exact description of the wave field and should not be confused with geometric ray theory. Whereas the reflectivity method includes automatically all multiples and also all conversions of PS and SP type inside the reflecting zone, these waves must be computed separately when the ray-theoretical method is used. In our computer programs for the ray theoretical method, we take into account only pure compressional rays, i.e. conversions are neglected. This is permitted for media with moderate variations of the velocities and the density. The elementary seismogram of a ray is the convolution of an excitation function with the impulse seismogram which can be calculated rather easily for excitation by a line source. We have used Garvin's method (GARVIN, 1956) which is essentially Cagniard's method (CAGNIARD, 1939) for two dimensions. Our way of computing point source seismograms is to multiply each elementary line-source seismogram by a

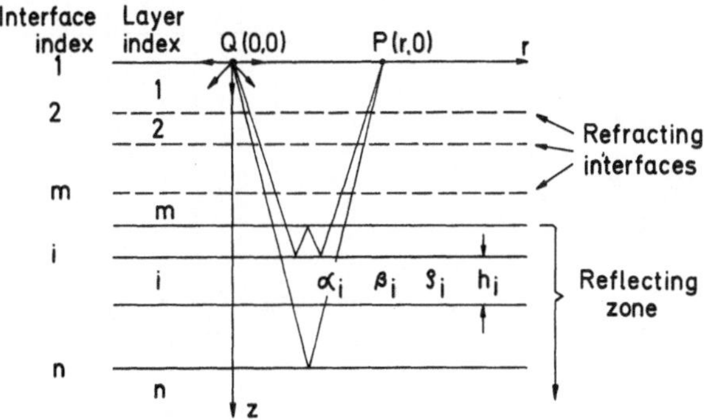

Fig. 1. Layered elastic medium with the explosive point source Q and the point of observation P

constant which follows from a comparison of the wave-front approximations of the line-source and the point-source case (MÜLLER, 1970).

A comparison of the reflectivity method and the ray-theoretical method gives the following results (FUCHS and MÜLLER, 1971). First of all, for very simple layered models with moderate contrasts of velocities and densities the agreement of both methods is very good. Furthermore, when the ray-theoretical method is used for more complicated models as for instance realistic crustal models, internal multiples must be neglected in order to avoid prohibitively long computing times. As a consequence, the ray-theoretical seismograms are not as accurate as those computed with the reflectivity method. However, the errors will seldom exceed the errors of the observed data, and the computing times are often less, sometimes considerably less, than the computing times with the reflectivity methods (FUCHS and MÜLLER, 1971, Fig. 8). Therefore, we propose to use the ray-theoretical method in order to obtain an overall picture of the complete refraction profile under investigation, and to use the reflectivity method for detailed studies of special features as for instance the amplitude variations in the neighborhood of travel-time cusps, diffraction phenomena, shadow zones, and in the case of strong variations of the velocity-depth distributions.

4.6.2 Reinterpretation of the Refraction Profile Hilders S

As a first example of how synthetic seismograms can be used for inversions in explosion seismology, we present

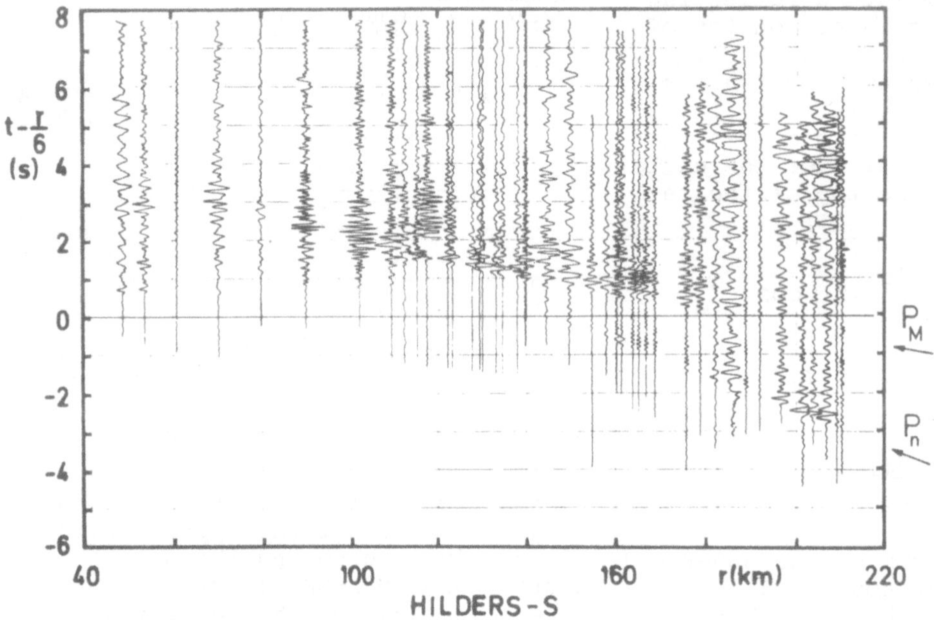

Fig. 2. Record section for the refraction profile Hilders-S in West Germany (KAMINSKI et al., 1967)

results computed in the course of a reinterpretation of the refraction profile Hilders S in West Germany. Travel-time interpretations of this profile have been given by FUCHS and LANDISMAN (1966a, b), by KAMINSKI et al. (1967), and by WANGEMANN (1970). The record section is given in Fig. 2. Although we did not know the magnifications of some of the traces, the instrumental characteristics, and the shape of the pulse from the source, we could improve the travel-time interpretation by modeling the most prominent amplitude characteristics of this section. These are firstly due to the Moho reflection P^M, extending from about 60 km to 190 km, and secondly due to the P_n wave from below the Moho which is visible only beyond 140 km. A zone of weak P_n arrivals between the critical point and larger distances can also be observed in other areas. For the interpretation of the P_n amplitudes, we used the amplitude ratio of P_n and P^M.

We started with the travel-time interpretation of KAMINSKI et al. (1967) (Fig. 3). The S velocities of this and all other models of this section follow from the P velocities under the assumption that Poisson's ratio is 0.25 everywhere. The Nafe-Drake relationship (quoted in TALWANI et al., 1959) was used to derive densities from P velocities. Furthermore, the explosive point source was assumed to radiate a spherical wave whose main frequency is 5 Hz. The synthetic seismograms for model MO and all other models of this section were computed with the reflectivity method. Comparing the synthetic with the observed record section in Fig. 3, we state the following discrepancies:

1. There are no clear observed arrivals that could correspond to the strong reflection from the discontinuity at a depth of about 21 km, having a reduced arrival time of about 1 s at 180 km.

2. The theoretical P_n wave has much too large amplitudes. Especially between 120 km and 140 km, where practically no arrivals should occur, the most prominent P_n amplitudes are found in the theoretical section. This discrepancy is also the main disadvantage of the interpretation which GIESE (1968a) has given for this profile.

Since we were mainly interested in the lower crust and the crust/mantle transition, we improved only the velocity-depth function below the low-velocity zone. We tried to keep the model as simple as possible and proceeded along the following lines. Firstly, we tried to approximate the lower crust between the low-velocity channel and the Moho by one layer with constant velocity gradient. Secondly, we started with the Moho as a first-order discontinuity. Only at a later stage should transition layers be included. Thirdly, we tried to explain

Fig. 3. P velocity-depth function, synthetic seismograms (vertical displacement), and travel-time curve according to the travel-time interpretation of KAMINSKI et al. (1967), compared with the observed record section. The synthetic seismograms and the travel-time curve include only the reflection response from the bottom of the low-velocity channel and from the layers below the channel. The numbers near the segments of the velocity-depth function with non-zero velocity gradient are the numbers of homogeneous layers used for the approximation of the corresponding depth range

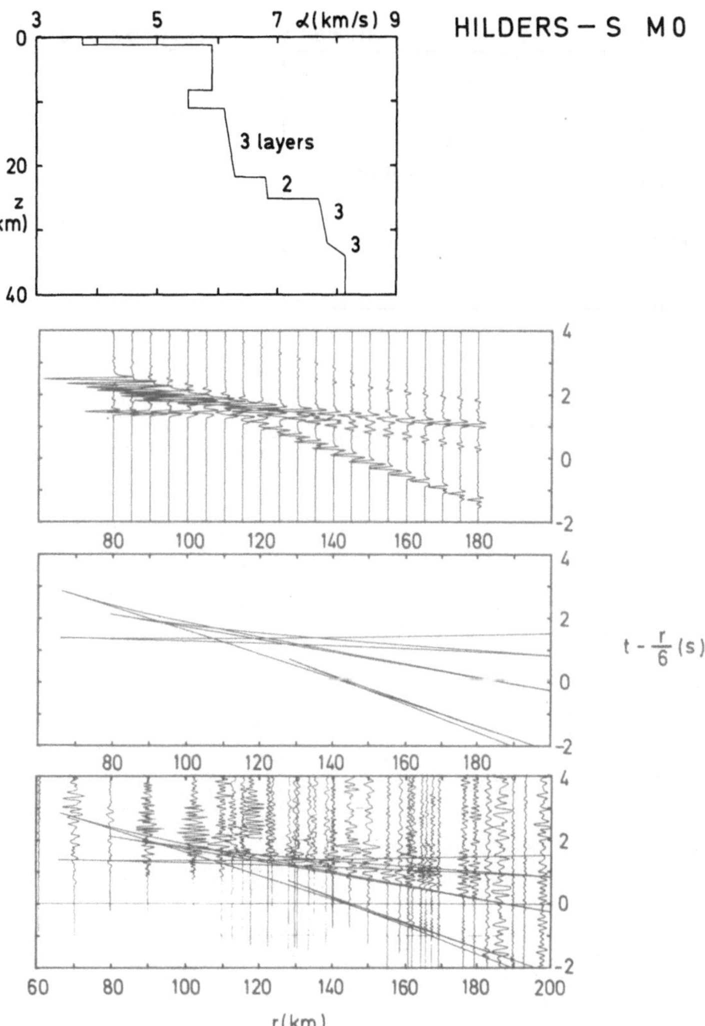

the amplitude variations of the P_n wave by a homogeneous layer immediately below the Moho, followed by an increase in velocity to 8.15-8.20 km/s, a value which can be derived from the record section.

Figs. 4-6 show three models according to these assumptions. The lower crust in these models is a constant-gradient layer, extending from a depth of 11 km to a depth of 27 km. The P velocity at the top is 6.11 km/s and 6.84 km/s at the bottom. Such a layer satisfactorily explains the disappearance of the Moho reflection beyond 180 km. The arrival after the Moho reflection at distances greater than 150 km is the reflection of the refracted wave in the gradient layer at the lower boundary of the low-velocity channel. For a less sharp boundary, the amplitudes of this arrival would be reduced.

The P_n phase of model M1, as visible in Fig. 4, is a superposition of two arrivals, the head wave from the layer below the Moho with a velocity of 7.90 km/s and the reflection from the interface at a depth of 35 km between this layer and the half-space with a velocity of 8.15 km/s. The reflection has much larger amplitudes than the head wave. From the observed record section, it is clearly seen that these amplitudes must be reduced. This can only be done by decreasing the velocity contrast at the interface at 35 km depth. Since we cannot reduce the half-space velocity of 8.15 km/s, we must increase the velocity of the layer below the Moho.

Fig. 5 shows the results for an increase of this layer velocity from 7.90 to 8.03 km/s. The P_n amplitudes are considerably reduced, compared with the P_n amplitudes of model M1. In addition, they have a minimum at a distance of about 130 km. If we investigate the ratio of the amplitudes of the P_n wave and the Moho reflection (the quantity which actually can be derived from the observed

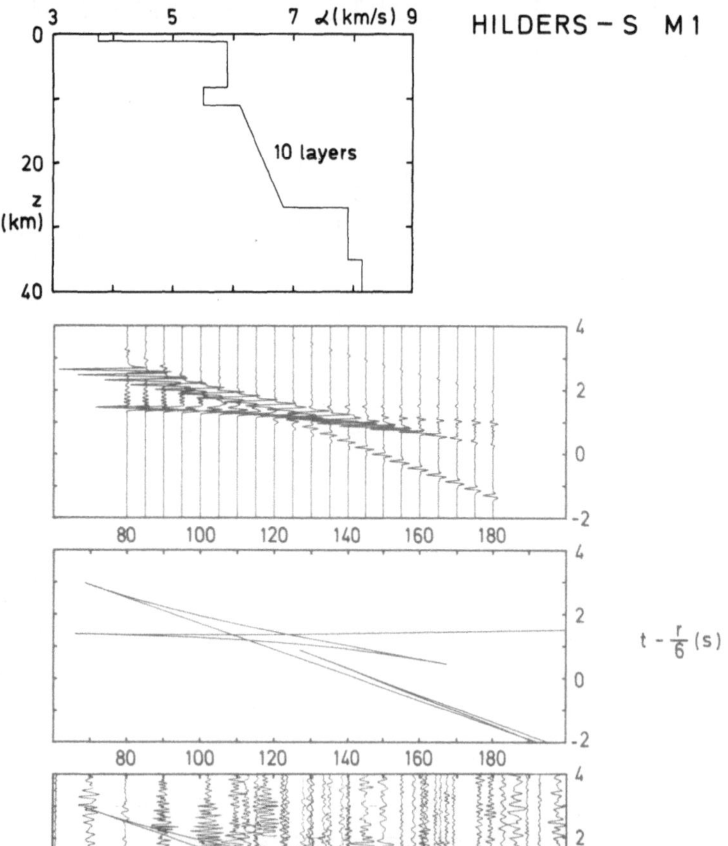

Fig. 4. The same as Fig. 3 for model M1

record section), we find very small values between 120 and 140 km and a value of 1 at distances around 180 km. This is in good agreement with the observations.

If instead of the first-order discontinuity at the depth of 35 km a transition layer is introduced, the amplitude minimum at the distance of 130 km can be made more pronounced, as is seen from Fig. 6. Here, we have replaced the interface by a transition of 2 km thickness. However, we cannot definitely say that model M3 is better than model M2, since the uncertainties in the observations are too large.

The structure below the Moho in model M3 (or model M2) is not the only one which can produce an amplitude minimum for P_n at 130 km distance. Actually, constant-gradient layers show the same effect (Fig. 7). For model M4, M5, and M6, the velocity increases linearly from 8.0, 7.9, and 7.8 km/s, respectively, to 8.2 km/s at a depth of 40 km. The P_n amplitude-distance curves were derived from synthetic seismograms by taking peak-to-peak amplitudes. Model M4 can be excluded as a possibility since its amplitude minimum is not pronounced enough. M6 has too large amplitudes and does not satisfactorily reproduce the travel times. Model M5, however, is indeed an alternative to M3. We prefer model M3 only because of the low value at the minimum of the amplitude-distance curve. These results suggest that a velocity of at least 7.90 km/s and more probably between 8.00 and 8.05 km/s must be assumed for the depth range immediately below the Moho.

Until now, in all models the Moho was a first-order discontinuity. We also performed computations for transition layers with a thickness of 2 and 4 km, respectively. A decision between these models (Fig. 8) can be made by comparing the amplitude-distance curves of their Moho reflections around the critical points. The amplitude maximum is raised and sharp-

Fig. 5. The same as Fig. 3 for model M2

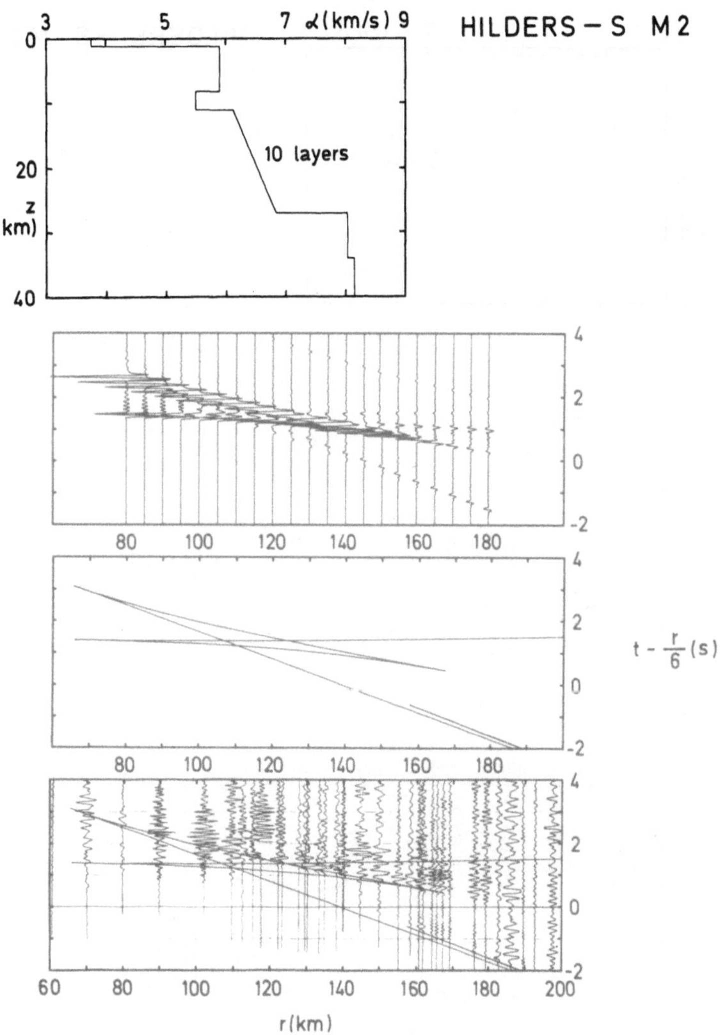

ened by increasing the thickness of the Moho zone. In addition, the subcritical reflections are strongly reduced in amplitude. Although we cannot compare these curves with an experimental amplitude-distance curve because we do not know the magnifications of all the traces, some very raw estimates of the amplitude decay to the left of the maximum do exclude thicknesses greater than 2 km. Therefore, we propose a transition at a depth of about 27 km with a thickness of 1 to 2 km and with an increase in velocity from about 6.8 to about 8.0 km/s for the Moho zone under the profile Hilders S.

Summarizing, our results for the lower crust and the transition to the upper mantle are as follows. Between the low-velocity layer and the Moho, there is a layer with constant velocity gradient and an increase in velocity from 6.1 to about 6.8 km/s. The Moho is a transition layer whose thickness is less than 2 km. Its depth is about 27 km. The velocity increases from 6.8 to 8.00-8.05 km/s. This relatively sharp transition makes a phase change rather unlikely. Just below the Moho, there is a homogeneous layer with a thickness of about 7 km, followed by a further increase in velocity to 8.15-8.20 km/s.

Looking at these results somewhat critically, we must say that the thick constant-gradient layer between the low-velocity channel and the Moho produces a refracted wave whose amplitudes are larger than the observed ones. This indicates a smaller gradient just below the channel than is assumed in the foregoing results. Then, the velocity-depth function between the channel and the Moho must necessarily consist of three portions whose gradients are smaller, greater, and again smaller than the original one. An example for such a model is shown in Fig. 9. However, in our opinion it is difficult to make a rational choice between this model of the lower crust and the constant gradient layer of model M3.

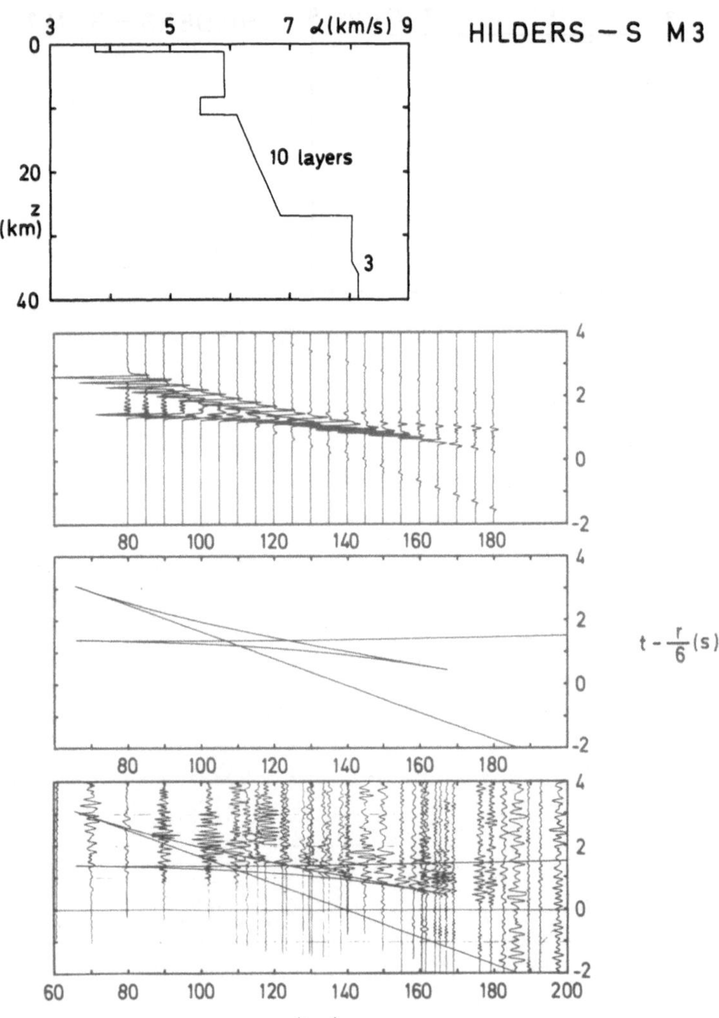

Fig. 6. The same as Fig. 3 for model M3

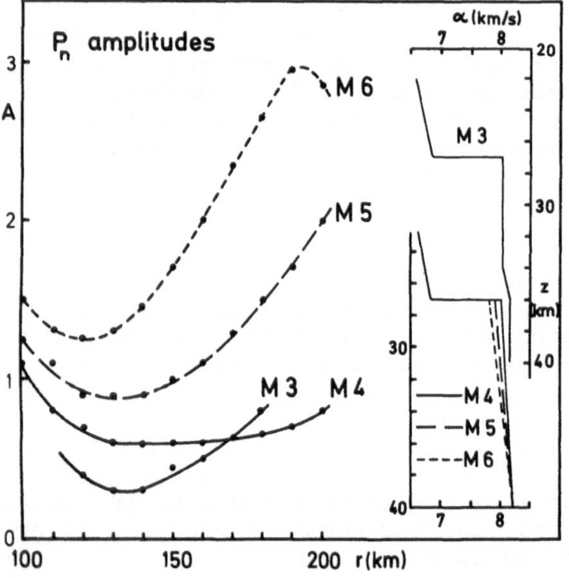

Fig. 7. Amplitude-distance curves for the P_n wave of models M3, M4, M5 and M6. The P velocity varies with depth as shown by the inserts

Fig. 8. Amplitude-distance curves for the Moho reflection around the critical points of three different models. The first-order discontinuity model corresponds to the crustal model M3. In the two other models, this first-order discontinuity is replaced by transition layers of thickness 2 km and 4 km, respectively. The amplitude units are different from those in Fig. 7

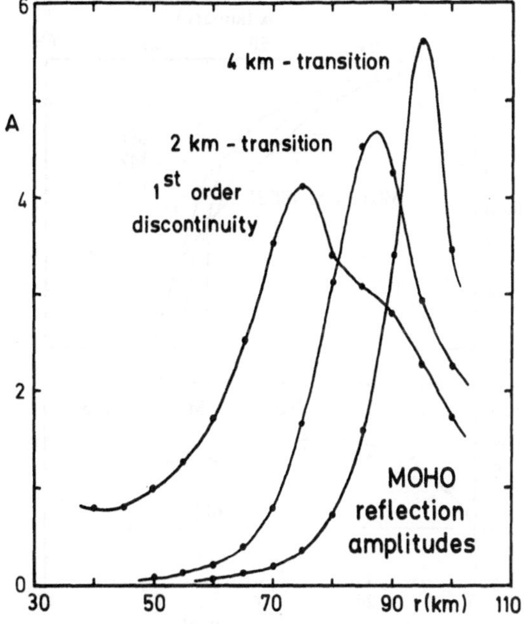

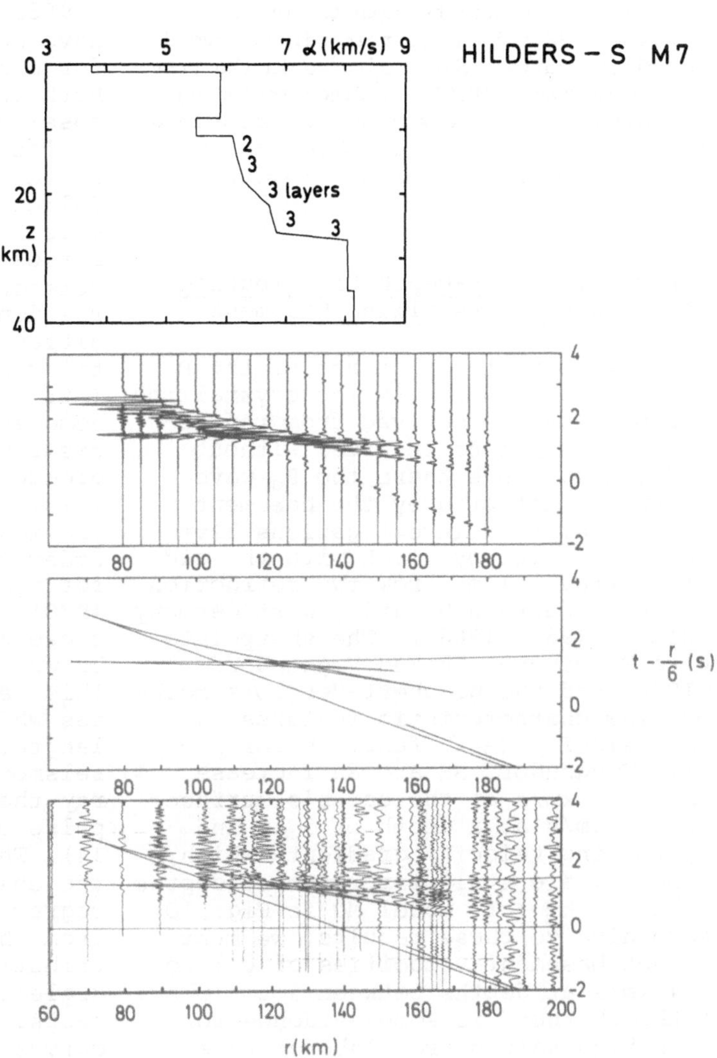

Fig. 9. The same as Fig. 3 for model M7

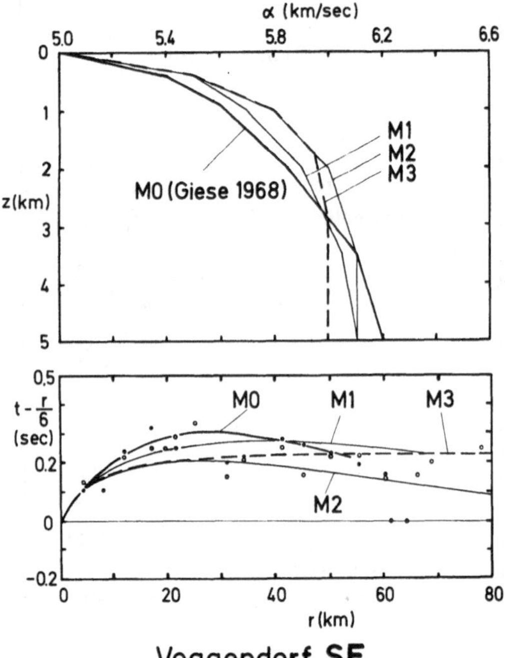

Fig. 10. P velocity-depth functions for the crystalline basement beneath the profile Voggendorf SE and the corresponding reduced traveltime curves. *Solid circles* in the traveltime diagram: GIESE's readings including topographic corrections which are not greater than 0.02 s (GIESE, 1968a). *Open circles:* our own readings from the record section (Fig. 11)

4.6.3 P Velocity-Depth Distribution of the Upper Crystalline Basement

Details about the increase of the P velocity with depth in the crystalline basement can be derived from refraction observations in areas without sediments, since there the P_g wave is solely affected by the basement structure. Such investigations have been performed by GIESE with the aid of P_g traveltimes from two refraction lines in eastern Bavaria, West Germany (GIESE, 1963, 1968a). The shotpoints were Böhmischbruck and Voggendorf; GIESE used the Wiechert-Herglotz method. The characteristic features of the velocity-depth function for profile Voggendorf SE are an increase from 5.0 km/s at the earth's surface to 6.2 km/s at a depth of 5 km and decreasing velocity gradients which, however, are still important at depths around 5 km (cf. model MO in Fig. 10). According to these results, we must expect basement velocities of 6.3 to 6.4 km/s at depths between 5 to 10 km. Although there is some evidence for such high values from laboratory experiments on granites under high pressures and temperatures (BIRCH, 1960; PRESS, 1966), most interpretations for refraction profiles in central Europe gave values between 5.8 and 6.0 km/s (German Research Group for Explosion Seismology, 1964; PRODEHL, 1964; FUCHS and LANDISMAN, 1966a, b; LANDISMAN and MUELLER, 1966; ANSORGE et al., 1970).

Looking at the record section of profile Voggendorf SE (Fig. 11) which has been prepared by WOLBER (1968), we would prefer a velocity of 6.0 km/s rather than 6.2 km/s and greater for the deeper parts of the basement. However, the scatter of the P_g arrival times in the lower part of Fig. 11 is so large that actually one cannot make a convincing choice between these two bounds. In this situation, we have tried to restrict the range of possible models by considering as an additional constraint the amplitude-distance curve of P_g which was derived from the observations by GIESE (1968a) and is shown in Fig. 12. We have searched, by trial and error, for a model which is compatible with both the travel-time and amplitude observarions.

The theoretical amplitude-distance curves in Fig. 12 were determined as follows. From the P velocities of the model under consideration, the S velocities were derived by assuming that Poisson's ratio is everywhere 0.25. The densities follow from the P velocities with the aid of the experimental relationship by Nafe and Drake (cf. TALWANI et al., 1959). This inhomogeneous velocity-density-depth distribution was approximated by a piecewise homogeneous distribution (Fig. 13). On the one hand, this approximation must not be too close in order to avoid too long computing times for synthetic seismograms (MÜLLER, 1970); on the other hand it must be close enough to reflect differences between different inhomogeneous models. This led us to choose layer thicknesses which are about half the wavelength. Then, theoretical P wave seismograms were computed with the ray-theoretical method for a source pulse of main frequency 10 Hz (Fig. 14). This frequency is derived from the observed P_g wavegroups. The seismograms include only the contributions from the primary reflections. The contribution from the multiples can be neglected as was shown by numerical tests. Theoretical amplitude-distance curves were then derived from the

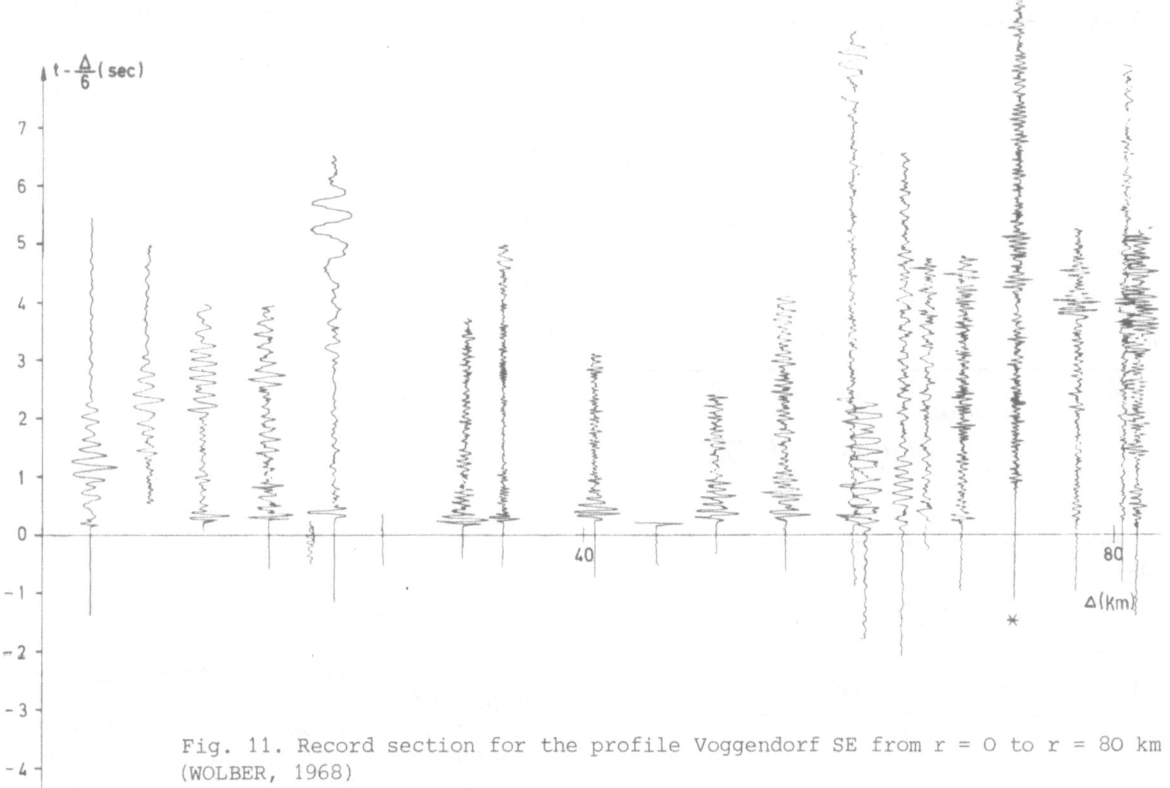

Fig. 11. Record section for the profile Voggendorf SE from r = 0 to r = 80 km (WOLBER, 1968)

synthetic record sections by taking peak-to-peak amplitudes. These quantities correspond to Giese's experimental amplitudes.

The amplitude-distance curve for model M0 in Fig. 12 deviates from the observed curve by a factor of about 2 in the distance range beyond r = 20 km. This is more than is allowed by the scatter of the observed data. In addition to this model, we have investigated seven similar models whose velocity-depth functions have different curvatures and reach velocities of 6.0 and 6.1 km/s, respectively, at 5 km depth. Three examples are given in Figs. 10, 12 and 13. Model M3 is the only one which satisfies both the travel-time and amplitude observations. Therefore, we consider M3 to be the best model for the upper crystalline basement beneath the profile Voggendorf SE.

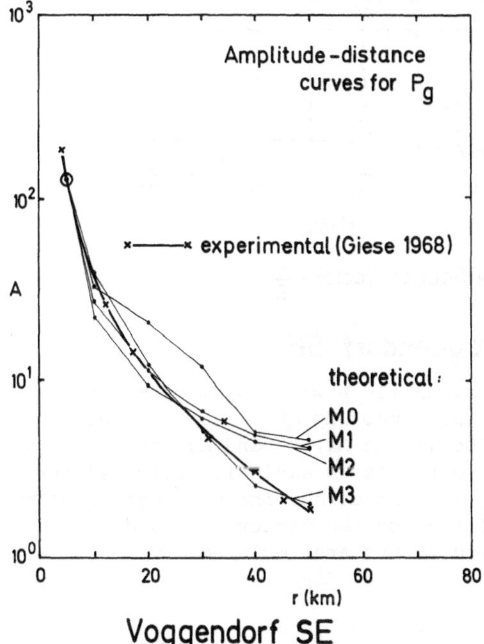

Fig. 12. Amplitude-distance curves for the vertical component of motion due to the P_g wave traveling through the basement. GIESE's observed curve and the theoretical curves have been normalized at r = 5 km, since only their forms can be compared. The theoretical curves correspond to the models shown in Figs. 10 and 13 and were taken from the seismograms in Fig. 14

4.6.4 Conclusions

We arrive at the following conclusions:

1. The reflectivity method and the ray-theoretical method can be used effectively in the quantitative interpretation of explosion seismological observations. In some cases, a com-

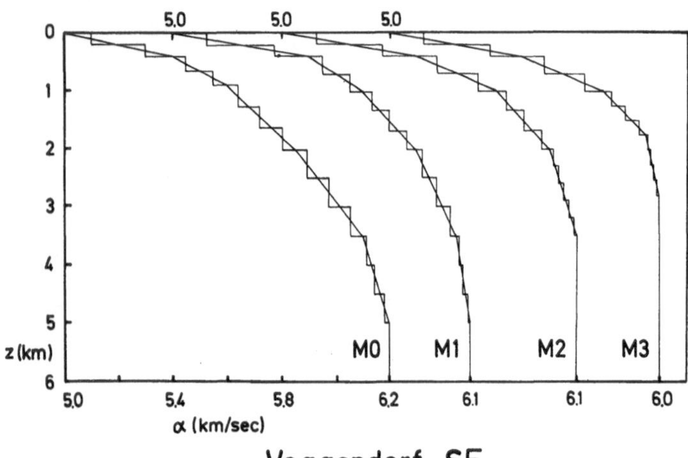

Fig. 13. The P velocity-depth functions of the models M0, M1, M2, and M3 and their discontinuous approximation for the computation of synthetic seismograms

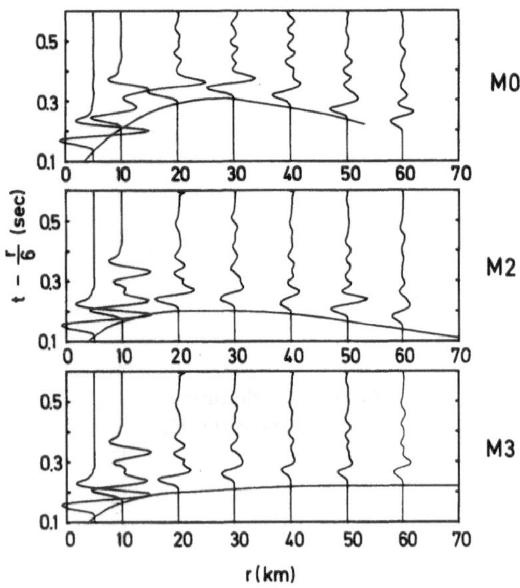

Amplification factor = $\frac{r}{5}$

Voggendorf SE

Fig. 14. Synthetic P wave seismograms for the vertical component at the earth's surface of the models M0, M2 and M3 (see Fig. 13). The source is an explosive point source radiating a 10 Hz pulse. The seismograms have been amplified by the factor r/5, and the travel-time curves are included

bined use will be the optimum inversion procedure.

2. It is an urgent need that amplitudes can be compared from trace to trace and not only within each trace. Therefore, instrumental characteristics and actual magnification values must be known in future measurements.

3. Prior to a comparison with synthetic seismograms, observed records should be passed through low-pass filters in order to reduce noise from small-scale inhomogeneities.

Acknowledgments. We are grateful to Jörg Ansorge, Werner Kaminski, and Stephan Müller for stimulating and helpful discussions. Our research was sponsored by the Deutsche Forschungsgemeinschaft (German Research Association). Computing facilities were made available by the Deutsches Rechenzentrum (German Computing Center) at Darmstadt, by the Gesellschaft für Kernforschung (Nuclear Research Center) at Karlsruhe, and by the university computing center at Karlsruhe.

4.7 Some Notes on Theoretical Travel Time and Amplitude Curves

U. Vetter and R. Meissner

ABSTRACT

Theoretical travel-time and amplitude curves based on geometrical ray optics are compared with field observations. Velocity-depth models can be improved by an iteration process if discrepancies between theoretical and observed curves are recognizable.

ZUSAMMENFASSUNG

Theoretische Laufzeit- und Amplitudenkurven, die auf der geometrischen Strahlenoptik beruhen, werden zu einem Vergleich mit seismischen Meßergebnissen benutzt. Geschwindigkeits-Tiefen-Modelle könnten durch einen Iterationsprozeß verbessert werden, falls die Diskrepanzen zwischen theoretischen und gemessenen Kurven erkennbar sind.

4.7.1 Introduction

Theoretical travel-time curves can be a great help for the interpretation of seismic data as is widely known from earthquake seismology using the Jeffreys-Bullen curves or from exploration seismics using synthetic seismograms. In either case theoretical travel-time curves from models are established and compared with the actually observed curves. In a similar way, theoretical amplitude curves of models can be calculated and compared with actually observed amplitude curves. This task is important for the calculation of the reflectivity of seismic boundaries, and in particular for the explanation of the reflected or the penetrating (diving) P^M wave from the Mohorovičić discontinuity (MD). This wave produces primarily very strong arrivals in seismograms of deep seismic soundings (DSS). In this paper we do not attempt to compare the observed amplitude curves in detail with theoretical curves because observed values are strongly influenced by local inhomogeneities near the surface or different sources of seismic energy. We will relate, however, the general appearance of the P^M wave in the seismograms to the theoretical amplitude curves.

4.7.2 Equations Used for Model Calculation

The basic equations for travel-time data are derived from ray optics (see, e.g. GRANT and WEST, 1965). We used these equations:

$$t = \sum_{\mu=o}^{n} t_\mu + \sum_{\nu=o}^{N} t_\nu \text{ and } x = \sum_{\mu=o}^{n} x_\mu + \sum_{\nu=o}^{N} x_\nu \qquad (1)$$

t = total travel time
x = total distance and
t_μ = travel time in a homogeneous, isotropic layer

$$t_\mu = \frac{2\,\Delta z_\mu}{V_\mu \cos i_\mu};\quad x_\mu = 2\,\Delta z_\mu \operatorname{tg} i_\mu \qquad (2)$$

$$\frac{\sin i_\mu}{\sin i_{\mu+1}} = \frac{V_\mu}{V_{\mu+1}}$$

Δz_μ = thickness of the μth layer
t_ν = traveltime in an isotropic layer with a constant velocity gradient with depth:

$$t_\nu = \frac{2}{a_\nu} \ln \frac{\operatorname{tg}\frac{i_\nu}{2}}{\operatorname{tg}\frac{i_{\nu a}}{2}};\quad x_\nu = \frac{2\,V_A}{a_\nu} \qquad (3)$$

$$(\cos i_{\nu a} - \cos i_\nu)$$

$$t_\nu\,(0^o) = \frac{2}{a_\nu}\ln\left(1 + \frac{a_\nu \cdot \Delta Z_\nu}{V_{\nu a}}\right);\quad x_\nu\,(0^o) = 0$$

i_ν = ray angle in layer ν; $i_\nu = \sin^{-1}\frac{V_\nu}{V_A}$

$i_{\nu a}$ = ray angle on top of layer ν;

$$i_{\nu a} = \sin^{-1} \frac{V_{\nu,a}}{V_A}$$

$a_\nu = \frac{\Delta V_\nu}{\Delta z}$ = velocity gradient in layer ν

V_A = apparent velocity $= \frac{V_\nu}{\sin i_\nu}$

$= \frac{V_\mu}{\sin i_\mu}$

Eqs. (1) to (3) are valid if no lateral velocity variation is present i.e. for V = V(z). Lateral changes can also be incorporated but this was not felt to be justified at the present stage of available seismic observations.

For the theoretical amplitudes an equation from GUTENBERG (1944) has been used which is also based on ray theory:

$$A\ (x,i_o) \sim \sqrt{\frac{\operatorname{tg} i_o}{x} \cdot \frac{\partial i_o}{\partial x}} \cdot \pi(R,D) \cdot f(i_o) \qquad (4)$$

with i_o = ray angle at the surface
$\pi(R,D)$ = product of all reflection and refraction coefficients R and D along the ray path
$f(i_o)$ = influence of the free surface.

As $f(i_o)$ for P waves can be approximated by $f(i_o) \sim \cos i_o$ for the Z-component, and as $\pi(R,D)$ is about 1 for the product of the reflection and refraction coefficients of the downgoing and upgoing ray, we use Eq. (4) in the form

$$A(x,i_o) \sim \sqrt{\frac{\sin i_o \cos i_o}{x} \cdot \frac{\partial i_o}{\partial x}} \cdot R_M \qquad (5)$$

$R_M \equiv$ rel. Amplitude

R_M = Refl. coefficient of the M-disc.

$R_M \approx$ 1 for overcritical angles and for diving waves.

For practical calculations of models $\frac{\partial i_o}{\partial x}$ has been replaced in Eq. (5) by $\frac{\Delta i_o}{\Delta x}$ and i_o was varied by 1, 2 or 5°. The succesful use of Eq. (5) and similar equations based on ray theory is limited to a slow to moderate change of elastic moduli over distances comparable to the wavelength λ. For simple harmonic waves we may write this condition in the from:

$$\frac{dV}{dz} << \omega;\ \omega = \frac{2\pi\ V}{\lambda} \quad \text{(e.g. GRANT and WEST, 1965)} \qquad (6)$$

Even if the velocity changes from 6 to 8 km/s in a 2-km-thick transition zone, $\frac{dV}{dz}$ is 1 compared to an ω of about 30 to 60/s for the generally observed frequency content of P^M waves.

Geometrical ray optics also limits the resolution of amplitude curves. If Δi_o is made too small some models result in very strong and small amplitude peaks which have no significance for the actual amplitude propagation or for a comparison with examples from field work. An averaging process or a Δi_o of 3° to 5° was found to be adequate. As much emphasis is placed on the amplitude peaks, absorption of seismic waves was not taken into account in the present study.

4.7.3 Examples from Models Comparable to Velocity-Depth Curves Obtained from DSS

Some simple velocity-depth models which are in general agreement with DSS results will be discussed. For both cases of Figs. 1 and 2 gradient zones at the bottom of the crust and first-order boundaries at the MD depth have been used. The reason for assuming a first-order discontinuity in the models was the fact that very often subcritical reflections with apparent velocities greater than that of the refractor velocity and with sufficient amplitudes have been observed, especially from the deeper crust where the theoretical amplitude curves do not show a strong gradient before the critical angle is reached. In other areas the appearance of near-vertical reflections is an indication for a first-order boundary at the MD. The gradient zone on top of this discontinuity, on the other hand, is also well-established in most areas as derived from a careful analysis of travel-time data (for instance MEISSNER, 1967a; GIESE, 1968a). Fig. 1, showing a crustal thickness of 50 km, represents a young mountain region or a depression with low-velocity material, i.e. low-density material, in the whole crust. This velocity distribution may be found in some parts of the Alps or in the Buchara Depression. In Asia the travel-time diagram shows only little deviations from a hyper-

Fig. 1. Schematic model, theoretical travel-time and amplitude curves of P^M-wave for the crust of young mountain areas and depressions (thickness of crust 50 km); gradient zone and first-order discontinuity at MD

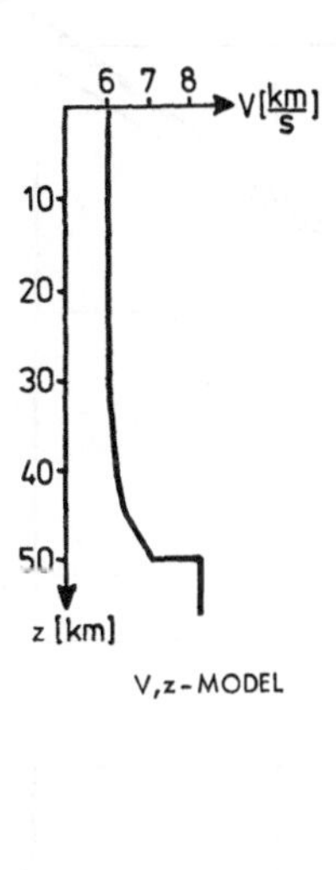

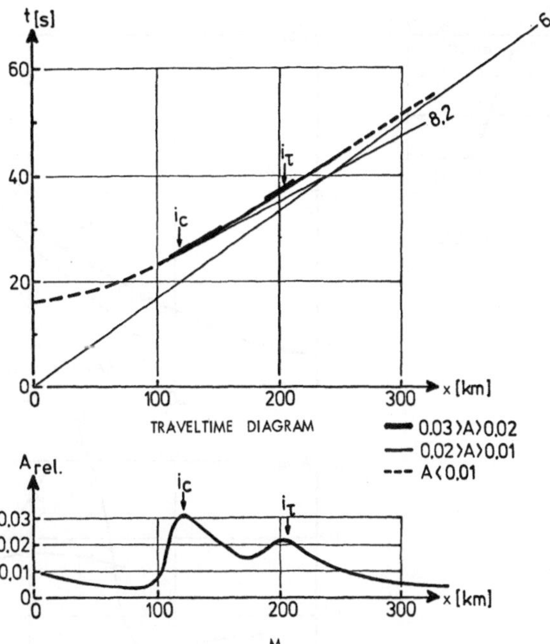

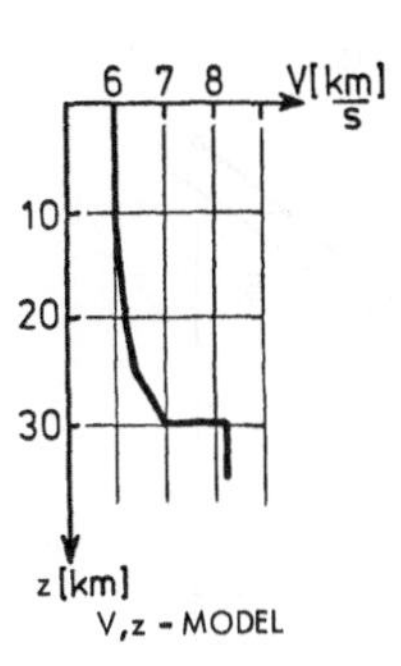

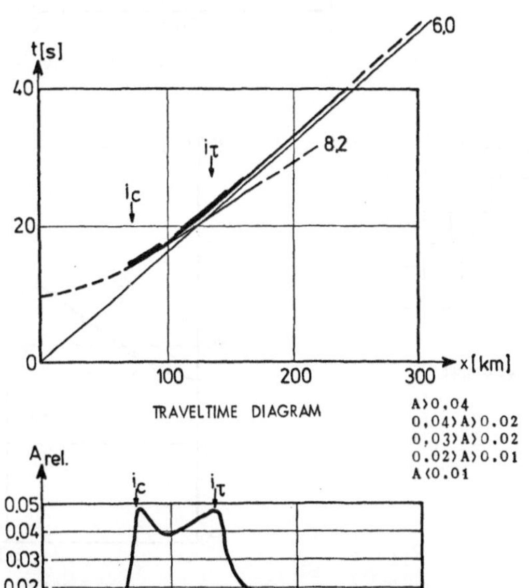

Fig. 2. Schematic model, theoretical travel-time and amplitude curves of P^M-wave for the crust of western Europe (thickness of crust: 30 km); gradient zone and first-order discontinuity at MD

bola. It is interesting to note that the amplitude curve for the P^M wave shows two maxima, one near the critical angle i_c, another in the region of the beginning of the diving wave (i_τ). The peaks are flat and the P^M wave should be observed between about 110 and 280 km under favorable conditions, which agrees roughly with the actual observations. Fig. 2 shows similar velocity gradients for a crustal thickness of 30 km which represents a low-velocity crust similar to that found in large parts of Western Europe. The amplitude curves again show two maxima, the first maximum shortly behind 70 km and the other at about 130 km distance. Both maxima are nearer to each other than for the 50 km crust. The P^M wave can be observed up to about 200 km under favorable conditions.

In addition to models with first-order boundaries some second-order discontinuity models have been investigated, which only show one amplitude maximum (at i_τ). The curves start at i_τ and have two branches for $i > i_\tau$

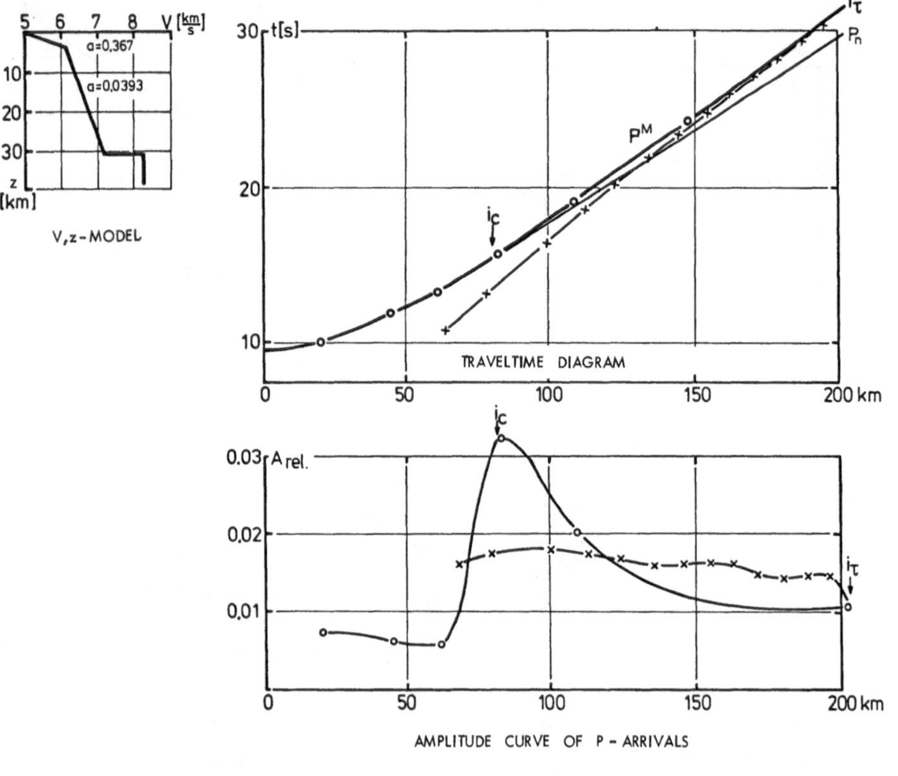

Fig. 3. Schematic model, theoretical travel-time and amplitude curves of P^M-wave for the crust as found in northern Europe; first-order discontinuities assumed

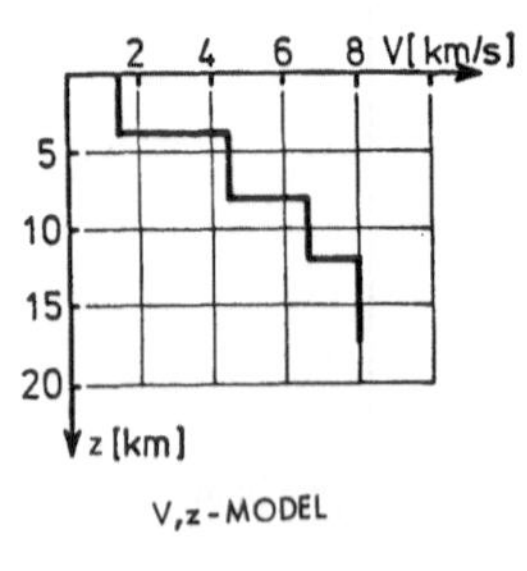

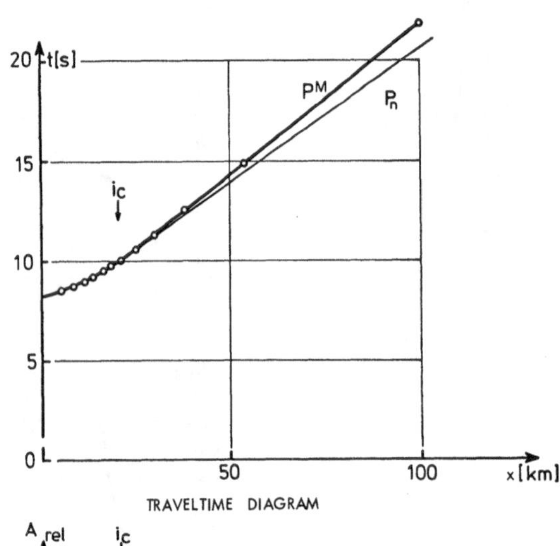

AMPLITUDE CURVE OF P^M- ARRIVALS

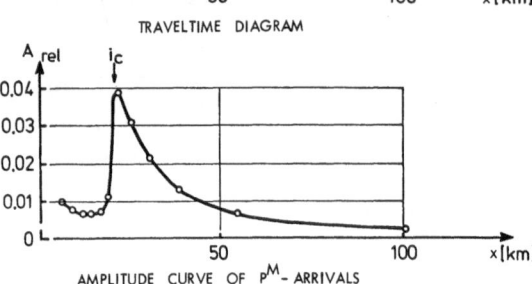

Fig. 4. Schematic model, theoretical travel-time and amplitude curves of P^M-wave for oceanic crust (thickness of crust: 12 km); first-order discontinuities assumed

and no branch for $i < i_\tau$ (MEISSNER, 1967b).

Crustal models with several first-order discontinuities have been investigated. Fig. 3 shows a model which was derived from DSS data in Northern Europe. The amplitude curve of the MD shows the maximum shortly before 100 km and stays at a rather high level up to a distance of 150 km. In the actual record section the observed P^M wave was very poor; only some P^M arrivals have been found at a distance between 100 and 160 km.

Figs. 4 and 5 show models of a crust with 12 km thickness as derived

Fig. 5. Schematic model, theoretical travel-time and amplitude curves of P^M-wave for oceanic crust (thickness of crust: 12 km); gradient zone and first-order discontinuities

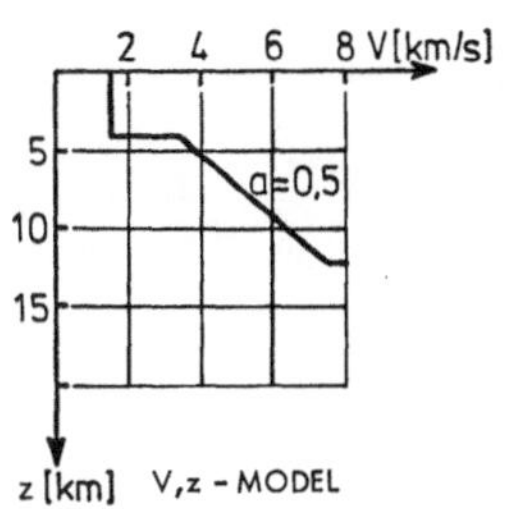

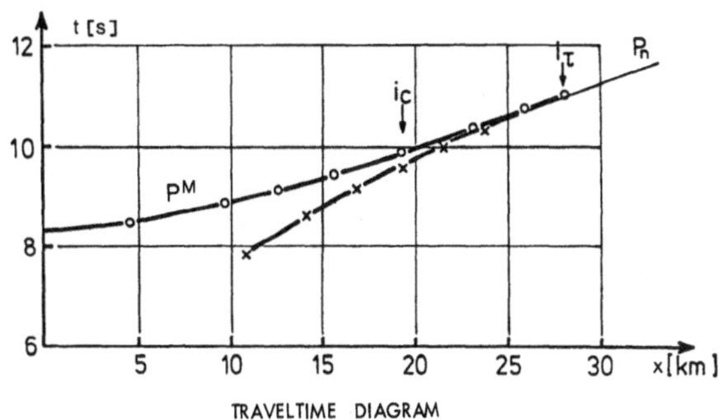

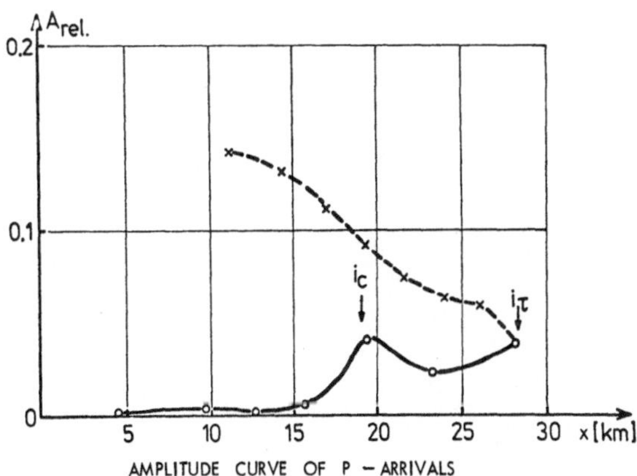

from marine seismic work over deep ocean floors. Both models, Fig. 4 with first-order discontinuities and Fig. 5 with a gradient zone show a pronounced maximum of the P^M wave amplitude in a rather limited distance range between 17 and 35 km. For some similar models the amplitude peak is still sharper and more limited so that it may occasionally be overlooked. Recent observations from MAYNARD and SUTTON (pers. comm.) and EWING and HOUTZ (1969) showed reliable P^M arrivals from the deep ocean floor between 18 and 33 km distance.

4.7.4 Practical Use of Model Curves

In the interpretation procedure first preliminary velocity-depth models are derived using GIESE's approach (GIESE, 1968a). Based on these velocity-depth models theoretical travel-time curves are calculated and checked with the observed travel-time curves. Models have to be revised, if differences between observed and theoretical curves exceed a certain time difference. This iteration process is important in order to enhance the reliability of velocity-depth functions. The theoretical amplitude curves, on the other hand, provide a clue as to where to look for large amplitudes in the seismogram and may be used to adjust the station distance in order to get the best possible information on reflected or diving waves from the MD or other interfaces in the crust. All observed amplitude curves obtained so far show a maximum of P^M arrivals before the critical angle and before the theoretical maximum. An explanation for this phenomenon may be a small scattering (diffraction), a deviation from ray optics, or an averaging process due to slightly different dips of lamellas in the gradient zone between crust and mantle. Also absorption, not taken into account in the present study, tends to shift amplitude maxima to shorter distances.

5. Main Features of Crustal Structure in Western Germany

P. Giese

As outlined in the previous chapter, the inversion of travel-time data into depth values encounters difficulties for different reasons. Therefore, methods of data evaluation are developed which allow a presentation of the main features of crustal structure without assuming detailed models.

In 5.1, the principal crustal structure is discussed as revealed from detailed seismic-reflection and refraction studies. Although all record sections differ from each other, some common features of the pattern of travel-time curves can be recognized. The main data of crustal structure can be derived from three wave groups, the P_g-, P^M-, and P_n-waves.

In 5.2 a number of different parameters is contoured, showing the lateral variations of crustal structure. Also the problem how to define the crust/mantle boundary, the well-known Mohorovičić discontinuity, is discussed.

A special method of a generalized interpretation of seismic-refraction data is presented by BAMFORD in 5.3. The main result of his time-term analysis is the statement that there exists a distinct anisotropy of upper-mantle velocity of about 0.5 km/s.

Finally in 5.4, a comparison between the main features of crustal and geological structures is attempted. As base for a tectonic division of the area under study, the different zones of the Hercynian orogene are used.

The problem of the primary thickness of the central zone of this orogene is discussed. Later, some peculiarities of the Rhenish Massif are studied in respect to the young volcanism. Finally, some considerations on the velocity distribution in the Rhine Graben area are given.

5.1 Models of Crustal Structure and Main Wave Groups

P. GIESE

ABSTRACT

From the seismic crustal studies carried out in the past 25 years it can be stated that the classic two-layer model is too simple and must be regarded as a rough approximation Seismic refraction and reflection measurements show that the earth's crust is composed of layers including numerous low-velocity layers. In horizontal direction, there are rapid changes of the elastic parameters. Thus, the picture of lamellae or sandwich-like structure has been proposed.

For a generalized description of the material, the three main wave groups P_g, P^M, and P_n have been used.

ZUSAMMENFASSUNG

Die krustenseismischen Untersuchungen der letzten 25 Jahre haben ergeben, daß das klassische Zwei-Schichten Modell nur als sehr grobe Annäherung betrachtet werden darf. Refraktions- und Reflexionsmessungen haben gezeigt, daß die Erdkruste aus einer Folge von Schichten aufgebaut ist, in der auch zahlreiche Bereiche mit geringer Geschwindigkeit auftreten. In lateraler Richtung ändern sich die Verhältnisse sehr rasch, so daß das Bild einer lamellen- oder sandwichartigen Struktur entsteht. Die Diskontinuitäten sind durch mehr oder minder breite Übergangszonen zu ersetzen.

Für eine generalisierende Beschreibung des Datenmaterials werden die drei Hauptgruppen P_g, P^M und P_n benutzt.

5.1.1 Introduction

When large-scale seismic refraction measurements for crustal studies were started 25 years ago, the already existing two-layer model, developed by seismologists from the study of near-earthquakes, was used as a basis for data evaluation. This model is based on the assumption that, in each of the two layers, the velocity is constant. At the Conrad discontinuity, separating these two layers, the velocities increase from 6.0 to 6.8 km/s. The well-known Mohorovičić discontinuity, with the typical velocity increase from 6.8 km/s to 8.2 km/s separates crust and mantle.

Detailed crustal investigations, based on refraction and reflection seismics, carried out in many countries during the past two decades, revealed that this simple model can no longer satisfactorily fit the data observed. GUTENBERG (1955) already suggested introducing a crustal low-velocity layer in order to interpret some phases hitherto not explained. This concept of crustal low-velocity layers - more or less neglected for a decade - was again proposed by several authors in the middle of the sixties (GIESE, 1966; MEISSNER, 1966; MUELLER and LANDISMAN, 1966; PAVLENKOVA, 1969).

Furthermore, the constant-velocity layers and the first-order discontinuities must be regarded as a first approximation of crustal velocity distribution. Evidenced by the results of velocity measurements on rock samples under pressure and temperature of crustal conditions, step-wise velocity functions were replaced by continuous ones.

The fine structure of the crust and its discontinuities and transition zones could be revealed by the study of near-vertical reflections (LIEBSCHER, 1964; MEISSNER, 1966; FUCHS, 1970).

The recent picture of crustal structure and composition is also strongly influenced by the results of modern petrography. From outcrops in shield areas, composition and structure of the crust can be deduced down to a depth of about 30 km. In special and rare cases, e.g. in the zone of Ivrea - situated in the inner arc of the Western Alps - even deeper-seated rocks are exposed today, showing what the lower crust looks like and how the transition from sialic to basic and ultrabasic rocks takes place (GIESE, 1968b).

The "stratification" of the crust must be seen in another way than that of the sedimentary cover. As demonstrated, e.g. by LIEBSCHER (1962, 1964), the crustal reflections cannot be traced over long distances but, in general, only over some hundreds of meters or, in rare cases, over a few kilometers. The rapid variations of the reflectivity in lateral direction are one of the main characteristic features of the crust. Thus, the picture of lamellae or sandwiches has been proposed in order to describe the fine structure of the crust (MEISSNER, 1966). KOSMINSKAYA and RIZNICHENKO (1964) already suggested the term "grainy-blocked" model, meaning that the crust is composed of large blocks as well as of small inhomogeneities.

The smallest inhomogeneity is the mineral grain, having on an average the size of a few millimeters. On the other side, there exist large bodies of magmatic rocks intruded into the overburden, showing, on an average, widely varying sizes. The largest crustal units, having a size of some 10 km, may be crustal blocks separated by fault zones. Within this wide spectrum (1 mm to 10 km), the wave lengths of seismic pulses are situated (100 m to 1 km). The resolving power of the deep-seismic sounding method is restricted by the shortest wave length recorded, that means the smallest detectable inhomogeneity is of 0.1 to 1 km in size.

Several authors have investigated the internal structure of discontinuities, especially that of the Mohorovičić discontinuity. Their results have been compiled e.g. by DAVYDOVA et al. (1972). In order to explain the complicated pattern of reflected and refracted waves, these authors propose a transition zone of variable width, composed of thin layers with high and low velocities, each having a thickness of about 0.1 to 1 km. In principle, however, there is no gap with respect to the thickness of crustal lower-velocity layers in the range between 10 km and 1 km (GIESE, 1972). Fig. 1 shows some examples of crustal velocity distribution proposed by different authors.

Summarizing the results, it can be stated that the principle of crustal velocity distribution, including the nature of discontinuities, is known with respect to the possible resolving power. On the other hand, it must be admitted that, in consequence of these results, it is practically impossible to obtain all travel-time data which would be necessary to determine completely the velocity field in the area under investigation. Therefore, in general, a simplified crustal model must be displayed and thus the question arises how to generalize the real structure without neglecting the main features of crustal velocity distribution.

The problem of simplification and generalization of deep-seismic sounding data is strongly connected with the problem how to correlate events in the record sections. The second part of this section deals with this problem.

5.1.2 Main Crustal Wave Groups

As mentioned in the introduction, field techniques used may differ with regard to the density of observations and the frequencies generated by the explosions, but a comparison of the wave pattern of both methods has shown that there are no principal differences in the main character of the waves observed. Evidently, in continuous profiling the resolving power is higher and the number of detectable intercrustal wave groups is greater than in point-wise technique. Any generalization must be based on features of wave fields that are present and characteristic for each system of observation. It is quite clear that this principle means a certain simplification of the real crustal structure.

In continental regions, three dominant wave groups are observed in general (Fig. 2) (KOSMINSKAYA and RIZNICHENKO, 1964; MEISSNER, 1967a; GIESE, 1968a; CHOUDHURY et al., 1971; GIESE and STEIN, 1971; PAVLENKOVA, 1973a).

1. P_g-wave ($P_o{}^k$-wave in Russian literature)
2. P^M- or P^MP-wave ($P^M{}_{refl}$-wave in Russian literature)
3. P_n-wave (P^M-wave in Russian literature)

5.1.3 The P_g-Wave (Upper Crust)

In general, the P_g-wave is associated with the crystalline basement. In extension of this meaning, the P_g-wave is considered here as a wave traveling in the upper crust including a

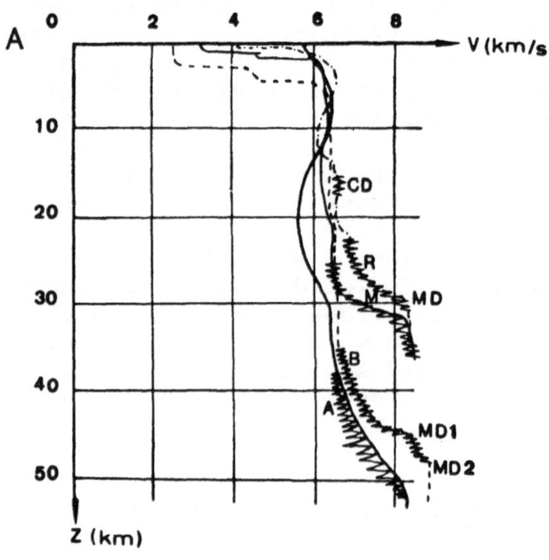
A
V (km/s)
Z (km)
CD
R
M
MD
B
A
MD1
MD2

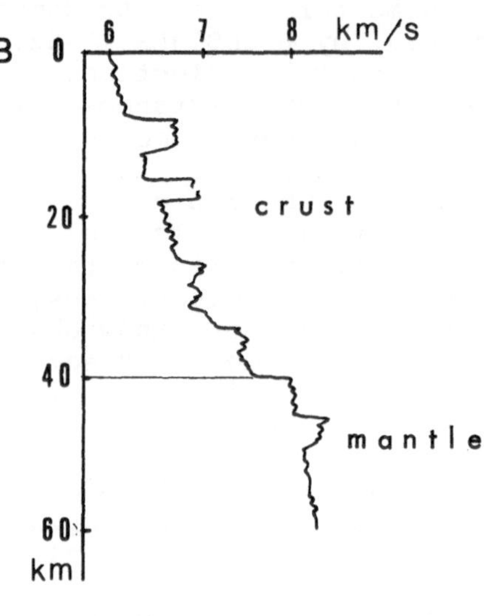
B
km/s
crust
mantle
km

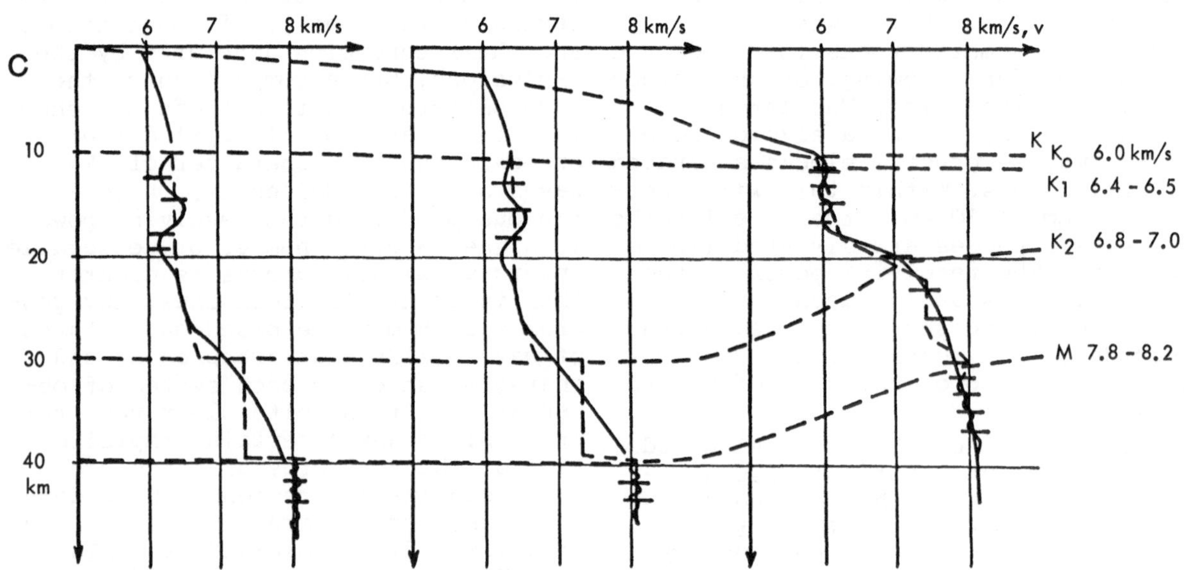
C
8 km/s
8 km/s, v
km
K
K_o 6.0 km/s
K_1 6.4 - 6.5
K_2 6.8 - 7.0
M 7.8 - 8.2

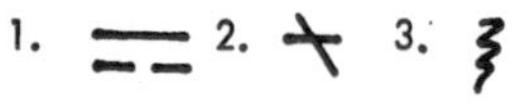
1.
2.
3.

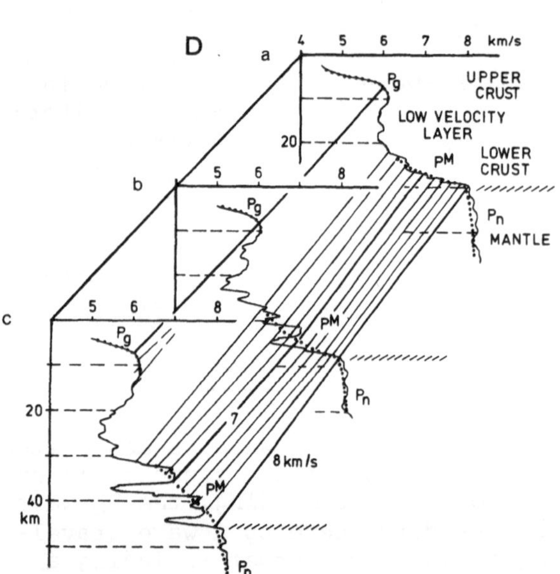
D
a
b
c
km/s
UPPER CRUST
LOW VELOCITY LAYER
LOWER CRUST
MANTLE
P_g
pM
P_n
8 km/s
km

◄ Fig. 1A-D. Examples of crustal velocity distributions proposed by different authors. (A) Distribution given by MEISSNER (1973) showing velocity-depth curves for different areas: *B* Buchara, *M* Molasse, *A* Alps, *R* Rhenish Massif, ⩘ Indication of lamella-type layering, *CD* Conrad discontinuity, *MD* Mohorovičić discontinuity. (B) Model of crustal velocity distribution from measurements on the territory of the USSR (KOSMINSKAYA et al., 1972). (C) Generalized crustal models of *a* shield areas, *b* platforms, *c* deep depressions (PAVLENKOVA, 1975), *1* velocity curves, *2* seismic boundaries, *3* zones characterized by thin layers. (D) Different crustal velocity distributions proposed by the author. Model *a*: a normal crustal velocity distribution for Central Europe with a weak velocity inversion and a well-expressed crust/mantle boundary. Model *b*: the transition type between the consolidated foreland and the Alps. The lower model *c*: for the Alpine type characterized by a thick crust and intensive velocity inversions in the middle as well as in the lower crust

---

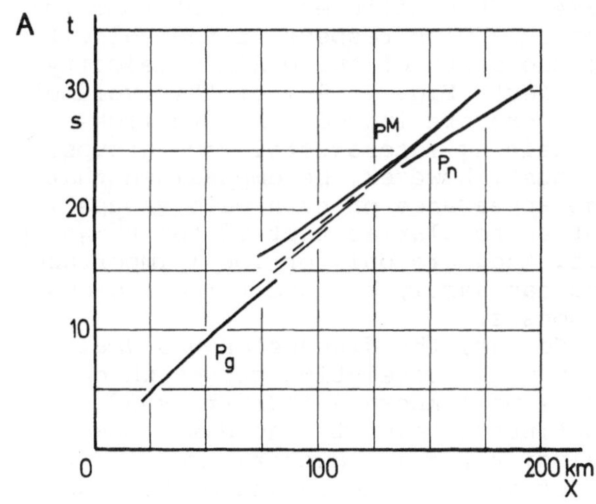

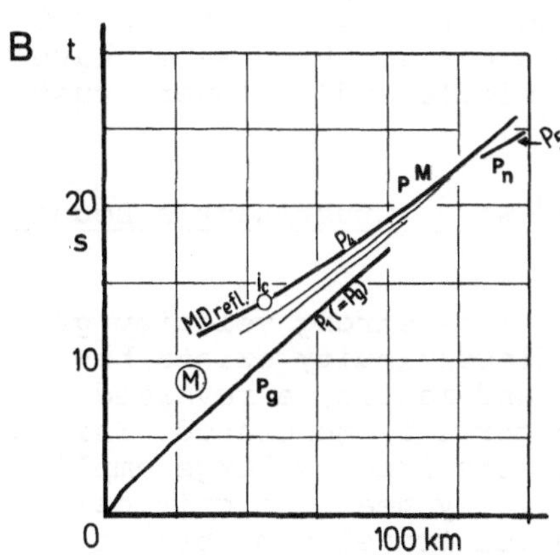

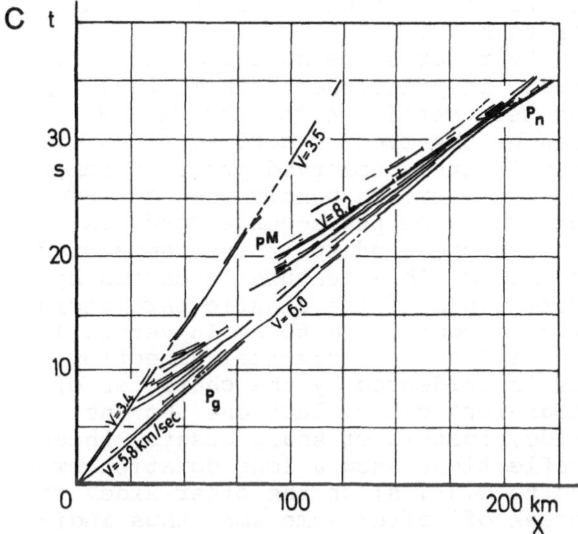

Fig. 2A-C. Travel-time diagrams with the main wave groups P_g, P^M, and P_n, observed in seismic crustal studies. (A) GIESE (1968a); (B) MEISSNER (1973); (C) PAVLENKOVA (1968). The main wave groups are indicated by thick lines. The thin lines show internal crustal wave groups of different nature (subcritical, overcritical, and refracted (head) waves)

possible sedimentary cover. Thus, the P_g-wave may be of complex nature, showing first and second arrivals. By this definition, the P_g-wave is composed of two groups, the sedimentary and the crystalline (basement) one. The nature of the sedimentary wave depends on the structure of and the velocity distribution in this material. But the nature of the pure (basement) P_g-group is complex, too. In continuous as well as in point-wise profiling, a phase-correlated branch can be lined up in the central distance range between 10-30 und 60-100 km. Beyond 100 km distance, the first P_g-phase becomes, in general, very weak and undetectable. There are many examples, especially in continuous profiling, showing that up to the point of intersection with the P_n-wave, more or less shorter travel-time branches occur, each delayed in respect to the preceding one. The amplitudes of a first phase become weaker with distance, and the next phase, delayed in respect to the first one, shows large amplitudes. At greater distances, the later phase becomes the first one, and the picture is repeated. This splitting feature may not be noticed if the spacing of detectors is too large. If regarding only first arrivals, the apparent velocity may increase with distance up to 6.4 km/s. At greater distances (100-130 km), the first arrivals may even show an apparent velocity between 6.4 and 7.0 km/s. In such cases, these waves have reached the middle and/or lower crust.

5.1.4 The P^M-Wave (Crust/Mantle Boundary)

If a more or less strong velocity gradient or a discontinuity exists between crust and mantle, a reversed travel-time curve is generated. This wave is characterized by large amplitudes, especially near the critical point. A convex travel-time branch may belong to reflected waves only or to penetrating waves (refracted) or to both.

As regards the question of correlation, in principle the same features as were mentioned for the P_g-wave can be observed. Record sections of continuously observed profiles demonstrate very frequently that the P^M-branch, when phase-correlated, is interrupted and split into separated segments. This feature is caused by variations of the elastic parameters within some 100 m to km in vertical as well as in horizontal direction, as is evidenced by the character of near-vertical reflections. On one side, instead of sharp onsets, these reflections show a long-duration sweep, up to 0.5-1 s; on the other side, they break off after some kms, thus indicating a sandwich-like structured crust. When representing these shorter and/or longer travel-time branches of the refracted or reflected waves by only one uninterrupted curve, the principle of group correlation is applied.

In the subcritical range, apparent velocity values greater than 8.2 km/s may be observed. The part beyond the critical point, the overcritical range, shows values that decrease at least down to about 7 km/s. But there are also cases showing a decrease of values of apparent velocity even down to the end-value of the P_g-curve, that means down to 6.5 to 6.0 km/s.

When the principle of group correlation is applied to the left part of the P^M-group, too high velocity values may result. Examples from measurements in the Alps and the Rhine Graben area show that the phase-correlated P^M-segments yield a maximum velocity about 0.5 km/s smaller than that of the group-correlated curve (GIESE, 1972; GIESE and PAVLENKOVA, 1974). This difference is of some importance with respect to the velocity at the depth of the maximum velocity gradient Fig. 4 (4.1 of this volume).

There are indeed many profiles containing intercrustal wave groups. It must, however, be emphasized that the appearance of such wave groups is not as regular as that of the P^M-group, i.e. they are only of local importance and can hardly be used for a regional synopsis.

So far, the discussion was dealing only with travel-time curves which - as is well-known - indicate sudden or continuous velocity increase. Low-velocity zones, however, are of equal importance. Criteria and methods how to detect velocity inversions are described in 4.3.

5.1.5 P_n-Wave (Upper Mantle)

The P_n-wave penetrates the upper mantle and is recorded as first arrival at larger distances (beyond 130-200 km) showing apparent velocities between 7.7 and 8.5 km/s. Recently, the P_n-wave has been explained to be not a Mintrop-wave (head wave), but a refracted one, penetrating into a medium with a weak positive velocity gradient. It is possible that the zone of the positive gradient in the uppermost mantle is only of some km width, and that the velocity gradient becomes negative in the underlying depth range. Evidently, the P_n-branch disappears more or less completely. In distances greater than 300-400 km, the phase-correlated P_n-branch is replaced by a new one but will be delayed in respect to the previous one, thus showing a behavior similar to the P_g-phase in the crust. This feature is not discussed here because this paper deals mainly with crustal structure.

5.2 Results of the Generalized Interpretation of the Deep-Seismic Sounding Data

P. GIESE

ABSTRACT

Based on the considerations of Chapter 4.2 of this volume, a number of basic and main parameters are contoured.

ZUSAMMENFASSUNG

Auf der Grundlage der Betrachtungen in Abschnitt 4.2 ist eine Reihe von Parametern in Form von Isolinienplänen dargestellt worden.

Methods are described in Chapter 4.2 that allow the presentation of basic travel-time data aiming to obtain some qualitative information on crustal structure.

Fig. 1 shows the contour map of the critical distance x_c of the P^M-group as given by GIESE and STEIN (1971). Although the resolving power of the quantity x_c is limited and its clear picking from the travel-time plot is sometimes difficult, this map already reflects variations of crustal thickness in the area under investigation. Fig. 2 shows the isolines of the quantity x_d, the cross-over-distance.

The determination of the travel time at a fixed distance is better defined than the reading of the critical distance x_c. A corresponding contour map is shown in Fig. 3. The similarity to the map of x_c is clearly visible.

As basic parameters, the intercept times for the velocities 6 and 7 km/s are contoured. Fig. 4 shows the intercept-time contours for v = 6 km/s, and Fig. 5 represents the corresponding depth values. Areas, e.g. Black Forest and Bavarian Forest, where crystalline rocks are outcropping, show depth values of about 4 km. That means that the rocks are deeply weathered here. In areas with sedimentary cover, the depth of the basement can be obtained by reducing the depth values by 1-3 km.

Fig. 6 shows the intercept-time contours of the velocity 7 km/s. The determination of the velocity 7 km/s from the slope of the reversed branch P^M is sometimes strongly influenced by the chosen correlation. Therefore, this value must be regarded with some caution. The same applies to the depth contours of velocity 7 km/s (Fig. 7). Because the velocity 7 km/s is derived from reversed segments, the general problem of apparent velocity exists (see 4.2). The velocity level 7 km/s may be situated within a sharp discontinuity or within a more or less wide transition zone.

The contour maps of Figs. 8 and 9 present data from the crust/mantle boundary. The definition of the crust/mantle boundary has been discussed in 4.2. The velocity at the level of the strongest velocity gradient varies more widely than previously assumed if a clear differentiation between the principle of phase and group correlation is taken into account. This quantity is mapped in Fig. 8. It must be admitted that the accuracy of the determination of the apparent velocity of the left part of the P^M-group is about ± 0.1 to 0.2 km/s, due to the uncertainty of correlation. Therefore, the lateral variations in Fig. 8 are mapped in intervals of 0.5 km/s.

For geological considerations the map of crustal thickness, shown in Fig. 9, may be the most important one. Compared with corresponding maps previously published (GIESE and STEIN, 1971; GIESE et al., 1973), the main features are unchanged. Smaller modifications due to new data have been introduced mainly in the southern Rhine Graben area and the Rhenish Massif including the Hessische Straße. Later reflection data have been used in order to complete the contours mainly in the northern part of NW Germany.

From the contour maps the following main features can be derived for the area under investigation. (1) In southern Germany the crustal thickness in-

Fig. 1. Contour map of the critical distance x_c of the P^M-wave group. Numbers and letters with points mean shotpoints. Numbers without points show the values of the critical distance which are plotted at half the distance. ▶
Shotpoints (•): see also Table 1 (3.1)

01 Eschenlohe	*09* Böhmischbruck	*17* Bransrode	*25* Vils
02 Hilders	*10* Voggendorf	*18* Lahr	*26* Lohne
03 Gersfeld Schw.A.	*11* Bischofsheim	*19* Merlebach	*27* Mauthaus Kronach
04 Grossenritte	*12* Romsthal	*20* Birresborn	*28* Ueffeln
05 Birkenau	*13* Dorheim	*21* Bermel	*29* Suhl
06 Adelebsen	*14* Mehrberg	*22* Dorndorf	
07 Gersfeld Nall.	*15* Büdingen	*23* Wilsenroth	
08 Kirchheimbolanden	*16* Taben Rodt	*24* Steinach	
BA Col des Bagenelles	*HE* Helgoland	*SB* Steinbrunn	*VS* Všetaty
BO Boubin	*MB* Merlebach	*SN* Saint Nabor	*WI* Wissembourg
HA Haslach			

creases from N to S towards the Alps. (2) A great crustal thickness also exists in the Bohemian Massif. (3) In the South German Triangle the crustal thickness is about 26-28 km. (4) The smallest crustal thickness of 20-22 km is found in the southern part of the Rhine Graben. (5) The crust becomes thicker E and W of the axis of the Rhine Graben. (6) The crustal thickness in the Rhenish Massif is between 26 and 29 km. (7) The Hessische Straße shows a crustal thickness by 2 to 3 km greater than the adjacent areas. (8) In the Northern Lowlands crustal thickness varies between 26 and 30 km. (9) A distinct crustal velocity inversion can be expected under the Rhine Graben. Its intensity decreases east- and westwards. (10) A low-velocity zone is also expressed in the Hessische Straße. (11) The velocity in the uppermost mantle varies between 7.5 and 8.2 km/s. Low values are found in the Rhine Graben area and regions showing young volcanism. (12) Distinct intercrustal discontinuities cannot be mapped over the whole area under investigation. In some regions very clear intercrustal reflections can be observed. A detailed discussion of the crustal structure and its relation to the geological structure follows in 5.4.

BAMFORD (1973, 5.3) presents maps showing qualitatively the variations of crustal thickness as well as absolute values. Although the data which are used by BAMFORD and by the author are different, the main features are the same.

For seismological purposes the mean crustal velocity may be of interest. This parameter is presented in the contour map of Fig. 10. The mean velocity was calculated from the critical-point part of the P^M-group by aid of the x^2, t^2 method. The resulting value has been reduced by 1.5% in order to get a better approximation.

The velocity inversion in the middle part of the crust is another main feature in crustal studies. Fig. 11 shows a contour map of the intensity of velocity inversion, which has already been published by GIESE and STEIN (1971). In the Rhine Graben area this map could be completed using the data of new measurements.

Fig. 12 shows a block diagram representing the geology after JACOBSHAGEN (1 of this volume) and the main features along some cross sections. For clarity only few quantities are displayed. Under the crust/mantle boundary the velocity value at the level of the strongest gradient is given. Dotted zones indicate intensive low-velocity layers. The velocity isoline v = 6 km/s can be regarded in a rough approximation as top of the basement or as level of unweathered crystalline rocks, respectively. Later the velocity isoline v = 7 km/s is shown.

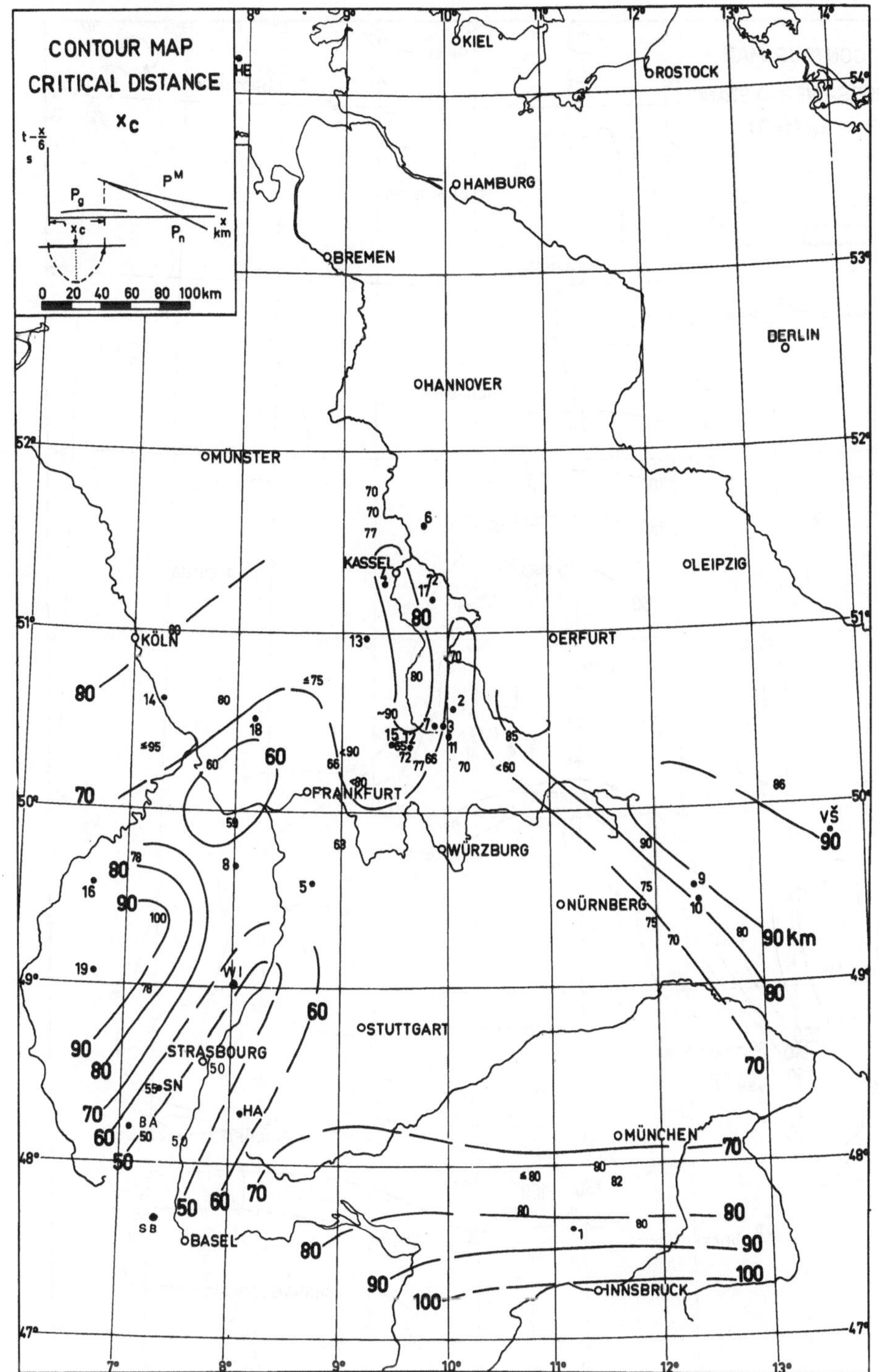
CONTOUR MAP
CRITICAL DISTANCE
x_c
KIEL
ROSTOCK
HAMBURG
BREMEN
BERLIN
HANNOVER
MÜNSTER
KASSEL
LEIPZIG
KÖLN
ERFURT
FRANKFURT
WÜRZBURG
NÜRNBERG
STUTTGART
STRASBOURG
MÜNCHEN
BASEL
INNSBRUCK
90 Km
0 20 40 60 80 100km

Fig. 1

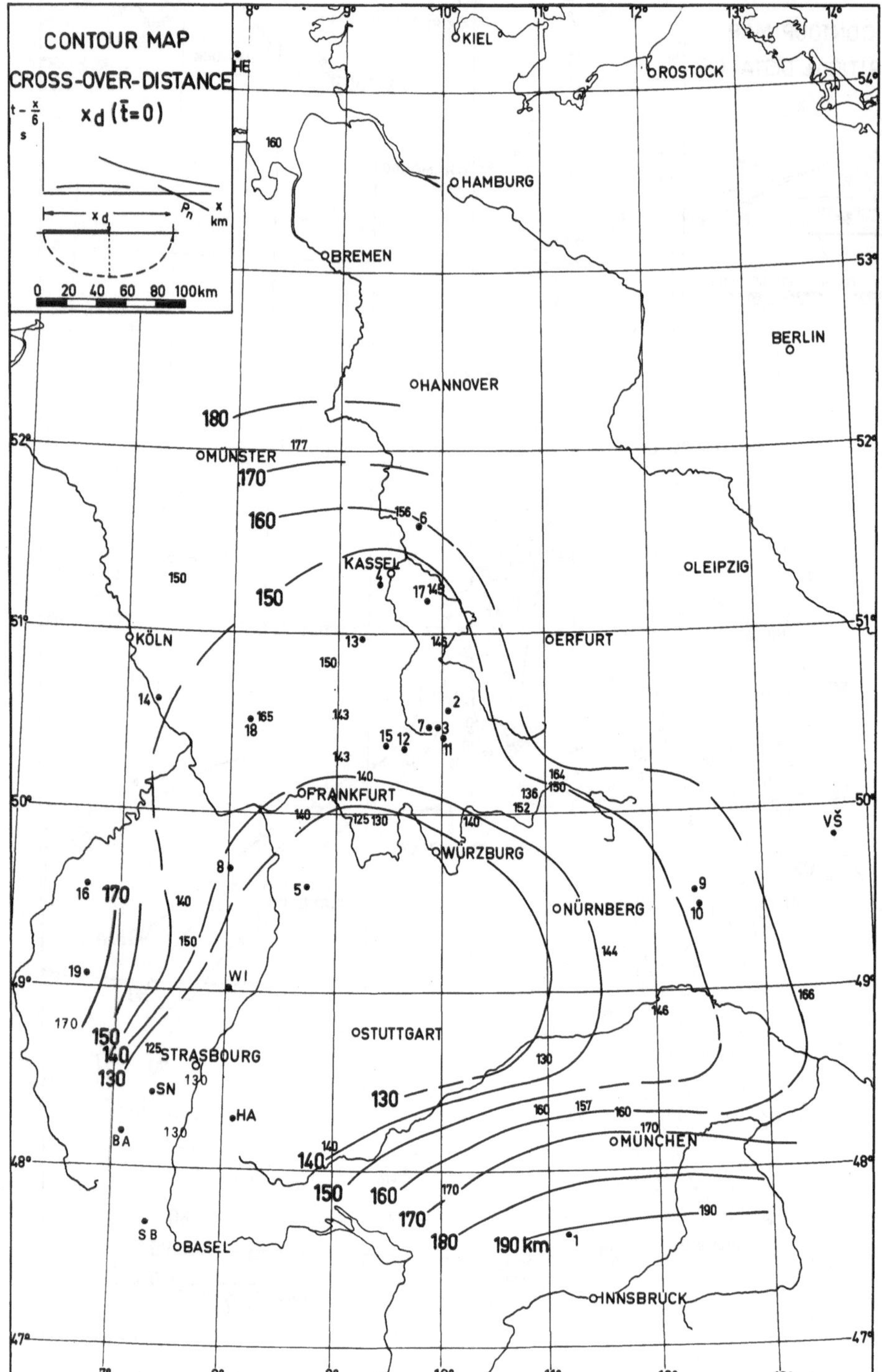

Fig. 2. Contour map of the cross-over distance between the P_n-wave group and v = 6 km/s (distance axis in reduced travel-time diagram). Further details see explanations to Fig. 1 (5.2)

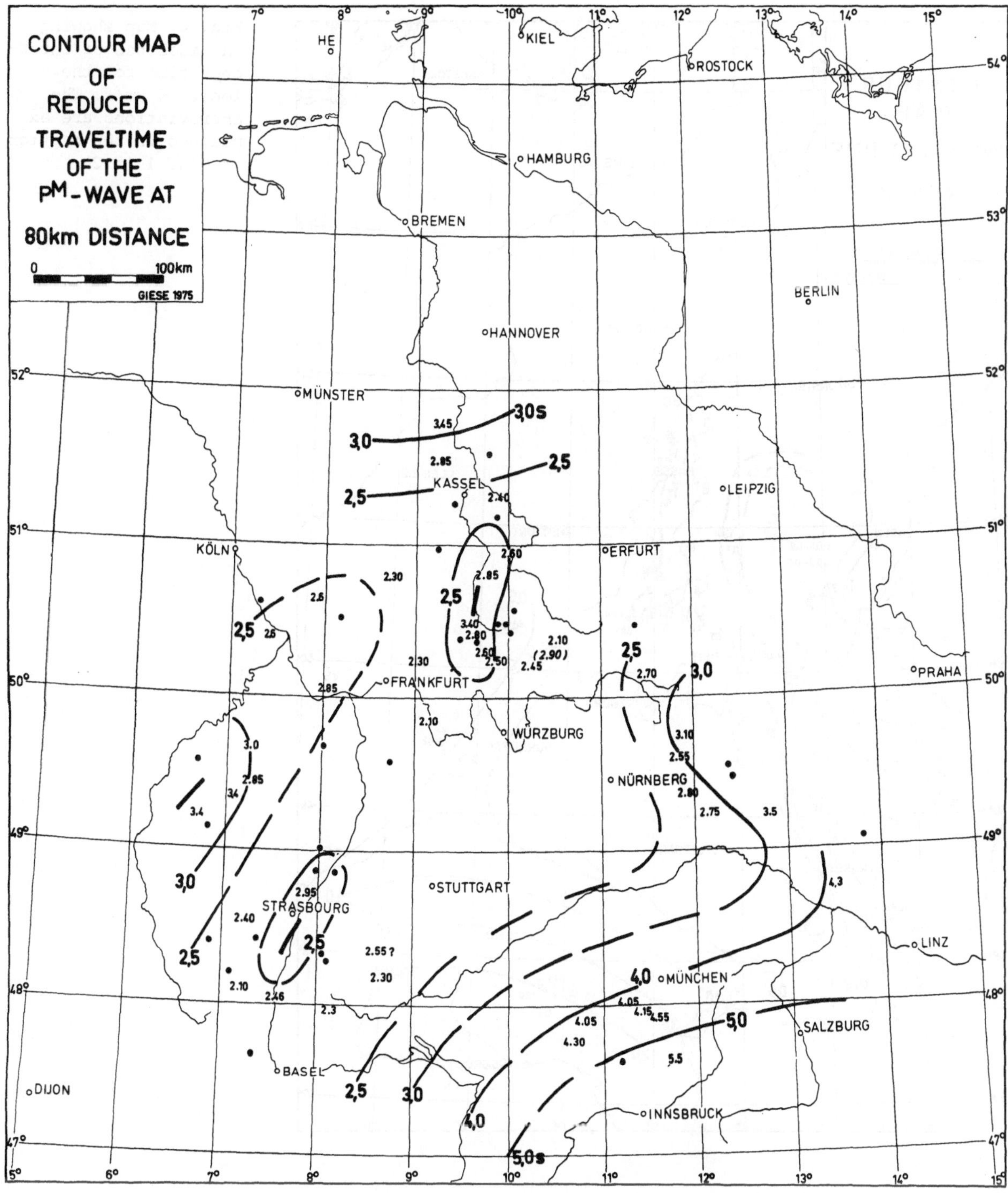

Fig. 3. Contour map of the reduced travel time of the P^M-wave group at 80 km distance. The corresponding values are plotted. The points indicate the position of shotpoints. Abbreviations see Fig. 1 (5.2)

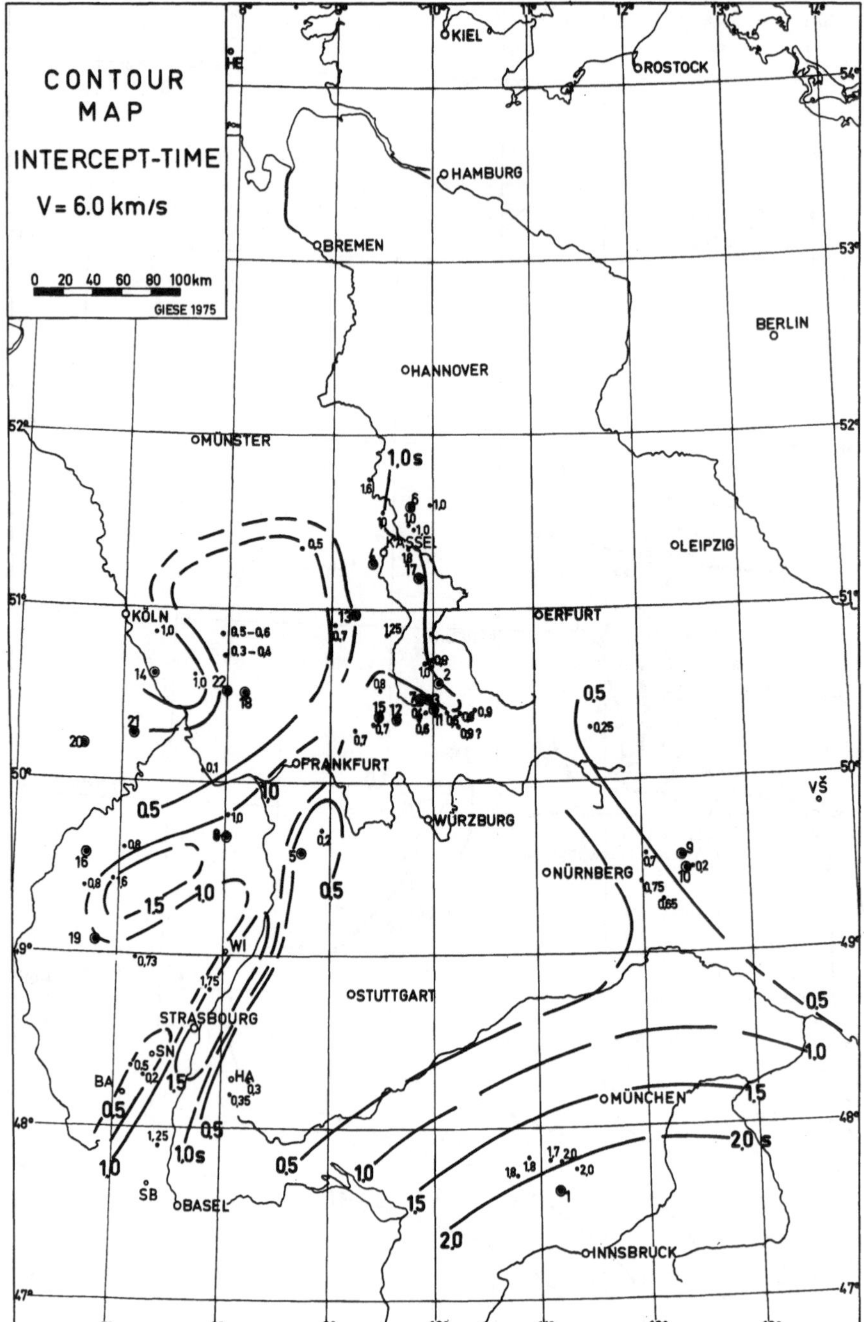

Fig. 4. Map showing the isolines of intercept time for the velocity 6 km/s. The abbreviations are explained in the caption of Fig. 1 (5.2)

Fig. 5. Contour map showing the depth of the velocity level 6 km/s. The abbreviations are explained in the caption of Fig. 1 (5.2)

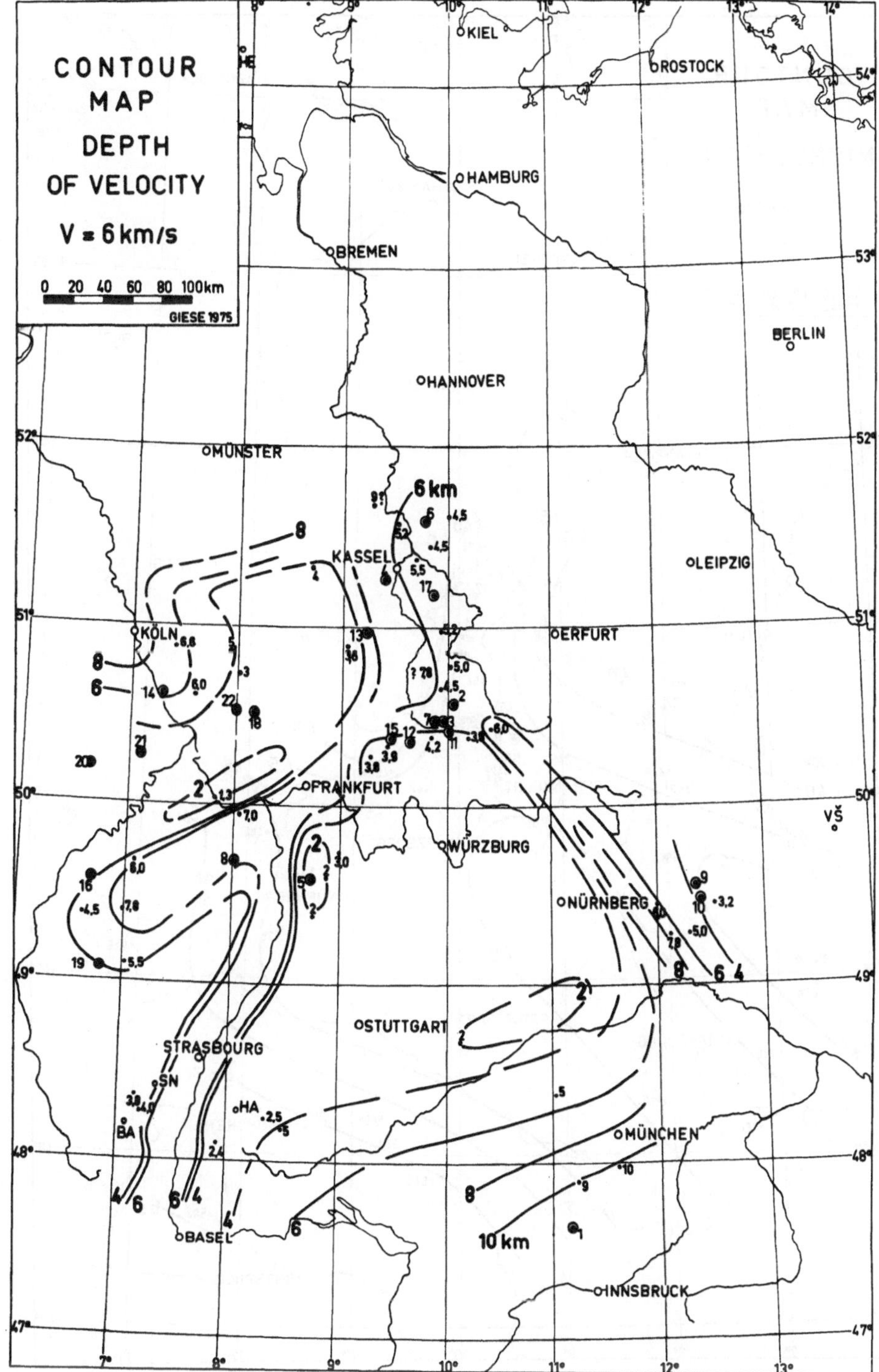

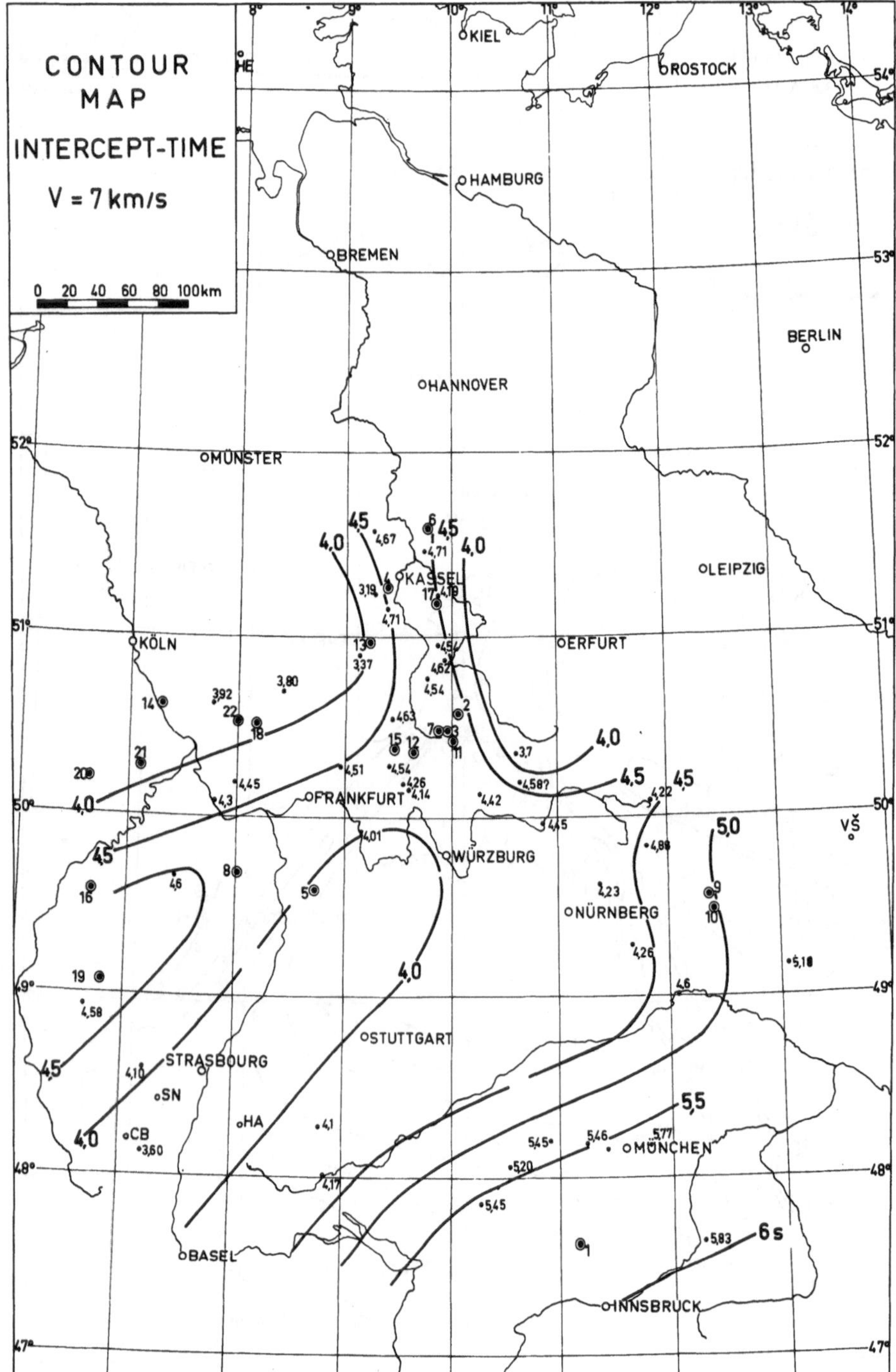

Fig. 6. Map showing the isolines of intercept-time 7 km/s. The abbreviations are explained in the caption of Fig. 1 (5.2)

Fig. 7. Contour map showing the depth of the velocity level 7 km/s. The abbreviations are explained in the caption of Fig. 1 (5.2)

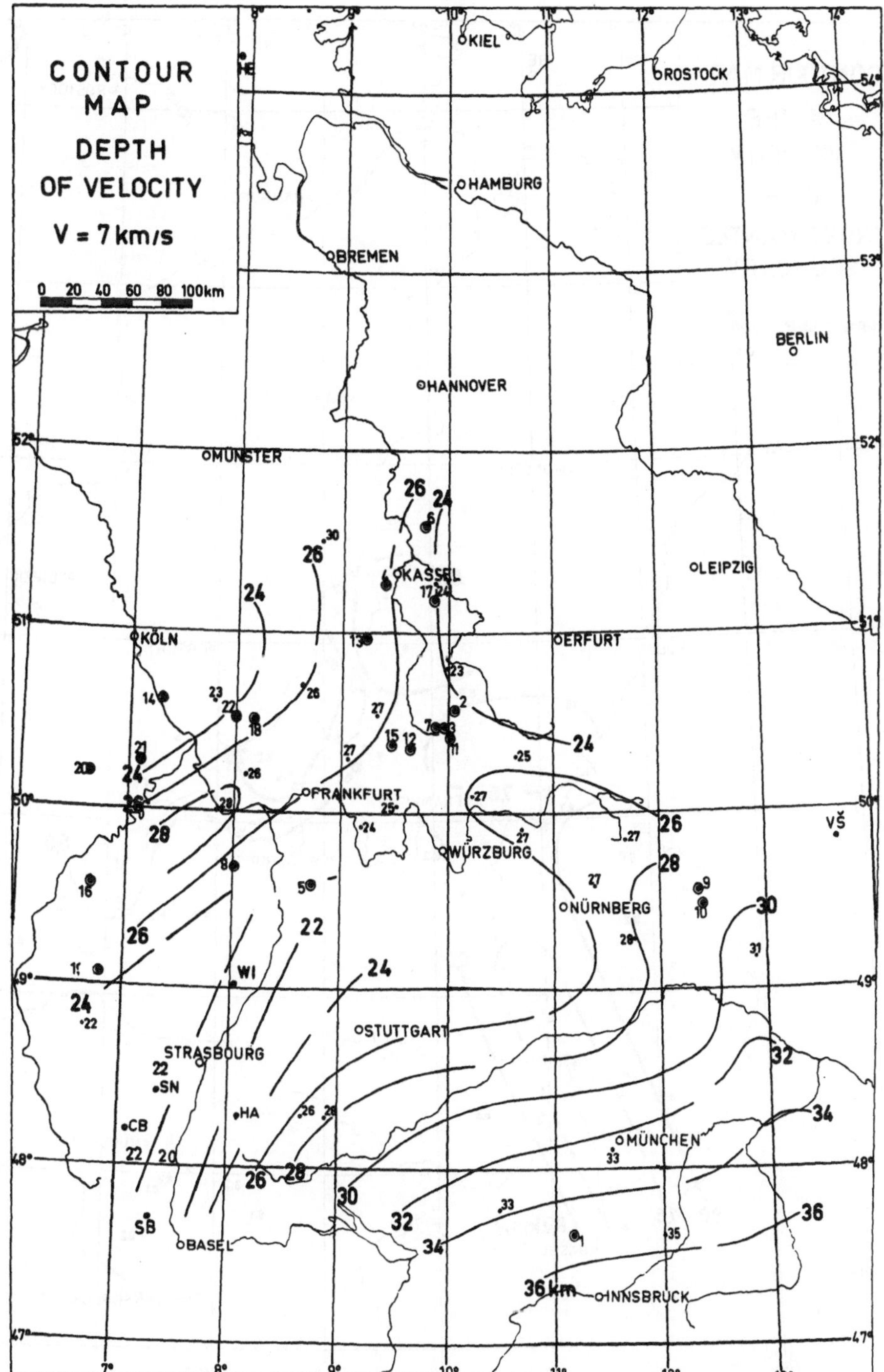

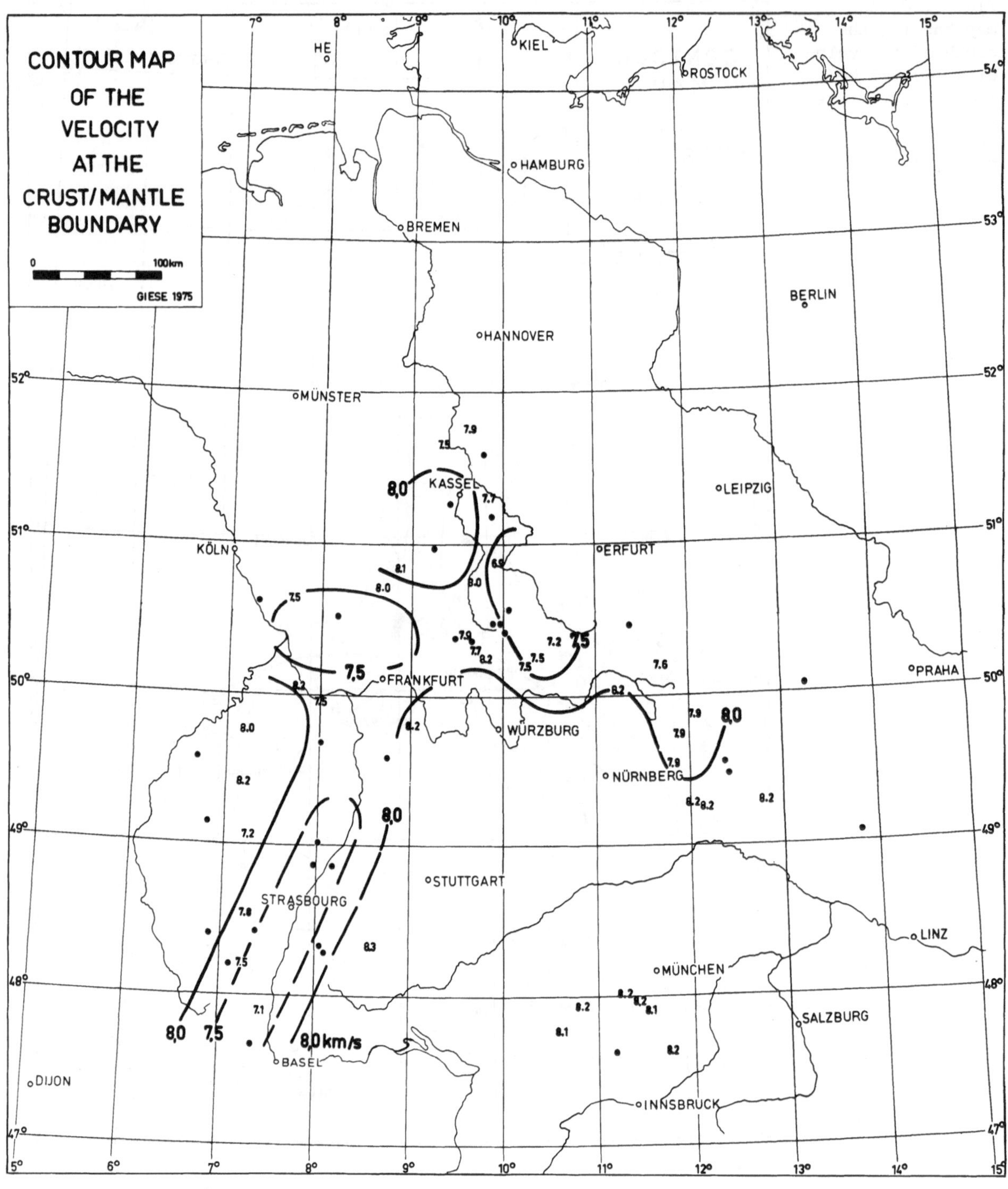

Fig. 8. Contour map showing the velocity at the depth of strongest-velocity gradient at the crust/mantle transition. Points indicate the position of the shotpoints

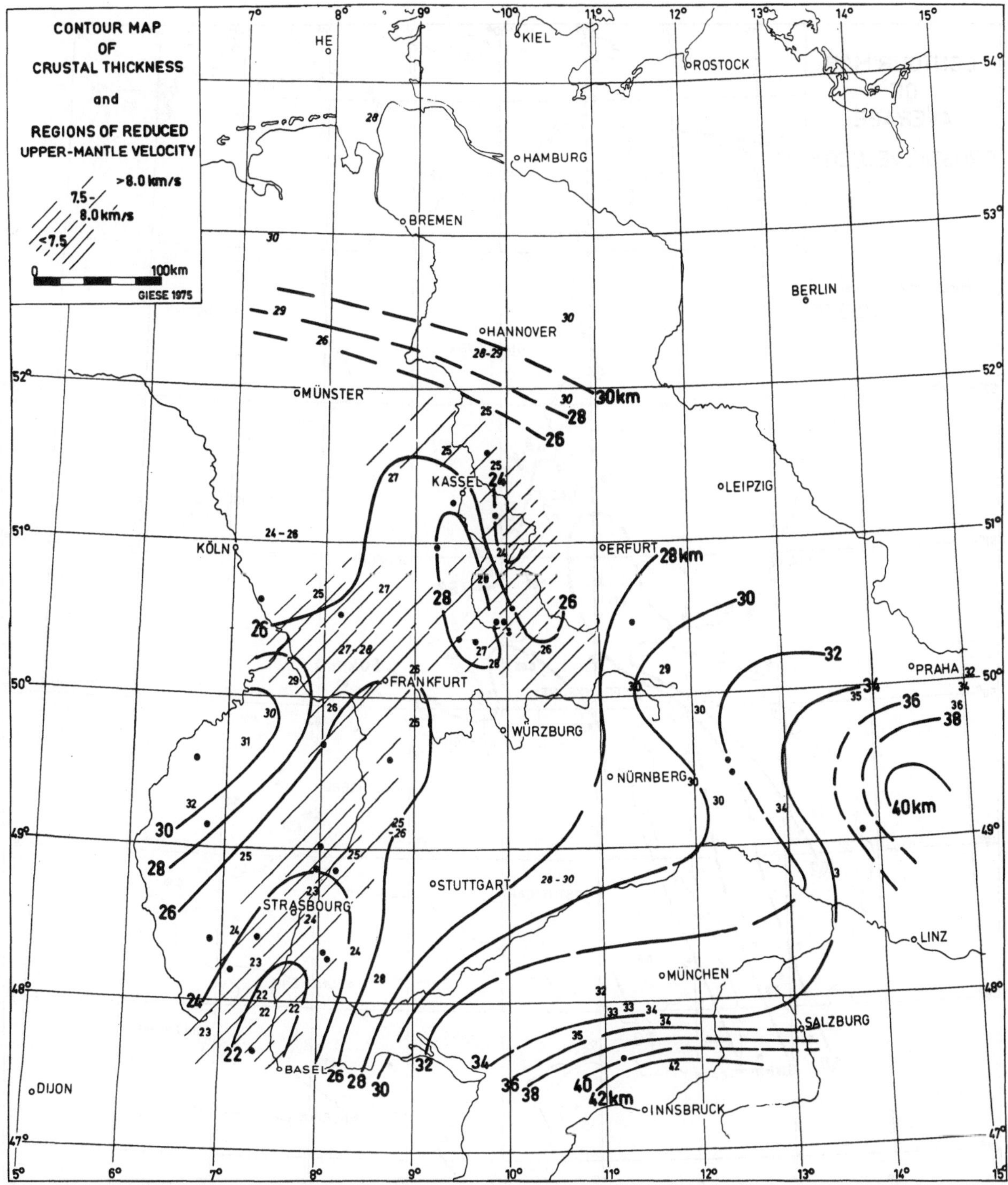

Fig. 9. Contour map of crustal thickness. The hatched area indicates upper-mantle velocities lower than 8 km/s. The numbers show the values obtained from the different profiles. *Widely hatched:* 7.5-8.0 km/s, *narrowly hatched:* < 7.5 km/s

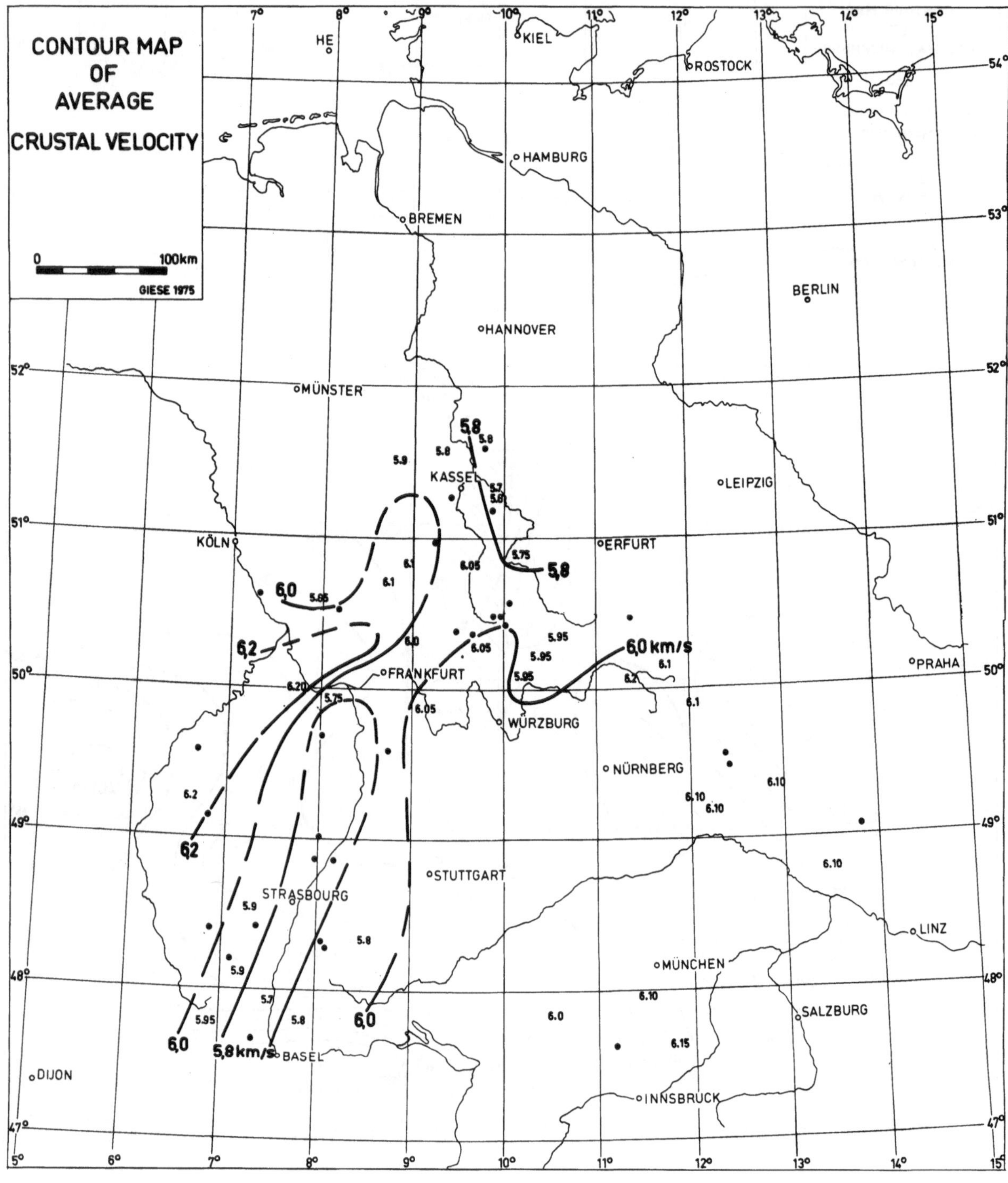

Fig. 10. Contour map of average crustal velocity. The numbers show the values obtained from different profiles

Fig. 11. Contour map showing the intensity of velocity inversion within the crust as defined by GIESE and STEIN (1971). The abbreviations are explained in the caption of Fig. 1 (5.2)

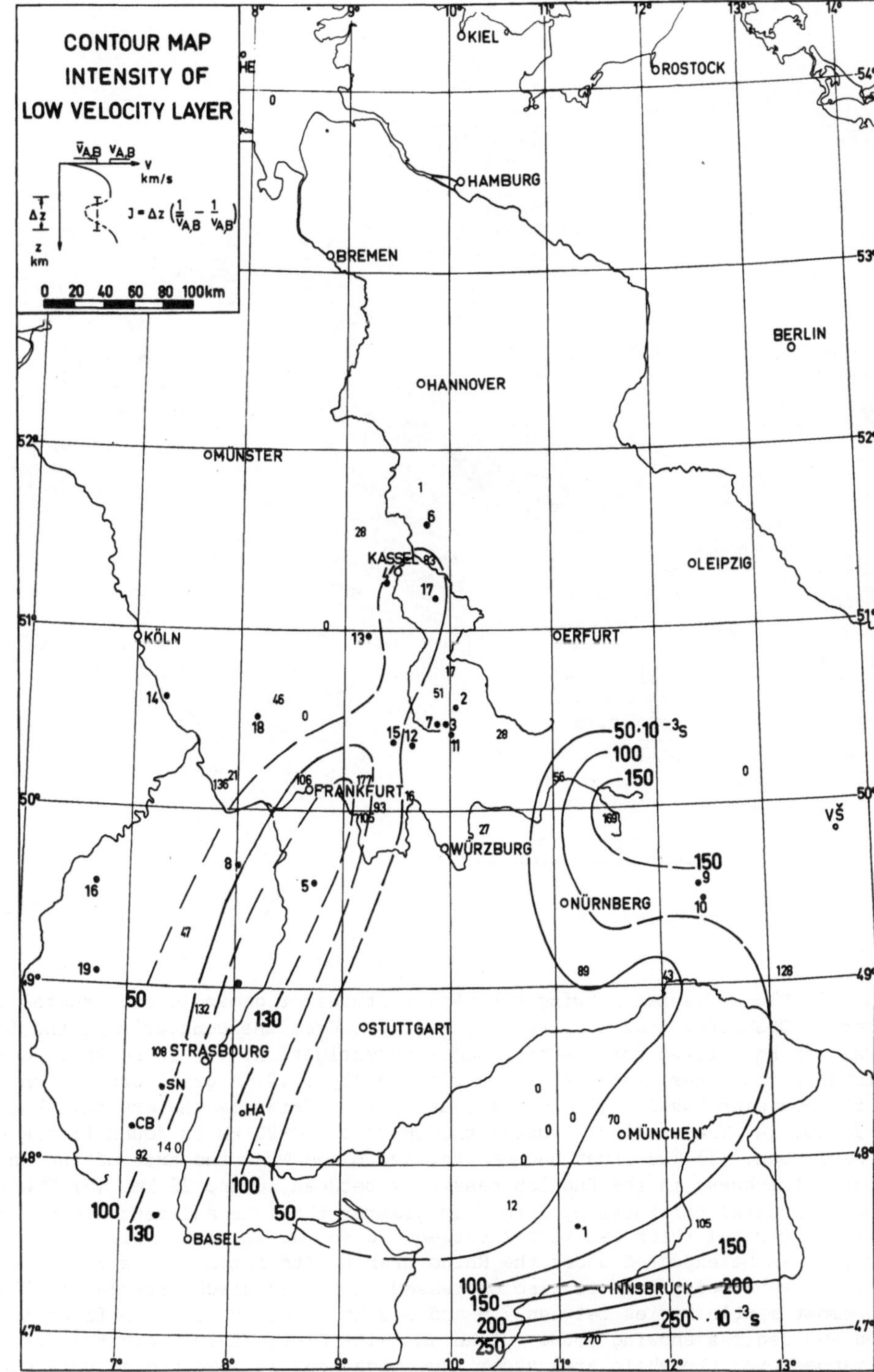

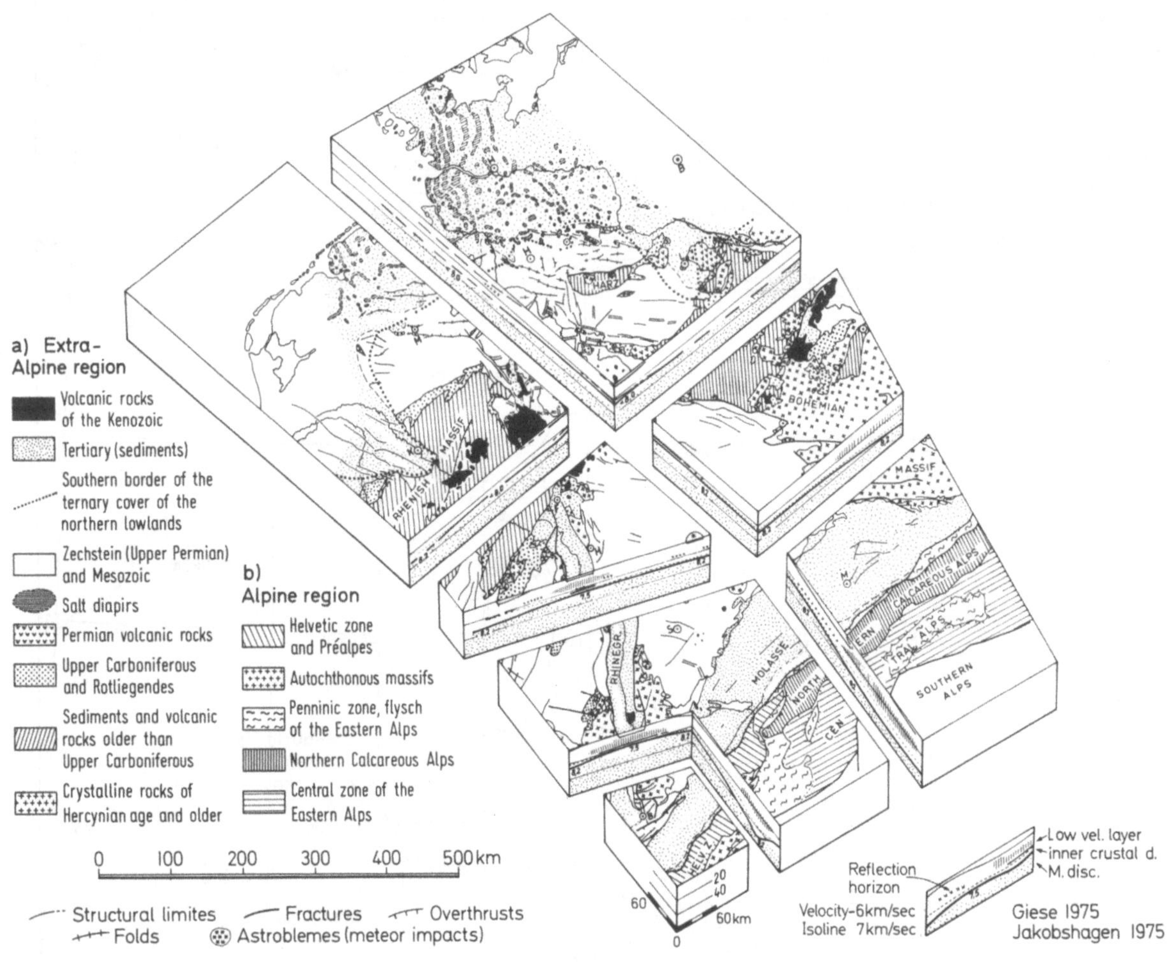

Fig. 12. Block diagram showing the main features of geologic and crustal structure (geology after JACOBSHAGEN, 1 of this volume, Fig. 2). From the contour maps the following main features can be derived for the area under investigation. (1) In southern Germany the crustal thickness increases from N to S towards the Alps. (2) A great crustal thickness also exists in the Bohemian Massif. (3) In the South German Triangle the crustal thickness is about 26-28 km. (4) The smallest crustal thickness of 20-22 km is found in the southern part of the Rhine Graben. (5) The crust becomes thicker E and W of the axis of the Rhine Graben. (6) The crustal thickness in the Rhenish Massif is between 26 and 29 km. (7) The Hessische Straße shows a crustal thickness by 2 to 3 km greater than the adjacent areas. (8) In the Northern Lowlands crustal thickness varies between 26 and 30 km. (9) A distinct crustal velocity inversion can be expected under the Rhine Graben. Its intensity decreases east- and westwards. (10) A low-velocity zone is also expressed in the Hessische Straße. (11) The velocity in the uppermost mantle varies between 7.5 and 8.2 km/s. Low values are found in the Rhine Graben area and regions showing young volcanism. (12) Distinct intercrustal discontinuities cannot be mapped over the whole area under investigation. In some region very clear intercrustal reflections can be observed. A detailed discussion of the crustal structure and its relation to the geological structure follows in 5.4

5.3 An Updated Time-Term Interpretation of P_n-Data from Quarry Blasts and Explosions in Western Germany

D. Bamford

ABSTRACT

An earlier time-term interpretation of P_n travel-time data from quarry blasts in western Germany, now supplemented by explosion data from the 1972 Rhine Graben experiment, has been updated and checked using a new MOZAIC version of the time-term method. As a result, an improved and extended map of Moho delay times is obtained; the data continues to require a considerable anisotropy of upper mantle P wave velocity. An overall velocity variation of 0.5 km/s is implied with a maximum velocity in a direction of 15° to 20° East of North.

ZUSAMMENFASSUNG

Eine frühere time-term-Analyse von P_n-Laufzeiten aus den Beobachtungen von Steinbruchsprengungen wurde mittels eines verfeinerten Verfahrens, der sog. MOZAIC-Version der time-term-Analyse, überprüft und verbessert. Außer den bereits vorliegenden Daten wurden die P_n-Daten, welche bei dem Rheingrabenexperiment im Jahre 1972 neu gewonnen wurden, in die Analyse mit einbezogen. Auf Grund der Berechnungsergebnisse wurde eine verbesserte und erweiterte Karte der Verzögerungszeiten der P_n-Welle am Registrierpunkt gezeichnet. Die Daten erfordern wiederum die Annahme einer beträchtlichen Anisotropie der P-Geschwindigkeit im oberen Mantel. Insgesamt ergibt sich eine Geschwindigkeitsvariation von 0.5 km/s, die Richtung der maximalen Geschwindigkeit liegt bei N 15°-20° E.

5.3.1 Introduction

A previous work (BAMFORD, 1973) presented a time-term interpretation of all P_n travel-time data available from the quarry blast recording program undertaken in western Germany as part of the Upper Mantle Project; this program is described in this volume in Chapter 3. The analysis demonstrated that the time-term method is indeed a powerful technique for the combined interpretation of refraction data from a network of profiles provided that the shot-station distribution satisfies certain conditions. The main result emerging from the interpretation, namely that a considerable anisotropy of upper mantle P wave velocity was required by the data, also demonstrated the power of the method in comparison with more subjective techniques which incorporate very restrictive assumptions, e.g. constant refractor velocity. However, no single interpretation should be regarded as final and conclusive and it is especially important to update interpretations as and when new data becomes available: not only will this add more details but also a fundamental test of any result is that it should not change significantly when new data is added. Thus, the availability of data from the 1972 Rhine Graben experiment (EDEL et al., 1975) offered the opportunity to expand, update and check the previous interpretation.

Furthermore, and for reasons that will be explained in the next section, the analytical basis of the time-term method has been changed; the new approach - henceforth called MOZAIC analysis - is significantly more powerful than that used in the previous interpretation. This paper introduces MOZAIC time-term analysis and presents some typical results of its application to the enlarged P_n data set.

5.3.2 An Introduction to MOZAIC Time-Term Analysis

5.3.2.1 The Meaning of "Time-Terms"

The fundamental principles of time-term analysis are well known (BAMFORD, 1973). Briefly, the theoretical travel time for a refracted wave traveling from site i to site j (Fig. 1) may be split into three independent parts or "terms" thus:

$$t_{ij} = a_i + a_j + \frac{D_{ij}}{V} \qquad (1)$$

The refractor velocity need not be constant; V can, for example, describe a uniform velocity, a vertical velocity gradient, horizontal velocity variations, velocity anisotropy and so on.

The terms a_i and a_j are the delay times, or "time-terms" at sites i and j; these terms describe the delay in travel-time caused by the presence of the overburden. If the depth to the refractor is h and the overburden velocity-depth function V(z) then the corresponding delay time, or time-term a is given by

$$a = \int_0^h \frac{[V^2 - V^2(z)]^{1/2}}{V \cdot V(z)} \cdot dz \qquad (2)$$

For a network of sites, at each of which there may be a shotpoint and/or recorder, a family of equations similar to Eq. (1) can be built up; for any observation involving site i, the time-term a_i will always be included in the corresponding theoretical equation and likewise for any other site. The same velocity function, i.e. the same coefficients describing the refractor velocity variations, will appear in each equation. For each family member, the measured travel time T_{ij}, and the distance D_{ij} will be observable quantities (but see BAMFORD, 1973, for further comments on D_{ij}) and thus, at least in principle, there remains only a fairly simple exercise in least-squares analysis to determine the unknowns in the system, that is the delay times and the coefficients of the velocity function, from the observations. However, in practice, the approach to this analysis depends very much on the observational network.

5.3.2.2 Some Approaches to Time-Term Analysis

Initially the time-term method was formulated (WILLMORE and BANCROFT, 1960) for the analysis of data from a particular type of refraction operation in which one shotpoint would be observed at several different recording points and each of these recording points would observe several shotpoints: the word 'site' in the previous section implied 'point'. A typical operation of this type would yield a large number of travel-time observations from a relatively small number of sites; the number of different time-terms occurring in the family of equations for the network would be equally small and thus, as only a small number of coefficients are required to describe typical velocity functions, the total number of unknowns would be very much less than the number of observations. The ensuing least-squares analysis is quite straightforward.

Reality is not always so simple. The quarry blast data in western Germany, for example, was collected over several years using a limited number of shotpoints and very many recording points, and the composite network consists of over a hundred criss-crossing profiles and fans. A single shotpoint may well have been observed at several hundred different recording points but only rarely and accidentally has a single recording point observed more thane one shotpoint. An analysis which, in these circumstances, allowed a separate time-term for each recording point would fail simply because the number of unknowns would approach the number of observations. In the earlier interpretation of the quarry blast data (BAMFORD, 1973), this weakness of the network was overcome by using an approach due to RAITT et al. (1969).

In essence, the assumption that individual delay times are completely independent is abandoned and it is assumed instead that the delay times within the network may be represented in terms of a regional delay time surface - usually taking the form of a linear polynominal combined with a double Fourier series. The number of

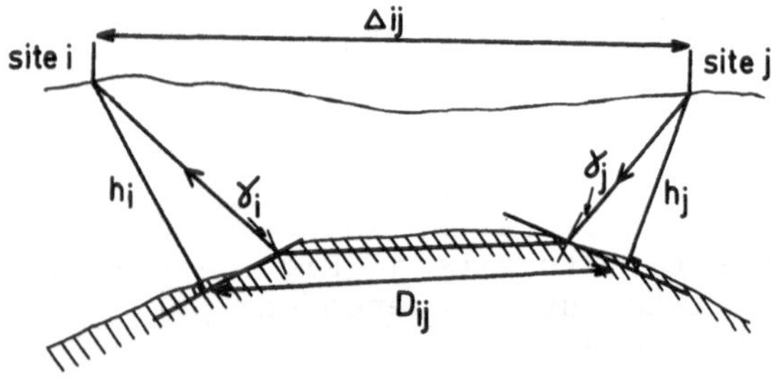

Fig. 1. Ray paths through a buried refractor. Δ horizontal observation distance, h depth to the refractor, γ critical angle

unknowns (now the coefficients of the delay time surface together with those of the velocity function) are considerably reduced and least-squares analysis is once again possible. However, this approach has at least two disadvantages: (1) the number of coefficients required increases, and the stability of solutions decreases, very rapidly as the order of the double Fourier series is increased and in practice one is restricted to a fairly smooth regional surface: for example, BAMFORD (1973) based his final map of Moho time-terms on a 4th order double Fourier series derived from well over 500 observations.(2) it is well known that in surface fitting with double Fourier series, the computed surface differs depending on the orientation of the reference axis; in particular genuine trends may not be accurately represented (WHITTEN, 1969).

Now delay times depend not only on depth-refractor but also on the overburden velocity-depth function [Eq. (2)] and, noting the extreme heterogeneity of the upper crust - evident, for example, in a geological map - it might be expected that delay times will typically not vary smoothly but will be subject to random and systematic fluctuations and will occasionally change discontinuously. Thus any analysis that constrained delay times to vary smoothly might give a reasonable average picture but leave some details unaccounted for; BAMFORD (1973) reached exactly this conclusion after examining statistical aspects of his solutions. Furthermore, if some of these 'details' happen to be part of a genuine trend in delay times, an accompanying analysis of the directional dependence of refractor velocity must be approached with great caution.

5.3.3 MOZAIC Analysis

Although the nature of the network itself rules out the application of the initial form of the time-term method to composite networks such as that available in western Germany, the nature of delay times themselves means that the regional-surface fitting approach will not be ideal, sacrificing the fundamental and reasonably realistic assumption that delay times at different sites are totally independent. How can this assumption be retained and yet the method adapted to deal with the problem in hand?

The step is in fact very simple and involves little more than a change in the implied meaning of 'site' from 'point' to 'area', coupled with an understanding of the behavior of refraction networks.

Locally there was no reason to restrict the initial formulation of the time-term method to imply 'points'. A small 'area' can be assumed to have a constant delay time to a particular refractor and it is possible to make an intelligent guess at the distribution of such 'areas' on the basis of gravity or magnetic anomaly, geological consistency, topography and so on. In principle, then, two or more recordings made some small distance apart may be assigned the same delay time, thus effectively reducing the number of unknowns in the system. At the same time, observations and sites should be interwoven in such a way as to build strength into the network. Two simple principles may be followed:

1. the more often a particular delay time is observed, the more accurately it is determined; in simple terms every site must have as many connections as possible to other sites, and

2. certain shot-station patterns determine velocity more effectively than others; for example, a reversed profile is more powerful than an unreversed one.

In this way, it is possible to construct, from a heterogeneous composite network, a powerful time-term network that will permit accurate determination of the unknowns remaining in it. Furthermore this network can always be modified, for example by splitting, amalgamating or redefining 'areas', to take account of local delay time anomalies or fluctuations without altering or at least unduly increasing the number of unknowns; no genuine variations in delay time need be unaccounted for. Thus not only will the statistical aspects of time-term solutions be improved in comparison with those of the surface-fitting approach but any results bearing on directional dependence of velocity can be accepted with confidence.

It might be expected that, in any typical example, the 'areas' defined would vary greatly in size, shape, distribution and density of data therein: working patterns are in fact so heterogeneous that they resemble a mozaic, hence MOZAIC time-term analysis.

Table 1. Comparison of various time-term solutions

Solution	Refractor velocity function	Variance (s^2)
Uniform velocity	V = constant	8.04×10^{-2}
Gradient velocity	V = v + k.z where v, k are constants, z is depth	8.04×10^{-2}
General velocity anisotropy	$V^2 = G^2 + A \cos 2\phi + B \sin 2\phi + C \cos 4\phi + D \sin 4\phi$ G,A,B,C,D constants; ϕ is azimuth in degrees east of north; offset distance 30 km	2.99×10^{-2}

5.3.4 Typical Results

Maps showing the distribution of profiles and fans obtained in the quarry blast program and the 1972 Rhine Graben experiment may be found in this volume (3 and 4.8 respectively). 762 P_n travel times have been measured by the author for these profiles; in most cases the availability of the data in record section form made P_n correlation very simple. The shot and station positions corresponding to these measurements were plotted to overlay the Geological Map of the Federal German Republic 1 : 1,000,000 (1973). On this scale adjacent stations can be clearly distinguished - the typical separation on the profiles is one to a few km - and yet the geological information is not too finely detailed. In this paper, the results of time-term analysis of a mozaic of 128 sites are described. These 'sites' ranged in size from a single point (e.g. a well-observed shot) to those with dimensions of a few (up to twenty) km. Apart from the upper limit on size, the sole factor in distinguishing sites (or combining recording-points) was common surface geology.

The actual conduct of a time-term analysis, including the various checks, tests and corrections, together with those aspects peculiar to the analysis of velocity anisotropy, is adequately described by BAMFORD (1973). Rigorous tests are applied before any single result is accepted; however, the results described below are quite typical of those obtained with this data.

Table 1 compares the variances obtained when three different velocity functions - constant velocity, velocity gradient and general velocity anisotropy - are included in time-term analysis of all available P_n data. The significant, indeed spectacular, improvement in variance resulting from the inclusion of velocity anisotropy parallels that obtained in the previous investigation and seems to confirm that velocity anisotropy is required by the data; furthermore, the statistics of the MOZAIC solution, as expected, are rather better.

The coefficients of the velocity anisotropy are given in Table 2. Broadly an overall velocity variation of approximately 0.5 km/s - minimum 7.8 km/s, maximum 8.3 km/s - is implied with the maximum in a direction of 15° to 20° east of north. Such values are typical of anisotropy solutions on this data and are similar to those of BAMFORD (1973).

The delay times resulting from this analysis are presented in Fig. 2. Following the discussion of the meaning of 'time terms' it is clearly both difficult and misleading to present a contour map of Moho delay times; instead the values are graded and plotted in position. In Fig. 2a, they are graded in groups of 0.30 s width; 0.30 s is roughly two to three times the typical error in a single delay time. Grading in Fig. 2b is on a slightly broader basis; the three grades could represent 'thin', 'normal' and 'thick' crust if one assumes constant crustal properties and multiplies delay times by a factor of about 10 to obtain depths. In this figure we see quite clearly the thickening of

Table 2. Coefficients of general anisotropy solution. ($V^2 = G^2 + A \cos 2\phi + B \sin 2\phi + C \cos 4\phi + D \sin 4\phi$)

G	=	8.06 ± 0.02	km/s
A	=	2.62 ± 0.33	$(km/s)^2$
B	=	3.35 ± 0.34	$(km/s)^2$
C	=	0.95 ± 0.21	$(km/s)^2$
D	=	-0.20 ± 0.20	$(km/s)^2$

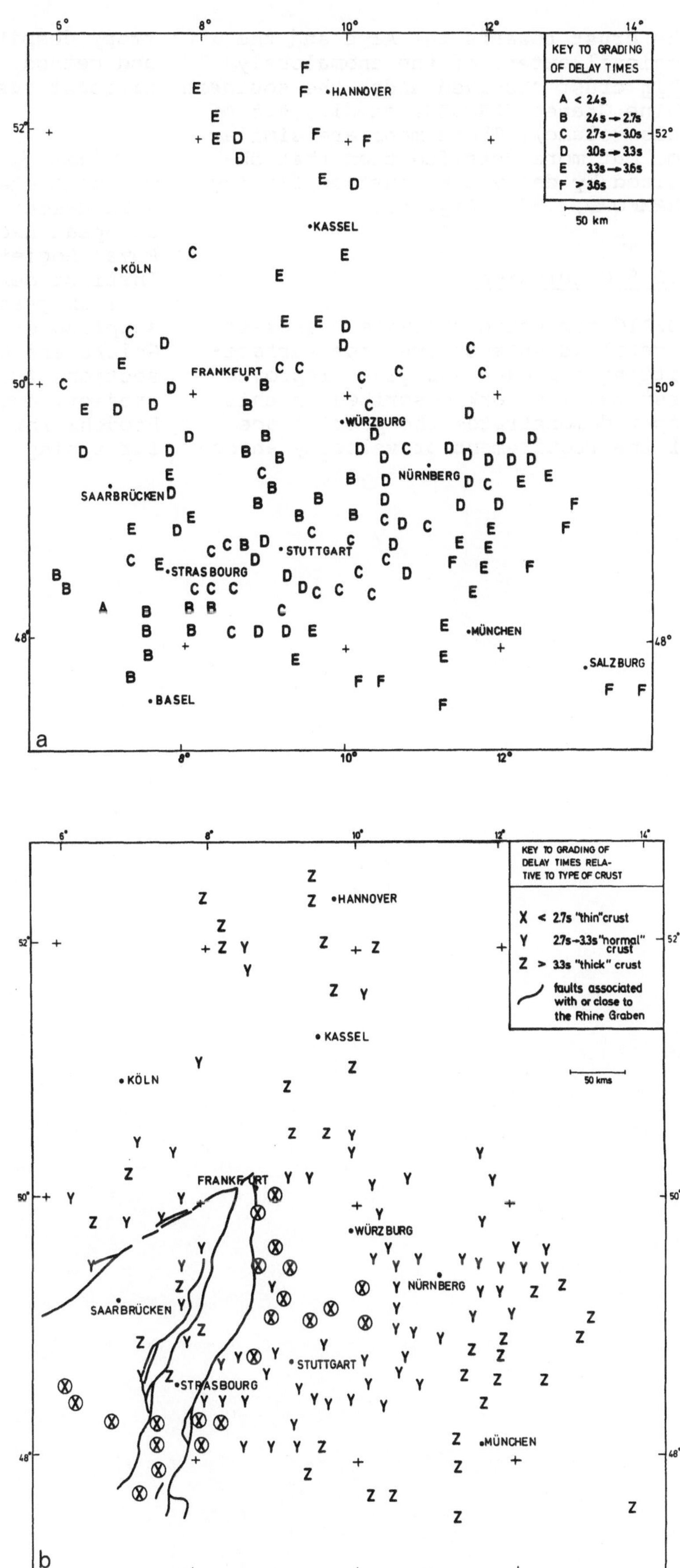

Fig. 2a and b. MOZAIC maps of Moho delay times: (a) graded every 0.30 s on simple base map; (b) graded to characterize 'thin', 'normal' or 'thick' crust; the 'thin' values have been circled to give them special emphasis. Note the relationship of crustal 'thinning' and the Rhine Graben

the crust towards the Alps and the regional extent of the anomalously thin crust observed under the southern Rhine Graben (PRODEHL et al., 6.8 of this volume). These maps are similar to, but more detailed than that obtained by delay time surface fitting (BAMFORD, 1973, Fig. 6).

5.3.5 Conclusion

MOZAIC time-term analysis possesses several advantages over the surface-fitting approach and gives improved results. The work described in this paper demonstrates the persistence of the requirement of velocity anisotropy despite changes in both data and method. This is an extremely significant result.

Acknowledgments. Much of this work was completed during the tenure of NATO Research Fellowship (under the European Exchange Programme of the Royal Society) at Geophysikalisches Institut der Universität Karlsruhe.

I am grateful to everyone who supplied data, especially Christoph Gelbke and Dieter Emter who provided sections from the 1972 Rhine Graben project, and to Karl Fuchs and Claus Prodehl for reading and Gisela Michael for typing this manuscript.

5.4 The Basic Features of Crustal Structure in Relation to the Main Geological Units

P. GIESE

ABSTRACT

In this chapter, crustal structure and geology will be discussed. Corresponding to the solving power of refraction seismics it can only be of use to consider somewhat larger units. In the first paragraph, the crustal structure of the individual Variscan zones will be described, the second part deals with the structure of the crust in the Rhine Graben. Finally, conditions at the border of the northern Alps will be briefly treated. At the end of each paragraph, a summary of the essential results will be given.

ZUSAMMENFASSUNG

In diesem Kapitel werden Krustenstruktur und Geologie diskutiert. Entsprechend dem Auflösungsvermögen der Refraktionsseismik ist es nur sinnvoll, größere Einheiten zu betrachten. Im ersten Abschnitt wird die Krustenstruktur der einzelnen variskischen Zonen beschrieben, im zweiten Teil wird auf den Bau der Erdkruste im Rheingraben eingegangen. Schließlich wird noch kurz auf die Verhältnisse am Nordalpenrand eingegangen. Am Ende jedes Abschnitts wird eine Zusammenfassung der wesentlichen Ergebnisse gegeben.

5.4.1 Introduction[1]

Studies dealing with a comparison of crustal structure with surface geology meet with many difficulties. One of them is due to the different amount and density of data in crustal structure and geology. Especially in central Europe a good deal of information on geological details is available, while the amount of data and the resolving power of deep-seismic sounding is considerably smaller. Geological maps are available on the scale of 1 : 25,000 for most areas, whereas anomalies of crustal thickness can be contoured easily in maps to the scale of 1 : 500,000 or 1 : 1,000,000.

This raises the question which geological units form an adequate basis of comparison with the resolving power of seismic crustal investigations. The tectonic structure of central Europe can be based, on one hand, on the different orogenic zones and, on the other hand, on blocks. Orogenic zones are characterized by similar age and behavior from the sedimentary, magmatic, and tectonic points of view. Blocks, on the other hand, are units separated by deep-reaching fracture zones. Due to different ages of formation, a relationship between these two types does not necessarily exist.

This means that central Europe outside the Alps, i.e. the region between the North Sea and the Alps, is subdivided into the zones of the Variscan orogene on the one side and blocks of younger origin on the other (JACOBSHAGEN, 1 of this volume; KNETSCH, 1963). Regarding their dimensions, these zones and blocks offer a basis corresponding to the resolving power of seismic crustal investigations.

The following discussions have to be confined to the main features of crustal structure. A detailed description of the sedimentary cover must be excluded from this paper because adequate data are not available. This is the task of reflection seismics carried out within the scope of oil and gas exploration. When determining the depth of the crystalline basement by the aid of the P_g-wave, it is necessary to deal with the question if and how far a clear boundary exists between sediments and crystalline basement.

It is not possible to contour, in the usual sense, the depth of the boundary between upper and lower crust, the so-called Conrad discontinuity in the whole region. The problematic nature of this question has been discussed in the previous chapter. In individual regions, the identification

[1] Geographical names mentioned in this section are summarized in Fig. 8.

of a clear boundary between upper and lower crust is feasible, as will be described in detail.

Even the task of contouring the Mohorovičić discontinuity - considered to be simple and a natural feature by outsiders - presents some difficulties in certain regions. Just these problematic cases, however, deserve special attention since they may supply some important information on special features of the lower crust and the uppermost part of the mantle. It is, at any rate, necessary to describe the way in which the crust/mantle boundary has been defined.

5.4.2 Hercynian System

5.4.2.1 Moldanubicum

The Moldanubic zone, the central and inner zone of the Hercynian orogene is situated in southern Germany and is exposed in the Black Forest and the Bavarian Forest. In the adjacent areas in the Vosges and the Bohemian Massif, parts of this zone are also outcropping (JACOBSHAGEN, 1 of this volume).

In the zone of the Moldanubicum, both the shotpoints of Böhmischbruck (09) and Voggendorf (10) are situated in the Bavarian Forest, whereas the shotpoints of Steinach (24) and Haslach (HA) lie in the Black Forest. Further on, the shotpoints of Col de Bagenelles (BA) and St. Nabor (SN) in the Vosges may be included, just as some shotpoints in the ČSSR, situated on the International Profile VII. The extremely important shotpoint of Eschenlohe (01), located at the northern border of the Alps, belongs in respect to its basement to the Moldanubic zone.

Using the records of Eschenlohe, PRODEHL (1964) has determined the depth of the basement in the Molasse area by aid of the P_g wave. BREYER (1956) has published the results of some short refraction lines observed in southern Germany in order to explore the thickness of the sedimentary cover. These results, completed by data from drillings, have been compiled by the German Research Group (1964) to a contour map showing the depth of the basement. In the area just discussed, the basement is well defined by the top of the strongly metamorphized rocks of Variscan and Prevariscan age. How far the P_g-wave penetrates into the crystalline rocks has been discussed by GIESE (1963), MEISSNER (1967a, b) and HOLUB (1974).

At the northern border of the Alps, south of Murnau, the sedimentary cover reaches a thickness of about 6 km. This value results from the profiles radiating northwards from the shotpoint Eschenlohe. The determination of the depth of the basement under the northern Calcareous Alps is the subject of a paper by SCHELIGA (6.9 of this volume), discussed below. Towards the north, in the direction of the Danube, the thickness of sediments decreases continuously and reaches no more than some hundred meters in the Swabian and Franconian Alb (MEISSNER, 1967a, b; BRAM and GIESE, 1968).

What statements can be made about the structure of the crust in the region of the Moldanubicum? In addition to the P_g and the P^M-onsets, there are numerous other events reflected and/or refracted within the crust. They cannot, however, be correlated over large distances with the same certainty as the P_g and P^M-waves. A regional mapping of the boundary between the upper and lower crust encounters difficulties because a typical and well-expressed reflected or strongly refracted phase is, in most cases, absent. The top of the crust/mantle transition, although difficult to determine, may serve as boundary between upper and lower crust.

An exception is the profile 09-200-01, starting from the shotpoint of Böhmischbruck and running toward SSW, with a record section showing a clear intercrustal phase (Δ = 90 km, $\bar{t}$ = 1.2 s) which can be followed over a distance of more than 30 km. MUELLER and LANDISMAN (1966) take it as an example for their P_c-phase. GIESE (1968a) interprets it as one of the possible groups appearing between the P_g and the P^M-branches. Whatever this phase may be called, it shows, at any rate, that, in the depth range between 8 and 12 km, there must be a distinct low-velocity zone. In the record sections of the neighboring profiles (09-240 and 10-135), this phase is less clearly expressed, thus indicating its limited importance. This anomaly refers to the SE border of the Franconian Alb, immediately west of Regensburg. A geological interpretation of this intercrustal interface must remain speculative at present. The local limitation suggests that there may possibly be a magmatic intrusion at a depth of about 12 km which spreads over some 10 km^2.

In the beginning of the sixties, an extensive reflection exploration work was carried out in the Molasse area. This opportunity was used in order to record reflections up to 12-15 s two-way travel time. These near-vertical reflections evaluated by LIEBSCHER (1962, 1964) give a detailed picture of the crust's structure and its lateral changes. In general it can be stated that there are depth ranges showing a greater number of reflections distributed over 1-2 s or even more than other intervals. In a histogram, presenting the number of reflections against travel time, the peaks may be correlated to discontinuities of refraction seismics. The crust/mantle boundary is detectable in general without difficulties in the whole area. The clearness of the upper/lower crustal boundary varies regionally, as well as the occurrence of interfaces within the upper crust.

The German part of the Bohemian Massif is relatively small. The evaluation of the profile Voggendorf-SE (10-135), situated completely within the Moldanubic zone, is facilitated by the absence of travel-time delays due to sediments. Down to a depth of 22 km, velocities do not exceed 6.35 km/s (GIESE, 1968a; HARCKE, 1972). In the depth range between 10 and 22 km, generally velocity inversions have to be assumed which, however, show a very heterogeneous structure, also in lateral direction, as a result of embedded bodies with slightly higher velocity (about 0.1-0.6 km/s).

The P^M-wave group indicates a transition zone between crust and mantle having a thickness of about 8-10 km, the top of which may be the beginning of the lower crust while the lower side can be regarded as the crust/mantle boundary. The observation of the International Profile VI in the ČSSR also results in 20-25 km thickness of the upper crust (BERANEK et al., 1971, 1973). Total crustal thickness amounts to about 32 km in the Bavarian Forest. Near Prague a maximum thickness of 42 km is reached.

The shots fired on the International Profile were observed in southern Germany and also in Austria. On the southeastern prolongation, a clearly perceptible P^M-wave could be observed, resulting in a crustal thickness of 32 km (MILLER and GEBRANDE, 6.11 of this volume). Surprisingly, the other profile observed, running in southern direction from the shotpoint of Boubin, contains no indication of any P^M arrivals but the intercrustal wave, yielding the 22 km discontinuity, is well developed (SCHÜTTE and MILLER, personal communication). Thus the nature of the crust/mantle boundary and its recognizability by refraction seismics change strongly within some 10 km in lateral direction.

The crustal thickness of the South German Triangle is derived from profiles radiating from Böhmischbruck (09-275, 09-240, 09-200-01), Eschenlohe (01-040, 01-020-09, 01-005, 01-345-02, 01-315-05, 01-290), and Hilders (02-140, 02-165-01, 02-215). In the region of the Swabian Alb and in the southern part of the Franconian Alb, the crust is 28-30 km thick (GIESE and STEIN, 1971; EMTER, 1971, 6.5 of this volume; AICHELE, personal communication). Generally, in this area the top of the upper mantle is placed 2-4 km higher than in the adjacent Bavarian Forest.

Southerly toward the Alps, the crust/mantle boundary dips and reaches a depth of about 38-40 km at the northern margin of the Alps. The increase of crustal thickness is mainly caused by a corresponding thickening of the sedimentary cover (PRODEHL, 1965; MEISSNER, 1967a, b; BRAM and GIESE, 1968; EMTER, 1971; GIESE and STEIN, 1971).

At the SW corner of the South German Triangle, in the Black Forest, the Moldanubicum is exposed again. Here, the profiles Steinach E and S (24-90, 24-170) prove a thin crust of only 24 km thickness (EMTER, 1971). A similar value has been found on the opposite side of the Rhine Graben in the Vosges (e.g. EMTER, 1971; EDEL et al., 1975).

The complex of questions regarding the crust in the vicinity of the Rhine Graben will be dealt with separately.

Since values for crustal thickness are available from various parts of the Moldanubicum, a brief comparative study is possible. The International Profile VI crosses the Prague Trough with nonmetamorphic early and late Paleozoic rocks (Barrandium). Here, the crustal thickness amounts to 35-38 km (BERANEK et al., 1971). Provided that, in Posthercynian time, no shifting of the crust/mantle boundary took place, the complete thickness of the crust as it existed at the end of the Paleozoic (apart from possibly eroded sediments) is here still present. At the western border of the Bohemian

Massif, thickness today has been determined at 32 km and in the region of the Franconian Alb, a further reduction to 28-30 km has been recorded.

It is, in this case, possible to assume that the decrease of crustal thickness in SW direction has its origin as early as in the Hercynian period. This idea cannot, however, be entirely sustained since, in the Bavarian Forest, Hercynian granites are exposed, which means that this part of the crystalline has been subject to erosion. The roof of the intruding granites may have been 5-10 km thick. But this is just the value which results as difference of recent crustal thickness in the Prague Trough (38 km) and in the Bavarian Forest (32 km).

In the SW part of the area under discussion, the Black Forest, a recent crustal thickness of only 24 km or even less has been found.

In summary, the following may be said about the crustal structure of the Moldanubicum.

In general, the crust/mantle boundary is well expressed, i.e. the transition from about 7 to 8 km/s takes place within a few km. In principle, there exists a lower crust, although it cannot be clearly separated from the upper crust in all places. The crustal thickness of about 40 km of the Bohemian Massif decreases towards SW to about 22 km in the southern Black Forest and the Vosges (Fig. 1).

5.4.2.2 Saxothuringian Zone

The Saxothuringian zone is adjacent to the north of the Moldanubic zone and can be followed from the Odenwald to the northern part of Bavaria, including the Oberpfälzer Forest, the Thuringian Forest and the SE part of the Harz Mountains (JACOGSHAGEN, 1 of this volume). This zone of the Hercynian orogene is interspersed with the younger structure of the upper Rhine Graben and its northern continuation, the Hessische Straße. Similar to the preceding chapters, however, the main attention, for the present, is paid to the Hercynian structure.

In this zone, profiles are located starting from the quarries of Merlebach (19), Taben Rodt (16), Kirchheimbolanden (08), Büdingen (15), Hilders (2), Böhmischbruck (09), and Mauthaus (27). The thickness of the Postvariscan sedimentary cover in the north Bavarian region is known from borehole data and some short refraction profiles (BREYER, 1956).

Large travel-time delays of the P_g-wave are characteristic of the Saar-Nahe Trough, an inner depression of the Hercynian orogene [Figs. 3 and 4 (5.2)]. In comparison with a sediment-free region, travel-time delays up to 1.5 s occur in the center of this depression, indicating that the thickness of unconsolidated sediments, including Rotliegendes and Upper Carboniferous, must be 4-5 km (v = 3000 m/s).

The Hessische Straße can be followed by a time delay of the P_g-wave of about 1 s with respect to the adjacent regions as shown in the contour map of Figs. 3 (5.2) and 4 (5.2).

As mentioned in the previous section, the structure of the crust is rather simple in the Moldanubic zone, as

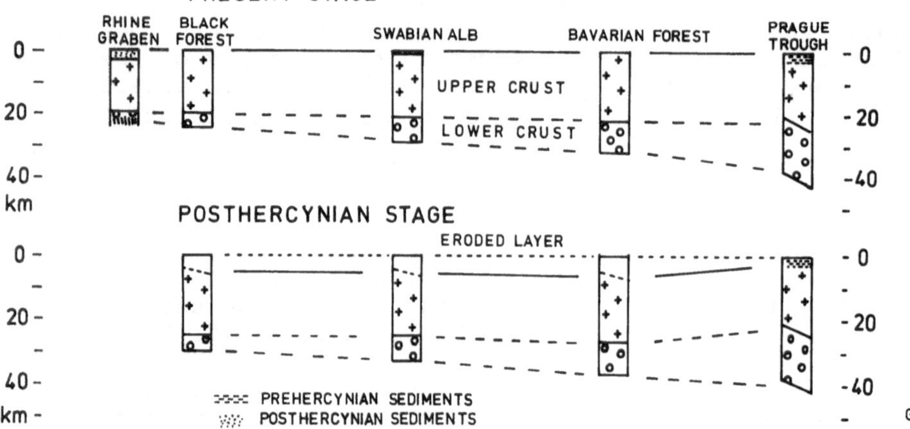

Fig. 1. The upper profile shows the recent crustal thickness along the Moldanubic zone, in the Bohemian Massif (BERANEK et al., 1971) and Bavarian Forest, in the Franconian Alb and the Black Forest. In the lower profile, the attempt of a reconstruction of crustal thickness at the end of the Hercynian orogenic period is given

indicated by the well-expressed P_g- and P^M-wave groups. This clearness and simplicity no longer exist in the region of the Saxothuringian zone; the crustal velocity distribution here is more complex. As an example of this, the wave pattern is discussed on profiles starting from the shotpoint of Hilders (02) in southern direction.

First, the profile 02-215, which runs from Hilders in SW direction (MEISSNER et al., 6.7 of this volume), will be studied. The record section shows two distinct wave groups in the range of 80-150 km distance. The first group (Δ = 80 km, $\bar{t}$ = 2 s) suggests an intercrustal discontinuity at a depth of about 22 km, while the second group may be regarded as P^M (Δ = 70, $\bar{t}$ = 3 s). For the crust/mantle boundary a depth of 29 km results (NE of the Spessart). The record section of the eastern adjacent profile (02-165-01) shows a P^M-group (Δ = 80 km, $\bar{t}$ = 2.5 s) resulting in a crustal thickness of 26 km and an intercrustal group (Δ = 100 km, $\bar{t}$ = 1.6 s), giving a depth of 22 km (FUCHS and LANDISMAN, 1966a, b).

On profile 02-140, the intercrustal group is clearly expressed whereas the P^M-group is rather suppressed (AICHELE, personal communication). The corresponding depth values of 22 and 24 km lie closely together. This trend where both phases approach each other and finally merge is best seen on profile 02-125-09. Due to the lack of a clearly separated P^M-wave, the crust/mantle boundary must consist of a transition zone of 5-10 km thickness, and the top of the upper mantle is characterized by a strongly reduced velocity (about 7.0 km/s). The crustal thickness is at least 24-25 km, but it may be questionable to regard this value as crustal thickness. It would, in this case, be more than 6 km smaller than on the International Profile VI, running through Thuringia (KNOTHE and SCHRÖDER, 1972). Obviously, the conditions of the crust/mantle transition on profile 02-215-09 may be similar to those of the profile Boubin-S; here, the crust/mantle boundary is less clearly developed than a somewhat higher intercrustal boundary zone (MILLER and SCHÜTTE, personal communication).

In the region of the Fichtelgebirge and Oberpfälzer Wald, the crust has a thickness of 28-30 km, as seen from the P^M-group on profiles 09-300-13 and 27-185-01 (AICHELE, personal communication).

A clear P^M-wave allowing a precise separation between crust and mantle can be detected on profile 05-040-02 which runs from the Odenwald in NE-direction to the Rhön. Between the Odenwald and the Spessart, the crust/mantle boundary is to be found at a depth of no more than 25 km (MEISSNER et al., 6.7 of this volume; STROBACH, 1963).

On the left side of the Rhine, a distinct increase of crustal thickness towards W can be stated. The record section of profile 16-195 (Taben Rodt) shows two well-expressed wave groups, resulting in discontinuities at depths of 22 km and 30-32 km (EDEL et al., 1975; PRODEHL et al., 6.8 of this volume). Although the observation material of profiles 08-00 and 16-080-02 is not too good, it can nevertheless not be overlooked that the P^M-phase splits into several individual phases. Similar observations were made on the profile in the northern Alps and the Rhine Graben area (GIESE, 1972; GIESE and PAVLENKOVA, 1974). This feature complicates a clear separation between upper and lower crust. South of Mainz crustal thickness, on the basis of profile 08-000, amounts to about 26 km (MEISSNER et al., 6.7 of this volume).

With its southern section, the Hessische Straße is just still a part of the Saxothuringian zone. In addition to its well-known sedimentary thickness it seems to also have a crust that is by 2 or 3 km thicker than under the flanks. It will be discussed in more detail in the next section which deals with the Rhenohercynian zone.

In summary it can be stated that crustal thickness and nature of the crust/mantle transition differ within the Saxothuringian zone. It is remarkable that an intercrustal boundary in the depth range of 20-22 km is clearly developed in the western part (profile 16-195) as well as in the eastern part of this zone (profiles south of 02).

The greatest crustal thickness of 32 km is reached in the depression of the Saar-Nahe Trough. In NE direction, thickness generally decreases and reaches values between 25 and 26 km between the Odenwald and the Thuringian Forest. The Hessische Straße is a rather special region which will be dealt with in the next section on the Rhenohercynian zone.

In the eastern part of the region under study, at the SW border of the Thuringian Forest, a wide transition zone (5-10 km) is obviosly situated between crust and mantle. This phenomenon is possibly of Postvariscan origin and will also be discussed in detail in the following section.

5.4.2.3 Rheno-Hercynian Zone

The Rheno-Hercynian zone covers, within the region under investigation, the Harz, the Rhenish Massif, Hunsrück, and Eifel. Between the Harz and the Rhenish Massif is situated the Hessische Straße, a depression that was active with varying intensity during the Mesozoic and the Tertiary (JACOBSHAGEN, 1 of this volume).

In the Rheno-Hercynian zone lie numerous quarries firing large explosions. This allowed the observation of a whole series of profiles in this region. Further on a common-depth point profile with its center between Koblenz and Bingen was carried out in the Rhenish Massif (BARTELSEN, 1970).

When discussing the crustal structure of the Rhenish Massif, the research program must be mentioned which served for the exploration of the siderite zone of the Siegerland, with its main activity in the fifties and the first half of the sixties (BOSUM et al., 1971). In the course of this program, not only geological investigations but also extensive seismic, gravimetric, magnetic, and geoelectric measurements were made which covered not only the shallow structures but also supplied information on the structure of the crust.

On all seismic refraction profiles, a wave has been recorded which may formally be called P_g, indicating at which depth velocities between 4.0 and 6.5 km/s are reached. In the Rhenish Massif and in the Harz, these values are due to clays and slates, sandstones, quartzites, and limestones existing in these regions. According to FRITSCH (1967) the velocity steadily increases from at least 4 or 5 to more than 6 km/s down to a depth of 8 km; a clear change of material cannot be ascertained by means of the velocity distribution. In the Rhenish Massif as well as in the Harz it is, therefore, not possible to separate clearly by aid of the velocity distribution the top of the hypothetic Prehercynian basement from the highly metamorphic rocks of Hercynian age. From the geological point of view it is even probable that, in the upper 10 km, a continuous transition from weakly to highly metamorphic rocks has to be expected.

In the central part of the Rhenish Massif, the region of Siegen, within the upper 5 km the velocity distribution reflects the NE-SW striking anticlines and synclines. The existence of a pluton which could be regarded as source of the ore deposits in this area could not be definitely proved down to a depth of 8 km (FRITSCH, 1967; BOSUM et al., 1971).

Within the region of the Hessische Straße, P_g-arrivals show a significant delay up to 1 s as compared to sediment-free zones. Maximum delays occur in the region between Fulda and Kassel, i.e. in a zone marked by an especially pronounced depression during the Zechstein and Buntsandstein times. In the center, the velocity of 6 km/s is reached at a depth of 6 to 7 km, whereas in the border zones of the Hessische Straße it occurs at a depth of 4 to 5 km. Due to the penetrating nature of P_g-waves, the top of the basement is situated shallower, by approximately 2 or 3 km. This means that an additional subsidence of about 1-2 km must have taken place. It shall be mentioned here that, in the vicinity of Kassel, the Buntsandstein alone reaches a thickness of 1000 m.

Regarding the velocity distribution in the deeper part of the crust, the common-depth profile in the southern part of the Rhenish Massif gives some detailed information (BARTELSEN, 1970). The thickness of the crust between Bingen and Koblenz amounts to 28-29 km. Two less prominent intercrustal discontinuities at 26 and 23 km depths are derived by BARTELSEN (1970). Between these discontinuities lie weak-velocity inversions. To name one of these discontinuities Conrad discontinuity seems hardly expedient since, in this case, a clear definition ought to exist which, however, cannot be given.

Similar to the Saxothuringian zone, the Rheno-Hercynian zone also shows a P^M-group of widely differing clearness. In the following the question will be discussed whether a relationship exists between the well-marked appearance of the P^M-group as well as its anomalous behavior and the geology of this region.

A clear P^M-group has been recorded on profile 13-240-20 from which a crustal thickness of 27 km results. The P_n-wave shows a velocity of 8.2 km/s. The reverse profile 13-240-20 shows the P^M-group as well. Thus the crust/mantle boundary in the NE part of the Rhenish Massif can be regarded as normal.

The situation is different on the refraction line between the quarries of Hilders (02) and Mehrberg (14) which passes the Westerwald, an area with young volcanism.

In the record section of profile 14-09-02, in a medium distance range of 80-150 km, a complex wave group occurs which probably has to be interpreted by two reversed segments. An intercrustal discontinuity lies at about 18 km depth. As the events of the second group can be indicated only with a tolerance of ± 0.1 s, the depth resulting varies between 25 and 27 km. Although depth data vary, the inclination of a P_n-type branch permits the statement that the velocity has to be smaller than 7.7 km/s in the region of the strongest gradient. Here also exists, therefore, an anomalous crust/mantle transition.

In this connection, the results of deep-reflection measurements in the Siegerland region, carried out 20-30 km N of profile 14-090-02, are of interest (BOSUM et al., 1971). On three profiles having a total length of 60 km, crustal reflections could be observed. The data were evaluated using the statistical methods proposed by DOHR (1957a, b) and LIEBSCHER (1962, 1964). The evaluation shows that no marked maxima exist in the histograms, such as LIEBSCHER (1962, 1964), for instance, has found in numerous South German regions covered by measurements. The general picture is that of a relatively constant reflection frequency with, on the whole, accidental variations. Two weak maxima are just noticeable on all three profiles, one at 1.5-2 s and another between 5 and 7.5 s two-way travel time. A relatively narrow reflection band as compared with the other reflections appears at 8 s travel time. An agreement of the results of reflection seismic with those of refraction seismics is signified by the fact that both reveal a transition zone between 18 and 25 km depth. Although the region under study around Siegen is situated about 20 km N of the refraction line 14-090-02, crustal structure, and especially crustal thickness, does not seem to show great variations in this area.

When dealing with the Saxothuringian zone, the profiles running S from the quarry of Hilders (02), have to be taken into account, while for the Rheno-Hercynian zone the profiles running W and N from this point are of interest.

The profile 02-265 does not show any discernible P^M-group, similar to profile 02-240-19, whereas this group is well developed on profile 02-220-05. In the record sections of the profiles radiating in NW direction (02-300 and 02-325) there are some indications of the P^M-group; since, however, these profiles cross the Hessische Straße causing a delay of P^M arrivals due to sedimentary fill, its correlation is complicated. On the record section of the profile 02-350-06, clear arrivals of a wave similar to P^M could be detected. The result, however, is a depth of only 23 km. A P_n-like branch shows a velocity of 7.7 km/s. These low values suggest here, too, a possible anomalous crust/mantle boundary. An inspection of the geological map reveals that the first part of this profile is passing the area N of the Rhön, showing young volcanism.

Before summarizing these comparative studies, the profiles of the shotpoints Dorheim (13) and Adelebsen (06) shall be investigated with regard to the development of the P^M-group.

On the profile 13-120-02 (09), distinct P^M arrivals have been recorded although they are delayed by the sedimentary fill of the Hessische Straße. An estimation of crustal thickness amounts to 28-29 km.

From the shotpoint Adelebsen 06, some profiles were observed which are still situated in the Rheno-Hercynian zone. The southwards running profile 06-170-02 shows a distinct reversed segment similar to a P^M-phase. The strongest velocity gradient has been found in the interval 6.8 to 7.2 km/s in a depth range of 23-24 km. From the clear P_n branch results a velocity of 7.9-8.1 km/s. This means small crustal thickness and normal velocity for the upper mantle.

On profile 06-260 running W from Adelebsen (06), the P^M-group is clearly pronounced while on the northern profile 06-350 the correlation of the existing P^M arrivals is complicated by the increasing thickness of sediments and the high noise level in this area.

Summarizing the main results of the qualitative comparison of the occurrence of the P^M-wave in the record sections, it can be stated that it is poorly developed in the Westerwald, in the Vogelsberg and in the Rhön, i.e. in regions with Tertiary volcanism. In the region between Hilders (02) and Adelebsen (06), the crust/mantle boundary extends over a wider depth range. In the sector SW to SE of the shotpoint of Hilders (02) which is free of young volcanism, the P^M-wave is well established in the record sections. Thus the conclusion may be drawn that a relationship exists between young volcanism and structure of the lower crust and the character of the crust/mantle boundary. In areas with extensive Tertiary volcanism, the P^M-group is distinctly suppressed, thus indicating a wide crust/mantle transition zone. In regions with weak volcanism, a P^M-like group is observed, but the thickness of the crust and the velocity at the depth of the strongest velocity gradient are reduced.

These considerations coincide with those of the previous section with regard to profile 02-125-09. The record section of this profile does not show any clear P^M arrivals. A glance at the geological map shows that the first 50 km of this profile reach as far as the "Heldburg dike zone", a region characterized by young basalt dikes.

In Fig. 2 the distribution of Tertiary volcanism in the area under discussion is shown. Besides, the figure reveals the regions where the P^M-group can be observed well, moderatly well or badly. It would be premature to look upon these relations as absolutely certain. The main purpose of these considerations is to draw attention to this problem.

In cooperation with commercial exploration, two seismic refraction profiles crossing each other perpendicularly could be observed between the Harz and the Rothaargebirge, the NE part of the Rhenish Massif (GRUBBE, 1969, 6.4 of this volume). For the point intersection of these two profiles GRUBBE (6.4 of this volume)

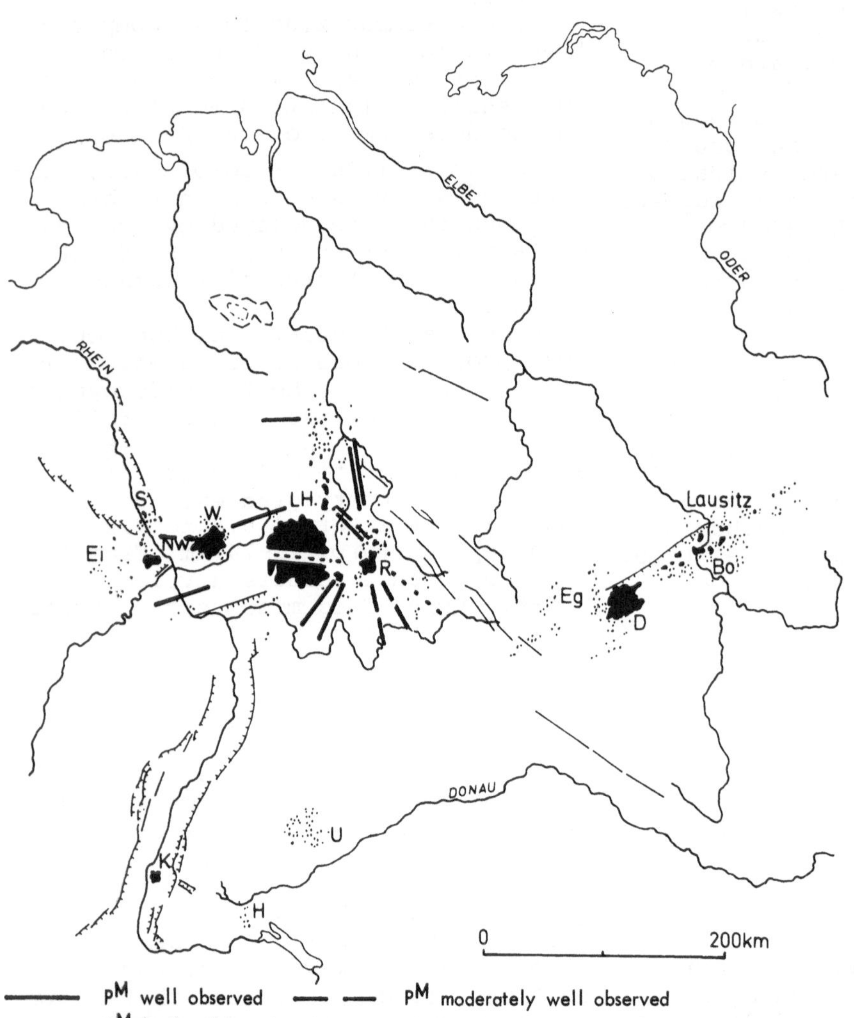

Fig. 2. Distribution of Tertiary volcanism and regions where the P^M-group can be observed well, moderately well, or badly. The areas indicated are related to the distance of ray vertices, i.e. half the observation distance

states a crustal thickness of 27 km while the crust/mantle boundary dips towards NW and SW. A reinterpretation by the author leads to results as to crustal thickness that are somewhat different from those given by GRUBBE. In the Rothaargebirge, a crustal thickness of 27 km has been determined, whereas for the region near Göttingen, a value of 27 km results. This value is in agreement with the results derived from the profiles radiating from the shotpoint of Adelebsen (06).

Some further information on the crustal structure of this zone is supplied by reflection recordings with long travel time, on which DOHR (1968) published a summarizing report. Here, regions S and N of the Harz are concerned. For the region under study at the southern border of the Harz, a crustal thickness of 27-32 km is given while, for the northern border, depth values of 23-25 km and 30-32 km are stated. The last-mentioned value eventually represents the depth of the crust/mantle boundary.

In the adjacent Thuringian trough, the NW part of which still belongs to the Saxothuringian zone, crustal thickness amounts to about 28-30 km, values corresponding to those of the Rhenish Massif (KNOTHE and SCHRÖDER, 1972).

Finally, the question will be briefly discussed whether the Hessische Straße, essentially of Postvariscan formation, influences in any way crustal structure. The largest crustal thickness of 28-30 km occurs in a N-S trending zone with its axis between Fulda and Kassel. East and west of this zone crustal thickness is by 2-3 km smaller, so that the depression of the Hessische Straße seems to be reflected also by increased crustal thickness. It must be mentioned, however, that, especially at the E border of the depression, values of small crustal thickness result from regions, where the crust/mantle boundary shows an anomalous behavior while in the depression itself a clearly developed P^M-group (profile 13-120-09) suggests a well-marked crust/mantle boundary. Therefore, it cannot be decided whether the larger crustal thickness in the depression has to be explained by a dipping crust or by an upward thrusting of the crust/mantle boundary in the adjacent regions to the east and west.

Within the regions of young volcanism, a time delay between the P_g and P^M-like branch indicates the existence of a crustal low velocity layer (e.g. 14-090-02). In areas without young volcanism, the velocity inversion seems to be weaker (e.g. 13-240-20).

In summary the crustal structure of the Rheno-Hercynian zone shows the following characteristics:

Crustal thickness in regions without Tertiary volcanism amounts to 29 km at the southern and to 25-27 km at the northern border. Intercrustal discontinuities are suggested but cannot be correlated over somewhat larger distances.

To an even greater extent than in the Saxothuringian zone, here Tertiary volcanism seems to have influenced the crust/mantle boundary and the lower crust in a way that led to a widening of the transition zone and to a marked shifting of the region of the strongest velocity gradient, situated normally between 7 and 8 km/s, into the range of 6.5-7.5 km/s.

An influence of the depression of the Hessische Straße can be seen only in its southern part, roughly between Göttingen and Fulda, where a crustal thickening of 2-3 km is observed. However it cannot be decided whether this is due to a real dipping of the entire crust or merely to an upward thrust of the crust/mantle boundary as a result of magmatic processes in the neighboring regions.

5.4.2.4 Subvariscan Foredeep and its Northern Foreland

To the north, in front of the Rheno-Hercynian zone, the Subvariscan zone extends as a molasse trough (JAGOBSHAGEN, 1 of this volume). Its northern boundary runs approximately along the line Osnabrück-Hannover; N of this zone, the foreland of the Hercynian orogene follows. This foreland, belonging to the Old Red Continent in the Devonian time, was part of the Caledonian orogenic system. Permian and Mesozoic sediments can reach a thickness of over 5000 m. The subsidence continued during the Tertiary, and 2000-m sediments were deposited in some places. Including Devonian and Carboniferous, the sedimentary cover of the Caledonian basement may reach 10,000 m or perhaps even more.

In the following, both zones will be treated together.

Contrary to Central and Southern Germany, relatively few crustal seismic data are available from the North German area.

Although crustal seismics scored their first great success in 1947 on the occasion of the Helgoland explosion and its observation on a profile through the North German lowlands, this region remained of secondary importance in the scope of activity during the decades that followed.

As the position map [Fig. 1 (3.1)] shows, no quarries are situated in the North German area where larger shots are fired. Only between Minden and Osnabrück, smaller shots fired in some quarries could be used for the observation of short refraction lines aiming to investigate the crustal structure of the Massif near Bramsche (THYSSEN et al., 1971). The North German lowlands are certainly covered in part by some seismic refraction profiles from the south but, unfortunately, at greater shotpoint distances. Thus, only P_n arrivals could be recorded, these events, however, being strongly disturbed by the high noise level in this area (HINZ et al., 6.3 of this volume).

The lack of crustal information of this area is somewhat mitigated by recording of deep reflections (HEHN, 1964; HADJEBI, 1966; DOHR, 1968), available from some areas. Further, a large-scale refraction program was carried out, ordered by the oil industry for the exploration of pre-Mesozoic formations. Many of these shots were recorded at greater distances for the purposes of crustal studies (GRUBBE, 1969, 6.4 of this volume; THYSSEN et al., 1971). The records have only partly been evaluated. The following statements, therefore, can only in general refer to the structure of the crust in the North German region.

As already mentioned, the Helgoland explosion in 1947 offered the first opportunity to record a long-distance refraction profile in a SSE direction. Results of this profile were published by WILLMORE (1949a, b), REICH (1950) and SCHULZE and FÖRTSCH (1950). The record section of the first part of this profile, compiled at a later date in the geophysical institute of Karlsruhe, is shown in the appendix.

On the seismic refraction profiles recorded in the North German lowlands, arrivals were recorded which, on the basis of their velocity, can be regarded as P_g-wave. Here, however, the question arises how this value can be interpreted from the geological point of view. Diagenetically strongly consolidated sediments, such as quartzites and limestones, can by all means have velocities up to 6 km/s for longitudinal waves, so that they do not differ from gneisses and granites. In the borehole Münsterland I, nonmetamorphic Devonian sediments were found at a depth of 6 km having a velocity of 6 km/s, as refraction recordings show (THYSSEN, 1964).

The crustal structure between Harz and Rhenish Massif is discussed by GRUBBE, 6.4 and HINZ et al., 6.3 of this volume. In the region between Göttingen and Hannover, a refracted wave with a velocity of 6.3 km/s in the depth range between 10 and 15 km was found. This velocity value may be correlated to high metamorphic Paleozoic sediments and/or to the crystalline rocks of the Caledonian basement. This boundary dips gently from S to N, reaching a depth of 15 km near Hannover. This result agrees well with reflection observations recorded south of Hannover showing clear phases with a two-way traveltime of 6.0 to 6.5 s (DOHR and MEISSNER, 1975). SCHULZE and FÖRTSCH (1950), using the data of the Helgoland profile, derived a Conrad-discontinuity (6.4 km/s) at a depth of 9-11 km, but this will presumably be the above-mentioned refraction horizon from paleozoic layers of the sedimentary cover or from the uppermost Caledonian basement. It must be kept in mind, however, that the results of the Helgoland profile are based on only relatively few records.

For the region S of Hannover, GRUBBE 6.4 and HINZ et al., 6.3 of this volume show crustal thicknesses of 25-30 km. These statements have been confirmed by more recent reflection measurements. DOHR and MEISSNER (1975) state two-way travel times of about 10 s; this corresponds to a crustal thickness of 28-29 km. As to the downdip, however, data differ. In accordance with the results presented in 6.4, the evaluation of all data from the region Göttingen-Hannover reveals an increase of crustal thickness from S to N.

Further data are available from commercial reflection exploration.

The evaluation of the deep reflections was carried out in the same manner as LIEBSCHER (1962, 1964) treated the data in Southern Germany (HEHN, 1964; HADJEBI, 1966; DOHR et al., 1967; DOHR, 1968). In general it can be stated that there must exist distinct essential differences in crustal structure between Northern and Southern Germany (DOHR, 1968). While in Southern

Germany the histograms show sharp maxima which can be well correlated also over regions lying somewhat further apart, the picture in Northern Germany is, in most cases, more complicated. The reason for this may partly be the fact that the real reflections are often superimposed by multiple reflections, but partly also by the different nature of the deep reflecting boundary zones. This idea is supported by the blurred appearance of the maxima in various regions, showing a considerable half value width.

From the eastern part of the Münsterländer Bucht, near Gütersloh, a reflection profile is published showing well-expressed deep reflections (HEHN, 1964). A wide band of reflections has to be attributed to a depth range of 18-25 km. This suggests a corresponding wide transition zone between crust and mantle. For the crustal thickness, a value of 25 km may be assumed.

At the northern margin of the Subvariscan foredeep in the region NW of Osnabrück, the anomaly near Bramsche is situated. This structure near Bramsche is characterized by a positive BOUGUER anomaly of 40-50 mgal and a positive magnetic anomaly of about 200 γ. The coal of the Carboniferous mined near Ibbenbüren shows a relatively high stage of coalification. This anomalous structure has been investigated in detail by the institute of geophysics in Münster (BROCKAMP, 1967; THYSSEN et al., 1971).

The body causing this anomaly has been identified as a layer with a P_g-wave velocity of 6.2 km/s; its top rises from a depth of 8 km in the north to 5 km in the center of the anomaly.

While BROCKAMP (1967) gives a crustal thickness of 28-30 km for the region Osnabrück-Ibbenbüren, DOHR et al. (1967) point out that, on the basis of the data of reflection measurements, the M-discontinuity lies at a depth of 26 km. THYSSEN et al. (1971) published the record section of a N-S profile passing this anomaly which shows a P^M-group yielding a crustal thickness of 25-26 km. It has to be mentioned, however, that the value of 26 km cannot be regarded as anomalously small because similar values could be found outside of the Bramsche anomaly, i.e. near Gütersloh.

THYSSEN et al. (1971) also cannot detect any premature P_n arrivals which would indicate a local uplift of the crust/mantle boundary in the region of Bramsche.

The investigations described above apply mainly to the region of the Subvariscan foredeep. Concerning the crustal structure of the foreland, i.e. the North German lowlands, some information can be derived from the Helgoland profile. From some regions, recordings of deep reflections are also available.

The record section of the Helgoland profile shows, in the distance range from 60-200 km, later arrivals with large amplitudes which may be interpreted as P^M-wave yielding a crustal thickness of 28-29 km for the region between Helgoland and Bremen. WILLMORE (1949a, b) and also SCHULZE and FÖRTSCH (1950) give a somewhat smaller value of 27.4 ± 1 km, but this small deviation is within the range of resolving power.

In this connection, attention should be drawn to an observation that can be derived from the record section of the Helgoland profile. At the critical distance of the P^M-wave between 60 and 70 km, the time difference between the P^M-wave and the P_g-group is relatively small, i.e. no more than 1 s. Normally, this value exceeds 1.5 s. The small difference between the two wave groups means that the thickness of the depth range between the velocity values 6 and 7.5-8 km/s cannot be more than about 15-16 km (6.0 km/s at a depth of 13 km, 7.5-8 km/s at a depth of 28-29 km). Although the boundary sediment against basement cannot be exactly determined, it may be expected that the pre-Caledonian crust in its present state has a thickness of only 15-20 km. Starting from the assumption that, at the end of the Paleozoic, the crust had about this thickness, the continuous subsidence during Mesozoic and Tertiary times because of isostatic reasons becomes understandable.

Concerning the crustal structure of the foreland, i.e. the North German lowlands, some clues can be gained by evaluation of deep reflections. Fig. 3 shows a compilation of frequency distributions of deep reflections in the region of eastern Emsland with the attempt of a correlation (DOHR et al., 1967).

Following this correlation, the two-way travel time between Ibbenbüren and Oberlanger-Tenge increases by 1.5 s for the crust/mantle reflections. This time can, essentially, be attributed to the sediments increasing in thickness towards N so that a crustal thickness of 29-31 km results for northern Emsland.

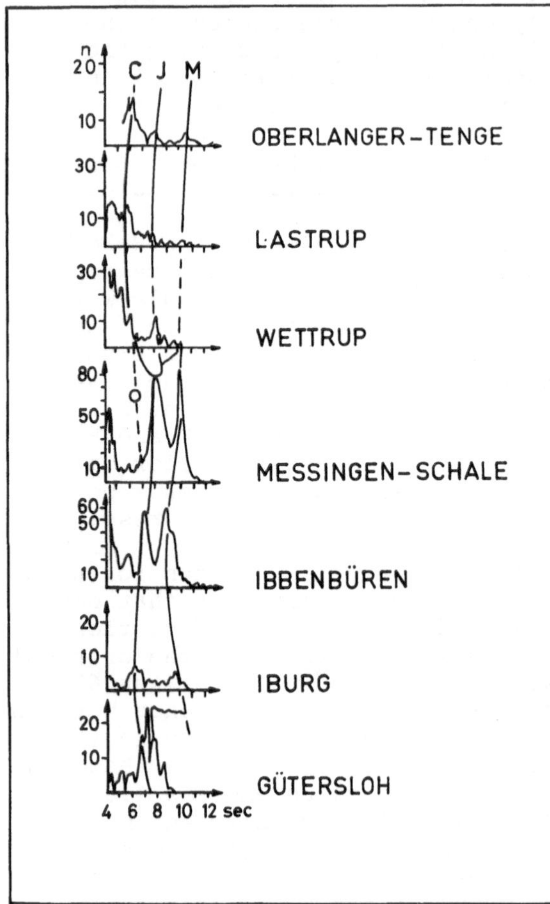

Fig. 3. Frequency distribution of deep reflections in the region of eastern Emsland with an attempt of a correlation between the several local exploring areas (DOHR et al., 1967). *C* Conrad discontinuity, *I* Intermediate discontinuity, *M* Mohorovičić discontinuity

A corresponding N-S profile for the region E of Hannover has been compiled by DOHR et al. in 1967; it also results in a crustal thickness of about 29-31 km between the northern border of the Harz and Wolfsburg.

Further results are available from the region of Schleswig-Holstein. Here, near-vertical as well as wide-angle reflection data exist. South of the Danish border, near Flensburg, both observations suggest a crustal thickness of 30-31 km (DOHR et al., 1967), the near-vertical reflections showing a downdip towards north. In central Holstein, the crust/mantle boundary is found at 26-27 km depth from near-vertical and wide-angle reflections. This value may confirm the dip of the M-discontinuity towards north.

A summary of the individual results discussed above permits the statement that crustal thickness increases from 25-26 km beneath the Variscan foredeep to about 30 km beneath the foreland. This increase of thickness is probably due to the Mesozoic and Tertiary sedimentary cover thickening towards north.

Whether at the coast of the North Sea and in Holstein crustal thickness decreases again over a wide area cannot be ascertained by aid of the few available data. In the North German lowlands, the sediment-free (pre-Caledonian) crust has but a thickness of 15-20 km. From the occurrence of clear reflections at 6-7 s it may be concluded that the boundary of upper crust against lower crust lies in a depth range of 18-22 km. In some regions this means, at the same time, the top of the crust/mantle transition zone.

The anomaly near Bramsche seems to be caused by the high position of intercrustal rocks. A higher position of the crust/mantle boundary obviously does not exist.

5.4.2.5 Comparative Study of Crustal Structure of the Variscan Orogene

After the individual zones of the Hercynian orogene have been dealt with, an attempt to give a summary comparison follows.

A crustal thickness of about 40 km may be assumed for the Moldanubic zone after the end of the orogenesis, i.e. in the beginning of the Permian period. This zone probably did not undergo an intensive tectonic compression during the Hercynian orogene as is shown, for instance, by the Barrandium. The essential orogenic phases are considered to be Prehercynian. The large crustal thickness can, however, be regarded as responsible for the fact that sialic material reached down to 20-30 km, melted there partially or even completely and rose after that to form granite plutons in the upper crust.

AUBOUIN (1965) ascribes the function of the hinterland to the Moldanubic. The question of how and when the anomalous crustal thickness came into existence must, for the present, be left undecided.

To the actual geosyncline belong the Saxothuringian and the Rheno-Hercynian zones. Here, a Post-Hercynian crustal thickness of about 30 km may be assumed. Both zones show strong compressions, but there exists no such nappe structure as in the Alpine mountains. It is, so far, not known in which depth range the compression took place. Corresponding to the moving direction of the orogenesis from the inside outwards, i.e. from S to N, the shearing-off planes probably rise to increasingly higher levels. In the Saxothuringian zone, crystalline from the upper and middle crust has already been included in the tectonics, as the gneiss block at Münchberg proves.

WUNDERLICH (1966), on the basis of fold stereometric studies, gives an original width of 310 km for the region between Wiesbaden and Mülheim which has now a width of 180 km. Although exact figures may be a matter in dispute, and there may be local differences, an approximation in the order of 100 km between the Moldanubic zone and the northern foreland has to be assumed.

In this connection, the question of the large crustal thickness of the Moldanubic zone shall be discussed once more. It has already been mentioned that, probably, deeper and deeper crustal regions were caught by shearing-off planes toward the south, so that it is quite imaginable that, finally, the Moldanubic zone as a whole moved like a block as hinterland toward the north and that, in this process, sialic crustal parts were subducted under it. In this way, the compression as well as the large crustal thickness of this zone can be explained. It must be expressly pointed out, however, that this can be no more than a vague hypothesis at the moment, which requires further discussion. The suggestion of such a model has its origin in analogous studies for the Southern Alps (GIESE, 1968a; GIESE et al., 1970).

5.4.3 Upper Rhine Graben

The upper Rhine Graben is the central segment of the system of tectonic rifts running through western Europe from N to S (e.g. ILLIES, 1970, 1974; JACOBSHAGEN, 1 of this volume). In the interest of trade and industry, the sedimentary filling of this graben has already been investigated quite intensively. Within the scope of the international Upper Mantle Project, the exploration of the graben was given a broad basis, and modern geoscientific methods were applied in order to investigate structure and development as well as its behavior in the present and the past. Three comprehensive publications on this research work have been published (ROTHE and SAUER, 1967; ILLIES and MÜLLER, 1970; ILLIES and FUCHS, 1974).

These monographs also contain contributions which deal with the crustal structure as it was determined by means of refraction seismic data. During the past decade, the opinion about the structure of the crust and the upper mantle in the region of the Upper Rhine Graben was subject to certain changes, caused by the increase of available data. An excellent survey of this example of a case history is given by PRODEHL (6.8 of this volume). In this paper, the most important reasons are mentioned that led to these different opinions. It also contains a summarizing presentation of an interpretation of all seismic refraction data available in this region as compiled by EDEL et al. (1975). Although, in the meantime, the different interpretations have been reconciled with regard to a number of questions, some points still require a discussion and supplementation. Therefore, the following problems will be discussed:

1. The structure of the lower crust and the crust/mantle boundary is of great importance for petrological studies. The wave groups supplying information on the velocities in these depth ranges are, however, of a complex nature and, therefore, the interpretation of data is, to a certain extent, ambiguous.

2. The question whether a low-velocity layer exists in the middle crust has, so far, always been answered in the affirmative. EDEL et al. (1975), however, are of the opinion that the question whether such a low-velocity layer exists cannot be decided definitely and they rather assume, therefore, a crust without velocity inversion. Just for geotectonic considerations, however, the existence of a low-velocity layer is of greatest importance (FUCHS, 1974; MÜLLER and RYBACH, 1974).

3. Finally, the crustal structure of the Rhine Graben will be compared with that of the neighboring regions to discuss what has to be considered normal and anomalous.

The discussion of these questions, especially those under (1) and (2) make it necessary that some details of the record sections be discussed.

In 6.8, a travel-time diagram showing the principal arrangement of the main branches is given. Although details of this draft may be controversial, it can generally be stated that the picture is more complex than, for instance, that of record sections of the Southern German region (see, for example, profiles 01-010-09 and 01-345-02). In the following will be discussed how far similarities or major deviations exist between the arrangement of the main travel-time branches in the record sections of the anomalous Rhine Graben and of the eastern and western adjacent areas.

In the record sections of the profiles in the Black Forest (24-090, 24-170) and the Vosges (BA-010, SN 170) the P_g-wave is clearly recognizable since here no sedimentary cover exists or its thickness can more or less be neglected. In the Rhine Graben, the situation is more complicated owing to the thick sediments reaching several km in thickness. No further investigation will be made as to whether the P_g-wave appears here as a reverse segment or a normal segment. At any rate, observations in a distance range between about 20 and 60 km show first arrivals which must be attributed to a wave that has travelled in the uppermost kilometers of the basement. Local premature or delayed arrivals suggest corresponding local uplifts or depressions in the Rhine Graben itself. Generally it may be said that the appearance of the P_g-wave may be called normal and does not differ from that in the neighboring regions, apart from the large time delay due to thick sediments. EDEL et al. (1975) study in detail the P_g onsets with respect to the thickness of the sediments by aid of a time-term analysis. It is remarkable, however, that the P_g-wave on most profiles disappears at a distance of about 60 km. This fact is, in principle, known and has often been described (e.g. GIESE, 1963; MEISSNER, 1966; HOLUB, 1974). Moreover, in undisturbed areas, the P_g-wave can often be observed over a distance of 80 km and even further (e.g. 01-020-09, 01-345-02, 02-165-01). The most striking wave group with large amplitudes appears in the second part of the record sections; it begins at about 60 km and can be followed up to 150 or 200 km. According to the terms used in 6.8 of this volume, branches 1 and 6 and partly also 2 and 3 belong to this group (Fig. 9, 6.8). These branches are situated in the area where, normally, the P^M-curve is found. Without doubt, this prominent wave group is complex and has to be interpreted by more than one travel-time curve. There is no absolutely definite answer, however, to the question how the arrivals have to be correlated into individual branches.

In order to eliminate strong lateral inhomogeneities, the following discussion will be confined to the record section of the profiles SB-010-WI and WI-190-SB which run in the graben and parallel to its axis. The right-hand part of branch 6 can be established quite well, partly even by phase correlation. In the direction towards the shot point, at a distance of 80 to 60 km, amplitudes become smaller and less clear. Possibly, branch 2 may be regarded as continuation of curve 6. Whether branch 2 is to be interpreted as over- or subcritical reflection shall be left undecided. As already mentioned, a second travel-time curve has to be introduced in order to satisfy further very late arrivals at a distance of 50-80 km (phase 1 in 6.8).

Since, in the case of these onsets, a clear and definite phase correlation is difficult, opinions on possible correlations may differ (Fig. 4a, b). EDEL et al. (1975) and 6.8 of this volume place curve 1 so that it cuts curve 6 at its left end or comes, at least, very close to it (Fig. 9, 6.8).

The author, however, is of the opinion that a less curved branch 1 provides a better answer to the observations (GIESE and PAVLENKOVA, 1974). In accordance with this interpretation, the left end of curve 6 and the right end of curve 1 are practically parallel and show an interval of 0.5 to 1 s (Fig. 4a, b). A similar interpretation was also proposed more recently by EDEL (1975).

Irrespective of these detailed questions, these waves suggest discontinuities or zones with strong velocity gradient at depths between 20 and 25 km. The question of the designation of this wave group and the corresponding discontinuities will be discussed later on.

The velocity depth functions resulting from these two versions differ in so far as in EDEL et al. (1975) no velocity inversion is supposed to lie between the two interfaces while, according to the version of the author, there exists an intensive velocity inversion between the two interfaces (Fig. 5). The more recent models by EDEL (1975) also contain such a velocity inversion in the lower crust.

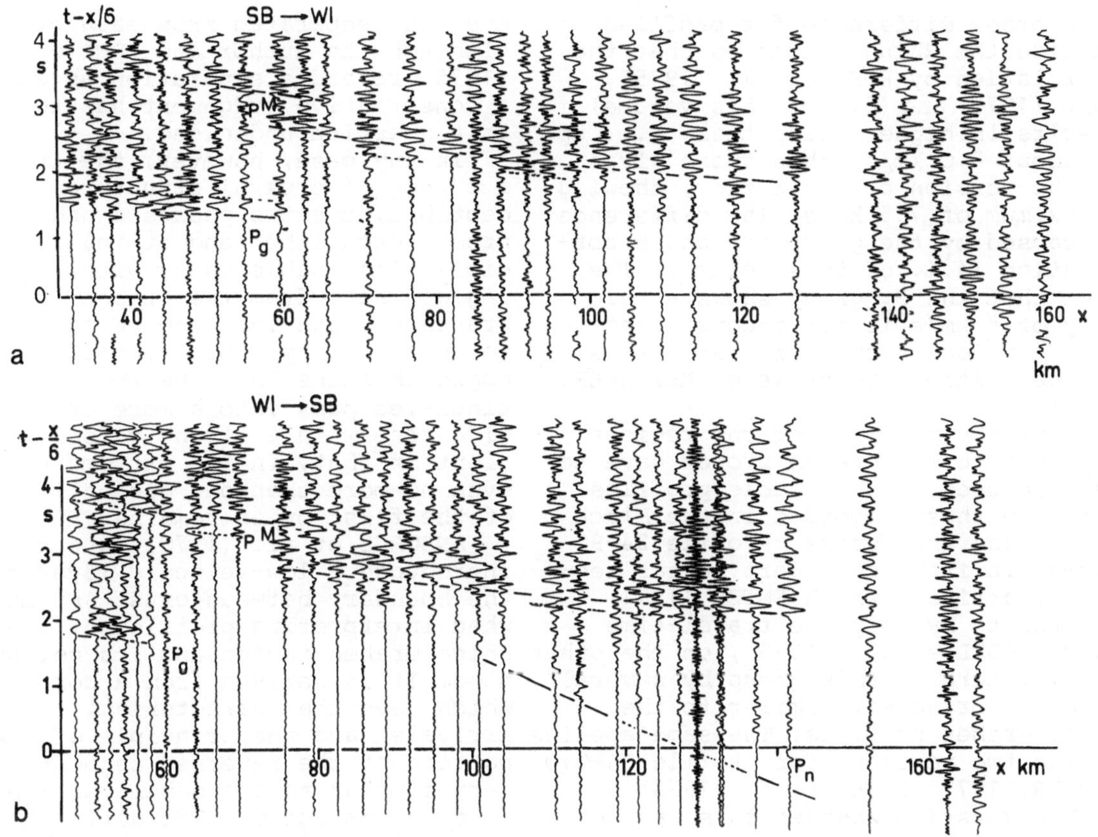

Fig. 4a and b. Record section of the profile (a) Steinbrunn (*SB*); Wissenbourg (*WI*) or (b) Wissenbourg (*WI*); Steinbrunn (*SB*) resp. (after EDEL et al., 1975). The correlation shown here differs from that given in EDEL et al. (1975). The important feature is the split nature of the P^M-group. The travel-time segments of the P^M group are approximately parallel

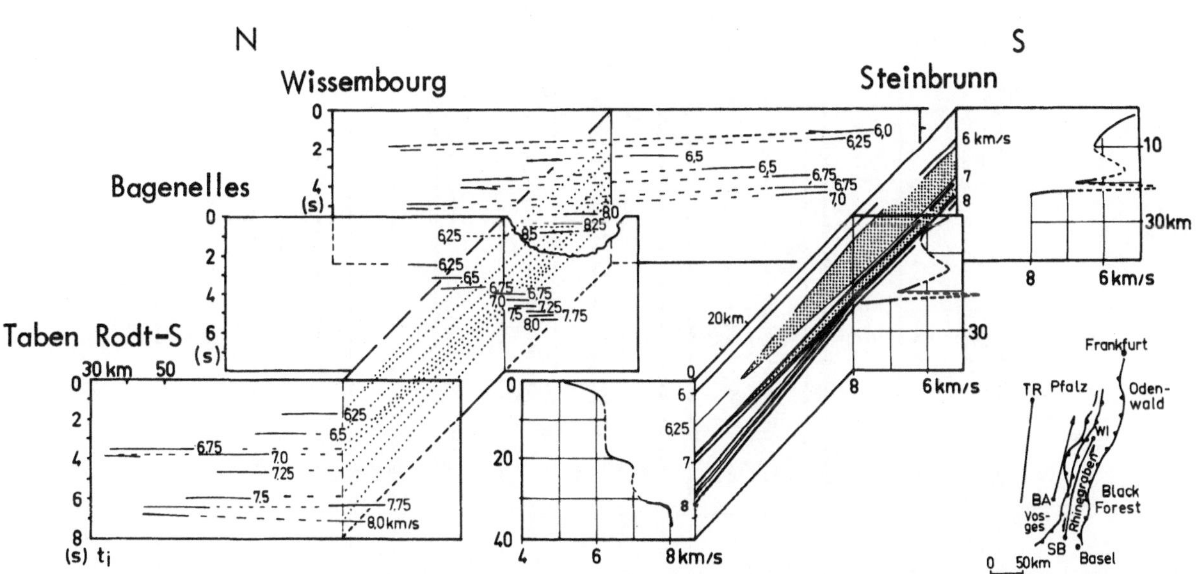

Fig. 5. Block diagram showing intercept time and depth cross sections of the profiles Steinbrunn-Wissembourg and reverse, Bagenelles-N, and 16-195 (Taben Rodt-S). The two profiles being perpendicular to the axis of the Rhine Graben are constructed from the parallel profiles

Another difference for profiles outside the Rhine Graben is that the correlation by EDEL et al. (1975), EDEL (1975) and (6.8 of this volume) results in higher velocities, i.e. up to about 8.0 km/s, than those given in the version of the author, showing a maximum of 7.5 km/s. The difference is caused by the group and phase correlation (4.1. of this volume). The fact that the P_n-group shows a velocity of 8 km/s is not inconsistent with a velocity smaller than 8 km/s at the critical point (e.g. MEISSNER, 1970).

Next the profiles of the regions adjacent to the graben proper have to be discussed. The suitable profiles here are those recorded in the Black Forest and the Vosges. Profile 24-90 shows, in its second part, clear secondary arrivals which EMTER (1971) interprets by only one travel-time curve. EDEL et al. (1975), on the other hand, interpret this group by several separate branches, similar to the Rhine Graben profiles. The same applies to the two versions for profile BA-010 (EMTER, 1971; EDEL et al., 1975).

The question whether this wave group is to be interpreted by one or several curves cannot be as clearly decided as in the case of the Rhine Graben profiles because the phases under study are less separated from each other. Although the author and EDEL et al. (1975) regard a separation as existing, a conservative judgement has to leave this question undecided.

In any case, however, the trend to a change of crustal structure is perceptible, the two phases 6 and 1 approach each other and it is, in principle, imaginable that, with increasing distance from the graben, both branches merge into one. With regard to the velocity-depth function this approach means that the two discontinuities pass into a more or less wide zone with a strong velocity gradient or that, in the extreme case, even no more than one discontinuity of the first order remains.

ANSORGE et al. (1970) have already introduced a low-velocity layer at the boundary between crust and mantle when interpreting profiles of the Rhine Graben region and its neighborhood. It is derived from a correlation which uses the characteristic later arrivals, and the principle of the pattern of the reversed segments is very similar to the scheme as identified by the author. The difference, as a matter of fact, is that these reversed segments are differently placed in the record sections (Fig. 6).

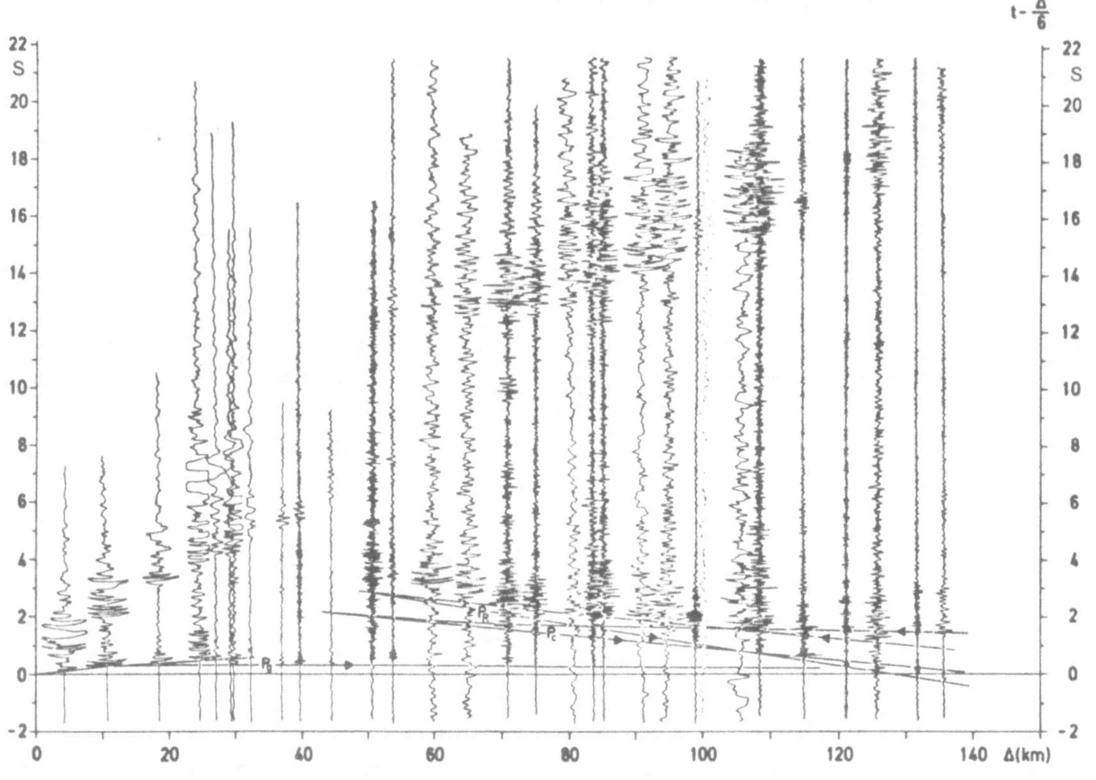

Fig. 6. Record section of the profile Bagenelles-N after ANSORGE et al. (1970). The P_c-phase should indicate the existence of a low-velocity layer in the middle part of the crust

In which way does this discussion help to determine the crust/mantle boundary? EDEL et al. (1975) and PRODEHL et al. (6.8 of this volume) can recognize phase 1 which they regard as the usual P^M-phase only on profiles or parts of profiles outside the graben but not in the graben itself. Thus, these authors consider the crust/mantle boundary outside the graben as given by a first-order discontinuity, while in the graben proper, the strongest gradient is encountered at a depth of about 21 km (phases 6 and 2), and they consider the whole depth range from 21 to 26 km as a zone of crust/mantle interaction.

The present author, on the other hand, is of the opinion that the profiles inside and directly outside the graben show qualitatively the same pattern of the corresponding phases and that merely quantitative differences exist. According to this idea, an intensive velocity inversion prevails in the lower crust, its intensity decreases in outward direction (GIESE and PAVLENKOVA, 1974). Since phase 1 (P^M) is interpreted by all authors concerned in the same manner, the determination of the crust/mantle boundary can be made in accordance with uniform points of view.

In agreement with EDEL et al. (1975), the present author also comes to the conclusion that, inside the graben, the velocity range of 7.5 to 8.2 km/s lies beneath the last well-marked discontinuity and gradually merges into the velocity of the upper mantle. In the opinion of the author, this velocity distribution also exists outside the graben (for instance BA-010), probably in a weaker form, while EDEL et al. (1975) derive here a first-order discontinuity for the crust/mantle boundary from their observations.

In summary, however, it can be said that both versions lead to the assumption of a crustal thickness between 22 and 25 km and to the opinion that no significant difference in crustal thickness exists between graben and graben shoulder.

In a similarly critical way, the question shall be discussed whether there exists a zone of low velocity in the middle crust, i.e. in the depth range between 6 and 18 km. A velocity inversion had been derived from the data of the Rhine Graben area by a number of authors (MUELLER et al., 1967, 1969, 1973; MEISSNER et al., 1970; ANSORGE et al., 1970; GIESE and STEIN, 1971; MEISSNER and VETTER, 1974). It is, therefore, surprising that EDEL et al. (1975) and PRODEHL et al. (6.8 of this volume) are of the opinion that the observations available do not permit a definite conclusion concerning the existence or nonexistence of a velocity inversion within the sialic upper crust. The data are compatible with both interpretations. Because this question is of great importance for the driving mechanism in graben formation (FUCHS, 1974; MUELLER and RYBACH, 1974), this problem will be discussed again under some additional aspects.

MUELLER and LANDISMAN (1966) derive the existence of a low-velocity layer from the occurrence of a so-called phase P_c. EDEL et al. (1975) point out that just the recent observations of 1972 do not reveal this phase. In the opinion of the author, this question has to be considered under somewhat more general aspects. A nonexistence of a low-velocity layer must not be concluded from the missing of the P_c-phase (GIESE, 1966, 1968).

A low-velocity zone can be deducted from various criteria. An anomalously small average velocity for the crust or small interval velocities derived from subcritical reflections suggest low-velocity layers (PAVLENKOVA, 1969). Another criterion used here is the observation that sometimes the far-distant ends of the P_g-phase and a later one form parallel parts of travel-time curves connected with a time shift between the corresponding phases.

In the case discussed, the far-distant ends of the P_g-phase and phase 6 according to EDEL et al. (1975) (Fig. 4a, b) have to be considered. From the slope of the right end of branch 6 a velocity of 6.2-6.3 km/s results, which is even partly established by phase correlation.

From a time-term analysis EDEL et al. (1975) derive a value of 6.00 ± 0.02 km/s. When using this value it is not necessary to introduce a low-velocity layer for the middle crust, because the straight line P_g (6.0 km/s) is tangent respectively asymptotic to curve 6; a shadow zone does not exist.

It should be remarked, however, that the value resulting from the time-term analysis must be considered as mean value for the velocities in the upper crust. The maximum velocity of the P_g-wave, reached in a depth range from 6 to 8 km, will certainly be greater by 0.2 to 0.3 km/s. From the profile BA-010 in the Vosges a maximum velocity of 6.1-6.2 km/s results for the P_g-wave. It can further be added that in areas free of sediments, as for instance in the Moldanubic zone of the Bavarian Forest, the P_g-wave shows velocities of about 6.2 km/s (GIESE, 1963). Essentially connected with the record sections of the reversed profile Steinbrunn-Wissenbourg, the following observation may serve as well as a qualitative indication for the existence of low-velocity layer in the middle crust. In the distance range of about 60-100 km, hardly any arrivals occur with higher amplitudes permitting correlation over larger distance intervals earlier than the phase, that are connected with the lower crust and the crust/mantle boundary. On the contrary, the record sections of the profiles 09-240, 09-275-08, 10-135 and 13-240 in southeast Germany also show numerous arrivals in the corresponding distance range. In the area covered by these profiles, a velocity inversion is not very clearly marked. This energy lack in the record sections of the Rhine Graben profiles is thus interpreted as qualitative indication for the existence of an intensive low-velocity layer in the middle crust, which prevents the occurence of refracted onsets. Therefore the author is of the opinion that, in the middle crust of the Rhine Graben area, a velocity inversion exists, its intensity decreasing from the graben proper towards east and west. Nevertheless the author agrees with EDEL et al. (1975) that structural details of upper and middle crust have no great influence on the propagation of the waves in the lower crust and in the crust/mantle boundary zone.

In the southern part of the Rhine Graben, the crust reaches its minimum thickness with only 22 km. Toward N, the crust/mantle boundary dips gently, a result that is also supported by evaluations of deep reflections (DEMNATI and DOHR, 1965). For the region between Karlsruhe and Heidelberg a crustal thickness of 24 km is obtained.

The frequency distributions, found by DEMNATI and DOHR (1965) in regions of the central Upper Rhine Graben show clear maxima at 7-7.8 s and 8.5-9.5 s. The medium traveltime difference of about 1.5 s corresponds, with an average velocity of 5.0 km/s, to a layer thickness of almost 4 km, i.e. to the difference between the two discontinuities at 18-19 and 22 km.

Somewhat surprising is the result, obvious to all scientists concerned, that crustal thickness is about the same for the graben shoulders (Black Forest and Vosges) on the one and the Rhine Graben itself on the other side. Also in these border regions, the crust is by 2-3 km thinner in the S than in the N. It is within the resolving power of refraction seismics to detect differences in thickness of 4-5 km.

While there is no essential difference in total crustal thickness of graben proper and shoulders this seems to be different for an intercrustal interface (DEMNATI and DOHR, 1965).

In the region of the Kraichgau, 8 out of 9 depth reflection registrations supplied arrivals with travel times at 5.2-5.4 s as well as at 7 and about 9 s. On the other hand, depth reflection observations from the graben region S of Karlsruhe are available showing maxima of the distribution curves at 7-8 s and about 9 s.

Supposing - and the picture of the reflections suggests this - that the reflections at 7 s in the Rhine valley correspond to those at 5.2-5.4 s in the Kraichgau, this results in a depth of 15 km for this discontinuity in the Kraichgau while, in the Rhine valley, the depth of this layer is supposed to be about 18-20 km. The difference of 3-5 km corresponds roughly to the dipping of the basement.

Consequently, the question arises whether the 15 km-discontinuity in the Kraichgau can also be detected on neighboring refraction profiles. Refraction profile 02-215 crosses through the Kraichgau but the distance range of observations is beyond 180 km. The record section of this profile shows a very well-marked intercrustal wave group which suggests a depth of about 20 km for a corresponding discontinuity in the region north of the Main.

In the profile of Fig. 7, an attempt has been made to combine these individual data. It has to be pointed out that this correlation has to be regarded as quite speculative. Should it be followed up all the same, this would be derived:

1. The Rhine Graben crust now consists mainly of the upper crust. If the discontinuity at a depth of 22-24 km is considered to be the crust/mantle boundary, the lower crust is but a few km thick, about 4-5 km. It has to be pointed out, however, that the uppermost mantle shows a somewhat

Fig. 7. Cross section between Würzburg (Unterfranken) and the Rhine Graben showing the reduction of crustal thickness towards the axis of the graben

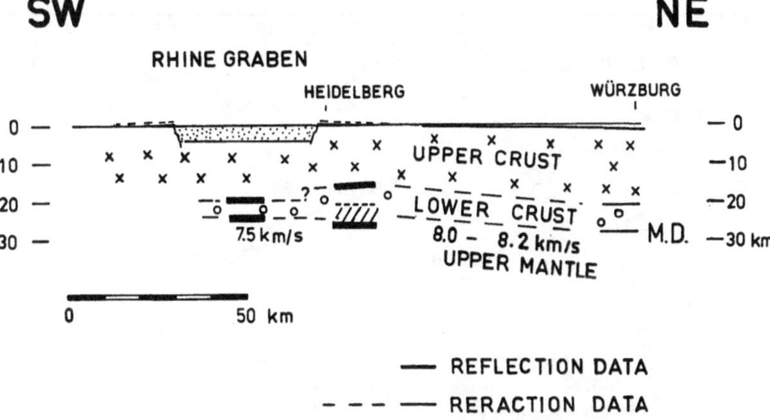

lower velocity and that the normal velocity values of 8.0-8.2 km/s are reached only 3-4 km below this boundary. In addition to a fall of the basement surface by 3-4 km, the process of rift formation has, obviously, caused a rising of the crust/mantle boundary. Thereby, regions of the lower crust have been transported into anomalous zones of the uppermost mantle.

2. In the region adjacent in the east, however, the thinning of the crust from E to W is mainly due to a reduction of the upper crust. The thickness of the lower crust of about 8 km remains constant.

It should be added that, on the basis of the evaluation of the Taben Rodt-S profile (16-195), an analogous conclusion would have to be drawn also for the western graben shoulder. Here, thickness of the upper crust amounts to 20 km, that of the lower crust 12 km.

5.4.4 The Alpine Area

That part of the Alps belonging to the Federal Republic covers the area between Lake Constance and Berchtesgaden. From N to S, this area comprises the Subalpine Molasse, the Helvetic zone, the Flysch zone and the Austroalpine unit.

The central shotpoint for this area is the quarry of Eschenlohe (01), situated at the northern border of the Alps. Within the scope of the German Research Society's program "Geodynamics of the Mediterranean Area", the crust of the Northern Alps has been investigated, using shots of Eschenlohe as well as those of some smaller quarries (ANGENHEISTER et al., 1972).

The crustal thickness under the northern margin of the Alps is about 40 km, and it increases southwards reaching values of 50-55 km under the axis of the Alps (PRODEHL, 1965; GIESE, 1968a; KOSCHYK, 1969; ANGENHEISTER et al., 1972; ZSCHAU and KOSCHYK, 6.10 of this volume).

The most interesting question with regard to the underground of the Northern Alps is whether the overthrust of Helvetic rocks, of Flysch and Austroalpine series over the autochthone Molasse, already claimed by RICHTER (1951) and, in the meantime, demonstrated by drillings, can be proved by aid of refraction seismics.

The profiles radiating from Eschenlohe into the Alpine area show, in the first 30 km at about $\bar{t}$ = 1.8 s, a clear phase that can be interpreted as reversed segment (e.g. profile 01-090). When the x^2, t^2 process is applied to the full length of this traveltime curve, a depth of 9-10 km results. A detailed investigation by aid of the z_{max} method shows, however, that the actual depth amounts to 6-7 km, a value that seems plausible for the thickness of the sedimentary cover. Average velocity is between 4 and 5 km/s. Considering the fact that, in the uppermost kilometer, a velocity of as much as over 6 km/s is reached (6.9 of this volume), the existence of a zone of low velocity in a depth range of about 3-6 km must be postulated, interpreted from the geological point of view as molasse (ANGENHEISTER et al., 1972). The formation with lower velocity (Flysch, Helvetic and Molasse) may extend at least 10 to 15 km to the south below the Calcareous Alps. Fig. 7 (4.5) shows a N-S cross-section presented by WILL (1975, 4.5) which can be regarded as typical for the Northern Alps. Under high-velocity material of the Northern Calcareous Alps, such as limestones and dolomites, low-velocity material follows which

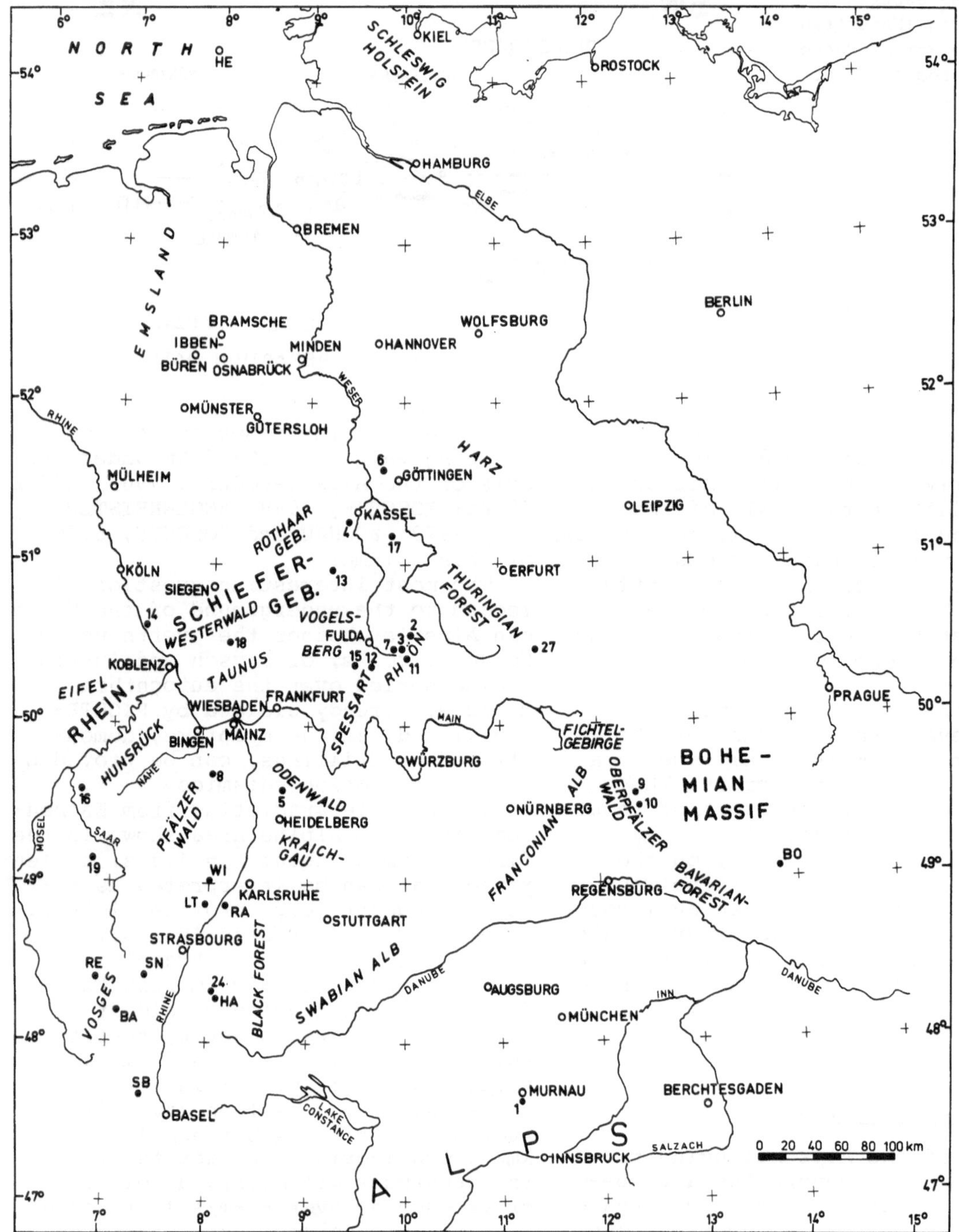

Fig. 8. Map containing geographical names mentioned in this section

can be associated with Flysch and Molasse series. It will also be pointed out that commercial seismic reflection work extends into the Alpine area and that the registrations of these measurements also reveal the overthrust feature of the Northern Alps (VEIT, 1963).

The intensive velocity inversion in the middle crustal range, proved especially by profile 01-090 for the northern border of the Alps, is particularly remarkable. Further on, an intensive inversion in the lower crust cannot be excluded (GIESE, 1972). Owing to the subsidence by 6-7 km, an increase of temperature will inevitably occur. An estimate should be formed whether the space of time derived from geological considerations has sufficed to bring about the necessary increase in temperature or whether a generally higher heat flow has to be postulated. This, however, requires a well-grounded knowledge of the connection between velocity and temperature (FIELITZ, 1971, 2.5).

In summary, it can therefore be stated that the reflection and refraction seismic measurements have, in principle, confirmed the geological ideas of the nappe structure of the Northern Alps. In the area of the Northern Alps, the sediment-free crust has a thickness of about 34 km. Velocity inversions in the middle and lower crust are clearly perceptible.

6. Interpretation of Regional Seismic Data

P. GIESE

This chapter deals with regional results. The content of the papers presented has been elaborated on within the framework of a thesis or as normal publications. Some of the contributions not only deal with regional crustal features, but also with methodical problems. The evaluation of the regional data is not based on homogeneous interpretation procedures, but it was one of the aims to show the different interpretation methods applied. On the other hand, the interpretation of the data has been performed under uniform principles in Chapter 5.

The first two papers (MEISSNER et al., 6.1; GLOCKE and MEISSNER, 6.2) deal with the crustal structure of the Rhenish Massif, the northern branch of the Variscan mountain system in central Europe. The area of the Rhenish Massif is covered by a network of refraction profiles, observed mainly by the aid of quarry blasts. The investigations are not yet finished. Along the axis of the Rhenish Massif a refraction program with reversed and overlapping profiles has been started. Another seismic investigation, including near-vertical observations, was started in 1973.

Based on the experience obtained in Southern Bavaria in 1964 a common-depth point profile was carried out in 1970 at the southern border of the Rhenish Massif north of Frankfurt (GLOCKE and MEISSNER, 6.2). It is noteworthy that the distinction between the upper and the lower crust is well established by clear overcritical reflections. The refraction measurements have been completed by near-vertical reflection measurements, which confirmed the sandwich-like structure of the transition zones between the upper and lower crust, as well as between the lower crust and the upper mantle.

GRUBBE (6.4) describes the crustal structure between the Rhenish Massif and the Harz. Here the geological situation is of some interest, because the northern prolongation of the Hessische Straße - a zone of subsidence during the Mesozoic and Tertiary - passes the area under investigation. From a technical point of view, it is remarkable that these crustal data could be obtained by observation of shots fired in the scope of exploration-refraction work. Although not all records are usable, the great volume of material enabled the elaboration of a crustal model. The type of travel-time diagram agrees well with that obtained by continuous profiling in the USSR.

Whereas in the central and southern parts of western Germany the main features of crustal structure are known, the North German Plain is only sparcely covered by long refraction profiles. Therefore, a profile with overlapping observation lines which extends from the region between the Harz and the Rhenish Massif to the North Sea, is of interest. Because of the high noise level, several processing procedures have been applied in order to improve the signal/noise ratio, whereby P_n-arrivals could be clearly established (HINZ et al., 6.3).

The crustal structure of SW Germany has been elaborated on by EMTER (6.5). Within the area under investigation, the occurrence of magnetic and high temperature-gradient anomalies is noteworthy. It is remarkable that on the profiles radiating from the Eschenlohe quarry and passing this area no P_n-arrivals could be observed, though the noise level was extremely low. The author discusses whether this is caused by absorption or by negative velocity gradient in the upper mantle. Finally, another problem is touched: the possible extension of the Rhine Graben structure towards the east, a question which is treated by PRODEHL et al. (6.8).

The Ries crater is located in the vicinity of the Swabian anomaly. Today there is general agreement that this phenomenon is caused by an impact of a meteor and not by volcanic process. The crustal structure of the Ries area and its vicinity has been investigated by a combined refraction and reflection program. Near-vertical reflections from the intermediate crustal boundaries and from the crust/mantle zone could be continuously observed over about 20 km length (ANGENHEISTER and POHL, 6.6).

The Rhine Graben between Basel and Frankfurt was a main object of an investigation, carried out in a close cooperation with French geophysical institutions. The paper of MEISSNER et al. (6.7) deals with the northern part in the Rhine-Main area. Here the anomalous structure is not so clearly developed as in the central and southern part.

The contribution of PRODEHL et al. (6.8) summarizes the main features and ideas of crustal structure of the central and southern Rhine Graben area, published within a period of eight years.

The crustal investigation of the Alps was another center of activity. This section comprises only investigations concerning the crustal structure of the German part of the Alps. One of the main shotpoints in the Eastern Alps is the Eschenlohe quarry at the northern border of the Alps. From this point profiles radiate into the foreland as well as into the Alps. The paper of SCHELIGA (6.9) deals only with the structure of the upper 10 km and is mainly concerned with the question of high-velocity material (Northern Calcareous Alps) overlying low velocity material (Molasse). The contribution of KOSCHYK and ZSCHAU (6.10) treats the profile Eschenlohe E. This record section shows one of the best P^M-curves ever observed in western Germany. The corresponding S^M-wave is also well recognizable. So the possibility was offered to evaluate not only P-waves but also S-waves. Some statements could be derived concerning the depth variation of Poisson's ratio.

The last paper by MILLER and GEBRANDE (6.11) deals with seismic-refraction measurements carried out in close cooperation with Czechoslovakian colleagues on the International Profile VII. Because of the great number of shots fired at the point of Boubin in the ČSSR, very closely spaced observations could be otained. The interpretation of the record section takes into consideration lateral variations of velocity.

6.1 An Interpretation of Wide-Angle Measurements in the Rhenish Massif

R. Meissner, H. Bartelsen, A. Glocke, and W. Kaminski

ABSTRACT

Results from a large wide-angle profile reveal accurate data on deep crustal layering. Small velocity reversals are found below 7 km and in the deeper crust. Between the Conrad and the Mohorovičić discontinuities another boundary is found which seems to be connected with the velocity-gradient zone at the base of the crust.

ZUSAMMENFASSUNG

Ergebnisse eines langen Weitwinkelprofils ermöglichen genaue Angaben über die Schichtung der tiefen Kruste. Schwache Geschwindigkeitsinversionen wurden unterhalb von 7 km und in der tieferen Kruste gefunden. Zwischen der Conrad- und der Mohorovičić-Diskontinuität wurde eine weitere Grenzfläche gefunden, die wahrscheinlich mit der Geschwindigkeitsgradientenzone an der Basis der Kruste zusammenhängt.

6.1.1 Introduction

Wide-angle profiles with a symmetrically arranged reflecting (or refracting) element (= common-depth point method) seem to provide the most accurate velocity-depth information of crustal structures, because dip effects can be controlled and easily eliminated. Based on the experience and results of wide-angle measurements in the Bavarian Molasse Basin in 1964 (German Research Group, 1966; MEISSNER, 1966, 1967a) another wide-angle profile was observed in 1968. It was the goal of this survey to investigate whether basic results of the previous study in southern Bavaria were the same in a geologically different region. This profile was observed in the Rhenish Massif where, along the profile, the crystalline basement is covered by rather uniform layers of Devonian slates.

6.1.2 Description of Test Site and Field Procedure

As shown in Fig. 1, a profile of 180 km in length in the direction ENE-WSW with its point of symmetry W of the Rhine river, was arranged near the southeastern boundary of the Devonian Rhenish Massif. Eight shots, 20 km apart, were observed continuously on two observation lines, 20 km long, and arranged symmetrically with regard to the center of the profile (Fig. 2). Only one of these shotpoints (SP 6) could not be drilled in Devonian slates but in thin Tertiary layers of sand and clay. In general, three holes of 15-25 m depth were drilled at each shotpoint. The charges varied between 100 kg for nearby observations and 225 kg for observations up to 180 km distance.

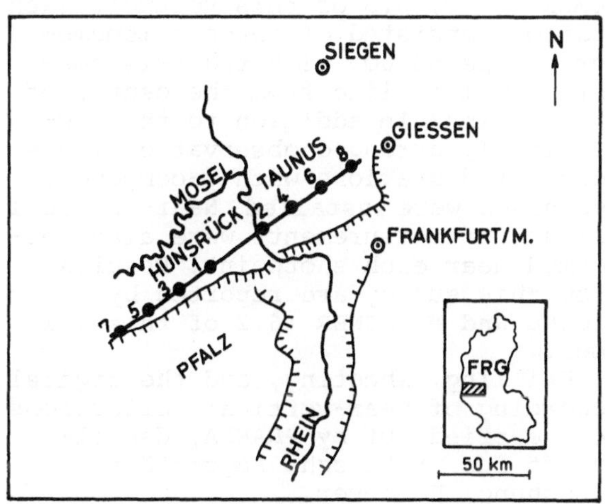

Fig. 1. Location map of the wide-angle profile

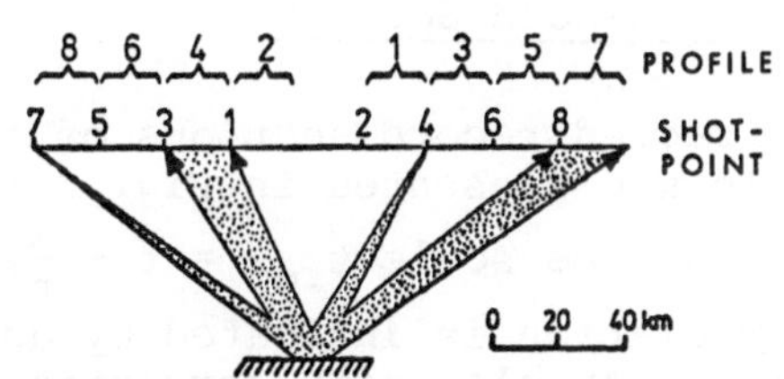

Fig. 2 Arrangement of shotpoints and observation lines

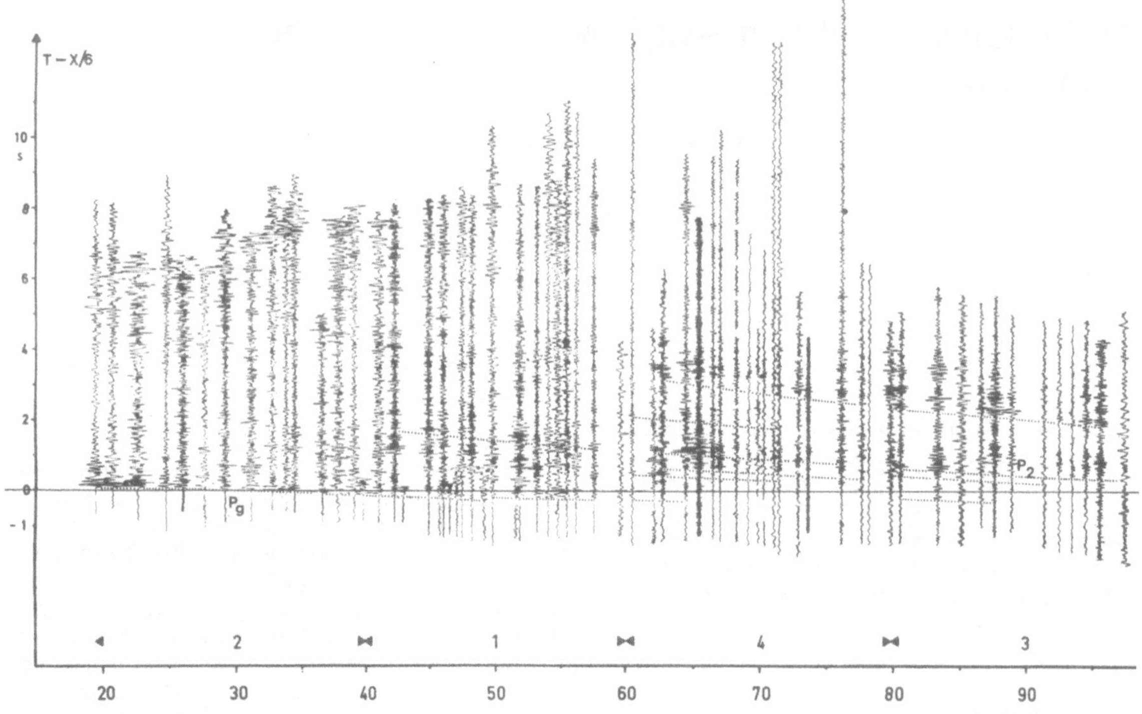

Fig. 3. Record section of the wide-angle profile 240-LO-060 with correlations

Seismometer spacing on the 20 km long observation lines was generally 300 m in order to obtain continuous phase correlation. All wide-angle stations were equipped with MARS 66 magnetic tape recording units (BERCKHEMER, 1970, 3.5 of this volume). Each station consisted of three seismometers, type FS 60, with two seismometers 300 m in line from the center of the station. In addition to the symmetrically arranged observation lines some fixed stations with 3-component geophones were installed. Near-vertical reflection measurements were also performed near each shotpoint. Results from this survey are reported by GLOCKE and MEISSNER (6.2 of this volume).

Drilling, shooting, and the digital recording of near-vertical reflections were carried out by PRAKLA, Gesellschaft für praktische Lagerstättenforschung, Hannover.

6.1.3 Record Section

Corrected record sections of the profile are presented in Fig. 3 in a reduced time scale $t_{red} = t - \frac{x}{6}$. The correlation is indicated by dotted lines. An elevation correction has been applied for shot and geophone positions.

In Fig. 3 the distances between shotpoints and geophones are referred to a common zero point. The eight different observation lines labeled with the same numbers as those of the corresponding shotpoints are marked underneath the seismograms. Even without any correlation plotted, the shape of the P_g- and P^M-branch can be clearly defined, and intermediate branches between the two can be recognized.

Four different travel-time branches of varying quality are plotted in Fig. 3. Some of them show offsets in Δt between the segments observed from different shotpoints which may be caused by weathering effects below the shotpoints and/or by radiation of different frequencies and small dispersion. The latter explanation certainly holds for SP 6 which could not be fired in a consolidated rock environment.

6.1.4 The P_g-Wave Branch

The first arrivals, P_g, are recorded clearly up to 60 km and permit an exact phase correlation. As their average velocity increases with increasing distance, regardless of the direction, these waves penetrate (= dive) considerably into the crystalline basement. Their early arrival times, compared to those of adjacent

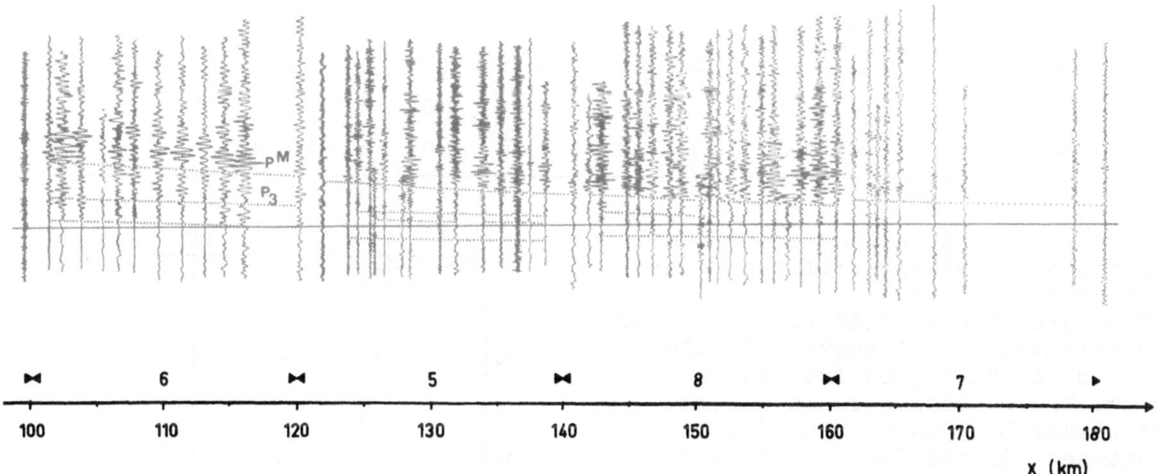

profiles, mean that the depth of the basement is small and the velocity in the Devonian slates near the surface is rather high. At a distance of 50 km a delay in the P_g-arrivals indicates a fault with a difference of about 200-300 m in height. This may well represent a rather deep fault zone as indicated from the vertical-reflection observations and from the observation of basaltic intrusions and extrusions in a strike direction perpendicular to the profile about 7 km ENE from Kaub.

As travel time and apparent velocity of the P_g-wave are a function of the depth of the basement and the velocity gradient within the basement, there is a certain ambiguity in the calculation of both values. The application of the Wiechert-Herglotz method for different averaging t(x) and v(x) functions results in different velocity-depth functions. All curves show that the velocity in the basement increases only down to approx. 7 km. Depth and undulations of the basement along the profile are plotted in Fig. 4 after a new method developed by BARTELSEN (1970).

6.1.5 The P_2, P_3, and P^M-Wave Branches

The indicated correlations for the branches following the P_g-wave branch are hyperbola-like. This means that they are reflections and/or diving waves in a strong gradient zone. The P^M-wave branch, in particular, is the

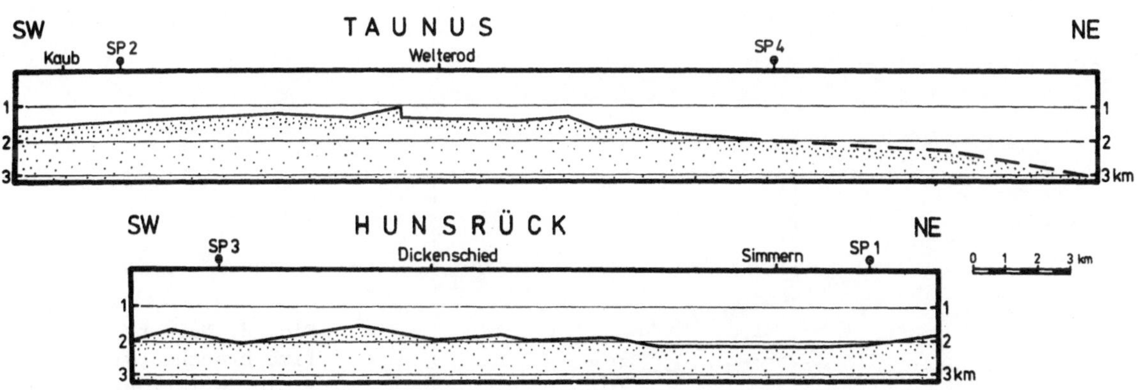

Fig. 4. Cross section of the crystalline basement along the profile

Table 1. Evaluation of travel-time branches.
$\tilde{V}$ = Root mean square velocity; from $t^2 - x^2$ diagram
V_{as} = Asymptotic velocity; largest possible velocity above interface
V_i = Interval velocity
The relatively large error in the V_i values should be noted

Branch	Discontinuity	t_o (s)	Depth Z (km)	$\tilde{V}$ (km/s)	V_{as} (km/s)	V_i (km/s)
P_g	basement	-	1....3	-	6.3 ± 0.2	-
P_2	Conrad D.	5.1 ± 0.2	15.7 ± 0.9	6.15 ± 0.01	6.3 ± 0.05	6.45 ± 1.2
P_3	Sub-Conrad D.	7.2 ± 0.3	22.5 ± 1.5	6.24 ± 0.01	6.6 ± 0.1	6.7 ± 1.3
P^M	Mohorovičić D.	9.2 ± 0.3	29.2 ± 1.7	6.35 ± 0.01	6.7 ± 0.2	

outstanding event in the record section and shows strong amplitudes between 60 and 160 km.

The P_2-branch related to the Conrad discontinuity (= CD) shows reliable arrivals in the region from 60 to 100 km. The P_3-branch is in general much weaker but may be observed clearly between 100 and 120 km; it is related to the Sub-Conrad discontinuity (= SCD) between the CD and MD. Both branches P_2 and P_3 can be correlated perfectly to strong near-vertical reflections as discussed by GLOCKE and MEISSNER (6.2 of this volume).

For the evaluation of the three branches P_2, P_3, and P^M, a $t^2 - x^2$ diagram and GIESE's method (GIESE, 1968a) have been used. In the $t^2 - x^2$ diagram, systematic deviations from a straight line have been found, especially for the P^M-branch. Similar to the Molasse survey in 1964, these deviations could be attributed mainly to refraction in a v(z) medium, to a transition from reflections to diving waves, and to an anisotropy. Interval velocities between discontinuities have been determined from the $t^2 - x^2$ diagram and DIX's method (DIX, 1952). For important data from the evaluation see Table 1.

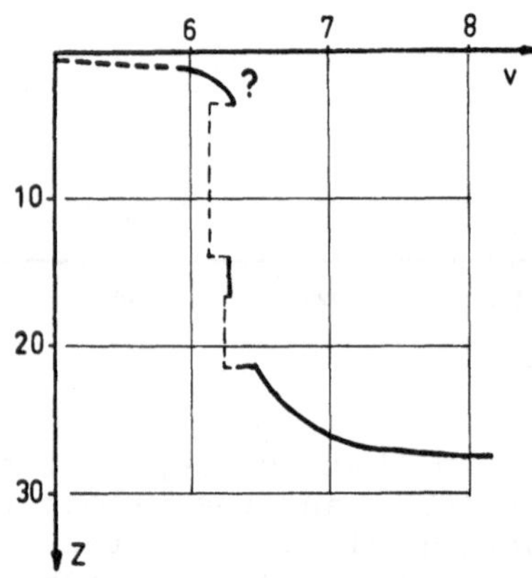

Fig. 5. Velocity-depth function after GIESE's method

6.1.6 Velocity-Depth Functions (= VDF)

Two velocity-depth functions are presented in Fig. 5 and 10. The first evaluation is based completely on the iteration method by GIESE (1968a). It uses the most reliable travel-time branches P_g, P_2, and P^M and shows some similarities to adjacent profiles (MEISSNER et al., 6.7 of this volume). The reflections from the near-vertical reflection survey are indicated. In Fig. 10 results from an additional

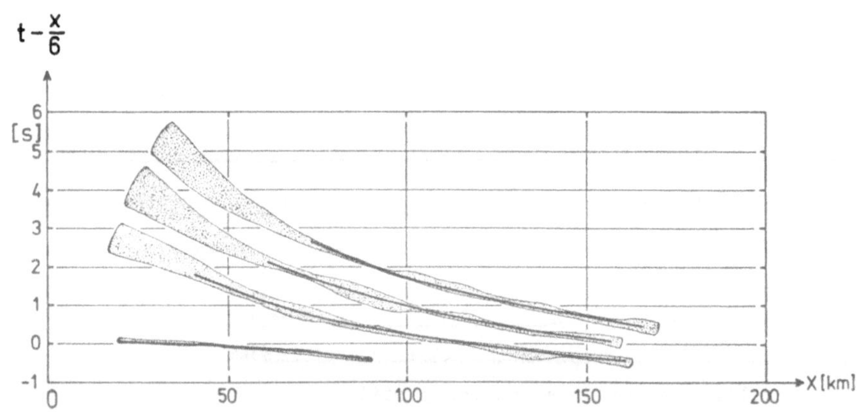

Fig. 6. Main travel-time branches with possible deviations indicated

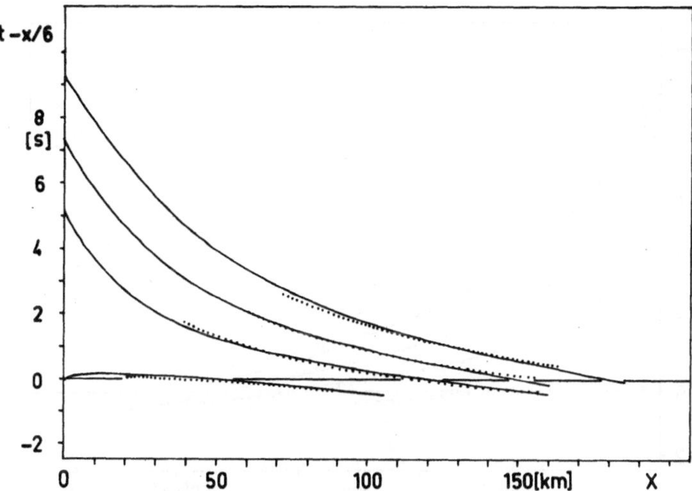

Fig. 7. Theoretical (——) and observed (....) travel-time branches

iteration process are shown, taking into account also the P_3-wave branch. This iteration process is based on a comparison between theoretical and observed travel-time curves and a systematic variation of velocity parameters as mentioned by VETTER and MEISSNER (4.7 of this volume).

First, the scatter of the correlation has been determined carefully from the record section. Each branch may have a certain deviation as shown in Fig. 6. The main wave groups are fitted best by assuming an additional low-velocity zone in the lower crust. In this case theoretical and correlated

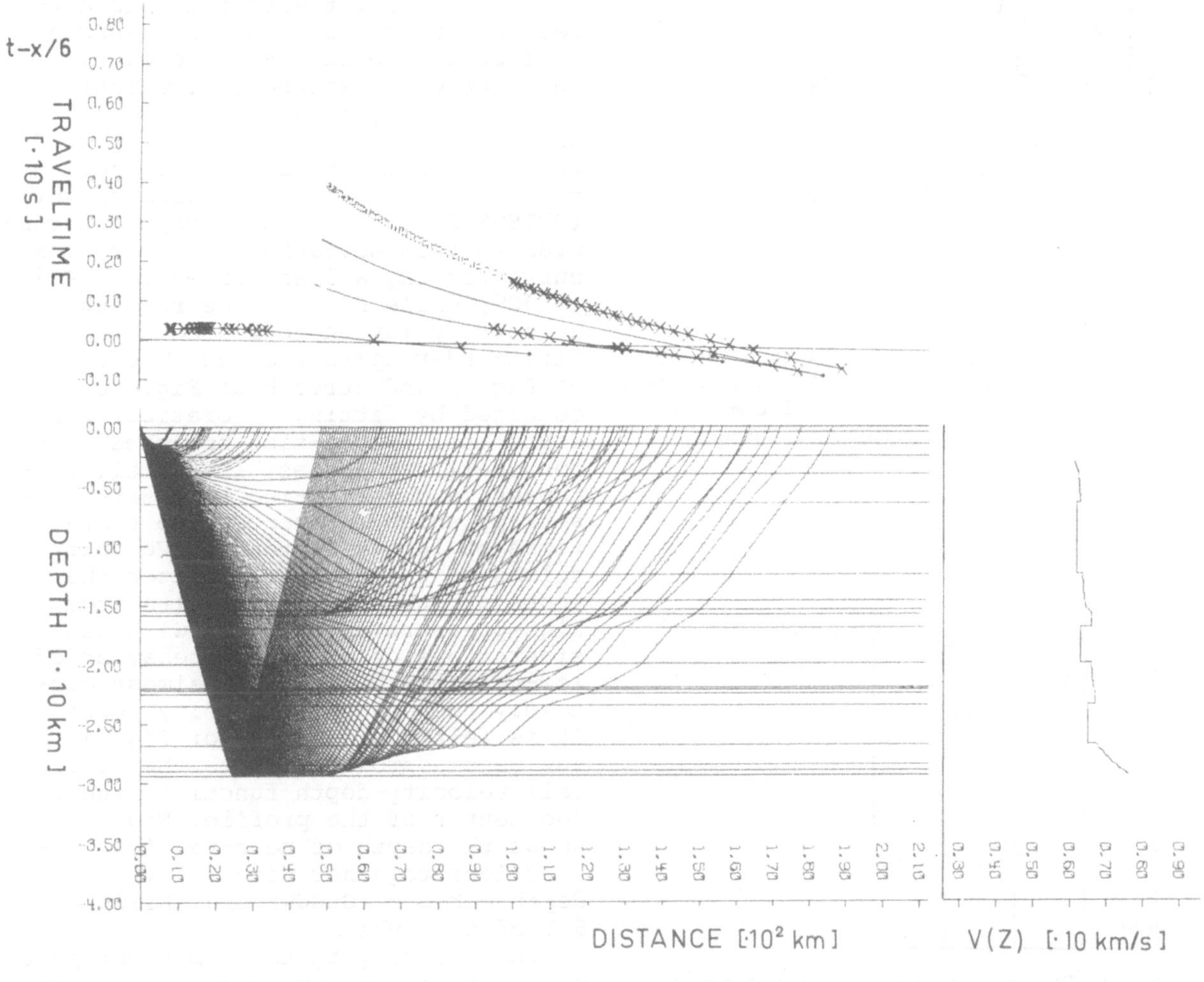

Fig. 8. Ray diagram and calculated travel-time branches □ reflected waves, x diving waves

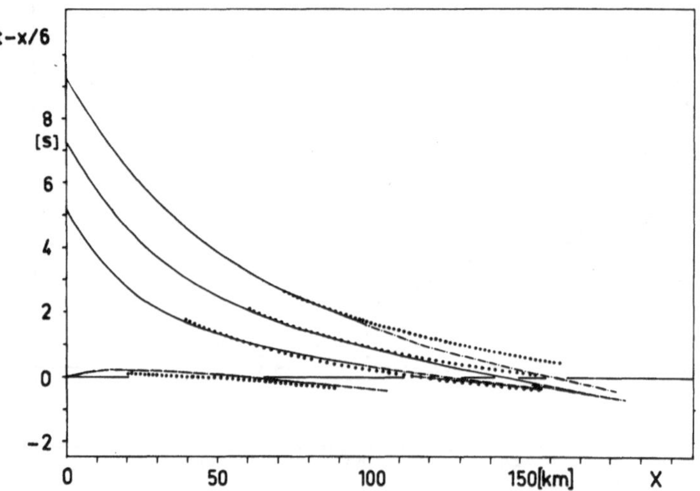

Fig. 9. Observed (····) and theoretical (——) travel-time branches

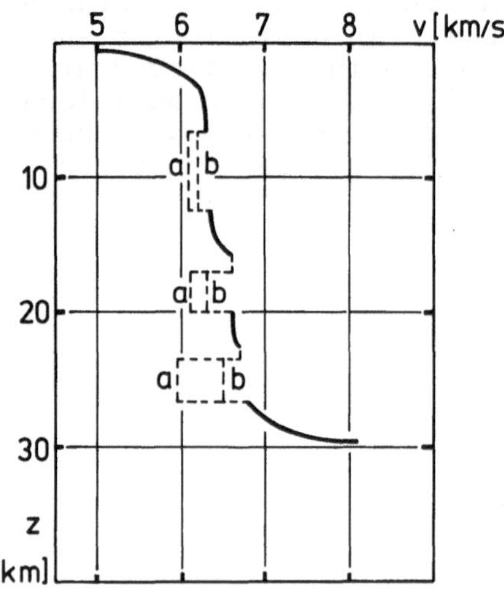

Fig. 10. Velocity-depth functions after applying GIESE'S method and an additional iteration process based on the travel-time curves shown in Figs. 8 and 9 (GIESE'S curve *b*)

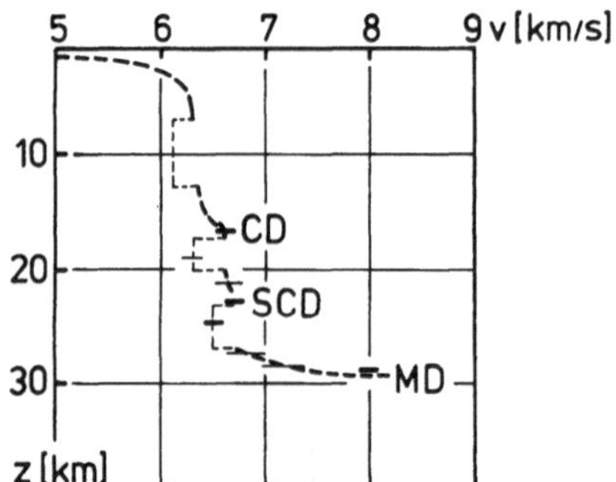

Fig. 11. Depth of near-vertical reflections and velocity-depth function

branches agree rather well for all distances as shown in Fig. 7. Travel-time curves and a ray diagram are shown in Fig. 8. The transition from reflected to the diving wave at the P^M-branch is clearly defined.

When fitting theoretical and correlated curves isotropy was assumed. It should be emphasized, however, that there are strong indications for an anisotropy, especially for the deeper crust. The $t^2 - x^2$ curve of P^M shows the same characteristic deviations as found at the Molasse survey (MEISSNER, 1967a). Anisotropy contributes to this deviation in the $t^2 - x^2$ curve for ray anlges larger than 40° to 50°, while below these ray angles anisotropy may be neglected in a first-order approximation. The curves of Fig. 9 and curve b of Fig. 10 are obtained by fitting theoretical and correlated travel-time branches only for small ray angles, i.e. small distances, and permitting a certain anisotropy which influences the travel times at larger distances. Velocity values of curve b agree better than those of curve a with results from high-pressure physics and with considerations of isostatic behavior of low-velocity zones which almost certainly represent low-density zones. It is felt that curve b of Fig. 10 is the best approximation to the (vertical) velocity-depth function beneath the center of the profile. Fig. 11 shows the depth of near-vertical reflections together with the velocity-depth curve b (GLOCKE and MEISSNER, 6.2 of this volume).

The velocity in the uppermost part of the mantle which is generally obtained from the P_n-wave could not be

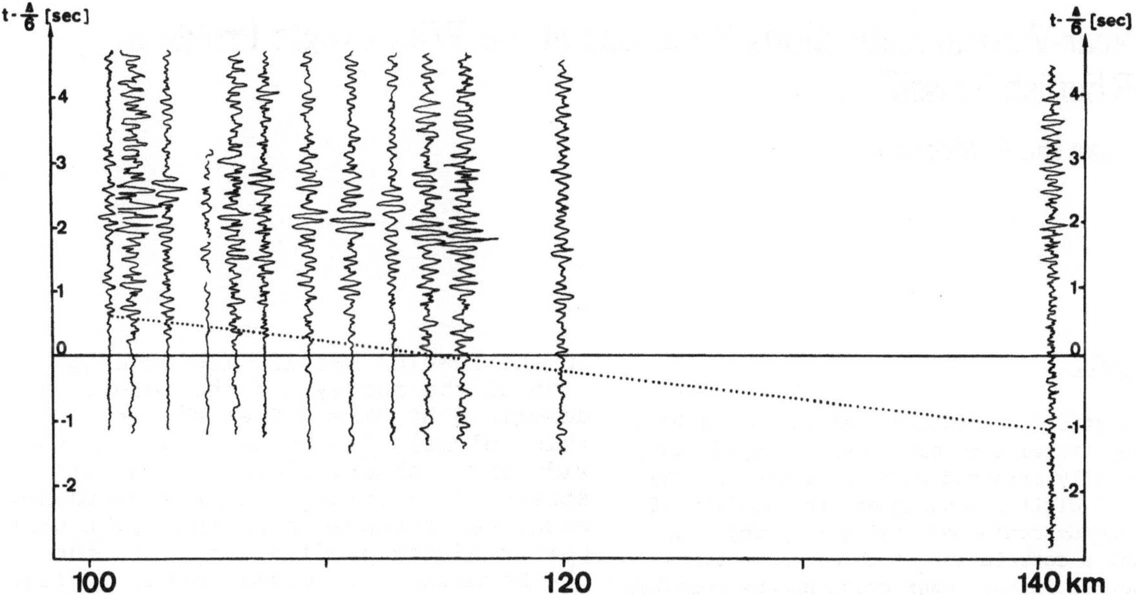

Fig. 12. Possible correlation of P_n-arrivals

derived with certainty. This may be caused by the fact that the frequencies of the radiated seismic energy from borehole shots were too high, the energy was too low, and the noise level on five rainy days of observation too high. Some indications of P_n may be observed from the low-frequency arrivals of shotpoint 6 recorded on line 6 and an additional station. A possible correlation is shown in Fig. 12 showing a P_n-velocity of 8.1 km/s.

6.1.7 Conclusions

Accurate velocity-depth functions have been obtained from the evaluation of record sections. Strong arrivals were obtained from the crystalline basement (P_g), from the CD (P_2) and MD (P^M), weak arrivals from the SCD (P_3) and from the upper mantle (P_n). Three low-velocity zones, with only small reversals, have been found. Basic results are similar to those of the Molasse survey of 1964 and to those of adjacent profiles.

Acknowledgments. Drilling and shooting facilities have been generously supplied by PRAKLA, Gesellschaft für praktische Lagerstättenforschung, Hannover. Field work and organization were carried out in team work by members and students of nearly all German geophysical institutes.

6.2 Near-Vertical Reflections Recorded at the Wide-Angle Profile in the Rhenish Massif

A. GLOCKE and R. MEISSNER

ABSTRACT

Near-vertical reflections between 4 and 10 s two-way travel time have been surveyed along a specially arranged wide-angle profile. By means of digital techniques, the quality of the records could be considerably improved so that a correlation of three characteristic boundaries in the lower crust became possible along the profile. The transition between crust and mantle is found to be partially step-wise.

ZUSAMMENFASSUNG

Entlang einem speziellen Weitwinkelprofil wurden Vertikalreflexionen mit Laufzeiten zwischen 4 und 10 Sekunden beobachtet. Mit Hilfe digitaler Techniken konnte die Qualität der Seismogramme so wesentlich verbessert werden, daß entlang dem Profil die Korrelation von drei charakteristischen Grenzflächen in der tiefen Kruste möglich wurde. Demnach erfolgt der Übergang zwischen Kruste und Mantel teilweise in Stufen.

6.2.1 Introduction

From Sept. 30 to Oct. 5, 1968, a large common-depth-point profile was surveyed in the Rhenish Massif. The organization of the survey and the results are described by MEISSNER et al. (6.1 of this volume). Simultaneously with the wide-angle observations, a central spread of 24 traces with six geophones each, was arranged near the eight shotpoints of the profile (Fig. 1). For the recording of seismic data, digital equipment from PRAKLA, Gesellschaft für praktische Lagerstättenforschung, Hannover, was used. Digital techniques of PRAKLA have also been applied for the evaluation of the recordings.

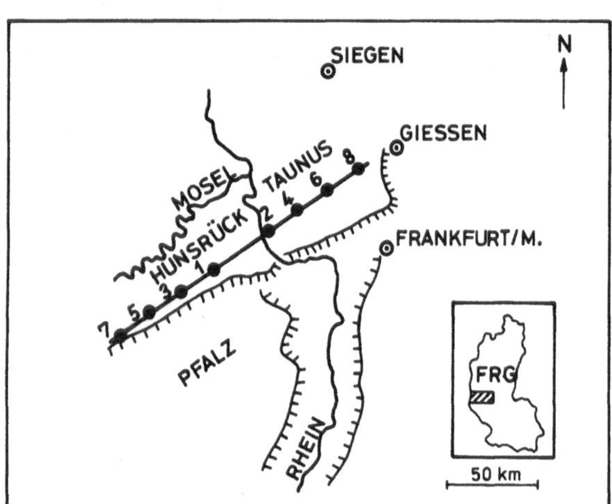

Fig, 1. Location map of the wide-angle profile

6.2.2 Techniques Used for Data Processing

Although the profile is situated in an area of extended Devonion slates and greywacke without a complicated weathered layer, the noise at some shotpoints was rather strong. In order to enhance the signal / noise ratio and to investigate the possibilities offered by digital techniques, a spike deconvolution and optimum filtering were applied. The operator for the spike deconvolution had a length of 80 point = 300 ms and was determined automatically from autocorrelation functions in overlapping time intervals of 3 s. The optimum filter was based on a discrimination of signal and statistical noise as obtained from autocorrelation functions of 12 adjacent traces. Filtering was done by a matched filter. The filter operator was calculated for 200-ms intervals with an effective overlap of 200 ms in order to avoid sudden changes of the operator. Finally, an automatic correction was applied to account for variations in the weathering layer and for those elevation irregularities which were not corrected in the normal static correction procedure.

Fig. 2 shows the obtained improvement of a field record (a) containing strong periodic noise originated by a 16 2/3 Hz railway powerline in the Rhine Valley. After deconvolution,

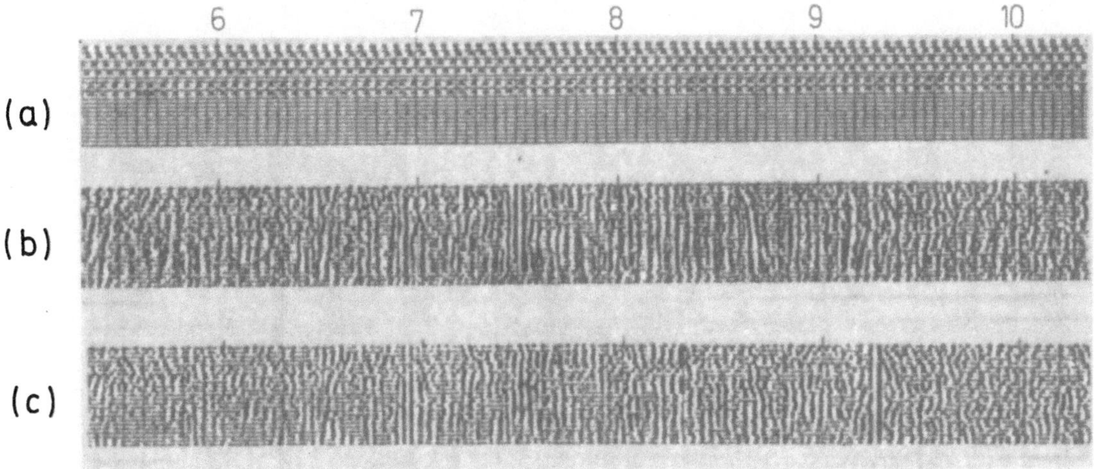

Fig. 2a-c. Example for the improvement of record quality by digital techniques. (a) Field record; (b) record after spike deconvolution applied; (c) record after spike deconvolution, optimum filter, and automatic corrections

this noise was completely eliminated and most signals were revealed (b) although they appear with a curved shape. After applying optimum filtering and automatic corrections (c), the reflected signals line up and can easily be detected.

6.2.3 Evaluation of Reflections

Seven out of eight recordings could be evaluated with the digital techniques mentioned above, whereas one record was already available as a good field recording. Results are shown in the variable-area records of Fig. 3. No reflections are found between 0 and 4 s and beyond 10 s. Between 4 and 10 s several arrivals can be detected, some of them showing a pronounced dip. These dips might represent the true dip of reflectors, a diffraction, or reflected refraction. Distinguishing between these possibilities is difficult because the next shotpoint is 20 km away. Reflected refraction, however, seems very improbable in this case as such arrivals should show up also before 4 and after 10 s and should not give a good correlation between the shotpoints. Strong reflections occur in general at about 5 s, at 7 s, and between 9 and 10 s. These three strong groups represent similar layering to that found in the Molasse in 1964 (MEISSNER, 1967a). They can be correlated to the Conrad discontinuity (= CD), the Sub Conrad discontinuity (= SCD), and the Mohorovičić discontinuity (= MD).

In Fig. 4 two histograms are shown, one for the western and another for the eastern part of the profile. Here, the three peaks related to CD, SCD, and MD can be clearly defined. Similar histograms were published by DÜRBAUM et al. (1967) from measurements in the Siegen area, 50 km north of shotpoint 8, and by DEMNATI and DOHR (1965) in the Rhine Graben and Kraichgau, about 130 km south of the profile. The measurements by DÜRBAUM et al. (1967) show certain correlations with the data obtained on the wide-angle profile. The CD peaks as the first strong reflections are found at the same travel time, later peaks show a poorer correlation, especially the MD arrivals. DOHR (1967) also finds a strong reflection at 5 s in the Kraichgau but not in the Rhine Graben. Small peaks at 8.5 to 9 s were attributed to the MD.

As the quality of reflections in the present survey was so surprisingly good, an attempt was made to correlate strong characteristic arrivals between the eight shotpoints. In the upper part of Fig. 5 a vertical travel-time plot is shown. The lower part indicates depths and positions of the reflectors, assuming that the dips of reflections match the dips of the reflectors in the vertical plane of the profile. The velocity data for the conversion of the two-way traveltime into depth, are listed by MEISSNER et al. (6.1 of this volume). The correlation for the CD, SCD, and MD seems to be well justified and matches very well with data from the wide-angle

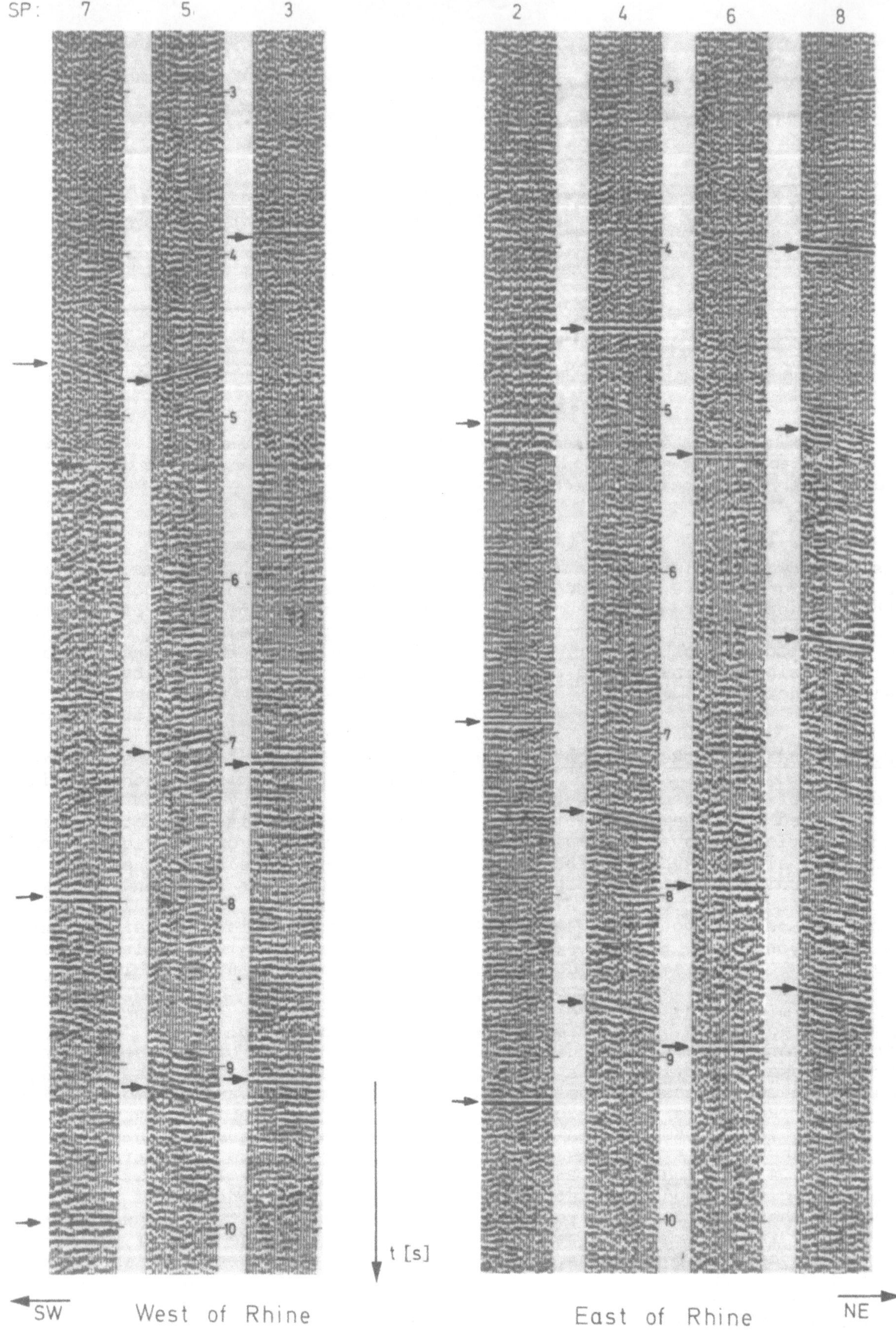

Fig. 3. Variable-area records of shotpoints 2 to 8. *Arrows:* strong reflections

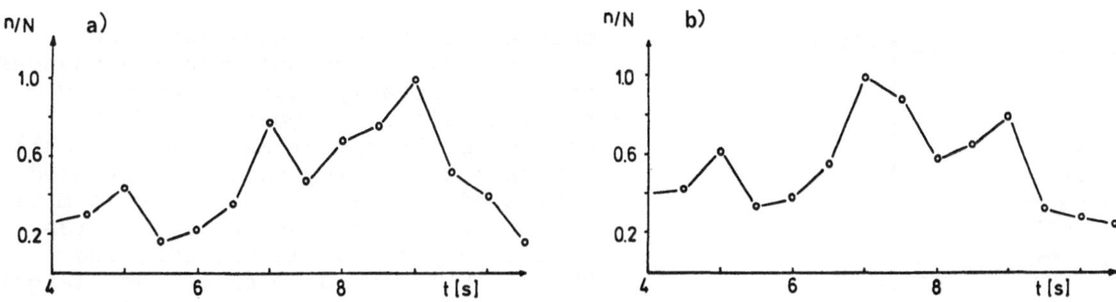

Fig. 4a and b. Histograms of reflections in the area of measurements (a) west of Rhine and (b) east of Rhine. Classification after DEMNATI and DOHR (1965)

measurements in the Rhenish Massif (MEISSNER et al., 6.1 of this volume). It should be stressed that the correlation of near-vertical reflections between the eight shotpoints was only possible after applying digital techniques with their strong improvement of the signal / noise ratio. It is felt that sometimes deconvolution and optimum filtering may be more powerful tools for reflections from deep crustal layers than their application to reflections in sedimentary areas. Deep crustal reflections always show a tendency to occur as a band of several oscillations, so that a pulse compression enhances the signal considerably.

6.2.4 Conclusions

No information on the actual length of reflecting elements can be obtained from Fig. 5. As indicated by the strong apparent dips of some elements, the CD, SCD, and MD certainly are no continuous first order boundaries.
In general, the concept of a lamellation of the lower crust as proposed by MEISSNER (1967a) seems to be supported. As in many other areas crustal reflections start at the CD and end at the MD, preferring the velocity-gradient zone at the base of the crust as seen in Fig. 6. The correlation between the depth of the near-vertical reflections and the structure of the

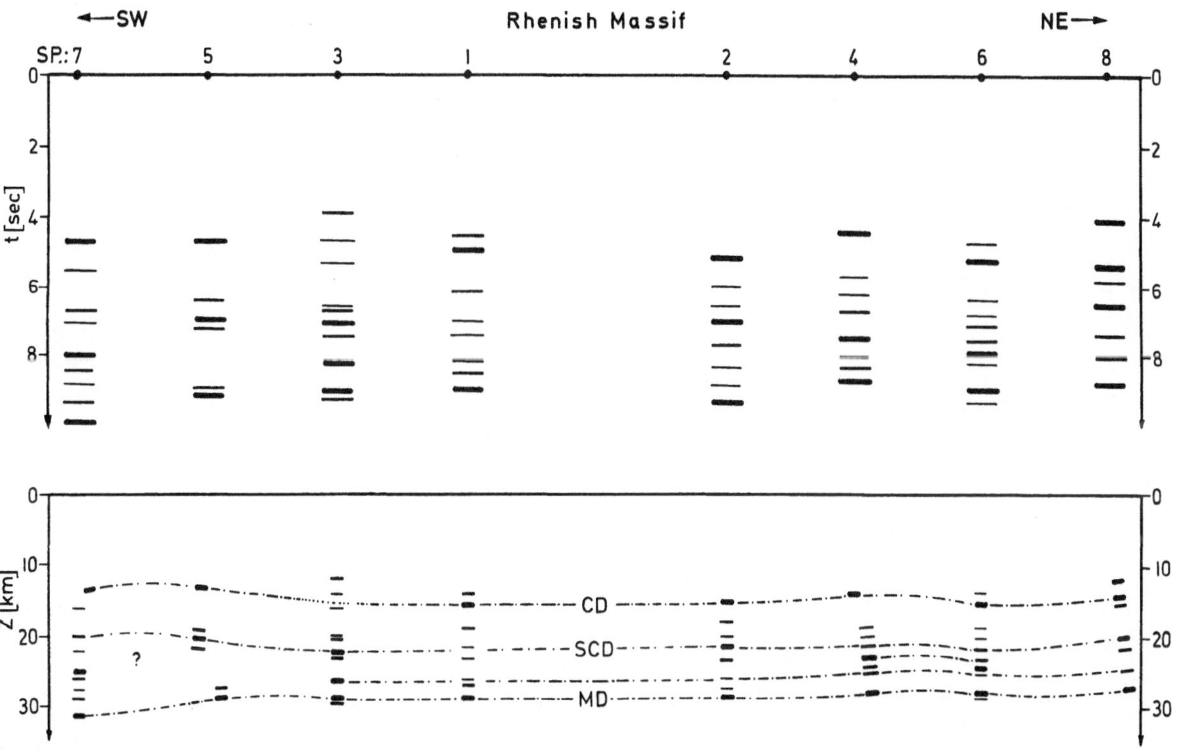

Fig. 5. Vertical traveltime and correlation of reflectors along the profile

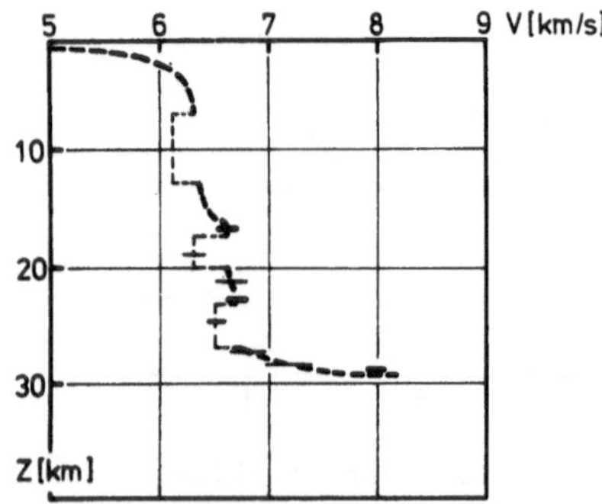

Fig. 6. Velocity-depth profile as derived by wide-angle measurements (MEISSNER et al., 6.1 of this volume) and reflectors obtained by near-vertical ray paths

velocity-depth function, as obtained from wide-angle reflections, is remarkably good. The correlated vertical reflections of CD, SCD, and MD indicate a nearly horizontal layering with relatively strong velocity changes and most probably sudden changes in the mineralogic composition. These changes might appear as large or small intrusions or a rhythmically arranged series of partial melts in the form of lenticular, sandwich-like structures. In the Rhenish Massif CD, SCD, and MD have been found to exist over larger distances than previously expected from experience in other areas. All three discontinuities have reflection coefficients comparable in magnitude. Thus, the transition from crust to mantle is partially stepwise along the profile and can be correlated over a distance of at least 180 km.

Acknowledment. The support of PRAKLA in supplying the digital facilities for recording and data processing is gratefully acknowledged.

6.3 Results of a Seismic-Refraction Profile from the Hoher Meissner to the North German Plain

E. Hinz, W. Kaminski, and A. Stein

ABSTRACT

Starting from two quarries at the Hoher Meissner and west of Göttingen an overlapping seismic-refraction profile runs northwards and ends in the North German Plain between Hannover and Bremen. The data along this profile are presented as record sections of normal and squared traces. The attempt was made to increase the signal-to-noise ratio by velocity stacking and by application of the modified method of correlation after SHIMSHONI and SMITH (1964). A simple crustal model, including time corrections for sediments, is derived. According to this the crustal thickness decreases from 28 to 26 km between the Solling and the Deister mountains.

ZUSAMMENFASSUNG

Ausgehend von zwei Steinbrüchen im Hohen Meissner und westlich Göttingen verläuft ein gestaffeltes Refraktionsprofil nordwärts und endet in der Norddeutschen Tiefebene zwischen Hannover und Bremen. Die Ergebnisse auf diesem Profil werden in Form von Montagen normaler und quadrierter Seismogrammspuren dargestellt. Es wird versucht, das Verhältnis zwischen Signalamplituden und Störamplituden durch Geschwindigkeitsstapelung und durch Anwendung des modifizierten Korrelationsverfahrens nach SHIMSHONI und SMITH (1964) zu verbessern. Ein einfaches Krustenmodell wird abgeleitet, wonach die Krustendicke zwischen Solling und Deister von 28 auf 26 km abnimmt.

6.3.1 Introduction

Deep seismic sounding in northern Germany requires high energy of the seismic sources because of the strong absorption of seismic body waves in the sediments with thicknesses of more than 5 km. Additionally the young sediments, that is the Tertiary and Quaternary layers in the North German Plain, cause a very high level of seismic noise and a corresponding reduction of the signal-to-noise ratio. The explosion at Helgoland in 1947, where 4000 metric tons of explosive were fired instantaneously, offered for the first time the possibility of crustal seismic observations in northern Germany. From 24 recording stations arranged along 3 profiles, a depth of 26 ± 2 km was derived for the Mohorovičić discontinuity (SCHULZE and FÖRTSCH, 1950).

CLOSS and HÄNEL (1964) compared these results with some new seismic refraction profiles and arrived at a modified model of the crustal structure for this region.

The progress in seismic instrumentation and the increasing number of deep-seismic sounding instruments in western Germany led to more detailed information from two overlapping profiles, with smaller gaps between the observation points. These are discussed subsequently.

6.3.2 The Position of the Profile

The strong explosions in two quarries near the villages of Bransrode (17) and Adelebsen (06) seemed to be suitable for the project. The code numbers in brackets refer to Section 3.3. of this Chapter: Fig. 1 (3.3), Tables 1 (3.3) and 2 (3.3). According to the tables, the charge weights in quarry No. 06 range from 3.0-9.0 metric tons of explosive and in quarry No. 17 from 14.0-24.0. Bransrode is situated in the Hoher Meissner about 26 km SE of Kassel. Adelebsen is situated in the southern Solling 15 km NW of Göttingen. The distance between the two points of explosion is about 43 km. The profile is the straight line between the two quarries and extends into the North German Plain. The exact azimuth of the overlapping profiles is 350°. They are denoted by 17-350 and 06-350.

6.3.3 Geology and Results of Former Seismic Projects

According to the surface geology the Buntsandstein formation is predominant along the profile between the Hoher Meißner and the northern Solling with some basaltic intrusions. This region corresponds to distances up to 75 km for the profile 17-350. The profile then runs across Jurassic and Cretaceous sediments in the region between Hils and Deister up to a distance of 115 km from the shotpoint near Bransrode and enters finally into Tertiary sediments of increasing thickness.

A review of the pre-Cretaceous sediments of the Northwest German Basin is given by NODOP (1962), who mapped the base of the Zechstein formation. Accordingly, the following depths are valid for profile 17-350:

Table 1

Shotpoint distance in km	100	130	150	200
Depth base of Zechstein in km	2.0	3.5	4.0	4.5

Between the center and the end of the profile 17-350 (shotpoint distance 190 km), the base of Zechstein dips northward by an average of 2°.

CLOSS and HÄNEL (1964) gave a reinterpretation of the Helgoland explosion data in combination with the profiles 06-260 and 06-080, evaluated by PLAUMANN (1961a). They deduced the following depths and velocities for the Solling, or more exactly, the region around the shotpoint near Adelebsen (06):

Table 2

Depth (km)	Velocity (km/s)	Geology
0 - 2.5	3.5	Sediments
2.5-14	5.9	Variscan basement
14 -28	6.55	Basic rocks
> 28	8.3	Ultrabasic rocks

More detailed and different values are given by GRUBBE (6.4 of this volume), where seismic shots of the oil industry for a seismic refraction survey were used.

The profile 06-170 from the shotpoint near Adelebsen to the south, the first 43 km of which represent a reversed profile for the profile 17-350, was evaluated by GIESE (1970). He found a continuous increase of velocity from 4.0 km/s to 6.2 km/s in the depth range up to 10 km, followed by a low-velocity layer with 6.1 km/s down to a depth of 15 km. In the depth range 20-30 km the velocity increases smoothly from 6.3 to 7.7 km/s without the existence of a sharp discontinuity between lower crust and upper mantle. FUCHS and LANDISMAN (1966a, b), however, found a sharp Conrad discontinuity as well as a sharp Mohorovičić discontinuity.

The discrepancies demonstrate that the crustal models depend considerably on the evaluation method.

6.3.4 Performance of the Seismic Field Survey

Most of the seismograms of both profiles were recorded on magnetic tape using the standardized deep seismic sounding equipment, described by BERCKHEMER (1970, 3.5). The distance between neighboring stations average of of 4 km. At the northern end of the profile it was partially reduced to 1 km to get a better phase correlation of the noisy seismograms. Almost every station had a geophone spread of 500-1000 m length with three vertical geophones. Further details of the profiles 06-350 and 17-350 as coordinates, altitudes, and exact shotpoint distances, are given in 3.3, Table 2.

6.3.5 Processing of the Seismic Data

The analog magnetic tapes were digitized on paper tape by a CDC 1700 electronic computer, whereby a low-pass filter was applied with a suppression of -6 db at 40 Hz and -70 db at 50 Hz. The Nyquist frequency was 50 Hz. The filtering was necessary since the seismograms are partially disturbed by electrical effects in the geophon cables. Several paper seismograms, which were recorded on the profile 17-350, using older equipment, were digitized on paper tape by a pencil-follower.

The paper tapes were used for further processing by an IBM 1620 computer and a Zuse 64 automatic plotter. The processing was made primarily with computer programs described in Arbeitsgruppe Digitalisierung (1968).

6.3.6 Record Sections

The record sections of the profiles 06-350 and 17-350, produced by the automatic plotter, were included in the seismogram album (3.4 of this volume). Both record sections showed first arrivals of good quality, the profile 06-350 up to about 105 km, and the profile 17-350 up to about 145 km distance from the relevant shotpoint. Later arrivals could not be correlated clearly because of interference of seismic waves and because the signal-to-noise ratio decreases with increased distance from the shotpoint. To improve first and later arrivals for larger distances, three methods have been applied on the seismic data. This is described in the following sections.

6.3.7 Squared Record Sections

One of the attempts to get a better correlation of the later arrivals consisted of plotting record sections of squared seismogram traces (Fig. 1 and Fig. 2). This process amplifies strong amplitudes and weakens small amplitudes. The squared record sections of the profile 17-350 for example, leads to improvements at a distance of about 45 km between 1 and 2 s of reduced traveltime and at a distance of about 80 km between 2 and 3 s. Remarkable are the large amplitudes which can be correlated for reduced travel times greater than 5 s in distances up to 40 km. The strong curvature of the latter travel-time branch leads to the supposition, that this phase can be interpreted as subcritical P^M-reflections; this was confirmed later by the velocity-depth function.

The disadvantage of squared seismogram sections is that small first arrivals may be completely suppressed, as was the case for the profile 17-350 in the distance ranges 35-65 km and 130-150 km. Therefore squared record sections should be used for the correlation of seismic waves only in connection with the original seismogram section. For comparison, a part of the original record section is represented together with the travel-time curves in Fig. 3. In this way, the travel-time branches which are represented by thick-heavy lines in Figs. 1 and 2 could be correlated. The reduced travel times and apparent velocities v_a which were found for the first arrivals are listed in Table 3.

6.3.8 Velocity-Stacking

To improve the weak P_n-arrivals, a velocity-stacking was applied by stacking traces with a given time lag, depending linearly on the distance, i.e., the traces are stacked with a given velocity.

For a distance interval in which the travel-time curve to be investigated is expected to be approximately

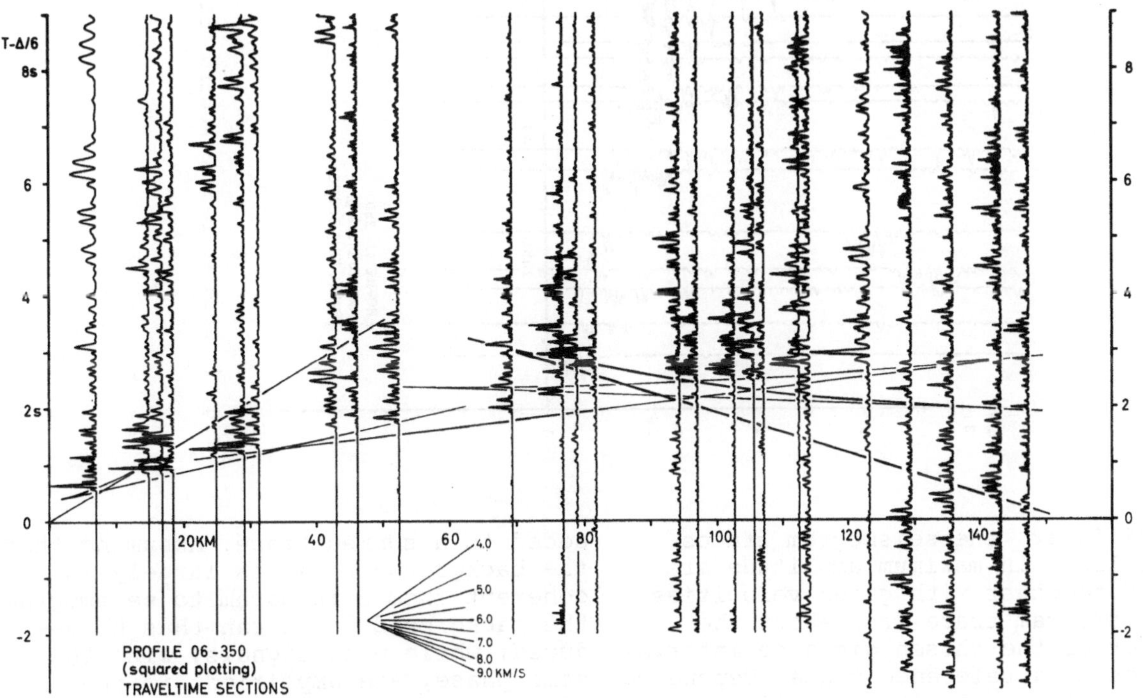

Fig. 1. Squared record section of the profile 06-350 with the travel-time curve

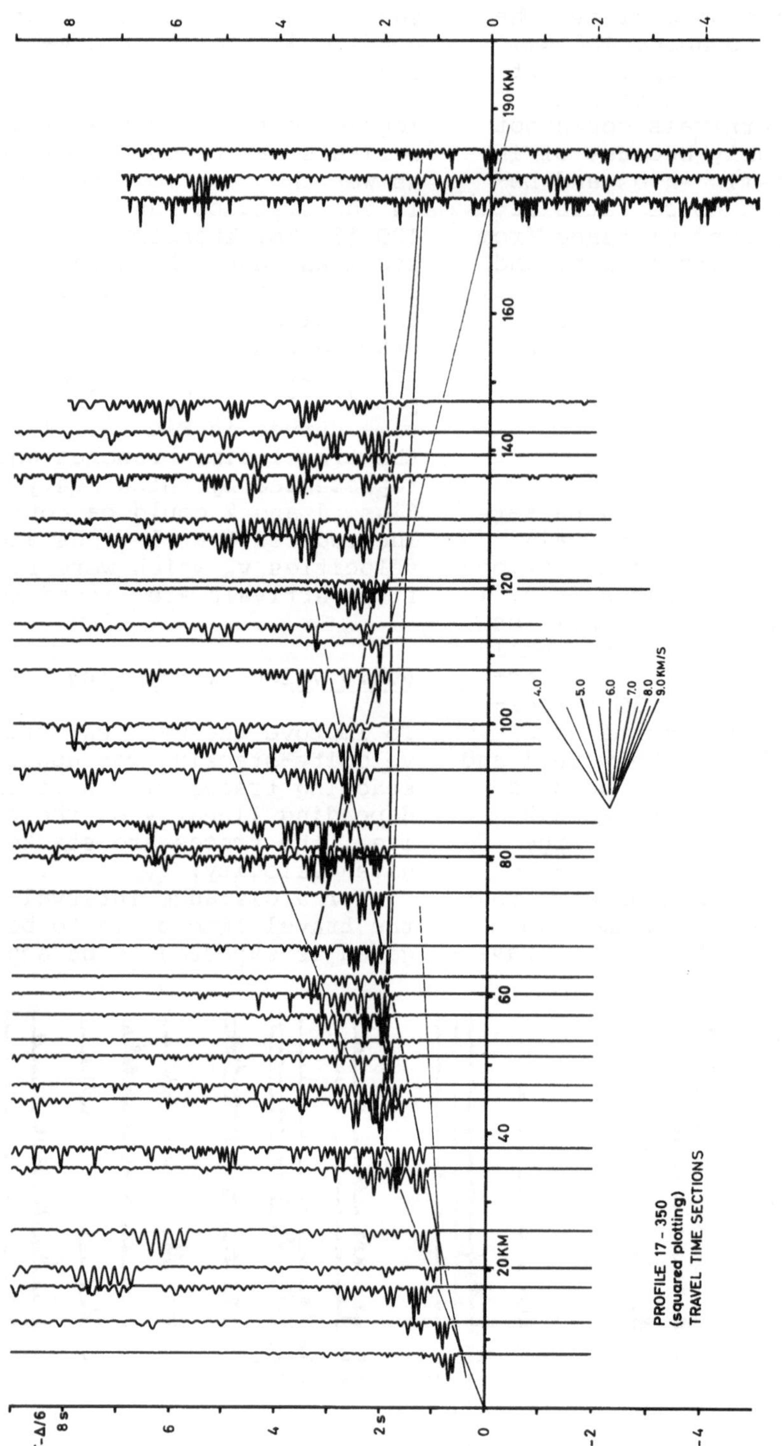

Fig. 2. Squared record section of the profile 17-350 with the travel-time curve

rectilinear, the seismogram traces normalized in maximum amplitude are added together with given velocities. A seismogram trace from about the center of the chosen distance interval serves as a reference trace. Depending on the given velocity for all other traces the corresponding samples are added to a summed trace. Assuming that the background noise is largely incoherent from seismogram to seismogram, the background noise can thus be reduced, while weak signals, with the same phase, are amplified, if the given velocity is equal to the apparent velocity.

Fig. 3. Record section of the profile 17-350, distance range 0-75 km

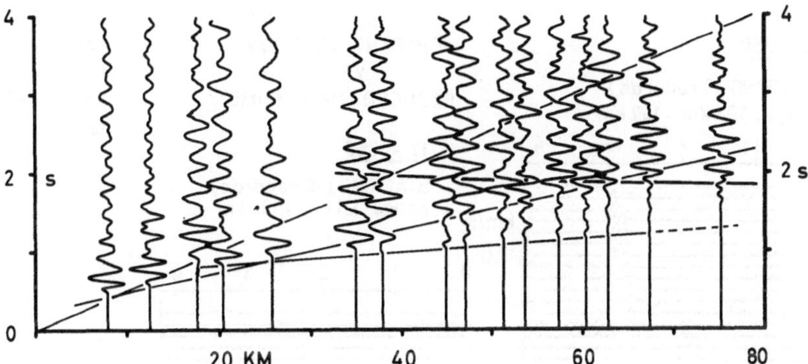

In a seismogram section of the summed traces of all given velocities, the appropiate apparent velocity is stressed by an amplitude maximum of the corresponding arrival.

In Fig. 4 such velocity-stacking is carried out for both profiles where the vertical axis denotes the pregiven velocity in steps of 0.05 km/s, and the horizontal axis represents the reduced traveltime of the central trace. The summed traces are given on a squared scale. For profile 06-350 the stacking was made in the 103-147 km distance range, and for profile 17-350 in the 121-153 km range.

Looking at the results of stacking on the right side of Fig. 4 it is obvious that for profile 17-350 almost no amplitudes occur before 1.3 s of reduced travel time. This can be traced back to the better signal-to-noise ratio for profile 17-350 as compared to profile 06.350 in the relevant distance interval. Larger amplitudes appear for about 1.3 s in the range 7.5-8.4 km/s. The amplitudes for 8.9 km/s and a reduced travel time of 2.3 s are very strongly emphasized. This velocity is, however, too large; obviously, neighboring phases have been included.

The following consideration should be used to determine to what extent the stacking of samples gives false apparent velocities, if a neighboring phase has been included instead of the true one. Let the time error be $\pm\Delta\tau$, which is equal to the period (here about 0.25 s, resulting from frequency analysis). If the interval between two adjacent seismograms is Δx (averaging about 4 km), the error estimation results in:

$$\frac{1}{v_a^*} = \frac{1}{v_a} \pm \frac{\Delta\tau}{\Delta x} \quad , \tag{1}$$

v_a is the true apparent velocity, and v_a^* the false determined one. The negative sign denotes the upper v_a^*, the positive the lower. Instead of an apparent velocity of 8.9 km/s, one of 5.7 km/s results, which coincides with the P_c-velocity in Fig. 2. The strong arrivals in Fig. 2 agree in travel time with those of the summed traces on the right side of Fig. 4.

The result of the stacking for profile 17-350 shows that the first arrivals cannot be associated definitely with the velocities. To explain this, we consider the P^M-curve in Fig. 2, which is clearly visible. The continuation of this curve to greater distances gives, with the guaranteed first arrivals of the original record section, a conjectural refraction curve of the P_n-wave at about 140 km and approx. 1.3 s, which has an apparent velocity of approx. 7.6 km/s. In reality, the stacking which corresponds

Table 3

Profile	Distance (km)	Reduced travel time (s)	v_a (km/s)	Wave-type
06-350	10- 30	0.6-1.2	5.1-5.2	
"	30- 80	1.2-1.9	5.6	P_g
"	80-100	1.8	6.2	P_c
"	100-120	1.8-1.1	7.6	P_n
17-350	0- 25	0.5-0.8	5.2	
"	25- 65	0.8-1.4	5.7	P_g
"	80-120	1.9	6.0	P_c
"	125-145	1.6-1.0	7.6	P_n

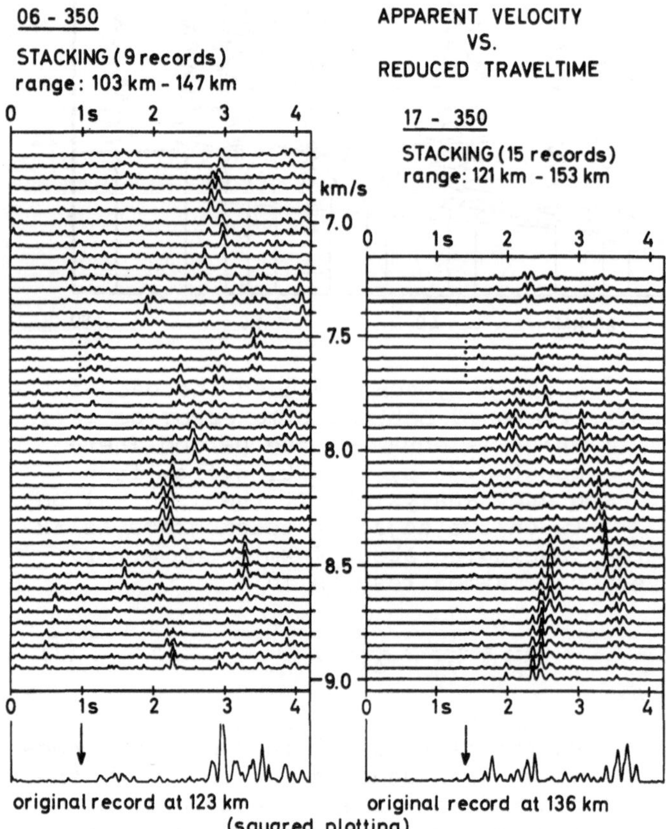

Fig. 4. An attempt of velocity stacking for parts of the profiles 06-350 and 17-350; detailed explanation in the text

to the velocity of 7.6 km/s causes an isolated group of amplitudes (dots on the right side of Fig. 4) which, due to the relatively small size, cannot be taken as representative. For profile 06-350 it is, by contrast, much easier to establish a strong first arrival on the summed trace at 7.6 km/s (dots on the left side of Fig. 4) so that for profile 06-350 this apparent velocity can be taken as certain.

6.3.9 Application of a Modified Correlation Method after SHIMSHONI and SMITH

Since for profile 17-350 the travel-time curve of the first arrivals in the distance interval 121-135 km cannot be uniquely determined by means of stacking, use should be made of the three-component method developed by SHIMSHONI and SMITH (1964). The method is described mathematically as follows:

$$M_j = \sum_{i=-n}^{+n} H_{i+j} \cdot V_{i+j} \qquad (2a)$$

$$\overline{V}_k = V_k \cdot M_k \qquad (2b)$$

$$\overline{H}_k = H_k \cdot M_k \qquad (2c)$$

$$k, j = 0, 1, 2 \ldots\ldots$$

the indexed quantities H represent the horizontal record, and the indexed quantities V the vertical. The quantities M_j may be understood as not normalized correlation coefficients of the jth point in the interval j-n to j+n. The products $V_k \cdot M_k$ represent the samples of the improved vertical trace, where the M_k may be understood as weighting coefficients according to the degree of correlation. The same holds for horizontal traces.

The modification of this method in the case under consideration is the restriction to vertical components. For gaps of 400 m or more between neighbored geophones the background noise may be considered as nearly incoherent (ARASCHMID, 1962). The correlation coefficients M_j would then by almost zero, but increase markedly for any seismic signals of sufficient amplitude. This consideration is also valid for geophone spreads perpendicular or oblique to the direction of the seismic main profile. Two vertical traces $V^{(1)}$ and $V^{(2)}$ with shot-point distances of x_1 and x_2 can be correlated according to the following equations:

$$M_j = \sum_{i=-n}^{+n} V^{(1)}_{i+j} \cdot V^{(2)}_{i+j+m} \qquad (3a)$$

$$m \approx \frac{x_1 - x_2}{v_a \cdot \Delta t} \qquad (3b)$$

Hereby v_a is the estimated apparent velocity of the seismic wave, the signal-to-noise ratio of which is to be improved, and Δt is the sampling rate. m is rounded to an integer.

Investigations show that a window of $2n \cdot \Delta t = 0.5$ s, about twice the period of the signal, leads to the best results, where the sampling rate is $\Delta t = 0.01$ s. Comparable to the velocity-stacking, a time lag must be considered in Eq. (3a) which is represented by the quantity m of Eq. (3b). Since the method is not very sensitive to errors in the chosen apparent velocity, it is sufficient to take an approximate value of 8 km/s for the investigation of the P_n- wave. Fig. 5 shows the correlation record section thus derived of profile 17-350. For comparison with the velocity-stacking, the correlation procedure was applied to adjacent traces at any given time for the same distance interval. Up to a distance of about 145 km, the noise was strongly reduced. Hence it was possible to identify first arrivals in the interval 135-145 km, which appear here as weak signals. The already-mentioned apparent velocity of 7.6 km/s for the P_n-wave could be confirmed.

6.3.10 Travel-Time Corrections

The emerging seismic waves are substantially delayed by the thickness of the sediments, increasing towards the north. According to the paper by NODOP (1962), mentioned above, the base of the Zechstein formation dips 2° northward on the average and near Hannover the depth of the base is about 3.5 km. To correct the travel time it is sufficient to use an average velocity for the overburden. SCHULZE and FÖRTSCH (1950) give a value of

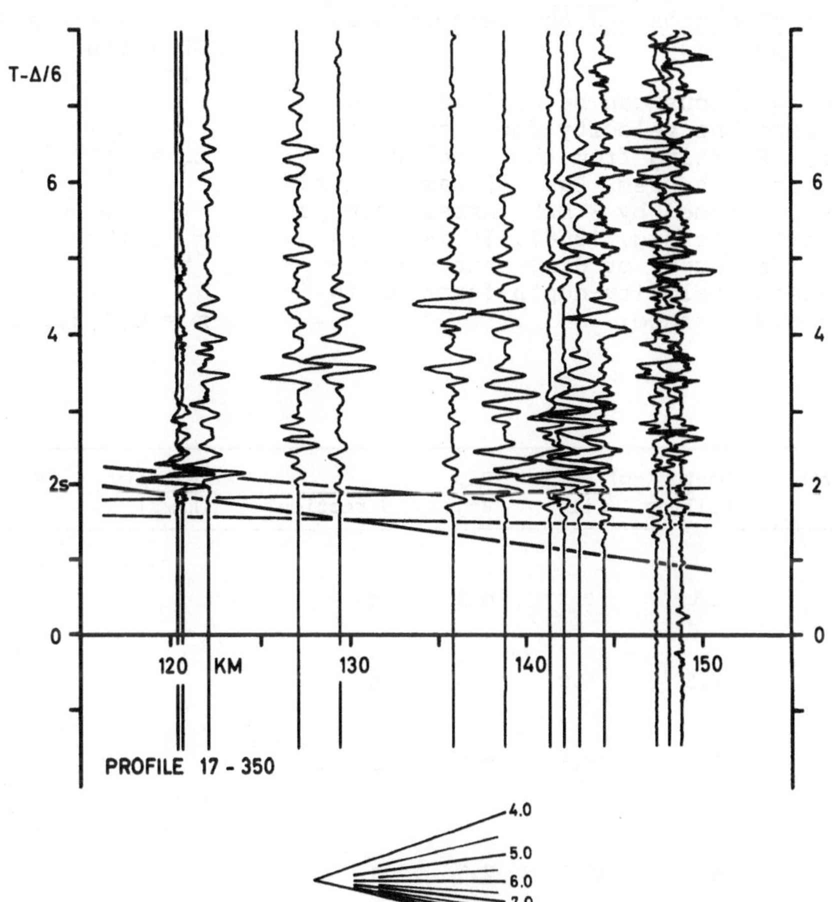

Fig. 5. Correlation record section of the profile 17-350; detailed explanation in the text

4.2 km/s, which is compatible with the values deduced from KREY and NODOP (1970) and GRUBBE (6.4). The corrections were performed approximately using an equation for the horizontal 3-layer case:

$$\frac{\Delta t}{\Delta z} = \sqrt{\frac{1}{v_1^2} - \frac{1}{v_3^2}} - \sqrt{\frac{1}{v_2^2} - \frac{1}{v_3^2}} \qquad (4)$$

where v_1 represents the average velocity down to the base of the Zechstein formation, v_2 the average velocity in the basement, and v_3 the apparent velocity of the wave, the traveltime of which is to be corrected. Δt is the delay caused by the sediments, and Δz the concerning sedimentary thickness down to the base of the Zechstein formation. Using the velocity values given in Table 4, a thickness of 1 km of the upper layer delays the P_n-phase by 0.096 s. An error estimation for the horizontal layer Eq. (4) shows that the influence of dip of the layers is negligible in the case under consideration.

6.3.11 Determination of the Velocity-Depth-Function and the Seismic Model thus Derived

Two equations can be applied to determine the velocity-depth function v(z). Starting from the Wiechert-Herglotz theorem, Eq. (5) was derived and programed by STEIN (Arbeitsgruppe Digitalisierung, 1968). It is valid for continuous or discontinuous increasing velocity-depth function in the wider sence:

$$z_n = \frac{1}{\pi} \sum_{m=0}^{n-1} \left((x_{m+1} - x_m) \cdot \ln \left(v_n/v_m + \sqrt{(v_n/v_m)^2 - 1} \right) \right) \qquad (5)$$

$$n = 1,2,3,\ldots, N-1; \; x_0 = 0; \; z_0 = 0$$

Hereby z_n is the depth, in which the velocity has the value v_n. The quantities x_m are the distances from the shotpoint in which travel times t_m are taken from the travel-time curve corrected for the sediments. The apparent velocities v_m are calculated approximately from adjacent values of x_m. This means

$$v_m = \frac{x_{m+1} - x_m}{t_{m+1} - t_{m-1}} \qquad (6)$$

The increasing indices correspond to the travel-time curve being followed continuously from the origin including cusps. This means that for overcritical reflections or corresponding diving waves the differences $x_{m+1} - x_m$ have negative values.

The second method is given by GIESE (1968a) and can be described by the following equations:

$$z_m = \frac{x_{m+1}}{2} \cdot \sqrt{\bar{v}_m \cdot \frac{t_{m+1}}{x_{m+1}} - 1} \qquad (7a)$$

$$\bar{v}_m = \sqrt{v_m \cdot \frac{x_{m+1}}{t_{m+1}}} \qquad (7b)$$

$$m = 0,1,2,\ldots, N-1$$

Table 4

Layer	Depth (km) Solling	Hils	Dip Angle	direction	Velocity (km/s)	Geology
1					4.2-4.6	Mesozoic, predominant Buntsandstein
	1	?	0.5	S		
2					5.2	Upper Palaeozoic, predom. Zechstein
	4.5	5.0	1.0	N		
3					5.7	Variscan basement
	16.0	15.5	0.5	S		
4					6.3	Intermediate, basic to ultrabasic rocks
	28	26.5	2.5	S		Crust-mantle boundary
5					7.8-7.9	Upper mantle

Hereby v_m can be taken according to Eq. (6).

Both methods were applied to the seismic profiles and the results are presented in Fig. 6, the results agreeing sufficiently well with each other. But in depth ranges with discontinuities of velocity or with steep velocity gradients, the velocity function may be ambiguous, as in the case under consideration. This ambiguity does not appear if a theoretical travel-time curve of a given model is used to reconstruct the model, using a computer program according to Eq. (5), as demonstrated by STEIN (1972).

Theoretical travel-time curves seem to fulfil conditions which cannot be formulated explicitely and which are normally not fulfilled for travel-time curves resulting from seismic field surveys. The condition of increasing apparent velocity is necessary but not sufficient. Other reasons for the discrepancies are correlation errors for the later arrivals and the dip of discontinuities or of equal velocity lines, which are not compatible with the Wiechert-Herglotz theorem in principle.

The ambiguities occurring around the discontinuities in Fig. 6 have been smoothed out. Therefore, the error from the depth determination for both equations can be estimated to ± 1 km. A seismic model for the area between Hoher Meißner and Hils can be derived from the v(z)-representation of profiles 06-350 and 17-350 considering that the distance of the vertex of the seismic ray from the shotpoint is half the recording distance. The result is given by the full line in Fig. 7. The velocities noted below the boundaries are valid in the upper range of the corresponding gradient layer. The broken line inserted to about 100 km from Bransrode corresponds to the base of the Zechstein formation (NODOP, 1962) which was used for the sediment time correction, assuming an average velocity of 4.2 km/s. The interval velocities of compressional waves in the sediments range between 1.6 km/s and 5 km/s. For comparison, Fig. 7 also shows the depth to the crystalline basement obtained from seismic refraction measurements (STEIN and DRUIVENGA, 1970), geoelectric measurements (BLOHM and HOMILIUS,

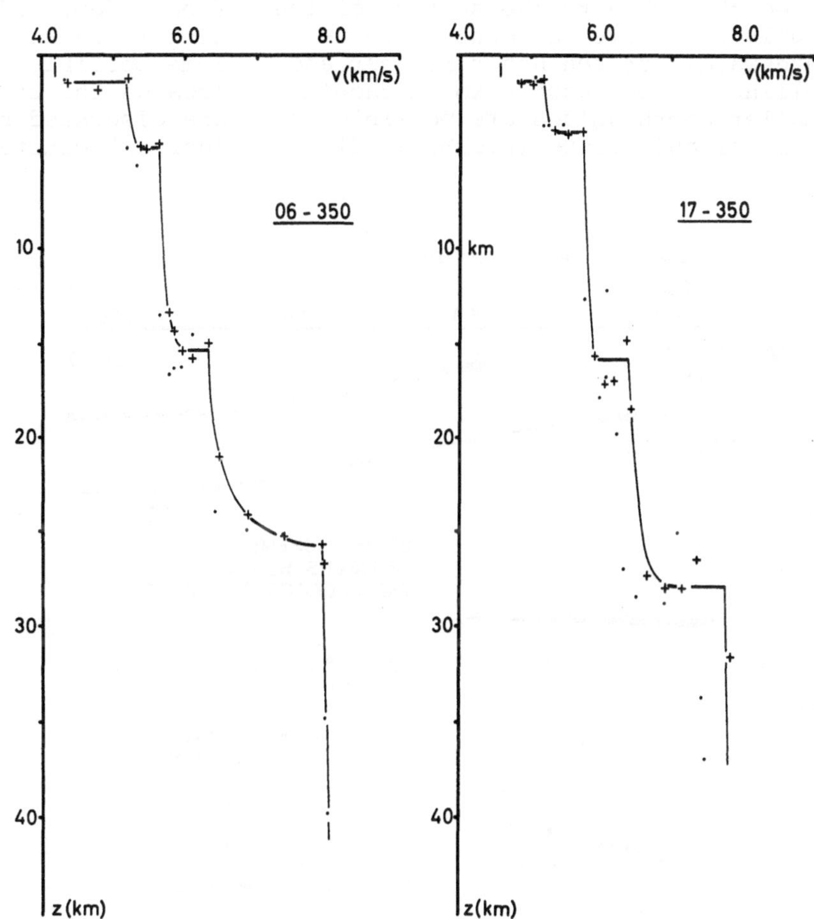

Fig. 6. Velocity-depth functions for the profiles 06-350 and 17-350. + After Wiechert-Herglotz formula; • after Giese formula

1971) and magnetotelluric measurements (LOSECKE, 1968), marked as SD, BH, and LO.

The results given in the vertical cross section of Fig. 7 are listed additionally in Table 4.

6.3.12 Discussion

The compressional wave velocities in the uppermost layer in the region between the shotpoints range from 4.2 km/s to 4.6 km/s, mainly determined by the Buntsandstein formation, while the velocity value of 5.2 km/s for the following layer was mainly determined by the Zechstein formation. For the third layer, the travel-time curve does not indicate any velocity inversion, such as found by FUCHS and LANDISMAN (1966a, b) and GIESE (1968a) for profile 06-170, starting southward from the shotpoint near Adelebsen. With a predominant velocity of 5.7 km/s this layer is supposed to be associated with the Variscan basement. The surface of this layer lies at a depth between 4 and 5 km. It should be mentioned that BLOHM and HOMILIUS (1971) using geoelectric measurements, derived the depth to the surface of the highly resistive basement at Reinhardswald, between Hoher Meißner and Solling, to be about 6 km, although smaller depth values are possible because of anisotropy (BLOHM, 1972).

The velocity of the third layer increases gradually downward, reaching granite-like velocities of about 5.9 km/s. At a depth of about 16 km the velocity increases discontinuously or with a thin transition zone to 6.3 km/s, the associated rocks being intermediate. The discontinuity or transition zone respectively dips less than 0.5^{o} towards the south.

It is not clear if this discontinuity should be taken as the Conrad discontinuity, as generally higher velocity values are expected for it, such as those found by FUCHS and LANDISMAN (1966a, b) for the south profile, mentioned above.

It is evident on the v(z)-curve of profile 06-350 (Fig. 6, on the left) that, within the lower crust, the velocity increases continuously with the depth, up to the final velocity of 7.8-7.9 km/s at about 28 km depth (the upper mantle). Certainly the transition from intermediate or basic to ultrabasic rocks is correspondingly continuous. The relation between rock type and wave velocity is discussed by PRESS (1966).

As can be seen from Fig. 7, the Mohorovičić discontinuity ascends, from a depth of 28 km beneath the Solling to 26 km beneath the Deister. Some possible reasons for this rise towards the basin of northern Germany are discussed by STEIN (1972). If a "normal" crustal thickness of about

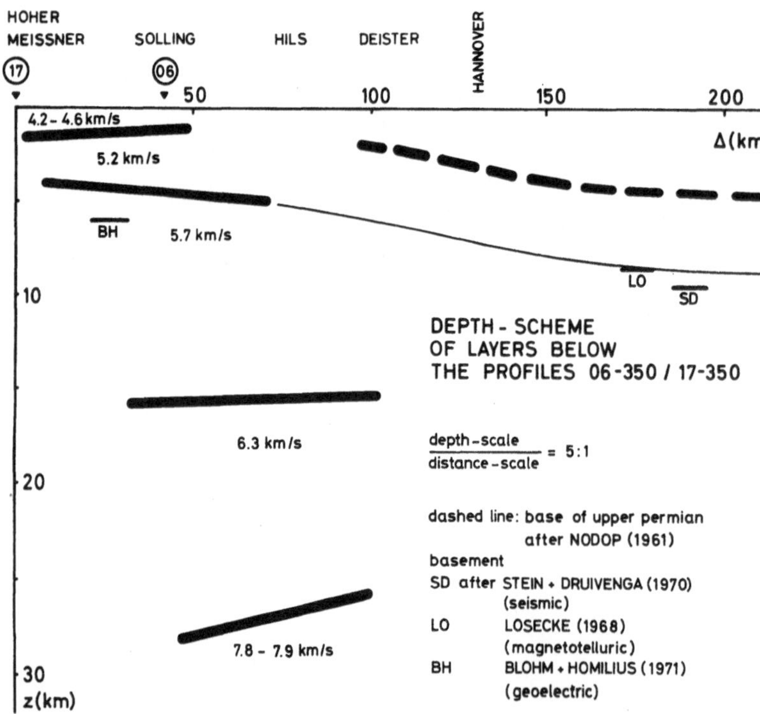

Fig. 7. Vertical cross section along the seismic refraction profile from the Hoher Meißner to the North German Plain; detailed explanation in the text

30 km is assumed at the end of the Variscan folding, the post-Variscan sedimentation, which resulted in deposits about 10 km thick in the basin of northern Germany, corresponds to a hypothetic depression of the Mohorovičić-discontinuity to a depth of about 40 km. The discrepancy of about 14 km between this estimation and the depth derived from deep seismic sounding, can only be explained by assuming an uplift of the crust-mantle boundary by phase transformation, opposite to the sinking of the same interface by sedimentation. This consideration should be valid at least qualitatively.

Such a process is also necessary from the standpoint of gravity to achieve an isostatic equilibrium, despite the comparable low average density of the basin sediments. STEIN and DRUIVENGA (1970) found on a refraction profile running from Versmold to Geesthacht, which crosses profile 17-350 at a distance of 190 km, a refractor at 9.4 km depth with a velocity of 6.2 km/s (denoted SD in Fig. 7). This locates the basement. LOSECKE (1968), using magnetotelluric methods derived a depth of the basement of about 8.4 km at Marwede near Celle (designated LO in Fig. 7). If one connects the top of the Variscan basement in the Solling with these scarce indications, then one obtains the thin line inserted in Fig. 7 which may be taken as a rough draft of the deeper range of the basin.

Extensive and detailed seismic refraction investigations have been performed by the oil industry down to depths of 10 km, the results of which have not been published to date.

6.4 Seismic-Refraction Measurements along Two Crossing Profiles in Northern Germany and Their Interpretation by a Ray-Tracing Method

K. GRUBBE

ABSTRACT

During exploration surveys performed by SEISMOS-GmbH, seismograms were obtained along two profiles 180 km long, one extending from Braunschweig to the Rothaargebirge, the other from Nienburg to Hoher Meißner, intersecting in the Solling area. The travel times are clearly influenced by different thicknesses of sediments and variations of the velocity distributions in the crust. Therefore interpretation was done by a ray-tracing method, which takes into account lateral velocity variations.

The results are given in two velocity cross sections along the profiles extending down to the Mohorovičić discontinuity. Apart from lateral variations, the sections are composed of a sedimentary cover of 3.5 and 4.9 km/s, the upper crust consisting of two interchanging layers of 5.4-5.8 km/s or 6.1 km/s and a layer with 6.3 km/s. The lower crust consists of one layer with 6.7 km/s. The Moho seems to dip to the SW and NW and at the intersection of the profiles lies at a depth of 27 km.

ZUSAMMENFASSUNG

Während eines refraktionsseismischen Meßprogrammes der SEISMOS-GmbH wurden zwei Profile von 180 km Länge beobachtet. Das erste liegt zwischen Braunschweig und dem Rothaargebirge, das zweite zwischen Nienburg und Hohem Meißner. Beide Profile kreuzen sich im Bereich des Sollings.

Aus den Laufzeiten lassen sich deutlich Unterschiede in der Sedimentbedeckung erkennen, aber auch die Geschwindigkeitsverteilung in der Kruste ist unterschiedlich. Daher wurde eine ray-tracing-Methode angewandt, welche laterale Geschwindigkeitsunterschiede zu berücksichtigen gestattet.

Die Auswertungsergebnisse sind in zwei Geschwindigkeits-Tiefen-Profilen zusammengefaßt. Abgesehen von lateralen Variationen folgt unter einer Sedimentbedeckung mit Geschwindigkeiten zwischen 3.5 und 4.9 km/s die obere Kruste, die sich aus zwei Schichten zusammensetzt: einer oberen mit 5.4-5.8 km/s bzw. 6.1 km/s und einer unteren Schicht mit 6.3 km/s Geschwindigkeit. Die untere Kruste besitzt eine mittlere Geschwindigkeit von 6.7 km/s. Die Mohorovičić-Diskontinuität taucht nach SW und NW ein und liegt am Schnittpunkt der beiden Profile in 27 km Tiefe.

6.4.1 Introduction

In 1967 and 1968 SEISMOS-GmbH carried out seismic-refraction work along two profiles in Northern Germany. The first profile extends from Braunschweig to the SW and in this paper will be called profile 1. The second profile extends from Nienburg to the SE, crosses profile 1 in the region of the Solling and will be called profile 2 (Fig. 1). The object was the mapping of pre-Permian horizons. Conventional shooting techniques were employed, charges up to 240 kg being set off in drill holes. By courtesy of SEISMOS-GmbH and the corresponding oil companies, the Institut für Angewandte Geophysik of the University of München was able to make additional measurements for an investigation of the lower crust in this region, using these commercial explosions.

6.4.2 Field Work

17 shotpoints were spaced at mean intervals of 9 km on profile 1, whereas there were 23 shotpoints at the same mean intervals on profile 2 (Fig. 1). At every point several shots with charges ranging in size from 48 to 240 km were fired. For these measurements 6 recording units of the type MARS 66 were used, each consisting of three vertical-component geophones, amplifiers, modulator, tape recorder, and time-signal receiver. This instrumentation has been described by BERCKHEMER (1970, 3.5 of this volume). To obtain continuous profiles of 200 km length the arrangment of the 6 recording units was different for the two profiles. On profile 1 they were placed successively on three spreads

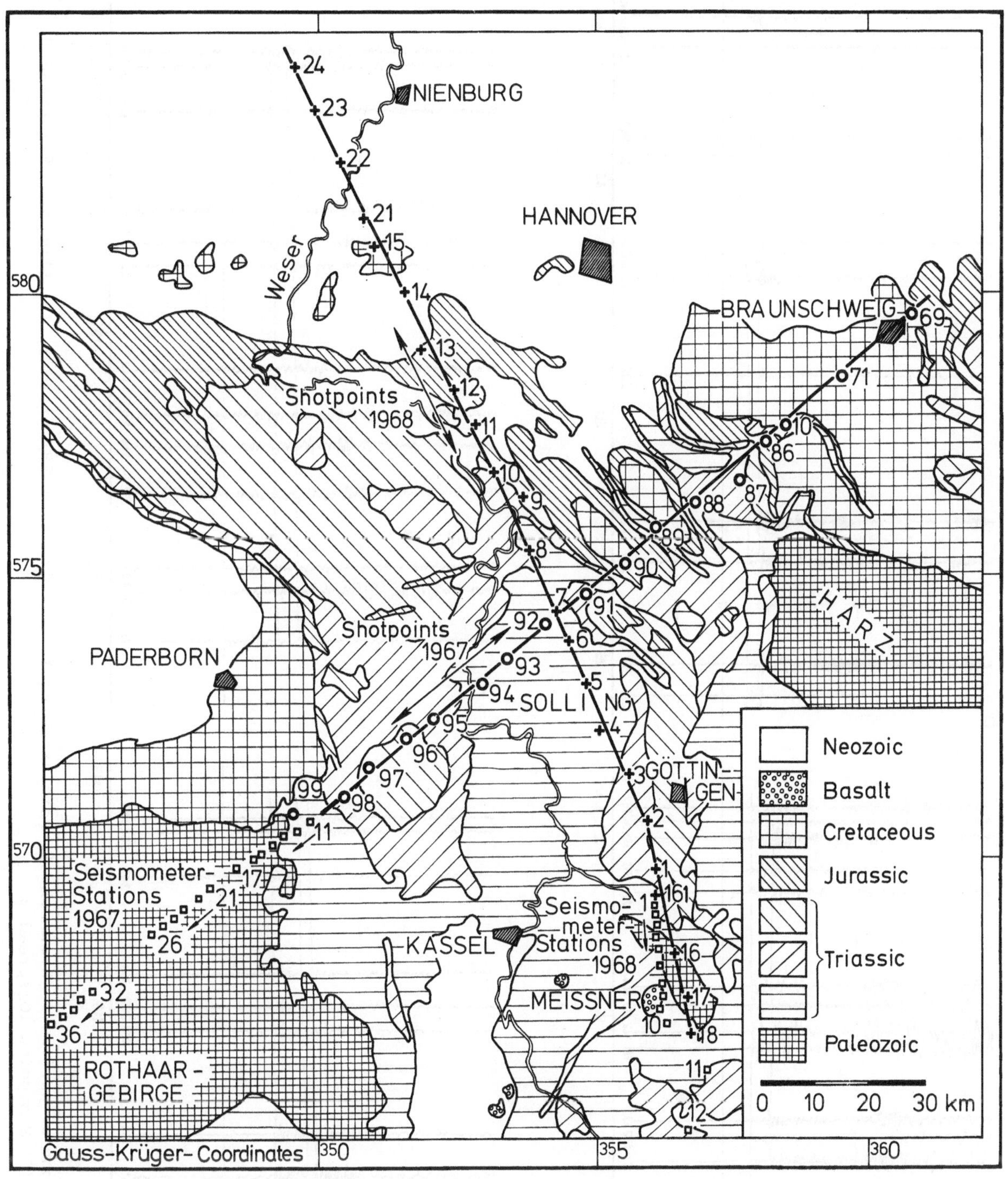

Fig. 1. Surface geology and location of the profiles. *Circles:* shotpoints in 1967 on profile 1; *crosses:* shotpoints in 1968 on profile 2; *squares:* recording stations, 3 geophones each

of 13 km length at the southwestern end of the profile, resulting in an average interval of 750 m between neighboring geophones for each spread. On profile 2 the recording units were changed between 12 stations with an approximate distance of 3 km between most of them.

In most cases the signal/noise ratio was low, although the measurements were done at night only. For distances larger than 50 km the minimum charge-to-yield usable seismograms was 120 kg. Thus, only one third of the recordings could be used for interpretation. Weak signal energy

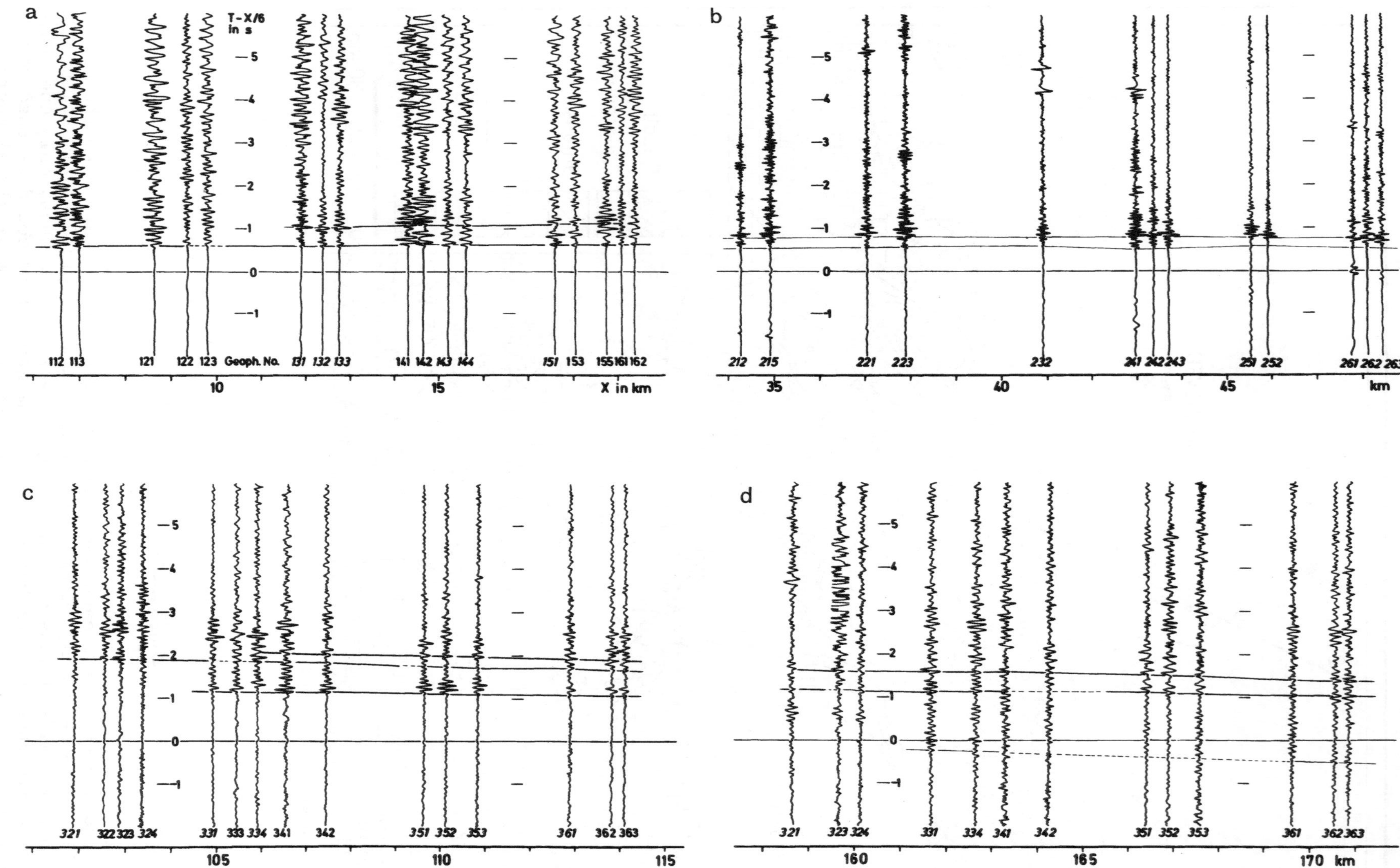

Fig. 2a-d. Seismogram sections of profile 1 (Braunschweig-Rothaargebirge). (a) Shotpoint 98/spread 1. (b) Shotpoint 97/spread 2. (c) Shotpoint 92/spread 3. (d) Shotpoint 10/spread 3

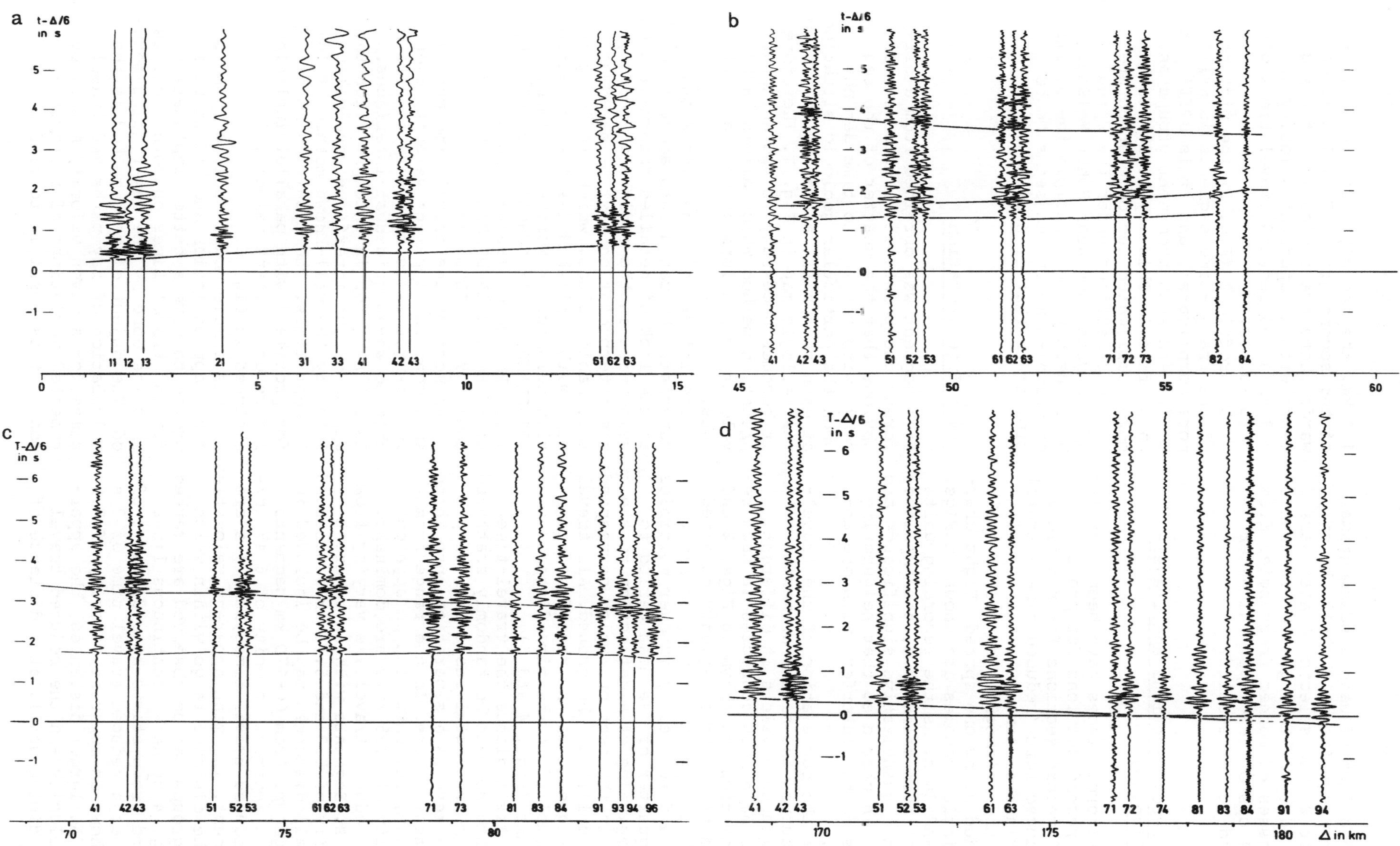

Fig. 3a-d. Seismogram sections of profile 2 (Nienburg-Hoher Meißner). (a) Shotpoint 161. (b) Shotpoint 5. (c) Shotpoint 8. (d) Shotpoint 24

caused a lack of usable seismograms in the range from 110 km-150 km on profile 2. The selected traces were improved by filtering with electronic bandpasses of corner frequencies from 2.5-24 cps or 6-16 cps, whichever yielded better results.

6.4.3 Record Sections and Travel-Time Diagrams

The filtered traces have been compiled in 24 record sections for profile 1 and 16 record sections for profile 2 in reduced form (reduced time = travel time minus distance/6). Each section shows the seismograms of one shotpoint recorded along one spread. Eight characteristic examples are shown in Figs. 2a-d and 3a-d. As the recording units had not been calibrated, the amplitudes have not quantitative significance and have been chosen so that neighboring traces do not interfere with one another.

For the correlation of the travel-time curves the minima or maxima of conspicuous wavelets in different traces have been connected in each record section, as shown in Figs. 2 and 3. The travel-time segments obtained have been approximated by straight lines and reduced by integer multiples of a quarter of the mean period to the beginning of a deflection to the right in the section, which corresponds to an uplift of the ground. All travel-time segments of the record sections are compiled in the time-distance diagrams (Figs. 4 and 5).

At first glance the travel-time segments seem to be randomly scattered, but a more careful inspection of profile 1 reveals a separation into three prominent groups. In the range from 20-70 km, segments with an apparent velocity of 6.25 km/s are dominant, their reduced traveltime varying from 0.5-1 s. Segments with an apparent velocity of 5.3 and 5.74 km/s observed at small distances may be included in this group. From 60-170 km segments with an apparent velocity of 6.47 km/s and reduced travel times near 1 s are prominent. The mean apparent velocities and the distance range within which both groups can be observed are marked in Fig. 4 by heavy continuous lines. A third group starts at a distance of 70 km at a reduced travel time of 3 s and shows an apparent velocity of 7.9 km/s. At larger distances, the apparent velocity and the reduced travel times dimish until, at a distance of 130 km, they nearly reach the values of the second group. From 130 km onwards there are indications for segments with an apparent velocity of 7.9 km/s and reduced travel times of 1 to -0.5 s.

On profile 2 (Fig. 5) the picture of the travel-time segments is still more complicated, and it is difficult to discern such prominent groups as on profile 1. This may be due to a more complicated geologic structure of the subsurface under the spread of the geophones near Hoher Meißner. Nevertheless, on principle the scheme of travel-time curves seems to be similar to that of profile 1.

6.4.4 Interpretation

Each segment was obtained from a different shotpoint. It is more or less obvious that the scatter of the segments with respect to time depends upon the particular geologic situation at each shotpoint, as can be seen from the geologic map in Fig. 1. Therefore it becomes necessary to use an interpretation method which allows lateral velocity variations as well as vertical ones.

The influence of sediments must especially be taken into account. Due to the lack of detailed information on velocities in the sedimentary cover, the interpretation must be based on reasonable values, which are in agreement with published results of refraction seismics in Northern Germany (BROCKAMP, 1931; SCHULZE and FÖRTSCH, 1950; PLAUMANN, 1961a; BEHNKE, 1961a, b; FUCHS and LANDISMAN, 1966b; BROKKAMP, 1967; GIESE, 1970).

As most of the travel-time curves show constant apparent velocity over long ranges it is necessary to relate them to waves guided at interfaces. As a first step in interpretation the depths of the interfaces were calculated by the well-known method of intercept times. The resulting model for profile 1 with parallel inclining layers of constant velocity is sketched in Fig. 6.

This simple model could only be a first approximation and had to be improved to give a better explanation for the travel times on both profiles. This was done by the ray-tracing method. The object of this method is the calculation of ray paths and travel times for a given velocity model. By repeated variations of a starting model it is possible to find a satis-

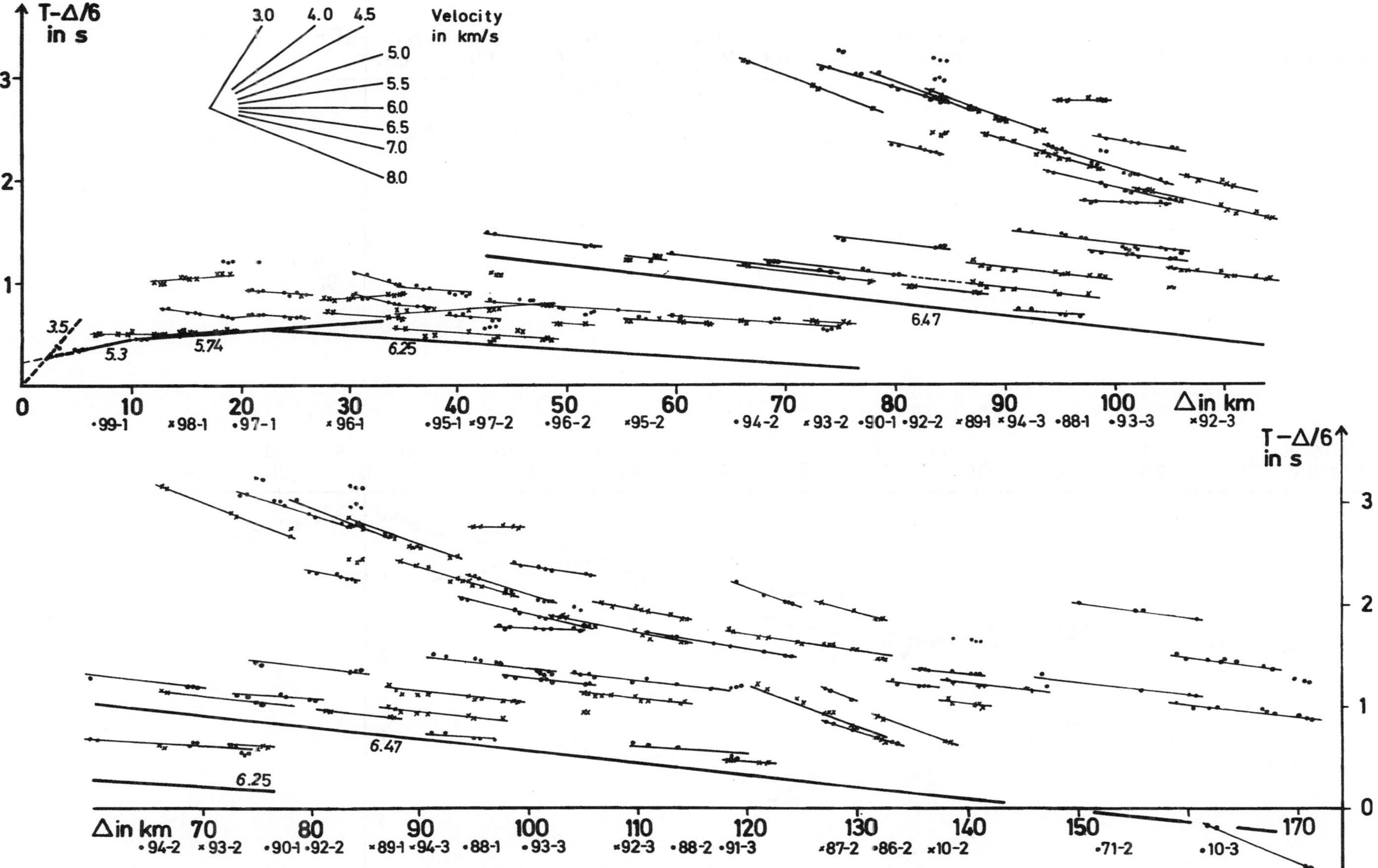

Fig. 4. Travel time-distance diagrams for profile 1 (Braunschweig-Rothaargebirge). The numbers at the heavy lines give the apparent velocities

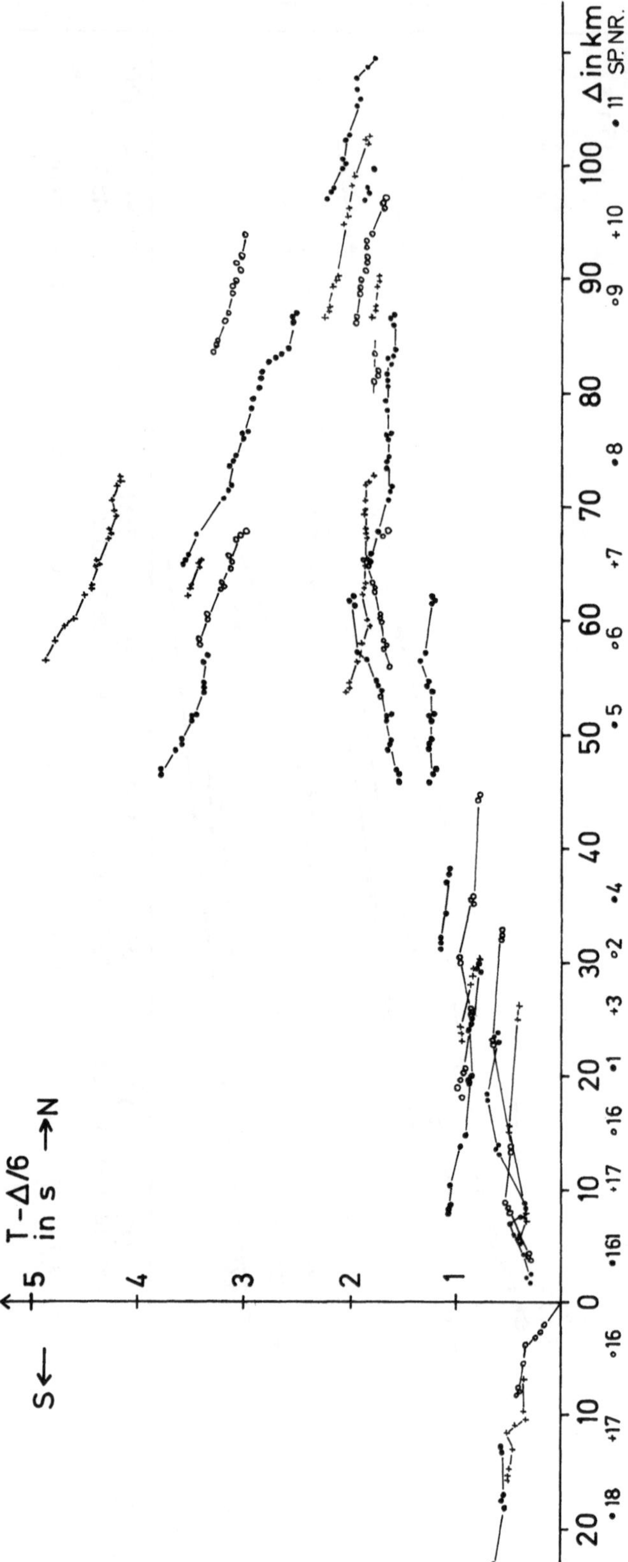

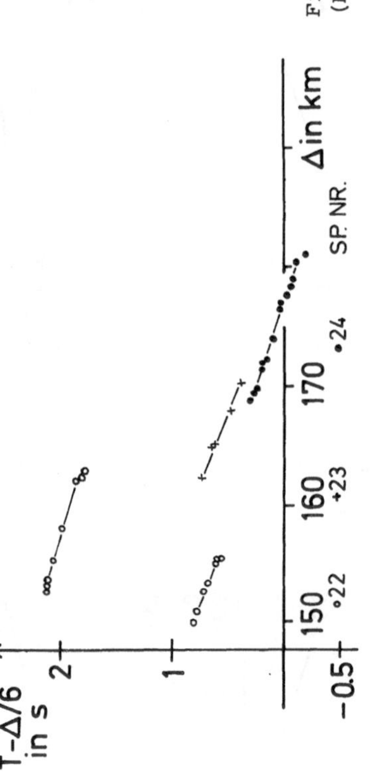

Fig. 5. Travel time-distance diagrams for profile 2 (Nienburg-Hoher Meißner)

factory fit between computed and observed travel times. The calculations were carried out with a program written by M. WILL. This program calculates and draws ray paths and travel times for a two-dimensional model. The model consists of areas with constant velocities or constant velocity gradients of arbitrary directions. The areas are bounded by polygonal lines. All calculations are based on the laws of refraction and reflection (WILL, 1975, 4.5).

The last of many models subjected to ray-tracing computation is represented by the two velocity cross sections in Fig. 7, the upper one for profile 1, the lower one for profile 2. They are identical at the intersection point in the northern Solling. The small gradients of 0.01 km/s/km in the layers of 6.1 km/s and 6.3 km/s have only computational significance. They are provided to allow simulation of travel times of guided waves by computing travel times along rays of slightly varying incident angles just penetrating a layer with a very small gradient.

Details of the sections are discussed with reference to the figures showing ray paths and travel times (Fig. 8a-d). The rays are traced starting at one reference point in each of the three geophone spreads on profile 1 (Fig. 8a-c) and the one spread on profile 2 (Fig. 8d). In essence, this corresponds to a reversal of shotpoints and receivers. The angles of incidence were chosen so that, together with the assumed surface velocities, they resulted in the observed apparent velocities at the reference points. Thus it is sufficient to compare the calculated travel times, indicated by crosses in the upper parts of the figures, with the observed travel times at the distance between reference point and shotpoint, indicated by dots, neglecting the apparent velocities.

All travel times are influenced by the sedimentary cover with a velocity of 3.5 km/s. There is a general thickening towards the North, an exception being the Solling saddle, as can be clearly seen in Fig. 8b, at a distance ('DELTA') of 60 km. Beneath a layer with 4.9 km/s, about 2 km thick, follows a thick layer with a velocity of 5.4 km/s at the top and a gradient of 0.05 km/s/km perpendicular to the upper interface. This layer is responsible for onsets at small distances on profile 1, spread 1, and the later arrivals (nearly 2 s reduced time) at distances between 50 and 70 km on profile 2. This layer is interspersed by high-velocity layers (6.1 km/s), which were necessary to explain the first arrivals between 30 and 80 km on both profiles. It is worth noting the different distributions of this high velocity layer in both sections. Beneath this 'mixed' layer there is with a downdip towards the North from 8 to 16 km depth a layer with 6.3 km/s, verified particularly on profile 1 by second arrivals between 40 and 110 km. The small distance of the critical reflection from this layer and relatively large travel times is evidence of the zone of low velocity (about 5.6 km/s) beneath the layer of 6.1 km/s on profile 1.

The lower crust is assumed to be formed by a layer of 6.7 km/s at a depth greater than 21 km. Arrivals from this layer could not be observed and ray-tracing calculations show that they would be concealed by the arrivals from other layers in a real seismogram with wavelets of a certain duration (see distance of 120 km). It should be mentioned that this is also true for reflections from the layers described above. The arrivals from the Moho are partly explained as reflected and refracted rays from an interface in 27 km depth, where the velocity increases to 8 km/s. For a best fit

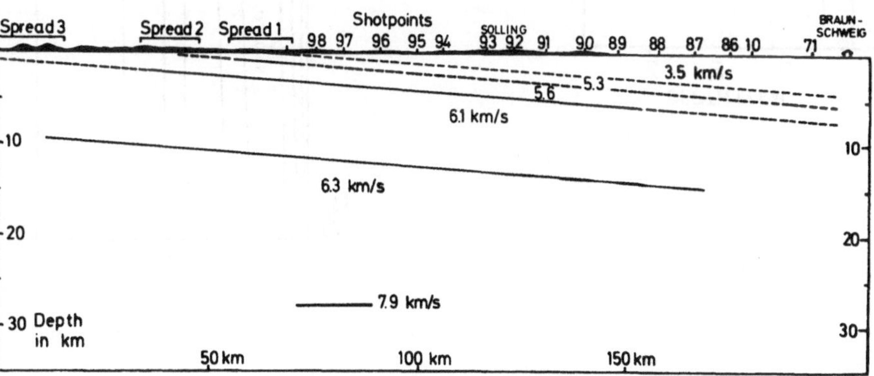

Fig. 6. First model for the velocity distribution beneath profile 1, found by direct inversion. The depth scale is exaggerated 2.5 times

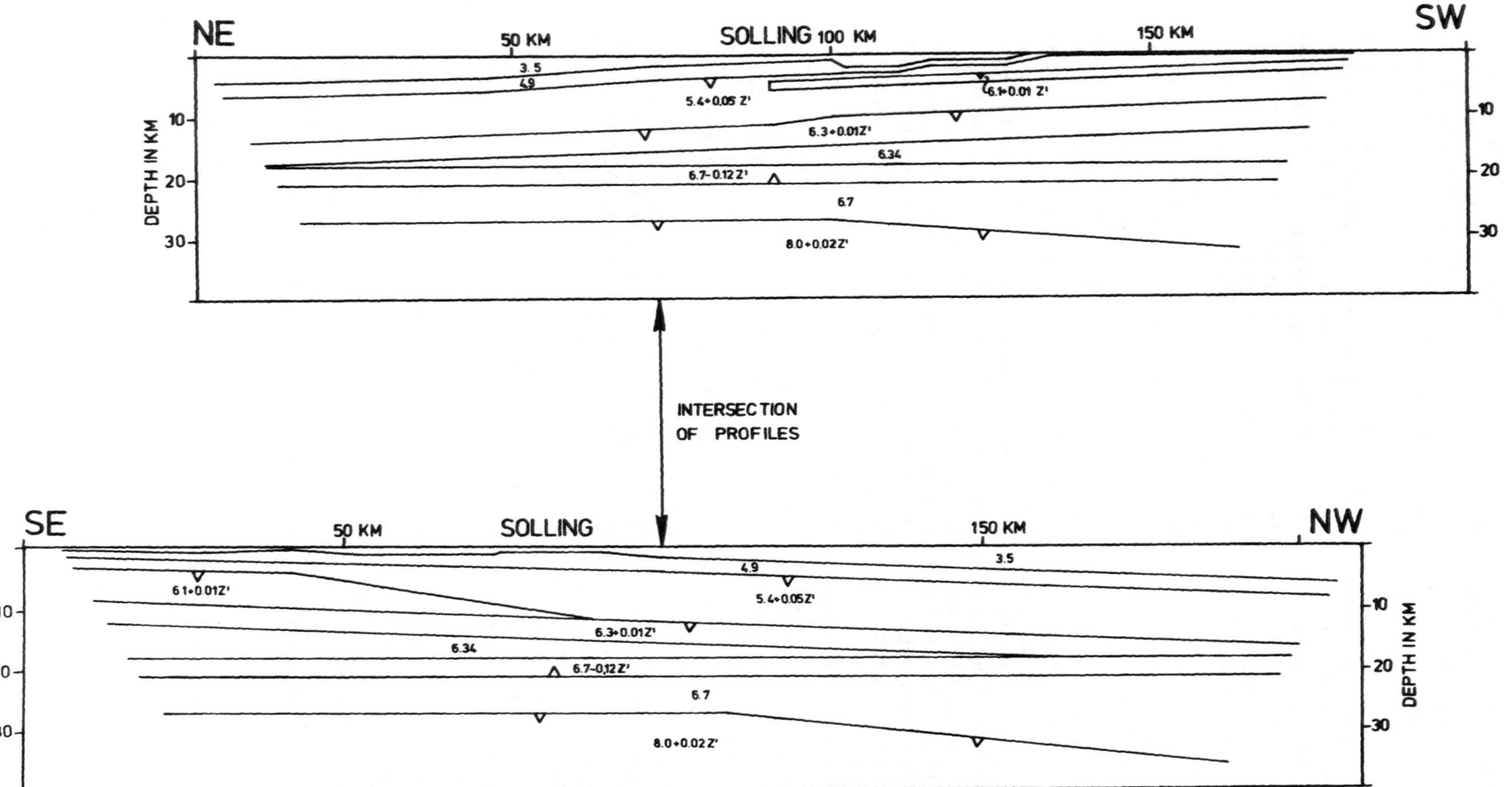

Fig. 7. Velocity model for profile 1 (*upper part*) and profile 2 (*lower part*) found by ray tracing. (Note the reversed orientation of profile 1 with respect to Fig. 6). The numbers in the layers give the constant velocity in km/s or the velocity in km/s at the interface marked by an triangle ± gradient in km/s/km perpendicular to these interfaces. z' is the distance from the marked interfaces

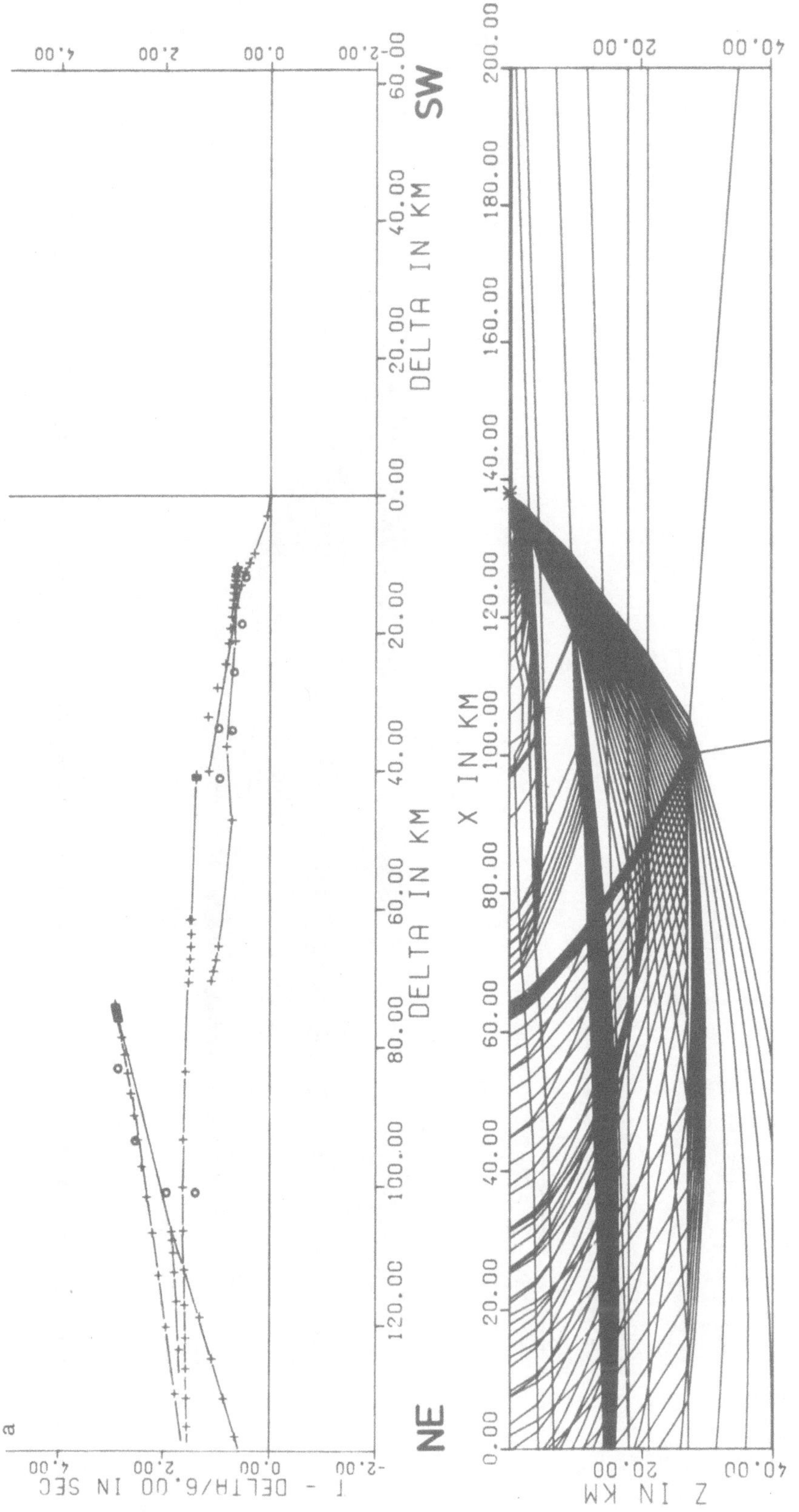

Fig. 8. (a) Ray paths and travel times for profile 1/spread 1. *Upper part:* crosses indicate calculated travel times, *circles:* observed travel times. *Lower part:* ray paths starting at a reference point (*asterix*) on the geophone spread

with all observed travel times, a dip of the Moho towards the southwestern end of profile 1 and the northwestern end of profile 2 had to be assumed. The dip on profile 2 did not suffice to explain the later arrivals. The remaining difference in travel times may, perhaps, be explained by a much thicker sedimentary cover or a different velocity distribution in the

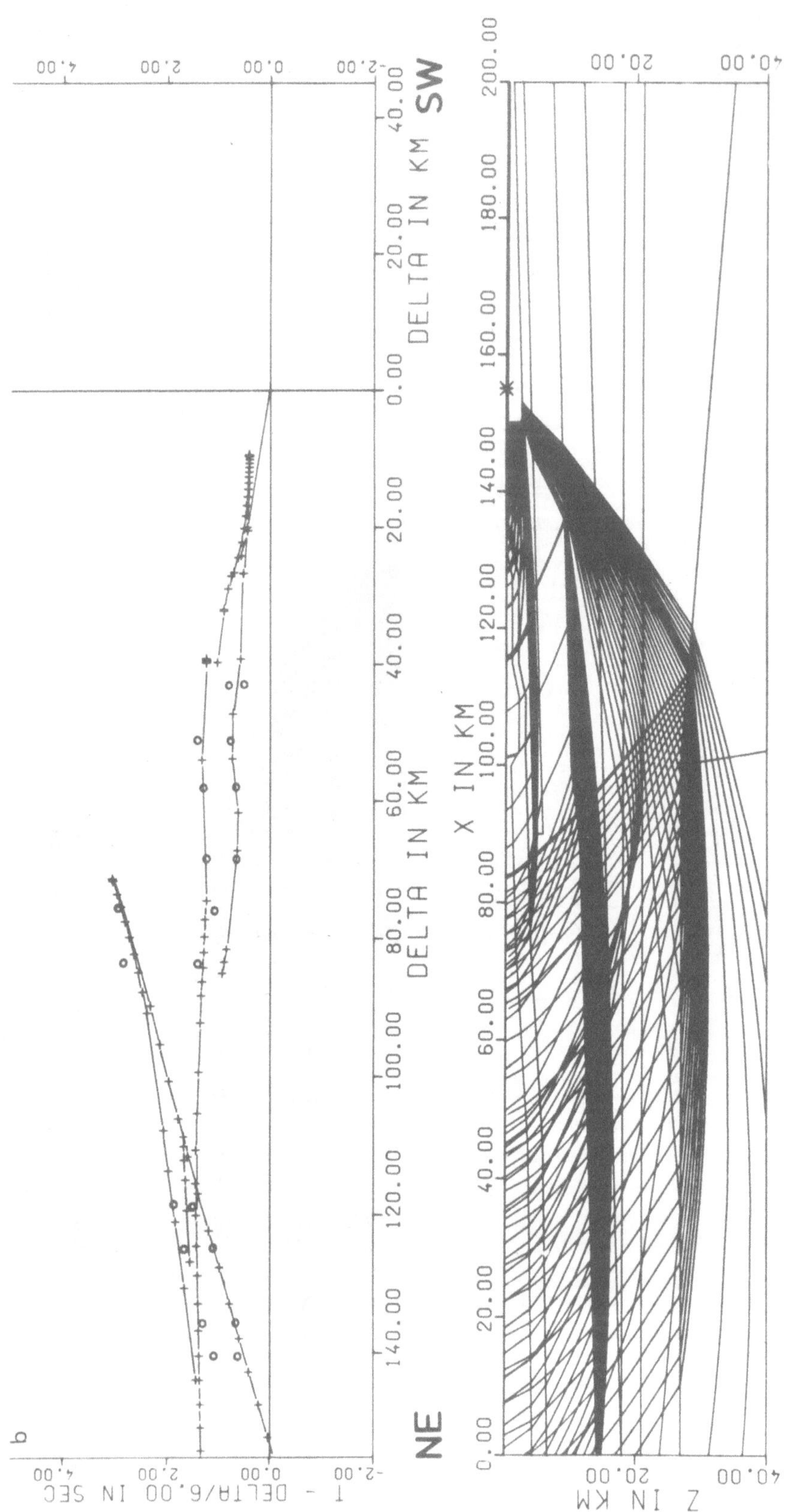

Fig. 8. (b) Ray paths and travel times for profile 1/spread 2. For explanation of symbols see Fig. 8a

lower or upper crust. The three dots on profile 2 at a distance of about 60 km represent only weak correlations and may be interpreted as subcritical reflections from the Moho.

6.4.5 Conclusions

A summary of the results is given by Fig. 9. Not every detail of the numerical model should be regarded as

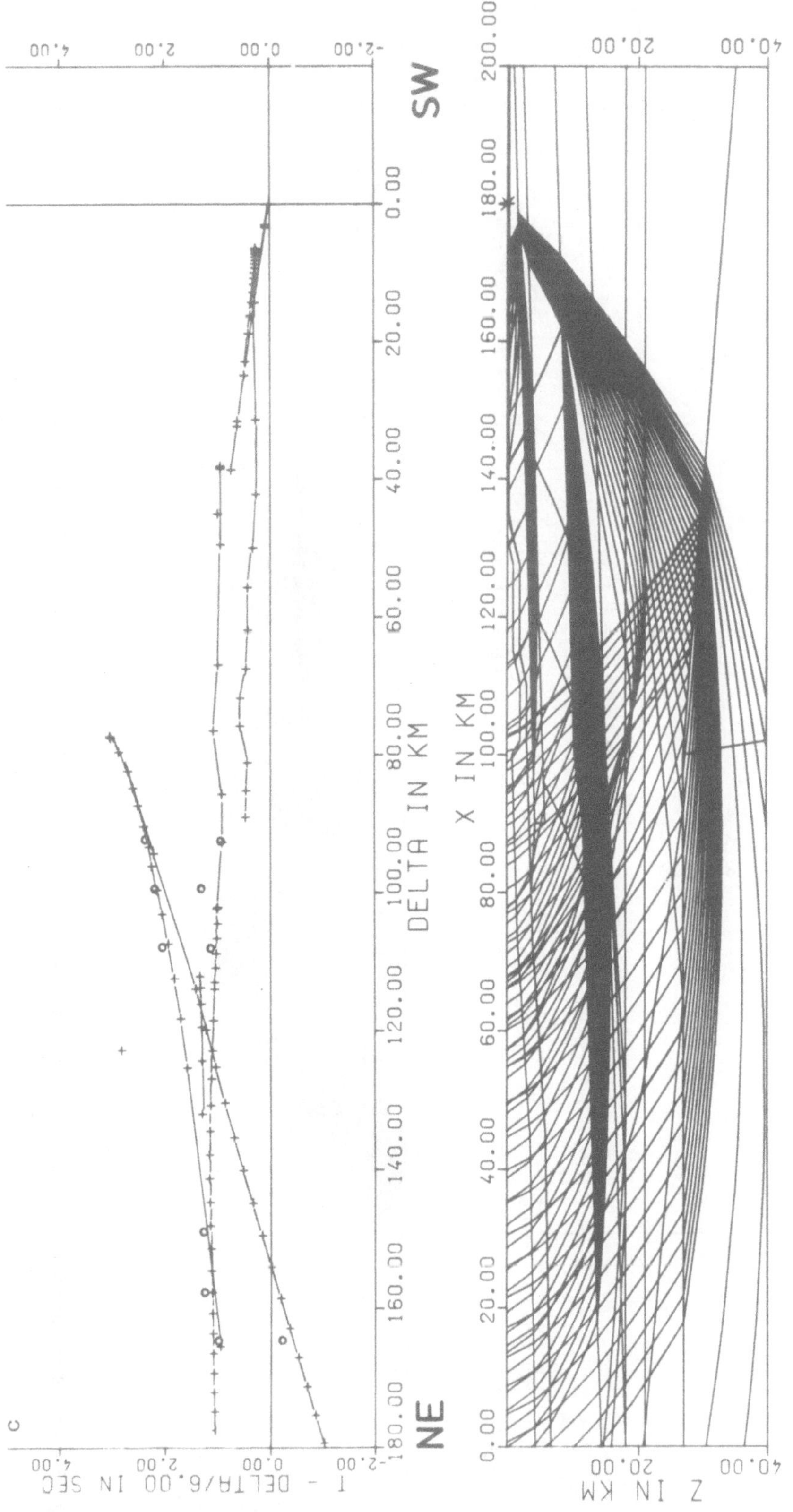

Fig. 8. (c) Ray paths and travel times for profile 1/spread 3. For explanation of symbols see Fig. 8a

real. For example there is a certain bandwidth for simultaneous variation of thicknesses and velocities without changing the travel times significantly. Furthermore the arrangement of the layers with 6.1 km/s and 5.4 km/s plus gradient should be looked at as an indication for a transition between two types of velocity distribution: an upper crust with somewhat over

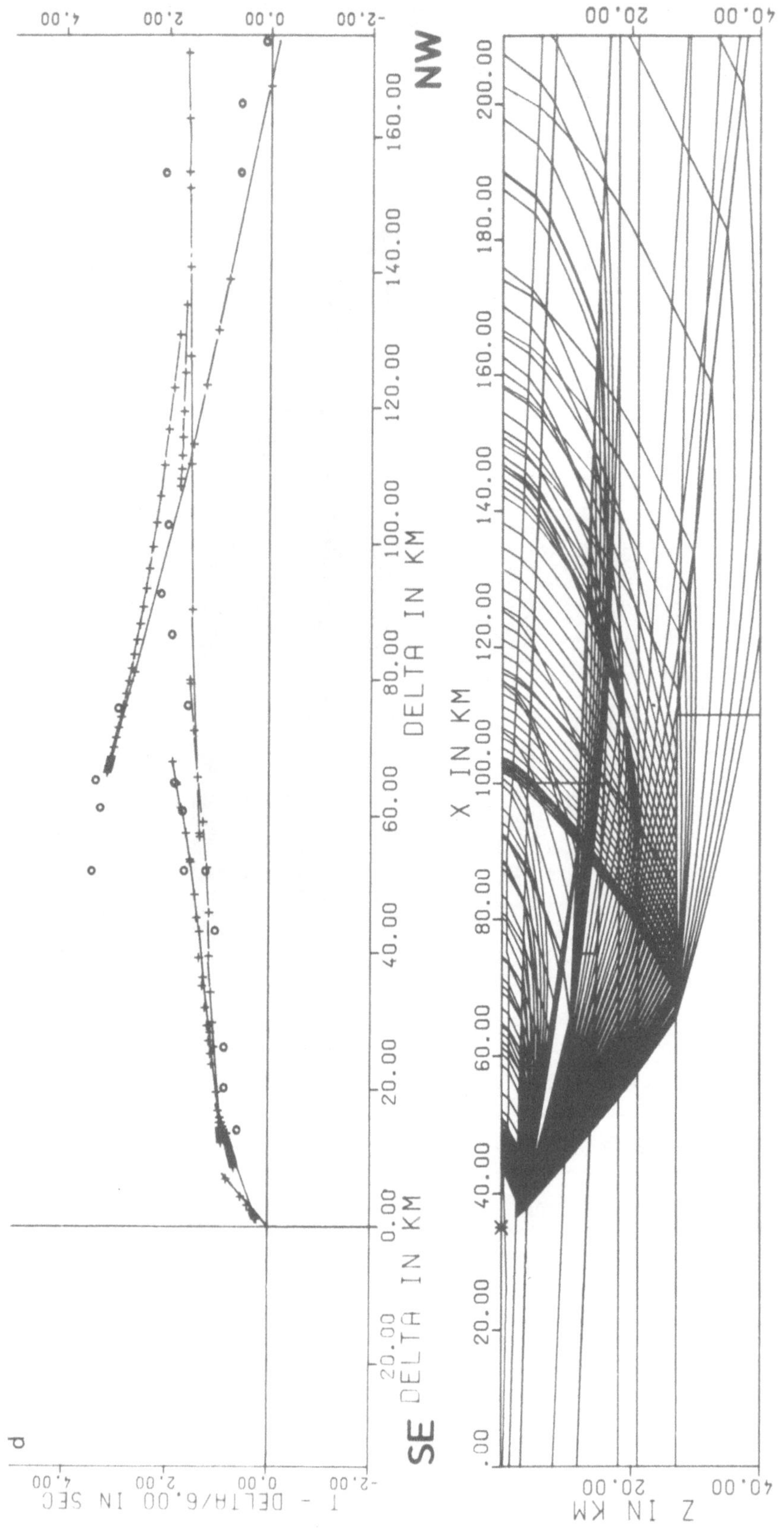

Fig. 8. (d) Ray paths and travel times for profile 2. For explanation of symbols see Fig. 8a

5.4 km/s in the northern part and an upper crust with 6.1 km/s in the southern part of the area.

However, as described above, the observed travel times and those calculated by ray tracing are in fairly good agreement. The model is compatible with surface geology and recent geophysical and geological knowledge of the area under investigation. Thus

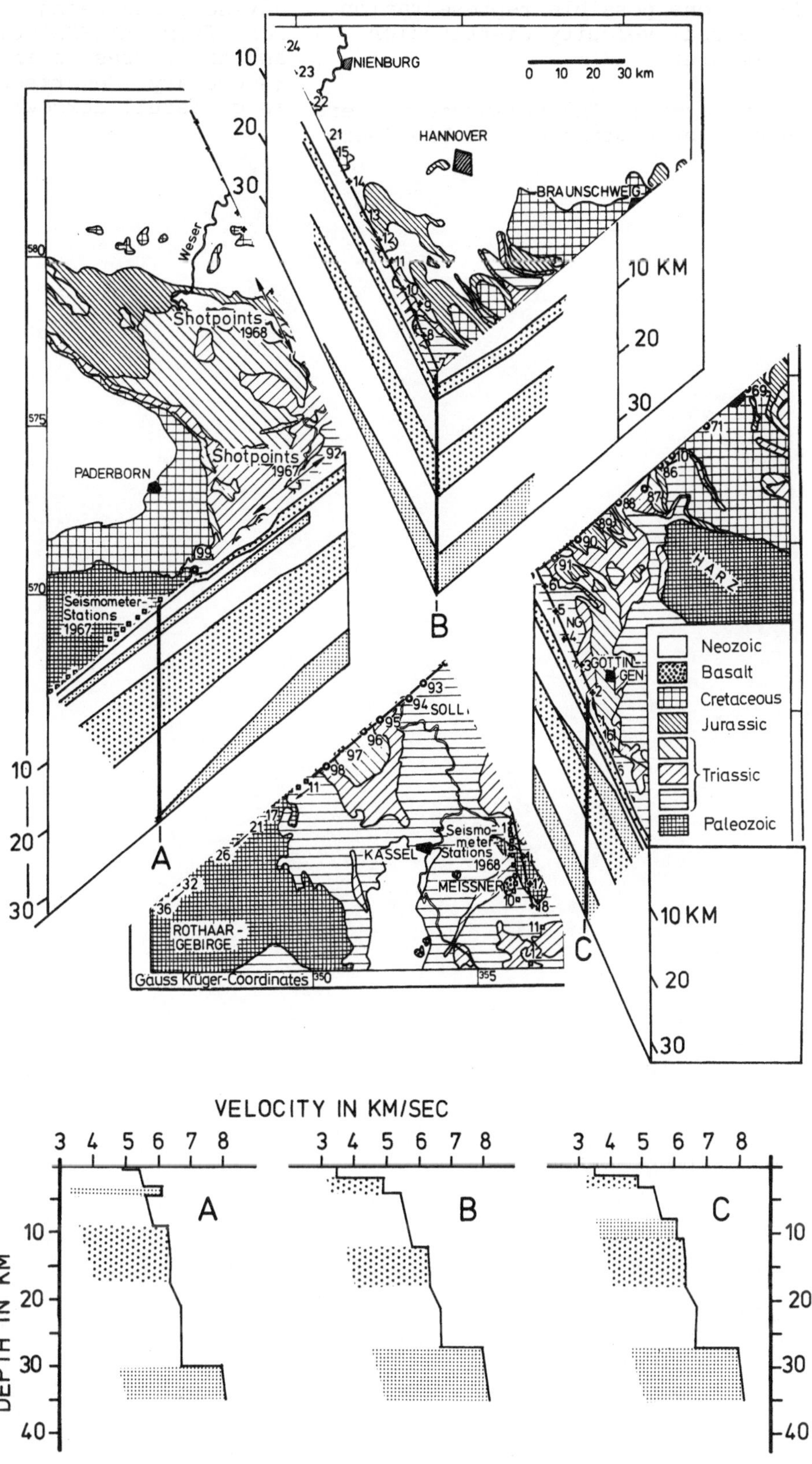

Fig. 9. Synoptical presentation of the derived velocity distribution beneath the profiles Braunschweig-Rothaargebirge and Nienburg-Hoher Meißner

we may conclude that the proposed model is one possible representation for the real velocity distribution in this area.

Acknowledgments. The measurements were supported by a grant of the DFG (German Research Society). Additional advisory and technical support by Dr. Dohr from PREUSSAG-GmbH, SEISMOS-GmbH and the corresponding oil companies (secretary: Gewerkschaft Elwerath) is gratefull acknowledged.

6.5 Seismic Results from Southwestern Germany

D. EMTER

ABSTRACT

This paper gives a short summary of the essential results of deep-seismic sounding in southwestern Germany up to 1970. Due to the location of shotpoints the crustal structure under the Swabian Jura can only be extrapolated. Observations of waves refracted or reflected at the crust-mantle boundary show a certain anisotropy of P-wave velocity in the uppermost part of the earth's mantle, a strong absorption of P-wave energy in a geothermal anomalous region of the Jura and only a gentle dip of the M-discontinuity from the Rhine Graben to the east. The most striking result is the observation of P-arrivals with a velocity of 8.50 km/s at distances between 260 and 400 km on profile 09-240. They replace the normal P_n-onsets and are attributed to a discontinuity or a zone of strong velocity gradient within the upper mantle in a depth between 50 and 60 km.

ZUSAMMENFASSUNG

Es wird ein kurzer Überblick über die wesentlichsten Ergebnisse sprengseismischer Untersuchungen in Südwestdeutschland bis 1970 gegeben. Wegen der ungünstigen Lage der Schußpunkte ist die Ableitung eines für den Schwäbischen Jura gültigen Krustenmodells nur durch Extrapolation möglich. Die Interpretation von Wellen, die an der Kruste-Mantel-Grenze refraktiert oder reflektiert werden, ergibt eine Anisotropie der P-Wellengeschwindigkeit im obersten Erdmantel, eine verstärkte Absorption von P-Wellen in einem Gebiet des Schwäbischen Jura mit anomaler geothermischer Tiefenstufe und ein nur sehr flaches Abfallen der M-Diskontinuität vom Rheingraben nach Osten hin. Das überraschendste Ergebnis ist das Auftreten von P-Einsätzen mit einer Geschwindigkeit von 8.50 km/s im Entfernungsbereich von 260 bis 400 km auf dem Langprofil 09-240. Diese Einsätze lösen die eigentlichen P_n-Einsätze ab und werden einer Diskontinuität oder Gradientenzone im oberen Mantel im Tiefenbereich zwischen 50 und 60 km zugeordnet.

Southwestern Germany is crossed by some seismic-refraction profiles and fans. These were partially fixed by the available shotpoints and reverse-shotpoints and used to study the propagation of seismic waves in regions interesting for their geologic and tectonic structure, such as the Swabian Volcano (CLOOS, 1941) with the geothermal anomaly sourrounding it (CARLÉ, 1958) and the seismic active zone along the 9°-meridian (SCHNEIDER, 1967). Fig. 1 gives a general view of the location of seismic measurements in the area under study. A common disadvantage of most of the profiles is the fact that they reach the interesting region in such far distances from the shotpoints that they give more information about the upper mantle than about the crust. A detailed discussion of the results of seismic investigations in southwestern Germany is given by EMTER (1971).

The profile Eschenlohe-NW (01-315) crosses two geologic units, the Molasse Basin and the South German Triangle including the Swabian and Franconian Jura and the Triassic region, before it reaches the rift system of the Rhine Graben in the Odenwald. In past years, the mostly older measurements on that profile have been completed by observations in some interesting distance intervals with modern equipments of the type MARS 66. As mentioned above, only for the first 130-150 km of this profile, i.e. for the Molasse Basin, can the velocity distribution within the crust be deduced. The velocity-depth function of the earth's crust in the Molasse was determined by many authors in the past with differing results (PRODEHL, 1965; LANDISMAN and MUELLER, 1966; MEISSNER, 1967a, b; GIESE, 1968; BRAM and GIESE, 1968). Therefore in this paper only some characteristics of the profile 01-315 are mentioned which may contribute some information on this problem.

The travel-time curve a (P_g) with a velocity of about 6.0 km/s shows no significant curvature and dies out be-

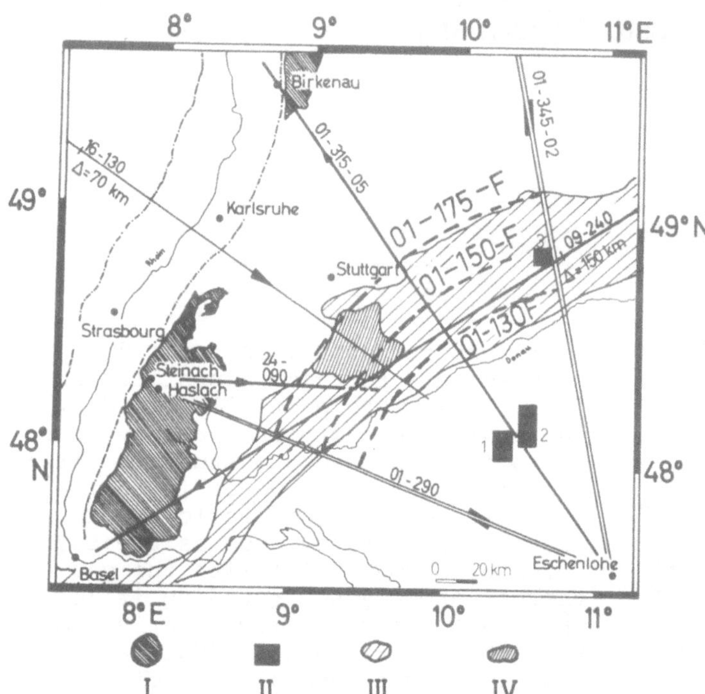

Fig. 1. Location map of seismic refraction profiles and fans in southwestern Germany. Profiles mentioned in this paper are marked by thick lines. *I* Crystalline basement (Black Forest and Odenwald). *II* Areas with reflection measurements near the profiles. *III* Swabian Jura. *IV* Region of Tertiary volcanism

tween 80 and 90 km distance. The reversed travel-time curve c (P^M) with an asymptotic velocity of 6.8 km/s can be explained with satisfying accuracy as a reflection from the Mohorovičić discontinuity rising from the Alps to the Molasse. The dip of the Moho is derived from the high apparent velocity of P_n (d). Phase c can be observed again as a very clear multiple event at distances longer than 180 km. This fact is typical of most profiles crossing the South German Triangle. Some events between the curves a and c can be correlated on nearly linear travel-time curves showing velocities of 6.3 and 6.8 km/s and relatively short critical distances. In order to explain these events by waves which are refracted at interfaces within the crust, one or two velocity-reversals have to be introduced between 8 and 20 km depth. Also agreement should be obtained with the deep-seismic reflection times of 4-4.5 and 7.5-8 s, observed by LIEBSCHER (1964) near the refraction profile (areas 1 and 2 in Fig. 1).

The P_n-arrivals, which can be observed very clearly on profile 01-315 up to distances of more than 150 km, can be traced at least on two of the fans in distances of 130 and 150 km round the shotpoint Eschenlohe. In Fig. 2 the reduced and corrected P_n-travel times on 01-150-F from NE (profile 01-345) to SE (profile 01-290) are compared with measured Bouguer anomalies (dashed line after GERKE, 1957) on the arc having a radius of the offset distance of about 120 km from Eschenlohe. The observed delay of the P_n-arrivals of nearly 1 s from NE to SW cannot be explained only by a dip of the M-discontinuity. An estimate of the thickening of the crust on the fan after the delay-time method (GARDNER, 1939) and after a formalism of MEISSNER (1970) independently leads to a difference of crustal thickness of about 10 km from NE to SW. In Fig. 2 the crustal thickness thus calculated is compared with a simple crustal model computed from gravity values. Similar results have been obtained for 01-130-F. Therefore at least half of the observed time delay must be caused by a decrease of seismic wave velocity in the upper mantle below the Molasse from NE to SW as was assumed by PRODEHL (1965).

If we consider the complete system of the fans (01-130-F, 01-150-F, 01-175-F, see Fig. 1) it is surprising to see that P_n-arrivals cannot be observed in certain regions. It is impossible to explain this fact by the radiation pattern of the quarry-blast. This could be proved by comparative amplitude measurements on some of the fans. An example is shown in Fig. 3. Fig. 4 shows tentatively how the appearance of the P_n-phases qualitatively correlates with the area of the Swabian Anomaly formed by the region of Tertiary volcanism in the Jura and

Fig. 2. Reduced and corrected P_n-travel-times (•) on 01-150-F compared with measured Bouguer anomalies (*dashed line*) and computed gravity values (⊙). M_t: M-discontinuity from P_n-travel-times (*dash-dotted line*). M_{gr}: M-discontinuity of the gravity model (*full line*)

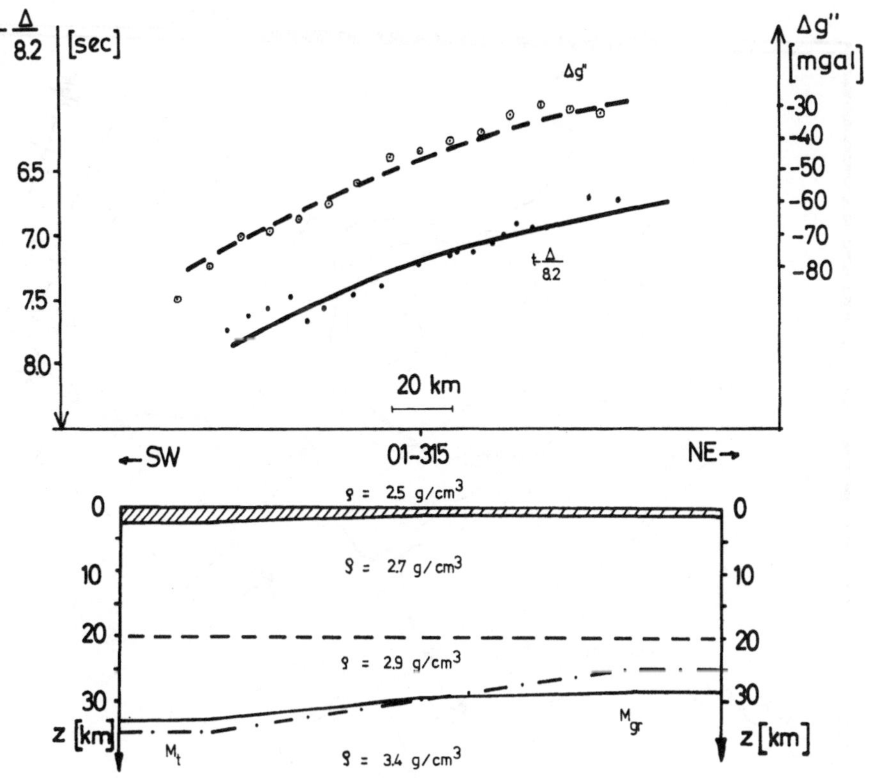

the geothermal anomaly surrounding it. It is obvious that this zone produces the seismic shadow for P_n-waves. This may be due to a strong absorption of the seismic waves in the upper mantle or in the crust, caused by a stronger "low-velocity zone", for instance. Moreover the anomaly is crossed by the profiles 09-240 and 16-130. On both profiles P_n-onsets can be observed throughout this region showing, however, a strong decrease in amplitude with increasing distance. On profile 09-240, along the section crossing the anomaly, the amplitudes of P_n-arrivals were measured using seismograms from the same explosion recorded by calibrated equipment (MARS 66). Assuming that the crust does not strongly change its composition within the anomaly, a determination of the absorption of compressional waves in the upper mantle is possible. For the present preliminary interpretation, the amplitude of the P_n-onset was correlated to its dominant frequency. An exact interpretation should deal with the amplitude spectrum of the signal. In the diagram of Fig. 5 the product of distance and P_n-amplitude is plotted in a logarithmic scale against the distance. The geometrical spreading factor Δ^{-1} is eliminated by this procedure. According to the equation

$$A = A_0 \Delta^{-1} \exp(-\alpha\Delta)$$

an absorption coefficient of $\alpha = 0.027$ km^{-1} can be determined from the slope of the regression line in Fig. 5. Assuming that only absorption in the upper mantle effects the strong decrease in P_n-amplitudes, a quality factor $Q = 100 \pm 30$ is obtained. A similar value is found for the part of profile 16-130 that crosses the anomaly. The quality Q observed for the upper mantle in the region of the Swabian Anomaly seems very low compared with Q-values reported in literature (RYALL and STUART, 1963: Q = 520; PASECHNIK, 1966: Q = 250; WOLBER,

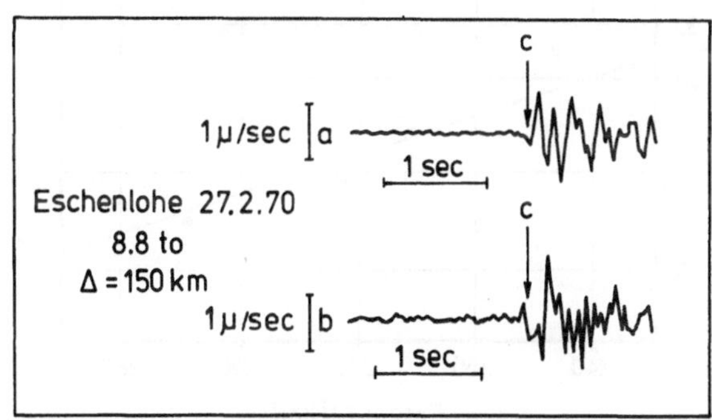

Fig. 3. Amplitude comparison between two stations of 01-150-F: *a* near profile 01-290; *b* near profile 01-345-02

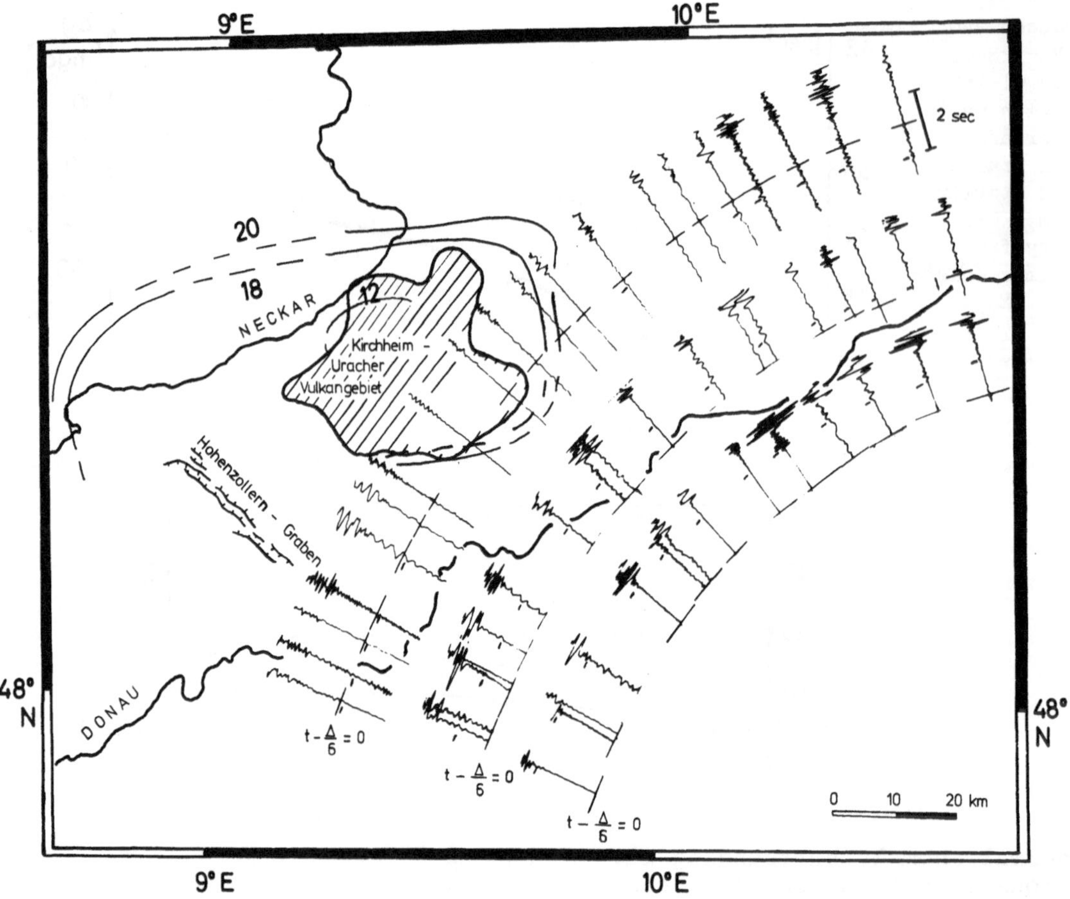

Fig. 4. Map with reduced seismograms plotted at the offset distances of three Eschenlohe fans (01-130-F, 01-150-F, 01-175-F). P_n-events are marked by bars. Moreover the map shows the borders of the Swabian Volcano and some isolines of the reciprocal geothermal gradient (m/°C). (After CARLÉ, 1958)

1968: Q = 250). That the strong decrease of the P_n-amplitudes might be caused by changing geometrical conditions for the radiation of seismic rays, i.e. by a velocity gradient or a velocity reversal in the uppermost

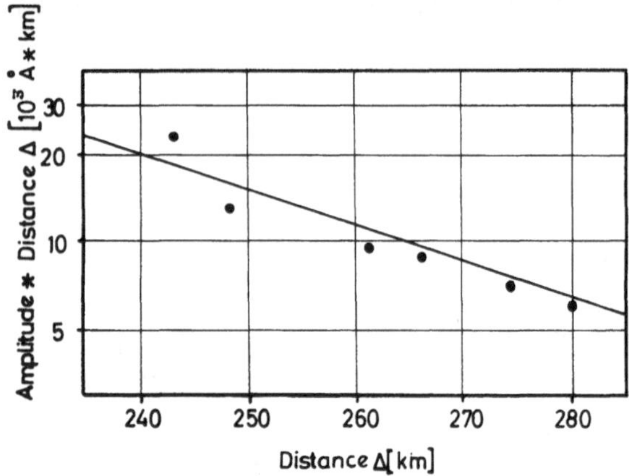

Fig. 5. Amplitude-distance relation for a section of profile 09-240 (shot on 20.4.68, charge 5.5 t)

part of the earth's mantle cannot, however, be excluded. In any case, it can be assumed that the crust and the upper mantle in the region of the Swabian Anomaly have slightly different physical properties than the crust and the upper mantle in the eastern part of the Jura for instance.

Information on the eastward extension of the Rhine Graben structure was expected from the measurements on profile 24-090, which were started in 1970. Because of the low seismic effect of the ripple firing no refracted arrivals can be observed except the P_g-wave. But, beginning at a critical distance of about 50-60 km strong events can be correlated to a reversed travel-time curve. The travel-times of these arrivals are identical with those observed by LAUER and PETERSCHMITT (1970) for reflections on some profiles in the Vosges. A mean velocity of 6.10 km/s for the overlying medium and a reflector depth of 25 ± 1 km can be determined from the T^2, Δ^2 diagram of these onsets as shown in Fig. 6. These values are nearly identical with those calculated by LAUER and PETERSCHMITT (1970) for the Vosges-profiles and confirm that the reflecting surface (now to be interpreted as the crust-mantle boundary, see PRODEHL et al., 6.8 of this volume) has a relatively shallow depth up to 60-80 km east of the border of the Rhine Graben itself (see also Fig. 8, PRODEHL et al., 6.8 of this volume).

The profile 09-240 running from the shotpoint Böhmischbruck in the strike of the Swabian Jura is, with a length of 400 km, the longest refraction profile in western Germany. For its first part only, up to distances of about 150-180 km, can the structure of the crust be deduced. This was partly done by GIESE (1968). Another correlation of the travel-time curve c (P^M), however, with an asymptotic velocity of 6.8 km/s does not lead to such a thick gradient zone between the crust and the upper mantle below the Jura as derived by GIESE (1968). Phase c is now explained with satisfying accuracy as reflection at the Mohorovičić discontinuity which can therefore be considered as "sharp" in comparison to the wave-length of the reflected waves. Clear arrivals between the curves a and c, which can be correlated on linear travel-time curves with velocities of 6.4 and 6.8 km/s must be explained, too. If these events are caused by waves refracted at interfaces within the crust, the calculated models (see also right hand side of crustal section in Fig. 7, PRODEHL et al., 6.8 of this volume) have to contain two reversals of the compressional velocity at a depth of 7 to 11 km and 16 to 19 km in order to explain the low critical distances of these waves and to obtain coincidence with the reflection times of 3.6-3.8 s and 6.0-6.8 s observed by ANGENHEISTER and POHL (1971) in the Ries region (see cross hatched area 3 in Fig. 1).

The amplitudes of P_n-signals decrease on profile 09-240 when approaching the region of the Swabian Anomaly. P_n-phases therefore cannot be observed at distances longer than 300 km. They are replaced by arrivals, called here P_n', which can be correlated with a high apparent velocity of 8.50 km/s (after all corrections applied) up to distances of 400 km from the shotpoint (see record section in Fig. 7). Behind the real P_n-arrivals, these events can be traced backwards with time delay to distances of about 250-270 km, and can therefore be explained by a wave refracted at an interface within the upper mantle. An explanation of the high velocity by a rise of the M-discontinuity or an intermediate layer claims improbably large dipping angles. The P_n'-wave may correspond to a wave reported for in-

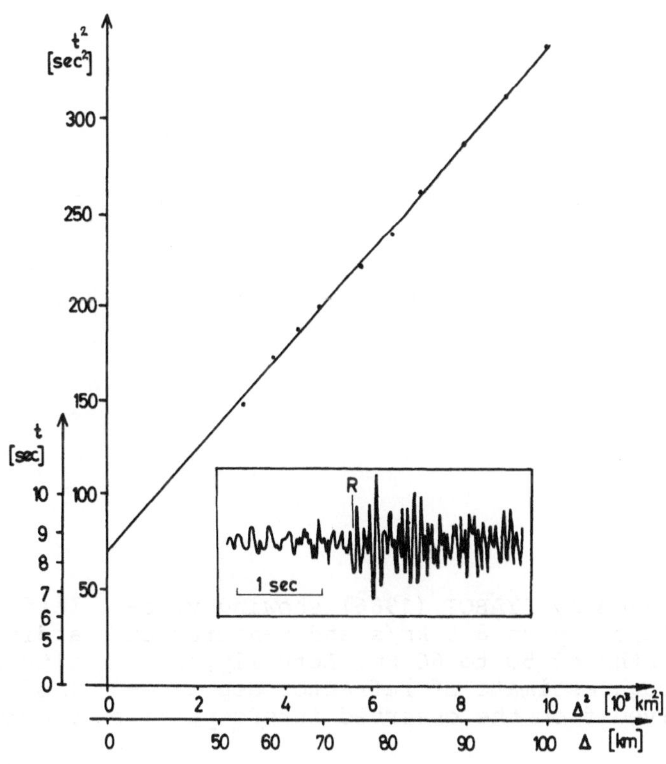

Fig. 6. T^2, Δ^2 diagram of the strong events observed on profile 24-090 with a seismogram sample

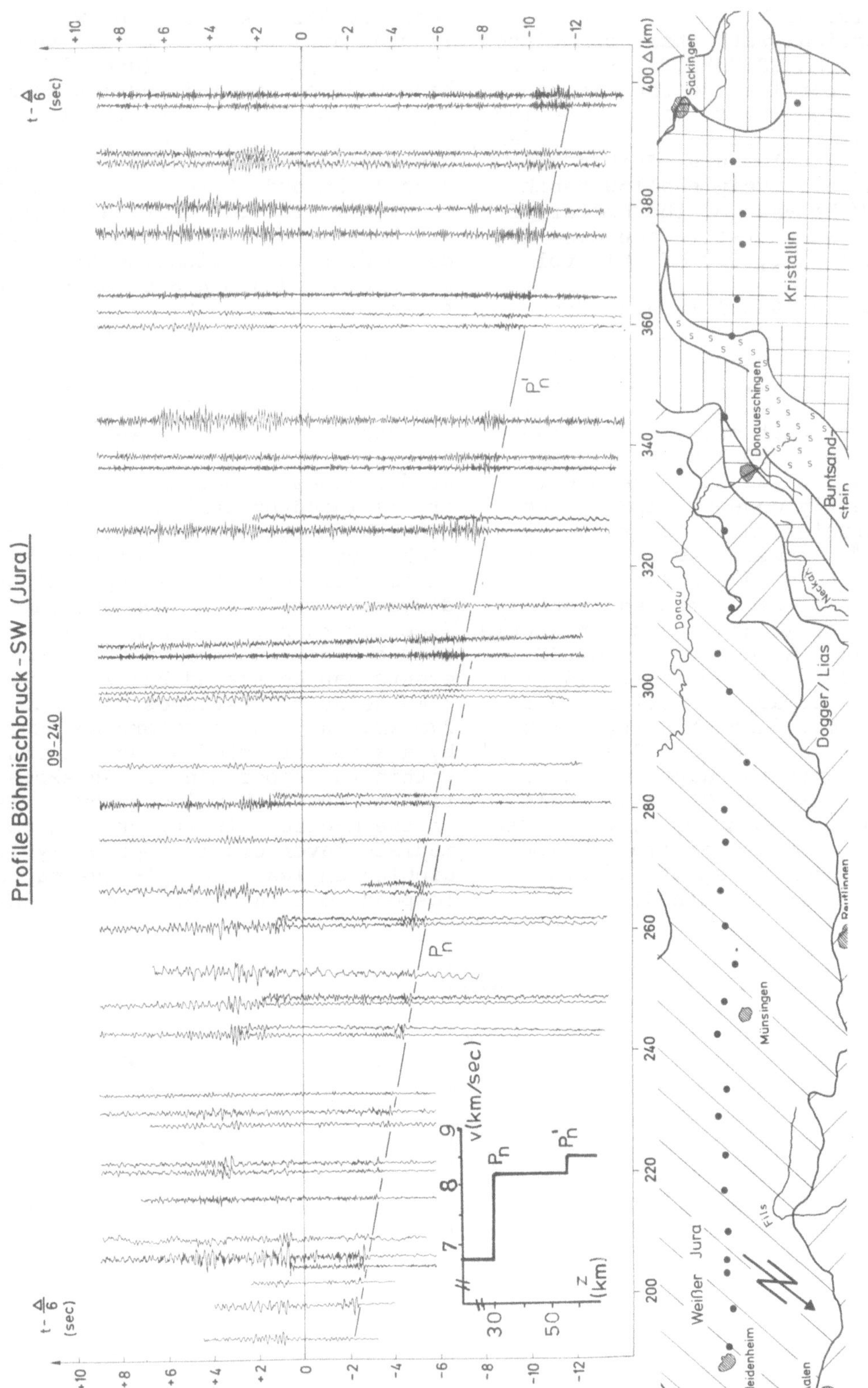

Fig. 7. Record section of the end of profile 09-240 with correlated arrivals of waves refracted within the upper mantle and with a first model of P-velocity distribution

stance by RYABOI (1966) showing velocities up to 8.8 km/s and penetration depths of 50 to 60 km. Actually, a rough estimate of refractor depth derived from the observed intercept time lead to values of 53-57 km depth for a discontinuity of zone of strong velocity gradient within the upper mantle (model in Fig. 7). For this estimate a crustal structure is supposed as derived for the eastern part of the Jura (see Fig. 7, PRODEHL et al., 6.8 of this volume).

It should be noted that only for such a strong velocity contrast, i.e. 8.50 km/s of P_n'-velocity to 8.20 km/s of P_n-velosity, is the critical distance of the P_n'-wave compatible to the observed value of about 260 km. Since, for its last 100 km, the profile touches the region of the Rhine Graben it cannot be excluded that the observed P_n'-velocity is influenced by on overlying inclined structure. This would not change the fact that the observed wave must have been refracted within the upper mantle, but could reduce the velocity contrast mentioned above. Furthermore if one takes into account that the P_n-velocity on profile 09-240 shows, after all corrections applied, a slight increase in the distance range between 230 and 280 km, the velocity contrast is also reduced. In both cases the models should contain a low-velocity zone in the region between the M discontinuity and 55 km in order to explain the observed critical distance of the P_n'-wave. As comparative observations, on long-distance profiles are rare in central Europe (ANSORGE and MUELLER, 1971) more exact models for the lower lithosphere in southwestern Germany can only be derived if more information is available about the real velocity of the P_n'-phase, by more profiles or by a reversed profile.

6.6 Results of Seismic Investigations in the Ries Crater Area (Southern Germany)

G. ANGENHEISTER and J. POHL

ABSTRACT

A review is given of the results obtained by seismic investigations on the structure of the Ries meteorite crater (diameter 25 km) and of the earth's crust in the crater area. In 1948, 1949 and 1952 refraction measurements along numerous 3 to 4 km long profiles were made in the crater area. In 1967 two 40-km-long refraction profiles across the crater were recorded using explosions in quarries. In 1968 a reflection profile from the center of the crater to the west was measured. The reflections were recorded digitally up to 14 s in order to obtain deep crustal reflections. Together with the reflection profile, two approximately 40 km long refraction profiles were recorded.

From these measurements the following picture of the crater structure, in accordance with gravimetric, magnetic and geological data, could be obtained. A central basin of about 11 km in diameter is surrounded by the so-called crystalline inner rim, which reaches to the surface in several places. Tertiary lake sediments fill the inner basin with a maximum thickness of about 350 m. Below the lake sediments a layer of glass-containing impact breccia (suevite) with variable thickness, up to 300-400 m, is found. The next layer, which consists probably of the brecciated crystalline crater filling, may be several km thick. At the center of the crater the P-wave velocity gradually increases and attains the value of the undisturbed crystalline basement which is outside the crater only at a depth of 2-2.5 km. The western boundary of the crater is indicated on the reflection profile by an abrupt disappearance of the reflections from the undisturbed Mesozoic layers and from the surface of the crystalline basement outside the crater.

Numerous deep crustal reflections were recorded on the reflection profile, some of which can be traced over several km. The main reflection zones have travel times of about 3-4 s, 6-7 s and 9-10 s (Mohorovičić zone).

ZUSAMMENFASSUNG

Es wird ein Überblick gegeben über die aus seismischen Messungen gewonnenen Ergebnisse über die Struktur des Ries-Kraters sowie der Erdkruste in der Umgebung des Kraters. Es wurden Reflexions- und Refraktions-Messungen durchgeführt.

Aus den Messungen ergibt sich in Übereinstimmung mit gravimetrischen, geoelektrischen, magnetischen und geologischen Daten folgende Struktur für den Bereich des Kraters: Ein zentrales Becken mit einem Durchmesser von etwa 11 km wird von dem sog. inneren kristallinen Wall umgeben, der an verschiedenen Stellen die Erdoberfläche erreicht. Das Becken ist mit tertiären See-Sedimenten, die maximal 350 m mächtig werden können, gefüllt. Unter den See-Sedimenten folgt eine Schicht aus Impakt-Breccien (Suevit) mit variabler Mächtigkeit, die bis zu 400 m betragen kann. Die nächste Schicht besteht aus zertrümmertem Kristallin, das im Zentrum einige km mächtig ist. Dies wird geschlossen aus der im Zentrum des Kraters allmählich zunehmenden Geschwindigkeit der P-Wellen, die den Wert für das nicht zerstörte Kristallin außerhalb des Kraters erst in einer Tiefe von mehr als 2,5 km erreicht. Die westliche Grenze des Kraters ist auf dem Reflexions-Profil gekennzeichnet durch das Verschwinden von Reflexionen von den mesozoischen Schichten und von der Oberfläche des Kristallins außerhalb des Kraters.

Bei den Reflexions-Messungen wurden zahlreiche Reflexionen aus der tieferen Kruste registriert, die z.T. über einige km hinweg korreliert werden könne. Die Hauptreflexionen haben Laufzeiten von 3-4 s, 6-7 s und 9-10 s (Mohorovičić-Zone).

6.6.1 Introduction

The Ries Crater (Nördlinger Ries) is a flat circular basin with a diameter of about 25 km situated in Southern Germany 120 km NW of München. It was formed 15 million years ago in a region where the crystalline basement is covered with 500-700 m of Mesozoic sediments. Outside the crater large masses of ejecta are found. Inside the crater, lake sediments have subsequently been deposited and now form the flat surface of the crater basin. In the past ten years good evidence for an impact origin of the crater has been obtained. A monograph containing results of mineralogical, geological, and geophysical investigations has been published (Bayer. Geolog. Landesamt, 1969). Geophysical surveys (gravimetric, geoelectric, magnetic, and seismic) have been carried out within and around the crater in order to obtain information about its structure. In this article a summary of the results of the seismic investigations is given. Though the measurements were made mainly with the aim of obtaining detailed information about the crater itself, some results about the structure of the earth's crust around the crater area were also deduced.

The first seismic measurements in the Ries Crater were made by REICH in 1948, 1949, and 1952 and interpreted in accordance with the theory of the volcanic origin of the crater. From the numerous short refraction profiles which were recorded inside and outside the crater, the following picture of its structure was obtained (Figs. 1 and 9 and REICH and HORRIX, 1955). The central part of the crater is filled with lake sediments with a maximum thickness of about 350 m (v_p = 1800 m/s). Below the lake sediments a layer with a higher velocity (v_p = 3000 m/s) and a mean thickness of about 300 m was found, which was thought, at that time, to consist of some kind of volcanic breccia. Today it is considered to be a layer of glass containing impact breccia (suevite). The next layer ($v_p \approx$ 4600 m/s) was considered to be the crystalline basement. The velocity in the upper part of this layer is however much

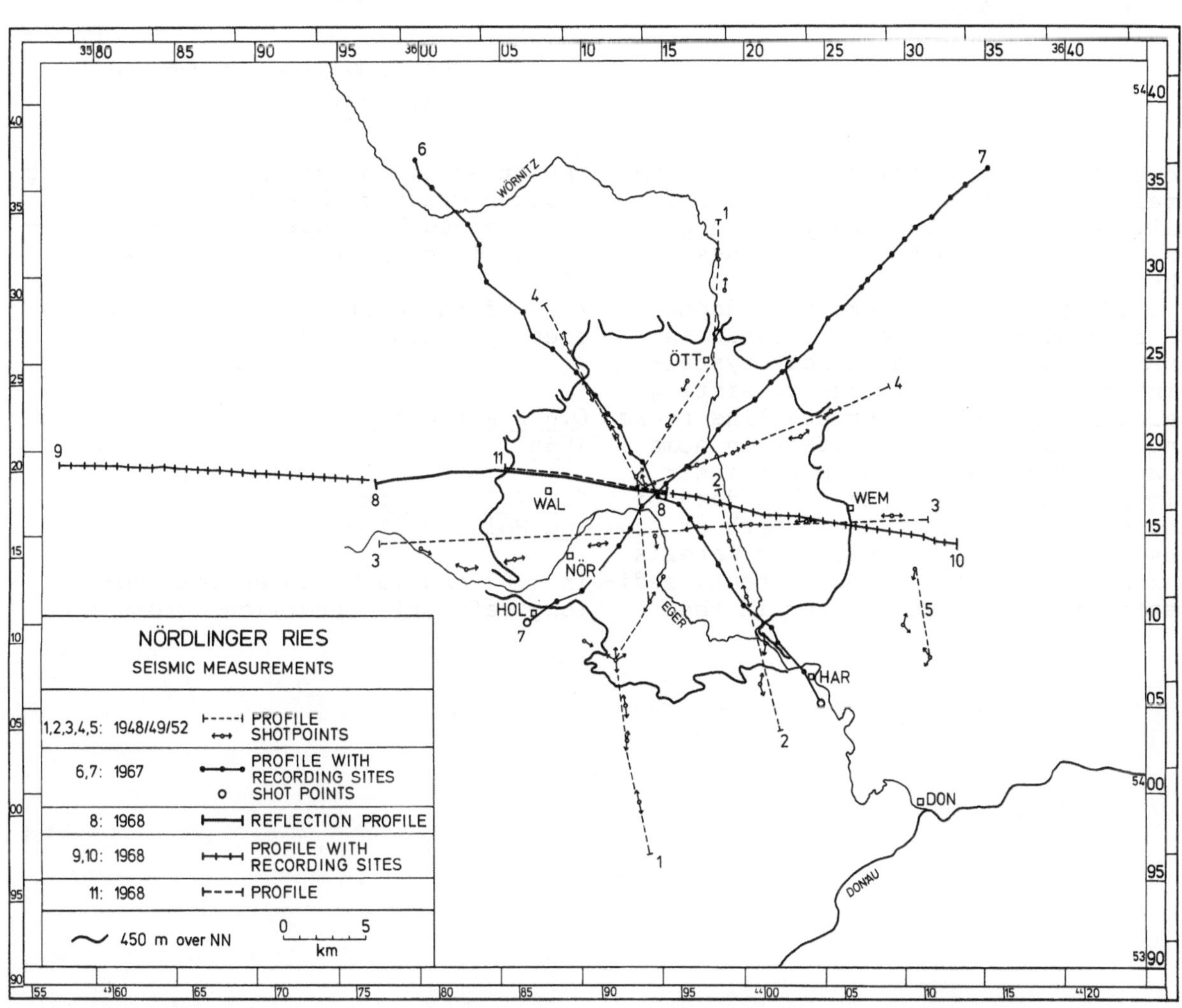

Fig, 1. Seismic measurements in the Ries Crater area. Profiles: *1-5* Refraction measurements 1948, 1949, and 1952; *6* Refraction profile Harburg 1967; *7* Refraction profile Holheim, 1967; *8* Reflection profile 1968; *9-10* Common-depth point profiles 1968. The common-depth points are located at both ends of profile 8, on which the shotpoints are located. *11* Short refraction profile 1968

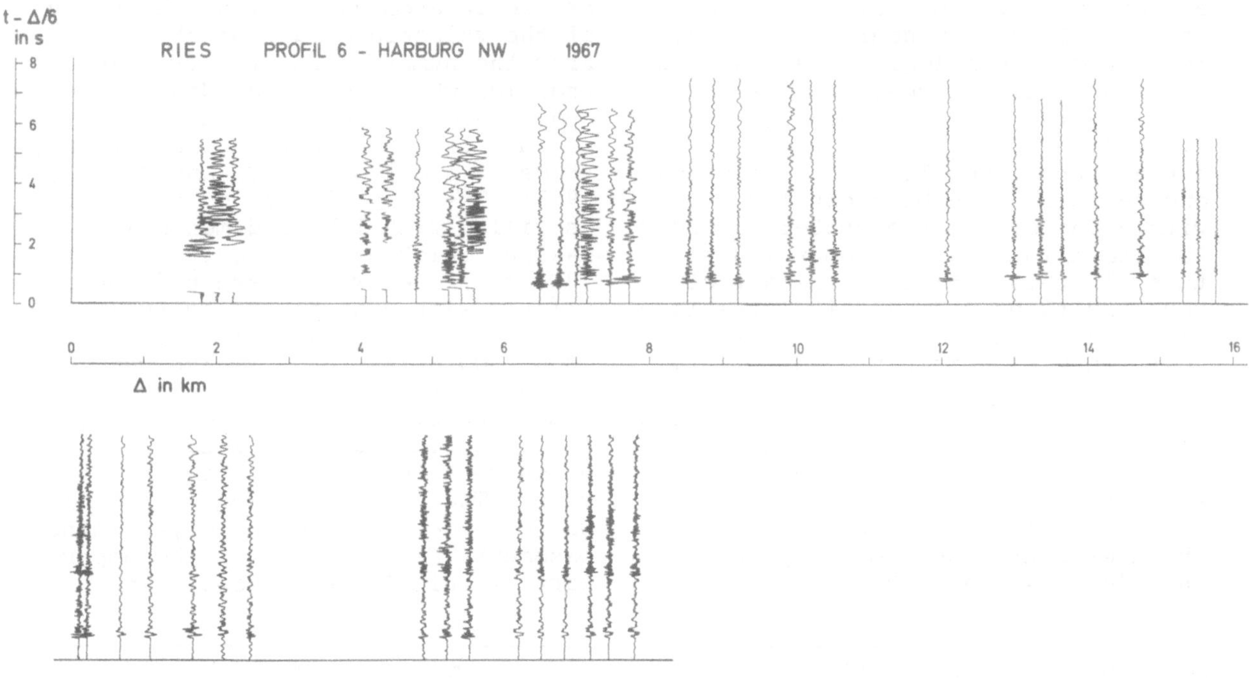

Fig. 2. Profile 6 Harburg NW, record section

lower than the velocity which is generally found in the undestroyed crystalline basement and today it is generally assumed that below the suevite the crater must be filled with brecciated crystalline rocks to a depth of several km. The central basin is surrounded by the so-called crystalline rim (inner rim) also consisting mainly of brecciated crystalline rocks, and extending to the surface in some places. The space between the inner rim and the topographical boundary of the crater is filled to a depth of several 100 m with brecciated material which contains to a great extent, sediments of the Mesozoic cover in the impact area. No information regarding a depth greater than 500-600 m could be obtained by these measurements because of the shortness of the refraction lines, but the concentric structure of the crater to this depth was clearly established.

6.6.2 Refraction Measurements 1967-1968

6.6.2.1 Refraction Profiles Harburg and Holheim (Profiles 6 and 7)

In 1967 and 1968 two approx. 40-km-long refraction profiles across the

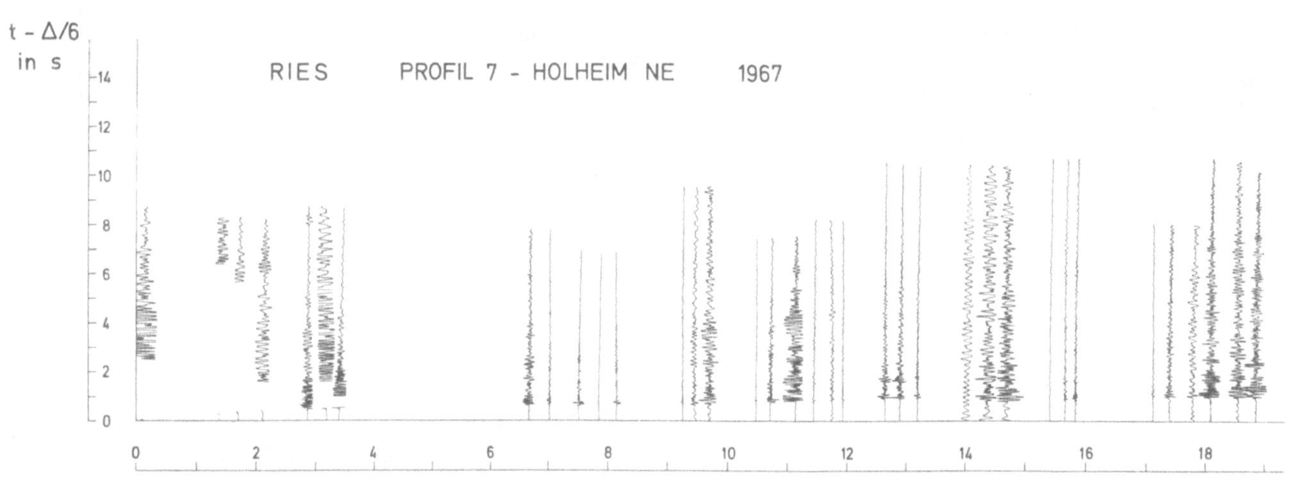

Fig. 3. Profile 7 Holheim NE, record section

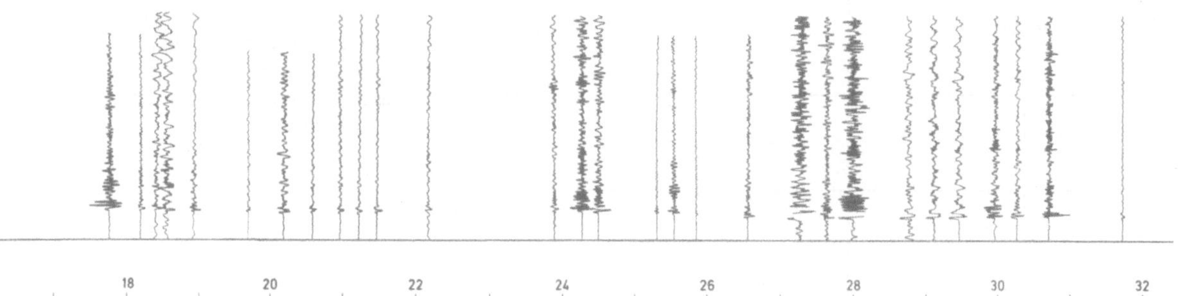

Ries Crater (6 and 7 in Fig. 1) were recorded using explosions in quarries near Harburg and Holheim. Record sections are shown in Figs. 2 and 3, and travel-time curves for the first arrivals are shown in Figs. 4 and 5.

For a preliminary interpretation it was assumed that the first arrivals beyond a distance of about 3 km are due to a refracted head wave coming outside the crater from the crystalline basement and inside the crater from a hypothetical boundary in the crater filling. A wave-front method adapted to refraction profiles with only one shotpoint (PRODEHL, 1965) was used to construct the corresponding refracting surface. The mean velocities for the layers above the refracting surface were known from the measurements of REICH and HORRIX (1955) and from later measurements described in this article.

The depth of the crystalline basement near the shotpoint was also known approximately from the other refraction measurements in the area and from geological observations. Small possible variations do not change the general features of the constructed surface.

Results are shown in Figs. 4 and 5 for various assumed velocities below the refracting surface. Velocities of about 5.75 to 6 km/s are reached at much greater depth inside than outside the crater. This is a first indication that the crater must be filled to a great depth with material of a lower velocity than in the undestroyed crystalline rocks outside the crater. It should be emphasized however that inside the crater the indicated surfaces for the different velocities in Figs. 4 and 5 are certainly not seismic boundary surfaces and that the used wave-front method can only give some indications about the depth at which certain velocities may be reached. On the part of the profile outside the crater the apparent velocity is nearly constant, and the use of the wave-front method seems to be justified. A next step in the interpretation will be to find more detailed models by using all the information recently obtained about the velocity-depth distribution in the crater, especially those described in the next section.

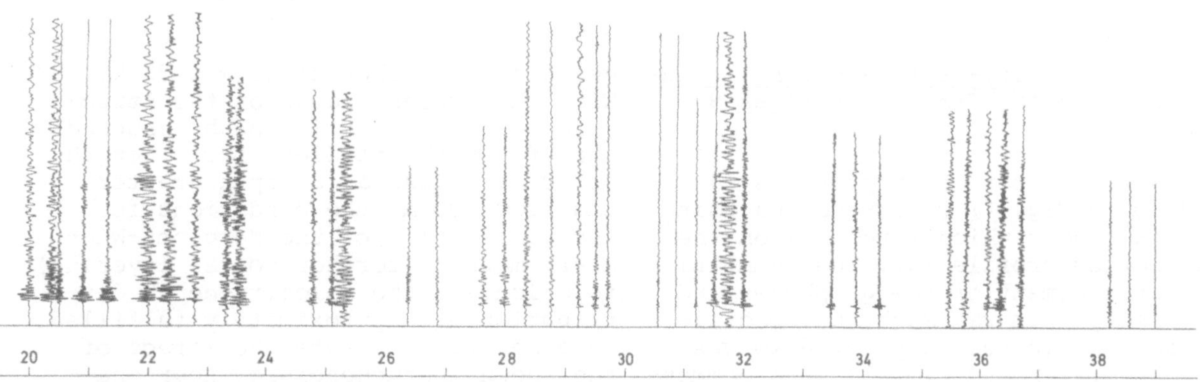

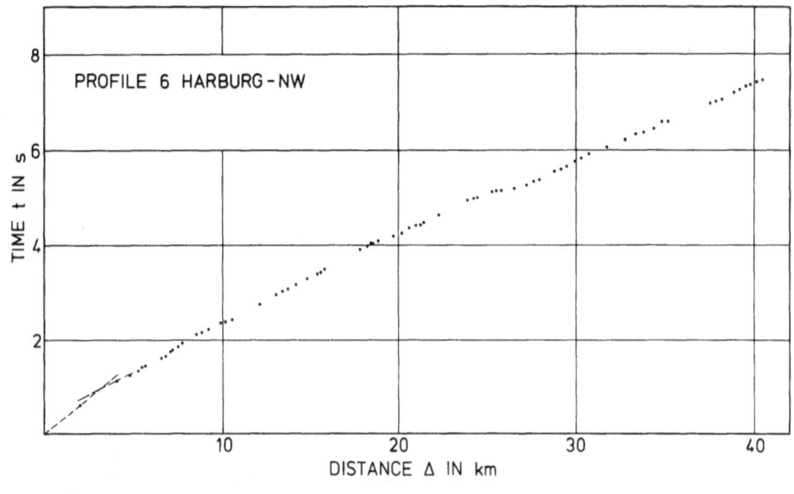

Fig. 4. Profile 6 Harburg NW. Travel-time curves of the first arrivals and constructed refracting surfaces for different assumed velocities in the lower layer

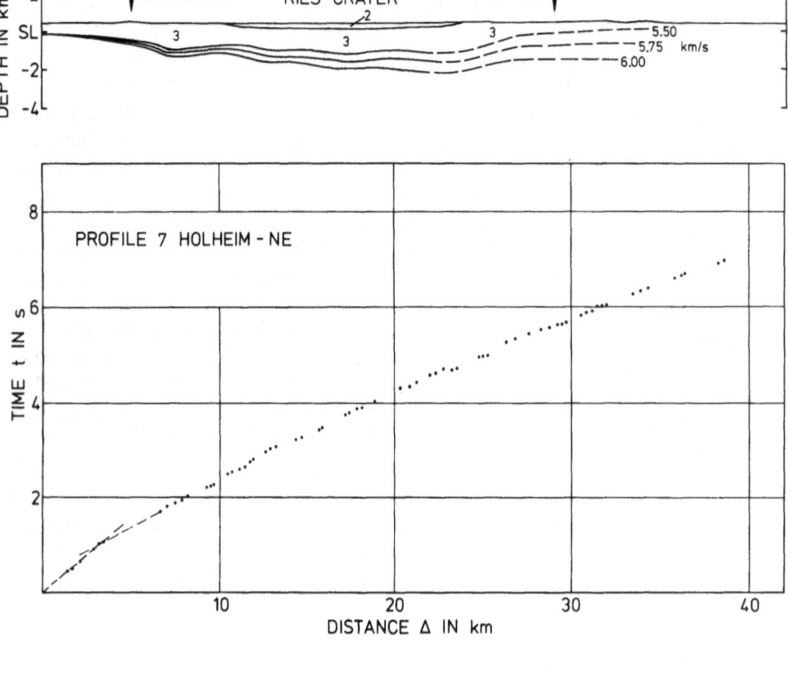

Fig. 5. Profile 7 Holheim NE. Travel-time curves of the first arrivals and constructed refracting surfaces for different assumed velocities in the lower layer

6.6.2.2 Velocity-Depth Sounding Inside and Outside the Crater (Profiles 9 and 10)

In 1968 two refraction profiles (9 and 10 in Fig. 1) were recorded using special shots which were made on the reflection profile (8). The symmetrical arrangement of the geophones and the shotpoints was such that profile 9 had a common-depth point 8 km west of the crater and profile 10 a common-depth point at the center of the crater. The maximum distance between geophones and shotpoints was 38 km. With the present ideas on the structure of the crater it can be assumed that for both profiles, for a length of 10-12 km and at a depth greater than 600-700 m, the conditions for the application of the Herglotz-Wiechert method (for horizontal layering) to calculate the velocity-depth distribution are approximately fulfilled. In order to eliminate the effect of the known inhomogeneities near the surface, the travel-time curves (travel times and distances) for the first

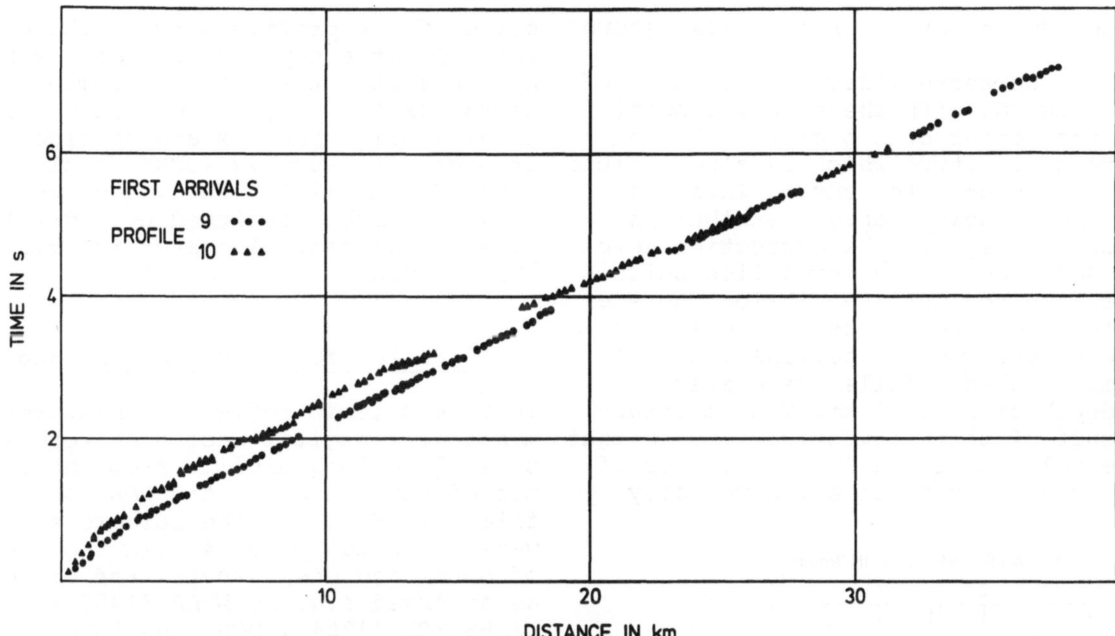

Fig. 6. Travel-time curves for the first arrivals for profile 9 (common-depth point outside the crater) and for profile 10 (common-depth point at the center of the crater). The partial travel-time curves have been reduced to a common shotpoint

arrivals were reduced to a horizontal plane lying at a depth of 300 m below sea level for profile 9 and at 400 m below sea level for profile 10, using the detailed knowledge of the seismic velocities and the layering in the upper 600 to 700 m inside and outside the crater obtained by the refraction and reflection measurements in 1948, 1949, 1952, and 1968 (WILL, 1970). The uncorrected and the corrected travel-time curves are shown in Figs. 6 and 7, and the velocity-depth curves in Fig. 8.

The corrected curve for profile 9, outside the crater, is approximately a straight line with an apparent velocity of 5500 m/s. This value is indicated in Fig. 8 by a vertically dashed line. The velocity of 5350 m/s, indicated by a point at a depth of 50 m below sea level (surface of the crystalline basement below the sounding point), has been deduced from the uncorrected travel-time curve, taking into account a small inclination of the basement in the area. As both travel-time curves have almost no curvature, homogeneous layers should be expected. There are, however, some arguments for a velocity increase in the basement. At shotpoint distances from 16-38 km the seismic waves traverse the low-velocity filling of the crater on paths of increasing length. This yields a retardation for the first arrivals and a smaller apparent

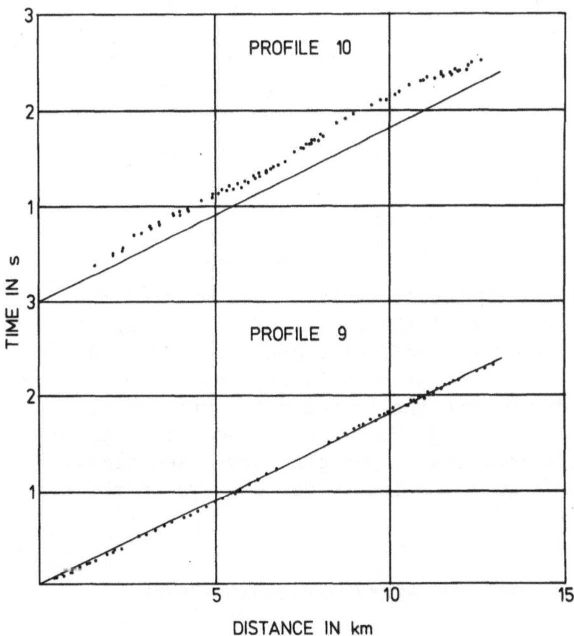

Fig. 7. Corrected travel-time curves for profile 9 at 300 m below sea level and for profile 10 at 400 m below sea level

velocity in the travel-time curves thus masking a velocity increase with depth. The velocity increase is corroborated by the fact that also in profile 7 and 8 (fixed shotpoint), Figs. 4 and 5) higher apparent velocities of 6 km/s are found at a shotpoint distance of 30 to 40 km. A probable extrapolated curve is also shown in Fig. 8.

The corrected travel-time curve of profile 10, with the sounding point at the center of the crater, has a strong curvature which is also visible in the uncorrected curve. This leads to the velocity-depth distribution shown in Fig. 8. The correction level of 400 m below sea level lies below the lake sediments and below the suevite layer and is assumed to lie in the brecciated crystalline material which probably fills the crater to a depth of several km. The important result of these investigations is that the velocity curve for the center of the crater approaches the velocity curve of the profile outside the crater only at a depth of at least 2.5 km. This must therefore be a minimum depth for the brecciated crater filling. Similar results are obtained from gravity surveys (JUNG et al., 1969; KAHLE, 1970). The velocity-depth curve has been confirmed by a drill hole in the central crater in 1973 (Bayer. Geol. Landesamt, 1974).

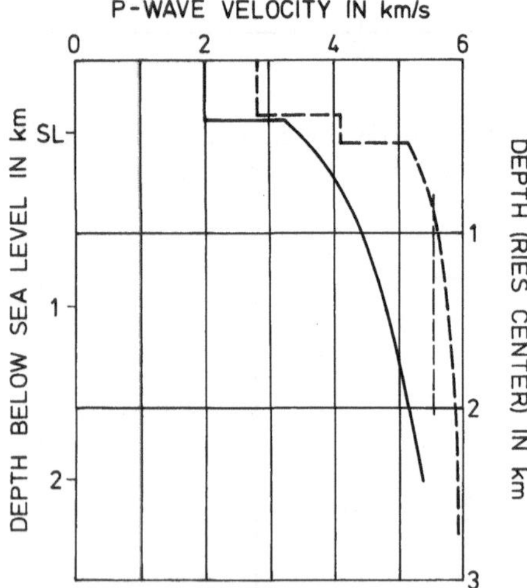

Fig. 8. Velocity-depth curves outside the crater and at the center of the crater obtained from profiles 9 and 10. *Dashed vertical line:* constant velocity (v_p = 5500 m/s) deduced from the travel-time curve of profile 9 in Fig. 7. *Dashed curved line:* probable real velocity-depth curve (see text). *Solid curve:* velocity-depth curve in the crater filling

6.6.3 Reflection Measurements 1968

In 1968 seismic reflection measurements with digital recording were made on a 17-km-long profile from the center of the crater to the west (profile 8 in Fig. 1). The seismic signals were recorded up to 14 s in the hope of obtaining deep crustal reflections as observed e.g. by DOHR (1957a), LIEBSCHER (1964), DOHR and FUCHS (1967). The field measurements were made by PRAKLA-SEISMOS, Hannover, in collaboration with the Institut für Angewandte Geophysik, München. The profile was recorded with 3-fold coverage, a mean shotpoint distance of 240 m, a spread length of 1380 m and a distance of 60 m between the seismometer groups. Generally charges of 12 kg of explosive (Seismogelit) were used. The digital data processing (static and dynamic corrections, stacking, filtering, deconvolution) was also made by PRAKLA-SEISMOS, Hannover.

6.6.3.1 Near-Surface Reflections

For reflection times less than 4 s special record sections were made in order to obtain optimum information about the structure of the Ries Crater to a depth of about 2 to 3 km. The best results were obtained with

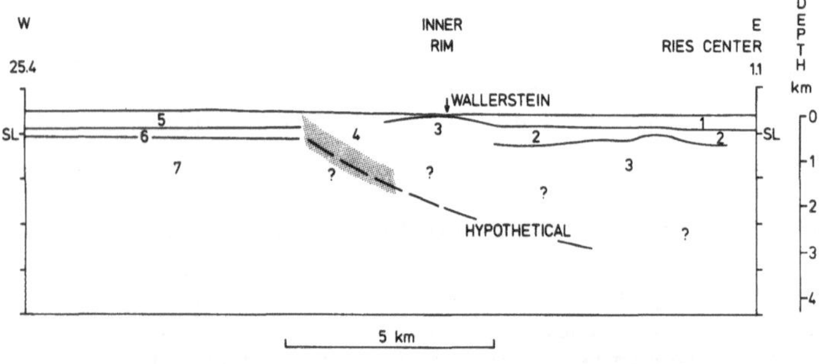

Fig. 9. Cross section of the Ries Crater along the seismic reflection profile. The hypothetical line indicating the lower boundary of the crater has been drawn by analogy with other meteorite craters

a frequency filter 25-90 Hz and spike deconvolution. The results have been published by ANGENHEISTER and POHL (1969). They have been incorporated in the cross section of Fig. 9. Outside the crater the undisturbed crystalline basement, which is about 50 m below sea level, is covered by Mesozoic sediments (Buntsandstein?, Muschelkalk?, Keuper, Lias, Dogger). The western boundary of the crater is clearly indicated on the reflection profile by an abrupt disappearance of the reflections coming from the undisturbed Mesozoic layers and from the surface of the crystalline basement outside the crater. This boundary coincides approximately with the topographical boundary of the Ries basin on the surface. East of this boundary an irregular zone with weak reflections dipping eastwards possibly indicates the lower boundary of the crater in this zone as shown in Fig. 9 by a shadowed strip. Apart from this reflecting zone no good reflections can be seen in the area between the outer and the inner rim, which is probably filled with brecciated material consisting of Mesozoic rocks and of rocks from the crystalline basement. The inner rim (crystalline rim) is mostly made of brecciated crystalline material as shown by the higher seismic velocities. In the central part of the crater the very flat lower boundary of the lake sediments gave strong reflections. Below the Tertiary sediments follows the layer of suevite of variable thickness, as determined by reflection and refraction measurements (profile 11, WILL, 1970). In the brecciated crystalline crater filling which is expected under the suevite layer waves are not reflected, nor can a lower boundary of the crater be seen in the reflection records. This may be due to a continuous variation of elasticity and density with depth in the crater filling and a gradual transition to the undisturbed crystalline basement.

6.6.3.2 Deep Crustal Reflections

The complete processing of the 14 s records was made with a sampling interval of 4 ms and different frequency filters. The best results were obtained with a constant band pass frequency filter (15-70 Hz) and an optimum filter, but the record sections obtained with only constant frequency filtering (15-70 Hz) show reflections of nearly the same quality. The reflections can also be seen in the unfiltered record sections, which, however, contain much low-frequency surface noise. The frequency of the reflected signals from inside the crust is about 18 to 20 Hz. Some deconvolution experiments show no improvements of the results. The record section is presented in Fig. 10.

In Fig. 11 all the elements from the record section were plotted, which could eventually be interpreted as seismic signals (reflected or refracted or surface waves) in order to improve their visibility. Some of these arrivals have a very low apparent velocity, especially in the center of the crater. They are probably partly refracted waves and reflected surface waves.

In the record section (Fig. 10) and in the drawing of Fig. 11 three zones can be distinguished. In the western part of the profile numerous reflections, that can be traced over several km, are clearly visible. This part of the profile lies outside the Ries Crater (Figs. 1 and 9). In the central part of the profile only very few reflections can be seen. We think that in this part of the profile, which is situated between the inner and the outer rim of the crater, most of the seismic energy is absorbed and diffused in the unconsolidated brecciated material which fills the crater near the border. In the eastern part of the profile, which ends at the center of the crater, numerous reflections are again seen, but the reflecions are of poorer quality than those outside the crater. Much oblique noise is also visible in this part of the record section.

In the vertical direction (time-axis) three main reflection zones with reflection times of about 3-4 s, 6-7 s and 9-10 s can be distinguished. There are nearly no reflections beyond 11 s. The depth of the reflecting elements can be calculated with the use of velocity-depth distributions deduced from refraction measurements.

For the western part of the profile (SP 15.1 to SP 25.1, outside the crater) a statistical distribution of the reflections and the depths of the reflecting elements were calculated. In Fig. 12 the results are compiled, together with velocity-depth curves and the corresponding travel-time curves for vertical reflections of different authors for neighboring regions (Fig. 13). Only the best re-

Fig. 10

◀ Fig. 10. Seismic reflection profile. Record section showing the deep crustal reflections. Digital recording and digital data processing by PRAKLA-SEISMOS, Hannover. 3-fold coverage, shotpoint distance 240 m, spread lengths 1380 m, distance between the geophone groups 60 m, 12 kg explosive charges (Seismogelit), static and dynamic corrections, stacking, constant band pass frequency filter (15-70 Hz) and optimum filter, sampling interval 4 ms

flections, which could be traced over at least 690 m in the record section, were used for this compilation. This corresponds to one geophone spread of 1380 m in the field. Thus the results can be compared with similar results in other areas, where single film records were used for statistical interpretation (e.g., LIEBSCHER, 1964; DOHR and FUCHS, 1967). The reflection-

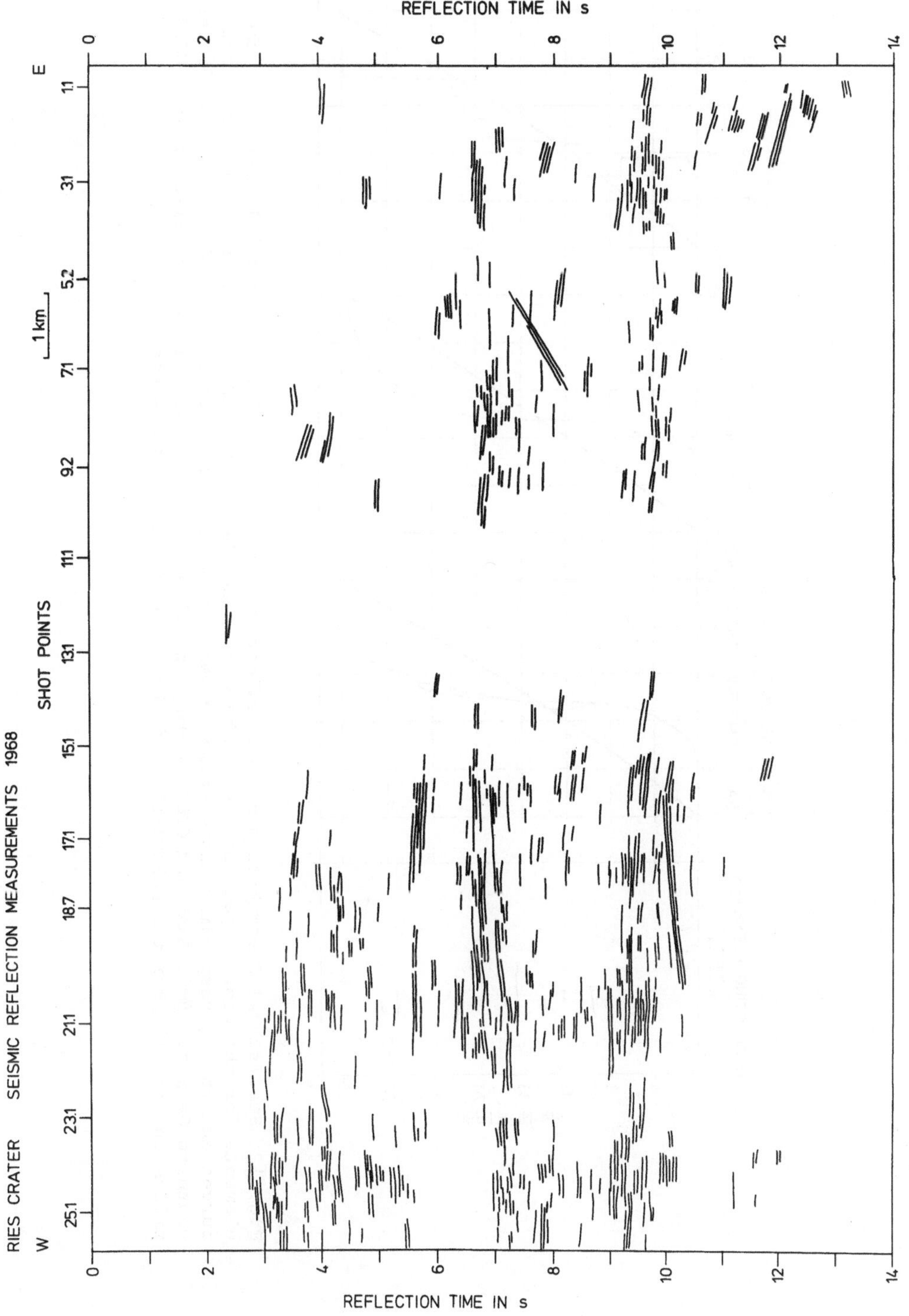

Fig. 11. Seismic reflection profile. Simplified diagram showing important events on the record section. 0 s corresponds to 400 m above sea level

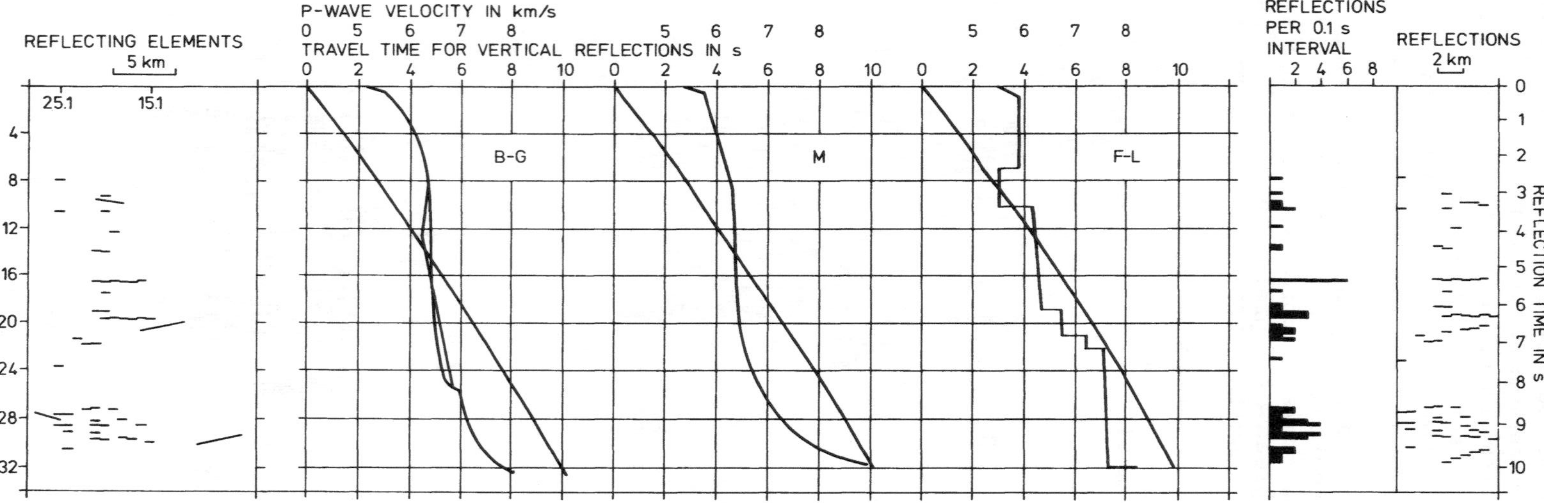

Fig. 12. Statistical distribution of the reflections (time interval 0.1 s) and calculation of the depth of the reflecting elements for the western part of the profile, outside the crater (SP 15.1 to SP 25.1). Depth below sea level. Velocity-depth curves and the corresponding travel-time curves for vertical reflections are also shown for different refraction profiles in southern Germany. *B-G:* BRAM and GIESE (1968), profile C; *M:* MEISSNER (1967a) profile C; *F-L:* FUCHS and LANDISMAN (1966b) profile A in Fig. 13. For the calculation of the depth the curve M was used

Fig. 13. Location of the Ries Crater, of the seismic reflection profile and of different seismic refraction profiles in southern Germany. *A:* FUCHS and LANDISMAN (1966b); *B:* EMTER (1971); *C:* MEISSNER (1967a, b), BRAM and GIESE (1968); *D:* reflection profile

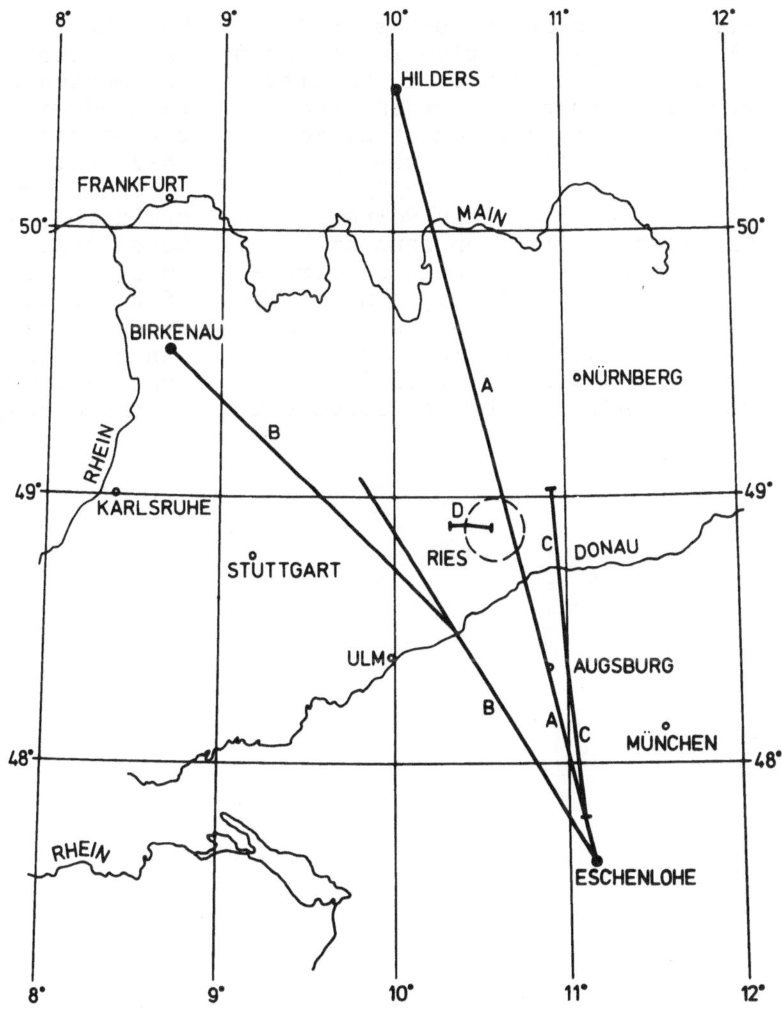

time interval for the statistical treatment is 0.1 s. It can be seen that the statistical treatment of sharp reflections extending over a length of several km and coming from inclined reflecting horizons yields broadened peaks. This can give a false idea of the structure of the reflecting horizons and about the thickness of the reflecting zones, in Fig. 12 especially for the Mohorovičić zone (27-30 km). The depth of the reflecting elements in Fig. 12 has been calculated with the travel-time curve for vertical reflections deduced from the velocity-depth curve M of MEISSNER (1967). The travel-time curves calculated from the two other velocity-depth distributions are also shown in Fig. 12 and give nearly the same results for the depth of the reflecting elements.

For some of the horizons the depth was calculated taking into account a possible inclination of the reflecting elements in the vertical plane through the reflection profile. The corresponding ray paths were also calculated with the velocity-depth curve of MEISSNER and with the assumption that the rays were continously refracted up to the reflecting horizon. As can be seen, rather high inclinations of the horizons can be expected in the crust.

With Fig. 12 a comparison of the results of the reflection and of some refraction measurements (velocity-depth distribution) in southern Germany can be made. The refraction profiles are shown in Fig. 13. They are all located in the South German Triangle where in general the layering of the crust seems to be very flat. Only the refraction profiles nearest to the Ries Crater have been included in Fig. 13. The curve B-G (BRAM and GIESE, 1968) and the curve M (MEISSNER, 1967a) in Fig. 12 have been obtained with profile C, the curve F-L (FUCHS and LANDISMAN, 1966b) has been obtained with profile A. The curve F-M is representative for a region about 100 km north of the Ries Crater, whereas the two other curves, B-G and M, are representative for a region about 60 to 80 km south of the Ries Crater. This must be kept in mind when the results are compared. Results for profile B have been published by EMTER et al. (1970) and EMTER (1971, 6.5 of this volume).

The first reflection zone at a depth of about 10 km could be correlated with a low-velocity zone in the upper crust. For the curves B-G or F-L the second reflection zone can be associated with the velocity increase as indicated by the curve F-L, but there is no equivalent increase in the two other curves. The third reflection zone clearly must be correlated with the velocity increase in the Mohorovičić zone at a depth of 28-29 km.

Until now no more detailed interpretation can be given, but it is hoped that from a careful and detailed analysis of the spectral properties of the reflections some information about the structure of the reflectors can be obtained. Special attention must be given to laminar models such as those described by FUCHS (1969a).

6.7 Results from Deep-Seismic Sounding in the Rhine-Main Area

R. MEISSNER, H. BERCKHEMER, and A. GLOCKE

ABSTRACT

Recent results from refraction, wide-angle, and near vertical reflection observations around Frankfurt/Main are compiled and discussed. Except for those of the Rhine Graben area, velocity-depth functions are similar and reveal two or three characteristic boundaries in the crust.

ZUSAMMENFASSUNG

Ergebnisse von Refraktions- und Reflexionsmessungen im Raum um Frankfurt/Main werden zusammengestellt und diskutiert. Sieht man von den Geschwindigkeitsfunktionen aus der Rheingraben-Region ab, so sind sie untereinander ähnlich und lassen zwei oder drei charakteristische Grenzen in der Kruste erkennen.

6.7.1 Introduction

Six refraction profiles have been observed in the Rhine-Main area around Frankfurt in the past eight years. Although these profiles are located close together, they cross geologically different areas (Fig. 1).

Profile 13-240-20, not yet completed, is a refraction profile in the central part of the Rhenish Massif; profiles 240-LO-060 is a wide-angle profile with a common reflecting (or refracting) element. It is situated in the southern part of the Rhenish Massif. Profile 03-250 crosses the Hessische Straße before entering the Rhenish Massif. It was observed using blasts from two quarries near Rhomsthal (12) and Gersfeld (03) in the Rhön. Profile 02-240-19 crosses the Rhine Graben and was observed using

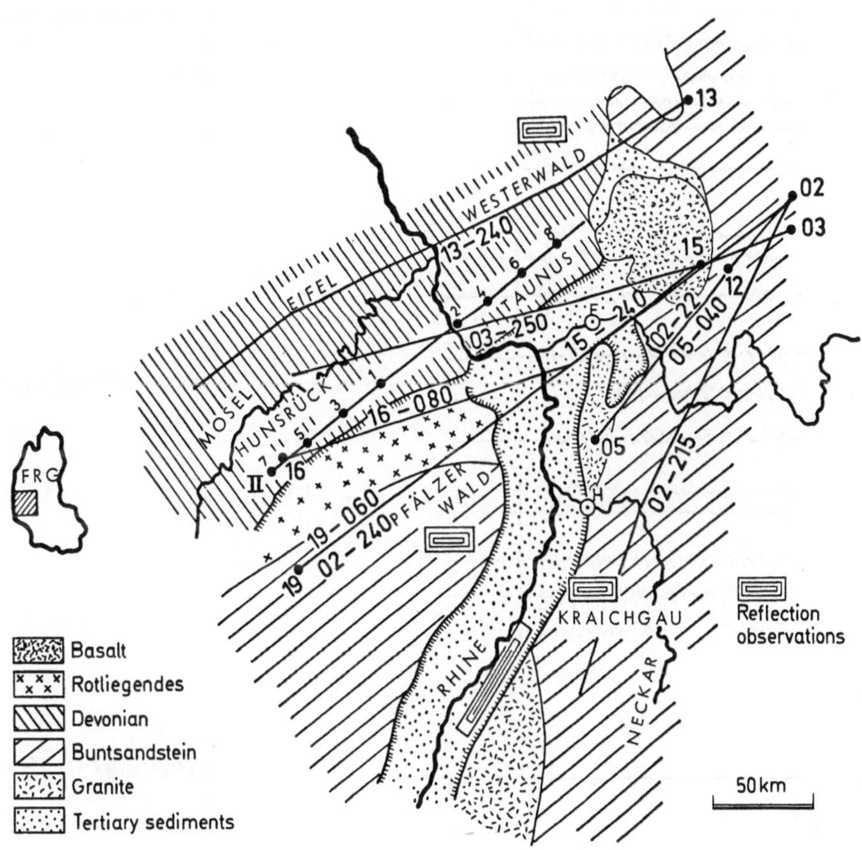

Fig. 1. Location map of profiles and geologic units

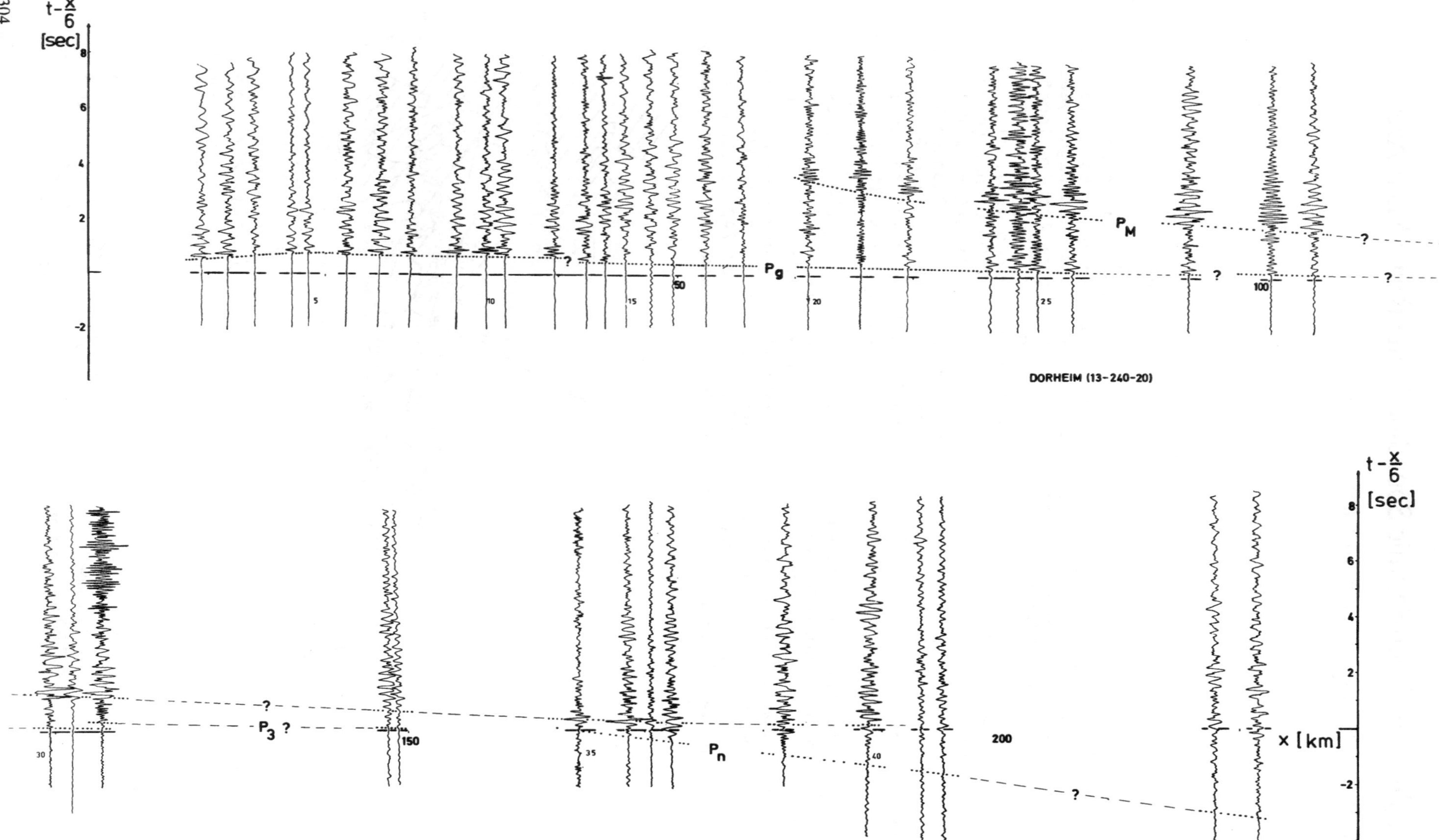

Fig. 2. Record section of profile 13-240-20 (shotpoint Dorheim)

three quarry blasts at both ends. Profile 02-220-05 with reverse shots in Hilders, Rhön (02) , and Birkenau (05) crosses the Odenwald. Profile 02-215 from Hilders (02) runs through the Spessart and the Kraichgau depression.

In addition to these six profiles near-vertical reflection observations are available from the Siegen area north of profile 240-LO-060, from the eight shotpoints of profile 240-LO-060, from the Pfälzer Wald, from the Kraichgau, and from the Rhine Graben (Fig. 1).

Results from the refraction profile and profile 240-LO-060, in particular velocity-depth functions which were obtained by applying the methods of Wiechert-Herglotz and GIESE (1968a) are compared. Furthermore, results from the near-vertical reflection observations are compared to each other and related to the results of the velocity-depth function.

6.7.2 Record Sections and Correlations

In general, the record sections show four or five wave groups, recognizable over greater distance ranges:

1. the P_g-wave is observed up to distances of 120 km and shows a small convex curvature;

2. The P_2- and/or P_3-wave mostly can be detected behind the P_g-wave branch and might be related to the Conrad discontinuity (=CD);

3. The P^M-wave is reflected from the Mohorovičić discontinuity (MD) and/or is a diving wave bottoming the velocity-gradient zone on top of the MD.

4. The P_n-wave is propagating in the uppermost mantle, sometimes apparently at some depth below the MD.

All record sections are presented in a reduced time scale with $t_{red} = t - \frac{x}{6}$. Correlations are indicated by dotted lines.

6.7.2.1 Profile 13-24P-20

Fig. 2 shows the record section of profile 13-240-20 with correlation. The P_g-curve with a slightly convex shape ends at 120 km with a velocity of 6.3 km/s. The travel-time anomaly at 105 km might be connected with a fault zone. There are only poor indications for the existence of a P_2-wave from the CD. The P^M-wave shows strong amplitudes between 60 and 100 km distance. The P_n-wave can also be clearly defined in the right-hand side of Fig. 2. Its extrapolation to shorter distances leads to an apparent intersection with the P^M-travel-time branch. This observation, which is rather often observed in deep-seismic sounding (MEISSNER, 1967b), indicates either a small velocity gradient in the uppermost part of the mantle and/or a kind of lamellation at the MD. In either case the P_n-wave travels at a depth greater than that of the reflected or diving P^M-wave.

6.7.2.2 Profile 240-LO-060 (Wide-Angle Profile)

This profile and its results are discussed in detail in 6.1 and 6.2 of this volume. The quality of the P^M-wave is excellent. Only very poor P_n-arrivals could be detected.

6.7.2.3 Profiles 03-250 and 12-260

Record sections of profile 03-250 and 12-260 have been compiled and interpreted by HÄNEL (1963). The reinterpretation (Fig. 3) takes into account some previously omitted arrivals of the P^M-wave branch and an extended P_g-branch. This evaluation seems to be in agreement with results from recently observed adjacent profiles. The convex curvature of the P_g-wave is clearly defined. The P_2-wave is rather poor and its correlation doubtful in some parts. The correlation of the P^M-wave is also doubtful, especially on the profile from Gersfeld (03-250).

Amplitudes of the P^M-wave, however, are rather strong between 60 and 150 km distance, and the indicated correlation may be justified. There are some indications that another concave travel-time branch follows the P^M-branch suggesting a complicated transition zone between crust and mantle. The P_n-wave is clearly defined. Its backward extrapolation to smaller distances again seems to cross the P^M-wave branch on both record sections.

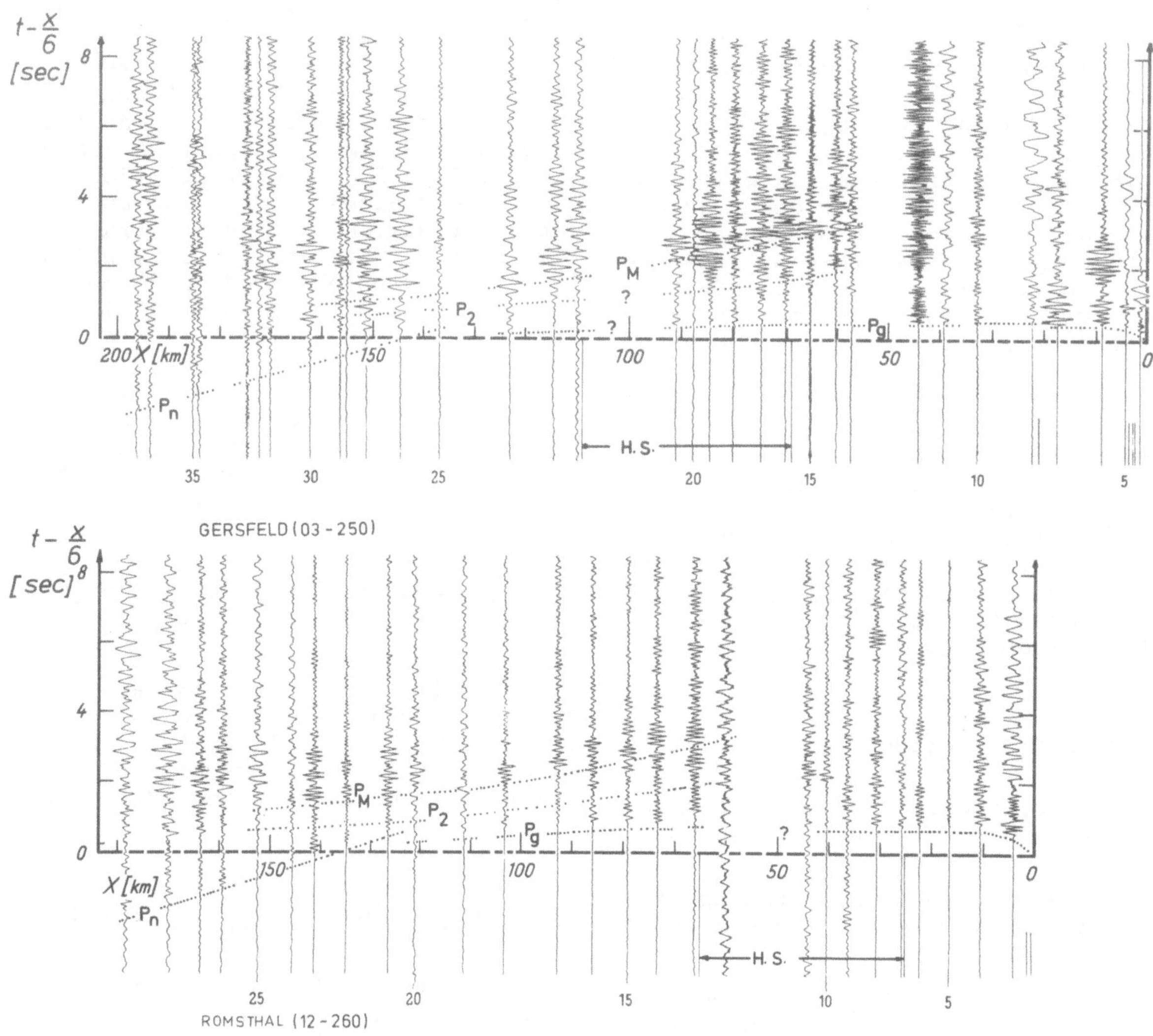

Fig. 3. Record sections of profiles 03-250 (shotpoint Gersfeld) and 12-260 (shotpoint Romsthal) compiled by HÄNEL and STEIN

6.7.2.4 Profiles 02-240-19, 15-240-08, 16-080-02 (Rhine Graben)

These profiles and their results are discussed in detail by MEISSNER et al. (1970). This interpretation uses observations from the shotpoints Büdingen (15) and Hilders (02) in the NE and from the shotpoint Taben Rodt (16) in the SW. The line could not be covered by equally spaced seismometers inside the graben zone because of a rather high noise level. Record sections are shown in Figs. 4, 5 and 6. The P_g-arrivals from Hilders (02) show a step wise displacement at the eastern flank of the graben. The arrivals from the other shots show high apparent velocities when approaching the graben border. This may be interpreted by a lifting of the graben shoulders which influences the whole crust. Intermediate layers in the deeper parts of the crust inside the graben show a synclinal shape with apparent velocities from 6.8-7.2 km/s. P^M-waves are rather sparse, so it does not seem justified to use them for the interpretation. P_n-arrivals are observed at the western end of the profile. They are missing beneath the graben and on the eastern flank.

The interpretation of the intermediate traveltime branches results in pillow-shaped boundaries in the deeper crust beneath the graben. The lower boundary of the "graben pillow" which could not be derived by seismic methods has been estimated by gravity models (Fig. 7). The eastern end of the pillow could not be mapped with certainty as the energy from the quar-

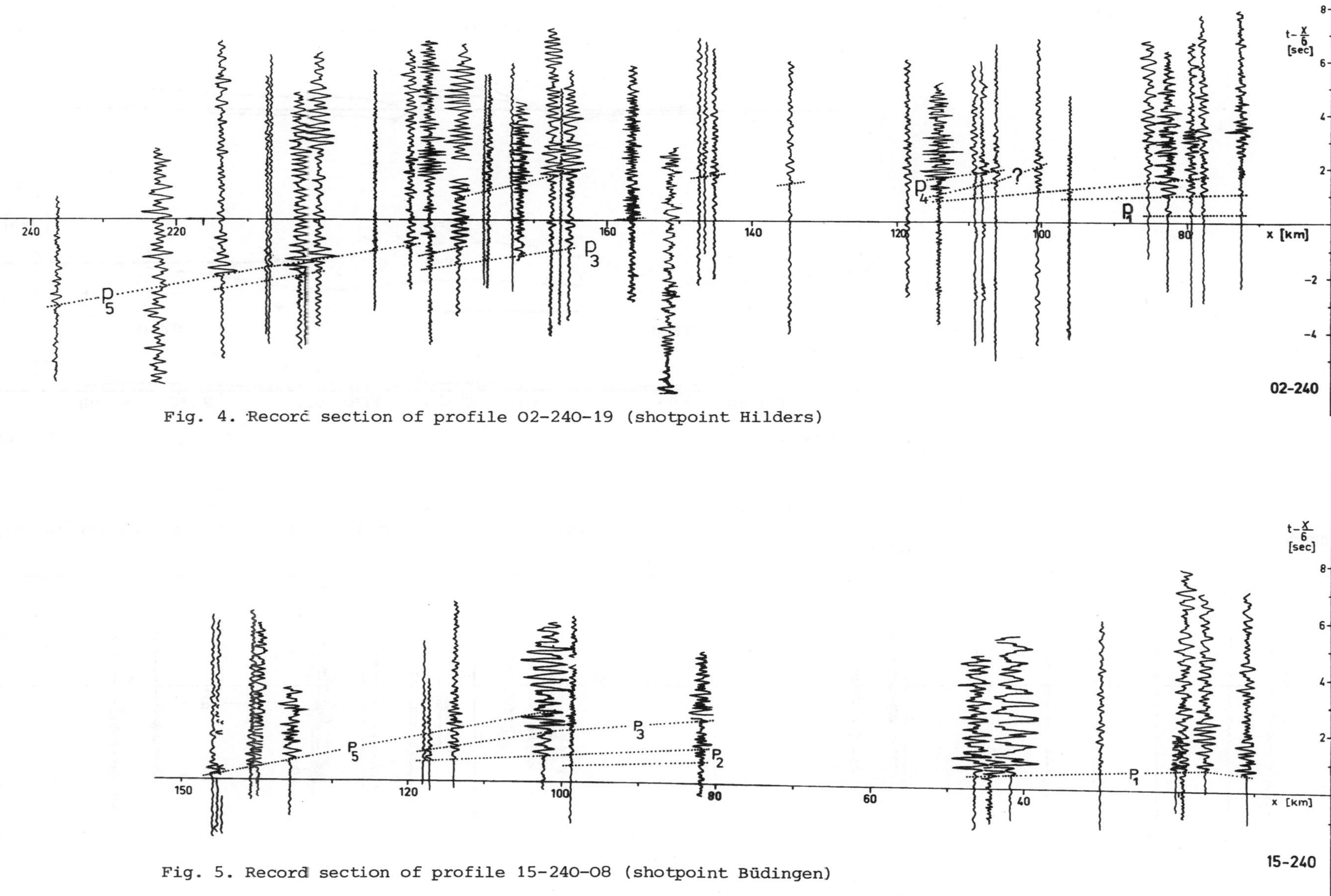

Fig. 4. Record section of profile 02-240-19 (shotpoint Hilders)

Fig. 5. Record section of profile 15-240-08 (shotpoint Büdingen)

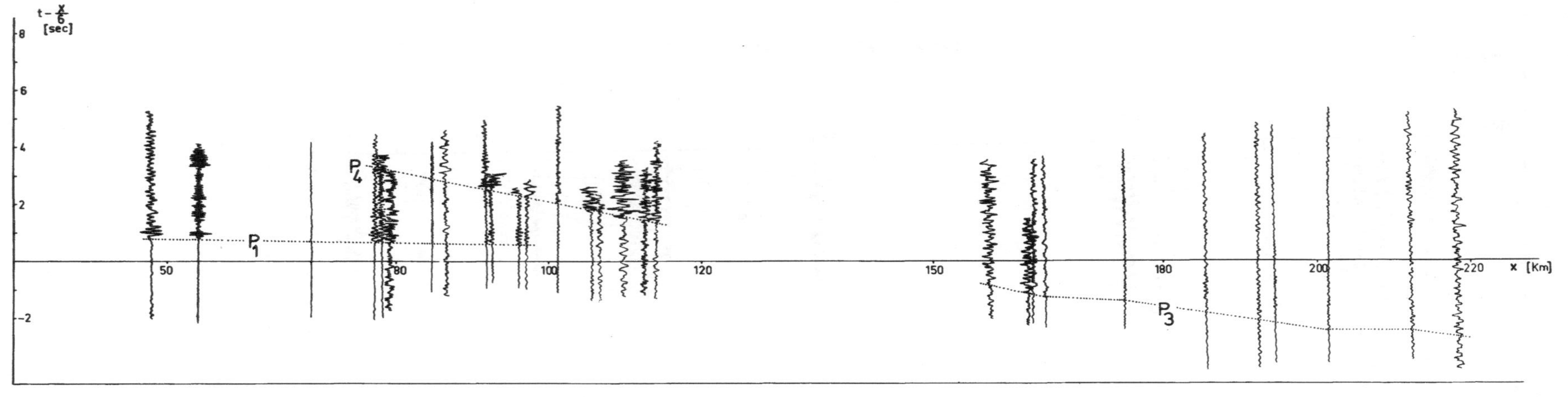

Fig. 6. Record section of profile 16-080-02 (shotpoint Taben Rodt)

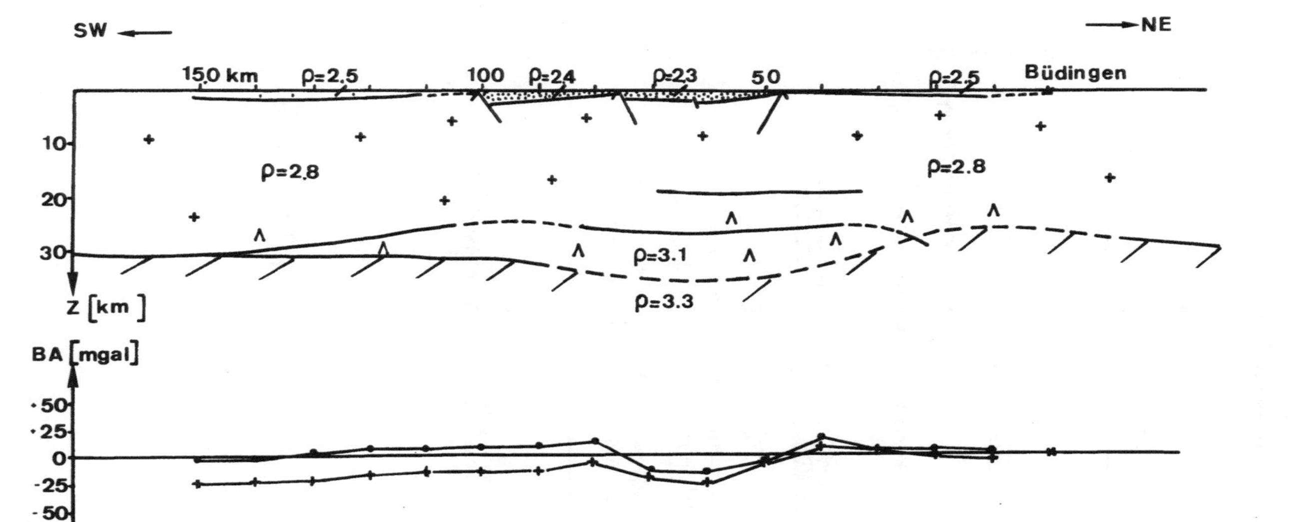

Fig. 7. The Rhine Graben Pillow derived from seismic and gravity calculations

ry Taben Rodt (16) was too weak. The profiles discussed next will show more reliable results for the eastern flank of the graben.

6.7.2.5 Profiles 02-220-05 and 05-040-02

Record sections of the reversed line Hilders (02) - Birkenau (05) have been compiled and interpreted by STROBACH (1963). The present reinterpretation (Fig. 8) shows only minor differences, mainly with regard to the P_2-branch. While those P_2-events which occur slightly before the concave P^M-branch are interpreted as a convex P^M-branch by STROBACH, the correlation of these arrivals may be better when connecting them with a concave line of correlation. This means that they belong to another, rather strong gradient zone in the middle part of the crust. This interpretation is supported by a similar correlation in the recently observed adjacent profile 02-215.

6.7.2.6 Profile 02-215

A record section of this profile has been compiled and interpreted by WILDE (1969) and MEISSNER et al. (1970). The southern end of this profile runs along the eastern shoulder of the Rhine Graben. The Mohorovičić discontinuity (MD) shows some undulations but no indications of a pillow-shaped structure as found below the graben in its central part (ANSORGE et al., 1970) (Fig. 9). Results of this profile indicate that the "Rhine Graben pillow" does not extend to the area under the eastern graben shoulder.

6.7.3 Velocity-Depth Functions

Velocity-depth functions of all profiles are shown in Figs. 10 and 11. Except for the traversing profiles the Rhine Graben velocity-depth functions all look rather similar. The depth of penetration of the P_g-wave is smaller on profile 240-LO-060 than on the other ones. The velocity-gradient zone at the base of the crust is very similar for profiles 13-240-20 and 240-LO-060 as well as for 02-220-05, 05-240-02, and 02-215. For the profiles 03-250 and 12-260 which cross the Hessische Straße, the velocity gradient seems to be steeper in the beginning of the gradient zone. A velocity of 7.1 km/s is found at 21 km depth compared to 26 km on the other profiles. This observation may be compared with the velocity-depth functions in the Rhine Graben where only a rough interpretation using first-order boundaries could be perfomed. The top of the pillow-shaped boundary with a 7.2 km/s velocity is at a depth of 28 km. The relatively high velocity of 7.1 km/s at 21 km depth below the Hessische Straße may be an indication of similar, though weaker, processes as found in the Rhine Graben where a tensional stress may lead to a mixing or a definite rise of mantle material into the transition zone at the base of the crust.

6.7.4 Results from Near-Vertical Reflection Measurements

Results of near-vertical reflection measurements have been published previously. DEMNATI and DOHR (1965) discuss the survey in the Rhine Graben and Kraichgau, SCHULZ (1957) gives results from the Pfälzerwald, DÜRBAUM et al. (1967) deal with the measurements around Siegen, and 6.2 of this volume discusses results from profile 240-LO-060. Data were evaluated statistically after methods proposed by DEMNATI and DOHR (1965). Some are shown in Fig. 12. The results can be summarized as follows:

1. In the Rhenish Massif, in the Siegerland, in the Kraichgau, and in the Pfälzerwald a band of strong reflections starts at about 5 s (two-way travel time). These reflections seem to represent the Conrad discontinuity. This group is missing in the Rhine Graben.

2. In all areas groups of reflections are found between 7 and 8 and between 8.5 and 9 s. The first group may indicate the beginning of the strong lamellation and gradient zone, the second group seems to mark its end and the top of the more homogeneous mantle. No strong reflections are found beyond 10 s. A lower boundary of the Rhine Graben pillow could not be detected by seismic-refraction or reflection investigations.

6.7.5 Contour Map of the Depth to the Mohorovičić Discontinuity in the Rhine-Main Area

Earlier papers dealing with contour maps of the depths to the MD and CD

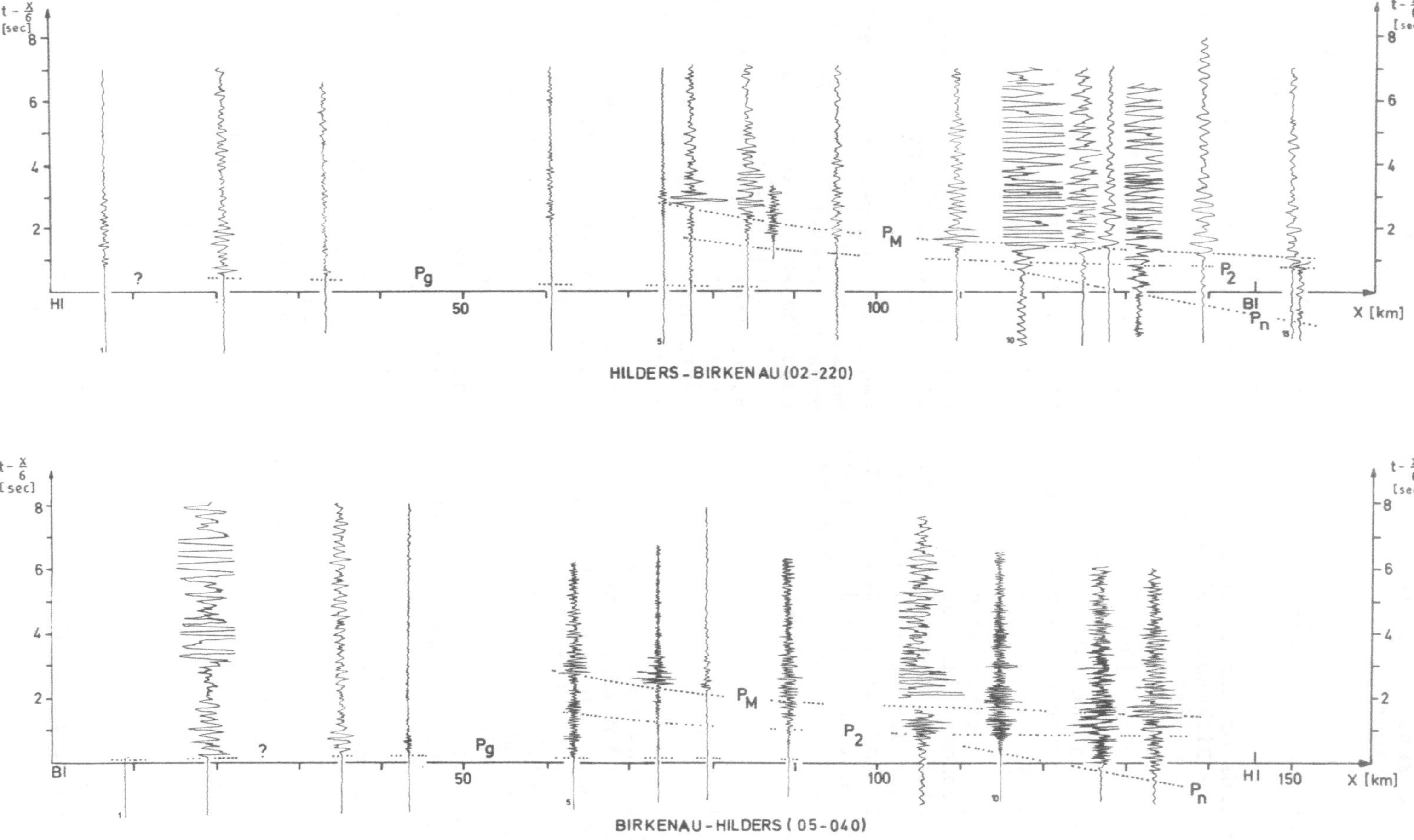

Fig. 8. Record sections of profile 02-220-05 (shotpoint Hilders) and 05-040-02 (shotpoint Birkenau) compiled by STROBACH (1963)

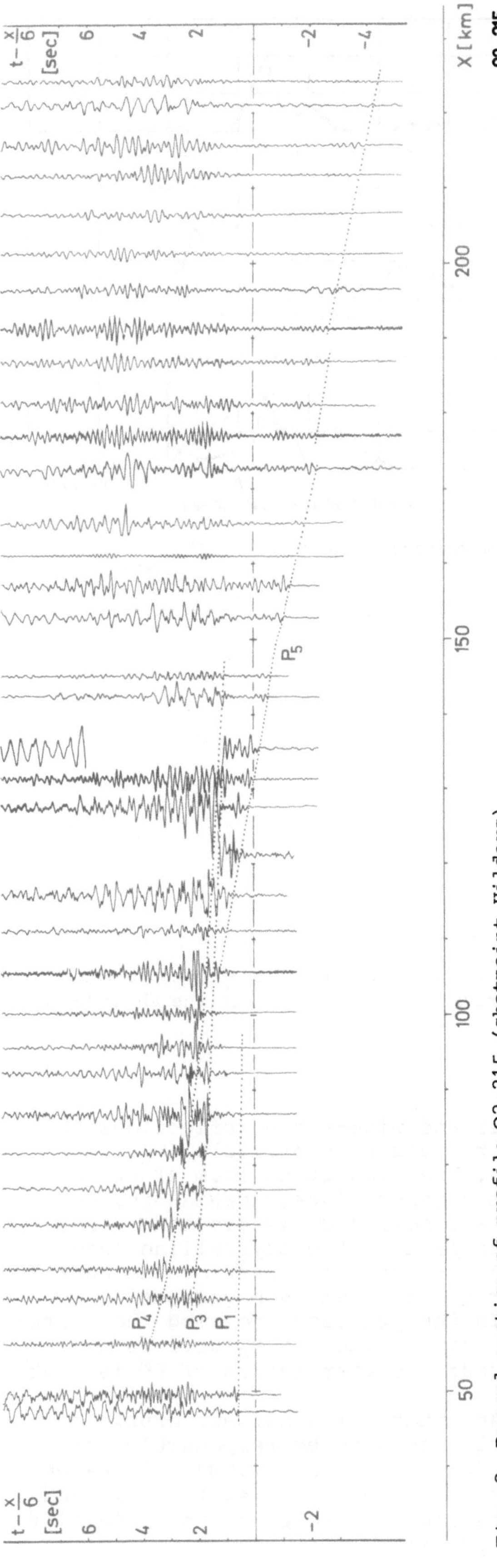

Fig. 9. Record section of profile 02-215 (shotpoint Hilders)

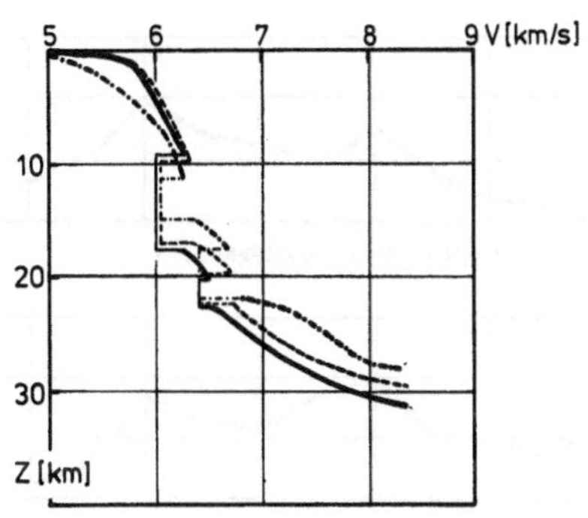

Fig. 10. Velocity-depth functions of profiles 13-240-20 (——), 240-LO-060 (----), and 03-250 (-.-.-.-)

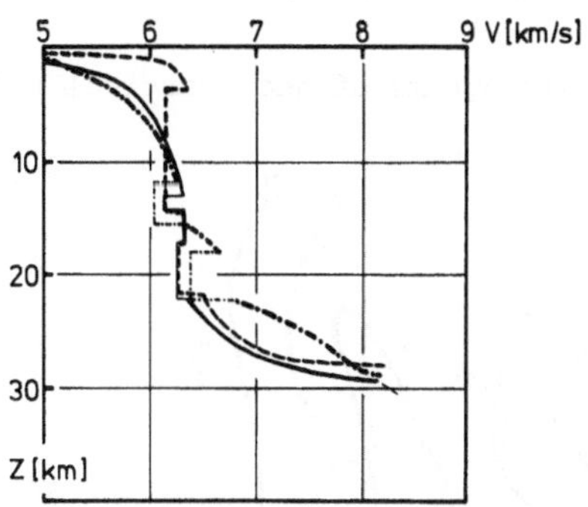

Fig. 11. Velocity-depth functions of profiles 03-250 (-.-.-), 02-220-05 (----), and 02-215 (——)

are based either on rather sparse P_n-data (German Research Group, 1964) or on P^M-data of seismic refraction profiles (GIESE, 1969). Based on the velocity-depth functions of the profiles described above and the depth from the peaks of the near-vertical reflection histograms, a new attempt is made to map the MD in the area under investigation. The map in Fig. 13 shows uncertainties mostly for the Rhine Graben. The top of the pillow has been tentatively correlated to the MD. As mentioned earlier, minor differences between depth data from the P^M- and P_n-waves and also between results from the near-vertical reflection measurements and the refraction data exist. The most reliable data and sometimes an average depth have been selected for the contour map of the depth to the MD. The map shows that the lifting of both shoulders of the graben extend down to the top of the upper mantle, indicating a deep-seated energy source for the formation of the Graben.

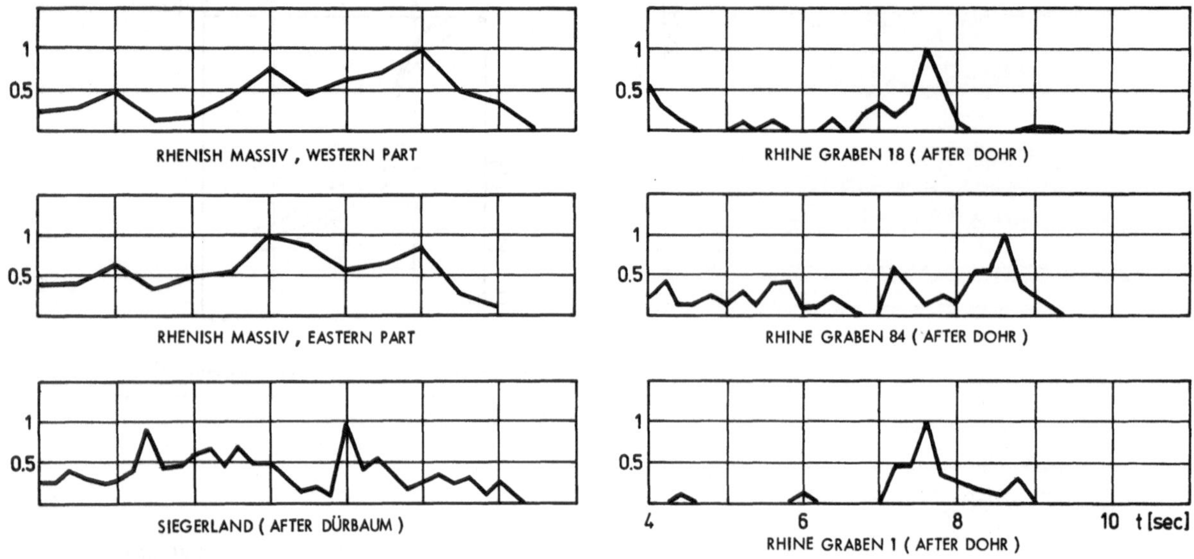

Fig. 12. Histograms of near-vertical reflections from various surveys

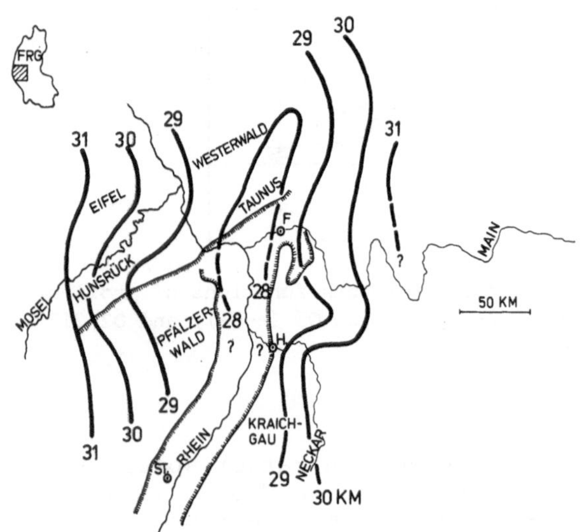

Fig. 13. Contour map of the depth to the Mohorovičić discontinuity in the Rhine-Main area

6.7.6 Conclusions

The following conclusions concern the area under investigation outside the Rhine Graben:

1. Velocity-depth functions show a strong similarity.
2. The structure of the deeper crust apparently is not strongly related to the different near-surface geology.
3. The Conrad discontinuity is indicated by the beginning of strongly reflected arrivals and is identified by refracted or diving waves.
4. The Mohorovičić discontinuity is found to coincide with the lower part of the zone of strong reflections and is identified by diving waves and wide-angle reflections (P^M) and by refracted arrivals (P_n).
5. The transition from crust to mantle takes place stepwise indicating a lamellated mixture zone. The upper part of the crystalline basement and the upper part of the mantle seem to be rather homogeneous.

In the profiles crossing the northern Rhine Graben no P_n-wave is observed; the correlation of P^M is doubtful.

An interaction between crust and mantle seems to be responsible for the "Rhine Graben pillow". Its extension to the west and east is limited by reliable P_n-waves on both sides of the graben.

6.8 Explosion-Seismology Research in the Central and Southern Rhine Graben - A Case History[1]

C. PRODEHL, J. ANSORGE, J. B. EDEL, D. EMTER, K. FUCHS, S. MUELLER, and E. PETERSCHMITT

ABSTRACT

A case history is presented summarizing different stages of explosion-seismic investigation in the central and southern Rhine Graben. Discrepancies and common aspects of the interpretation of data originating from quarry blasts, specially arranged explosions, and reflection measurements are outlined. Until 1966 few data were available. In 1967 a more detailed investigation was started under the auspices of the International Upper Mantle Project. Shortly thereafter a crustal model was proposed which showed a pronounced velocity inversion in the upper crust and a high-velocity layer (7.7 km/s) within the lower crust extending from about 25 to 40 km depth, this lense-shaped body had been termed the "rift cushion". With some modifications this model was supported by additional measurements, and detailed crustal sections of the P-wave velocity distribution were presented for the central and southern part of the Rhine Graben during the subsequent years. Significant changes of the existing model had to be applied by the results of a specially designed seismic-refraction experiment in 1972. From a reversed profile in the graben proper the true velocity in the uppermost mantle could be determined to be 8.0-8.1 km/s at a depth between 25 and 27 km. Thus the surface of the "rift cushion" was reinterpreted as top of the mantle. A velocity inversion in the upper crust is not necessarily required to interpret the data, but its possible existence cannot be excluded.

ZUSAMMENFASSUNG

Eine Reihe von Modellen wird zusammengestellt, die im Verlauf fortschreitender Untersuchungen über die Struktur von Kruste und oberstem Mantel im mittleren und südlichen Oberrheingraben in den zurückliegenden Jahren veröffentlicht wurden und die jeweils dem neuesten Stand der Forschung angepaßt waren. Dabei werden die gemeinsamen Aspekte und noch offene Fragen diskutiert. - Bis 1966, dem Beginn eingehender Untersuchungen im Rahmen des "International Upper Mantle Project", waren nur wenige Daten vorhanden. 1967 wurde das erste auf neueren Beobachtungen beruhende Modell veröffentlicht, das sich durch eine ausgeprägte Geschwindigkeitsinversion in der oberen Kruste und eine Schicht hoher Geschwindigkeit (7,7 km/s) in der unteren Kruste (25-40 km Tiefe) auszeichnete. Diese linsenförmige Einlagerung wurde als "Riftkissen" bezeichnet. Bis auf einige Änderungen konnte dieses Modell durch Zusatzmessungen bestätigt werden, und für den mittleren und südlichen Rheingraben wurden in den folgenden Jahren detaillierte Krustenschnitte abgeleitet. Die Ergebnisse der Auswertung einer 1972 durchgeführten refraktionsseismischen Spezialuntersuchung erforderten einige wesentliche Abänderungen des bestehenden Rheingrabenmodells. Aus dem gegengeschossenen Profil im Grabeninneren folgt für den obersten Mantel eine wahre Geschwindigkeit von 8,0-8,1 km/s in einer Tiefe von 25-27 km, und die Oberfläche des "Riftkissens" wurde uminterpretiert als Grenzfläche zwischen Kruste und Mantel. Eine Geschwindigkeitsumkehr in der oberen Kruste folgt nicht zwangsläufig aus den Daten, kann aber auch nicht ausgeschlossen werden.

6.8.1 Introduction

The study of the crustal structure in the Rhine Graben area was started as early as 1948, when some mobile stations recorded the large explosions near Haslach in the Black Forest on a short profile towards Strasbourg. A detailed research in the graben itself, however, was not started prior to 1966, when an international program was initiated aiming at a collection and synoptic interpretation of all available data in order to investigate the deep structure of the Rhine Graben which is a classical example of a continental rift valley in central Europe (ILLIES, 1970; MUELLER, 1970).

The details and the reliability of the different models of crustal struc-

---

[1] Contribution No. 74, Geophysikalisches Institut, Universität Karlsruhe.
Contribution No. 69, Institute of Geophysics, Swiss Federal Institute of Technology, Zürich.

ture depend largely on the number and quality of data which are available when results are published. It is the purpose of this paper to present the sequence of models in the form of a case history and to draw attention to the common aspects of the models presented at different stages of the Rhine Graben investigations.

With the exception of the very first studies prior to 1966 the paper will only deal with interpretations concerning the central and southern part of the Rhine Graben area.

Results from seismic investigations in the northern part of the Rhine Graben area are presented by MEISSNER et al. (6.7); the reader is also referred to MEISSNER and BERCKHEMER (1967), MEISSNER et al. (1970), GIESE and STEIN (1971), as well as MEISSNER and VETTER (1974).

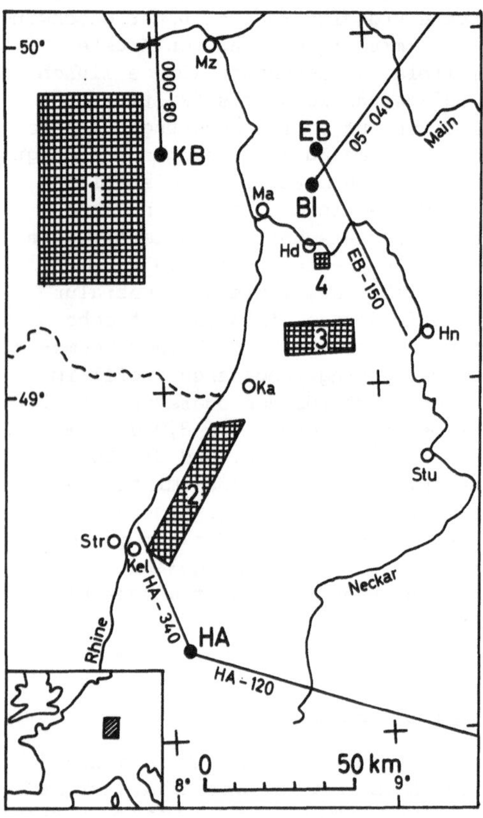

Fig. 1. Location map of seismic reflection and refraction surveys up to 1966. Explanations:
05-040 Seismic-refraction lines with shotpoints
Each profile is designated by: shotpoint code - azimuth
• Shotpoints: *BI*(05) Birkenau, *EB* Erlenbach *HA* Haslach, *KB* (08) Kirchheimbolanden
▦ Areas of reflection surveys: *1* Pfälzer Wald, *2* Upper Rhine Graben, *3* Kraichgau, *4* Near Heidelberg
o Cities: *Hd* Heidelberg, *Hn* Heilbronn, *Ka* Karlsruhe, *Kel* Kehl, *Ma* Mannheim, *Mz* Mainz, *Str* Strasbourg, *Stu* Stuttgart

6.8.2 Seismic-Refraction Investigations Prior to 1966

In 1948, the large explosions near Haslach in the Black Forest furnished the first information on crustal structure in and around the Rhine Graben (REICH et al., 1948; ROTHÉ and PETERSCHMITT, 1950; FÖRTSCH, 1951). The main results came from a profile running towards ESE (profile HA-120 in Fig. 1), but some stations were also installed along a profile towards Strasbourg (HA-340 in Fig. 1). Fig. 2 shows the crustal cross-section by ROTHÉ and PETERSCHMITT (1950) which reaches into the Rhine Graben proper. Two major discontinuities are indicated: a discontinuity at a depth of about 15 km, separating the "granitic" from the "basaltic" layer, and the Mohorovičić discontinuity (M-discontinuity) at about 25 km depth on top of the "peridotitic" layer.

VON ZUR MÜHLEN (1956) reported on seismic-refraction measurements carried out between 1950 and 1954 using quarry blasts. One of the profiles was recorded from a shotpoint near Erlenbach/Odenwald extending into the Kraichgau (profile EB-150 in Fig. 1) and crossing the so-called Neckar-Tauber gravity anomaly. Below a layer of "granitic" or metamorphic composition (5.5. km/s) VON ZUR MÜHLEN (1956) found a "gabbroic" layer (6.5 km/s) which rises from about 8 km depth near Michelbach/Odenwald to 2 km near Heilbronn. Because of the short recording distance of only 65 km no information was obtained about the depth of the M-discontinuity.

Some more insight into the structure of the crust by seismic-refraction profiles was obtained from the reversed profile Birkenau-Hilders (profile 05-040 in Fig. 1) which was compiled by STROBACH (1963), and from some profiles crossing the northern end of the Rhine Graben (e.g., profile 08-000 in Fig. 1) and interpreted by PLAUMANN (1961b) and HÄNEL (1964).

All results available in Germany up to 1963 were summarized by the German Research Group (1964). The depth contours for the Variscan mountain system did not cover the Rhine Graben area, but the contour maps of the Conrad and the Mohorovičić dis-

Fig. 2. Geologic cross section derived from seismic and gravimetric results presented by ROTHÉ and PETERSCHMITT (1950, Fig. 14)

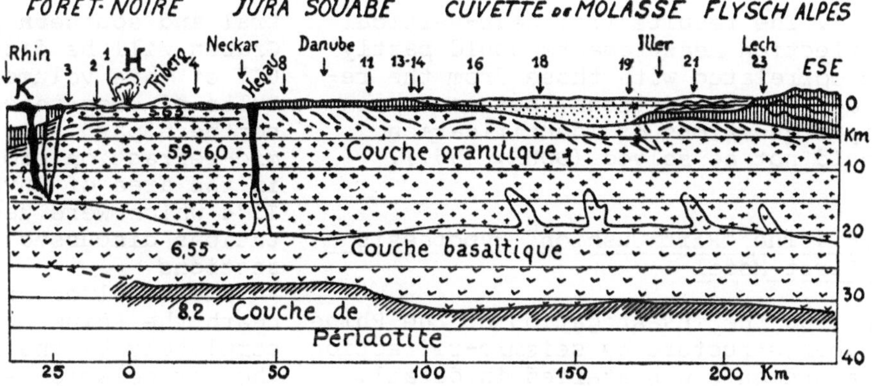

continuities include the Rhine Graben, indicating a mean depth of 16 to 18 km for the Conrad discontinuity, and a total crustal thickness of about 28 km beneath the Rhine Graben and its eastern flank. The M-discontinuity shows a slightly elevated ridge extending from SW to NE with its axis approximately along the line Basel-Stuttgart-Nürnberg.

Until 1966 no additional detailed seismic-refraction measurements in the region of the Rhine Graben had been carried out except for some profiles which reach this area only at distances beyond 130 km such as the profiles from Eschenlohe or Hilders. BEAUFILS (1967) interpreted a NE-profile from Lago Bianco (Gotthard Massif) which traverses the graben area at a distance greater than 140 km, and from the P_n-arrivals obtained a crustal thickness of 24-26 km under the Vosges.

6.8.3 Seismic-Reflection Surveys Prior to 1966

Additional information on the crustal structure in the Pfälzer Bergland, the graben area between Karlsruhe and Kehl, and in the Kraichgau (Fig. 1) became available from seismic reflection surveys (SCHULZ, 1957; DOHR, 1957a, b, 1959; DEMNATI and DOHR, 1965). These results were summarized by DOHR and MENZEL (1966) and DOHR (1967).

In the Pfälzer Bergland reflections with travel times of about 4.0 and 5.4 s were observed from which the depths to the respective reflectors were determined to be 10.4 and 13.4 km (SCHULZ, 1957). For interpretation, histograms of the reflection times were prepared, i.e., the number of observed reflections per time interval of 0.2 s were plotted versus travel time (DEMNATI and DOHR, 1965). From the recorded "deep" reflections in the area between Karlsruhe and Kehl with travel times between 7 and 9 s a depth of 18 to 21 km was determined for the Conrad discontinuity and 24-25 km for a deeper interface (DOHR, 1959). A special seismic-reflection experiment in the Kraichgau gave reflections at 5.2-5.4, 7.0, and 8.7 s resulting in depths of about 15, 21, and 27 km (DEMNATI and DOHR, 1965).

The statistical evaluation of commercial seismic-reflection measurements in the Rhine Graben near Heidelberg in 1966 showed a maximum of well-recorded reflections at travel times of 9.2-9.3 s, while no reflections with travel times between 7.0 and 7.5 s were picked up. According to DOHR (1967) in the area between Karlsruhe and Kehl, however, the reflections with travel times around 7 s were the best recorded ones. The author discussed in detail the problem of correlation from one area with another and concluded that, in agreement with the results of the seismic-refraction investigations (German Research Group, 1964), the reflections with travel times of about 9 s probably originate at the Mohorovičić discontinuity.

In summary, until the beginning of a more detailed research in 1966 the investigations of the crustal structure under the Rhine Graben area had led to the following results:

1. A general uplift of the Mohorovičić discontinuity with a minimum depth of 25 to 28 km and a P_n-velocity of 8.2 km/s (ROTHÉ and PETERSCHMITT, 1950; German Research Group, 1964).

2. A mean depth of the Conrad discontinuity of 16 km beneath the graben flanks and a depression to 18 km along the graben axis, with a velocity for the lower crust of 6.55 km/s (ROTHÉ and PETERSCHMITT, 1950; German Research Group, 1964).

3. The results from near-vertical reflection measurements could partly be correlated with those from the refraction observations, but some principle questions of identification could not be solved (DOHR, 1967).

6.8.4 The "Rift Cushion" Hypothesis - a First Model

In 1966 the investigation of the Rhine Graben structure by seismic-refraction measurements was started in detail. As already mentioned, only observations and results concerning the central and southern part of the Rhine Graben will be discussed here, while 6.7 of this volume reports on the results of deep-seismic sounding in its northern part. Fig. 3 shows all seismic-refraction profiles recorded within the area of investigation up to 1974 which were used for the interpretations discussed in the following sections.

The model of crustal structure beneath the Rhine Graben was changed completely by the interpretation of the new seismic-refraction measurements compared to the previous model described above. Suggesting that the

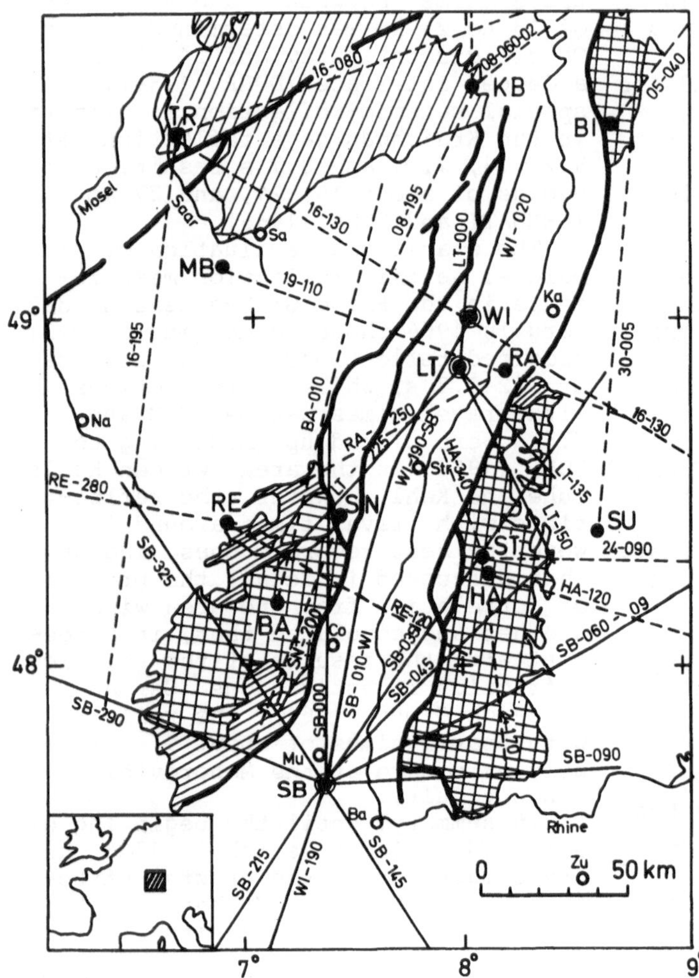

Fig. 3. Location map of seismic-refraction profiles in the central and southern Rhine Graben up to 1974. Explanations: ⊚—— profiles recorded in September 1972; •----- other profiles. Each profile is designated by: shotpoint code - azimuth - (reverse shotpoint code) ▬ faults; ▦ crystalline outcrops; ▨ Paleozoic outcrops;
⊚ • Shotpoints: *BI* Birkenau (05)* , *BA* Col des Bagenelles, *HA* Haslach, *KB* Kirchheimbolanden (08), *LT* Leutenheim, *MB* Merlebach, *RE* Raon l'Étape, *RA* Rastatt, *SN* Saint Nabor, *ST* Steinach (24), *SB* Steinbrunn, *SU* Sulz (30), *TR* Taben Rodt (16), *WI* Wissenbourg.
o Cities: *Ba* Basel, *Co* Colmar, *Ka* Karlsruhe, *Mu* Mulhouse, *Na* Nancy, *Sa* Saarbrücken, *Str* Strasbourg, *Zu* Zürich

* The numbers refer to the designations used for shotpoints (quarries) within western Germany (BAMFORD, 1973; GIESE and STEIN, 1971; STEIN and SCHRÖDER, 3.1)

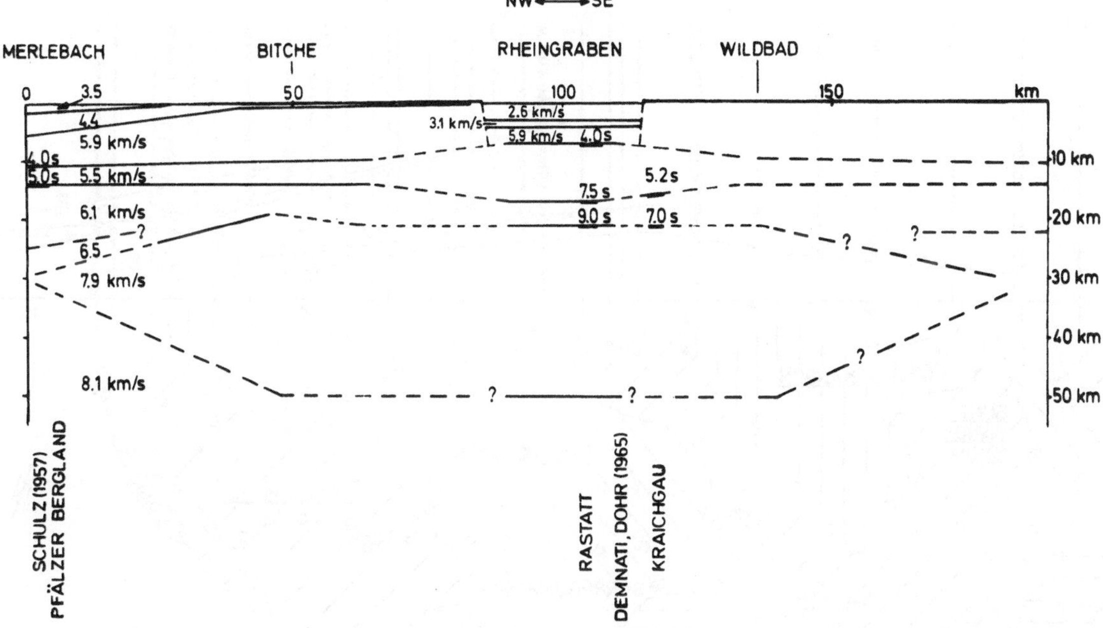

Fig. 4. Model of the crust and upper mantle structure in the central part of the region of the Rhine Graben, derived by MUELLER et al. (1967, Fig. 5)

Rhine Graben is the central part of a rift system, the existence of a high-velocity intermediate layer in the lower crust had been postulated by MUELLER and PETERSCHMITT (1966).

Based on profiles recorded in 1966 and 1967 (profiles 16-130 and 19-110), a model of the structure of the crust for the central part of the Rhine Graben was presented by MUELLER et al. (1967) including in the interpretation the profiles from Haslach (HA-120, HA-340). The resulting crustal model is shown in Fig. 4. The main features are as follows: The crystalline basement (5.9 km/s) is underlain by a sialic low-velocity zone at about 10 km depth, a zone which has first been described by MUELLER and LANDISMAN (1966) and LANDISMAN and MUELLER(1966). In that depth range the velocity is supposed to decrease from 5.9 to 5.5 km/s. This low-velocity zone is thicker beneath the Rhine Graben - reaching to a depth of about 17 km - than under the adjacent areas to the east and west. The next discontinuity is reached at 21 km depth separating crustal material with velocities between 6.1 and 6.5 km/s from a layer with a velocity of 7.9 km/s. The correlation chosen for this so-called phase P_R with an apparent velocity of 9.4 km/s and an assumed true velocity of 8.2 km/s would have produced a gravity anomaly much too high. Therefore, an intermediate layer was introduced above the classical Mohorovičić discontinuity. The true P-wave velocity of 7.9 km/s within this layer could only be estimated since no reversed profiles were available at that time.

It was demonstrated by MUELLER et al. (1967) and CLOSS and PLAUMANN (1967) that the sedimentary filling of the graben accounts for most of the observed negative gravity anomaly. Thus, the earlier postulated depression in the Conrad discontinuity (German Research Group, 1964) could not be real. Therefore, the lower crust must be isostatically compensated and the layer at 21 km depth could not be interpreted as the Mohorovičić discontinuity.

Arrivals with a velocity of 8.1 km/s observed along the profile 16-130 east of the Rhine Graben were interpreted as P_n-phases guided at the lower boundary of this intermediate layer. From the P_n-observations the depth to the base of the intermediate layer was estimated to be about 50 km. A characteristic splitting of P_n-arrivals observed on the profiles Lago Bianco-NW, Lac Negre-N, Eschenlohe-Haslach, and Eschenlohe-Birkenau at the distances where these profiles intersect the Rhine Graben region was explained by MUELLER et al. (1967) as

PROFILE

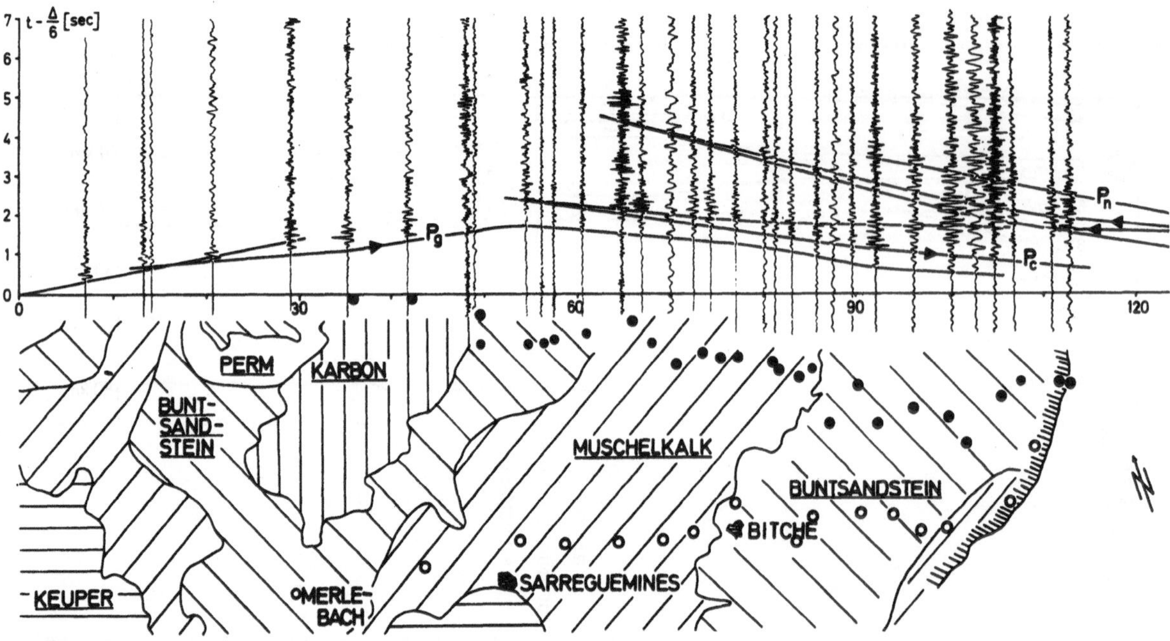

Fig. 5. Record section of the profile 16-130 with correlations of ANSORGE et al. (1970, Fig. 3). The geologic map below indicates the station locations for profiles 16-130 (*black circles*) and 19-110 (*open circles*)

additional evidence for the existence of this high-velocity intermediate layer. The interpretation was also guided by the results from seismic-reflection measurements described above, and from phase-velocity measurements of seismic surface waves along the lines Stuttgart-Oropa (western Alps), Stuttgart-Besançon, and Stuttgart-Strasbourg (SEIDL et al., 1966).

6.8.5 A Detailed Crustal Model of the Central Rhine Graben Based on the "Rift Cushion" Hypothesis

The observation of new seismic-refraction profiles (BA-010, SN-200, 24-170) and additional recordings on existing profiles (profile 16-130 and profiles from Eschenlohe) in the area of the Rhine Graben as well as additional deep seismic-reflection measurements near Rastatt (shotpoint RA in Fig. 3) (DOHR, 1970) permitted the derivation of a more detailed model by MUELLER et al. (1969) and ANSORGE et al. (1970). This model is an alteration of the older model shown in Fig. 4 adjusted in details according to modifications of the earlier signal correlations. Four main wave groups are correlated (Fig. 5): P_g corresponds to a wave through the crystalline basement with velocities of 5.9-6.0 km/s, P_c (6.7-6.9 km/s) denotes a wave refracted by the abrupt increase of velocity at the base of the sialic low-velocity zone, and P_R is attributed to the already-mentioned intermediate layer with a true velocity of 7.6-7.7 km/s. A P_n-velocity of 8.2 km/s could only be observed at greater distances east of the Rhine Graben.

Fig. 6 shows the crustal cross-section through the central part of the Rhine Graben area as presented by MUELLER et al. (1969, Fig. 5) and ANSORGE et al. (1970, Fig. 4a). The main features of this model are a pronounced reversal of the compressional velocity in the sialic upper crust with a lamellar structure underneath in the middle crust, including an additional slight velocity inversion. In the lower crust a cushion-shaped intermediate layer is found between a depth of 25 and 40 km with velocities between 7.6 and 7.7 km/s which extends about 100 km to the east and west of the Rhine Graben axis. According to MUELLER et al. (1969) the existence of this high-velocity layer supported the view of the Rhine Graben being a rift system. The "rift cushion" houses the suspected driving mechanism of the rifting process.

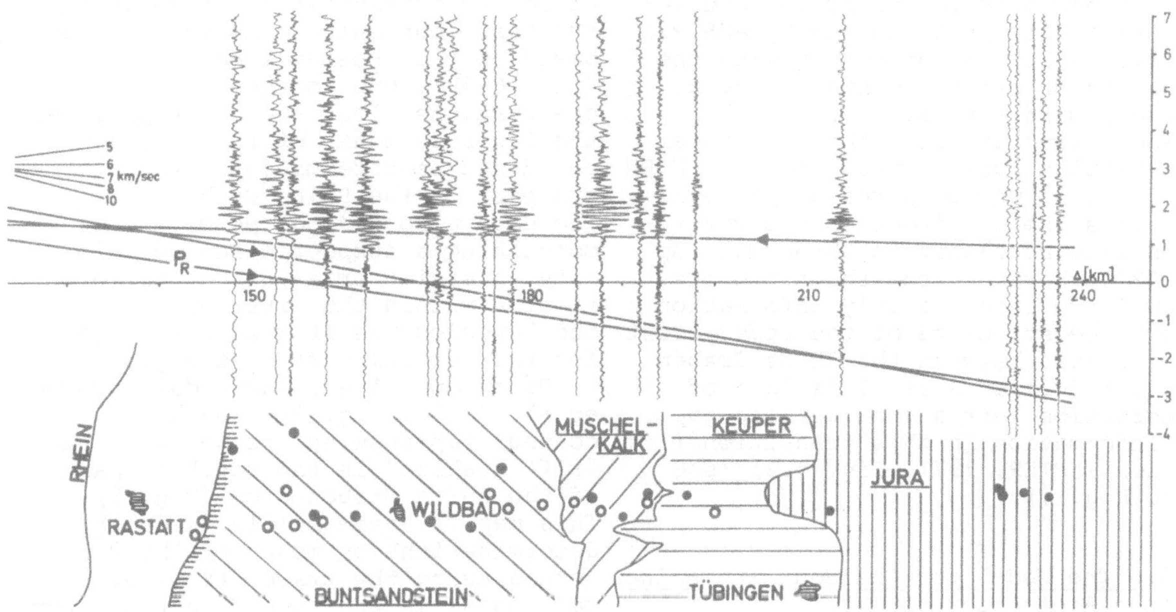

As can be seen in Fig. 6, the authors assume a marked asymmetry in the crustal section of the Rhine Graben which is mainly derived from near-vertical reflection measurements in and around the graben area (LIEBSCHER, 1964; DOHR, 1970). This interference is discussed in detail in connection with other asymmetric features of the Rhine Graben, e.g., the Quaternary and Tertiary graben filling, the geothermal field, the distribution of focal depths determined for shallow earthquakes, or measurements of teleseismic P travel-time delays at stations in the Rhine Graben area.

The evidence for the existence of a lower boundary of the intermediate

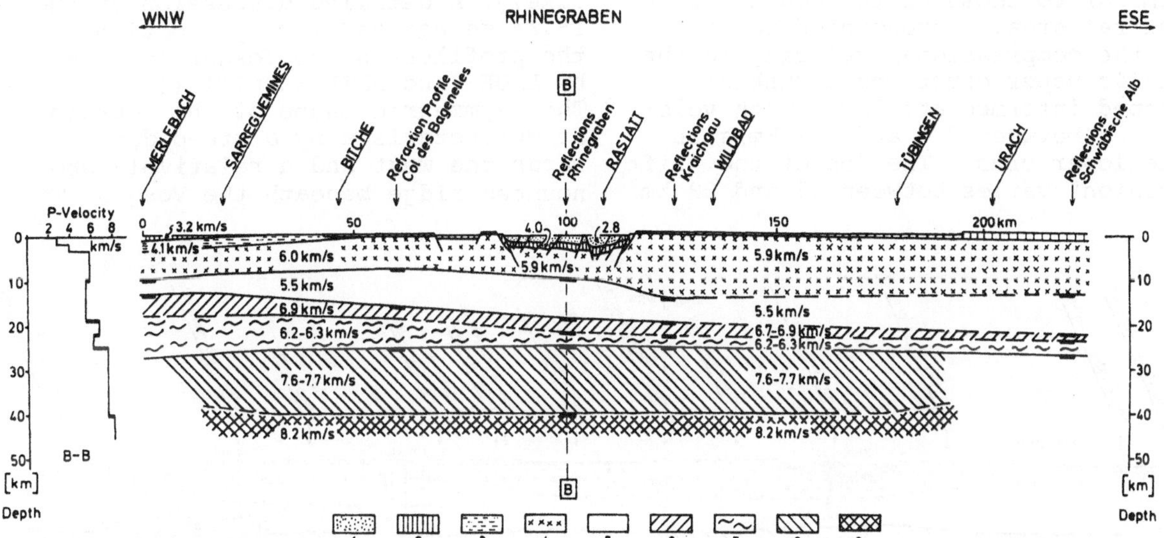

Fig. 6. Crustal section through the central part of the Rhine Graben area by MUELLER et al. (1969, Fig. 5) and ANSORGE et al. (1970, Fig. 4a). The P velocity-cross section B-B is shown on the left. Reflecting horizons are marked by heavy bars. *1* Quaternary and Tertiary; *2* Jurassic, Triassic, and Permian; *3* Carboniferous and Devonian; *4* crystalline basement; *5* sialic low-velocity channel; *6* intermediate crustal layer; *7* intermediate low-velocity zone; *8* rift cushion; *9* upper mantle

layer at a depth of about 40 km in the center of the Rhine Graben based on the new seismic refraction and reflection data is discussed by ANSORGE et al. (1970) in connection with the analysis of other geophysical data. The eastward extension of the Rhine Graben structure into the area under the Swabian Jura is presented by EMTER (1971, 6.5) who prepared a synthesis of all seismic refraction and reflection data available in SW-Germany up to 1970. However, the short profile 24-090 provided the only information about the structure of the upper crust in the area between the Rhine Graben and the Swabian Jura. This lack of observation left a considerable gap of information in this transitional region (EMTER, 1971, Fig. 7.2) (see also Fig. 7).

6.8.6 The Model for the Southern Rhine Graben Area Prior to 1972

MUELLER et al. (1973, Fig. 6) (see also Rhine Graben Research Group, 1964, Fig. 10) presented another crustal section through the southern part of the Rhine Graben (Fig. 7). In addition to the profiles mentioned above, profiles 16-195, RE-290 provided new information. To the east, the cross-section is based on the above quoted results by EMTER (1971, 6.5). The main features of the model for the southern part of the Rhine Graben are similar to those of the model for the central area: a pronounced reversal of the compressional velocity in the sialic upper crust and a cushion-shaped intermediate layer with velocities between 7.6 and 7.7 km/s in the lower crust. The top of the "rift cushion" varies between 22 and 29 km depth with a fairly flat portion of 25 km beneath the graben proper. The structure extends at least 90 km to the east, but only 50 km to the west, based on the observations along profile 16-195. Due to the lack of P_n-observations along the short profiles, the lower boundary of the "cushion" could not be determined in the southern part of the Rhine Graben area. The data also did not permit studies concerning a suspected second velocity inversion or a further differentiation within the crust on top of the "cushion" as it was postulated for the central part of the graben.

Based on all available data ANSORGE et al. (1972, Fig. 2) presented a contour map showing the depth to the "rift cushion" in the southern part of the Rhine Graben area (Fig. 8). This map is based mainly on depth determinations by means of the T^2, Δ^2 method using the travel times of the main phase in the record sections for the distance range from 50 to 100 km. This main phase was interpreted as the overcritical reflection from the top of the "cushion". For all seismograms in which the respective phase was recognized and used for depth calculations the location of the corresponding depth value is marked by a dot at half the shotpoint-station distance. For the overburden an average velocity of 6.1 km/s was assumed except for the profiles from Taben-Rodt for which 6.3 km/s seemed to be more likely. A detailed discussion of the T^2,Δ^2 method and its application to the profiles in the Vosges is given by LAUER and PETERSCHMITT (1970). The asymmetric shape of the "cushion" is characterized by a steep dip towards the west and a relatively pronounced ridge beneath the Vosges. To-

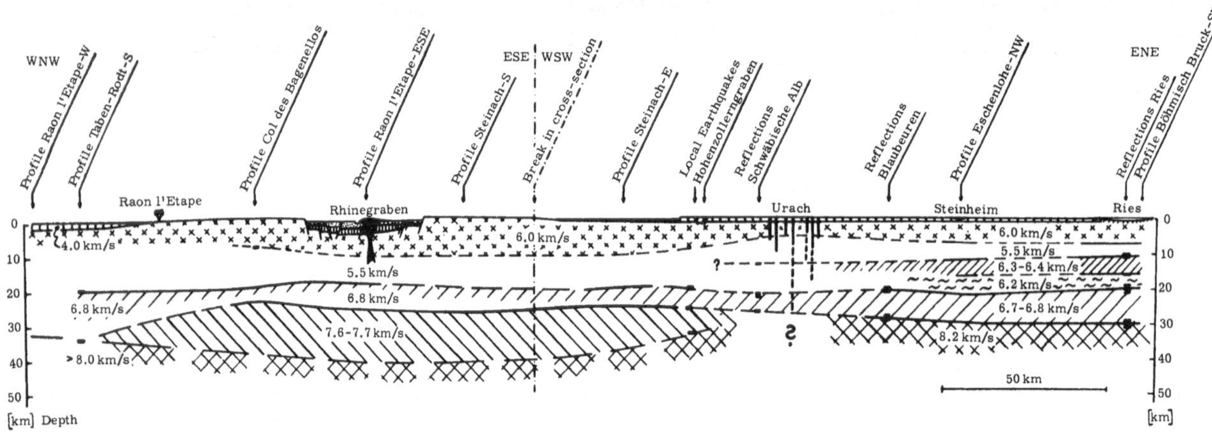

Fig. 7. Crustal section through the southern part of the Rhine Graben area by MUELLER et al. (1973, Fig. 6). Explanations see Fig. 6

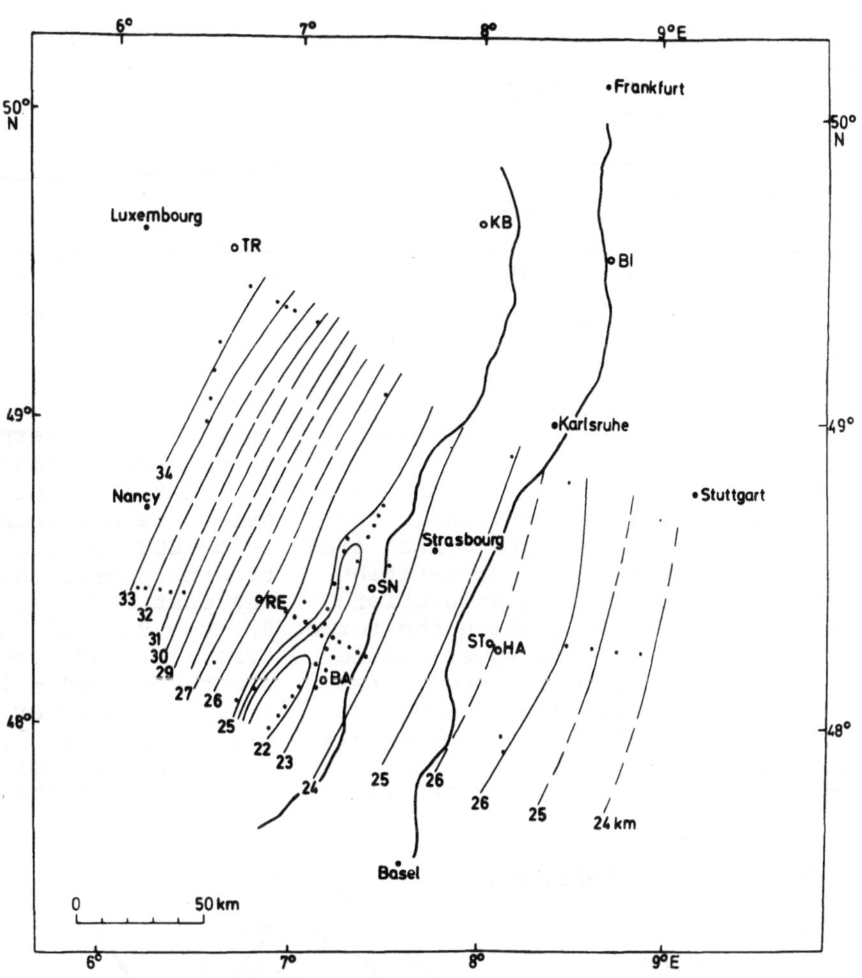

Fig. 8. Contour map of the top of the "rift cushion" in the southern part of the Rhine Graben area by ANSORGE et al. (1972, Fig. 2). *Thick lines:* boundaries of the graben proper. *Points:* the positions for which depths were calculated from observations. o Shotpoints (see Fig. 3)

wards the east the dip of the surface of the "cushion" is considerably less. Its flat top lying at a depth of about 25 km extends 90 km eastward from the graben axis.

6.8.7 The 1972 Seismic-Refraction Experiment in the Southern Rhine Graben - General Results

So far all refraction observations depended solely on quarry blasts as energy sources. The limited amount of explosions and often unfavorable geographical locations of these blasts largely prevented the layout of profiles which were best suited to the tectonic and geologic problems of the Rhine Graben. Because of the lack of sufficient energy most profiles could not be observed up to distances where arrivals from below the Mohorovičić discontinuity may be recorded. Reversed profiles were impossible to obtain, because of the unsuitable locations of quarries. Therefore, in September 1972 a specially designed seismic-refraction experiment was performed in the southern part of the Rhine Graben with profiles extending from the shotpoints Steinbrunn (SB), Wissembourg (WI), and Leutenheim (LT) indicated by solid lines in Fig. 3. The experiment was described in detail by the Rhine Graben Research Group (1974). The most important profile is the one recorded between Steinbrunn and Wissembourg in the graben proper (SB-010-WI and WI-190-SB), because it is the first and the only reversed seismic-refraction profile in the area of investigation which allows the determination of the true upper-mantle velocity.

STREICHER (1974) investigated in detail the velocity-depth structure of the sediments within the graben

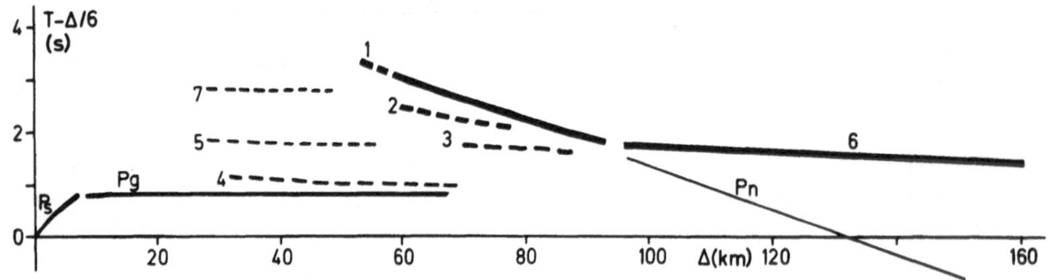

Fig. 9. Basic travel-time diagram of the profiles recorded in the Rhine Graben area as defined by EDEL et al. (1975, Fig. 7). Explanations see text

along the seismic-refraction lines of 1972.

The more comprehensive and better-quality seismic data obtained in the southern part of the Rhine Graben area required a complete reinterpretation of all the profiles available prior to September 1972 upon which the former models of the crust and upper mantle in this region had been based. First results for the area of the southern Rhine Graben using all available data were published by the Rhine Graben Research Group (1974). They are based only on the interpretation of the correlated main phases (P_g, P^M, P_n). A more detailed interpretation of the same data has been prepared subsequently by EDEL et al. (1975). Generally up to nine phases can be correlated in the most recent interpretation (Fig. 9): the P_g- and P_n-phases in first arrivals, reflected and multiply reflected phases - labeled by numbers 1 to 7 - in later arrivals (EDEL et al., 1975, Fig. 7).

The density of the P_g-arrivals in the region of the Vosges, the southern

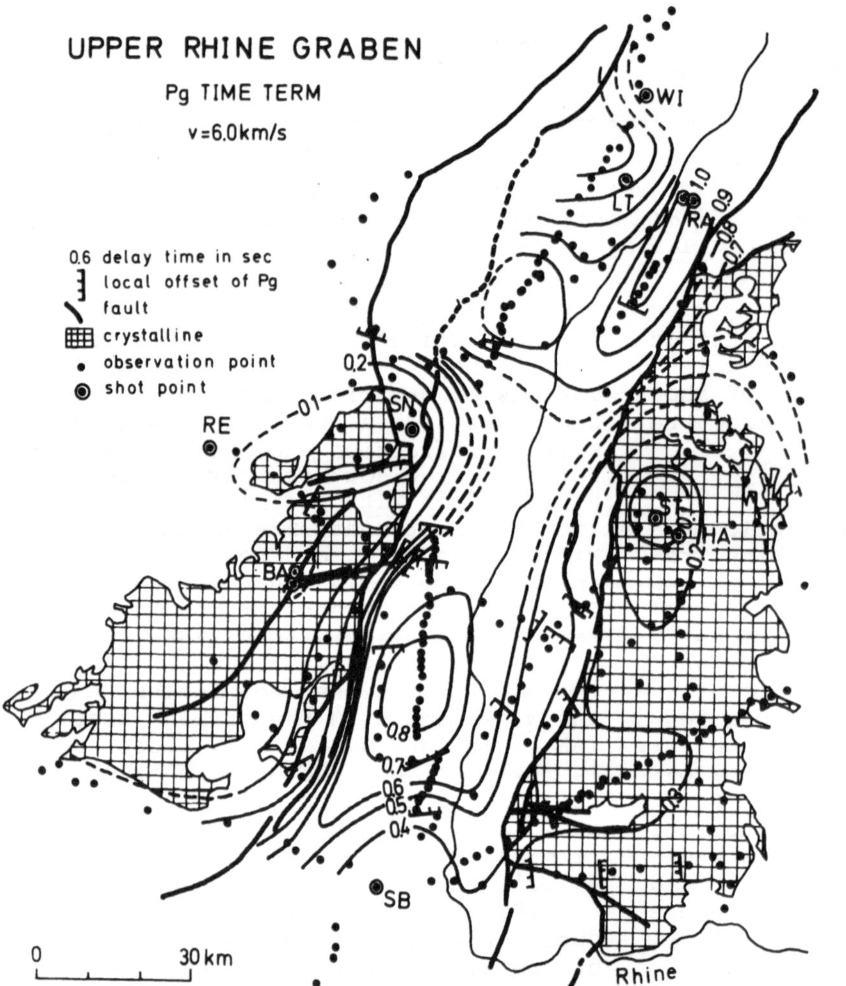

Fig. 10. Delay time map to the 6.0 km/s surface for the upper Rhine Graben based on fourth order Fourier series solutions by EDEL et al. (1975, Fig. 8). Shotpoint codes see Fig. 3

Rhine Graben, and the Black Forest allowed the application of a modified version of the so-called "time-term approach" (BAMFORD, 1973) which gave a "refractor"-velocity of 6.00 ± 0.02 km/s (EDEL et al., 1975, Fig. 8). The contour lines of this delay time map (Fig. 10) in the graben represent in a first approximation the variations to the depth of the basement. The small variable values of the delay times in the areas of outcropping basement may be due to the fact that the influence of weathering and fracturing changes from one area of the basement to the other. Sudden delays in the P_g-travel-time curves possibly caused by vertical displacements of the basement are also indicated in the map.

The P_n-phase can be observed on several profiles, especially on the reverse profile between Wissembourg and Steinbrunn in the graben proper. Although the scattering of P_n-arrivals causes difficulties in the exact determination of the apparent velocities, the data indicate clearly that the true velocity within the uppermost mantle cannot be less than 8.0-8.1 km/s. With this value the apparent P_n-velocities on the other profiles which are less or equal to 8.0 km/s are interpreted to be caused by a dip of the crust/mantle boundary along the corresponding profiles. The depths calculated from the P_n-arrivals for the M-discontinuity lie between 25 and 29 km (Fig. 11) as shown by the Rhine Graben Research Group (1974, Fig. 8).

Because of these results based on reversed P_n-observations, the hypothesis of the "cushion" defined as an intermediate layer in the lower crust above the Mohorovičić discontinuity with a true velocity of 7.6-7.7 km/s had to be abandoned by the Rhine Graben Research Group (1974) and EDEL et al. (1975), and the most prominent phase of the reflected signals (phase 1 in Fig. 9) was interpreted as P^M, i.e. the reflected wave from the crust/mantle boundary (Figs. 11 and 12). Some differences can be seen in the two contour maps of Figs. 11 and 12: While the preliminary contour map of Fig. 11 published by the Rhine Graben Research Group (1974) is based on phases interpreted as P^M-arrivals in the distance interval from 50 to 150 km (phases 1 and 6 in Fig. 9), the more detailed interpretation by EDEL et al. (1975) limits the pure P^M-phase to the distance range from 50 to 100 km (phase 1 in Fig. 9) and raises the

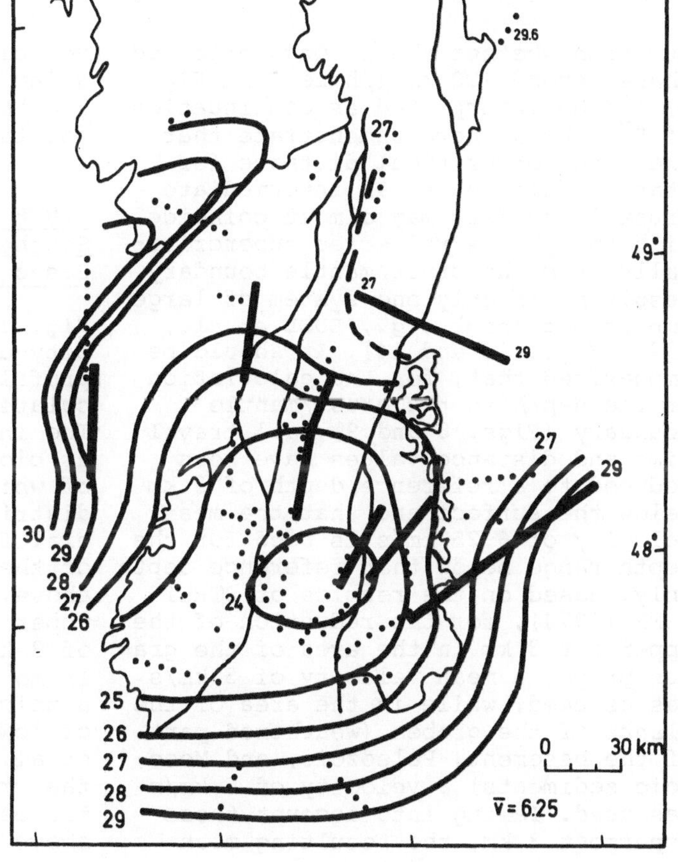

Fig. 11. Contour map of the depth to the crust/mantle boundary for the area of the southern Rhine Graben, using a mean velocity of 6.25 km/s below a reference depth of 3 km, by the Rhine Graben Research Group (1974, Fig. 8). The mean velocity of 6.25 km/s below 3 km depth corresponds to a mean velocity of 5.8 km/s of the total crust below the graben proper resp. 6.05 km/s below the flanks of the graben. *Points:* locations for which depths were calculated from the P^M-data. *Solid lines:* the distance in which P_n-phases were recorded

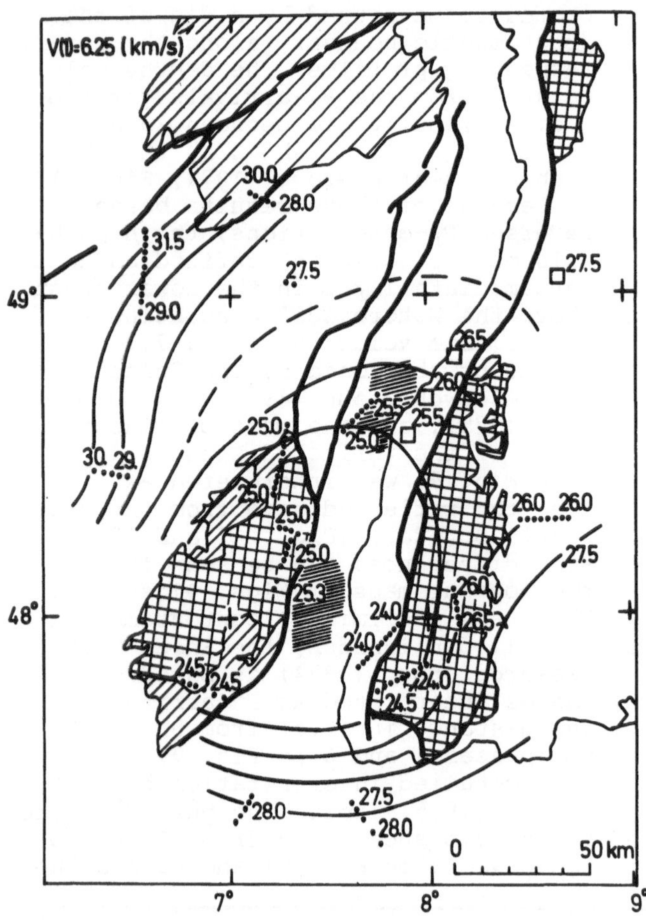

Fig. 12. Contour map of the depth to the crust/mantle boundary for the area of the southern Rhine Graben, using a mean velocity of 6.25 km/s below a reference depth of 3 km, by EDEL et al. (1975, Fig. 10) based on phase 1 (see Fig. 9). The mean velocity of 6.25 km/s below 3 km depth corresponds to a mean velocity of 5.8 km/s of the total crust below the graben proper resp. 6.05 km/s below the flanks of the graben. Contour interval 1 km. 28.8 depth in km; ..... positions for which depths were calculated from seismic-refraction data; □ positions for which depths were calculated from seismic-reflection data. ▨ areas where phase 1 is not observed. Further explanations see Fig. 3

question whether the strong reflected phase beyond 100 km (phase 6 in Fig. 9) can be interpreted as continuation of P^M. The authors demonstrate that the range of critical distance for phases reflected at an intermediate crustal boundary may almost coincide with the phases reflected supercritically from the crust/mantle boundary resulting in only one system of large amplitudes (see, e.g., EDEL et al., 1975, Figs. 13 and 14). It should be emphasized that, for the calculation of the depth to the crust/mantle boundary (Figs. 8 and 9), all travel time and distance values have been reduced to a reference depth of 3 km below the surface and that the mean velocity of 6.25 km/s is used for the depth range below that reference depth only. Based on the results of STREICHER (1974), for the reduction of the uppermost 3 km in the area of the graben proper a mean velocity of 3 km/s was assumed, while in the area of the flanks of the graben (weathered part of the basement, Paleozoic, and Mesozoic sediments) a velocity of 5 km/s was used. Taking into account these uppermost 3 km, the resulting mean velocity of the total crust of the Rhine Graben area from the surface is 5.8 km/s for the graben and 6.05 km/s for the flanks.

6.8.8 A Detailed Crustal Model of the Southern and Central Rhine Graben Based on the 1972 Experiment

Fig. 13 presents a map showing velocity-depth functions for selected profiles in the Rhine Graben area calculated by EDEL et al. (1975, Fig. 15). The inlets with the models are placed as close as possible into the regions of which they are typical. The model distributions show the following general features: Based on the results of the time-term analysis mentioned above, the authors assume for the upper crust below the reference depth of 3 km an average velocity of 6.0 km/s. In most cases the upper crust has a thickness of about 15 km. No zone of low velocity was included by EDEL et al. (1975) in the upper part of the crust which had been described in the earlier publications discussed above (MUELLER et al., 1967, 1969,

Fig. 13. Map showing velocity-depth distributions for different profiles of the central and southern Rhine Graben by EDEL et al. (1975, Fig. 15). The inlets with the models are placed as close as possible into the regions for which they are typical. Position of profiles and further explanations see Fig. 3

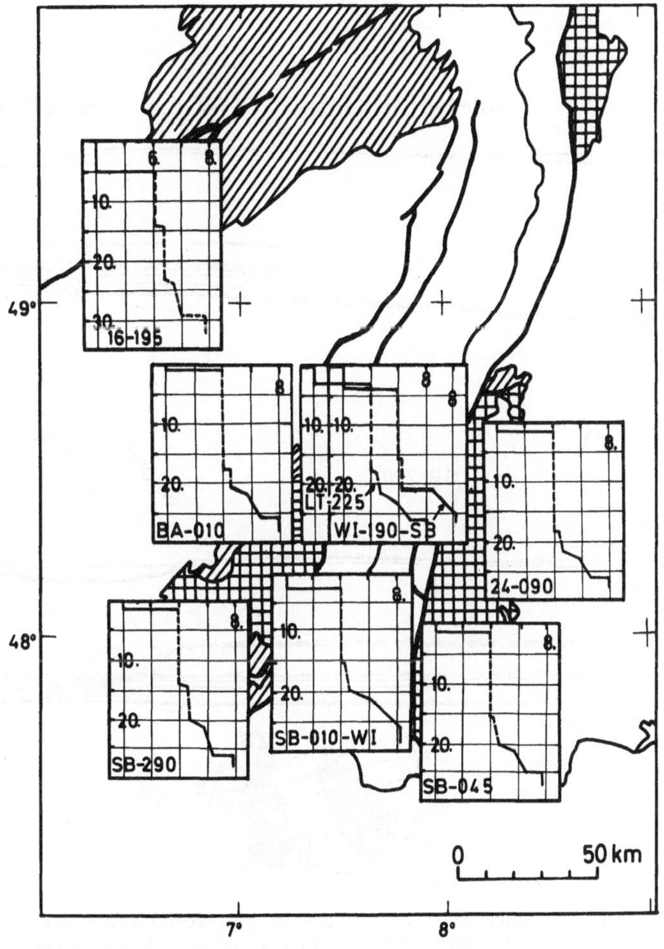

1973; ANSORGE et al., 1970, 1972) and which was based on a phase termed P_C (see Fig. 5). The new observations of 1972, however, furnished no further information concerning the phase P_C. The explosions within the graben proper generated strong multiple reflections and reverberations within the sediments which completely mask the distance range important for the detection of P_C. Therefore, no fine structure was included in the upper part of the crustal models shown in Fig. 13. Instead, only its average velocity is taken into account. A more complicated velocity-depth distribution in this depth range, however, does not affect the model derived for the structure of the middle and lower crust (EDEL et al., 1975, Fig. 14).

The depth range below 15 km can be subdivided into two parts: the middle crust with a mean velocity of 6.3-6.5 km/s and the lower crust with 6.8-7.2 km/s, separated by a transition zone in which the velocity increases more or less rapidly with increasing depth. All models for the regions outside the graben proper are characterized by a first-order discontinuity at 25-26 km which is interpreted as Mohorovičić discontinuity. Only on profile 16-195 in the northwest is its depth greater. In the graben proper, the strongest gradient is encountered in a depth range of about 21 km, and EDEL et al. (1975) consider the whole depth range between 21 and 26 km as a zone of crust-mantle interaction. This zone seems to be confined rather strictly to the graben proper since the first-order discontinuity at 25-26 km depth appears as soon as profiles originating in the graben enter the flanks at about 60-80 km recording distance (LT-225 and SB-035 in Fig. 3).

These features can also be seen in the crustal sections (Fig. 14) by EDEL et al. (1975, Fig. 16). The crust-mantle boundary rises from the west towards the Rhine Graben proper and dips gently to the east on the other side of the graben. Outside the graben the crust-mantle boundary forms a first-order discontinuity, into which the lines of equal velocity from 7.4 to 8.0 km/s converge. Beneath the graben proper these lines of equal velocity are rising steeply to shal-

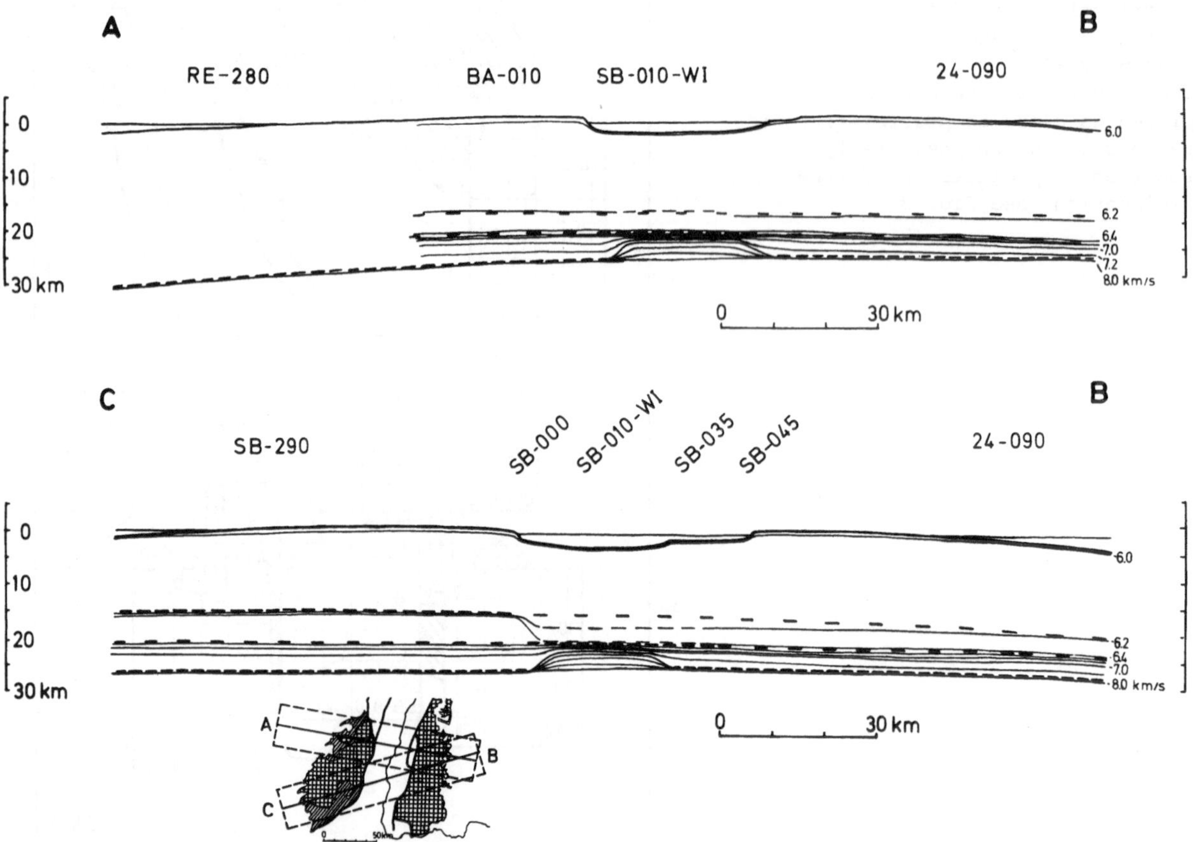

Fig. 14. Crustal sections through the southern part of the Rhine Graben by EDEL et al. (1975, Fig. 16). *Thin lines:* lines of equal velocity, contour interval 0.2 km/s. *Thick continuous line:* surface of the crystalline basement. *Thick dashed lines:* mean depth of the main crustal boundaries

lower depth forming a new zone of strongest velocity gradient at a depth of only 21 km. This zone has its continuation into the area outside the graben proper forming the border between middle and lower crust, but it is no longer the zone of strongest gradient.

6.8.9 Discussion and Comparison of the Different Models

The model presented by EDEL et al. (1975) shows a surprisingly good agreement with the first model published by ROTHÉ and PETERSCHMITT (1950) who found for the "granitic" layer beneath the Rhine Graben a velocity of 5.9-6.0 km/s with a lower boundary at 15 km depth. For the M-discontinuity they deduced a depth of 25 km leading again to almost identical results. Also the depth values reported by BEAUFILS (1967) are in fairly good agreement with the most recent observations.

GIESE and STEIN (1971) have published several contour maps compiling all data from seismic-refraction measurements in the Federal Republic of Germany up to 1968. Their map displaying the depth of the strongest velocity gradient between crust and mantle (GIESE and STEIN, 1971, Fig. 12) corresponds also relatively well with the depth to the crust/mantle boundary calculated by the Rhine Graben Research Group (1974) and EDEL et al. (1975) for the Rhine Graben area.

The crustal structures of the cited authors (ROTHÉ and PETERSCHMITT, 1950; BEAUFILS, 1967; GIESE and STEIN, 1971) are much less differentiated than the more recent model presented by ANSORGE et al. (1970, 1972), EMTER (1971), LAUER and PETERSCHMITT (1970), and MUELLER et al. (1967, 1969, 1973). Thus, a comparison with the newly developed structure by the Rhine Graben Research Group (1974) and EDEL et al. (1975) (Fig. 13) raises the question which aspects of previous detailed crustal concept (Figs. 6 and 7) can be confirmed and which have to be modified.

EDEL et al. (1975) discuss the ambiguity of the observations with regard to the existence or non-existence of a velocity inversion within the sialic upper crust. They demonstrate with an example that the observations available do not permit a definite conclusion, but are compatible with both interpretations. These results give rise to further discussions about the role of the low-velocity zone in the upper crust as part of the driving mechanism for graben formation (FUCHS, 1974; MUELLER and RYBACH, 1974).

The lower crust as defined by EDEL et al. (1975) is thicker than that shown by the previous interpretations (see, e.g., Figs. 6 and 7), its top lying at depths of 14 to 18 km in the models of Fig. 13. It is thus seen that the boundary between the upper and lower crust in this newly proposed model lies at about the same depth as the bottom of the low-velocity zone (see Figs. 6 and 7). The subdivision of the lower crust into two parts had already been shown in the crustal cross-section of Fig. 6. But when the section for the southern graben area (Fig. 7) was developed the data were not good enough to justify a similar detailed differentiation. However, comparing the correlated phases as well as the resulting depths and velocities, the top of the layer with 6.8-7.2 km/s (Fig. 13) by EDEL et al. (1975) corresponds to the layer with 6.7-6.9 km/s or 6.8 km/s, respectively (Figs. 6 and 7) by ANSORGE et al. (1970, 1972) and MUELLER et al. (1969, 1973). In agreement with the model shown in Fig. 6, in a more recent interpretation EDEL (1975, Fig. 16) concludes that the lower crust most probably contains a velocity inversion at about 20 km depth beneath the Vosges and the Black Forest which separates a middle and a lower crust.

As stated by the Rhine Graben Research Group (1974), the uplift of the crust/mantle boundary or the top of the "rift cushion", fairly well established in the interpretations prior to 1972, remains essentially unchanged. Also the general topography of this interface with its asymmetric steep slope to the west and northwest and its more gentle slope to the east was confirmed. However, the "rift cushion" hypothesis itself and, consequently, ideas regarding the function of this intermediate layer in the rifting process had to be changed considerably. The low velocity of 7.6-7.7 km/s in the "rift cushion" mentioned earlier was based mainly on the unreversed observations of apparent velocities along the profiles 16-130 and BA-010 (Fig. 3). Since the reversed line Steinbrunn-Wissembourg (SB-010-WI, WI-190-SB) now gave a true velocity of 8.0-8.1 km/s, any anomalous values of P_n-velocities in that area can now be explained in terms of an 8.0-8.1 km/s refractor with a varying topography.

As mentioned already in a previous section results of seismic reflection surveys were part of the information which was used in addition to the refraction observations. Since these data are of fairly good quality the new model had also to be checked against it. Of the velocity-depth distributions shown in Fig. 13 those of profiles LT-225, BA-010, and WI-190-SB are representative for the same Rhine Graben area between Rastatt and Kehl where the reflection histograms were obtained by DOHR (1967). The echoes around 7 s correspond to the strong velocity gradient in the lower crust at around 21 km depth. Reflections with echo times of 8.5 to 9.0 s can then be attributed to the shallow Mohorovičić discontinuity at about 25 km depth. The new crustal model of EDEL et al. (1975) leaves the echoes between 4 s and 5 s unexplained.

The lower boundary of the "rift cushion" in previous models for the area south of Karlsruhe (ANSORGE et al., 1970) was based on the observations of signal velocities of 8.1 km/s for the P_n-phase on the profile 16-130 beyond a distance of 210 km. Since this phase was interpreted now also as originating from the upper surface of the "cushion" (Rhine Graben Research Group, 1974) the existence of a lower boundary for the 8.1 km/s-layer is still an open question. There are some independent indications for further discontinuities of velocity below the uplifted crust/mantle boundary. DOHR (1970), during a special reflection survey in the central part of the Rhine Graben, observed echoes at 13.5 s and around 17.5 s which may indicate more or less abrupt velocity changes below the main interface. Recent long-range observations in France (HIRN et al., 1973) and similar observations by EMTER (1971) under the Swabian Jura about 80 to 100 km east of the graben axis have shown that there exist pronounced heterogeneities in the uppermost mantle. At present we cannot make definite statements about the deeper structure beneath the Mohorovičić discontinuity in the Rhine Graben area. Instead, the new situation has raised new questions. For example, there is no positive gravity anomaly associated with the uplift of the crust/mantle boundary. This fact poses an unsolved problem. Several possibilities to explain the absence of a gravity anomaly have been proposed by FUCHS (1974). In addition the possibility of a velocity anisotropy in the uppermost mantle should be considered as suggested by BAMFORD (1973, 5.3 of this volume).

6.8.10 Limits of the Seismic-Refraction Investigations in the Rhine Graben Area

As Fig. 3 illustrates, a great number of seismic-refraction profiles has been observed in the Rhine Graben and its surrounding. In spite of this great number of data the evaluation in terms of crustal structure turned out to be very difficult and was not always very satisfactory. WARREN et al. (1973) discuss in detail this general problem in connection with the interpretation of explosion seismology data obtained in the investigation of the crustal structure under the Large-Aperture Seismic Array (LASA) in Montana. Some of their discussions and conclusions can be equally applied to the Rhine Graben data.

Although characteristic phases can be mapped, the seismograms show inconsistent occurrences and changes in character. In 25 record sections prepared from the data and presented by EDEL et al. (1975), it is difficult to find two that look alike. These differences suggest a fine structure that is beyond the resolving power of the methods which can be used and which require certain assumptions not being true in all cases.

A number of reasons for these difficulties is evident and is discussed by EDEL et al. (1975). On the one hand, the geologic and tectonic structure of the area of investigation is extremely complex. Three main tectonic events have influenced the area: the Variscan orogeny mainly in NE-SW direction, the Alpine orogeny causing a general uplift and block faulting of the Variscan region, and the formation of the graben. In consequence the sedimentary cover shows a rather variable thickness: It reaches 6000 m in the "Saar-Nahe-Trog" in the NW and 2000-4000 m in the Rhine Graben, while in other areas (Vosges, Black Forest, Odenwald) the crystalline basement is exposed.

On the other hand, because of the high costs of seismic experiments, special explosions could be organized only in a few cases. In most other cases other energy sources, usually quarry blasts, had to be used. Therefore, the positions of the profiles were not well chosen with regard to the geologic and tectonic setting of the area, but more often were determined by the accidental location of suitable quarries.

The complexity of the seismic data obtainable may be a fundamental property of the crust that explains the contradiction between the great variation of structure and rock types shown on a detailed geologic map and the simple layered crustal structures shown in crustal sections. The rocks of the earth's crust tend to separate into zones or layers having similar average physical properties, but within each layer there are rocks with a wide range of properties and a variety of structure. Seismic waves sense these average properties that are difficult to obtain from geologic data. The fact that these averages can be stable and therefore meaningful is clearly demonstrated by the analysis of the P^M-data, e.g.: the corresponding boundary can be mapped with fair accuracy as demonstrated also by the coincidence of different interpretations shown in this paper. Nevertheless, the changes in character of the wide-angle reflections from one profile to the other indicate that the real model is a more complex one than that shown in the crustal sections.

Acknowledgments. The research of the 1972 seismic-refraction experiment in the southern Rhine Graben was a joint project of French, German, and Swiss geophysical institutes. It was enabled by the financial support of the Institut National d'Astronomie et Géophysique of France, the German Research Society, and the Fond National Swiss de la Recherche Scientifique.

6.9 P-Wave Velocities in the Northern Calcareous Alps Derived from Observations of Explosions in the Eschenlohe Quarry

G. SCHELIGA

ABSTRACT

The first-arrival data of six seismic-refraction profiles radiating from the Eschenlohe quarry into Northern Calcareous Alps are used to derive a contour map of the quantity x/t in steps of 0.1 km/s. An average velocity-depth function yielding a maximum depth of penetration of 5.5 km with a velocity of 6.6. km/s is compared with data from other investigations.

ZUSAMMENFASSUNG

Die ersten Einsätze von sechs refraktionsseismischen Profilen in den nördlichen Kalkalpen werden interpretiert und eine Karte mit Linien gleicher Werte x/t in Schritten von 0,1 km/s wird entworfen. Eine aus diesen Daten berechnete mittlere Geschwindigkeits-Tiefen-Funktion erreicht in einer maximalen Eindringtiefe von 5,5 km eine Geschwindigkeit von 6,6 km/s. Das Ergebnis wird mit anderen Untersuchungen verglichen.

Within the programs "The Deeper Crustal Structure of Central Europe" and "Upper Mantle", refraction measurements were carried out in the Northern Calcareous Alps. Using the explosions in the Eschenlohe quarry, six profiles radiating from this quarry were laid out in the Eastern Alps (Fig. 1).

The tectonic structure of the Northern Alps is very complicated and still partially open to discussion. Today it is generally accepted and proved (boreholes, windows) that the Flysch, Helvetic, and Subalpine sediments have been overridden for at least 10 to 20 km by the nappes of upper Austro-

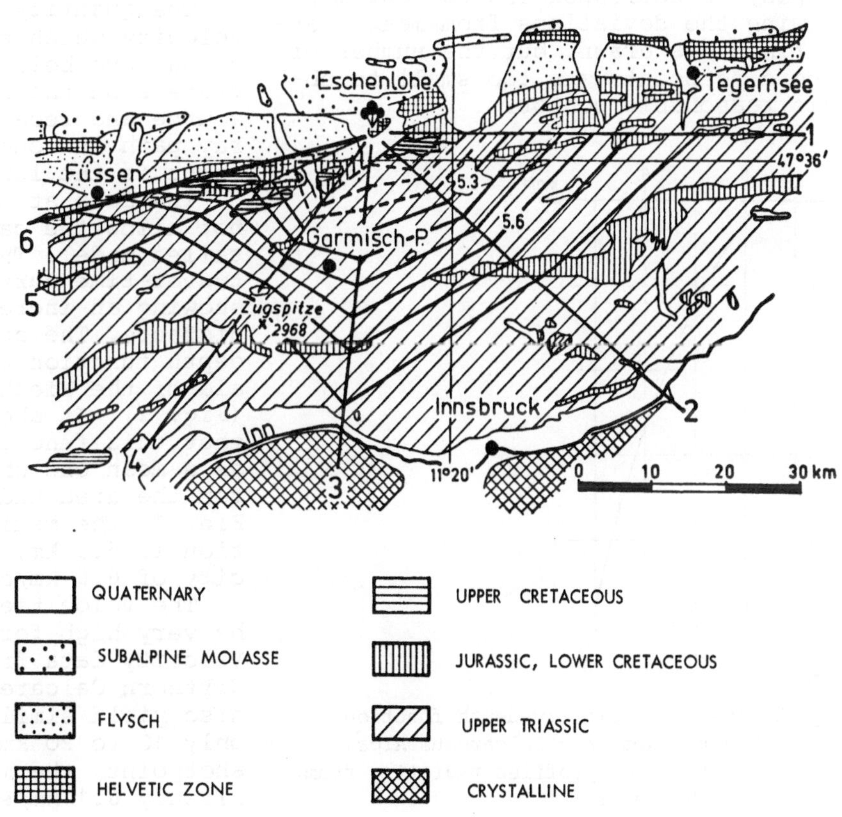

Fig. 1. Simplified geological map, position of refraction profiles and lines for x/t = constant in km/s

alpine units mainly composed of carbonate rocks with a total thickness of about 4000 m.

From experience it may be stated that the P-wave velocity in the Molasse is less than that in the Northern Calcareous Alps. From this model, a low-velocity layer in the sedimentary cover of the Variscan basement of the Northern Alps is to be expected. A low-velocity layer in the deeper basement starting at 10 km depth is well established by several authors (GIESE, 1968a; ANGENHEISTER et al., 1972).

This paper deals with the velocity distribution in the sedimentary cover in the Northern Alps.

For the object under study, only the first part of the record sections are of interest. As shown in Fig. 1, the first travel-time curves break off at 50 km distance due to vanishing arrivals. Thus a total of 58 recorded traces was available for the area covered by the profiles, that means on an average one station per 50 km^2. This low density of stations does not allow a detailed interpretation. Some attempts have been made to find out a suitable display of the data obtained.

First, a contour map of equal travel times was drawn, but the deviation from circles was not significant. In order to enlarge the resolution of display, a corresponding contour map showing the deviations from mean travel times was drawn, but the number of recording points was too small to obtain reasonable contours.

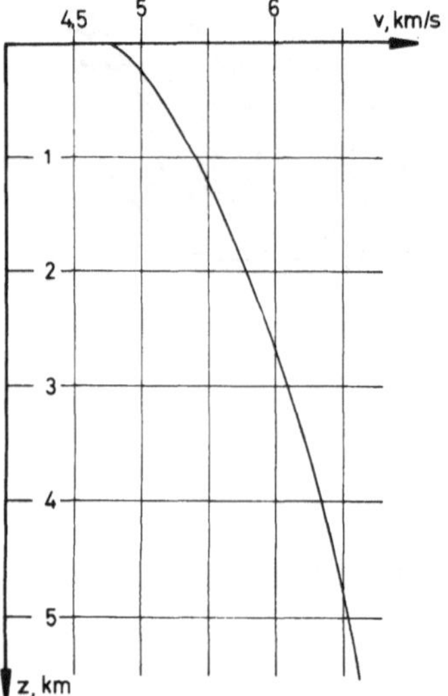

Fig. 2. The mean velocity-depth function for P-waves in the Northern Calcareous Alps, based on refraction profiles radiating from the Eschenlohe quarry

Then the observed travel-time curves were approximated individually by a polynomial of third degree to eliminate small deviations:

$$t = ax + bx^2 + cx^3$$

t: travel time
x: distance from shotpoint
a, b, c: constants which must be determined by least-squares approximation.

From the equation obtained by this procedure, the quotient x/t was calculated along the profiles.

Fig. 1 represents a contour map of equal x/t in steps of 0.1 km/s, showing that in southern direction relatively high velocity values (5.8 km/s) occur in rather short distances, so demonstrating lateral imhomogeneities of the velocity distribution in the uppermost crust. The dashed contour lines near the shotpoints are not well established due to rare data.

Furthermore it must be mentioned that all six travel-time curves show a point of inflexion at about 28 km distance, probably caused by the special geological situation of the quarry Eschenlohe. As shown in Fig. 1, the Eschenlohe quarry is located in the Helvetic zone outside the Northern Calcareous Alps, probably causing the complicated travel-time curves.

The quantity x/t represents a velocity which can be interpreted as a mean one belonging to a ray that travels on the shortest way between source and receiver. Because the real ray path is longer, this velocity value is the lowest possible one.

The data at present available do not allow the calculation of a detailed model. Therefore, an average travel-time curve was determined showing an increase of velocity with distance. The corresponding velocity-depth function was calculated by the aid of the Wiechert-Herglotz method (Fig. 2). It should be pointed out that this function must be regarded as a mean one which is representative for the area under study as shown in Fig. 1. The maximum depth of penetration is 5.5 km, here reading a velocity of 6.6 km/s.

The velocities obtained seem to be very high for sedimentary layers. Velocity data from other parts of the Northern Calcareous Alps, however, also yield similar high values. At only 10 to 20 km distance from the shotpoint, the mean velocity x/t is already 6.1 km/s, whereas for the true velocity the value of 6.5 km/s was found.

In some of the record sections here studied, later arrivals can be seen which may be interpreted as a reflection. Applying the t^2, x^2 method and taking into account the velocity-depth function of Fig. 2, a depth value of 10 km results. Then, the mean velocity for the depth range between 5.5 and 10 km is 5.5 km/s. It is evident that a smaller mean velocity for this zone lowers the depth. With a velocity of 4.3 km/s, a depth of 9 km is obtained. ZSCHAU and KOSCHYK (6.10 of this volume) have found similar depths for a reflection horizon by their interpretation of S-waves observed on the profile Eschenlohe-E (profile 01-090). The velocity value of 4.3 km/s was revealed in the Molasse at 2.5 km depth in the well Urmannsau I by well velocity survey (KRÖLL and WESSELY, 1967). This well is situated in the Northern Calcareous Alps E of Salzburg. Here, Molasse was found under limestones and dolomites on the Austroalpine units.

It is still open whether the maximum depth of 5.5 km as indicated by the first travel-time branch is identical with average thickness of the Calcareous sediments of the Northern Alps.

Further, there is a discrepancy concerning the depth of the basement under the Northern Alps. Whereas PRODEHL (1964) obtained a thickness of 5.7 km for the sedimentary cover at the northern border of the Alps, the interpretation of later arrivals yielded a value of 9-10 km for the depth of the basement. Further measurements should be carried out in order to resolve the problem.

6.10 Results of a Combined Evaluation of Longitudinal and Transverse Waves on a Seismic Profile along the Northern Margin of the Alps

J. ZSCHAU and K. KOSCHYK

ABSTRACT

Travel-time curves and velocity-depth functions of longitudinal and transverse waves on the profile Eschenlohe-E are discussed. A comparison is made between the velocity-depth functions of P- and S-waves. From the velocity of both wave types, Poisson's ratio, the ratio bulk modulus to shear modulus and the ratio bulk modulus to density are calculated as a function of depth.

ZUSAMMENFASSUNG

Laufzeitkurven und Zusammenhänge zwischen Geschwindigkeit und Tiefe für longitudinale und transversale Wellen werden für das Profil Eschenlohe-E diskutiert. Es wird ein Vergleich der Beziehung zwischen Geschwindigkeit und Tiefe für P- und S-Wellen angestellt. Aus den Geschwindigkeiten beider Wellentypen wird die Poissonsche Zahl, das Verhältnis Kompressionsmodul/Schubmodul und das Verhältnis Kompressionsmodul/Dichte in Abhängigkeit von der Tiefe berechnet.

6.10.1 Introduction

The present paper deals with a refraction survey along the profile Eschenlohe-E. This profile is 340 km long, beginning about 60 km SW of München at Eschenlohe and passing almost due east through the Northern Limestone Alps. Its direction therefore coincides with the strike of this alpine zone.

In analyzing the seismic records of this profile, travel times of transverse waves (S-waves) as well as those of longitudinal waves (P-waves) have been interpreted. Knowing the velocity-depth function for both kinds of seismic waves, one can obtain information on elastic parameters of rocks, these parameters being independent of density.

6.10.2 Interpretation of the P-wave Arrivals

6.10.2.1 Time-Distance Graph

The record section for longitudinal waves along the profile Eschenlohe-E is shown in the lower part of Fig. 1. As one can see from the correlation of the P-wave arrivals, the time-distance graph fits in the scheme of travel-time curves, which has been found for the Alps and its northern foreland (GIESE, 1968a). The well-marked later arrivals between 0 and 50 km distance are the only exception from this typical arrangement of the curves. They are considered to belong to a reflected wave, because the corresponding travel-time curve is almost a hyperbola. The first onsets in the seismograms between 0 and 52 km distance lie on a continuous curve a. Between 95 and 234 km they lie on a curve c (P^M). The fact that the a-curve does not meet the c-curve, indicates a low-velocity layer. A P_n-wave could not be detected definitely and therefore was not evaluated.

6.10.2.2 Velocity-Depth Function

The velocity-depth function has been computed by the Wiechert-Herglotz method. This method is applicable only for a continuous travel-time curve, and if the apparent velocity dX/dT increases monotonously along this curve. In the present case the first condition was not satisfied because of the zone with low velocities in the earth's crust. The corresponding gap in the travel-time curve could, however, be closed by calculating suitable models for the low-velocity layer and eliminating its effects on the observed travel-time curve.

Fig. 2 shows the velocity-depth functions obtained by the above method. The different velocity functions correspond to different models for the low-velocity layer. The velocity in-

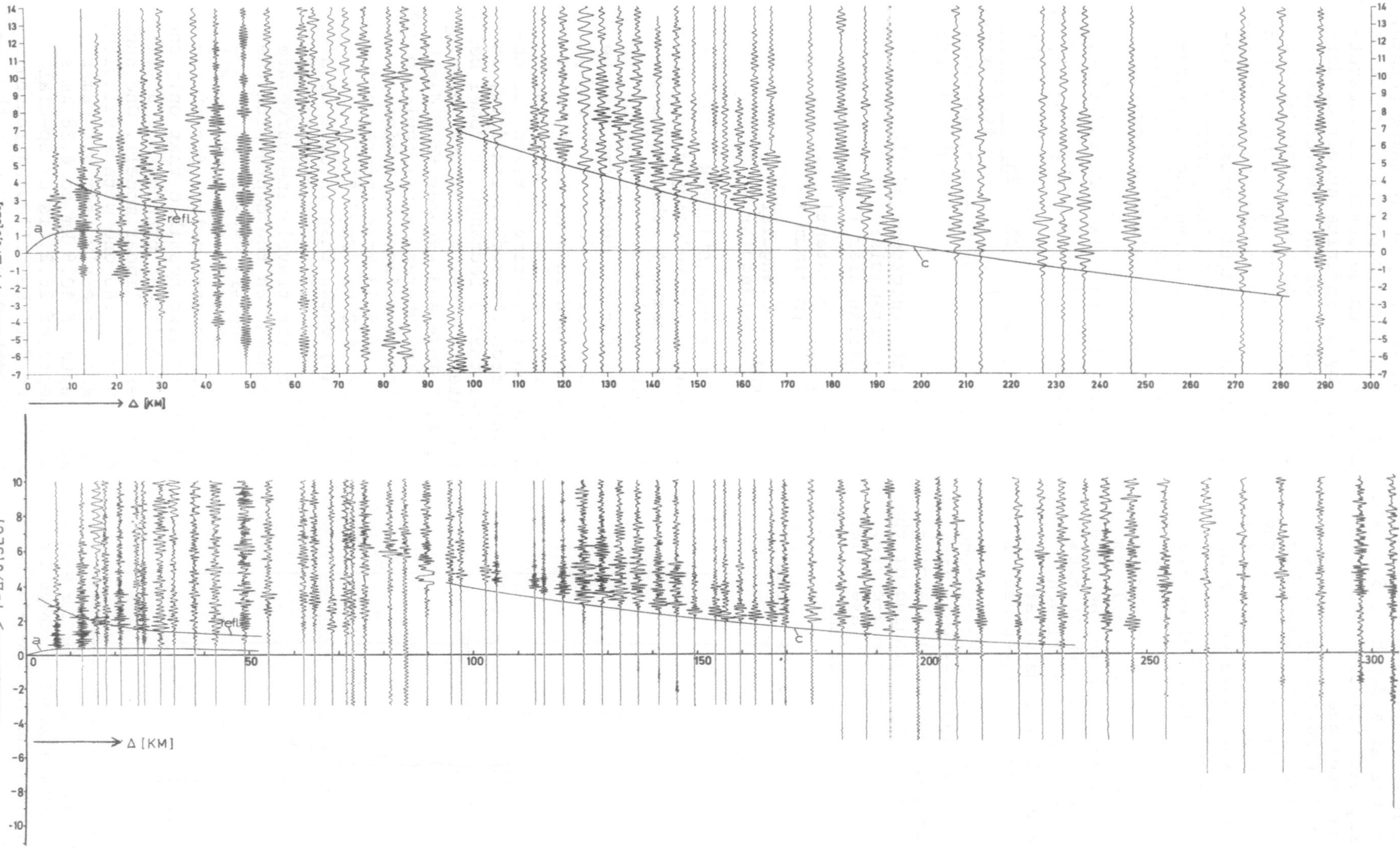

Fig. 1. Record section of the profile Eschenlohe-E (01-090), reduced with 3.5 km/s (*upper part*) and 6.0 km/s (*lower part*)

creases from 4.3 km/s at the surface up to 6.38 km/s at a depth of 5.75 km. These fairly high velocities are attributed to limestone.

The only information on the interior of the low-velocity layer has been obtained from the onsets of the reflections between X = 0 km and X = 50 km. These onsets have been evaluated using the $X^2 - T^2$-method. It follows that there is a reflecting horizon at a depth of 10.5 km and that the average velocity above this interface up to a depth of 5.75 km is 5.64 km/s. The material in the zone with this average velocity of 5.64 km/s is considered to be Flysch, Molasse, and Helvetic material, which underly the limestone layer near the surface.

For the entire low-velocity zone three parameters are available from the travel-time curve: (1) the distance ΔX between the beginning of the c-curve and the end of the a-curve; (2) the difference ΔT in travel time between these two points; and (3) the apparent velocity v_l at these points, being equal for both. These three parameters can be satisfied theoretically by an infinite number of different models of the low-velocity zone. In case of a constant velocity distribution (rectangular model) and in case of a triangular velocity distribution (triangular model) the relationship between the parameters of the travel-time curve and the parameters of the zone with low velocities is given by the following equations:

$$\left.\begin{array}{l}\text{rectangular}\\ \text{model}\end{array}\right\} : \quad \begin{array}{l}\Delta X = 2h_r v_r / \sqrt{v_l^2 - v_r^2}\\ \Delta T = 2h_r v_l / (v_r \sqrt{v_l^2 - v_r^2})\end{array} \tag{1}$$

$$\left.\begin{array}{l}\text{triangular}\\ \text{model}\end{array}\right\} : \quad \begin{array}{l}\Delta X = \dfrac{2h_t}{\sqrt{(v_l + v_t)/(v_l - v_t)}}\\ \Delta T = (2h_t/(v_l - v_t)) \cosh^{-1} (v_l/v_t)\end{array} \tag{1}$$

$\left.\begin{array}{l}h_r\\ h_t\end{array}\right\}$: the thickness of the $\left\{\begin{array}{l}\text{rectangular}\\ \text{triangular}\end{array}\right\}$ model

$\left.\begin{array}{l}v_r\\ v_t\end{array}\right\}$: the $\left\{\begin{array}{l}\text{constant}\\ \text{minimum}\end{array}\right\}$ velocity in the $\left\{\begin{array}{l}\text{rectangular}\\ \text{triangular}\end{array}\right\}$ model

It can be shown (SLICHTER, 1932) that the rectangular velocity distribution yields a maximum thickness for the low-velocity layer.

The different models can also be combined as it is done in this paper. By Eq. (1) and in consideration of the rectangular low-velocity layer above the reflecting horizon, as obtained from the reflection data, it follows that a rectangular model below the interface has a constant velocity of 6.23 km/s and extends to a depth of 28.7 km, whereas a triangular model below the interface has a minimum velocity of 5.93 km/s and reaches a depth of 26.2 km (Fig. 2).

Eq. (1) shows that in case of a triangular model the parameters ΔX and ΔT of the travel-time curve are not dependent on the depth with the minimum velocity. Hence, there is an infinite number of triangular models, yielding the same values for ΔX and ΔT. It is also possible that only one of the two linear branches of a triangle exists (linear model). One such case, namely the one with a velocity, increasing linearly from 5.93 km/s at a depth of 10.5 km to 6.38 km/s at a depth of 26.2, is of special interest for the following reason: the velocity of 5.93 km/s at the re-

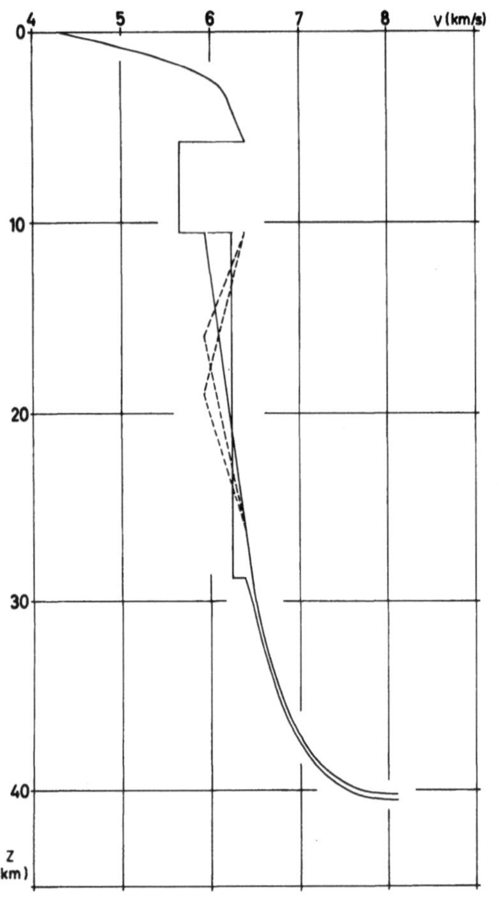

Fig. 2. Velocity-depth functions for P-waves on the profile Eschenlohe-E, corresponding to different models for the low-velocity layer

flecting interface corresponds with velocities found for the crystalline stratum at the northern margin of the Alps (PRODEHL, 1964). Hence, the reflecting horizon at a depth of 10.5 km may be the crystalline basement.

It is evident that a velocity distribution obtained for the lower earth's crust depends on the model chosen for the middle crust. However, the influence is not strong in the present case, as Fig. 2 demonstrates. For all velocity distributions plotted in Fig. 2 the velocity increases from 6.38 km/s immediately below the low-velocity layer up to 8.1 km/s at a depth of about 40.2 km.

Fig. 3 shows velocity-depth functions corresponding to different models for the low-velocity layer and to the following four different correlations of the P-wave arrivals: correlation by GIESE (1970) in 1967 (dashed-dotted line), two correlations by KOSCHYK (1969), one taking a b-curve into consideration (solid lines) and the other without a b-curve (dashed lines), correlation used in this paper. It can be seen that the influence of correlation errors on the velocity distribution is stronger than the influence of different models for the zone with low velocities. However, in all the curves plotted in Fig. 3 the velocity of 8.1 km/s is reached between 40 and 42 km. Above the low-velocity layer the velocity distribution found by GIESE in 1967 differs rather strongly from the others. This may result from the fact, that in 1967 a refraction survey along the first 70 km had not yet been carried out. Thus GIESE obtained the corresponding part of the velocity-depth function by extrapolating the results of other profiles.

6.10.3 Interpretation of the S-wave Arrivals

6.10.3.1 Time-Distance Graph

The time-distance graph for transverse waves (upper part of Fig. 1) is essentially similar to the one for longitudinal waves. At small distances from the shotpoint the first arrivals of the S-wave lie on a curve a, at larger distances on a curve c (S^M). The a-curve for S-waves is shorter than the a-curve for P-waves. It already ends at a distance of about 33 km. The c-curve for S-waves on the other hand is longer than the c-curve for P-waves. It extends from 97 km to about 280 km. Because the a-curve does not meet the c-curve, a low-velocity layer for S-waves is indicated. A S_n-wave was not detected. The arrivals of a reflected wave could be correlated up to a distance of 40 km.

6.10.3.2 Velocity-Depth Function

The velocity distribution v(z) has been calculated by means of polygonal models for v(z). For calculating the corresponding travel-time curve, the following equations have been used:

For $v(z_i) = v(z_{i-1})$:

$$X_i^s = 2(z_i - z_{i-1})\left[v(z_i)/\sqrt{v^2(z_s) - v^2(z_i)}\right]$$

$$T_i^s = X_i^s\left[v(z_s)/v^2(z_i)\right]$$

$$z_o < z_i < z_s,\ v(z_i) < v(z_s)$$

For $v(z_i) = v(z_{i-1})$: (2)

$$X_i^s = 2(z_i - z_{i-1})\left[\left(\sqrt{v^2(z_s) - v^2(z_{i-1})} - \sqrt{v^2(z_s) - v^2(z_i)}\right)/(v(z_i) - v(z_{i-1}))\right]$$

$$T_i^s = 2(z_i - z_{i-1})\left[\left(\cosh^{-1}(v(z_s)/v(z_{i-1})) - \cosh^{-1}(v(z_s)/v(z_i))\right)/(v(z_i) - v(z_{i-1}))\right]$$

$$X^s = \sum_{i=1}^{s} X_i^s,$$

$$T^s = \sum_{i=1}^{s} T_i^s$$

$$z_o < z_i \le z_s,\ v(z_i) \le v(z_s)$$

z_i : depth up to the lower margin of the i-th layer with a linear velocity distribution

z_s : depth of the deepest point of a ray

$v(z_i)$, $v(z_s)$: velocity belonging to the depth z_i, z_s

X_i^s : horizontal distance traversed in the i-th layer along the ray with the deepest point in z_s

T_i^s : travel time of a wave in the i-th layer along the ray with the deepest point in z_s

X^s : horizontal distance traversed along the whole ray with the deepest point in z_s

T^s : travel time of a wave along the whole ray with the deepest point in z_s

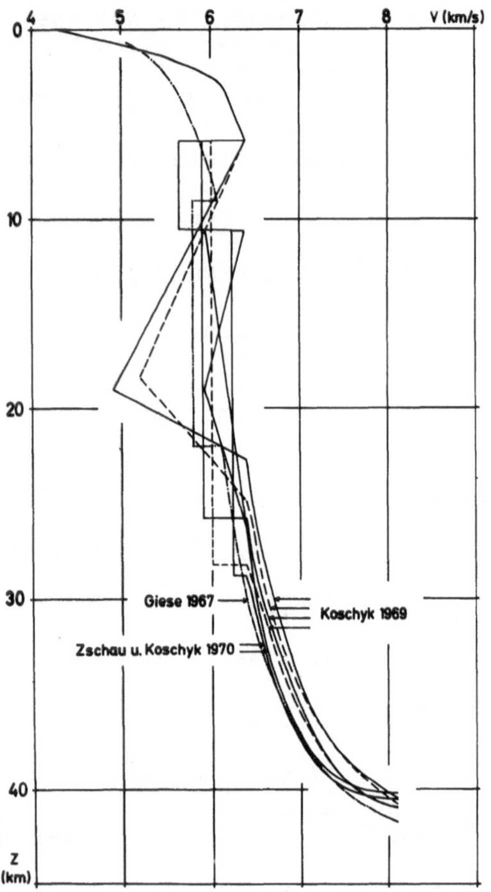

Fig. 3. Velocity-depth functions for P-waves on the profile Eschenlohe-E, corresponding to different correlations and different models for the low-velocity layer

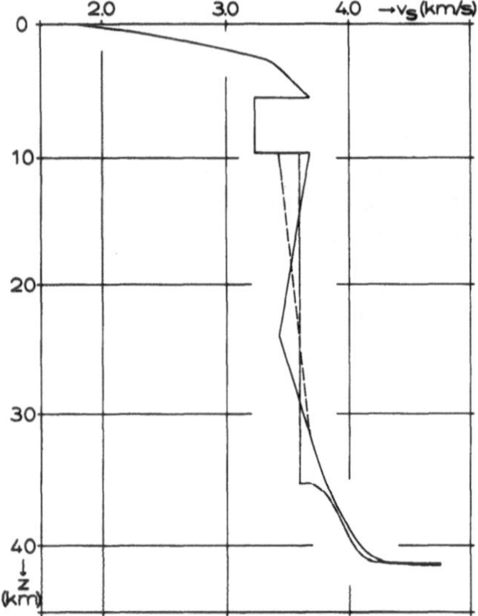

Fig. 4. Velocity-depth functions for S-waves on the profile Eschenlohe-E, corresponding to different models for the low-velocity layer

The number of 20 sections for the polygon v(z) has been adequate to obtain a travel-time curve which is almost the same as the observed one. Models for the low-velocity layer have been calculated for S-waves in the same way as for P-waves: the reflected wave has been taken into account and a combination of rectangular and triangular models has been chosen.

Fig. 4 shows the velocity-depth funtions obtained in the way described above, but smoothed later on. The velocity at the surface has a low value of 1.80 km/s. It increases rather rapidly up to 3.68 km/s at a depth of 5.6 km. The low-velocity layer follows beneath. By evaluating the arrivals of the reflected wave by the $X^2 - T^2$-method, a horizon was found in this layer at a depth of 9.7 km. The difference between this value and the one obtained from longitudinal waves is about 850 m. This may be a result of errors in correlation. Between 5.6 and 9.7 km in depth the average velocity was found to be 3.23 km/s. The velocity of the rectangular model below the reflecting interface is 3.59 km/s, the minimum velocity of the triangular model 3.42 km/s. The rectangular model extends to 35.4 km in depth, the triangular model to a depth of 31.8 km. In the lower crust the velocity distributions, corresponding to the different models, do not differ greatly. For all models presented in Fig. 4 the velocity increases strongly at about a depth of 41 km and reaches the value of 4.75 km/s at about 41.5 km.

6.10.4 Conclusions from the Velocity Distributions of Both Kinds of Waves

For facilitating a comparison between the velocity-depth functions of both wave types, the velocities of P-waves have been devided by $\sqrt{3}$ (Fig. 5). The value $\sqrt{3}$ has been chosen, because most solid bodies have a ratio v_P/v_S of P-wave to S-wave velocities of approximately $\sqrt{3}$. Fig. 5 shows that after thus normalizing the P-wave velocities, both velocity distributions are quite similar immediately below the reflecting interface. Near the surface and in deeper ranges of the earth's crust the values of $v_P/\sqrt{3}$ are greater than those of v_S. The low-velocity layer for transverse waves extends to a greater depth than for longitudinal waves. The constant S-wave velocity in the rectangular model and the minimum S-wave velocity in the triangular models are equal to the corresponding normalized P-wave velocities. It is apparent that velocities, which are characteristic of the uppermost mantle, are reached approx. 1.3 km deeper for S-waves than for P-waves. This may be connected with correlation errors.

For characterizing the elastic properties of rock, the ratio k/ρ of the bulk modulus k to the density ρ is often used. One can calculate this ratio from the velocities of the longitudinal and the transverse waves by the following equation:

$$k/\rho = v_P^2-(4/3)v_S^2 = (k/\mu)v_S^2 \qquad (3)$$

The depth-dependence of the ratio k/ρ determined from the velocity-depth functions of the P- and S-waves is shown in the right part of Fig. 6. This ratio increases from 14.1 km^2/s^2 at the surface to 23.4 km^2/s^2 at a depth of 4 km. In the interior of the low-velocity layer the values for k/ρ are between 17.9 km^2/s^2 and 24.3 km^2/s^2. The low-velocity layer is a zone with low (k/ρ)-values. Below this layer the ratio k/ρ increases up to 37.4 km^2/s^2 at a depth of 40 km.

A measure of the differences in the velocity distributions of both wave kinds is given by the ratio v_S/v_P. As Eq. (4) shows, this ratio does not depend on the density ρ of the material transversed by the waves. It is only dependent on the elastic properties of this material.

$$v_P = \sqrt{(k+(4/3)\mu)/\rho}\,, \quad v_S=\sqrt{\mu/\rho} \qquad (4)$$
$$v_P/v_S = \sqrt{k/\mu+4/3} = \sqrt{2(1-\sigma)/(1-2\sigma)}$$

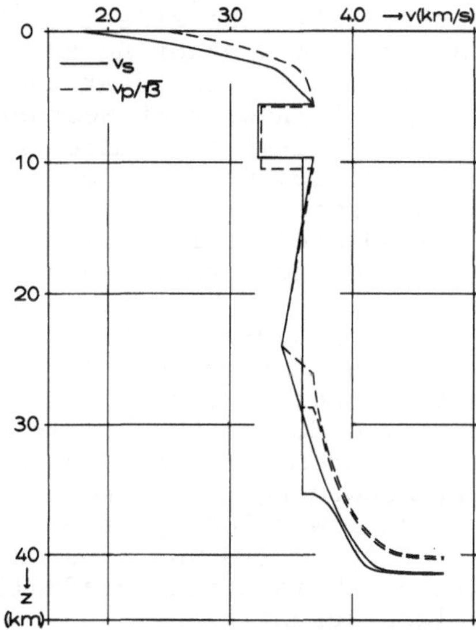

Fig. 5. Variation of S-wave and normalized P-wave velocities with depth, corresponding to a rectangular and a triangular model for the low-velocity layer

k: bulk modulus
μ: shear modulus
ρ: density
σ: Piosson's ratio

The relationship between the (v_S/v_P)-ratio and the depth is given in the left part of Fig. 6. Poisson's ratio σ and the ratio k/μ of bulk to shear modulus, corresponding to a (v_S/v_P)-value, may be taken from two scales below the (v_S/v_P)-graph. The ratio $(v_S/v_P)\sqrt{3}$ increases from approx. 0.73 at the surface to 1 at a depth of 5.6 km. This corresponds to a decrease in Poisson's ratio σ from approx. 0.39 to 0.25 and a decrease in the (k/μ)-ratio from 4.35 to 1.67

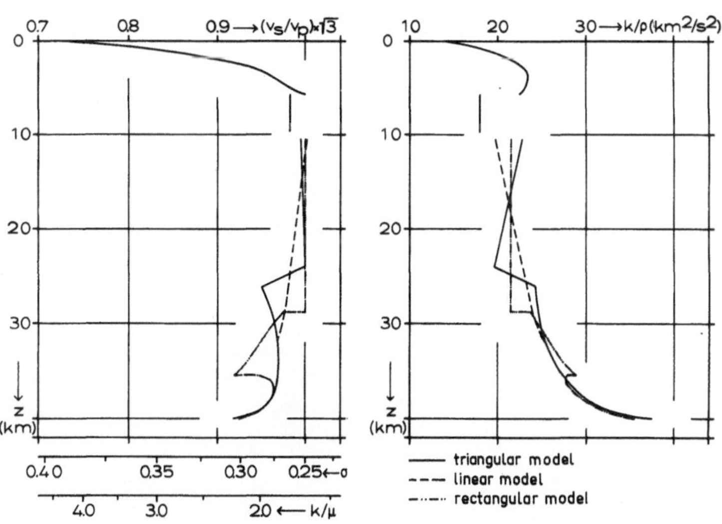

Fig. 6. Variation of some elastic properties of the crust with depth, deduced from the velocity distributions of both longitudinal and transverse waves. The data have been computed for the following three different velocity models for the low-velocity layer: *a* triangular model with $z(v_t)$ = 24 km; *b* linear model with $z(v_t)$ = 10.5 km for P-waves and $z(v_t)$ = 9.7 km for S-waves; *c* rectangular model

respectively. The effect most likely results from weathering and the great pore volume of rock near the surface. It is well known that weathering and pore volume have more influence on the rigidity of a rock than on its incompressibility.

Immediately below the reflecting interface the normalized velocity ratio $(v_S/v_P)\sqrt{3}$ has values near 1. At greater depths this ratio decreases by maximally 8%, whereas Poisson's ratio σ increases from 0.25 up to approx. 0.30 and the (k/μ)-ratio increases from 1.67 up to a maximum of 2.2. According to the model chosen for the low-velocity layer, this variation of $(v_S/v_P)\sqrt{3}$, σ and k/μ with depth is either gradually or more or less abruptly.

As ultrasonic investigations showed Poisson's ratio is rather high for basic and ultrabasic rock (VOLAROVICH, 1965). HUGHES and MAURETTE (1957a) have measured the variation of seismic velocities with pressure and temperature in three different gabbros. Poisson's ratios were between 0.28 and 0.33. On the other hand Poisson's ratio found for five different granites were smaller. They lay between 0.22 and 0.30 (HUGHES and MAURETTE, 1956). Thus, the observation that the velocity ratio (v_S/v_P) decreases with depth and according to this Poisson's ratio and the (k/μ)-ratio increase with depth, can be explained by the fraction of basic and ultrabasic rock in the earth's crust, which increases with depth. The fact that the low-velocity layer for S-waves ends at a greater depth than for P-waves, can be explained likewise, because this fact can be a result of Poisson's ratio, increasing with depth.

From Fig. 6 it may also be seen that the low-velocity layer in the middle crust is not at all a layer of abnormal Poisson's ratio. This is a strong argument against the hypothesis that high temperature or even partial melting in the presence of water in the middle crust is responsible for the low velocities in this range of depth. Temperature near the melting point as well as partial melting have a greater effect on the S-wave- than on the P-wave velocities, and therefore should increase Poisson's ratio. Of course, this conclusion regarding the nature of the low-velocity layer in the area studied cannot be generalized ad hoc. It is of interest, therefore, to make the same type of investigation of the low-velocity layer in other regions.

Acknowledgment. The authors are grateful to H. Gebrande (München) for many helpful discussions.

6.11 Crustal Structure in Southeastern Bavaria Derived from Seismic-Refraction Measurements by Ray-Tracing Methods

H. MILLER and H. GEBRANDE

ABSTRACT

In 1970 refraction seismic measurements were carried out along the International profile VII. Shotpoints were situated in Czechoslovakia, but shots were also observed between the Bohemian Massif and the northern Limestone Alps. Two slightly different models for the crustal structure along this profile are presented. They were obtained by model calculations using ray-tracing techniques. Additional data from boreholes and deep reflections faciliated interpretation and helped in reducing the ambiguity of models.

Both models show characteristic differences in crustal structure along the profile: total thickness of the crust underneath the Molasse basin is smaller, whereas the thickness of the lower crust (with p-wave velocities between 6.4 and 8 km/s) is larger than underneath the Bohemian Massif.

ZUSAMMENFASSUNG

1970 wurden refraktionsseismische Messungen auf dem Internationalen Profil VII durchgeführt. Die Sprengungen, die in der Tschechoslovakei durchgeführt wurden, wurden auch zwischen der Böhmischen Masse und den nördlichen Kalkalpen beobachtet. Zwei nur wenig unterschiedliche Modelle wurden nach zwei verschiedenen Verfahren der Modellrechnung für zweidimensional inhomogene Medien berechnet. Die Berücksichtigung von Bohrergebnissen und von Laufzeiten von Tiefenreflexionen ermöglichte es, die Vieldeutigkeit der Modelle einzuschränken.

Beide Modelle lassen charakteristische Unterschiede der Krustenstruktur längs des Profils erkennen: Die Gesamtmächtigkeit der Kruste ist unterhalb des Molasse-Beckens geringer, die Mächtigkeit der Unterkruste (v_p etwa zwischen 6,4 und 8 km/sec) hingegen größer als unter der Böhmischen Masse.

6.11.1 Introduction

The International Profile VII extends from the northern border of Poland approximately 40 km south of Kaliningrad to the SW. It crosses Poland, Czechoslovakia, the southeastern part of Bavaria and ends in the Northern Limestone Alps of Austria (Fig. 1). Measurements along this profile and its localization had been proposed at a meeting of the European Seismological Commission in Leningrad and by the Working Group on Generalization of Explosion Seismology Data (BERANEK et al., 1973).

In this paper, results from observations along the German and Austrian part of the profile and from shotpoint No. 1 only will be presented. Starting in the Bohemian Massif, the profile crosses the Molasse basin and reaches the Northern Limestone Alps

Table 1. Technical data for the shots at the shotpoint 1. Coordinates: $\lambda = 13^{o}\ 51.70'$; $\phi = 49^{o}\ 2.88'$

Shot no.		Date	Time	Charge (kg)	Holes Number	Depth (m)
11	A	8.6.70	20 00 04.2	500	3	30
12	B	10.6.70	20 00 03.43	600	3	35
13	C	18.6.70	22 00 02.89	600	3	35
14	D	22.6.70	22 00 03.26	800	4	35
15	E	24.6.70	22 00 02.59	800	4	35
16	F	16.7.70	22 00 02.94	800	4	40
17	G	20.7.70	22 00 01.98	800	4	40
18	H	22.7.70	22 00 02.99	800	6	30
19	I	5.8.70	22 00 02.67	730	6	30

south of the Chiemsee, turning slightly to the West there.

6.11.2 Measurements

Measurements for shotpoint 1 were carried out during June, July and August 1970. Table 1 gives a list of shots together with their respective technical data after HOLUB (1972). Special care was taken in choosing quiet recording sites, although freedom in choosing was limited by an unusually close spacing of stations. Distance between geophones generally was one km or less, except for distances greater than 200 km. As shots were rather late in the evening, ground noise was relatively low and if it so happened that a recording proved to be of poor quality it was repeated at a later date.

Instruments used were the standard MARS 66 stations (BERCKHEMER, 1970) with three vertical component seismometers spaced approximately 500 m apart along the profile. All instruments were calibrated to the same standard. Records were played back using an analog band-pass filter to eliminate ground noise. The pass band generally was 4-20 Hz. As the dominating frequency of the signals lies between 12 and 16 Hz, they are centered very well in this pass band. Records were obtained on a calibrated 4-channel recorder with a time scale of 4 cm/s. From these records a seismogram section was assembled on a reduced scale (reduction velocity being 6 km/s). In order to facilitate phase correlation, care was exercised in selecting the seismograms incorporated into the sections. Due to the close spacing of stations many more seismograms were available than could be reasonably plotted. Therefore only the best and most "similar" looking seismograms were plotted. These record sections are shown on maps inside the back cover, each covering 50 km of profile. Seismograms are numbered consecutively, every tenth seismogram being numbered. The letters below each seismogram identify the shot, identical letters indicating that the respective seismograms were obtained from the same shot (see Table 1 for charge weight of each shot). Immediately below the seismograms an amplitude A* is given on a logarithmic scale. This amplitude A* is a normative correction factor which includes station calibration and amplification during recording and playback. Amplification during playback had been chosen such that the maximum amplitude (peak to peak) in each seismogram was approximately 2 cm (or 1 km in the scale of the abszissa). This gives a rather uniform appearance and facilitates phase correlation. To obtain the true amplitude of the velocity of ground motion in μ/s at a particular geophone site, the amplitudes of the seismograms measured in units of the abscissa of the seismogram section have to be multiplied by the appropri-

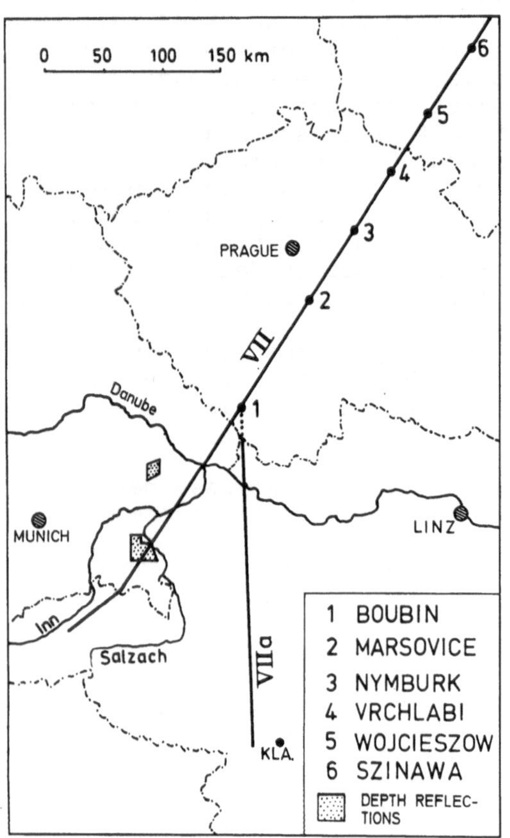

Fig. 1. Map showing the location of the International Profile VII with shotpoints indicated by numbers. *Dotted areas:* regions, where data for the lower crust and the Moho zone are available from deep reflections

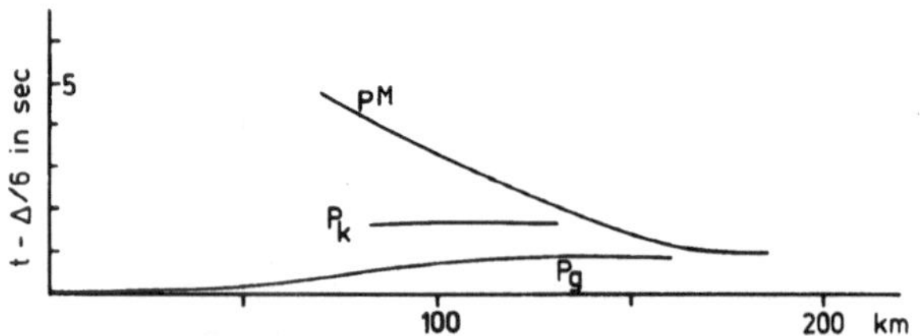

Fig. 2. Generalized pattern of travel-time curves obtained and used for interpretation. A P_n phase has not clearly been identified

ate factor A^* below. In the lowermost part of the sections the surface topography along the profile is given.

6.11.3 Interpretation

The record sections show 3 rather prominent wave groups. Fig. 2 gives a diagrammatic sketch of correlated travel-time curves. The P_g phase is readily observed up to a relatively large distance of some 160 km. It shows an apparent velocity of approximately 5.8-6.2 km/s with reduced traveltimes between 0 and 0.2 s up to a distance of 80 km; beyond this distance a marked increase in travel times indicates a downdip of the crystalline basement, which would be in good agreement with the geological situation, as at this distance the profile crosses the boundary between the Variscan Bohemian Massif and the young sediments of the Molasse.

First onsets generally become weaker, almost disappearing around 104 km, then their amplitudes get larger again between 115 and 140 km. It may very well be that P_g changes in character from a simple penetrating wave to a head wave at distances beyond 80 km.

The P^M phase is a very prominent one. It can first be detected at a distance of 72 km and a reduced travel time of 4.8 s. By its large amplitudes it may easily be traced to a distance of about 170 km. Sometimes two onsets with a small time difference may be observed in this group. As will be shown later, this can be explained by the fact that this P^M-phase is a combination of retrograde and prograde travel-time curve segments. Another phase marked by relatively large amplitudes between 100 and 140 km, reduced travel times of about 2 s and apparent velocities between 6 and 7 km/s may also be easily distinguished. This wave group will be called the P_k phase.

The P_n phase finally is very weak and may only be observed, if at all, at distances beyond 200 km. Recording conditions in this region however had been rather poor.

Interpretation was carried out by model calculations with the seismic ray-tracing method for two dimensional inhomogeneous media as described by GEBRANDE (4.4 of this volume). We feel that first we should describe the strategy of interpretation using this method. To obtain a first model for the structure one uses direct inversion techniques together with information from other sources if available. This generally will yield a first model, which, if put to the test by calculating travel times by ray tracing and comparing calculated times with observed ones, will often be found to be rather incorrect. Although the general pattern of travel-time curves will be satisfied, if direct inversion techniques have been used very carefully, deviations between computed and observed travel times in the presence of strong lateral variations may reach as much as one second. Other boundary conditions such as critical distance and length of individual travel-time segments will not be satisfied either. One then tries to obtain a better fit between calculated and observed travel times by changing the first model in a certain way. The plotted seismic rays have been found to be very helpful for determining changes towards a better model. By repeating this procedure of changing the model and calculating travel times one will eventually reach good agreement between observed and calculated data. We have as of yet not found a sure strategy for changing the model in such a way that successive models will yield better and better approximations to the observed data. However we found that applying few changes at once and starting from the surface downwards, one obtains results most quickly.

Assuming that one has arrived at a model that fits observed travel times and other data e.g. from seismic reflection work, there is still the problem of uniqueness inherent to most model calculations in geophysics. By experience with a number of model calculations for different areas we feel that given reasonably good and sufficient data one may be rather sure of the final model. By using the trial-and-error method in changing the models one finds that one is always pushed in a certain direction; that is, independent of the type of model one starts out with, one will end up with one particular class of model. This however does not imply that the final model will truly represent actual conditions - it will still be only a model, albeit a fairly good one.

For the interpretation of this profile, data on P-wave velocities for the Molasse were available from reflection seismic work and from bore-

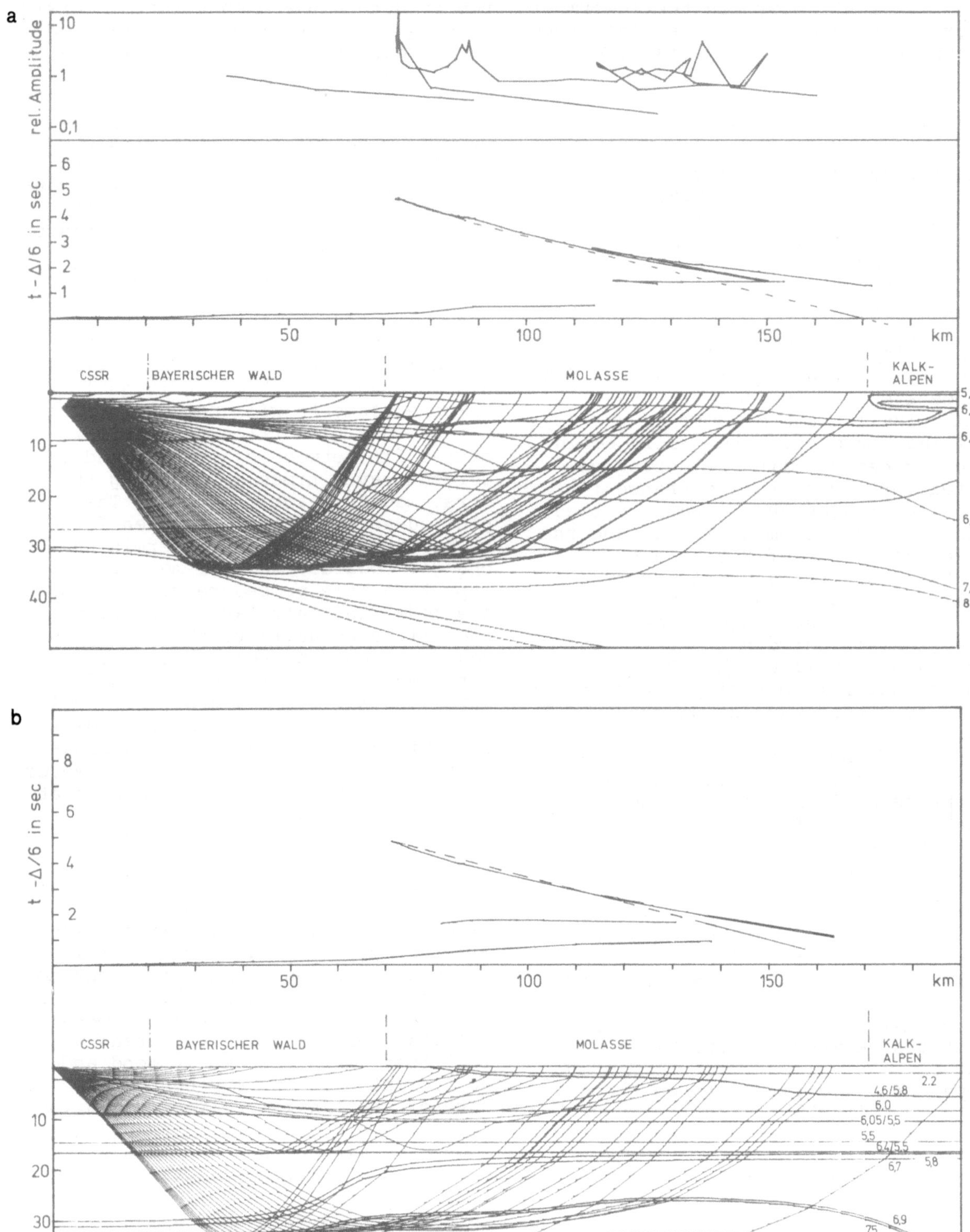

Fig. 3. (a) *Model A:* lines of equal velocity with numbers indicating the respective velocities, ray paths, calculated travel-time curves and geometric amplitudes. Seismic rays are plotted with increments of 0.5° for the radiation angle. (b) *Model B:* lines of equal velocity with numbers indicating the respective velocities, ray paths and calculated travel-time curves. Seismic rays are plotted with increments of 1° for the radiation angle

hole investigations, which also in a few places gave depth to the crystalline basement. For the deeper crust, data from deep reflections were available for the two areas indicated in Fig. 1 (LIEBSCHER, 1964).

As there were no observations for the first 20 km, the velocity distribution for the uppermost part of the crust beneath the shotpoint had to be inferred from the fact that the reduced travel time at a distance of 20 km was only 0.05 s. This of course is possible only if the velocity in the top few km is very close to 6 km/s. Data from the same shotpoint but in the opposite direction yield quite the same velocity distribution (BERANEK et al., 1974; HOLUB, 1974). As the mean velocity for the whole crust however is less than 6 km/s - this also can be derived from the apparent velocity and travel time of the P^M phase in the critical distance - at least one, and as we will show, in some parts even two low velocity zones must be introduced at greater depth.

Fig. 3a and 3b show two models for the area under investigation. Each of the models is the last and best-fitting of a whole series of models. The model of Fig. 3a is a model with smoothly varying velocity depth functions (called model A); the one of Fig. 3b is constructed from layers of varying thickness with velocities defined by interpolating vertically between the upper and lower lines of equal velocity (called model B). The difference between these two different types of models in mathematical terms is given in 4.4. This difference is probably seen most easily in Fig. 4a and 4b. These figures show in a semiperspective way the velocity-depth functions along the profile with a 10-km spacing between them. At the top of Fig. 3a a relative amplitude is plotted. We will not discuss this amplitude in this paper in any detail. Basically it is derived from ray divergence assuming radially symmetric energy radiation near the source. A detailed description may be found in GEBRANDE (1975). Lateral variations of velocity are indicated in Fig. 3a and b by the varying depth of the lines of equal velocity. These lines are marked by numbers which indicate the velocity for each line. Two numbers separated by a dash indicate a discontinuity, the second number giving the velocity of the lower layer in the case of constant velocity or the velocity at the top of the lower layer in the case of a velocity gra-

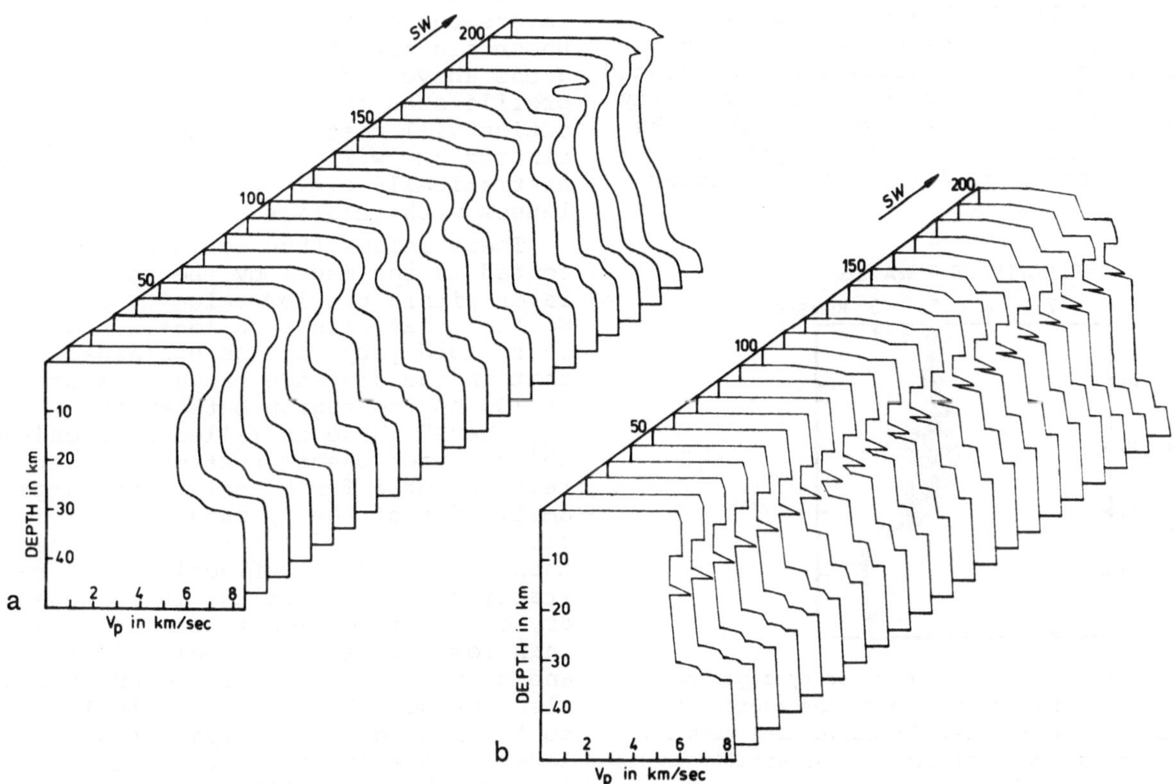

Fig. 4a and b. Distribution of velocity-depth curves of model A (a) and model B (b) along the profile in perspective view

dient. Computed travel times are plotted on a reduced scale.

The most striking features of the models may best be seen by following the lines of equal velocity of model B (Fig. 3b). At first there is the downdip of the line 4.6/5.8 km/s which starts at some 80 km and increases still further at some 140 km. This line indicates the crystalline basement which is covered by tertiary and recent sediments of the Molasse. In Fig. 3a (model A) the Molasse is shown to be overlain by the northern Limestone Alps with their rather high velocity. More detailed data on this feature have been obtained by WILL (1975, 4.5 of this volume) using special observations and a similar method of model calculation. We did not carry our model calculations to this distance because we felt that our data were too insufficient to give a detailed picture in this region. In model B this feature is completely omitted as memory space in our desk-top computer (HP 9810) is rather limited and we felt that we should better use the available space to get a more detailed model at smaller distances.

Another feature of the models is the low-velocity zone. Whereas model A shows one low velocity zone only, which underneath the shotpoint is some 17 km thick and decreases in thickness towards the SW, model B shows two low-velocity zones separated by a thin layer with maximum velocity of 6.4 km/s. This layer, which looks like a nose in the velocity-depth function (Fig. 4b), should be understood as a model representation of actual conditions only. It has been introduced into the model to explain the P_k phase. In model A the onsets of this phase are caused by reflected and refracted waves from the lower boundary of the low-velocity zone. It is impossible, however, with this model to explain all the onsets observed at distances between 90 and 120 km which clearly belong to the P_k phase. By introducing this thin layer it is possible to compute travel times which are in good agreement with the observed ones. It seems highly improbable for such a layer to exist as a continuous feature.

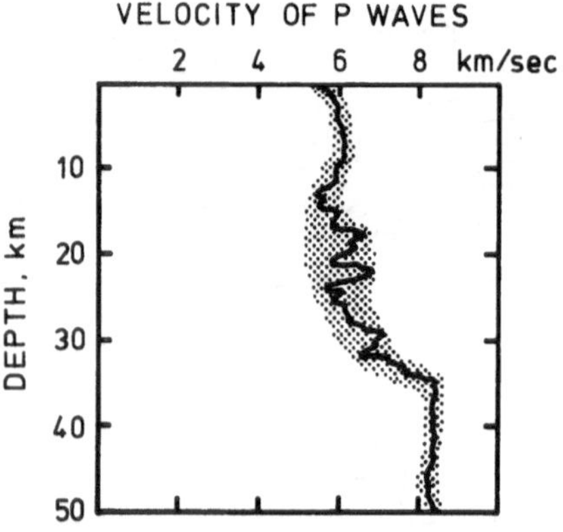

Fig. 5. Representative velocity depth curve. The dotted area represents the region of possible solutions. The solid line with its rather strong oscillations, especially in the upper part of the velocity inversion zone, is an example for a probably realistic velocity-depth curve

We prefer to think that a real velocity-depth function will resemble the one shown in Fig. 5. The dotted area in Fig. 5 defines a region in velocity-depth space, within which any particular velocity-depth function may vary. The dark line shows what we think a real velocity depth function for this profile should look like. It will oscillate rather strongly within the low velocity zone, probably reaching rather high velocities within small depth intervals - the integral effect still being that of a velocity inversion. The dotted area may also be understood as some sort of a confidence area, results for the region of the velocity inversions obviously being much more uncertain than the ones for the regions above and below it. Along the profile such thin zones with high velocities will most likely not be continuous but will be rather distributed statistically e.g. in the form of lenses with relatively high velocities.

These lenses then would be represented in the model by the one single layer within the low-velocity zone. We feel that with such lenses, the fact could very well be explained, that in the distance range of some 70 to 100 km the region between the P_g and the P^M phase is filled with energy, which causes onsets that may be correlated for a few km only. The larger amplitudes of the P_k phase in the distance range of 110 to 135 km may then be caused by reflections and refraction of waves at the lower boundary of the inversion zone. The very thin low velocity zone below 17 km and beyond distances of 100 km should also not be taken for real. It is due to the lack of memory space that this layer exists at all, but because it is very thin, it will in any case give very little contribution to the total travel times.

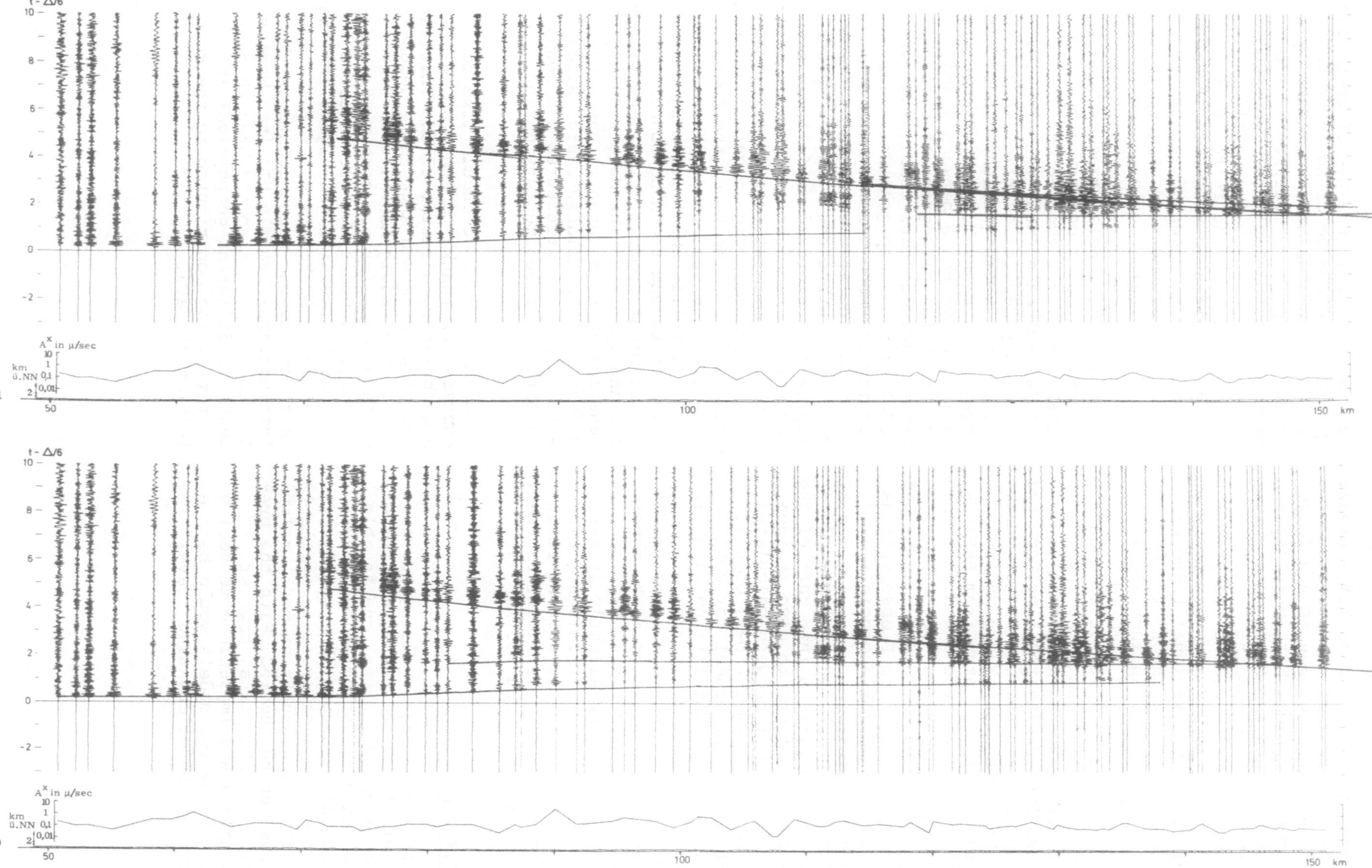

Fig. 6a and b. Seismogram section with travel-time curves calculated for model A (a) and model B (b)

Both models show a marked increase of the thickness of the lower crust in the distance between 50 and 70 km from the shotpoint. Within the same distance range, the Moho zone, which underneath the shotpoint is rather narrow, gradually widens and splits into two separate rather strong gradient zones. The regional trend of the Moho zone from the shotpoint to the SW first shows an increase in crustal thickness to a maximum of 35 km, then a decrease to 32 km underneath the Molasse with the upper gradient zone reaching much smaller depths. Approaching the Alps, crustal thickness again becomes larger. Data for the crustal thickness underneath the shotpoint are taken from BERANEK et al. (1974) and the downdip of the Moho zone, which is shown in Model A (Fig. 3a) only, follows from apparent velocities taken from P_n observations on this profile for shotpoints 2, 3, 5 and 6 (Fig. 1).

Figs. 6a and b show calculated travel times plotted into part of the seismogram sections to give an idea of the agreement between calculated and observed travel times. Generally there is good agreement for both models - each model fitting slightly better in particular parts than the other one. For seismograms No. 83-95 however, there still seems to remain a systematic error with observed travel times being slightly larger than calculated ones. This we think must be attributed to a local basin or trough in the crystalline basement between about 95 and 105 km.

6.11.4 Conclusion

We have shown that by using ray-tracing methods for computing the travel times of seismic waves, one is able to obtain a rather detailed picture of crustal structure, especially if sufficient data material is available, against which results of model calculations can be checked. With the presently available data, however, it obviously seems impossible to decide which of the two models would be the better one by calculating travel times. We also feel that the use of geometric amplitudes would be of no decisive help, although we have not actually tried to come to a decision by comparing computed amplitudes with observed ones. It would possibly in any case be splitting hairs to try to reach a decision, especially if one considers the suggested semistatistical nature of the velocity-depth distribution within the inversion zone.

Vertical reflection times calculated for our models agree very well with reflection times given by LIEBSCHER (1964) for two regions (at 90 and 145 km) on or near the profile. Especially reflection times calculated for the two steep gradient zones which together form the Moho zone agree well with two peaks between 9 and 11.4 s in the histograms of reflections observed, thus confirming our model of two steep gradient zones forming the Moho zone.

In particular the general pattern of crustal structure along our profile may be characterized as follows: a low velocity zone below 10 km is rather thick underneath the Bohemian Massif and becomes thinner underneath the Molasse. The thinning of the low-velocity zone goes hand in hand with a marked uplift and thickening of the lower crust and the Moho zone.

One might speculate that the widening and uplift of the Moho zone may genetically be related to the subsidence of the Molasse basin, since similar crustal structures are found beneath other sedimentary basins.

Acknowledgments. We should like to express our thanks to Prof. Dr. G. Angenheister, director of the Institut für Allgemeine und Angewandte Geophysik of the Ludwig-Maximilians-Universität München for his continuous and extensive support. We are indebted to our Czechoslovakian colleagues for the excellent collaboration, which allowed us to extend the measurements on the International Profile VII into Western Germany. We are also indebted to many teams of different geophysical institutes, who contributed to the collection of data.

7. Main Features of Crustal Structure in the Alps

P. GIESE and C. PRODEHL

ABSTRACT

This report presents a review of the main features of crustal structure in the Alps, based predominantly on published results. Some minor new details are included.

An intensive study of the structure of the Alps by explosion seismology was started about 20 years ago. Numerous profiles, partly reversed, have been recorded from various shotpoints in the Eastern, Western, and Southern Alps. Based on record sections available for the profiles, all material has been interpreted homogeneously. To avoid the complexity of correlation only the most important phases which could be identified more or less clearly were used: the P_g-, P^M-, and P_n-wave groups.

A first qualitative picture of the crustal structure in the Alps is presented in contour maps showing some basic parameters of the travel-time diagram as there are: the critical distance Δ_c of P^M,, the cross-over distance Δ_d between P_n and v = 6 km/s, and the reduced travel time $\bar{t}_{80}$ of P^M at the fixed distance 80 km. These maps reproduce qualitatively variations of crustal thickness under the Alps.

The methods applied for depth calculation include the possible existence of velocity gradients as well as low-velocity layers. The velocity distribution is displayed in the form of cross sections with lines of equal velocity.

The results are discussed in detail along traverses through the eastern part of the Alps, the border zone between Eastern and Western Alps, and the Western Alps including the Ivrea zone. The structure of the whole Alps is summarized in the block diagrams and in a contour map of the depth of the crust/mantle boundary.

The main results are as follows: The crust reaches a maximum thickness of 50-60 km under the axis of the Alps. The lower crust and/or the transition from crust to mantle increases in thickness from the foreland towards the axis of the Alps. In general an extensive velocity inversion exists under the Alps. In the Western Alps, the low-velocity zone extends even eastwards under the high-velocity material (7.2 km/s) of the Ivrea zone which is characterized by a strong gravity high. On the eastern side, this high-velocity material is linked to the lower crust and the mantle under the Po plain.

ZUSAMMENFASSUNG

Der vorliegende Beitrag gibt eine zusammenfassende Darstellung der wesentlichen Merkmale der Krustenstruktur unter den Alpen und basiert im wesentlichen auf veröffentlichten Ergebnissen. Einige Ergänzungen sind hinzugefügt.

Die intensive Untersuchung der Krustenstruktur der Alpen wurde vor etwa 20 Jahren begonnen. Zahlreiche, zum Teil gegengeschossene Profile wurden von verschiedenen Schußpunkten in Ost-, West- und Südalpen beobachtet. Auf Grund von vorliegenden Seismogramm-Montagen wurde das gesamte Beobachtungsmaterial einheitlich interpretiert. Dabei wurden nur die wichtigsten Phasen benutzt, die sich mehr oder weniger deutlich auf allen Profilen identifizieren ließen: die P_g-, P^M- und P_n-Wellengruppen.

Ein erstes qualitatives Bild über die Krustenstruktur unter den Alpen kann man aus Isolinienplänen gewinnen, in denen einige grundlegende Parameter kartiert sind, die dem Laufzeitdiagramm entnommen wurden: die kritische Entfernung Δ_c der P^M-Welle, die Überholentfernung, in der die P_n-Laufzeitkurve die Entfernungsachse ($\bar{v}$ = 6 km/s) kreuzt, und die reduzierte Laufzeit $\bar{t}_{80}$ der P^M-Welle bei der konstant gehaltenen Entfernung 80 km. Diese Pläne spiegeln bereits qualitativ die Änderungen in der Krustenmächtigkeit unter den Alpen wider.

Die Methoden, die zur Tiefenberechnung verwendet wurden, schließen die Erfassung von Geschwindigkeitsgradienten sowie von Zonen erniedrigter Geschwindigkeit ein. Die Geschwindigkeitsverteilung wird in Profilschnitten durch Kruste und obersten Mantel dargestellt, wobei Linien gleicher Geschwindigkeit in Abhängigkeit von der Tiefe aufgetragen sind.

Die Ergebnisse werden im einzelnen an drei Krusteneinschnitten diskutiert, die durch den östlichen Teil der Alpen, durch den Grenzbereich zwischen West- und Ostalpen und durch die Westalpen einschließlich der Ivreazone

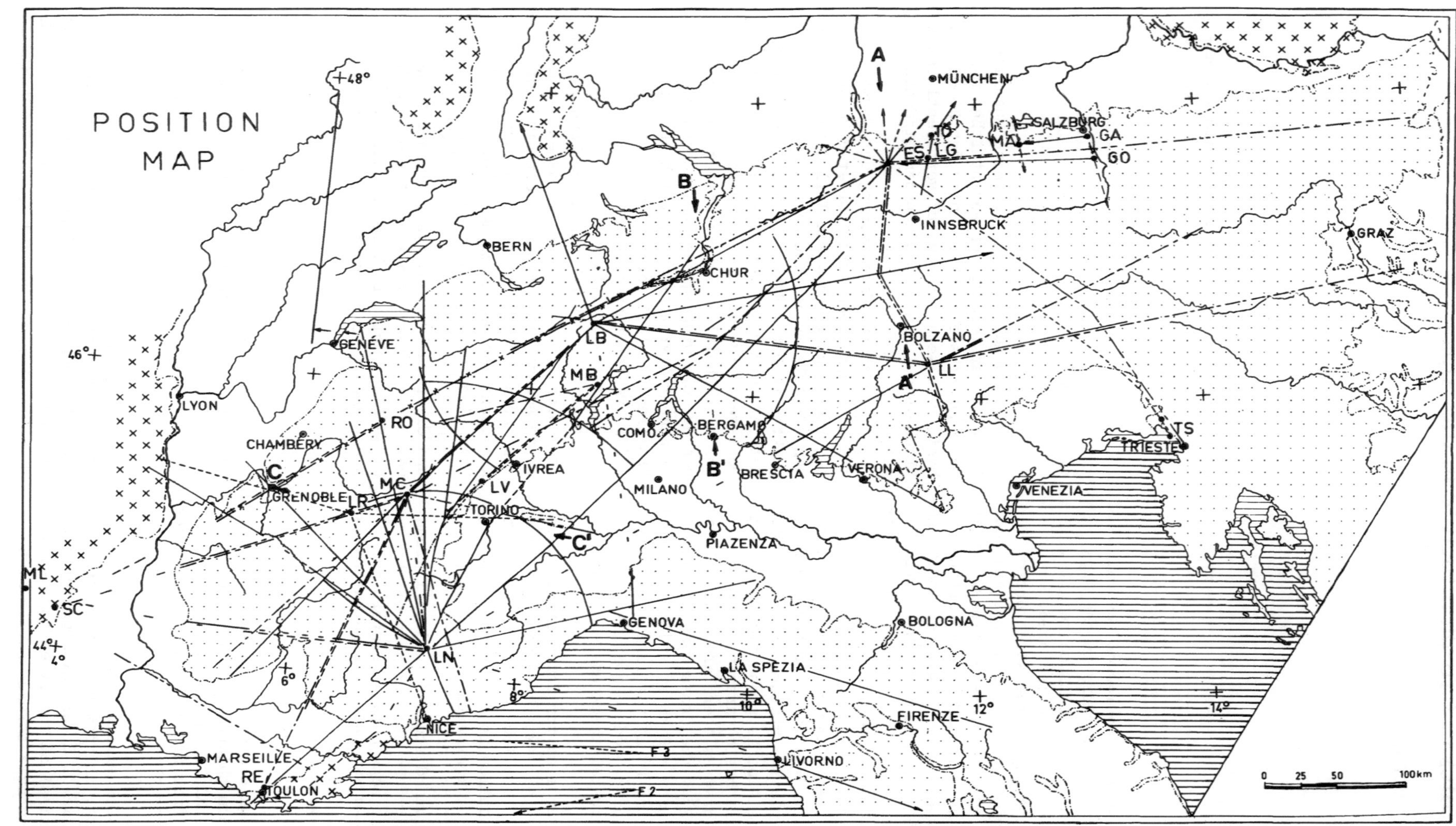

Fig. 1. Position map of shotpoints and seismic-refraction profiles. ● Shotpoints: *ES* Eschenlohe, *GA* Gartenau, *GO* Golling, *LB* Lago Bianco, *LG* Lenggries, *LL* Lago Lagorai, *LN* Lac Nègre, *LR* Lac Rond, *LV* Levone, *MA* Marquartstein, *MB* Monte Bavarione, *MC* Mont Cenis, *ML* Lozère, *RE* Le Revest, *RO* Roselend, *SC* Ste. Cècile d'Andorge, *TÖ* Tölz, *TS* Trieste. *Dotted:* areas of Alpine orogeny; *crosses:* Variscan massifs outside the Alps. *AA'* position of cross section shown in Fig. 13, *BB'* position of cross section shown in Fig. 19, *CC'* position of cross section shown in Fig. 27

laufen. Eine Zusammenschau der Struktur der gesamten Alpen vermitteln zwei Blockdiagramme und ein Tiefenlinienplan der Kruste/Mantel Grenze.

Die wichtigsten Ergebnisse sind: Die Kruste erreicht eine maximale Mächtigkeit von 50-60 km unter der Achse der Alpen. Ausgehend vom Vorland nimmt die Breite der Übergangszone zwischen Kruste und Mantel zur Alpenachse hin zu. Eine starke Geschwindigkeitserniedrigung im mittleren Krustenbereich ist unter den gesamten Alpen vorhanden. In den Westalpen erstreckt sich diese Zone erniedrigter Geschwindigkeit nach Osten bis unter den Ivreakörper und erreicht hier extrem niedrige Werte von 4-5 km/s. Der Ivreakörper selbst zeichnet sich durch anormale hohe Geschwindigkeiten (7,2 km/s) in geringer Tiefe (5-20 km) aus und ist durch ein intensives Schwerehoch charakterisiert. Nach Osten zu fällt er unter die Poebene ein und geht hier in die untere Kruste/Mantelgrenze des Poebenenblocks über.

7.1 Introduction

Essential geologic results have been obtained by the intensive investigation of the Alps. There is no mountain system on the whole earth which has been as deeply explored as the Alps. Although still today ideas on structure and development of the Alps differ in details, they agree, however, in that the observed phenomena at the surface are caused by processes generated by sources in the deeper crust and upper mantle. The investigation of such deeper structure cannot be carried out by pure geologic methods, the basic framework for any geotectonic theory is rather formed by the results of geophysical measurements.

For four decades it has been known that a characteristic geophysical feature of the Alps is a strong negative Bouguer anomaly of 150-250 mgal which indicates a thickening of the earth's crust in the sense of a mountain root. While the results of gravity measurements facilitate only integral statements, being ambiguous due to potential theory, seismic measurements allow the determination of velocity-depth distribution.

When the seismic investigations of crustal structure started in western Germany in the beginning of the fifties, measurements were also performed at the same time in the northern Alps. In the following two decades, with international cooperation, the Alpine area was covered with a dense network of seismic-refraction profiles. German geophysical institutions took part in all these experiments and the interpretation of the data. As Germany borders on the Alpine area, the crustal structure of this young orogene as well as the transition from the foreland into the orogene is of special interest. By this extension many new fruitful suggestions stimulated the interpretation of seismic data obtained in western Germany. The main results of Alpine explosion seismology are presented here.

7.2. Brief Historical Review on Alpine Explosion Seismology

7.2.1 Field Experiments

The first experiments of explosion seismology in the area of the Alps were initiated by H. Reich in the beginning of the fifties, recording commercial blasts of the Eschenlohe quarry situated at the northern margin of the Eastern Alps (Fig. 1). In the same region two special seismic projects were executed using borehole shots in 1954 (Tölz) and 1958 (Lenggries).

The explosions of the Eschenlohe quarry were a great help for the investigation of crustal structure of the whole Eastern Alps as well as their northern foreland for two decades.

During the meeting of the European Seismological Commission in Rome in 1954, following a suggestion by W. Hiller, a Sub-Commission for Alpine Explosions was established with the aim of organizing international cooperation for the investigation of the crustal structure of the Alps. At the same time the General Assembly of the IUGG adopted a resolution requesting the governments of the countries surrounding the Alps to contribute to the study of the deep Alpine structures.

The first large program with international cooperation was carried out in the Western Alps in 1956. A series of six large shots (0.5-10.1 t) was fired at Lac Rond in the Pelvoux Massif near Briançon and was observed on three profiles crossing the Alps. These were followed by a second series of six explosions (0.1-25 t) fired at Lac Nègre, Mercantour Massif, north of Nice, in 1958. These two experimental programs were accomplished thanks to the efforts of Mme Y. Labrouste.

One particular result of the 1958 experiments revealed the existence of high-velocity material (7.3-7.4 km/s) at shallow depth (8-10 km) under the Ivrea zone which is characterized by a very strong positive gravity anomaly (Fig. 2). The necessity of organizing new explosions was, therefore, evident. Thus two shotpoints were selected on the axis of the gravity anomaly and, in 1960, under the organization of L. Solaini and O. Vecchia, the experiments were carried out.

In the beginning of the sixties the main activity moved to the Eastern Alps. In 1961 and 1962 the efforts of C. Morelli enabled two series of explosions in Lago Lagorai, in the eastern Dolomites, SE of Bolzano, which were observed along five profiles (Fig. 1).

In order to connect the profiles of the Western and the Eastern Alps, in 1963 and 1964 A.E. Süsstrunk organized explosions in Lago Bianco, west of the St. Gotthard Pass. The position of this shotpoint allowed reversing some of the already existing profiles: from Eschenlohe, Lago Lagorai, and Lac Nègre. In addition some new profiles could be recorded.

To complete the observations in the Western Alps and their western foreland, a series of shots was organized in 1965 by Mme Y. Labrouste at various shotpoints: Mont Cenis and Roselend in the Alps, Le Revest near Toulon and Mont Lozère and Sainte Cécile d'Andorge in the Central Massif. From Mont Cenis the observations were completed in 1967.

Although the data obtained so far in the Western Alps gave very interesting results, their complex structure and particularly that of the Ivrea zone demanded more experiments. Again, in 1966, Mme Y. Labrouste organized a series of nine large explosions up to 42 t fired in Lac Nègre, the same shotpoint as in 1958. The number of shots and the big charges permitted observations up to 750 km in- and outside the Alps.

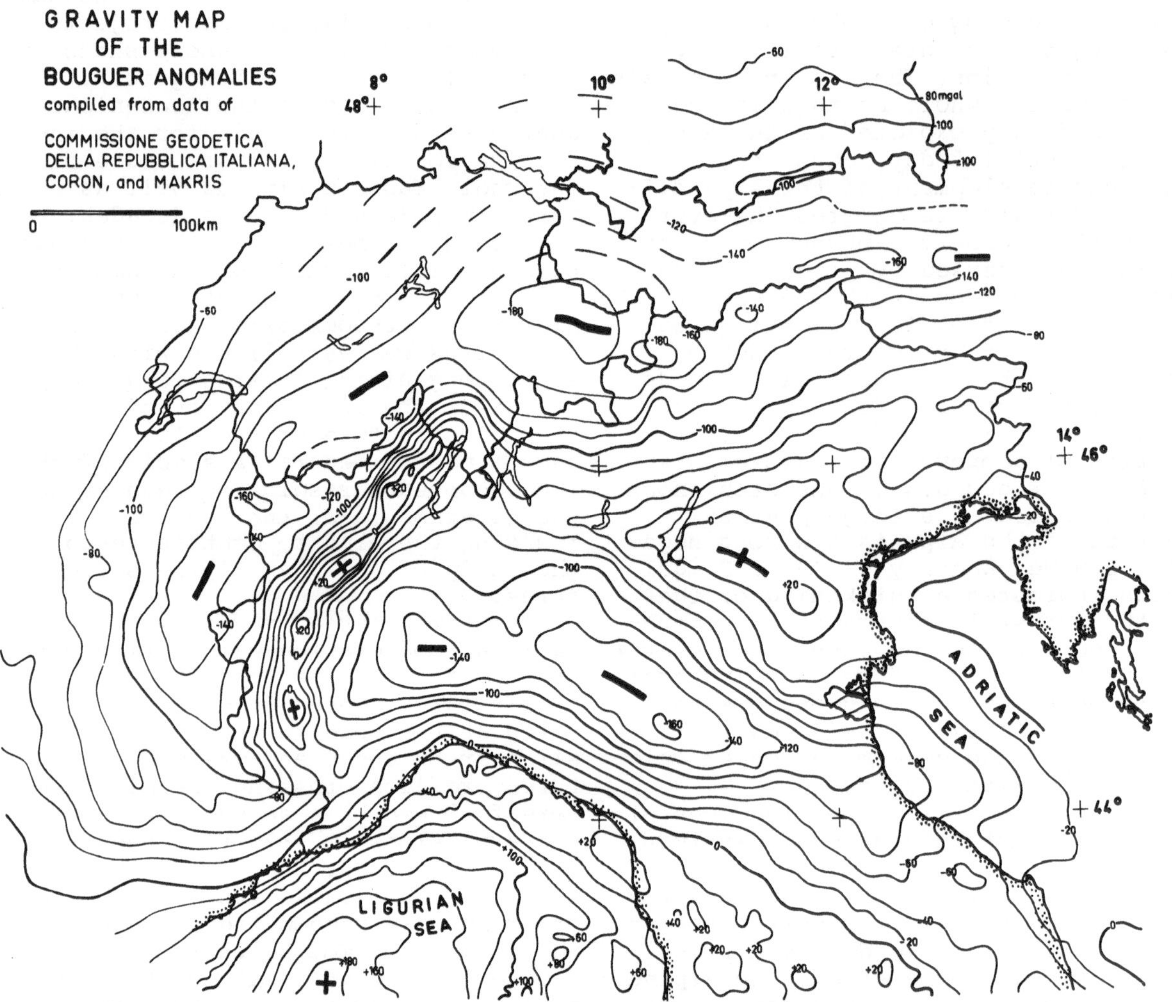

Fig. 2. Gravity map of the Bouguer anomalies for the Alps and adjacent areas

Due to the uninterrupted use of explosions at the Eschenlohe quarry, mentioned above, a fan of profiles up to 300 km distance were in the meantime obtained in the Eastern Alps. Within the program Geodynamics of the Mediterranean Area the reversed profile Eschenlohe-Lago Lagorai became the axis of the geotraverse Ia, running from the Molasse basin to the Po plain. During this project a special seismic-refraction survey was conducted in the Northern Calcareous Alps.

Fig. 1 presents all the seismic-refraction lines observed successfully up to 1974. It must be pointed out generally the shotpoints could not be chosen according to optimum structural and recording conditions, but were more or less bound to the accidental position of quarries and lakes suitable for seismic purposes. As a consequence, the position of profiles is very often neither parallel nor perpendicular to the tectonic alignments and in some areas profiles are completely lacking.

In the first half of the seventies the activity of the Western and Southern European groups was concentrated on projects outside the Alps. These different experiments stimulated new aspects of details of crustal and upper-mantle structure and led finally to a new experiment in the Alps in 1975.

Contrary to the hitherto existing observations it was felt that special drillhole shots had to be organized at positions aiming to solve open problems.

With 175 participating recording stations from Austria, France, Germany, Great Britain, Hungary, Italy and Switzerland an 800 km long profile was organized running in E-W direction from the Pannonian Basin to S of Genève. It was supplemented by some fans and smaller observation lines in Austria and Italy. Six shotpoints were distributed along the lines. This was the largest experiment ever carried out in the Eurasian continent. Aims of this project were the study of the structure of the crust and the lower lithosphere along the axis of the Central Alps, the detailed structure of the crust/mantle boundary and the transition from the Pannonian Basin to the Eastern Alps. In addition, the Italian group investigated the gravity high near Verona. The present report can only summarize the results obtained up to 1974.

7.2.2. Crustal Models

The first Eschenlohe, Tölz and Lenggries experiments aimed mainly at investigating the upper part of the crust under the Subalpine Molasse and the Northern Calcareous Alps (REICH, 1957, 1958a, b, 1960a, b).

Data and results of all the seismic experiments in the Western Alps up to 1960 were published in a monograph edited by CLOSS and LABROUSTE (1963). All the models derived by different authors agree with the existence of a root under the axis of the Western Alps in the form of a depression of the Mohorovičić discontinuity. Further on, at the inner arc of the Western Alps, in the Ivrea zone, a high P-wave velocity of about 7.2 km/s has been discovered at a shallow depth of only 5-10 km.

With respect to the deeper structure of this anomaly, two models were developed. CHOUDHURY et al. (1963) proposed a structure without any intermediate layer in the crust and a continuous transition between the high-velocity body of Ivrea and the upper mantle. In the other model, FUCHS et al. (1963) introduced a Conrad discontinuity and assumed a separation of the Ivrea body from the upper mantle by the conventional Mohorovičić discontinuity. In the model of FUCHS et al. (1963) the depth of the crust/mantle boundary was calculated on the hypothesis that the wave with large second arrivals (overcritical reflections) is a converted wave of the type sP_p. Both models agree more or less with gravity data.

The first models for crustal structure of the Eastern Alps were published by BEHNKE et al. (1962) and PETERSCHMITT et al. (1965) using the observations of the Lago Lagorai experiments. These interpretations, utilizing the usual methods, yielded a Conrad- and Mohorovičić discontinuity. From the data of the reverse profile Eschenlohe-Lago Lagorai, PRODEHL (1965) constructed a cross section of the Eastern and Southern Alps using the wave-front method. He introduced an additional layer (7.2-7.4 km/s) between the Conrad and the Mohorovičić discontinuities. Contrary to the interpretation of FUCHS et al. (1963), all models for the Eastern Alps were derived on the assumption that all waves are of P-type.

A new concept for crustal models in the Alps was started by GIESE

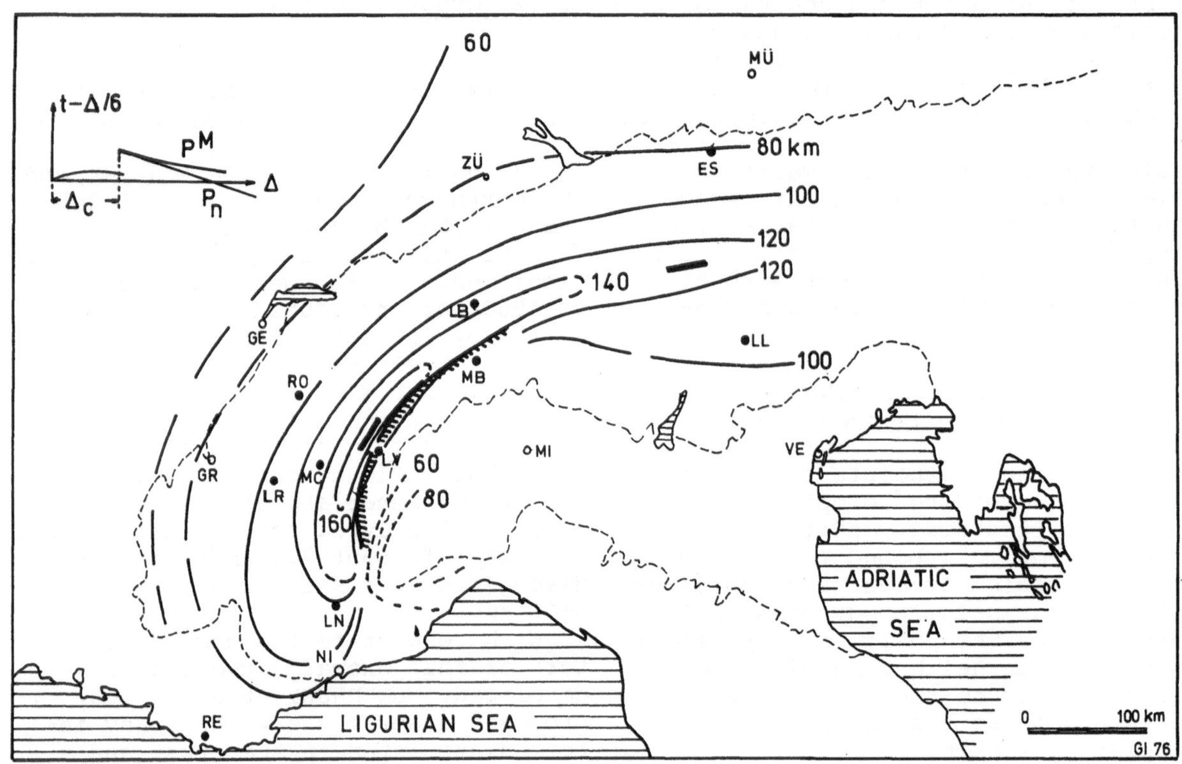

Fig. 3. Contour map of the critical distance Δ_c (P^M-wave group). In the Ivrea zone, dotted lines contour the corresponding critical distance for the wave group P^I or the P^M-group resp. of the Po plain. • Shotpoints: *ES* Eschenlohe, *LB* Lago Bianco, *LL* Lago Lagorai, *LR* Lac Rond, *LV* Levone, *MB* Monte Bavarione, *MC* Mont Cenis, *RE* Le Revest, *RO* Roselend. o Cities: *Ge* Genève, *Gr* Grenoble, *Mi* Milano, *Mü* München, *Ni* Nice, *Ve* Venezia, *Zü* Zürich

(1966, 1968a). Based on data from a number of refraction profiles in southern Bavaria, the Eastern, the Western, and the Southern Alps, he demonstrated that, for the first direct P-wave, there exists a curvature in the time-distance curve and that the overcritical Moho reflections and their asymptote may be separated by a variable intercept-time delay. Such a time delay indicates the possible existence of a velocity inversion which was not taken into account in the previous interpretations. Applying this new concept to the Alpine data, the existence of a distinct velocity inversion in the middle part of the crust, especially in the Central Alps, could be proved. This result became important for the Western Alps. Introducing also a strong velocity inversion under the high-velocity material of the Ivrea zone, it became possible to explain the typical high-amplitude phase by pure P-waves. Also the crust/mantle boundary is no more interpreted as a first-order discontinuity but is replaced by a more or less broad transition zone, reaching its maximum extension under the Central Alps.

In the light of these ideas, GIESE et al. (1967) presented a new model of the crustal structure of the Western Alps. The peculiarities of the structure of the Ivrea zone were discussed on a special symposium "Zone Ivrea-Verbano" held in Locarno in 1968 (ANSORGE, 1968; BERCKHEMER, 1968; GIESE, 1968b). Here the main features of the new model were generally accepted. The areal extension of this anomaly was investigated by GIESE et al. (1971).

CHOUDHURY et al. (1971) compiled all the data and elaborated a structural model for the Alps based on a unique concept of interpretation, which was presented first at the General Assembly of the IUGG in 1967. A special summary on the previously mentioned Geotraverse Ia through the Eastern Alps was published by ANGENHEISTER et al. (1972).

7.3 General Remarks on the Data Observed

In view of the complexity of the surficial geology of the Alps it has to

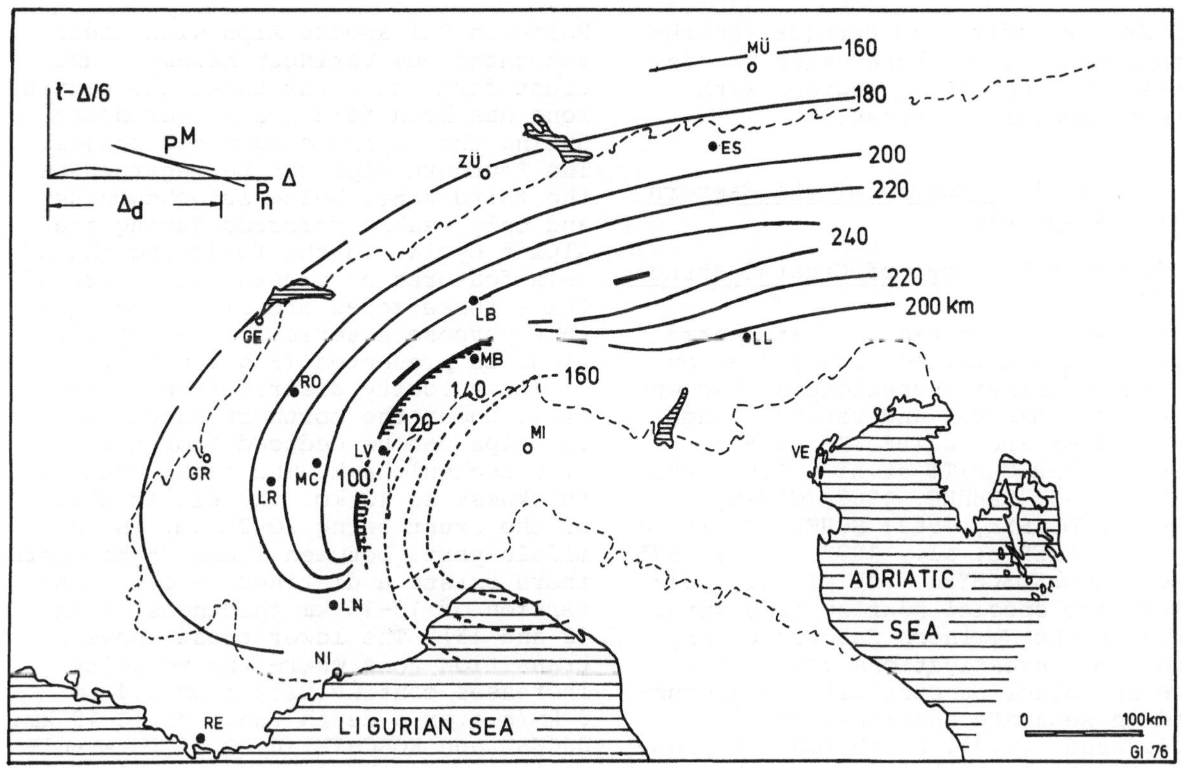

Fig. 4. Contour map of the cross-over distance Δ_d between the P_n-travel-time curve and v = 6 km/s (Δ-axis). Explanations see Fig. 3

be expected that the deep structure is also very complicated. In spite of this it turns out that some common features in the arrangement of the main phases can be recognized in the record sections of the majority of profiles. However, such a generalized pattern shows wide modifications due to the complexity of Alpine structure.

Similar to the northern foreland (GIESE, 5.2 of this volume) the following three main wave groups can be distinguished:

1. the P_g-group traveling through the upper 10 km of the crust,

2. the P^M-group bottoming the deeper crust and the crust/mantle boundary,

3. the P_n-group penetrating the uppermost mantle.

The meaning of the term "group" instead of "wave" indicates that the corresponding travel-time phases may be composed of several branches. Here the problem of phase and group correlation is touched (GIESE, 4.1 of this volume). Other wave groups can be recognized on one or the other record section. However, as they are not generally present, they cannot be taken into account for any generalizing interpretation, but can only be used for a detailed study of the corresponding profile.

In the zone of Ivrea, due to its anomalous structure, additional main wave groups are observed and must be considered.

Before starting any detailed regional interpretation and depth calculation, some basic parameters are presented which give qualitative information on the general features of any structures (PRODEHL, 1970a, b; BEHNKE, 1971; CHOUDHURY et al., 1971; GIESE and STEIN, 1971; GIESE, 5.2; PRODEHL, 8.2 of this volume).

The values of the critical distance of the P^M-wave (Fig. 3) and of the cross-over-distance of the P_n-wave with the line v = Δ/6 (Fig. 4) are approximately proportional to the thickness of the crust. The quantity $\bar{t}_{80}$ also reflects lateral variations of crustal thickness (Fig. 5). These maps also contain some values obtained from investigations of crustal structure of the Rhônegraben (SAPIN and HIRN, 1975) and of the Rhine Graben (EDEL et al., 1975; PRODEHL et al., 6.8 of this volume) bordering the Alps in the west and north. The quantities Δ_d, Δ_c, and $\bar{t}_{80}$ increase progressively from the foreland towards the axis of the Alps. In the region of Gran Paradiso and Bernina (southern Switzerland) there exist two depressions indicating regions with great crustal thickness.

7.4 Crustal Structure of the Eastern Part of the Alps

7.4.1 Main Features of Crustal Structure

The results reported here are based mainly on a detailed interpretation of the profiles Eschenlohe-E, Eschenlohe-Lago Lagorai and reverse, Lago Lagorai-NE and -E (BEHNKE et al., 1962; PETERSCHMITT et al., 1965; PRODEHL, 1965; BEHNKE, 1967; GIESE, 1968a; KOSCHYK, 1969; SCHELIGA, 1971; ANGENHEISTER et al., 1972; WILL, 1975; ZSCHAU and KOSCHYK, 6.10 of this volume). The special near-surface structure of the Northern Calcareous Alps and some investigations concerning the Engadiner Fenster will be discussed in separate sections (Sections 7.4.2 and 7.4.3). Furtheron the problem of the fine structure of the crust-mantle transition zone is also treated separately (Section 7.4.4).

A cross section through the eastern part of the Alps comprises three main tectonic zones. The allochthonous Northern Calcareous Alps with their autochthonous Variscan basement and crust form the first zone. The central zone has been folded and overthrust during the Alpine orogenetic phases. The Southern Alps can be regarded as the third zone, being autochthonous and only weakly deformed during the Alpine cycle. In the following the main features of crustal structure of these three zones are discussed. For this purpose a series of record sections is presented (Figs. 6-11).

The velocity distribution of the crust under the northern margin of the Alps can be deduced from the profile Eschenlohe-E. The total crustal thickness is 40 km, the sialic part of the crust being 20-25 km. In the middle part, between 7 and 25 km depth, there exists a distinct velocity inversion of 10-15 km thickness (Figs. 12 and 13). The lower crust shows a transition zone where the velocity increases continuously from 6.4 to 8 km/s over a depth range of 10-15 km. ZSCHAU and KOSCHYK (6.10 of this volume) discuss the records of this profile in detail and take also S-phases into consideration, while WILL (1975, 1976, 4.5 of this volume) primarily discusses the structure of the sedimentary cover.

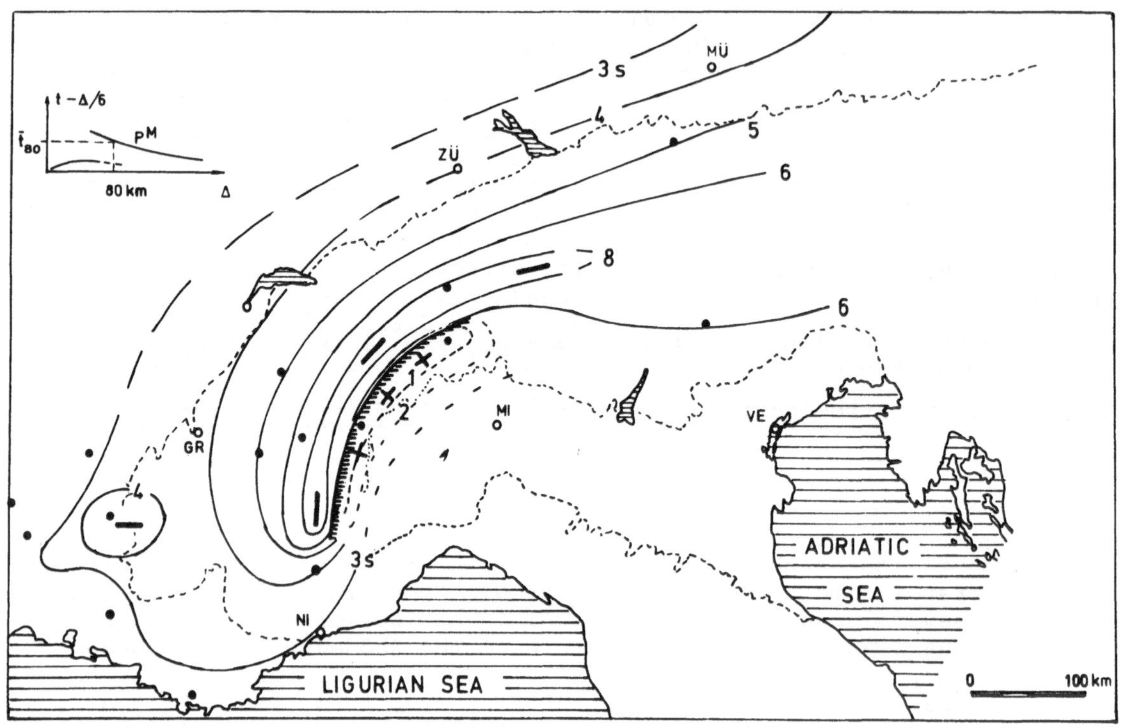

Fig. 5. Contour map of the reduced travel time $\bar{t}_{80}$ of the P^M-wave group at 80 km distance. In the Ivrea zone, *widely spaced dashed lines:* the corresponding time $\bar{t}_{80}$ for the wave group P^I or P^M resp. of the Po plain. Western border of the Ivrea body. *Narrow dashed lines:* border of Alps and Apennines

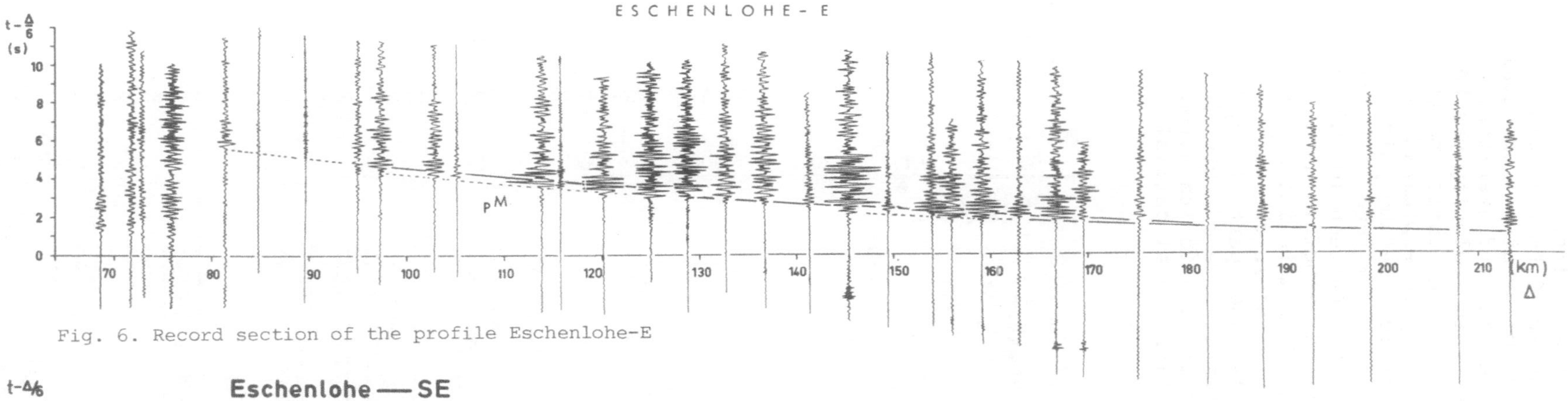

Fig. 6. Record section of the profile Eschenlohe-E

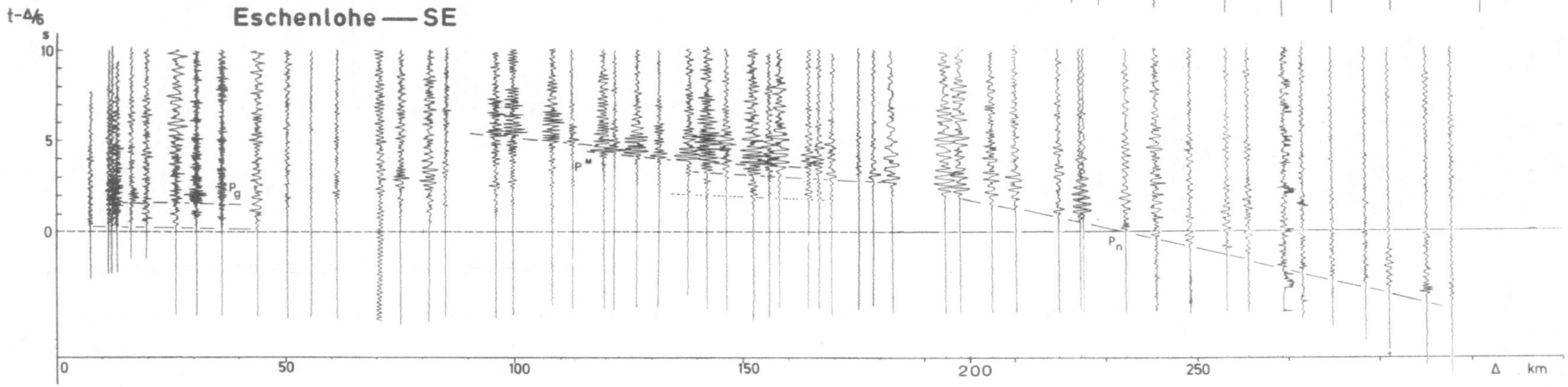

Fig. 7. Record section of the profile Eschenlohe-SE

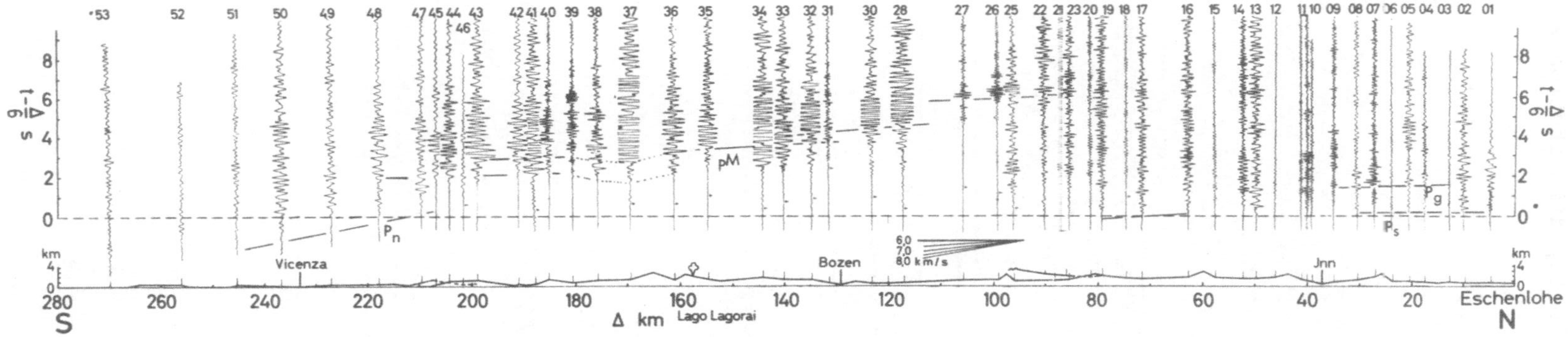

Fig. 8. Record section of the profile Eschenlohe-Lago Lagorai

The structure of the central Alps is derived from the reversed profile Eschenlohe-Lago Lagorai (Figs. 8 and 10). Here (Fig. 13) the total crustal thickness is about 45 km. The internal velocity distribution shows some distinct differences with respect to that of the Northern Alps.

Within the upper 10 km the velocity is 0.2-0.3 km/ greater than usually known for gneisses and granites. This anomalously high velocity can be seen in general in the central parts of the Alps and may be explained by the existence of meso-metamorphic schists of the Penninic zone with embedded basic rocks. Between 10 and 25 km depth, a strong velocity inversion is observed which is more intensive (down to 5 km/s) than in the northern foreland. In its upper part it may be explained by temperature effects, but in its lower part, due to its strong velocity reduction, local partial

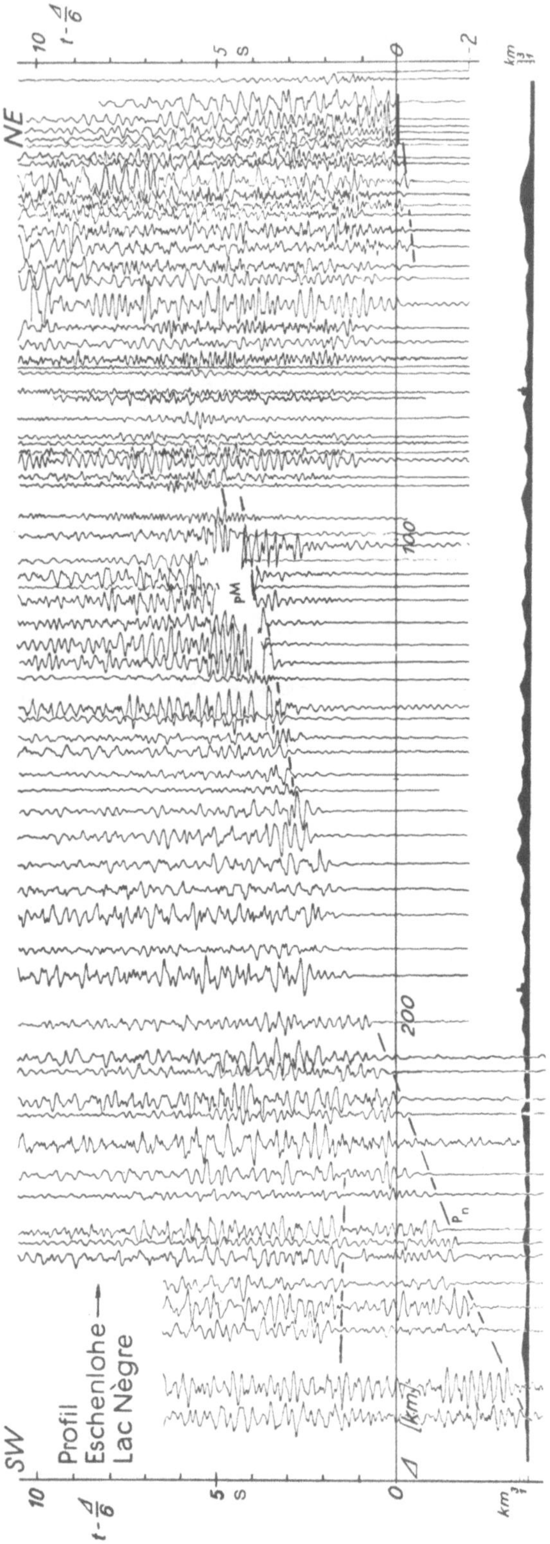

Fig. 9. Record section of the profile Eschenlohe-SW

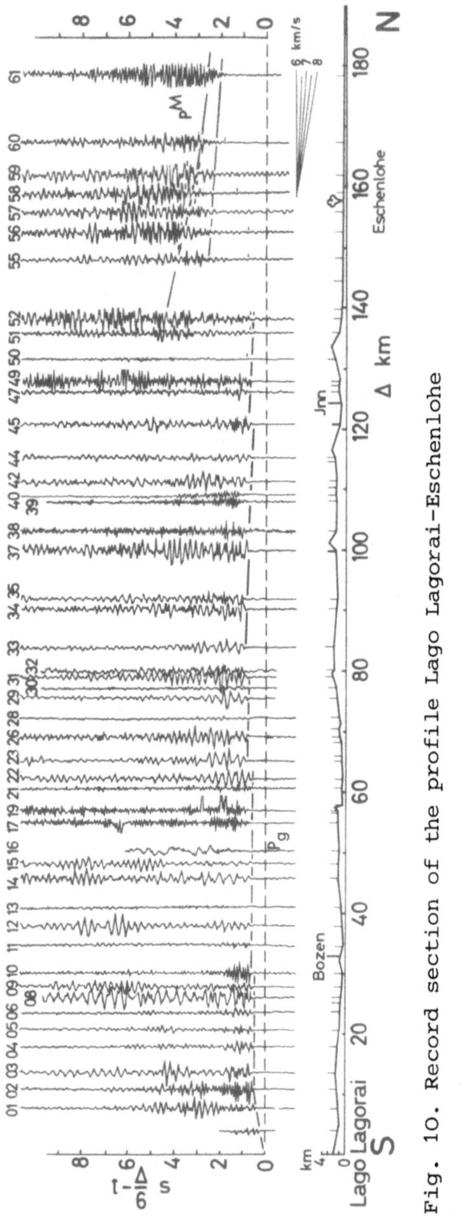

Fig. 10. Record section of the profile Lago Lagorai-Eschenlohe

melting must be taken into account which may be connected with the metamorphism of Pennine rocks visible in windows of the Tauern and the Engadine. In total, the sialic crust is 30-35 km thick. The lower crust shows a wide transition zone with a thickness of about 15 km. In general within this transition zone, the velocity increases up to upper-mantle values, but embedded velocity reversals may exist (see Section 7.4.4).

The main features of the crustal structure of the Southern Alps are deduced from the profile Lago Lagorai-E (Fig. 11). The results are surprising. While the shortening is only small at the surface, as indicated by the relatively calm tectonics, the total crustal thickness in the area of the Dolomites is about the same as in the strongly folded central Alps. The sialic part is about 30-35 km thick and a moderate velocity inversion exists. It must be pointed out that no remarkable change in crustal thickness is evident from the record sections when crossing the Periadriatic line, the tectonic boundary between Central and Southern Alps.

7.4.2 Northern Calcareous Alps

In 1970-1974, within the program Geodynamics of the Mediterranean Area, several seismic-refraction profiles were recorded in the Northern Calcareous Alps, mainly between Salzach and Inn (Fig. 14), using commercial borehole blasts in quarries which were fired with delays in the order of some 10 ms. Because of this delay technique the maximum recording distance was only 65 km, with one exception.

These profiles have been evaluated by WILL (1975, 1976, 4.5 of this volume). Most of the correlated traveltime curves show effects which must be explained by lateral variations of seismic velocities. For this reason WILL developed a method which allows the computing of travel paths and travel times of seismic signals in models with two-dimensional velocity distribution (WILL, 4.5 of this volume).

Remarkable are the extremely high P-velocities measured in the limestones and dolomites of the Northern Calcareous Alps which reach from 5.5 km/s at the surface to more than 7 km/s at the base. The average thickness of limestones and dolomites is about 3 km.

Because of this high-velocity layer at the surface only some of the short profiles allowed the gathering of information on the material underneath, where shotpoint and recording sites are arranged in a suitable position in order to record arrivals from layers at greater depths. For the model shown in Fig. 15 corresponding arrivals were observed at distances beyond 30 km which are delayed by about 2 s with respect to the first arrivals up to 12 km distance being generated by the top layer (see 4.5 of this volume, Fig. 7). These delayed arrivals were attributed to the basement which is separated from the high-velocity limestone/dolomite layer by a low-velocity zone of several km thickness. According to the velocity this layer can be interpreted as Molasse and/or Helvetic rocks and/or Flysch underlying the Northern Calcareous Alps at least under the northernmost 15 km.

7.4.3 Pennine Window of the Engadine (Engadiner Fenster)

For the structure and the development of the Alps the so-called windows - an eroded area of a nappe that displays the rocks beneath it - are of special interest. At the margin of such a window the problem can be studied, if it is possible to detect thrust planes by seismic methods. In the Central Alps this means distinguishing between high-metamorphic crystalline rocks of Upper-Austroalpine units and the metamorphic schists of Pennine age. In the area of the Engadiner Fenster detailed seismic investigations organized by the Institut für Angewandte Geophysik, München, could be carried out thanks to the support of the Tiroler Wasserkraftwerke A.G., Innsbruck (SCHELIGA, 1971).

Two short seismic-refraction profiles (Kauner- and Radurschl-Tal, Fig. 16) were observed. In spite of great technical difficulties some reflection measurements could be performed in a tunnel connecting both valleys (Fig. 16).

The 16-km-long refraction profile in the Kaunertal gives a velocity distribution for the crystalline rocks of the Ötztal-nappe, quite typical for gneisses and granites of the uppermost kilometers. The reflection re-

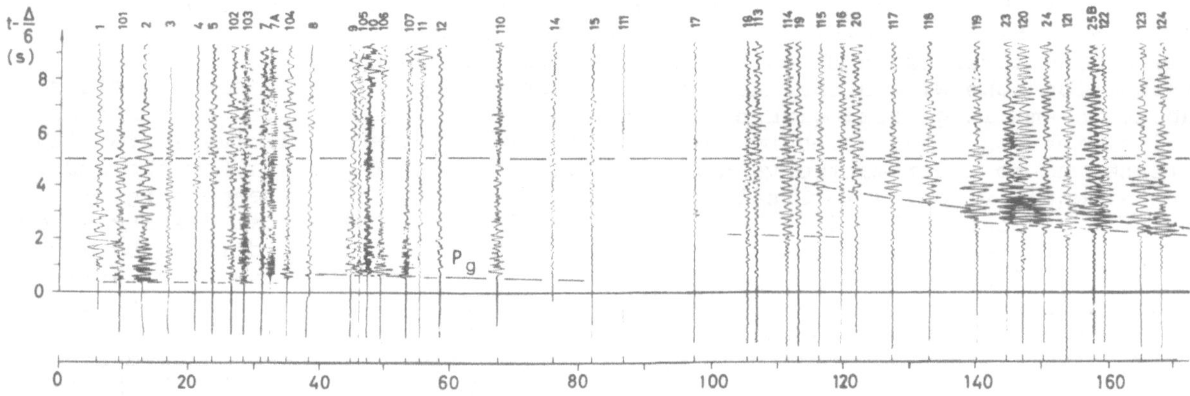

Fig. 11. Record section of the profile Lago Lagorai-E

cordings show only one clear event with a two-way travel time of 1.1 s resulting in a depth of about 3 km. Thus it does not seem unreasonable to associate this reflection with the thrust plane between Ötztal-nappe and Pennine rocks.

The situation is quite different in the Radurschl-Tal. Here the seismic-refraction profile (Fig. 17) crosses the outcropping contact between Ötztal-gneisses and Pennine schists (Bündner Schiefer). The travel times cannot be explained without assuming lateral inhomogeneity. Fig. 17 shows the model derived from the seismic observations including a geologic interpretation. The observed velocities, which are higher for Pennine rocks than for gneisses, are in agreement with the results from other observations in Pennine rocks. In the block diagram of Fig. 18 these results are summarized.

Finally it should be pointed out that the observations on the long-range profile from Eschenlohe to SW passing this area (Fig. 9) does not show any peculiarities in comparison to profiles outside.

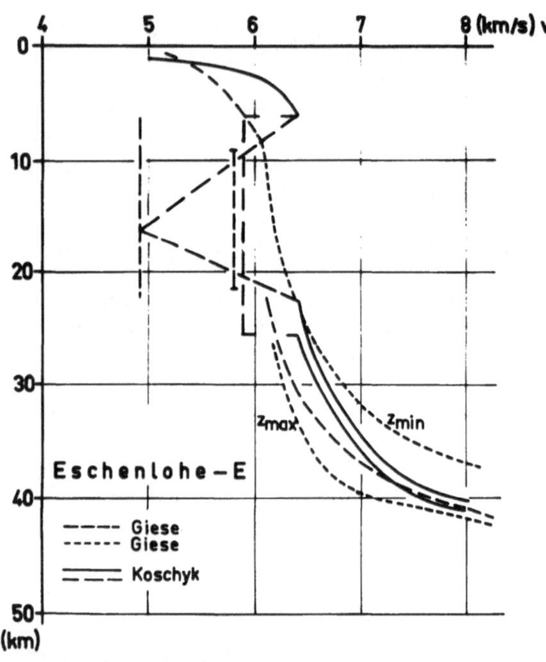

Fig. 12. Velocity-depth function for the profile Eschenlohe-E showing the range of possible solutions

7.4.4 A Possible Fine Structure of the Crust-Mantle Boundary

A detailed study of the P^M-wave group of record sections obtained in the Alpine area reveals that often it can hardly be explained by only one travel-time curve. GIESE (1972) has shown that the P^M-group splits into several branches showing a shingle-like arrangement, especially on record sections radiating from Eschenlohe. This effect is very clear on the profiles from Eschenlohe to the SE (Fig. 7), S (Fig. 8), and SW (Fig. 9). GIESE (1972) explains this feature by several narrow low-velocity zones in the lower crust characterized by only a few km in thickness, but a very intensive velocity decrease of the order of 1-3 km/s. Examples of such velocity-depth functions are shown with the cross section of Fig. 13. A similar structure is also proposed by MEISSNER (1967a) who

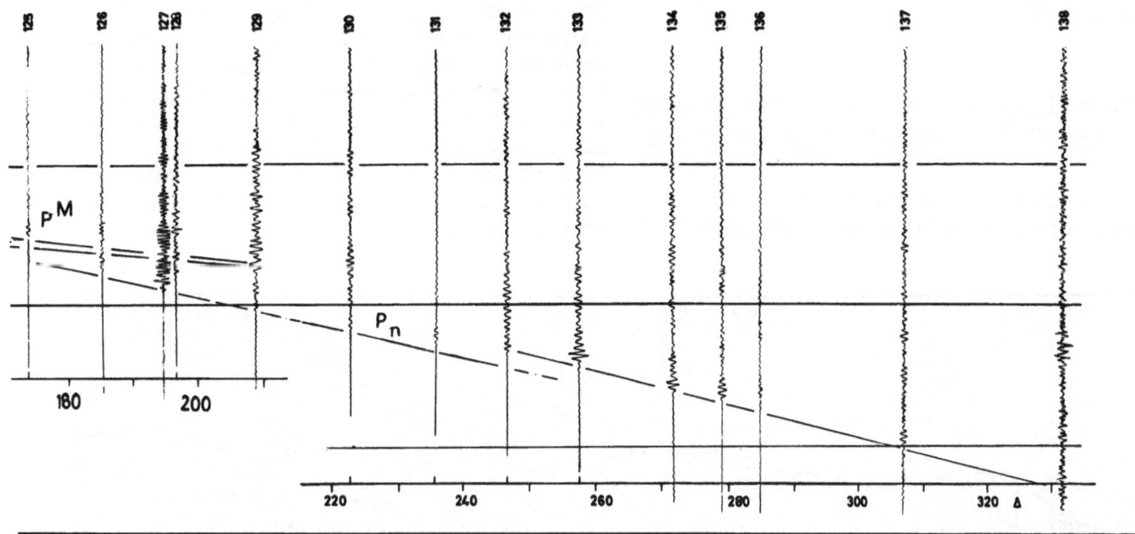

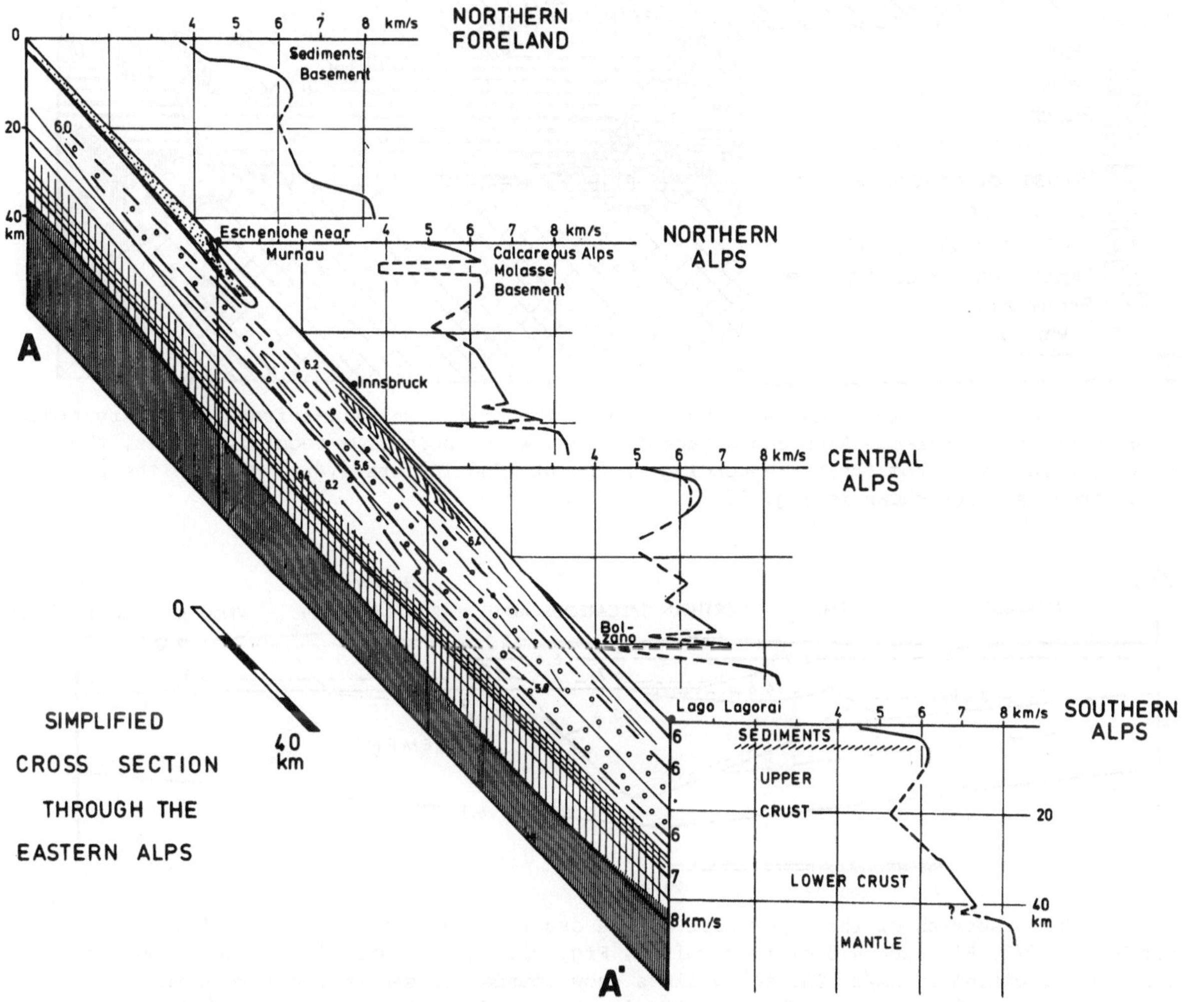

Fig. 13. Simplified N-S cross section through the eastern part of the Alps (*AA'* in Fig. 1), showing contours of equal velocity and typical velocity-depth functions. *Small circles:* the position of the low-velocity zone

derives a laminated structure of the crust/mantle transition from near-vertical reflections.

7.5 Crustal Structure along the Border Zone between Eastern and Western Alps

Fig. 19 shows a N-S section crossing the Alps approximately at the border zone between the Western and Eastern Alps. This section is based mainly on the evaluation of the profiles Lago Bianco-Eschenlohe (Fig. 20) and reverse (Fig. 9), Lago Bianco-ENE, Lago Bianco-Lago Lagorai (Fig. 21) and reverse (Fig. 22), Lago Bianco-SE (Fig. 23), fan observations at 140 km east of Lago Bianco, and Lago Bianco-NNW (see, e.g., BEHNKE, 1967, 1969, 1971).

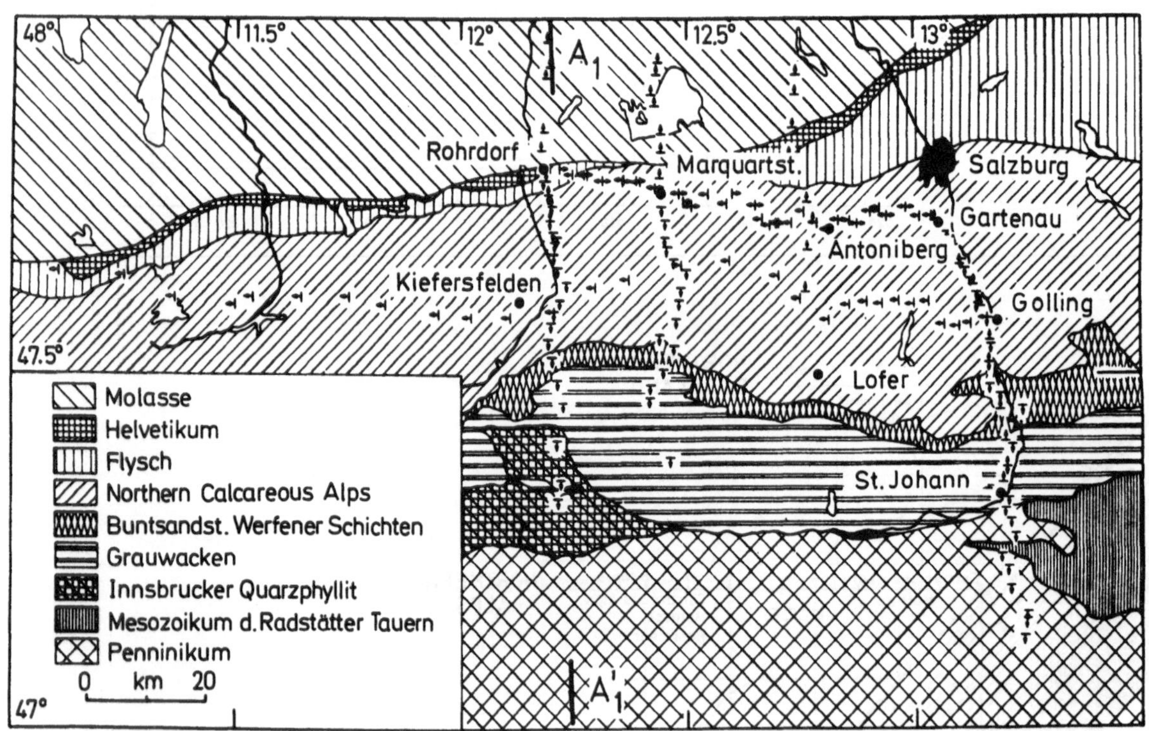

Fig. 14. Simplified geologic map showing the position of the seismic-refraction observations in the Northern Calcareous Alps and adjacent areas. ● Shotpoints, ↦ recording sites, the arrow pointing away from the corresponding shotpoint. The letters A_1A_1' indicate the position of the cross section shown in Fig. 15

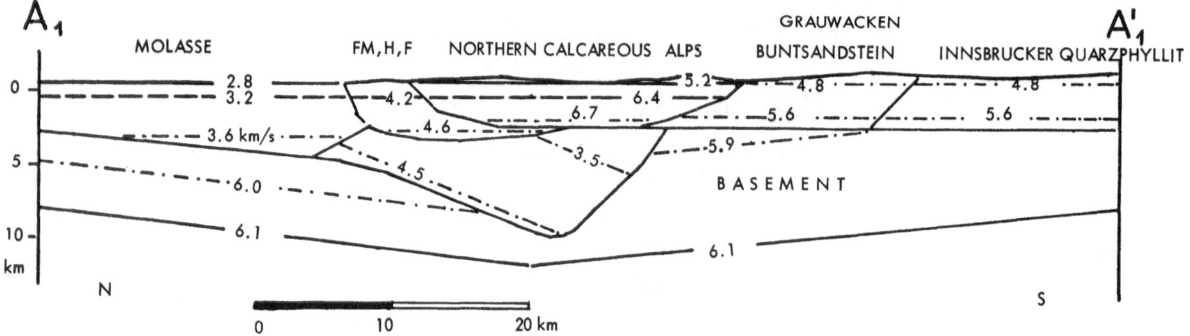

Fig. 15. Cross section of the upper part of the crust through the Northern Calcareous Alps. From WILL (1975, Fig. 28, 4.5 of this volume, Fig. 7). The numbers within the cross section indicate velocities in km/s. The solid lines show boundaries separating layers which can be interpreted as geologic units. The velocity is changed discontinuously at such boundaries as well as at internal boundaries (*dashed lines*). *Dashed-dotted velocity lines:* not boundaries, but their interval distance indicates the size of the velocity gradient within the corresponding layer. *FM* folded Molasse, *H* Helvetic series, *F* Flysch

The structure of the foreland is similar to that in southern Bavaria. Towards the central part of the Alps, crustal thickness increases considerably from 35 to 55 km. In the Central Alps (Graubünden), the pillow with higher velocities in the upper crust is even more distinct than beneath the Eastern Alps. The low-velocity zone is located between 15 and 35 km depth. Thus the thickness of the upper crust is about 35 km and the total crustal thickness is 50-60 km.

When moving from the Central to the Southern Alps - in contrast to the Eastern Alps - crustal thickness decreases drastically to about 30 km. The thickness of the sialic crust is here only 20 km, less than under the Dolomites, and the velocity inversion is weak.

7.6 Crustal Structure of the Western Part of the Alps

7.6.1 Foreland and Central Zone

For the discussion of the crustal structure of the Western Alps, three main areas must be distinguished: the foreland, the Central Alps, and the anomalous zone of Ivrea which belongs from the geological point of view to the unit of the Southern Alps. The foreland is bordered by the Rhône Gra-

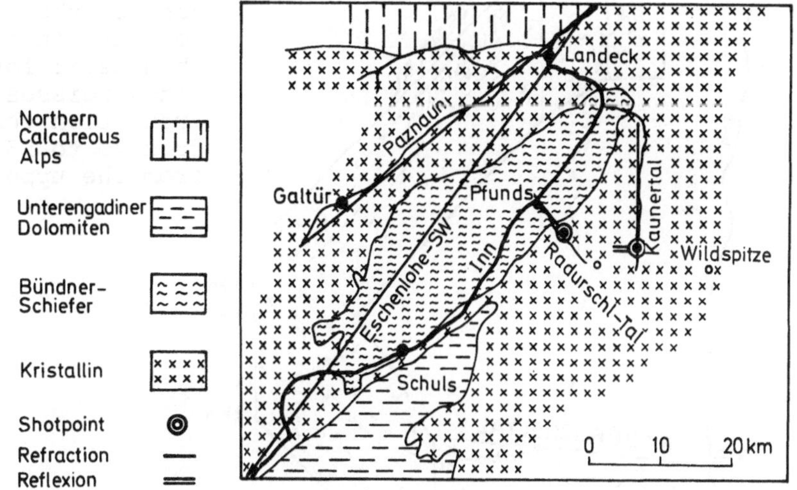

Fig. 16. Position map of profiles and shotpoints in the northeastern part of the Engadiner Fenster

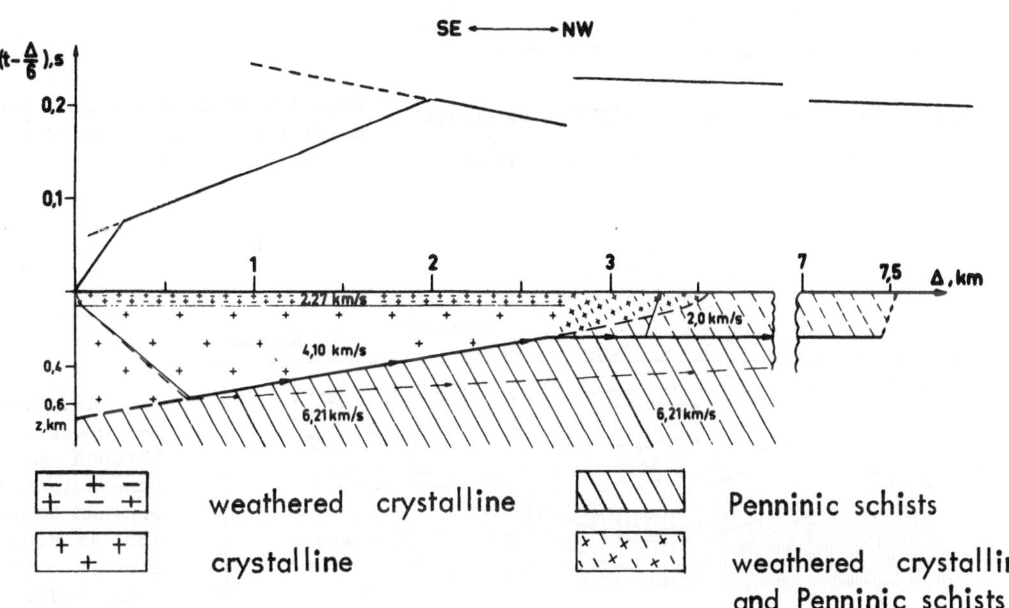

Fig. 17. Travel-time observations and cross section along the Radurschl-Tal crossing the boundary Upper Austroalpine unit versus Penninic zone. (After SCHELIGA, 1971)

ben in the west and by the Swiss Jura and the Swiss Molasse Basin in the northwest, and it is separated from the Central Alps by the granitic autochthonous massifs, as there are the Aar Massif, the Belledonne and Pelvoux Massifs, and the Mercantour Massif.

Several profiles have been observed here, in the so-called zone externe of southeastern France, a part of which was discussed in detail by GIESE et al (1967). Two examples, the record sections of the profiles Lac Nègre-NW and Roselend-NE, are shown in Figs. 24 and 25. The average velocity of the crust amounts to slightly more than 6 km/s, its thickness being about 40 km. A velocity inversion covering a depth range of approx. 15 km exists all over the foreland of the Western Alps, but its position in depth varies.

Only few complete profiles were recorded which deal only with the structure of the central part of the Western Alps, namely the profiles from Lago Bianco to the east and towards Mont Cenis (Fig. 26). Fig. 27 shows a cross section between Grenoble and Torino. As typical for the central zone of the Alps, the increase of velocity in the uppermost km is greater than usual for regions with outcropping gneisses and granites. A broad transition zone reaching approximately from 40 to 55 km depth is separated from the upper crust by a very dis-

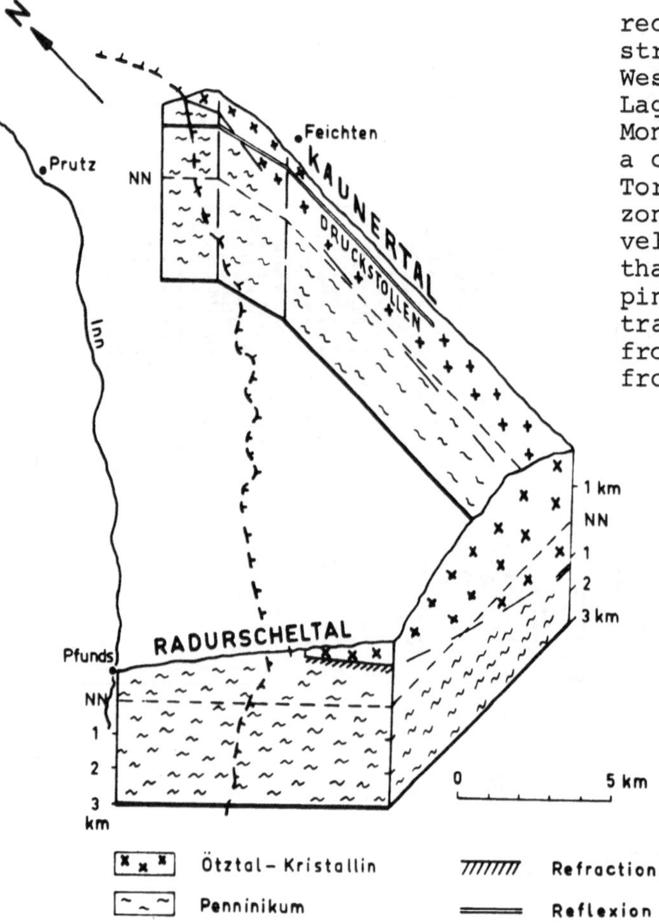

Fig. 18. Block diagram of the northeastern Engadiner Fenster between Kaunertal and Inn

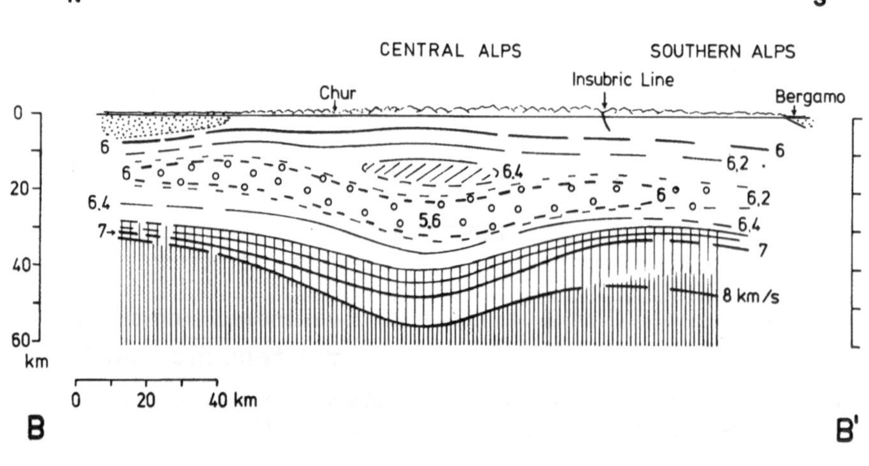

Fig. 19. Simplified N-S cross section through the border zone between Eastern and Western Alps (*BB'* in Fig. 1) showing contours of equal velocity. *Small circles:* Position of the low-velocity zone

tinct low-velocity zone of more than 20 km thickness.

Similar average crustal velocity of 6.25 km/s and total crustal thickness of about 55 km are derived from the relatively poor P^M phase of the profile Mont Cenis-Lago Bianco (Fig. 28). From the observations of this profile, a discontinuity showing a velocity increase up to 7 km/s is found at about 25 km depth. This layer is separated from the mantle by a distinct velocity inversion. The relation to the high-velocity body of Ivrea, adjacent to the southeast, is an open question.

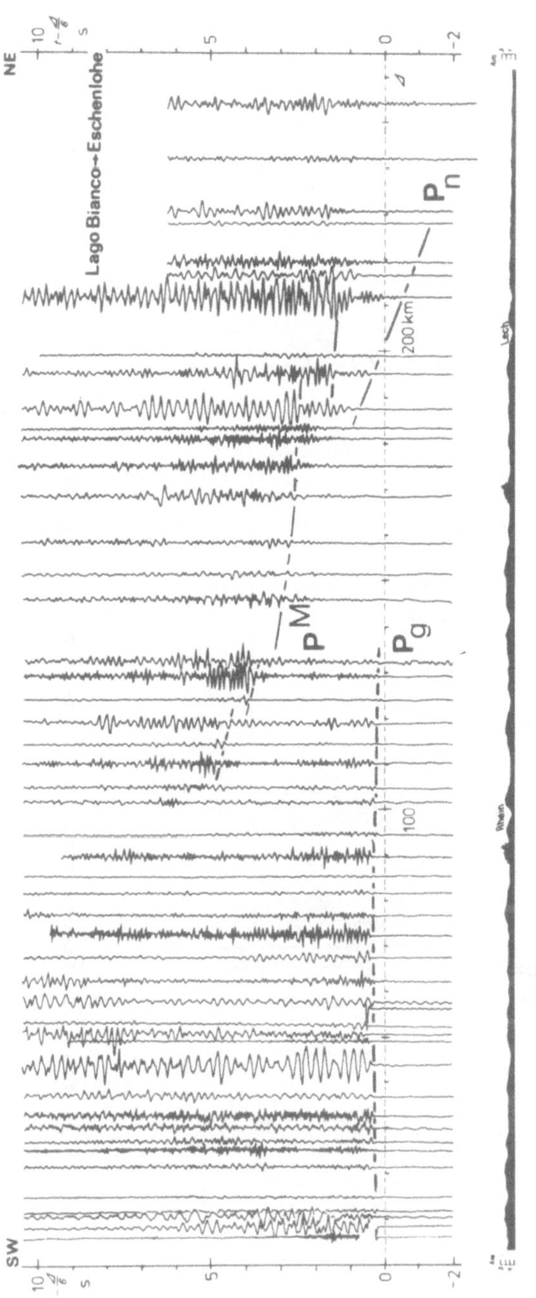

Fig. 20. Record section of the profile Lago Bianco-NE

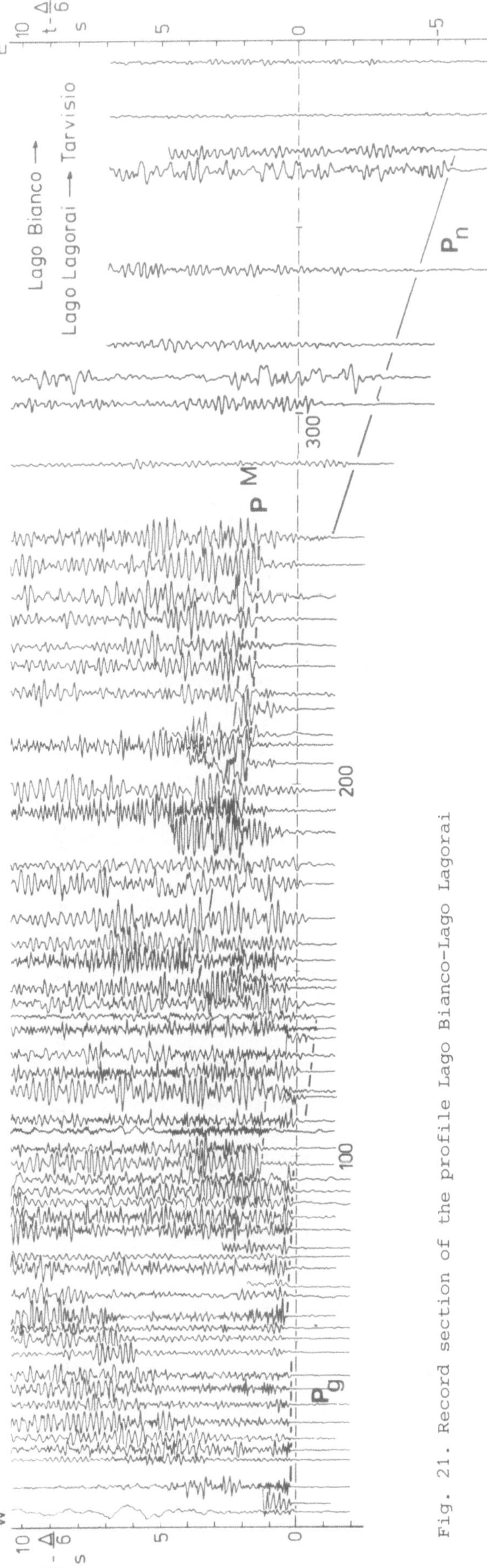

Fig. 21. Record section of the profile Lago Bianco-Lago Lagorai

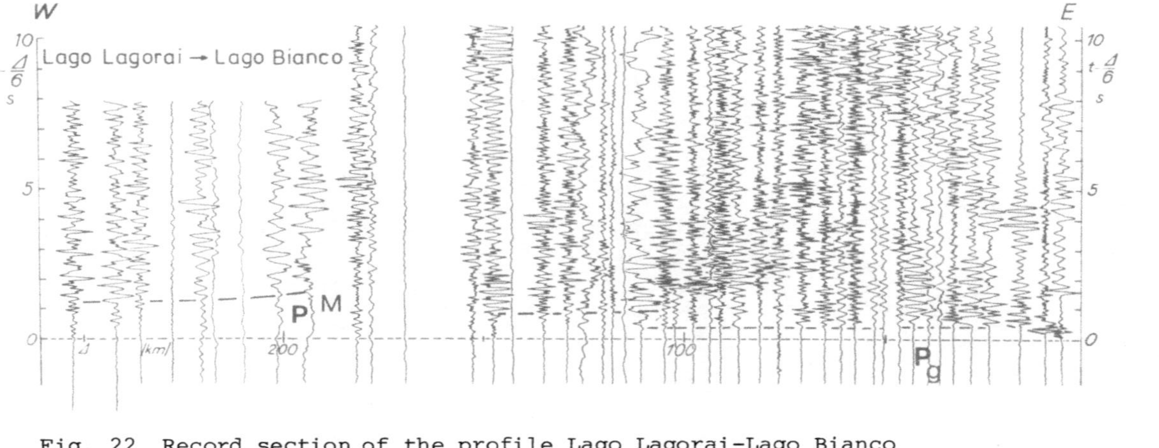

Fig. 22. Record section of the profile Lago Lagorai-Lago Bianco

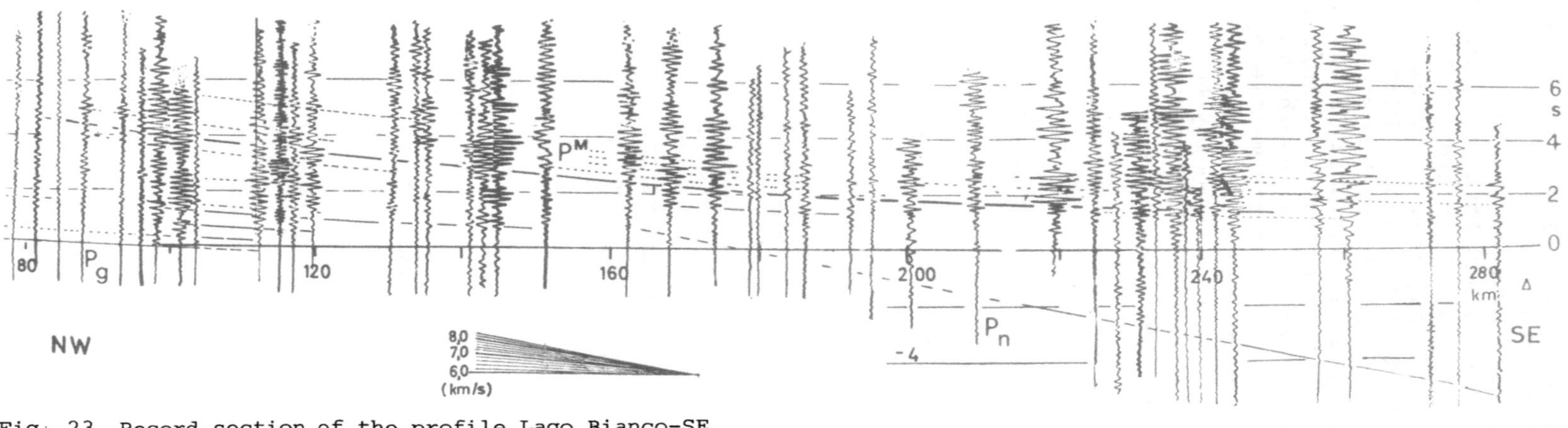

Fig. 23. Record section of the profile Lago Bianco-SE

Fig. 24. Record section of the profile Lac Nègre-NW

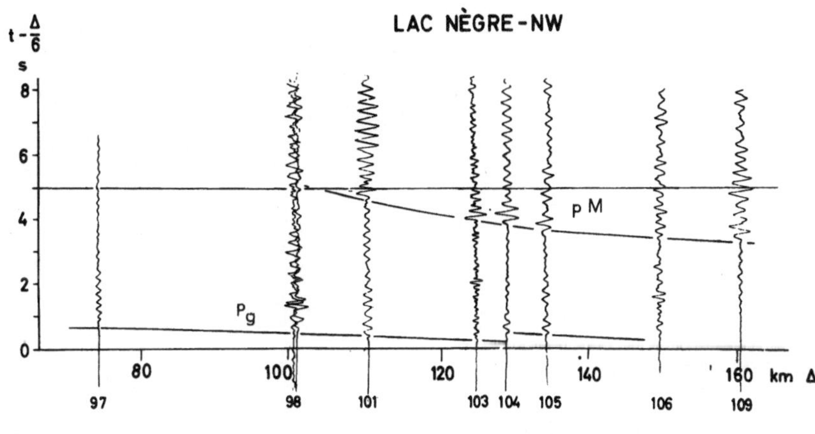

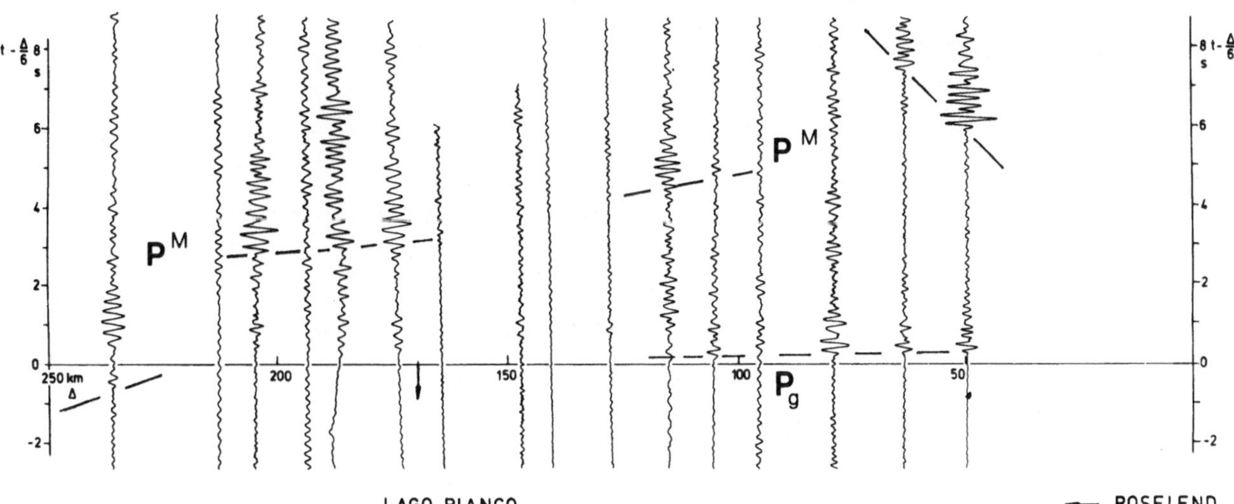

Fig. 25. Record section of the profile Roselend-NE

7.6.2 The Zone of Ivrea

The inner arc of the Western Alps is characterized by a strong gravity high which is presented in the contour map of Bouguer anomalies (Fig. 2) as well as in that of isostatic anomalies. This gravity high extends from Locarno in the north via Ivrea to Cuneo in the south. As NIGGLI (1946) has already pointed out, this sharp anomaly must be mainly caused by high-density material in the depth range between 0 and 20 km. Deeply seated high-density rocks cannot significantly influence the sharp form of this anomaly.

The seismic-refraction observations on the profiles crossing this anomalous zone of Ivrea show some particular features which will be discussed briefly. As a characteristic example, the profile Lac Nègre-Ivrea is shown in Fig. 29. From the first arrivals beyond a distance of 70 km, a velocity of about 7.4 km/s is derived. In addition to this phase - named P_i in Fig. 29 - a retrograde travel-time curve P^I, based on clear secondary arrivals, is observed to which P_i is tangent at about 40 km distance. Both phases indicate the existence of high-velocity material at only 5-10 km depth (Fig. 27). The record section also shows the typical P^M-phase, but at a distance of only 115 km with an unusually large reduced travel time of 6 s which can be explained only by assuming an intensive low-velocity zone (4-5 km/s) beneath the Ivrea body at a depth between 20 and 40-50 km. As second example the profile Mont Cenis-SSE is shown in Fig. 30. Here also, the P^M-group can be recognized following the phases P^I and P_i with a considerable intercept-time delay.

For the southern part of the zone of Ivrea, GIESE et al. (1971) have studied the areal extension of this

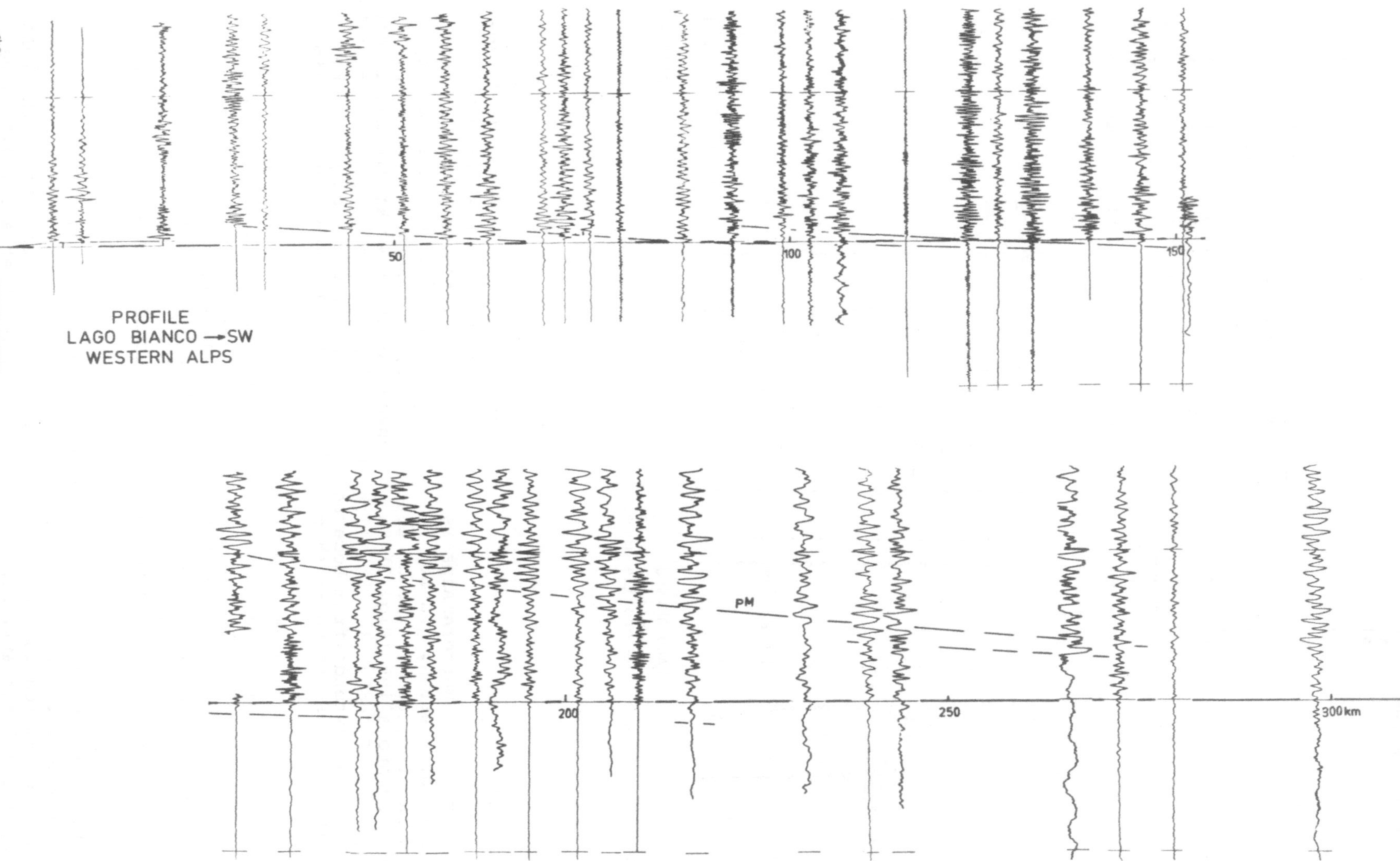

Fig. 26. Record section of the profile Lago Bianco-SW

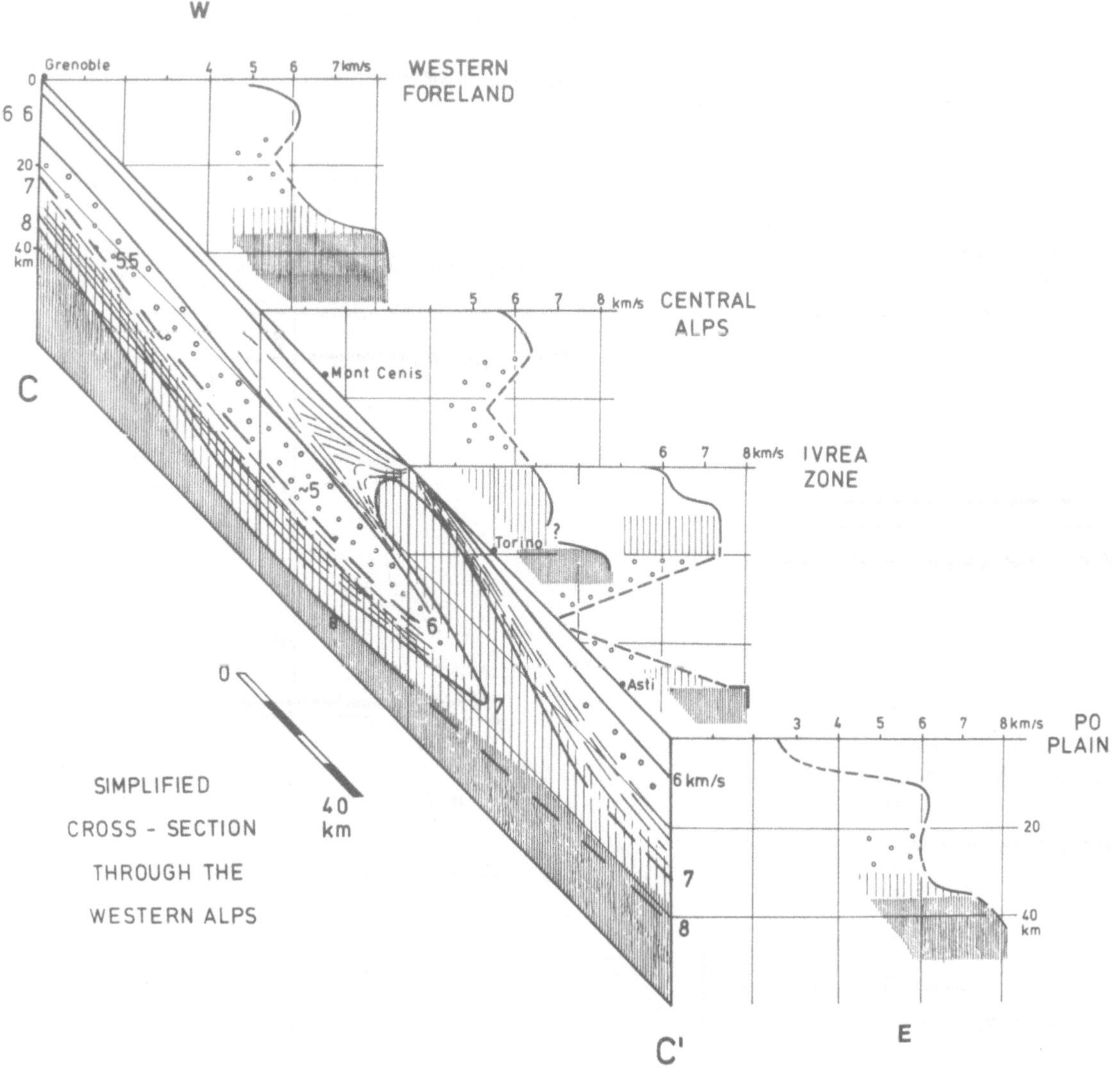

Fig. 27. Simplified W-E cross section through the western part of the Alps (*CC'* in Fig. 1) showing contours of equal velocity and typical velocity-depth functions. *Small circles:* Position of the low-velocity zone

anomalous body investigating systematically the areal distribution of the wave groups P^I and P_i, related with the existence of a high-velocity body, and of the wave groups P^M and P_n, connected with the upper mantle (Fig. 31 and Table 1). If both pairs of wave groups exist, the material of the body of Ivrea and its eastward continuation is clearly separated from the rocks of the transition zone crust/mantle. Those parts of profiles where this separation can be recognized are marked by thick lines in Fig. 31 (crosshatched area). The distance interval of the travel-time curve under discussion is related to the distance interval of the corresponding ray vertices. Therefore, in Fig. 31, the interesting distance ranges are plotted at half their original value. The 200 km fan, for instance, is plotted at 100 km distance. On the eastern side of the gravity high of Ivrea, the material characterized by the wave groups P^I, P_i dips and passes into the transition zone crust/mantle under the western part of the Po plain. Here, a separation, based on travel-time diagrams, is no more possible. Those parts of the profiles are marked by thin lines in Fig. 31 (area hatched by horizontal lines).

On the profile Mont Cenis-NE (Fig. 32), crossing the northern part of the zone of Ivrea, phases P^I and P^M can be correlated, too, proving that the structure north of Ivrea is similar

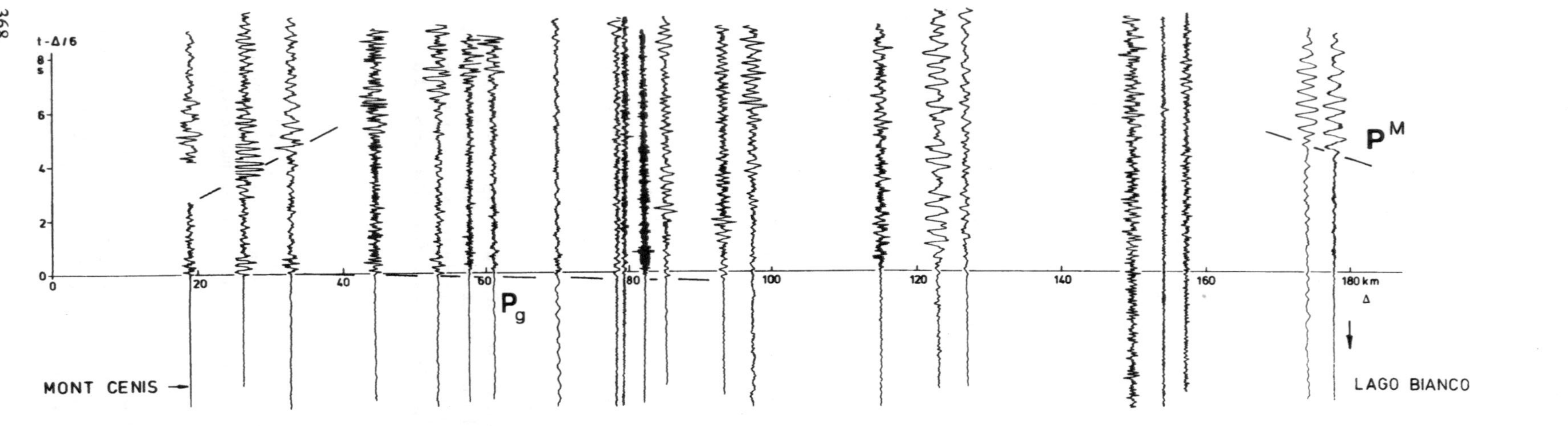

Fig. 28. Record section of the profile Mont Cenis-Lago Bianco

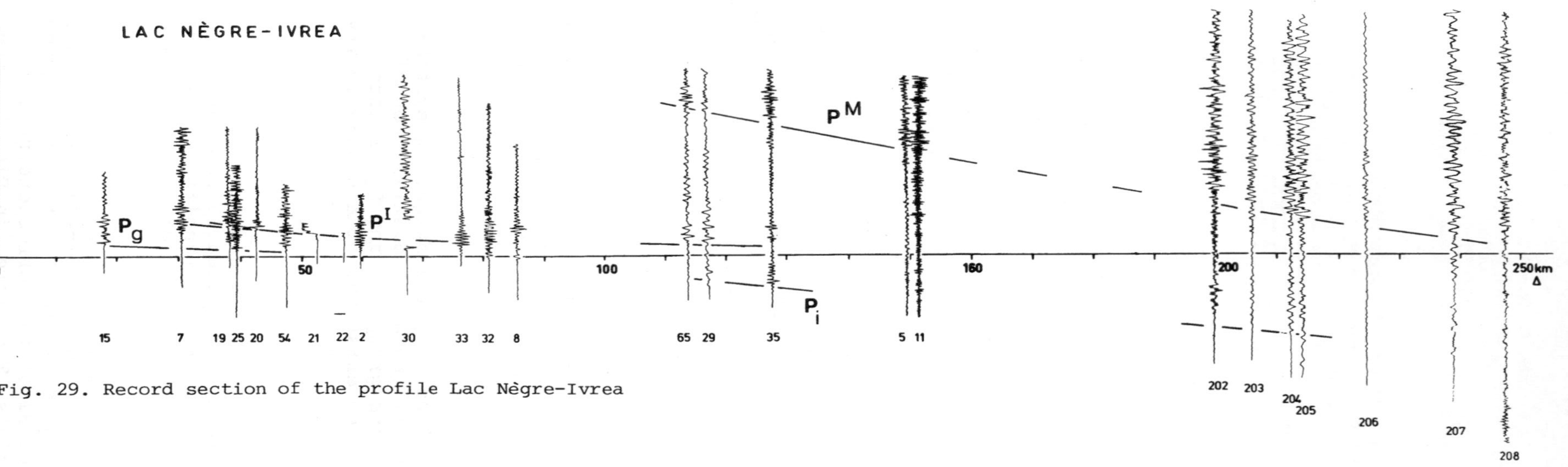

Fig. 29. Record section of the profile Lac Nègre-Ivrea

to that under the southern Ivrea zone.

Fig. 27 demonstrates these features in the crustal cross section. The well-expressed low-velocity zone of the Western Alps and the Ivrea zone extends to Torino and disappears eastward unter the Po plain. The high-velocity material of the Ivrea body has rather a direct link to the upper mantle at its eastern flank, i.e. to the upper mantle under the western Po plain.

This anomalous crustal structure can only be explained by regarding the development of the Alpine orogene. The Ivrea zone with its high-velocity and high-density material belonging primarily to the lower crust, forms the northwestern and western margin

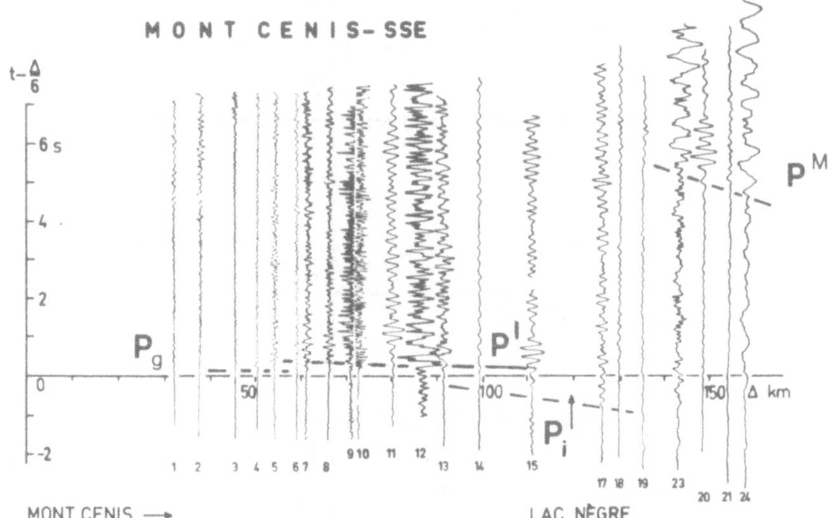

Fig. 30. Record section of the profile Mont Cenis-SSE

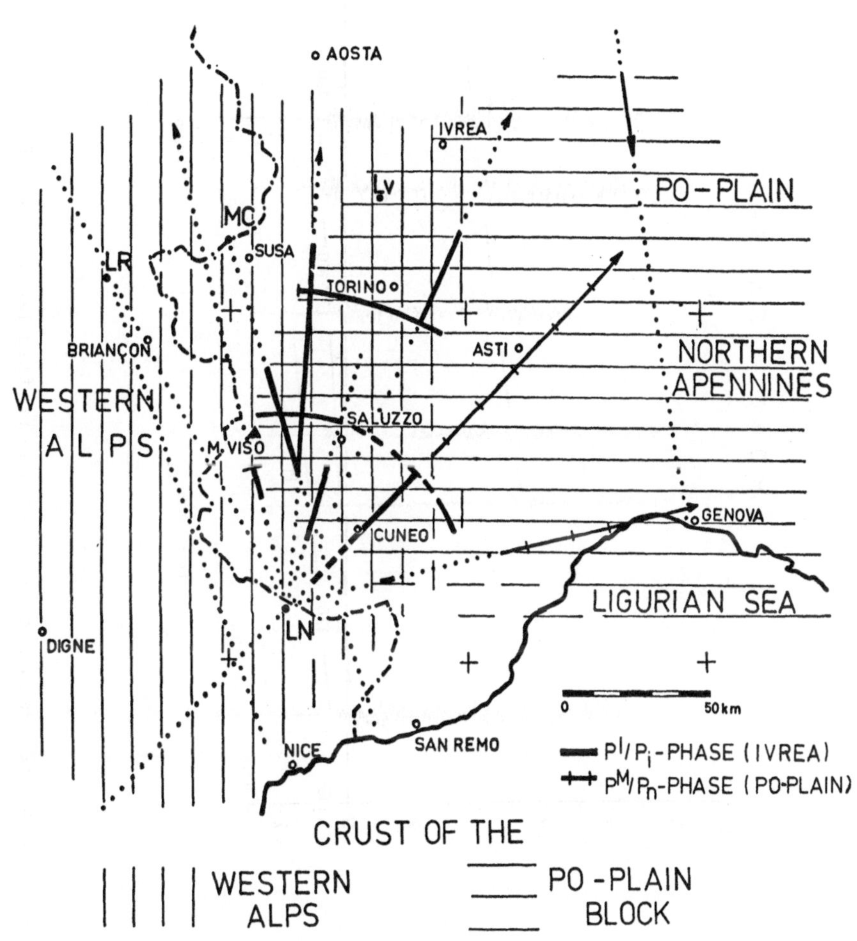

Fig. 31. Areal distribution of the phases P^I/P_i (Ivrea) and P^M/P_n (Po plain) demonstrating the link of the crust/mantle transition zone of the Po-plain block (*hatched by horizontal lines*) with the Ivrea body overlying upper-crustal material of the Alpine geosynclinal trough (*cross-hatched area*), east of the normal crust of the Western Alps (*hatched by vertical lines*)

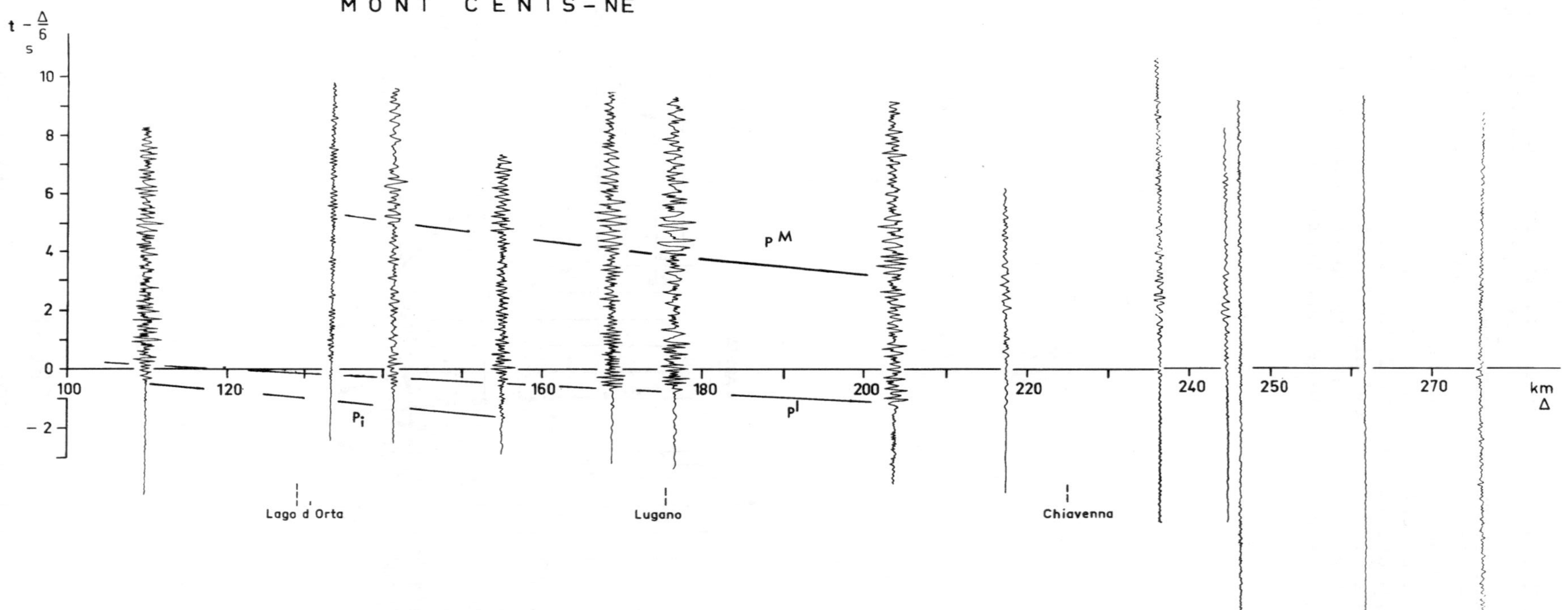

Fig. 32. Record section of the profile Mont Cenis-NE

Table 1. Distances at which the wave groups P^I, P_i, and P^M, P_n are recorded

I. Wave groups P^I, P_i

Profile	Wave group	Distance (km)
Mont Cenis-SSE	P_i	90-160-?
Mont Cenis-Le Revest	-	-
Lac Rond-Nice	-	-
Lac Nègre-Lac Rond	-	-
Lac Nègre-Lac Léman	P_i	70-84
Lac Nègre-Aosta	P_i	67-200-?
Lac Nègre-Saluzzo	P^I	30-85-?
Lac Nègre-Lago Maggiore	P_i	?-200-260
Lac Nègre-Asti	P^I	18(?)-58-120
Lac Nègre-Genova	P^I	70- > 140
Lago Bianco-S	-	-
fan Lac Nègre 120 km	-	-
fan Lac Nègre 200 km	-	-

II. Wave groups P^M, P_n

Profiles	Wave group	Distance (km)
Mont Cenis-SSE	P^M	130-160-?
Mont Cenis-Le Revest	P^M	120-200(?)
Lac Rond-Nice	P^M	?-148-?
Lac Nègre-Lac Rond	P^M	93-155-?
Lac Nègre-Lac Léman	P^M	95-150-?
Lac Nègre-Aosta	P^M	100-183
Lac Nègre-Saluzzo	P^M ?	85
Lac Nègre-Lago Maggiore	P^M	?-200-250
Lac Nègre-Asti	P^M	?-135-280
Lac Nègre-Genova	P^M	70(?)
	P_n	148-?
Lago Bianco-S	P_n	?-182-220-?
fan Lac Nègre 120 km	-	-
fan Lac Nègre 200 km	-	-

of the backland of the Alps, the Po-plain block. When, during the Alpine orogeny, the foreland and the backland had approached each other to the present position, the lower-crustal material of the western margin of the Po-plain block emerged to the position of the present Ivrea zone or Ivrea body respectively, while upper-crustal material of the Alpine geosynclinal trough subsided and shifted underneath. The subsided crust can be associated with the low-velocity zone beneath the high-velocity material of the Ivrea zone. Due to the very strong velocity reduction, partial melting of this material must be assumed (GIESE et al., 1968).

7.7 Compilation of the Main Results

Using reverse and crossing profiles, cross sections with lines of equal velocity can be constructed. To elaborate the main crustal features, some lines have been chosen crossing perpendicularly the axis of the Alps and containing a maximum number of reverse and crossing profiles. The velocity-depth cross-sections have then been compiled along these lines which are arranged to a block diagram and shown in Figs. 33 and 34, together with the gravity contours or the main tectonic features resp. On the basis of the seismic-cross sections MAKRIS (1971) calculated gravity models and could demonstrate good agreement of these calculated models with the observed gravity data when matching the seismic results.

An examination of Figs. 33 and 34 shows the following main features:

1. The crust becomes thicker from 30 km under the northern and western foreland towards the axis of the Alps and reaches values of 50-60 km. The crust becomes thicker also when moving from the Mediterranean coast northward to the Western Alps.

2. Regarding the two lines v = 7 km/s and v = 8 km/s it can be stated that the crust/mantle transition zone thickens in the same direction from 2-5 to about 10 km.

3. Under the axis of the Alps, at depths between 10 and 30 km, i.e. in the lower part of the upper crust, a marked velocity inversion exists which becomes more accentuated under the axis of the Alps with minimum velocities between 5.0 and 5.5 km/s. It is interesting to note that near the Mediterranean coast, this inversion scarcely exists.

4. At the inner arc of the Western Alps, in the area of the zone of Ivrea, the crust shows a very complicated structure: a layer with extremely low velocities (4-5 km/s) underlies the high-velocity material (7.2-7.4 km/s) of the body of Ivrea which is linked to the east with the lower crust and upper mantle under the Po plain.

A contour map of total crustal thickness is approximated in Fig. 35, showing the depth of the strongest velocity gradient in the velocity

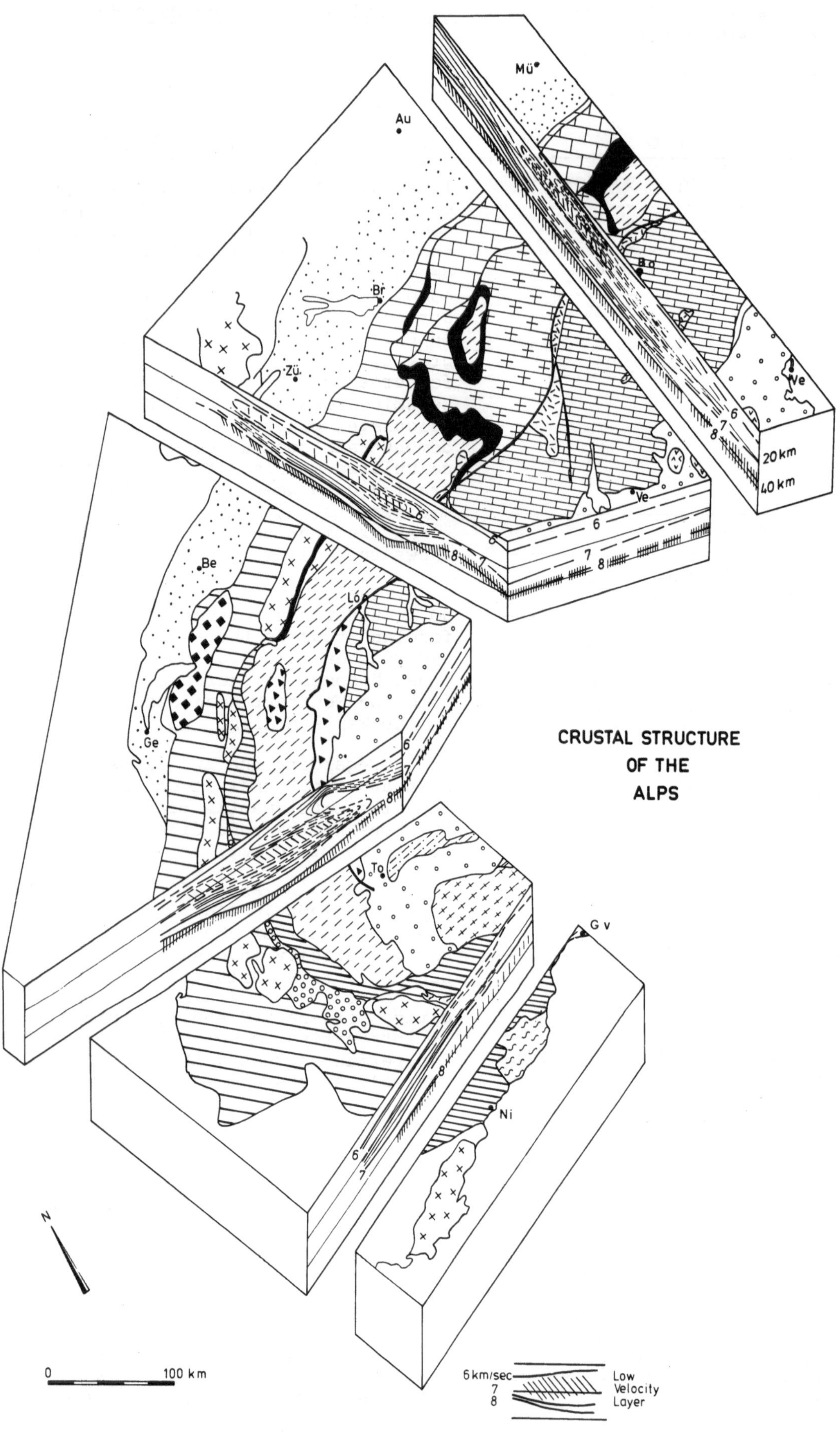
CRUSTAL STRUCTURE
OF THE
ALPS
Mü
Au
Br
Zü
Ba
Ve
Be
Lo
Ge
To
Gv
Ni
20 km
40 km
N
0
100 km
6 km/sec
7
8
Low
Velocity
Layer

◄ Fig. 33. Block diagram of the Alps showing the main features of crustal structure together with the main tectonic units. Explanaions:
Au Augsburg, *Be* Bern, *Bo* Bolzano, *Br* Bregenz, *Ge* Genève, *Gv* Genova, *Lo* Locarno, *Mü* München, *Ni* Nice, *To* Torino, *Ve* Venezia, *Zü* Zürich

- Molasse basin
- Helvetic and Dauphiné zone
- Pre-Alps nappes (penninic to ultrahelvetic domain)
- Ubaye-Embrunais nappes
- Helminthoid nappes (penninic domain)
- Briançonnais zone
- Penninic zone
- External crystalline massifs
- Ivrea zone (Southern Alps)
- Northern Calcareous Alps (upper Austro-Alpine nappes)
- Crystalline of the central zone in the Eastern Alps (Austro-Alpine nappes)
- Lower Austro-Alpine unit
- Southern Alps
- Tertiary igneous rocks
- Basic volcanic rocks
- External Ligurian nappes
- Post-geosynclinal sediments
- Young sediments of the Po plain
- Insubric line

---

range 7.5-8.2 km/s, i.e. the crust/mantle boundary. The special feature of the zone of Ivrea is indicated by two sets of contours: Solid lines indicate the crust/mantle boundary when approaching from the west, long-dashed lines show the crust/mantle boundary of the Po plain and its continuation to the west into the surface of the body of Ivrea.

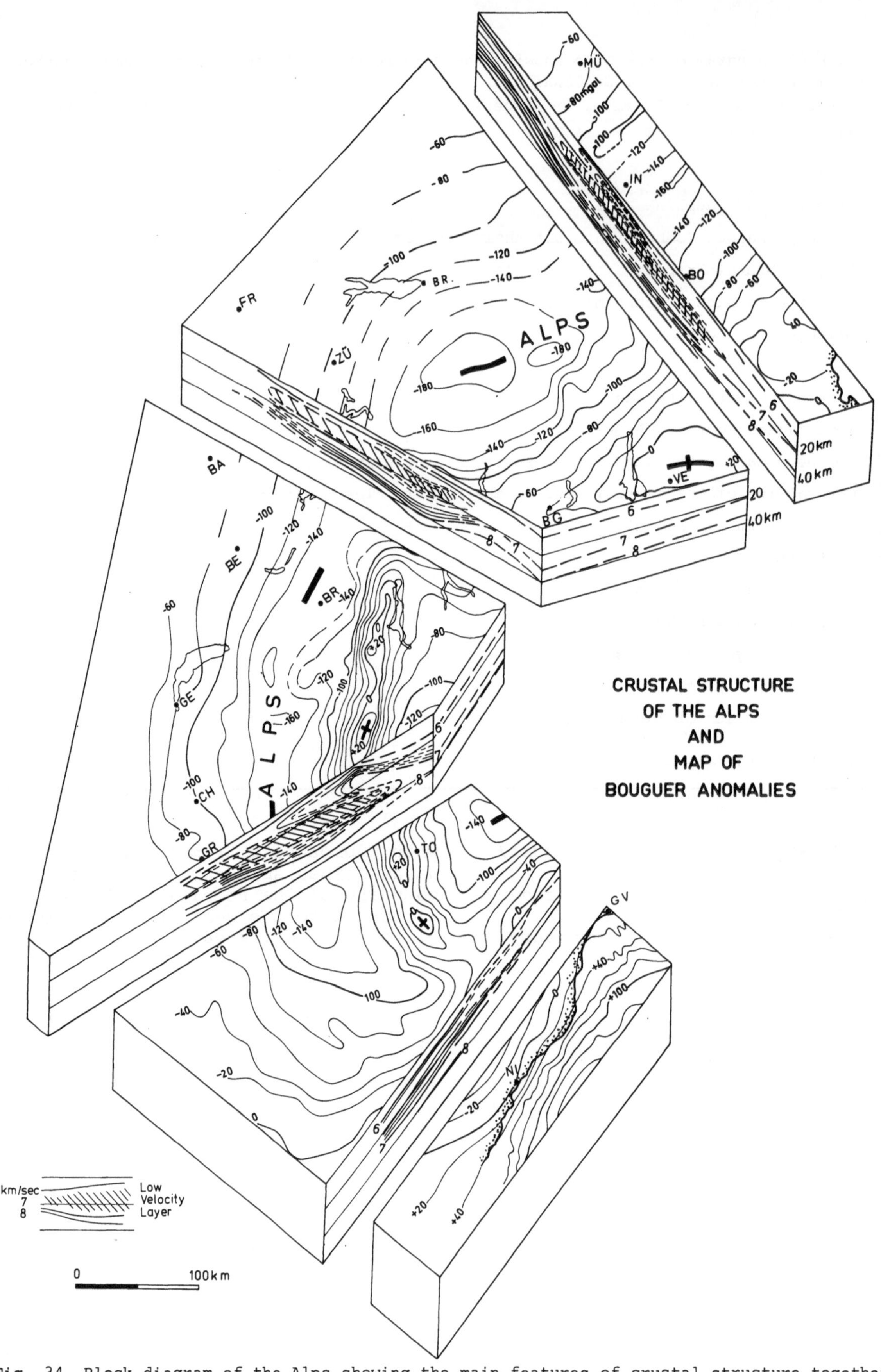

Fig. 34. Block diagram of the Alps showing the main features of crustal structure together with the Bouguer anomalies. Explanations: *BA* Basel, *BE* Bern, *BG* Bergamo, *BO* Bolzano, *BR* Bregenz, *CH* Chamonix, *FR* Freiburg, *GE* Genève, *GR* Grenoble, *GV* Genova, *IN* Innsbruck, *MÜ* München, *NI* Nice, *TO* Torino, *VE* Venezia, *ZÜ* Zürich

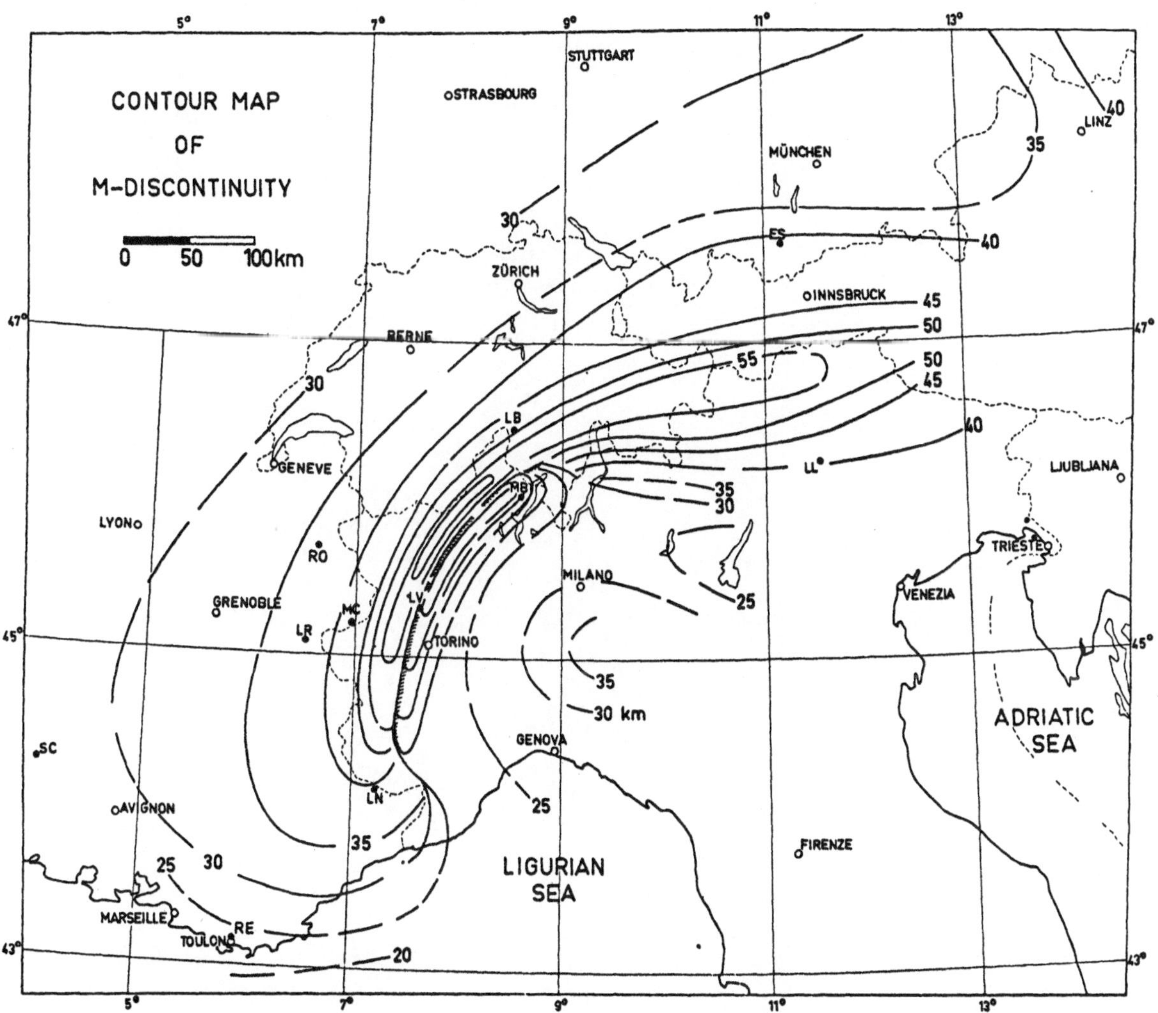

Fig. 35. Contour map of the depth of strongest velocity gradient in the velocity range 7.5-8.2 km/s which is regarded as crust-mantle boundary. Explanations see Fig. 3

8. Some Aspects of Comparing Seismic and Gravimetric Results from Selected Areas

R. MEISSNER

ABSTRACT

A comparison between crustal thicknesses, discontinuities, velocity-depth structures, and composition shows distinct differences for certain geologic units. Compared to the rather uniform oceanic lithosphere, the continental crust and upper mantle have a more complicated structure. However, many similarities are found when comparing similar tectonic areas such as young mountain belts, old shield areas, or certain transition areas. A possible connection exists between the age and the crustal thickness in stable areas.

ZUSAMMENFASSUNG

Ein Vergleich zwischen Krustenmächtigkeiten, Diskontinuitäten, Geschwindigkeits-Tiefen-Profilen und Materialzusammensetzung zeigt für bestimmte geologische Großstrukturen ausgeprägte Unterschiede. Im Vergleich zu der recht einheitlichen ozeanischen Lithosphäre haben kontinentale Kruste und oberer Erdmantel einen viel komplizierteren Aufbau. Trotzdem finden sich bei einem Vergleich ähnlicher tektonischer Strukturen wie junger Faltengebirge, alter Schilde und Übergangszonen viele Ähnlichkeiten. Es scheint ein Zusammenhang zwischen Alter und Krustenmächtigkeit stabiler Gebiete zu bestehen.

8.1 Introduction

The ultimate goal of deep-seismic sounding (DSS), besides the determination of detailed velocity-depth functions and regional structure of interfaces, must be a correlation of DSS data to geologic and tectonic units and to different evolutionary stages of the earth's crust and mantle. The comparison of DSS data should involve oceanic, continental, and transitional regions as well as different continental units like shields, platforms, old and new orogenic regions and the present margins of continents. The following three papers in this Chapter should be regarded as only a beginning of a planned greater study.

8.2 The Oceanic Lithosphere

The new concept of plate tectonics seems to explain the forming and evolution of oceanic crust quite satisfactorily (for review see BULLARD, 1969, or MENARD, 1969). Together with the underlying mantle, the oceanic crust is formed along ridges. These ridges or rift zones may represent the present stage of an evolution which has started as a simple intercontinental or interoceanic graben (DIETZ, 1961; HESS, 1962; TALWANI et al., 1965). In these areas no definite upper mantle velocity is found, but intermediate velocities between 7.3 and 7.7 km/s, sometimes interpreted as mantle-crust mix and sometimes as young and not completely differentiated mantle material in a high temperature state. In fact, velocities within the a.m. range are recorded from the oceanic ridges or from graben zones - as for instance from the Utah graben (COOK, 1967), the Baikal graben (ARTEMJEV and ARTYUSHKOV, 1971) and from the Afar triangle (BERCKHEMER, personal communication).

The normal oceanic lithosphere shows a crustal thickness of 5-7 km, a Mohorovičić discontinuity (MD) at 10-12 km depth with "normal" mantle velocities of 8.0 to 8.2 km/s. No continuous or strong reflections are reported from the deep ocean basins. This lack of good reflection data can only partly be explained by geometric reasons (VETTER and MEISSNER, 4.7 of this volume). There seems to be a basic difference between oceanic and continental lithosphere. The uppermost part of the lithosphere, the crust, is well known to be quite different. The oceanic crust has no granitic layer with velocities of 5.8 to 6.3 km/s, but consists of basalts and gabbros underlain by a mantle most probably composed of peridotite, at least in its upper part. This is indicated by lots of peridotite rocks dredged from the rift zones, by long-range refraction experiments, by topographic changes away from the ridges and by density and velocity comparisons (FORSYTH and PRESS, 1971). Further down at a depth of 30-40 km there seems to be a general increase in seismic velocity as reported by HALES (1969, HALES et al. (1970), and ZVEREV (1970). Due to FORSYTH and PRESS (1971) these data might indicate a contribution of eclogite in the lower zone of the lithosphere.

In general, according to geophysical, geologic, and mineralogic data the evolutionary behavior of the oceanic lithosphere appears to be quite uniform and simple compared to the different and more complicated continental regions.

8.3 The Continental Crust

Large parts of the continental crust are older than their oceanic counterparts, i.e. older than 200 million years. Most complexes have suffered repeated differentiation and crystallization processes; the oldest and most stable areas of shields and platforms have been consolidated more than a billion years ago. We must suppose that by repeated processes of orogeny, sedimentation, erosion, and metamorphism granitic layers have been formed by melting and remelting of sediments and basalts. By seismic observations velocities of 5.8 to 6.3 km/s are found virtually in all parts of the continental crust. In many areas such as in shield areas of the USSR (PAVLENKOVA, 1969; KOSMINSKAYA, 1971) and Canada (CLOWES et al., 1968) and in areas of the Hercynic orogeny like the German Mittelgebirge a boundary inside the crust - the Conrad discontinuity (CD) - can be clearly defined. It seems to represent the lower part of a predominantly sialic layer, often about 10 km deep. This CD is not clearly defined in young orogenic belts like the Alps (BERCKHEMER, 1968; GIESE, 1968b), the Pamir (BELJAJEVSKY et al., 1967), and the Rocky Mountains (PRODEHL, 8.2 of this volume). In these regions apparently large parts of sialic material have accumulated, probably by the mechanism of subduction and compression at plate boundaries. In the Alps and the southern and middle Rocky Mountains a pronounced low-velocity layer has been detected with velocities sometimes as low as 5 km/s as is the case under the Ivrea body (GIESE, 1968b, 1970), where a high-velocity layer with V_p = 7.2 to 7.4 km/s is found at very shallow depth. Similar "spans" from the mantle, which might have moved upward during a period of compression, may be present in the northernmost Andes as indicated by some suspicious gravity highs along the mountain belt (Instituto Geografico, 1959).

Reflections from the MD in the wide-angle range are very often the most pronounced arrivals in the records. They seem to be best developed at the margins of young mountain belts as observed near the Alps, the Sierra Nevada, the Rocky Mountains (PRODEHL, 8.2 of this volume), and the Tien Shan (MEISSNER, 1967). Under the young orogenic belts the transition zone between crust and mantle with velocities increasing from 6.6-7.2 to 7.8-8.2 km/s is extremely large. More than 10 km have been reported by CHOUDHURY et al. (1971) and PRODEHL (8.2 of this volume) for the Alps and the Rocky Mountains.

Low-velocity zones in the crust also seem to prefer young orogenic belts with much sialic material and probably higher heat flow than in the surrounding zones. Data from platform areas like USSR and Canada do not show low-velocity layers in the crust. This may be due to lower temperatures and less sialic material than in young orogenic belts. Below the CD (or the equivalent RIEL discontinuity in Canada) intermediate velocities of 6.5 to 6.8 km/s are found in most areas which might not present a gabbroic but more a dioritic or noritic material. It is interesting that the reflectivity of near-vertically traveling seismic waves increases considerably below the CD. Both the measurements in the Rhenish massif (MEISSNER et al., 6.1 of this volume) and in the Bavarian molasse area (MEISSNER, 1967a) show a discontinuity between CD and MD which seems to be quite consistent. The increased reflectivity in the lower crust was also observed in the Canadian Shield and the USSR and has been interpreted as an increased inhomogeneity or lamellation. It may be connected with differentiation or crystallization seams, intrusions, phase transitions, or with mechanical layering due to lateral movements of layers during periods of compression. Whatever the reasons may be, the upper 10 km of the crystalline basement mostly do not show this kind of layering.

In shield areas velocities of 7.0 to 7.2 km/s are often reported for the base of the crust (LITVINENKO, 1968). Recently similar velocity values have also been found at the bottom of oceanic crusts (MAYNARD, 1970) which tentatively have been explained as a general phenomenon.

The depth of the MD in continental areas varies between 20 and 70 km. Except for the young orogenic belts there seems to be a general rule for stable regions that older crusts have greater crustal thicknesses than younger ones. These observations are confirmed in North America and Eurasia where the shields show MD depths of 40-45 km whereas the later-accumulated parts of the continents have only 25 to 30 km crustal thickness. The explanation for this observation might be found in a different temperature and a different temperature gradient in the period of crustal accreation. Higher temperatures and stronger gradients than today have prevailed at the time of forming of the old continental nuclei, a larger part of the mantle may have been involved in the differentiation of a thicker and layered crust.

The following papers contain some details of comparisons between selected areas of continental crust where continuous observations or a network of seismic profiles provide a reliable background for such a task. It will be shown that travel-time diagrams in some areas of the USSR, USA, and central Europe are remarkably similar, indicating similar mechanism of their origin whereas basic differences in other parts are indications for a different or stronger activity during the process of mountain building or accretion of new lithosphere.

8.1 Comparison of Wide-Angle Measurements in the USSR and the Federal Republic of Germany

R. MEISSNER

ABSTRACT

Many analogies have been found between results of wide-angle observations in western Germany and similar investigations in the USSR. The pattern of travel-time curves and the wave form of characteristic arrivals show a strong similarity between both parts of Eurasia. There is in general a small positive velocity gradient in the upper 10 km of the crust and a strong gradient at the base of the crust which may consist of a lamellation of layers with different velocities. Characteristic differences have been found mostly for structures of the lower crust.

ZUSAMMENFASSUNG

Die Ergebnisse von Weitwinkelbeobachtungen in der Sowjetunion und in der Bundesrepublik zeigen viele Gemeinsamkeiten. Das System der Laufzeitkurven und die Wellenform charakteristischer Einsätze zeigen für die beiden Untersuchungsgebiete eine große Ähnlichkeit. Es existiert allgemein ein schwacher positiver Geschwindigkeitsgradient in den oberen 10 km der Kruste und ein starker Gradient an der Basis der Kruste, der aus einer Wechsellagerung von Lamellen mit unterschiedlichen Geschwindigkeiten bestehen könnte.

8.1.1 Introduction

In 1964 the first large wide-angle measurements with a common-depth point were performed in the Bavarian Molasse Basin (German Research Group, 1966; MEISSNER, 1966, 1967a). By applying this method, dip effects of discontinuities in the earth's crust can be easily observed and eliminated; reliable velocity-depth curves, however, can be obtained only for the center of the profile. In large parts of the USSR many profiles with great observation density for deep-seismic sounding have been carried out, using overtaking recording techniques. They also yield very accurate data regarding dip effects and velocity-depth functions along the profile. A comparison of data from both kinds of observations will reveal many similarities and some basic differences for different geotectonic units.

8.1.2 The Shape of Arrivals and the Different Travel-Time Branches

In general, the shape of refracted and reflected events and their frequencies are very similar in most parts of Eurasia. Fig. 1 shows a com-

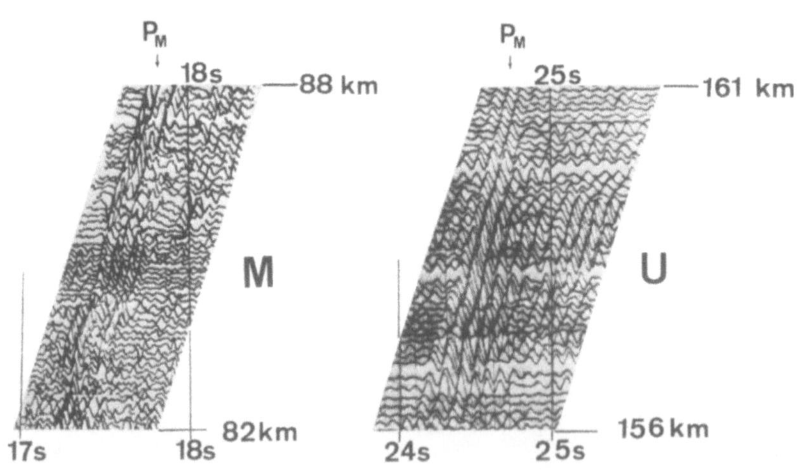

Fig. 1. P^M-wavelets, reflected and/or diving waves from the Moho discontinuity in the Buchara region and the Molasse Basin

parison between arrivals of P-branches representing reflected and diving waves from the base of the crust. The two areas, the Bavarian Molasse basin and the Buchara-Khiva region, both represent large sediment troughs in front of folded mountain ranges such as the Alps and the Tien Shan, respectively. In these troughs and in large parts of the Russian Platform the P^M-branches are the most powerful arrivals of the records and show many oscillations with frequencies around 10 Hz.

Also the form of other travel-time branches are very similar. Fig. 2 shows a comparison between travel-time patterns from the Buchara region, from the Molasse area, and from the Rhenish Massif. As in many other areas five characteristic branches are observed: first the P_g (or P_1) wave from 0 to about 100 km is recorded showing a small convex curvature indicating a small positive velocity gradient from about 5.9 to about 6.3 km/s in the upper part of the crystalline basement. The maximum penetration of these rays is about 10 km. The next branch, P_2, mostly shows a certain offset with regard to the P_1 branch. As its asymptotic velocity is mostly about the same as that of the P_1 branch, a zone with smaller velocities below about 10 km depth may be deduced. The P_2 branch, however, is not as strong and reliable as the P_g (= P_1) or the P^M (= P_4) branch. The P_2 and the P_3 waves both come from the middle part of the crust and often differ in curvature and amplitude. This observation shows, according to STEINHART and MEYER (1961) and GIESE (1968a), that the middle part of continental crusts may be heterogeneous but does not show characteristic, continuous boundaries like the crystalline basement or the crust/mantle transition. The Conrad discontinuity found in many parts of western Europe (MEISSNER et al., 6.1 of this volume) cannot be correlated with certainty to the eastern part of Europe. While in western Europe the Conrad discontinuity mostly resembles the P_2-wave branch and shows asymptotic velocities between 6.3 and 6.5 km/s (which is also valid for the Buchara-profile), most parts of the shield area in eastern Europe and Asia show higher asymptotic velocities in the middle part of the crust. These asymptotic velocities seem to be a good indicator for the maximum velocities on top of the interface. Velocities of 6.3 to 6.5 km/s can be correlated to sialic or dioritic but not to gabbroic or granulitic material.

8.1.3 Some Implications from Travel-Time Observations

The observation of the asymptotic velocity and the initial apparent velocity of a given travel-time branch plays a major role for the strong P^M (or P_4) arrivals. As mentioned before, they represent reflected and/or diving waves from the base of the crust. It is known, for instance, that the true velocity of the M-discontinuity varies between 7.8 and 8.4 km/s. If an apparent velocity greater than 9 km/s is observed in the beginning of different P^M wave branches and if dip effects can be eliminated, a part of this branch must consist of subcritical reflections corresponding to a first-order boundary. This may be seen from fundamental relations of ray optics, especially from the first line of the following formula connecting the apparent velocity of the P^M branch, V_{aM},

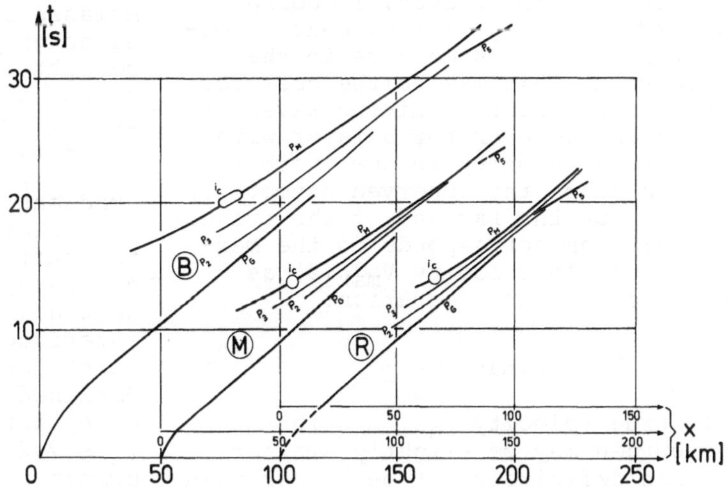

Fig. 2. Travel-time pattern of deep-seismic sounding in the Buchara region *B*, in the Molasse Basin *M*, and in the Rhenish Massif *R*.

$P_1 = P_g$	corresponds to branch	a
P_2	" " "	b
P_3	" " "	b
$P_4 = P^M$	" " "	c
$P_5 = P_n$	" " "	d

all (GIESE, 1968a)

with the true velocity V_n of the refracting interface, the uppermost part of the mantle:

$$V_{aM}=\frac{V_n}{\sin i_n}\begin{cases} >V_n \text{ for } \sin i_n < 1 \text{ or } i_{n-1} < i_c \text{ (subcrit. refl.)} \\ =V_n \text{ for } \sin i_n = 1 \text{ or } i_{n-1} = i_c \text{ (crit.refl.)} \\ <V_n \text{ for } i_{n-1} > i_c \text{ (overcritical refl.)} \end{cases} \quad (1)$$

with i_n = ray angle below the interface n

i_{n-1} = ray angle on top of the interface n.

Eq. (1) describes the three parts of the P^M travel-time branch. Subcritical reflections with $V_{aM} > V_n$ in ray optics imply the existence of a first-order boundary. On the far end side of the travel-time branch, on the other hand, parts with $V_{aM} < V_n$ are found representing overcritical reflections and - if there is a velocity gradient on top of the first-order interface - a diving wave. So, with decreasing apparent velocities along the travel-time branch a minimum apparent velocity is approached which is the maximum true velocity of any layer on top of the reflecting interface. This is called the asymptotic case and may be described by Eq. (2)

$$\frac{V_n}{\sin i_n} = \frac{V_{n-1}}{\sin i_{n-1}} \rightarrow V_{n-1} = V_{asy} \quad \text{for } i_{n-1} \rightarrow 90^\circ \quad (2)$$

In Eq. (2) V_{asy} is the minimum observable apparent velocity on the far end of a travel-time branch. It corresponds to $\sin i_{n-1} \rightarrow 1$ i.e. nearly horizontal ray path somewhere in the overburden. This may be the case for the turning point of diving waves in a gradient zone or for a layer with a discrete velocity in the overburden. In both cases the observed asymptotic velocity on the far end of the travel-time diagram corresponds to the maximum possible velocity $V_{max}(z)$ as seen from the next equation:

$$V_{asy} \geq V_{n-1,max} \quad (3)$$

This true velocity $V_{n-1,max}$ in the overburden may be slightly smaller but can definitely not be larger than the observed apparent velocity V_{asy} at the far end of the travel-time branch. Hence, the actual observation of asymptotic velocities of 6.3 to 6.5 km/s for the Buchara, Molasse, and Rhenish Massif areas means that there is no thick gabbroic or granulitic layer with velocities of 6.5 km/s or larger in these areas. On the Russian platform and other old shield areas, on the other hand, asymptotic velocities of the P^M branch are much higher, showing that the middle and lower crust are basically different. In this way the observed asymptotic velocities are an important tool for comparing true velocities in the lower crust.

Additional information on the kind of velocity transition in the lower crust comes from a more continuous evaluation of the P^M branch. According to GIESE (1965), the depth of a reflection from a first-order boundary is given by

$$z = \frac{x}{2}\sqrt{\frac{V_a \cdot t}{x} - 1} \quad (4)$$

in a first-order approximation. Here, different values of t, x and V_a along the P^M branch should yield about the same depth values for a first-order M-discontinuity. If the plotted depth values show a systematic shifting to shallower depths with decreasing V_a there must definitely be a gradient zone involved (GIESE, 1968a; MEISSNER, 1967b). The determination of V_a as a function of z is particularly reliable for common-depth point profiles with only small dip effects. When plotting $z(V_a)$ according to Eq. (4) for the Buchara region, a significant shifting of constructed depth points with decreasing V_a values is observed (Fig. 3). A similar, though weaker, shifting of the points is observed for the Molasse area and the Rhenish Massif. Although these depth points are not correct for a diving wave in a gradient layer and may be improved by iteration, the large deviation of the curve for velocity values smaller than 8.5 km/s shows that there is a gradient layer on top of the interface where rays are refracted back to the surface. The large apparent velocities $V_a > 9$ km/s on the other hand show a (rather strong) first-order interface at the bottom of the gradient layer. Similar observations are obtained from shield areas with the exception of apparent velocities with $V_a \approx 7.2$ km/s at the far end of the P^M branch.

Another observation obtained from travel times shown in Fig. 2 is the fact that the backward extrapolation of the P_n (or P_5) branch does not run asymptotically to the P^M (or P_4) branch but intersects it. As shown by MEISSNER (1967b), this means that the mantle wave P_n belongs to a greater depth than that of the reflected wave P^M and that there exists no plane interface common to both P^M and P_n. There is either a small positive gradient in the uppermost part of the mantle, or the reflected waves come from a lamellation zone at the base of the crust, or both. The concept of lamellation was developed from the observation of both vertically reflected waves and wide-angle events based on the Molasse survey and seems to be valid in other areas as well (GLOCKE and MEISSNER, 6.2 of this volume).

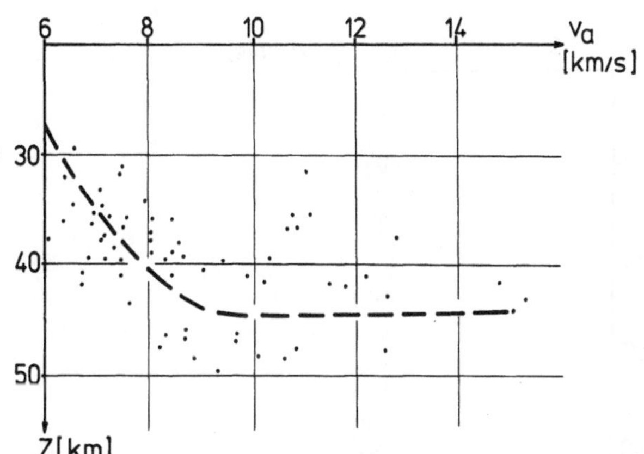

Fig. 3. Apparent velocities of the P^M-branch plotted versus apparent depths from reflection EQ. (4); (Buchara region)

8.1.4 Velocity-Depth Relations

The comparison of velocity-depth profiles of different areas may give the most reliable information regarding the material of the crust. Details of obtaining velocity-depth profiles have been given by GIESE (1968a) and MEISSNER (1967a). Figs. 4 and 5 show some velocity-depth profiles from different areas of Eurasia. For all these curves gradient layers in the basement have been obtained from the curvature of the P_g wave branch; gradients in the deeper crust have been derived from the curvature of the P^M wave branch (sometimes also from the P_3 wave branch). Low-velocity zones are obtained either from interval velocities due to DIX (1955), or (mostly) from differences between the average velocity obtained from the P^M branch and the other constructed parts of the V(z)curve. In this way strong low-velocity zones have been observed with certainty for large parts of the Alps, weaker low-velocity zones are found in Western Germany, and (up to now) no reliable low-velocity zones have been found for the Russian platform. Often apparent low-velocity

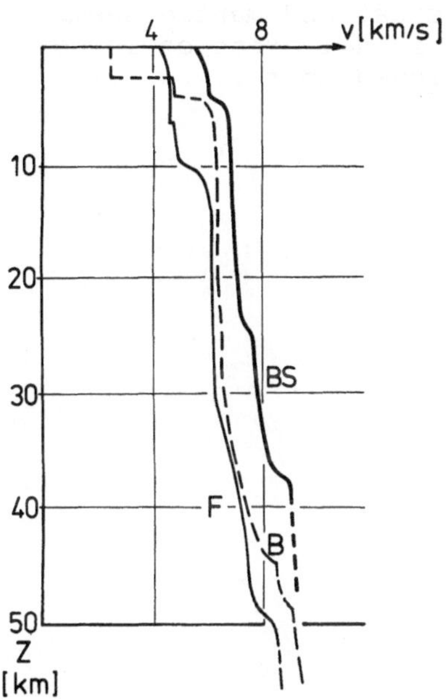

Fig. 4. Velocity-depth curves for the Baltic shield *BS*, the Buchara region *B*, and the Fergana Degression *F*

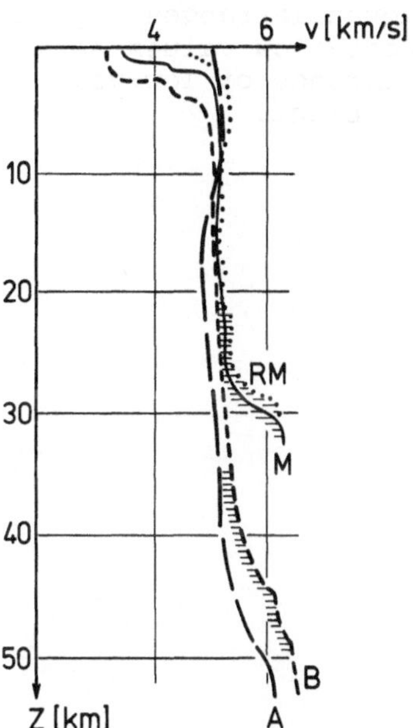

Fig. 5. Velocity-depth curves for the Alps *A*, the Molasse *M*, the Rhenish Massif *RM*, and Buchara region *B*

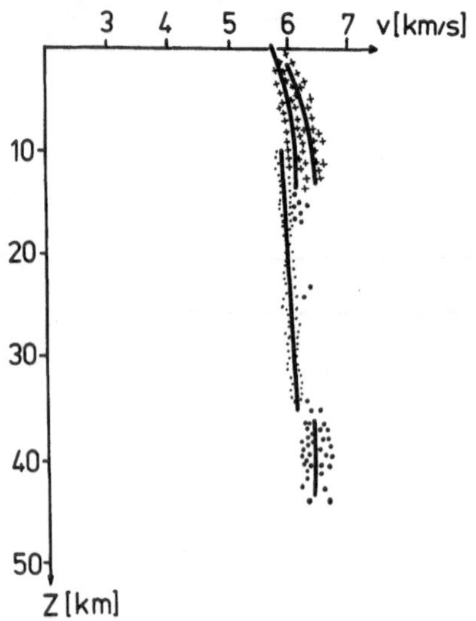

Fig. 6. Velocity-depth data of different regions in the Ukraine after PAVLENKOVA (1969)

parts may simply represent an anisotropy as suggested by the work of PAVLENKOVA (1969) (Fig. 6). Here, data from refraction and wide-angle observations always show higher velocities than those obtained from near-vertical reflections. A similar observation has been obtained from the $t^2 - x^2$ diagrams of the common-depth point profile in the Molasse area and seems also to hold for the Rhenish Massif region. Low-velocity zones in the crust depend on temperature and chemistry. Large sialic quantities as they are found in troughs, geosynclines, and mountain ranges together with higher heat-flow values seem to favor the occurrence of low-velocity zones in the crust.

8.1.5 Conclusions

1. The old shield area in eastern Europe, a very old and stable part of continental crust with small heat-flow values, show high-velocity material in the lower and middle part of the crust. It must represent high-density layers. No low-velocity layers have been detected so far.

2. Both margins of folded mountain belts under investigation, the Buchara region and the Bavarian Molasse, show pronounced velocity-gradient zones at the base of the crust. In the Buchara region depth and transition zone to the mantle are larger. The actual depth of the horizontally traveling P_n wave is larger than that of P^M. Both areas show no pronounced gabbroic lower crust and no Conrad discontinuity. Low-velocity zones in the crust are not significant and certainly much smaller than those of the adjacent mountain belts.

3. In the Rhenish Massif as well as in other areas of the Hercynian orogeny a Conrad discontinuity is found at depths of 14 to 18 km. Also, a weak Sub-Conrad boundary is present between CD and MD. Otherwise, travel-time patterns are similar to those of Buchara and Bavaria including small zones with velocity reversals.

4. A lamellation of the transition zone between crust and mantle seems to be present in all areas of investigation mentioned so far.

8.2 Comparison of Seismic-Refraction Studies in Central Europe and the Western United States

C. Prodehl

ABSTRACT

The results of a reinterpretation of 64 seismic-refraction profiles recorded in the western United States are summarized and compared with the crustal structure of central Europe. A basic travel-time diagram can be derived, showing features similar to those found for European profiles. Contour maps for several parameters: the critical distance of the phase reflected from the crust/mantle boundary, the average P_n-velocity, the depth to the strongest velocity gradient, e.g., as well as a fence diagram for the area of investigation are presented and compared with respective presentations for central Europe.

ZUSAMMENFASSUNG

Die Ergebnisse einer Reinterpretation von 64 Profilen, die in den westlichen Vereinigten Staaten beobachtet wurden, werden zusammengefaßt und mit der Krustenstruktur Mitteleuropas verglichen. Für alle Profile ergibt sich eine charakteristische Anordnung von Laufzeitkurven, die den Ergebnissen in Mitteleuropa ähnelt. Verschiedene Parameter wie z.B. die kritische Entfernung der von der Krusten-Mantel-Grenze reflektierten Phase, die durchschnittliche P_n-Geschwindigkeit und die Tiefenlage des stärksten Geschwindigkeitsgradienten in der Kruste, wurden kartiert und die Ergebnisse in einem Blockbild zusammengefaßt. Die Ergebnisse werden mit Daten und Interpretationen für Mitteleuropa verglichen.

8.2.1 Introduction

A detailed study of crustal and upper-mantle structure in the western United States by explosion seismology was started in 1960, organized and carried out by the U.S. Geological Survey. From 1961 to 1963, a network of 64 profiles was recorded in California and Nevada and adjacent areas (Fig. 1). The reinterpretation of many of these data (PRODEHL, 1970a, b) is based on similar principles and the same method as was used by GIESE (1968a, 4.3) for the reinterpretation of seismic-refraction data in central Europe. So it is possible to show up a direct comparison of the main results for central Europe and the western United States.

8.2.2 The Travel-Time Diagram

The comparison of the travel-time diagrams found on profiles in central Europe with those on profiles in the western United States (see, e.g., Fig. 2) shows that, generally, the same features are evident. In most cases the first arrivals align on two travel-time curves. The first one can be traced up to distances of 100-150 km (curve a) showing velocities of 5.9-6.3 km/s and is correlated with the basement. As far as can be seen, in the western United States this curve a is formed by a single phase comparable to the observations in the Variscan basement in central Europe.

The second travel-time curve is observed mostly at distances greater than 150-200 km (curve d) showing velocities of 7.6-8.2 km/s and is correlated with the upper mantle. On the record sections of the profiles in the Sierra Nevada, the Rocky Mountains, and the Colorado Plateau as well as the profiles in the central Alps, this curve d crosses the distance-axis at greater distances than on profiles in other areas indicating a relatively thick crust. Generally the amplitudes of the P_n-wave increase with increasing distance.

In secondary arrivals one or two dominant phases can be correlated by travel-time curves across 50-200 km in distance, their velocity decreasing with increasing distance. The most significant one (curve c) is observed between 80 and 200 km distance from the corresponding shotpoint and is correlated with the crust/mantle boundary zone. This phase c, characterized

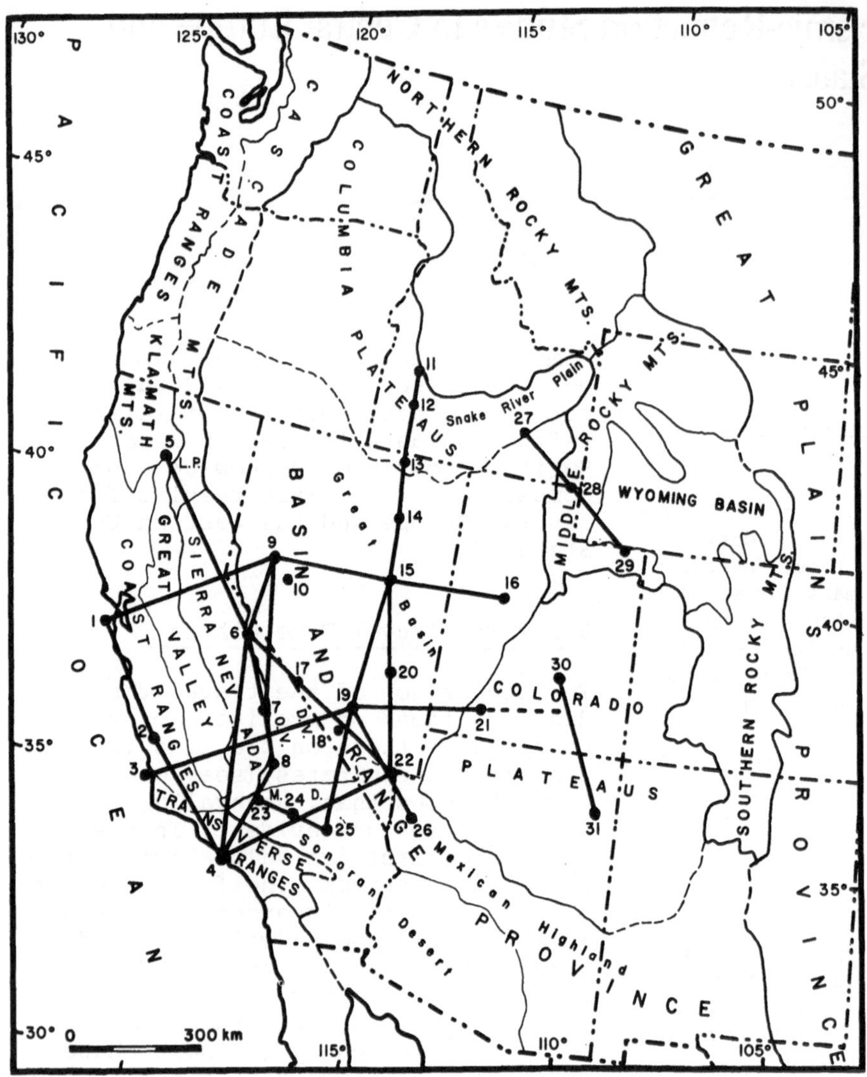

Fig. 1. Physical divisions of the western United States and location of seismic profiles. After the map of the physical divisions of the United States by FENNEMAN and JOHNSON (1946). *D.V.* Death Valley, *L.P.* Lassen Peak National Park, *M.D.* Mojave Desert, *O.V.* Owens Valley. Shotpoints: *1* San Francisco, *2* Camp Roberts, *3* San Luis Obispo, *4* Santa Monica Bay, *5* Shasta Lake, *6* Mono Lake, *7* Independence, *8* China Lake *9* Fallon, *10* SHOAL, *11* Boise, *12* Strike Reservoir, *13* Mountain City, *14* Elko, *15* Eureka, *16* Delta, *17* Lido Junction, *18* Lathrop Wells, *19* Nevada Test Site, *20* Hiko, *21* Navajo Lake, *22* Lake Mead, *23* Mojave, *24* Barstow, *25* Ludlow, *26* Kingman, *27* American Falls Reservoir, *28* Bear Lake, *29* Flaming Gorge Reservoir, *30* Hanksville, *31* Chinle

by large amplitudes in nearly all record sections, is the most evident similarity to profiles in central Europe. It seems to be best developed on profiles at the margin of or outside the Alps and not so well expressed on profiles within the central Alps. A similar, comparably weak, development of phase c can be seen on the profiles in the Sierra Nevada and the Rocky Mountains, in contrast to most profiles in the Basin and Range province, where this phase appears extremely strong and with large amplitudes. However, as published record sections of profiles in the Great Plains (STEWART, 1968; HEALY and WARREN, 1969) and the southern Rocky Mountains (HEALY and WARREN, 1969) indicate this conclusion probably cannot be generalized.

On some profiles in the western United States, at distances greater than 150 km, a travel-time curve parallel to curve d can possibly be traced through distinct secondary arrivals which follow the first arrivals at a time interval of about 0.5 s. Similar arrivals are found on profiles from Eschenlohe to the Bohemian crystalline massif (PRODEHL, 1965; GIESE and STEIN, 1971) and on profiles across the Rhine Graben (MEISSNER and BERCKHEMER, 1967; MUELLER et al., 1967). The interpretation remains open, the proposed one for the Rhine Graben probably cannot be transferred to the profiles in the Bohemian crystalline massif and the western United States, but rather may be correlated with structure below the crust/mantle boundary.

The scattered existence of more or less well correlated phases between phases a and c up to 160-200 km distance on most profiles in the western United States corresponds with the general weak occurrence of those phases on profiles in central Europe. However, in the western United States some areas exist where a phase b can be correlated very well indicating a defined boundary zone within the crust. In some cases also a curve d(b) is found which is tangent to the retrograde curve b at the point of "critical" reflection. Those areas are, for example, the northern part of the Basin and Range province in Nevada and Utah, the Snake River Plain, and the middle Rocky Mountains. In central Europe, only on profiles crossing the zone of Ivrea phases exist - named c(i) and d(i) (see, e.g., GIESE, 1968a, b) - which may be comparable with the phases b and d(b) in the western United States.

8.2.3 Basic Data

Similar to the maps published by CHOUDHURY et al. (1971), GIESE and STEIN (1971), GIESE (5.2), and GIESE and PRODEHL (7 of this volume) for central Europe, some basic parameters (Figs. 3 and 4) were mapped for the area of investigation of the western United States shown in Fig. 1. In Fig. 3, the distance Δ_d, at which the P_n-travel-time curve (curve d) crosses the distance-axis of any record section, as well as the "critical" distance Δ_c, at which the "refracted" P_n-travel-time curve and the "reflected" curve correlated with the crust/mantle transition zone (curve c) are tangent to each other, represent in a first approximation the variation of total crustal thickness. The comparison of these contour maps with the corresponding maps for central Europe shows that the maximum values of $\Delta_d \geq 200$ km and $\Delta_c = 120$-140 km in the western United States are comparable with the results found for the Alps, while the minimum values of $\Delta_d = 120$-140 km and $\Delta_c = 60$-80 km in the western United States correspond with values in western Germany outside the Alps.

The values of the reduced travel time $\overline{T}_c$ read off at the distance of "critical" reflection for the phase reflected at the crust/mantle boundary represent approximately the intensity of a velocity inversion, if any exists. For the western United States these reduced travel times were corrected by the corresponding reduced P_g-travel time at the same distance to eliminate the travel-time delays caused by sedimentary layers (Fig. 3). Values of $\overline{T}_c - \overline{T}_{a,c} = 3$-$4$ s, which are found in central Nevada and across southern Nevada and adjacent parts of California, can be seen on the corresponding map for western Germany in southeastern Bavaria and adjacent parts of Tyrol. Times of 1-2 s in other areas of the western United States are comparable with values found for the main part of western Germany.

A contour map for the reduced travel time of curve c at a fixed distance avoids the definition of the critical distance and seems to be more independent of subjective interpretational influence. While GIESE (4.2 of this volume) proposes a fixed distance of 80 km for central Europe where the crustal thickness is 30-35 km on the average, for the western United States a fixed distance of 100 km seems to be more suitable, because on many profiles, curve c (phase P^M) is not observed at distances smaller than 100 km, due to the greater crustal thickness of 45-50 km in average.

Fig. 4 shows the resulting contour map for the reduced travel time of curve c at the constant distance 100 km: $\overline{T}_{c,100}$, corrected by the corresponding reduced travel time of curve a: $\overline{T}_{a,100}$, covering the western United States including the southern Rocky Mountains. The values shown by crosses are the observed ones, plotted at half the critical distance Δ_c from the corresponding shotpoint. The largest values are observed in the area of the Rocky Mountains, indicating great crustal thickness as well as low mean crustal velocity. Another maximum is obtained in the area of the Sierra Nevada.

The P_n-velocity (Fig. 3) (based on curve d) differs significantly between

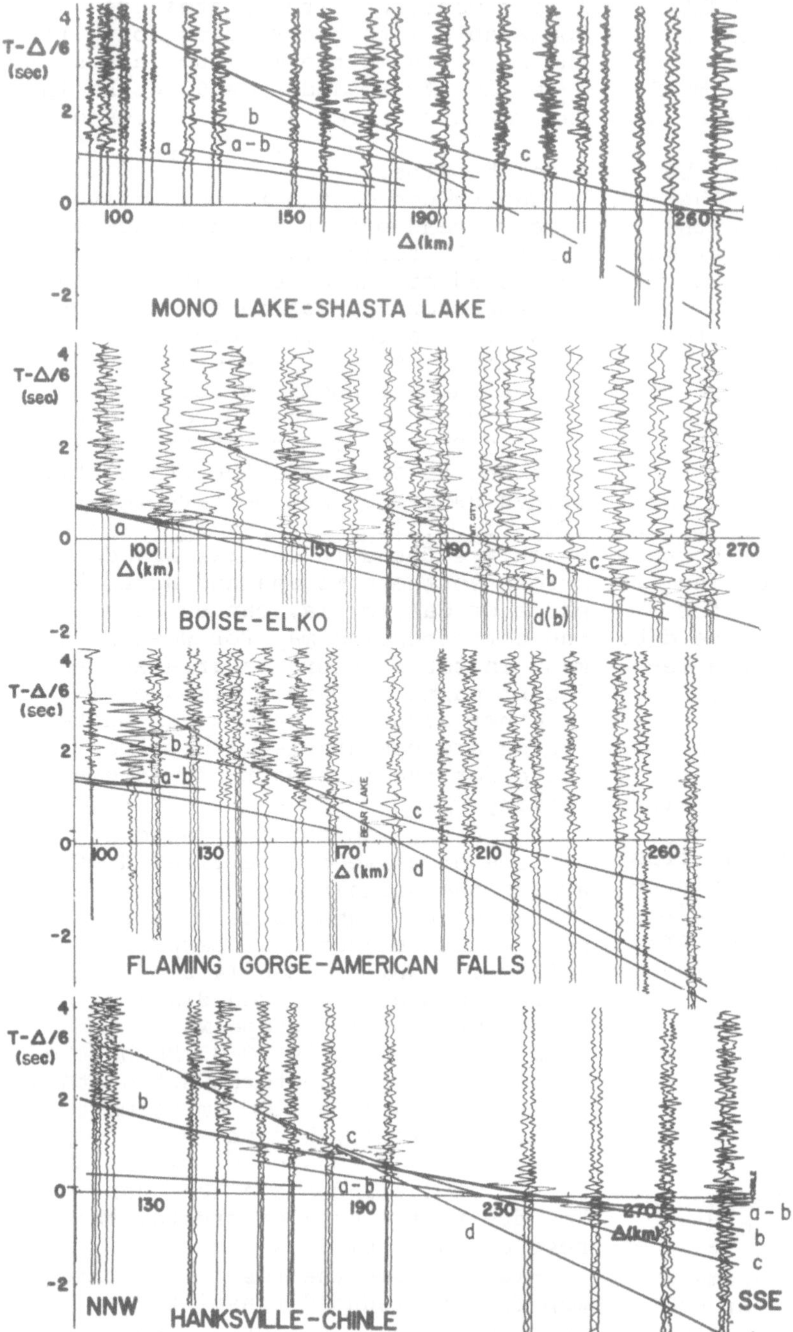

Fig. 2. (a) Part of the record sections of the profiles Mono Lake-Shasta Lake (Sierra Nevada), Boise-Elko (Snake River Plain), Flaming Gorge Res.-American Falls Res. (middle Rocky Mountains) and Hanksville-Chinle (Colorado Plateau)

the Alps including their foreland and the western United States. For the Alps and southern Germany velocities are generally greater than 8.0 km/s. Under the western United States, the P_n-velocity is generally lower than 8.0 km/s. Only under the middle Rocky Mountains, the Mojave Desert, and the Coast Ranges of California does the P_n-velocity equal or exceed 8.0 km/s. Lowest values of 7.6 km/s are present under central and southern Utah.

8.2.4 Crustal Structure

Although the tectonic relations of the western United States are not directly comparable with those of the Alps, some significant differences and similarities in crustal structure can be identified and discussed. The most comparable parameter is the total crustal thickness, i.e., the depth to the strongest velocity gradient ($z(\Delta_c)$ in Fig. 3). The Alps, the Sier-

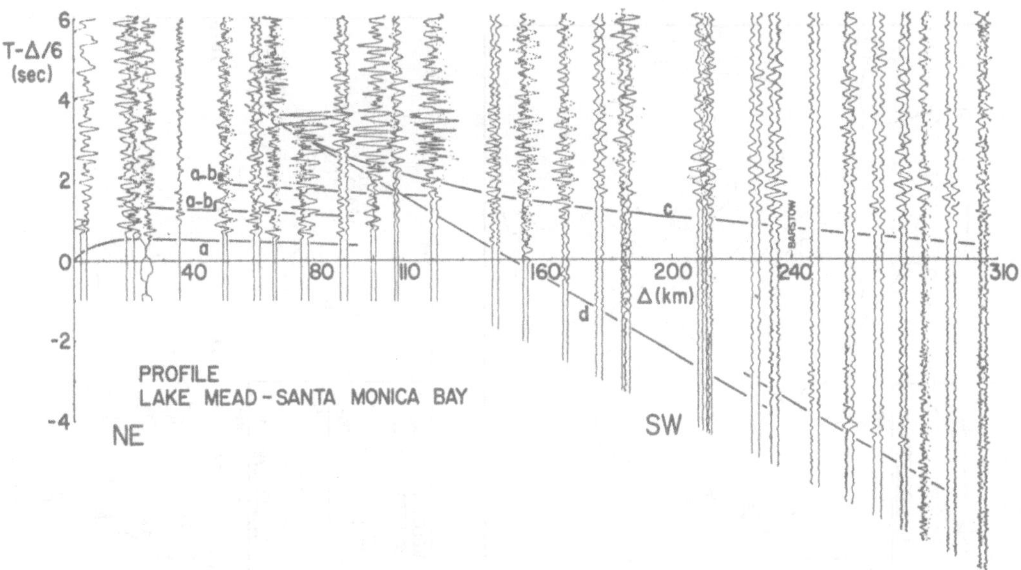

Fig. 2. (b) Record section of the profile Lake Mead-Santa Monica Bay (southern Basin and Range province)

ra Nevada, and the Rocky Mountains are underlain by a thick crust (more than 40 km), compared with surrounding areas. However, under the Colorado Plateau, the crust is just as thick as under the adjacent middle Rocky Mountains. The total crustal thickness is also about the same under the southern Rocky Mountains (see Fig. 6) and the adjacent Great Plains of eastern Colorado (PAKISER, 1965; PRODEHL and PAKISER, 1976). The crustal structure of the Colorado Plateau may be comparable with the structure of the Great Plains of eastern Colorado (JACKSON et al., 1963) and southern Missouri (STEWART, 1968). However, the crustal thickness of more than 40 km under the Colorado Plateau and the Great Plains is significantly greater than that found for the "normal" crust in central Europe which is 30 km thick. The Great Valley and the Coast Ranges of California may be comparable with the Po Plain and the Apennines. The crustal thickness of about 25-30 km (GIESE et al., 1968) agrees approximately with the results found for the Coast Ranges of central California (24-26 km).

Corresponding to the depth $z(\Delta_c)$, at which the velocity-gradient reaches its maximum value, the velocity $v(\Delta_c)$ is mapped. For some areas as well in central Europe as in the western United States significantly lower velocity values result than those shown in the contour maps of the P_n-velocity. These differences may be explained by an increasing velocity, but decreasing velocity gradient below the depth of strongest velocity gradient.

The fence diagram (Fig. 5) summarizes the results of the velocity-depth determinations combined in 15 crustal cross sections showing lines of equal velocity, similar to the fence diagrams shown by CHOUDHURY et al. (1971) and GIESE and STEIN (1971) for the Alps and western Germany. Under the Coast Ranges of California, a nearly sharp crust/mantle boundary is found (west of the San Andreas fault), the average crustal velocity between 10 and 24 km depth is 6.3-6.4 km/s, and the total crustal thickness is about 26 km beneath its central part, but increases under the Transverse Ranges. Under the central Sierra Nevada the mean crustal velocity is higher than under the Basin and Range province, the velocity increasing slightly from 6.2 to 6.6 km/s between 5 and 35 km depth. A low-velocity zone is not present. The transition zone between crust and mantle is up to 10 km thick, its base rising southward from 42 to 33 km. On many profiles in the Basin and Range province a velocity inversion within the upper crust is present, the velocity decreasing from about 6.2 to 6.1 or 6.0 km/s. In the north and east of the province an intermediate boundary zone between upper and lower crust where the velocity gradient is very steep is present which disappears toward south and west resulting in a low crustal mean velocity of 6.2 km/s in these parts of the province. The

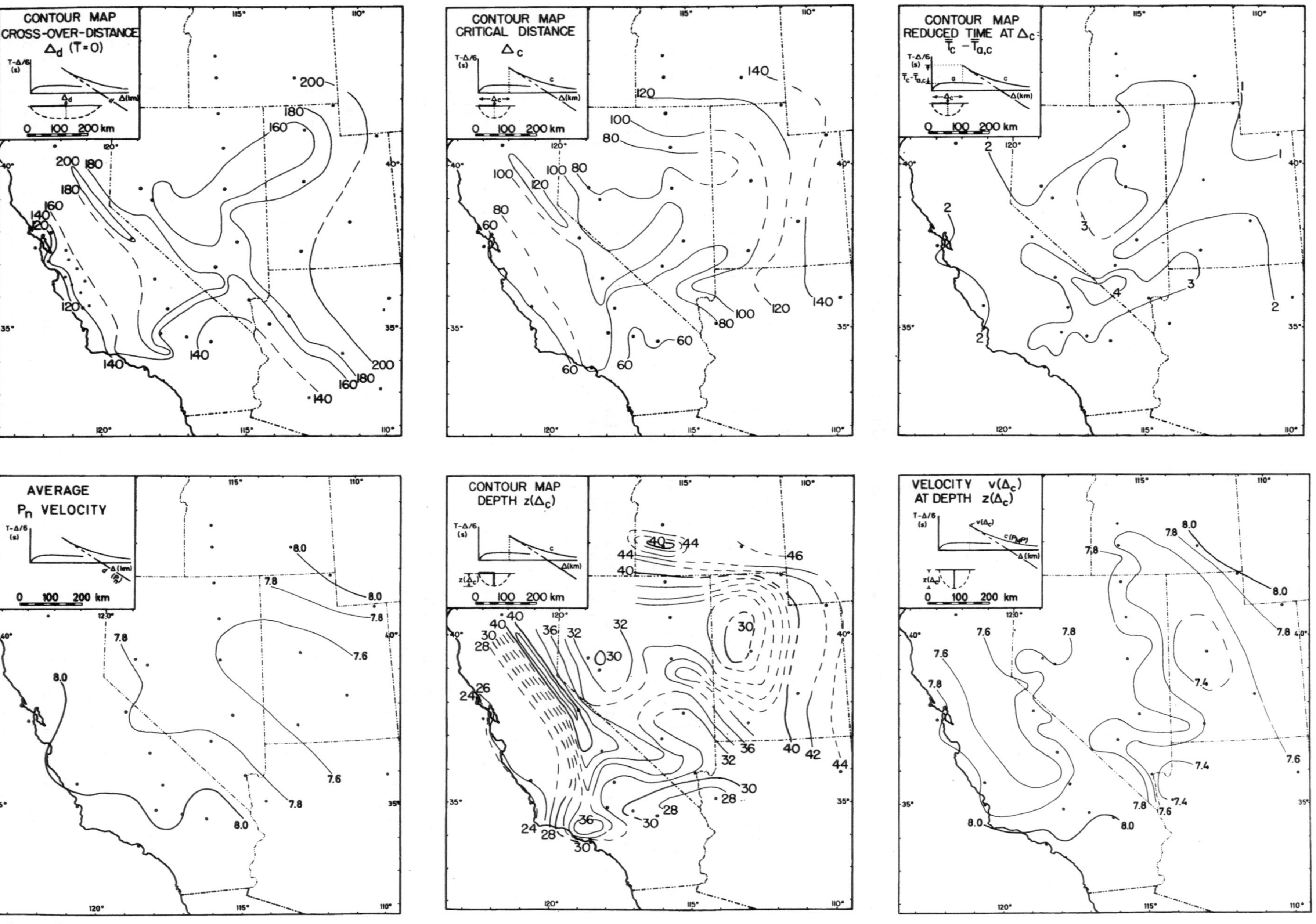

CONTOUR MAP
CROSS-OVER-DISTANCE
Δ_d (T=0)
CONTOUR MAP
CRITICAL DISTANCE
Δ_c
CONTOUR MAP
REDUCED TIME AT Δ_c:
$\bar{T}_c - \bar{T}_{a,c}$
AVERAGE
P_n VELOCITY
CONTOUR MAP
DEPTH $z(\Delta_c)$
VELOCITY $v(\Delta_c)$
AT DEPTH $z(\Delta_c)$
0 100 200 km

Fig. 3

Fig. 3. Contour maps of cross-over distance Δ_d, "critical" distance Δ_c, reduced travel time $\overline{T}_c - \overline{T}_{a,c}$, average P_n velocity, depth of strongest velocity gradient $z(\Delta_c)$, and velocity $v(\Delta_c)$ at depth $z(\Delta_c)$ for California and Nevada and adjacent areas. The values Δ_c, $\overline{T}_c - \overline{T}_{a,c}$, $z(\Delta_c)$, and $v(\Delta_c)$ are plotted at half the distance Δ_c, the values Δ_d at half the distance Δ_d, and the P_n velocity values half way between corresponding reversed shotpoints. Shotpoints according to Fig. 1 are marked by dots. In addition, for the map of Δ_d P_n velocity values of other seismic-refraction surveys are used which else are not included in this investigation. The corresponding shotpoints are also shown in the map of Δ_d

crust/mantle boundary in the entire Basin and Range province is very sharp, the average crustal thickness being 32-34 km, showing minima near Fallon in Nevada, near Delta in northwestern Utah, and in the south in southern California toward northwestern Arizona. Upper crustal material may be missing under the southern Cascade Mountains and the western Snake River Plain, a low-velocity zone being present under the Lassen Peak National Park area (velocity-decrease from 6.6 to 6.0 km/s at 8-10 km depth). The total crustal thickness exceeds 40 km. On the profiles in the Colorado Plateau and in the middle Rocky Mountains, the existence of an intermediate boundary zone within the crust is indicated, and the total crustal thickness varies between 41 and 46 km. A low-velocity zone within the crust apparently is present under the middle Rocky Mountains at a depth of approximately 17 km, the velocity decreasing from 6.4 to 5.8 km/s.

The transition zone between crust and mantle in which the velocity increases rapidly from 6.6-7.0 to 7.8-8.0 km/s extends under the central Alps over a depth range of more than 10 km and becomes thinner at the northern and western margin (CHOUDHURY et al., 1971). Under the Rocky Mountains and under the Sierra Nevada, a comparable thickness of 10 km is found for this transition zone which is greater than under the neighboring Basin and Range province and other surrounding areas of the Sierra Nevada. A main difference in crustal structure seems to be the existence of a well-defined low-velocity zone under the Alps where the velocity decreases from 6.0-6.2 to 5.5-5.6 km/s between 10 and 25 km depth (CHOUDHURY et al., 1971) and under western Germany where a velocity decrease of 0.1-0.45 km/s at a depth of about 10-15 km is reported (GIESE and STEIN, 1971), while it is, in general, not present under the area of investigation of the western United States. An exception seem to be the Rocky Mountains. The narrow low-velocity zone found under the middle Rocky Mountains evidently continues under the southern Rocky Mountains (see Fig. 6). Here the velocity in the upper crust down to a depth of 30-35 km is 6 km/s in average, in contrast to the adjacent areas of the Colorado Plateaus and the Great Plains, where the average velocity in the crust is higher. However, the velocity inversions found within the upper crust

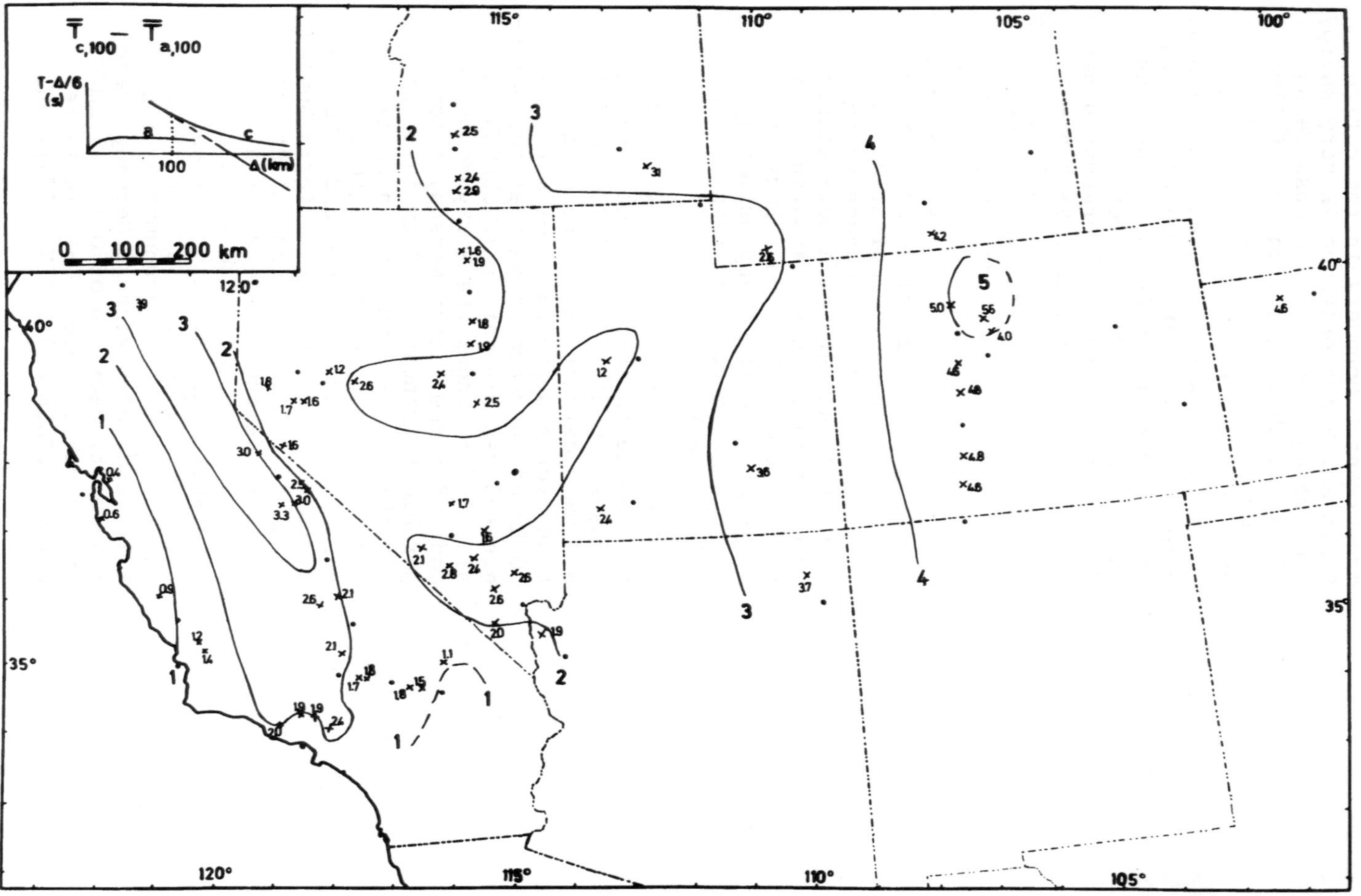

Fig. 4. Contour map of the reduced travel time $\overline{T}_{c,100} - \overline{T}_{a,100}$ for the western United States. The reduced travel time $\overline{T}_{c,100}$ of the P^M-phase (curve *c*) at the fixed distance 100 km is corrected by the corresponding travel time $\overline{T}_{a,100}$ of the P_g-wave (curve *a*). The contour interval is 1 second. The values shown by crosses are the observed ones and are plotted at half the critical distance Δ_c from the corresponding shotpoint

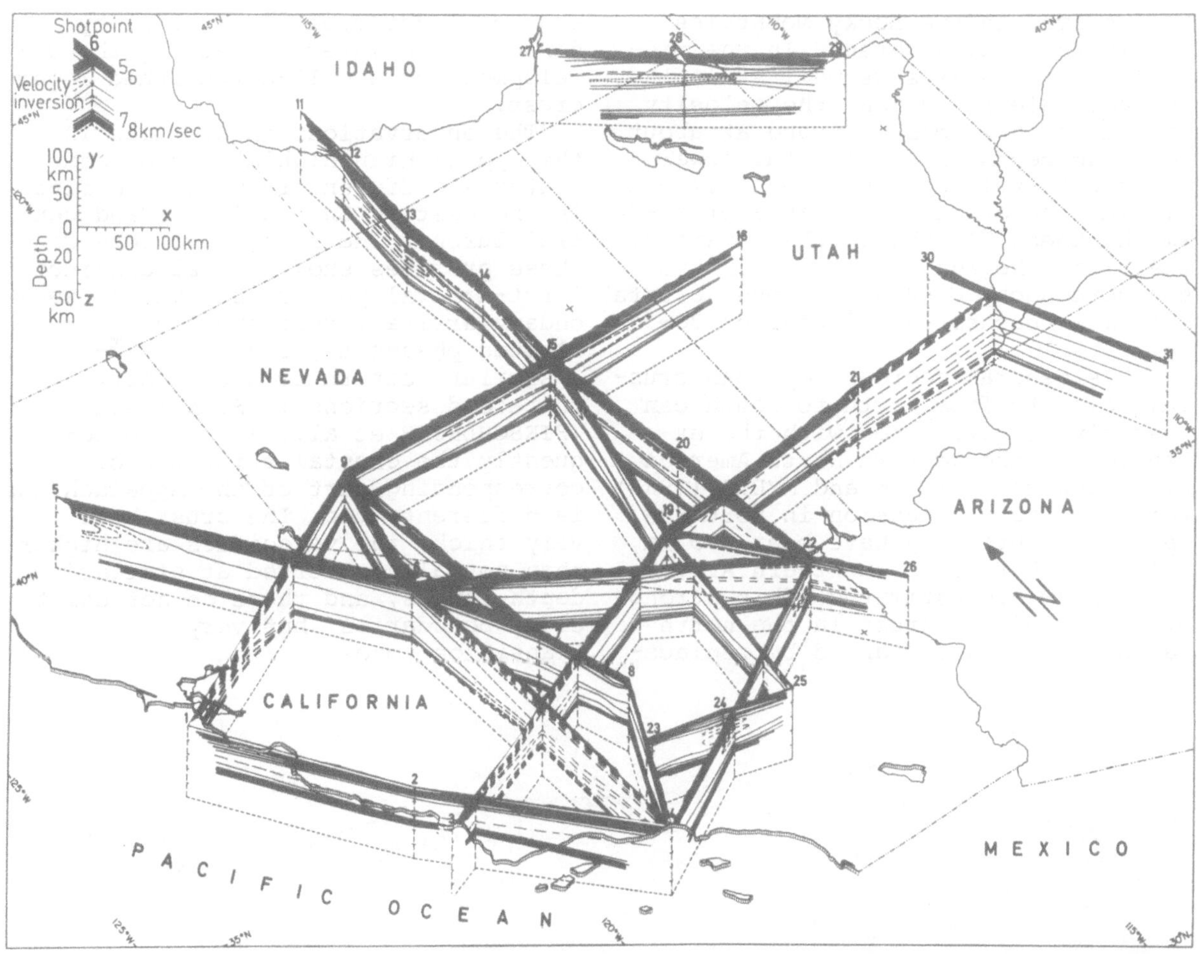

Fig. 5. Fence diagram showing the crustal structure under California and Nevada and adjacent areas. The diagram is viewed from an angle of 45° from the Pacific Ocean toward northeast, approximately parallel to the line from Los Angeles to Salt Lake City. The depth z is exaggerated 2:1 versus the horizontal direction (SW to NE). The scale of the surface altitude corresponds to the scale of the depth z. The contour interval of the lines of equal velocity is 0.2 km/s. Velocity lines less than 5 km/s are not shown. *Dashed lines*: uncertain results. The depth scales under the shotpoints are divided into 10 km intervals. The shotpoints are numbered according to Fig. 1

of the southern Rocky Mountains are not as strong as beneath the central Alps where velocities as low as 5.5 km/s are reported. The nonexistence of a low-velocity zone under the Sierra Nevada indicates that the crustal structure here differs significantly from that under the central Alps.

8.2.5 Some Characteristic Features of the Velocity-Depth Structure

Fig. 6 shows some selected crustal cross sections through the United States including the southern Rocky Mountains and the Appalachians. Also, for each cross section one or two velocity-depth functions are plotted which may be typical for the corresponding province. The map seems to indicate a zonal division from north to south: in the Cascade Mountains as well as in the Columbia Plateaus in the north, the upper crust is very thin or even may not exist. In the adjacent northern Basin and Range province, upper and lower crust are equally well developed, have similar thickness, and can well be separated from each other. In the southern Basin and Range province, however, apparently the lower crust has almost disappeared, the upper crust reaching into a depth of about 25 km with velocities of 6.0-6.2 km/s, while the lower crust is not more than a 5 km thick transition zone between crust and mantle. A similar increase of the sialic crust from north to south can be observed in the Rocky Mountains and in the Sierra Nevada. In the middle Rocky Mountains as well as in the northern Sierra Nevada, the velocity contour of 6.4 km/s is found at about 20-25 km depth, i.e. also the lower crust with velocities between 6.4 and 7.0 km/s covers a depth range of 20-25 km thickness. In the southern part of the Sierra Nevada as well as in the southern Rocky Mountains, upper-crustal material reaches into depths of 30-35 km.

This general decrease of mean crustal velocity from north to south can possibly be correlated with the extension of the western North America. According to HAMILTON and MYERS (1966) total Cenozoic extension in northern Nevada and Utah may have been 300 km, while further south (northern Arizona) only 50 km are estimated. This strong extension of the crust in the north may have been accompanied by replacement of light crustal material by dense mantle material causing the development of a well-established lower crust.

The observations on profiles in the Appalachian Highlands show some features different from the profiles in the western United States and central Europe. The record sections of these profiles show several distinct first-arrival phases, but hardly secondary arrivals corresponding to reflected phases as, e.g., the P^M-phase. A similar pattern is also observed on record sections in Scandinavia (HIRSCHLEBER et al., 1975). Consequently the crustal structure of the corresponding part of the Appalachians is different, too. The crust here is very thick, the velocities are higher than normally observed at similar depth ranges, and there do not exist clear boundaries, but very broad transition zones.

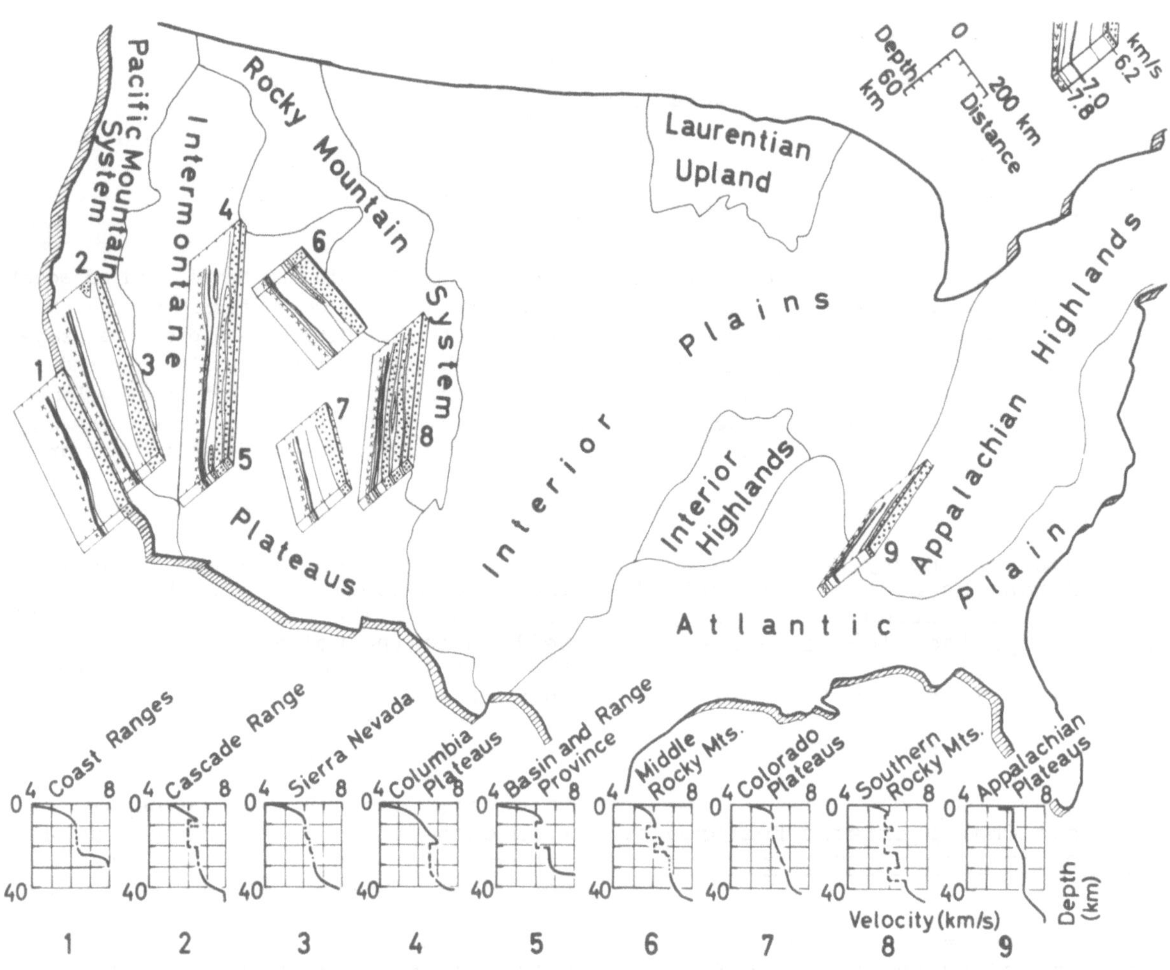

Fig. 6. Map of the United States showing main physical divisions (after FENNEMAN and JOHNSON, 1946), selected crustal cross sections with lines of equal velocity, and typical velocity-depth functions. In the crustal cross sections: *Dotted areas:* depth ranges with velocities < 6.2 km/s; *crosses:* uppermost part of the mantle with velocities ≥ 7.4-7.6 km/s

Looking at the crustal thickness of different mountainous areas of the United States, it seems questionable if the term "mountain root" is justified. Beneath the Alps, crust/mantle boundary as well as upper-crustal material reach into greater depths than in neighboring areas. Beneath North America, however, only one of both features is relevant: the total crust of the Sierra Nevada is thicker than that of the surrounding provinces; a low-velocity zone within the crust, however, does not exist and the thickness of the "upper" crust is normal. In the southern Rocky Mountains, total crustal thickness is similar to that of the neighboring areas, but the upper crust with an average velocity of about 6 km/s reaches into depths of about 35 km in contrast to velocity values found at similar depth beneath the adjacent Colorado Plateaus in the west and the Great Plains in the east. The Appalachians, finally, are apparently quite similar in crustal structure, thickness, and velocity behavior to the Great Plains as far as observations are available to decide on this question.

Acknowledgments. The study was enabled by a grant of the Deutsche Forschungsgemeinschaft and a succeeding appointment as a Visiting Scientist at the National Center for Earthquake Research of the U.S. Geological Survey in Menlo Park, California.

8.3 Investigations on Isostatic Balance in Different Parts of Eurasia Based on Seismic and Gravity Data

R. MEISSNER and U. VETTER

ABSTRACT

Density-depth functions are derived from velocity-depth profiles obtained from deep-seismic soundings. Characteristic functions from eastern and western Eurasia are compared. In general, isostatic balance between east and west is achieved by lighter sialic material and shallower crustal depths in the west compared to heavier material and deeper crustal blocks in the Russian shield. Some geotectonic implications are discussed.

ZUSAMMENFASSUNG

Es werden Geschwindigkeits-Tiefen-Profile in Dichte-Tiefen-Profile umgerechnet und Vergleiche für verschiedene Gebiete Eurasiens durchgeführt. Ein isostatischer Ausgleich zwischen östlichem und westlichem Teil Eurasiens wird nicht durch PRATT-AIRY-Modelle, sondern durch leichteres sialisches Material und geringere Krustentiefe im Westen im Vergleich zu schwererem Material und größerer Tiefe im russischen Schild erreicht. Einige geotektonische Folgerungen werden diskutiert.

8.3.1 Introduction

Contributions to the problem of isostatic behavior of crustal blocks have come from various fields of geophysics. Gravity as the tool for isostatic studies has been supported by deep-seismic soundings (DSS), by thermal, magnetotelluric, and geodetic methods in order to find depths, densities, and thicknesses of crustal blocks. In this paper we deal mainly with seismic methods. They seem to furnish important data for density-depth relations, and hence for details of isostatic balance between different crustal sections.

In the last five years some new techniques and interpretation procedures have been established in deep-seismic sounding (DSS), for instance by MEISSNER (1967a), SOLLOGUB (1969a, b), PAVLENKOVA (1969), GIESE (1970). Together with an increasing density of seismic observation lines, a fairly good knowledge on velocity-depth functions has been developed for various parts of Eurasia (MEISSNER, 1967b; GIESE, 1968a; KOSMINSKAYA, 1969). A transformation of velocity values into density values has been attempted by several authors (NAFE and DRAKE, 1957; WOOLLARD, 1962, 1968, 1969a, b; DORTMAN and MAGID, 1968; KOSMINSKAYA et al., 1969). These velocity-density functions are based on high-pressure studies in laboratories using many available rock samples (see for instance NAFE and DRAKE, 1957; CHRISTENSEN, 1965; BIRCH, 1960, 1969; DORTMAN and MAGID, 1968). The V,ρ values obtained show a certain scattering. Even average, V,ρ curves from different authors are slightly different. The latest results after DORTMAN and MAGID (1968) and WOOLLARD (1969a, b) are shown in Fig. 1. Especially for densities smaller than 5 g/cm^3 some uncertainties remain. The average curve D - based on studies of DORTMAN and MAGID and WOOLLARD'S curve W will be considered for velocity-density transformations mentioned below.

Regarding the reliability of present velocity-depth profiles, their transformation into density-depth profiles seems to be the largest uncertainty of the seismic methods at the present stage of investigation. The pure gravity approach to isostatic compensation, on the other hand, has also serious shortcomings: Bouguer anomalies do not only depend on the density distribution below a datum level but also on the lateral extension of a subsurface density anomaly and several other factors. Moreover, calculations of isostatic anomalies are in general based on the AIRY concept (which might not be valid in some areas) and furthermore need a good knowledge on the compensation depth. This depth definitely cannot be obtained from the pure gravity approach but should be taken from seis-

Fig. 1. Velocity-density relations after laboratory studies as compiled by WOOLLARD (1962, 1969a, b) (*W*) and DORTMAN and MAGID (1968) (*D*)

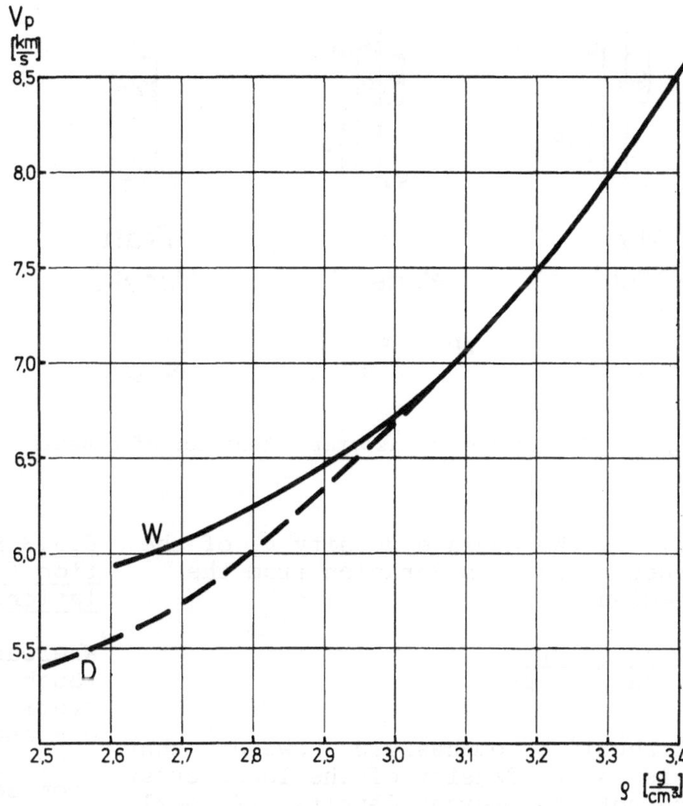

mic crustal studies. The compensation depth may often coincide with the Mohorovičić discontinuity (MD), but density inhomogeneities in the upper mantle also contribute to the observed gravity distribution.

In fact, a combination of seismic and gravity studies has been used very successfully for crustal problems by WOOLLARD (1969), DEMENITZKAYA and BELJAJEVSKY (1969), and others. WOOLLARD, by comparing the depth of the MD, obtained from seismic crustal studies, with the average elevation and assuming that AIRY'S concept of isostasy holds, comes to the conclusion that the established average crustal density used for the calculation of the Bouguer anomalies is too low. This observation seems to be supported by laboratory studies of rocks under high pressure.

The relation between crustal depth, average density, and elevation is useless, however, in those large areas of Eurasia which do not have any considerable elevation. These regions, on the other hand, show remarkable differences in crustal depth. If the crustal composition were similar, western Europe with its shallow MD would be much heavier than the old shield areas in the east. Thus, a careful compilation of present velocity-depth profiles, a conversion into density-depth profiles, and a comparison with gravity data seems to be necessary for attacking problems of isostatic balance.

8.3.2 Simple Models for an Isostatic Behavior of Crustal Blocks

In Fig. 2 five simple models for isostatic compensation of crustal blocks are presented. Models 1 and 3 show the AIRY and PRATT concepts with different heights of crustal blocks. It is felt that model 1 (AIRY) is valid for young mountain regions, and model 2 is mostly a good approximation for older areas with elevation differences. For model 2, if ΔR and Δh are known, the gradient in a $\Delta h, \Delta R$ graph gives the density ρ_2 in the lower crust under the assumption that values of ρ_o and ρ_M are fairly well known:

$$\frac{\Delta R}{\Delta h} = \frac{\rho_o}{\rho_M - \rho_2}$$

ΔR = depth difference of crustal blocks
Δh = elevation difference of crustal blocks
ρ = density being calculated

This procedure is very similar to that used by WOOLLARD (1962) who applies this method mostly to model 1. He cal-

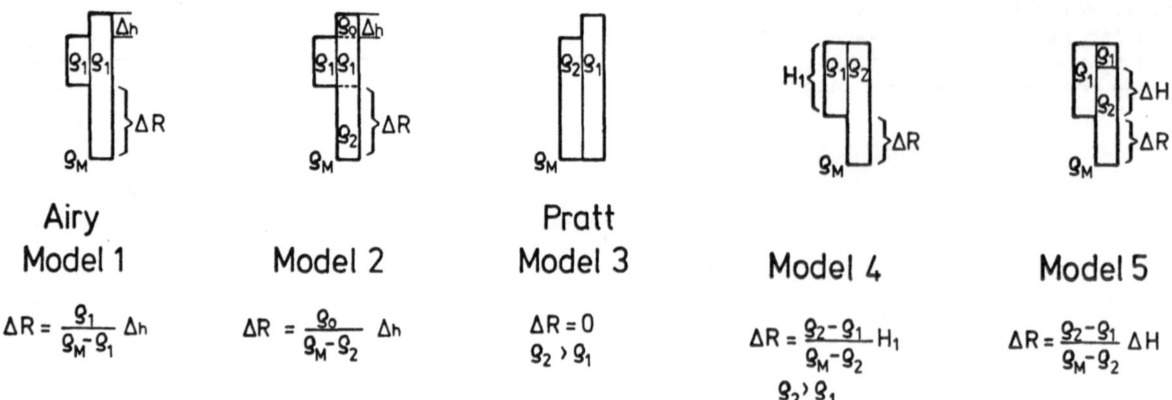

Fig. 2. Five models for isostatic behavior of crustal blocks

culates the average density ρ_1 of the crust under consideration from the gradient:

$$\frac{\Delta R}{\Delta h} = \frac{\rho_1}{\rho_M - \rho_1}$$

It should be emphasized, however, that this is the density of the lower crust and not the average density, if model 2 is valid. Thus, considering a general increase of density with depth, it is not surprising that the density values of 2.8 to 2.9 g/cm^3 calculated by WOOLLARD are always greater than those of the crystalline basement in the upper part of the crust. Nevertheless, by calculating the gradient $\frac{\rho_1}{\rho_M - \rho_1}$ or $\frac{\rho_0}{\rho_M - \rho_2}$ on an areal basis valuable information on ρ_1 and ρ_2 is obtained.

Models 4 and 5 in Fig. 2 show differences in crustal depth without elevation changes. These models might be applied for large areas of eastern and western Europe. Conditions such as these are actually observed from $\Delta h, \Delta R$ plots of the Russian Platform where the gradient $\frac{\Delta R}{\Delta h}$ tends toward infinity (DEMENITZKAYA and BELJAJEVSKY, 1969). Selecting between models 4 and 5, number 5 must be preferred because the seismic P_g wave which penetrates into the upper 10 km of the crust shows very similar characteristics in eastern and western Europe, indicating similar density in the upper 10 km. The deeper crust, on the other hand, must be denser in eastern Europe than in the west in order to create an isostatic balance between both parts of Europe. In order to find more details about the density-depth function recent seismic data will be analyzed.

8.3.3 Seismic Velocity-Depth Information for Estimating Density-Depth Relations

Whereas model 5 of Fig. 2 gives a rough estimate on the possible isostatic behavior of large parts of eastern and western Europe, more exact density-depth relations may be derived from seismic velocity-depth profiles. It has been mentioned already that the crustal depth of the Russian Platform exceeds that of western Europe. It is mostly about 40 to 45 km, compared to only 30 km in western Europe. The velocities in the deeper crust of eastern Europe, on the other hand, are much higher than in the western part.

In the following some velocity-depth profiles will be compared to density-depth profiles obtained from the velocity-density relations of Fig. 1. In Fig. 3a a V,z curve from western Germany is compared with one of the Baltic Shield. As most velocity-depth curves of western Europe (with the exception of the young Alpine Belt) are rather similar, the function presented in Fig. 3a may be used as a standard curve. Also the density-depth curves of both regions are compared (Fig. 3b). As the areas between intersections of the curves of both regions in Fig. 3b are about equal we may state that the crusts of both regions are approximately in isostatic balance. Parts of the Baltic Shield might be slightly heavier, parts of the Baltic Sea slightly lighter. More data from the Russian and West Siberian Platforms and western Europe all give similar information: the western part of the European continent has obtained approximately an isostatic balance with regard to its older neighbor in the east. Here, the deeper

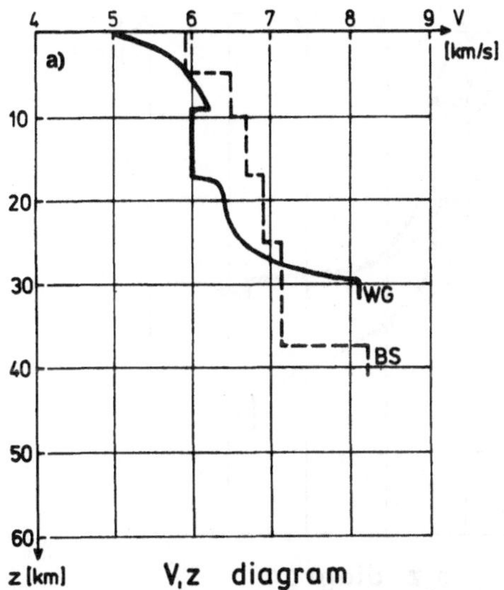

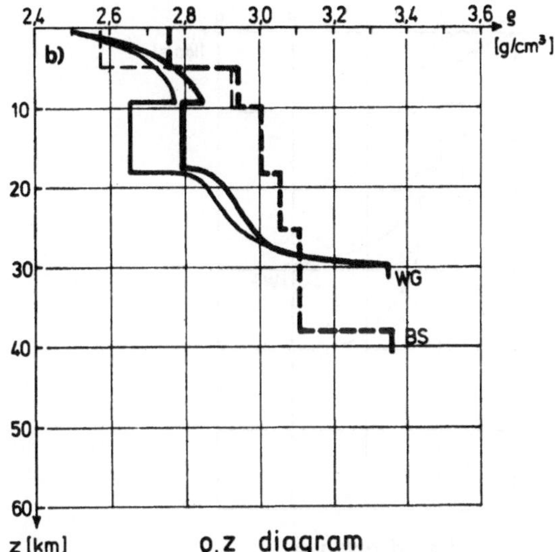

Fig. 3. (a) Velocity-depth curves from western Germany (*WG*) and the Baltic Shield (*BS*). (b) Density-depth curves from western Germany (*WG*) and the Baltic Shield (*BS*). Relations after WOOLLARD (1969a, b) and DORTMAN and MAGID (1968)

crust definitely consists of heavier, probably more gabbroic or granulitic material than in the west. Isostatic anomalies in both regions differ only by 10 mgal. In general, they are about equal for the eastern and western part of Europe. This may be taken as evidence that there is no large error and no considerable difference in the conversion of velocity into density values and that the density-depth relation obtained from velocity-depth curves are correct for both parts of Europe. The V,ρ relation of DORTMAN and MAGID seems to provide a slightly better correlation to the gravity field in Eurasia than WOOLLARD's curve.

While western Europe as a whole seems to be well adapted to the old shield area, young mountain belts and depressions show a strong deviation from isostasy. In Fig. 4a and b an example from the Fergana depression is compared to the standard curve. No intersection of the ρ,z curves takes place. The curve of Fergana indicates very light, probably sialic material in the whole crust. This information provides the clue for the strong isostatic anomaly of -195 mgal observed in the Fergana region.

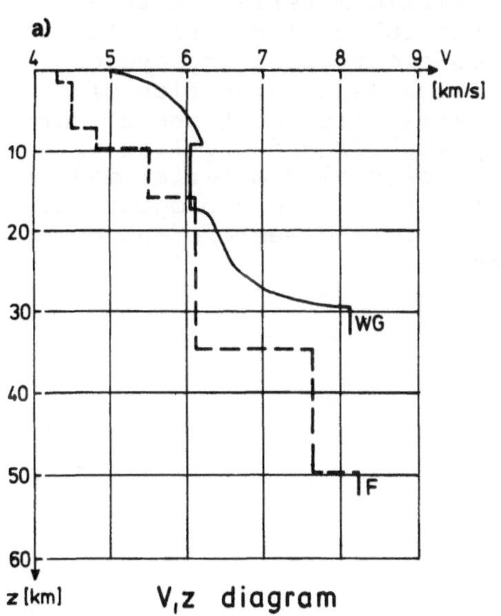

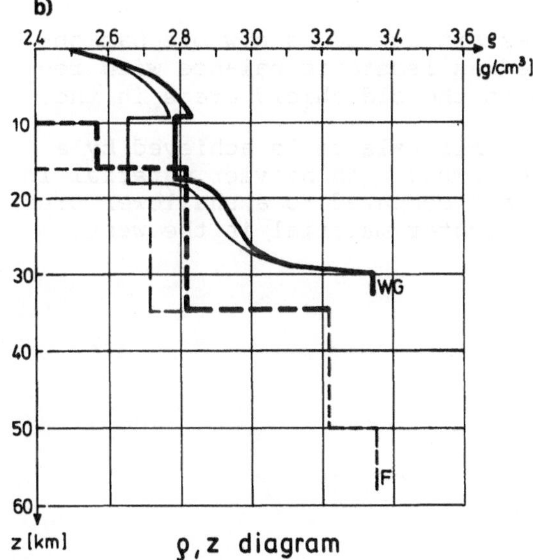

Fig. 4. (a) V,z curves from western Germany (*WG*) and Fergana Basin (*F*). (b) ρ,z curves from western Germany (*WG*) and Fergana Basin (*F*) further explanations Fig. 3b

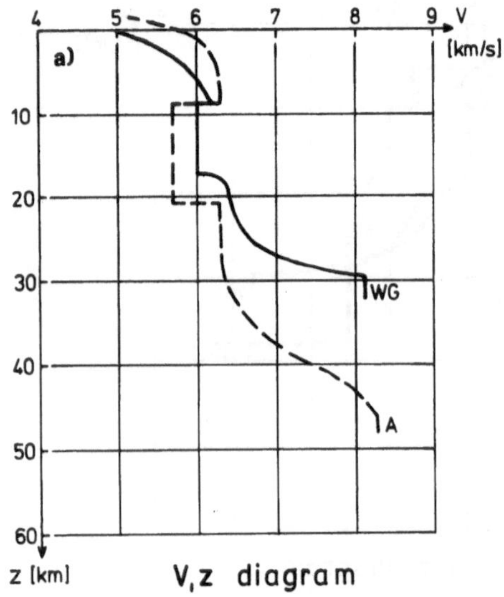

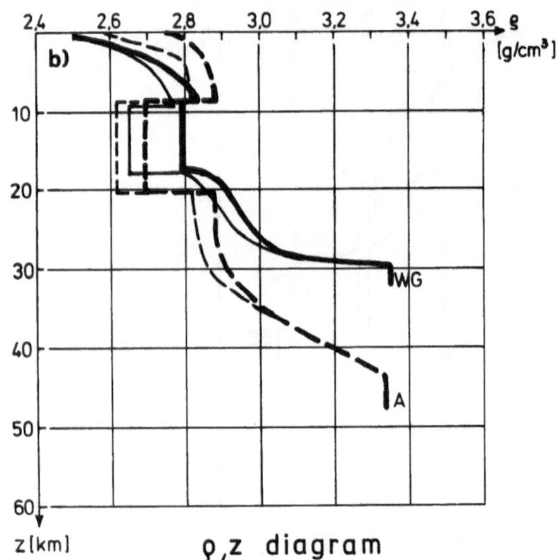

Fig. 5. (a) V,z curves from western Germany (*WG*) and Lago Lagorai, Alps (*A*). (b) ρ,z curves from western Germany (*WG*) and Lago Lagorai, Alps (*A*) further explanations Fig. 3b

Fig. 5a and b give examples for the central Alps. Here, the elevation difference plays an important role for the isostatic behavior. However, the curves of Fig. 5a and b suggest that an isostatic balance is still missing. The deeper part of the crust consists of lighter material compared to the crust in the shield area. Minor seismicity may be an indication that isostatic adjustments in the Alps are still going on. The isostatic anomalies in this area show an extended low of -20 mgal in the average.

8.3.4 Conclusions

1. Western Europe as a whole has obtained an isostatic balance with regard to the old shield areas in the east.

2. This balance is achieved by a deeper crust with heavier material in the east compared to a shallower crust with lighter material in the west.

3. For young mountain regions and depressions a considerable deviation from isostasy could be found by comparing density-depth profiles.

4. The lighter material in the deep crust of western Europe must be more sialic, the heavier material in the shield area more gabbroic or granulitic.

5. This indicates the growing of the old continental nuclei by the formation of geosynclines around them, filling them mainly with light, sialic debris. Later, isostatic balance is obtained.

6. Most of the growth of the old nuclei which have accumulated belt with more sialic material in their vicinity has taken place prior to the present sea-floor spreading cycle. Processes such as a subduction of oceanic plates below continental margins stripping off and buckling up large parts of sialic material may also account for the old orogenic processes which formed western and central Europe.

9. Acknowledgments

The realization of a crustal study program such as described in this volume calls for the participation and collaboration of many persons as well as the support of many institutions. Pages would be filled by naming everybody and by describing all contributions.

A program of this size can be started and successfully carried out only if adequate financial support is available. In the first place, therefore, the continuous granting of funds for this crustal study program by the German Research Society (Deutsche Forschungsgemeinschaft) must be gratefully acknowledged. Investigations were furthered in three priority programs, The Deep Structure of Central Europe, Upper Mantle Project, and Geodynamics of the Mediterranean Area, as well as in several individual normal programs. The funds granted for these programs covered means for employing scientific and technical personnel, for procuring equipment and expendable supplies and for financing field expeditions. In this connection, Dr. F. Goerlich and Dr. W. Heitz who, on behalf of the German Research Society, were responsible for the support of these programs, must be given credit for their valuable assistance.

For planning and execution the geophysical institutes of the universities and the corresponding departments of some government agencies of the Federal Republic of Germany were responsible. These institutions also contributed considerable means to the realization of this project.

Finally, the Volkswagen Foundation must be mentioned here for helping to procure modern magnetic-tape recording equipment.

The program could also not have been carried out without the courtesy and cooperation of several quarry companies whose explosions were used. Valuable assistance was also given by several radio stations and a number of other institutions.

All interpreters of the seismic data are indebted to the great number of participants who contributed in many ways by their personal efforts to the success of the field operations.

Some of the experiments of this program were carried out within the scope of common projects with French, Italian, and Swiss geophysical institutions. In the course of years, a close cooperation with the colleagues of these countries developed. On behalf of all German groups, we should like to thank them for this pleasant and fruitful teamwork.

In compiling this volume, support was given by the technical staff of the Geophysical Institute of the Freie Universität Berlin and the Geophysical Institute of the Universität Karlsruhe. We should like to thank especially Mrs. E. Gebert, Berlin, and Miss I. Hörnchen, Karlsruhe, for their help in typing and reviewing the English version. Assistance in typing and compiling was also given by Mrs. M. Steinhage. A great deal of photographic and drawing work was done by Mrs. R. Schmidt, and drawings were also prepared by Mrs. I. Stark.

10. References

AHORNER, L.: Herdmechanismen rheinischer Erdbeben und der seismotektonische Beanspruchungsplan im nordwestlichen Mitteleuropa. Sonderveröffentlichung. Geol. Inst. Univ. Köln 13, 105-130 (1967).

AHORNER, L.: Erdbeben und jüngste Tektonik im Braunkohlenrevier der niederrheinischen Bucht. Z. dt. geol. Ges. 11, 150-160 (1968).

AHORNER, L., MURAWSKI, H., SCHNEIDER, G.: Die Verbreitung von schadenverursachenden Erdbeben auf dem Gebiet der Bundesrepublik Deutschland. Z. Geophys. 36, 313-343 (1970).

ALPIN, L.M., BERDICHEVSKII, M.N., VEDRINTSEV, G.A., ZAGARMISTRA, A.M.: Dipole methods for measuring earth conductivity, 302 p. New York: Consultants Bureau 1966.

ANDERSON, D.L., SAMMIS, C.: Partial melting in the upper mantle. Phys. Earth Planet. Interiors 3, 41-50 (1970).

ANDERSON, O.L., SCHREIBER, E., LIEBERMANN, R.C., SOGA, N.: Some elastic constant data on minerals relevant to geophysics. Rev. Geophys. 6, 491-524 (1968).

ANGENHEISTER, G.: Struktur der tieferen Erdkruste im nördlichen Alpenvorland nach Ergebnissen der Refraktions- und Reflexions-Seismik. Beih. Geol. Jb. 80, 31-41 (1969).

ANGENHEISTER, G., BÖGEL, H., GEBRANDE, H., GIESE, P., SCHMIDT-THOME, P., ZEIL, W.: Recent investigations of surficial and deeper crustal structure of the eastern and southern Alps. Geol. Rdsch. 61, 349-395 (1972).

ANGENHEISTER, G., BÖGEL, H., MORTEANI, G.: Die Ostalpen im Bereich einer Geotraverse vom Chiemsee bis Vicenza. N. Jb. Geol. Paläont. Abh. 148, 50-137 (1975).

ANGENHEISTER, G., POHL, J.: Die seismischen Messungen im Ries von 1948-1969. Geologica Bavarica 61, 304-326 (1969).

ANGENHEISTER, G., POHL, J.: Deep crustal reflections on a 17 km digital reflection profile in South Germany (Nördlinger Ries). Proc. 12th Gen. Ass. Europ. Seismol. Comm. (Luxembourg 1970), Obs. Royal de Belgique, Comm. Sér. A-N° 13, Sér. Geophys. 101, 173-176 (1971).

ANSORGE, J.: Die Struktur der Erdkruste an der Westflanke der Zone von Ivrea. Schweiz. Min. Petr. Mitt. 48, 247-254 (1968).

ANSORGE, J.: Die Feinstruktur des obersten Erdmantels unter Europa und dem mittleren Nordamerika. Diss. Univ. Karlsruhe, 111p., 1975.

ANSORGE, J., EMTER, D., FUCHS, K., LAUER, J.P., MUELLER, S., PETERSCHMITT, E.: Structure of the crust and upper mantle in the rift system around the Rhinegraben. In: Graben Problems (H. ILLIES, S. MUELLER, eds.), pp. 190-197. Stuttgart: Schweizerbart 1970.

ANSORGE, J., EMTER, D., FUCHS, K., LAUER, J.P., MUELLER, S., PETERSCHMITT, E., PRODEHL, C.: Refraktionsseismische und reflektionsseismische Untersuchungen im Gebiet des Oberrheingrabens. 5. Koll. Internat. Rheingraben-Forschungsgruppe, Karlsruhe, 5 p., 1972.

ANSORGE, J., MAYER-ROSA, D.: Evidence of velocity reversals within the upper mantle in Europe from bodywave observations. Proc. 10th Gen. Ass. Europ. Seismol. Comm. (Leningrad 1968), Acad. Sciences USSR, Moscow, pp. 49-55, 1970.

ANSORGE, J., MUELLER, S.: The fine structure of the upper mantle in Europe and in North America. Proc. 12th Gen. Ass. Europ. Seismol. Comm. (Luxembourg 1970), Obs. Royal de Belgique, Comm. Sér. A-N° 13, Sér. Geophys. 101, 196-197 (1971).

ARASCHMID, A.: Über die Bündelung von Geophonen in der Refraktionsseismik. Z. Geophys. 28, 185-208 (1962).

Arbeitsgruppe Digitalisierung im Forschungskollegium Physik des Erdkörpers: Katalog von Rechenprogrammen für digitale Großrechner aus dem Gebiet der Geophysik - Kurzbeschreibungen, 72 p. Unpublished report, Karlsruhe 1968.

ARTEMJEV, M.E., ARTYUSHKOV, E.V.: Structure and isostasy of the Baical rift and the mechanism of rifting. J. Geophys. Res. 76, 1197-1211 (1971).

ASFA (STEIN, A., ed.): Bericht über sprengseismische Untersuchungen 1968, 69 p. Niedersächs. Landesamt f. Bodenforschung, Hannover 1969.

ASFA (STEIN, A., ed.): Bericht über sprengseismische Untersuchungen 1969, 46 p. Niedersächs. Landesamt f. Bodenforschung, Hannover 1970.

ASFA (STEIN, A., ed.): Bericht über sprengseismische Untersuchungen 1970, 49 p. Niedersächs. Landesamt f. Bodenforschung, Hannover 1971.

ASFA (PRODEHL, C., SCHRÖDER, H., eds.): Bericht über sprengseismische Untersuchungen 1971 und 1972, 63 p. Niedersächs. Landesamt f. Bodenforschung, Hannover 1974.

ATANASOFF, J.V., HART, P.J.: Dynamical determination of the elastic constants and their temperature coefficients for quartz. Phys. Rev. 59, 85-96 (1941).

AUBOUIN, J.: Geosynclines, 335 p. Amsterdam-London-New York: Elsevier Publ. Comp. 1965.

AUBRAT, J., GIESE, P., PASCAL, G., PERRIER, G., PUNTOUS, R., RECQ, M., SIMONIN, A.: Structure de la croûte terrestre dans les Alpes occidentales et la vallée du Rhône. C.R. Acad. Sci. 265, 533-536 (1967).

BALTENBERGER, P., LABROUSTE, Y., PERRIER, G., RECQ, M.: Courbes d'égale profondeur de la surface de Mohorovicic dans le Sud-Est de la France. C.R. Acad. Sci. 266, 1530-1533 (1968).

BAMFORD, D.: Refraction data in western Germany - a time-term interpretation. Z. Geophys. 39, 907-927 (1973).

BANKS, R.J.: Geomagnetic variations and the electrical conductivity of the upper mantle. Geophys. J.R. astr. Soc. 17, 457-487 (1969).

BARTELSEN, H.: Deutung der seismischen Weitwinkelmessungen in der Rheinischen Masse unter Verwendung neuer Kontroll- und Auswerteverfahren, 90 p. Dipl.-Arbeit. Univ. Frankfurt 1970.

BARTENSTEIN, H.: Present status of the Paleozoic palaeogeography of northern Germany and adjacent parts of North-West Europe. In: Geology of Shelf Seas (D.T. DONOVAN, ed.), pp. 31-54. Edinburgh-London: Oliver and Boyd 1968.

BATEMAN, H.: The solution of the integral equation connecting the velocity of propagation of an earthquake-wave in the interior of the earth with the times which the disturbance takes to travel to the different stations on the earth's surface. Phil. Mag. 19, 576-587 (1910). Also in Phys. Z. 11, 96-99 (1910).

Bayerisches Geologisches Landesamt: Das Ries. Geologica Bavarica 61, 478 p. (1969).

Bayerisches Geologisches Landesamt: Die Forschungsbohrung Nördlingen 1973. Geologica Bavarica 72, 98 p. (1974).

BEAUFILS, Y.: Experience du Lac Blanc. Proc. 9th Gen. Ass. Europ. Seismol. Comm. (Copenhagen 1966), pp. 257-263. Kobenhavn: Akademisk Forlag 1967.

BEHNKE, C.: Bericht über die Auswertung refraktionsseismischer Messungen "Westprofil Thaiden". Frankfurter DFG-Kolloquium, 7 p. Hannover: Niedersächs. Landesamt. Bodenf. 1961a.

BEHNKE, C.: Bericht über die Auswertung refraktionsseismischer Messungen "Profil Grossenritte". Frankfurter DFG-Kolloquium, 3 p. Hannover: Niedersächs. Landesamt Bodenf. 1961b.

BEHNKE, C.: Über Geschwindigkeiten seismischer Wellen im zentralen Teil der Ostalpen. Proc. 9th Gen. Ass. Europ. Seismol. Comm. (Copenhagen 1966), pp. 97-102. Kobenhavn: Akademisk Forlag 1967.

BEHNKE, C.: Meßdaten seismischer Untersuchungen in den Alpen 1959-1969. Bericht, 71 p. Hannover: Niedersächs. Landesamt Bodenf. 1969.

BEHNKE, C.: Explosion seismology data generalization - examples from the Eastern Alps. Proc. 12th Gen. Ass. Europ. Seismol. Comm. (Luxembourg 1970), Obs. Royal de Belgique, Comm. Sér. A-N° 13, Sér. Géophys. 101, 177-181 (1971).

BEHNKE, C., GIESE, P., PRODEHL, C., de VISINTINI, G.: Seismic refraction investigations in the Dolomites for the earth's crust in the Eastern Alpine area. Boll. Geofis. teor. ed. appl. 4, 110-132 (1962).

BELJAJEVSKY, N.A., BORISOV, A.A., VOLVOVSKY, I.S.: Tiefbau des Territoriums der UdSSR (russ.). Sovjetskaya Geologya 11, 56-84 (1967).

BERÁNEK, B., DUDEK, A.: The results of deep seismic sounding in Czechoslovakia. Z. Geophys. 38, 415-427 (1972).

BERÁNEK, B., MAYEROVA, M., ZOUNKOVÁ, M., GUTERCH, A., MATERZOK, R., PAJCHEL, J.: Results of deep seismic soundings along International Profile VII in Czechoslovakia and Poland. Studia geoph. et geod. 17, 205-217 (1973).

BERÁNEK, B., MAYEROVA, M., ZOUNKOVÁ, M., ZATOPEK, A., HOLUB, K., ANGENHEISTER, G., GEBRANDE, H., MILLER, H.: Results from deep seismic sounding along the International Profile VII in Czechoslovakia and Federal Republic of Germany. Abstr. 14th Gen. Ass. Europ. Seismol. Comm., Trieste, ESC 15, 1974.

BERÁNEK, B., ZOUNKOVÁ, M., HOLUB, K.: Results of deep seismic sounding in Czechoslovakia. In: Upper Mantle Project Programme in Czechoslovakia 1962-1970 (M. PICK, J. PICHA, eds.), pp. 94-115. Praha: Geophys. Academia 1971.

BERCKHEMER, H. (German Research Group for Explosion Seismology): Topographie des Ivrea Körpers, abgeleitet aus seismischen und gravimetrischen Daten. Schweizer. Min. Petr. Mitt. 48, 235-246 (1968).

BERCKHEMER, H.: Direct evidence for the composition of the lower crust and the Moho. Tectonophysics 8, 97-105 (1969).

BERCKHEMER, H.: MARS 66; Eine Magnetbandapparatur für seismische Tiefensondierung. Z. Geophys. 36, 501-518 (1970).

BERRY, M.J.: Depth uncertainties from seismic frist-arrival refraction studies. J. Geophys. Res. 76, 6464-6468 (1971).

BERRY, M.J., WEST, G.F.: An interpretation of the first-arrival data of the Lake Superior experiment by the time-term method. Bull. Seism. Soc. Am. 56, 141-171 (1966).

BESSONOVA, E.N., FISHMAN, V.M., RYABOI, V.Z., SITNIKOVA, G.A.: The Tau method for inversion of travel times - I. Deep seismic sounding data. Geophys. J.R. Astr. Soc. 36, 377-398 (1974).

BHATTACHARYYA, B.K.: Propagation of transient electromagnetic waves in a medium of finite conductivity. Geophysics 12, 75-88 (1957).

BINDER, K.: Beteiligung des Landeserdbebendienstes Baden-Württemberg und des Geophysikalischen Institutes Stuttgart an den Schwerpunktprogrammen der Deutschen Forschungsgemeinschaft zur Erforschung des tieferen Untergrundes in Mitteleuropa. Festschrift Wilhelm Hiller, Inst. Geophysik Univ. Stuttgart, pp. 164-169, 1969.

BIRCH, F.: Elasticity of igneous rocks at high temperatures and pressures. Geol. Soc. Am. Bull. 54, 263-286 (1943).

BIRCH, F.: Interpretation of the seismic structure of the crust in the light of experimental studies of wave velocities in rocks. In: Contributions in Geophysics in Honor of Beno Gutenberg (E. INGERSON, ed.), pp. 158-170. London: Pergamon Press 1958.

BIRCH, F.: The velocity of compressional waves in rocks to 10 kb, part 1. J. Geophys. Res. 65, 1083-1102 (1960).

BIRCH, F.: The velocity of compressional waves in rocks to 10 kb, part 2, J. Geophys. Res. 66, 2199-2224 (1961).

BIRCH, F.: Density and composition of the upper mantle: First approximations as an olivine layer. In: The Earth's Crust and Upper Mantle (J. HART, ed.). Geophys. Monogr., Am. Geophys. Un., Washington, D.C. 13, 18-36 (1969).

BIRCH, F.: Interpretations of the low-velocity zone. Phys. Earth Planet. Interiors 3, 178-181 (1970).

BIRCH, F., CLARK, S.: The thermal conductivity of rocks and its dependence upon temperature and composition. Am. J. Sci. 238, 529-558, 613-635 (1940).

BLOHM, E.K.: Geoelektrische Messungen am Wamberger Sattel. Bericht E 1217. Archiv BfB/NLfB, Hannover 1970.

BLOHM, E.K.: Die Methode der geoelektrischen Tiefensondierungen mit großen Elektrodenentfernungen. Diss. TU Clausthal 1972.

BLOHM, E.K., FLATHE, H.: Geoelectrical deep sounding in the Rhinegraben. In: Graben Problems (H. ILLIES, S. MUELLER, eds.), pp. 239-242. Stuttgart: Schweizerbart 1970.

BLOHM, E.K., HOMILIUS, J.: Geoelektrische Tiefensondierungen im Reinhardswald und Bramwald. Bericht E 1273 Archiv BfB/NLfB, Hannover 1971.

BÖGEL, H.: Zur Literatur über die "Periadriatische Naht". Verh. geol. Bundesanstalt, Heft 2/3, 163-199 (1975).

BOIGK, H.: Gedanken zur Entwicklung des Niedersächsischen Tektogens. Geol. Jb. 85, 861-900 (1968).

BOMMERT, R.: Die Kammersprengung, ein Mittel zur Leistungssteigerung bei der Rohstoffgewinnung über Tage. Silikattechnik 4, 173-176 (1953).

BONJER, K.-P., FUCHS, K.: Crustal structure in south-west Germany from spectral transfer ratios of longperiod body waves. In: Graben Problems (H. ILLIES, S. MUELLER, eds.), pp. 198-202. Stuttgart: Schweizerbart 1970.

BORCHERT, H.: Geosynklinale Lagerstätten, was dazu gehört und was nicht dazu gehört, sowie deren Beziehung zu Geotektonik und Magmatismus. In: Lagerstättenkunde, Freiberger Forschungshefte C 79, 7-61 (1960).

BORTFELD, R.: A method of dip corrections for expanding spread velocity measurements. Geofisica Pura ed Applicata 38, 32-44 (1957).

BORTFELD, R.: Moderne Seismik dringt in größere Tiefen vor. Erdöl u. Kohle 24, 289-298 (1971).

BOSUM, W.: Interpretation magnetischer Anomalien durch dreidimensionale Modellkörper zur Klärung geologischer Probleme. Geol. Jb. 83, 667-680 (1965).

BOSUM, W., DÜRBAUM, H.J., FENCHEL, W., FRITSCH, J., LUSZNAT, M., NICKEL, H., PLAUMANN, S., SCHERP, A., STADLER, G., VOGLER, H.: Geologisch-lagerstättenkundliche und geophysikalische Untersuchungen im Siegerländer-Wieder Spateisensteinbezirk. Beih. geol. Jb. 90, 139 p. (1971).

BOSUM, W., HAHN, A.: Interpretation der Flugmagnetometervermessung des Oberrheingrabens.

In: Graben Problems (H. ILLIES, S. MUELLER, eds.), pp. 219-223. Stuttgart: Schweizerbart 1970.

BOSUM, W., ULRICH, H.J.: Die Flugmagnetometervermessung des Oberrheingrabens und ihre Interpretation. Geol. Rdsch. 59, 83-106 (1969).

BRADLEY, R.S., JAMIL, A.K., MUNRO, D.C.: The electrical conductivity of olivine at high temperatures and pressures. Geochim. Cosmochim. Acta 28, 1669-1678 (1964).

BRAM, K.: Die Geschwindigkeitsverteilung für P-Wellen innerhalb der Erdkruste entlang eines Profiles in Bayern zwischen Weißenburg und Weilheim, abgeleitet aus Refraktionsbeobachtungen im Sommer 1964. Dipl.-Arbeit, 62 p. Univ. München 1967.

BRAM, K., GIESE, P.: Die Geschwindigkeitsverteilung der P-Welle in der Erdkruste im Raum Augsburg (Süd-Deutschland) - Ergebnisse und Vergleich zweier seismischer Messungen. Z. Geophys. 34, 611-626 (1968).

BREYER, F.: Ergebnisse seismischer Messungen auf der süddeutschen Großscholle besonders im Hinblick auf die Oberfläche des Varistikums. Z. dt. geol. Ges. 108, 21-36 (1956).

BREYER, F., DOHR, G.: Betrachtungen über den Bau der Gefalteten Molasse im westlichen Bayern mit Beziehung auf das Molasse-Vorland und die angrenzenden Teile der Alpen auf Grund geophysikalischer Untersuchungen. Erdöl und Kohle 12, 315-323 (1959).

BRIDGMAN, P.W.: The thermal conductivity and compressibility of several rocks under high pressures. Am. J. Sci. 7, 81 (1924).

BRINKMANN, R.: Die Mitteldeutsche Schwelle. Geol. Rdsch. 36, 56-66 (1948).

BROCKAMP, B.: Seismische Beobachtungen bei Steinbruchsprengungen. Z. Geophys. 7, 295-317 (1931).

BROCKAMP, B.: Kurzbericht über die im Gebiet von Osnabrück durchgeführten seismischen Arbeiten des Instituts für Reine und Angewandte Geophysik der Universität Münster. Veröffentl. d. Dtsch. Geod. Komm., Reihe B, Heft 153, 12 p. (1967).

BROCKAMP, B., WOELCKEN, K.: Bemerkungen zu den Beobachtungen bei Steinbruchsprengungen. Z. Geophys. 5, 163-171 (1929).

BULLARD, E.: The origin of oceans. Sci. Am. 221(3), 66-75 (1969).

BURCKHARDT, H.: Some physical aspects of seismic scaling laws for underwater explosions. Geophys. Prospect. 12, 192-214 (1964).

BURG, J.P.: Three-dimensional filtering with an array of seismometers. Geophysics 29, 693-713 (1964).

CAGNIARD, L.: Réflexion et réfraction des ondes séismiques progressives, 255 p. Paris: Gauthier-Villars 1939.

CARLÉ, W.: Kohlensäure und Herdlage im Uracher Vulkangebiet und seiner weiteren Umgebung. Z. dt. geol. Ges. 110, 71-101 (1958).

CERVENY, V.: On some kinematic and dynamic properties of reflected and head waves in the case of a layered overburden. Geofysikálni Sborník, Academia, Prague 14, 105-179 (1966) (1967).

CHARLIER, Ch.: Deuxième rapport sur l'explosion d'Heligoland. Obs. Roy. Belgique à Uccle, Publ. Serv. Seism. et Gravim., 27 p. Brussels 1947.

CHOUDHURY, M., GIESE, P., de VISINTINI, G.: Crustal structure of the Alps - some general features from explosion seismology. Boll. Geofis. Teor. ed Appl. 13, 211-240 (1971).

CHOUDHURY, M., LABROUSTE, Y., PERRIER, G.: Essai d'interprétation no. 2. In: Recherches séismologiques dans les Alpes occidentales au moyen de grandes explosions en 1956, 1958 et 1960 (H. CLOSS, Y. LABROUSTE, eds.). Mémoire Collectif, Année Géophysique Internationale, C.N.R.S., Sér. III, Fasc. 2, 176-200, Paris (1963).

CHRISTENSEN, N.I.: Compressional wave velocities in metamorphic rocks. J. Geophys. Res. 70, 6147-6164 (1965).

CHRISTOV, V.K.: Die Gaußschen und geographischen Koordinaten auf dem Ellipsoid von Krassowsky, 254 p. Berlin: Verlag Technik 1955.

CLARK, S.P.: Heat flow in the Austrian Alps. Geoph. J.R. astr. Soc. 6, 54-63 (1961).

CLASEN, G.: Salzstockrandbestimmungen mit Hilfe gravimetrischer und seismischer Methoden. Erdöl u. Kohle 11, 2-5 (1958).

CLOOS, H.: Bau und Tätigkeit von Tuffschloten, Untersuchungen am Schwäbischen Vulkan. Geol. Rdsch. 32, 709-800 (1941).

CLOSS, H.: Geophysik und tektonische Richtungen im weiteren Unterelbegebiet. Jb. d. Reichsamt f. Bodenf. 62, 117-154, 1941 (1944).

CLOSS, H., BEHNKE, C.: Fortschritte der Anwendung seismischer Methoden in der Erforschung der Erdkruste. Geol. Rdsch. 51, 315-330 (1961).

CLOSS, H., HÄNEL, R.: Eine neue Darstellung der Ergebnisse der Helgolandsprengung. Bericht NLfB-Archiv (Seismik), Hannover 1964.

CLOSS, H., LABROUSTE, Y. (eds.): Recherches séismologiques dans les Alpes occidentales au moyen de grandes explosions en 1956, 1958 et 1960. Mémoire Collectif, Année Géophysique Internationale, C.N.R.S. Sér. III, Fasc. 2, 241 p., Paris (1963).

CLOSS, H., PLAUMANN, S.: On the gravity of the upper Rhinegraben. In: The Rhinegraben Progress Report 1967 (J.P. ROTHÉ, K. SAUER, eds.). Abh. Geol. Landesamt Baden-Württemberg 6, 92-93 (1967).

CLOSS, H., PLAUMANN, S.: Gedanken zur Tektonik der Kruste im Oberrheingraben auf Grund von Schweremessungen. Geol. Jb. 85, 371-382 (1968).

CLOWES, R.M., KANASEWICH, E.R., CUMMING, G.L.: Deep crustal seismic reflections at near vertical incidence. Geophysics 33, 441-451 (1968).

COLOMBI, B., SCARASCIA, S.: Sulla interpretazione dei profili sismici crostali. Calcolo diretto della funzione velocità-profondita. Revista Italiana di Geofisica 22, 213-226 (1973).

COOK, K.L.: Active rift systems in the Basin and Range province. Abs. of Pap. UMC, Symp. World Rift Syst., Gen. Ass. IUGG, 1 p., Zürich 1967.

DAVYDOVA, N.K., KOSMINSKAYA, I.P., KAPUSTIAN, N.K., MICHOTA, G.G.: Models of the earth's crust and M-boundary. Z. Geophys. 38, 369-393 (1972).

DEMENITZKAYA, R.M., BELJAJEVSKY, N.A.: The relation between the earth's crust, surface relief, and gravity field in the USSR. In: The earth's crust and upper mantle. (P.J. HART, ed.), pp. 312-319. Geophys. Monogr. 13, Am. Geophys. Un., Washington, D.C. 1969.

DEMNATI, A., DOHR, G.: Reflexionsseismische Tiefensondierungen im Bereich des Oberrheintalgrabens und des Kraichgaues. Z. Geophys. 31, 229-245 (1965).

DEPPERMANN, K.: Die Abhängigkeit des scheinbaren Widerstandes vom Sondenabstand bei der Vierpunktmethode. Geophys. Prospect 2, 262-273 (1954).

DEPPERMANN, K.: Zur Eliminierung der Störspannungen bei geoelektrischen Widerstandsmessungen. Geol. Jb. 85, 901-918 (1968).

DIETZ, R.S.: Continent and ocean basin evolution by spreading of the sea floor. Nature 190, 854-857 (1961).

DIX, C.H.: Seismic Prospecting for Oil, 414 p. New York: Harper and Brothers 1952.

DIX, C.H.: Seismic velocities from surface measurements. Geophysics 20, 17-26 (1955).

DOHR, G.: Zur reflexionsseismischen Erfassung sehr tiefer Unstetigkeitsflächen. Erdöl u. Kohle 10, 278-281 (1957a).

DOHR, G.: Ein Beitrag der Reflexionsseismik zur Erforschung des tieferen Untergrundes. Geol. Rdsch. 46, 17-26 (1957b).

DOHR, G.: Über die Beobachtungen von Reflexionen aus dem tieferen Untergrund im Rahmen routinemäßiger reflexionsseismischer Messungen. Z. Geophys. 25, 280-300 (1959).

DOHR, G.: Beobachtungen von Tiefenreflexionen im Oberrheingraben. In: The Rhinegraben Progress Report 1967 (J.P. ROTHÉ, K. SAUER, eds.). Abh. Geol. Landesamt Baden-Württemberg 6, 94-95 (1967).

DOHR, G.: Ergebnisse reflexionsseismischer Messungen zur Untersuchung des Baues der Erdkruste in der Bundesrepublik. Proc. 8th Gen. Ass. Europ. Seismol. Comm. (Budapest 1964), pp. 59-81. Budapest: Akadémiai Kiadó 1968.

DOHR, G.: Reflexionsseismische Messungen im Oberrheingraben mit digitaler Aufzeichnungstechnik und Bearbeitung. In: Graben Problems (H. ILLIES, S. MUELLER, eds.), pp. 207-218. Stuttgart: Schweizerbart 1970.

DOHR, G., FUCHS, K.: Statistical evaluation of deep crustal reflections in Germany. Geophysics 32, 951-967 (1967).

DOHR, G., HADJEBI, B., HEHN, K.: Beobachtungen von Tiefenreflexionen in Norddeutschland. Proc. 9th Gen. Ass. Europ. Seismol. Comm. (Copenhagen 1966), pp. 87-96. Kobenhavn: Akademisk Forlag 1967.

DOHR, G., MEISSNER, R.: Deep crustal reflections in Europe. Geophysics 40, 25-39 (1975).

DOHR, G., MENZEL, H.: Die Reflexionsseismik im Oberrheingraben - ein Überblick, 5 p. Wiesloch: DFG-Kolloquium Oberrheingraben 1966.

DORTMANN, N.B., MAGID, M.: New data on the velocity of compressional waves in crystalline rocks. Sovjetsk. Geologya 5, 123-129 (1968).

DUBA, A., NICHOLLS, I.A.: The influence of oxidation state on the electrical conductivity of olivine. Earth Planet. Sci. Letters 18, 59-64 (1973).

DÜRBAUM, H.-J., FRITSCH, J., NICKEL, H.: Deep-seismic sounding in the eastern part of the Rhenish Massif. Proc. 9th Gen. Ass. Europ. Seismol. Comm. (Copenhagen 1966), pp. 265-271. Kobenhavn: Akademisk Forlag 1967.

EBERLE, D.: Aeromagnetische Karte der Bundesrepublik Deutschland 1 : 1,000,000. - Hannover: Bundesanst. f. Bodenf. 1973.

EBERLE, D.: Interpretation der aeromagnetischen Anomalien im Gebiet der Schwäbischen und südlichen Fränkischen Alb. Abstr., 35. Jtag. Dt. Geophys. Ges., Stuttgart, 5.33, 1975.

EDEL, J.B.: Structure de la croûte terrestre sous le fossé Rhénan et ses bordures, 207 p. Thèse, Univ. Strasbourg 1975.

EDEL, J.B., FUCHS, K., GELBKE, C., PRODEHL, C.: Deep structure of the southern Rhinegraben area from seismic-refraction investigations. Z. Geophys. 41, 333-356 (1975).

EMTER, D.: Refraktionsseismische Untersuchungen im Gebiet des Steinheimer Beckens. Festschrift Wilhelm Hiller, pp. 170-186. Inst. Geophysik Univ. Stuttgart 1969.

EMTER, D.: Ergebnisse seismischer Untersuchungen der Erdkruste und des oberen Erdmantels in Südwestdeutschland, 108 p. Diss., Univ. Stuttgart 1971.

EMTER, D., SCHNEIDER, G., ZÜRN, W.: The 1969/70 Swabian Jura earthquakes. Proc. 12th Ass. Gen. Comm. Seism. Europ. (Luxembourg 1970), Obs. Royal de Belgique, Comm. Sér. A-N° 13, Sér. Géophys. 101, 129-133 (1971).

Erläuterungen zur Geologischen Karte von Bayern 1:500 000, 2 ed., 344 p. Bayer. Geol. Landesamt, München 1964.

EWING, J., HOUTZ, R.: Mantle reflections in airgun-sonebuoy profiles. J. Geophys. Res. 74, 6706-6709 (1969).

FENNEMAN, N.M., JOHNSON, D.W.: Physical divisions of the United States. U.S. Geol. Survey Map, 1:7 000 000, 1946.

FIELITZ, K.: Untersuchungen zur Temperaturabhängigkeit von Kompressions- und Scherwellengeschwindigkeiten in Gesteinen unter erhöhtem Druck, 111p. Diss., Techn. Univ. Clausthal 1971.

FLATHE, H.: The determination of the electrical resistivity of the crust within the region of the Rhinegraben. In: The Rhinegraben Progress Report 1967 (J.P. ROTHÉ, K. SAUER, eds.). Abh. Geol. Landesamt Baden-Württemberg 6, 96-98 (1967).

FLATHE, H.: Comment on "The automatic fitting of a resistivity sounding by geometrical progression of depths". Geophys. Prospect. 22, (1), (1974).

FLATHE, H., HOMILIUS, H.: Probleme bei der Deutung geoelektrischer Sondierungskurven. In: Die Wassererschließung (H. SCHNEIDER, ed.), pp. 224-231. Essen 1973.

FORSYTH, D.F., PRESS, F.: Geophysical test of petrologic models of the spreading lithosphere. J. Geophys. Res. 76, 7963-7979 (1971).

FÖRTSCH, O.: Analyse der seismischen Registrierungen der Großsprengung bei Haslach im Schwarzwald am 28. April 1948. Geol. Jb. 66, 65-80 (1951).

FRITSCH, J.: Ein Beitrag zur Lösung des seismischen Refraktionsproblems, 62 p. Diss., FU Berlin 1967.

FUCHS, K.: The reflection of spherical waves from transition zones with arbitrary depth-dependent elastic moduli and density. J. Phys. Earth 16 (Special Issue), 27-41 (1968).

FUCHS, K.: On the properties of deep crustal reflectors. Z. Geophys. 35, 133-149 (1969a).

FUCHS, K.: The method of stationary phase as a diagnostic aid in estimating the field pattern of body waves reflected from transition zones. Z. Geophys. 35, 431-434 (1969b).

FUCHS, K.: On the determination of velocity depth distributions of elastic waves from the dynamic characteristics of the reflected wave field. Z. Geophys. 36, 531-548 (1970).

FUCHS, K.: Geophysical contributions to Taphrogenesis. In: Approaches to Taphrogenesis (H. ILLIES, K. FUCHS, eds.), pp. 420-432. Stuttgart: Schweizerbart 1974.

FUCHS, K., LANDISMAN, M.: Results of a reinterpretation of the N-S refraction line Adelebsen-Hilders South in West-Germany. Z. Geophys. 32, 121-123 (1966a).

FUCHS, K., LANDISMAN, M.: Detailed crustal investigation along a north-south section through the central part of Western Germany. In: The earth beneath the continents (J.S. STEINHART, T.J. SMITH, eds.), pp. 433-452. Geophys. Monogr. 10, Am. Geophys. Un., Washington, D.C. 1966b.

FUCHS, K., MÜLLER, G.: Computation of synthetic seismograms with the reflectivity method and comparison with observations. Geophys. J.R. Astr. Soc. 23, 417-433 (1971).

FUCHS, K., MUELLER, S., PETERSCHMITT, E., ROTHÉ, J.P., STEIN, A., STROBACH, K.: Krustenstruktur der Westalpen nach refraktionsseismischen Messungen. Gerl. Beitr. z. Geophys. 72, 149-169 (1963).

GARDNER, L.W.: An arcal plan of mapping subsurface structure by refraction shooting. Geophysics 4, 247-259 (1939).

GARVIN, W.W.: Exact transient solution of the buried line source problem. Proc. R. Soc. London A, 234, 528-541 (1956).

GEBRANDE, H.: Ein Beitrag zur Theorie thermischer Konvektion im Erdmantel mit besonderer Berücksichtigung der Möglichkeit eines Nachweises mit Methoden der Seismologie, 159 p. Diss., Univ. München 1975.

Geologische Karte von Nordwestdeutschland 1:300 000. - Amt f. Bodenforsch.: Hannover 1951.

Geologische Übersichtskarte von Südwestdeutschland 1:600 000. - Geol. Landesamt Baden-Württemberg: Freiburg i. Br. 1953.

Geologische Übersichtskarte von Hessen 1:300 000. - Hess. Landesamt f. Bodenforsch.: Wiesbaden 1960.

Geologische Übersichtskarte von Nordrhein-Westfalen 1:500 000. - 3. ed., Geol. Landesamt Nordrhein-Westf.: Krefeld 1963.

Geologische Karte von Bayern 1:500 000. - 2. ed., Bayer. Geol. Landesamt: München 1964.

Geologische Karte der Bundesrepublik Deutschland 1:1 000 000. - Bundesanstalt für Bodenforschung: Hannover 1973.

GERKE, K.: Die Karte der Bouguer-Isanomalen 1:1 000 000 von Westdeutschland. Dt. Geod. Komm. bei d. Bayer. Akad. Wiss., Reihe B, Heft 46, Teil I, 13 p. München 1957.

GERKE, K., WATERMANN, H.: Übersichtskarten der Schwere und der mittleren Höhen von Westdeutschland 1:4 000 000. Dt. Geod. Komm. bei d. Bayer. Akad. Wiss., Reihe B, Heft 46, Teil III, Frankfurt/Main: Verlag des Instituts für Angewandte Geodäsie 1960.

German Research Group for Explosion Seismology: Crustal structure in Western Germany. Z. Geophys. 30, 209-234 (1964).

German Research Group for Explosion Seismology: Seismic wide-angle measurement in the Bavarian Molasse Basin. Geophys. Prospect. 14, 1-6 (1966).

GERVER, M., MARKUSHEVICH, V.: Determination of a seismic wave velocity from the traveltime curve. Geophys. J.R. Astr. Soc. 11, 165-173 (1966).

GERVER, M., MARKUSHEVICH, V.: On the characteristic properties of travel-time curves. Geophys. J.R. Astr. Soc. 13, 241-246 (1967).

GIESE, P.: Die Geschwindigkeitsverteilung im obersten Bereich des Kristallins, abgeleitet aus Refraktionsbeobachtungen auf dem Profil Böhmischbruck-Eschenlohe. Z. Geophys. 29, 197-214 (1963).

GIESE, P.: Beispiele zur Geschwindigkeitsverteilung im obersten Bereich der Erdkruste. Bad Kreuznacher DFG-Kolloquium, 9N6, 5 p. Univ. Mainz 1964.

GIESE, P.: Ergebnisse der bisherigen seismischen Messungen in den Alpen und Erörterung einiger damit zusammenhängender Probleme, pp. 271-290. Max Richter Festschrift, Clausthal-Zellerfeld 1965.

GIESE, P.: Neue Gesichtspunkte zur Gliederung der Erdkruste auf Grund refraktionsseismischer Messungen. Z. Geophys. 32, 488-491 (1966).

GIESE, P.: Versuch einer Gliederung der Erdkruste im nördlichen Alpenvorland, in den Ostalpen und in Teilen der Westalpen mit Hilfe charakteristischer Refraktions-Laufzeit-Kurven sowie einer geologischen Deutung. Geophys. Abh. Inst. Meteorol. u. Geophys., FU Berlin 1(2), 202 p. (1968a).

GIESE, P.: Die Struktur der Erdkruste im Bereich der Ivrea-Zone. Schweiz. Min. Petr. Mitt. 48, 261-284 (1968b).

GIESE, P.: The determination of the velocity-depth distribution for separated traveltime segments, 20 p. Unpublished report, 1969.

GIESE, P.: The structure of the earth's crust in central Europe. Proc. 10th Gen. Ass. Europ. Seismol. Comm. (Leningrad 1968), pp. 387-403. Moscow: Acad. Sci. USSR 1970.

GIESE, P.: The special structure of the P^MP traveltime curve. Z. Geophys. 38, 395-405 (1972).

GIESE, P., GÜNTHER, K., REUTTER, K.-J.: Vergleichende geologische und geophysikalische Betrachtungen der W-Alpen und des N-Apennin. Z. dt. geol. Ges. 120, 151-195, 1968 (1970).

GIESE, P., MORELLI, C., PRODEHL, C., VECCHIA, O.: Crust and upper mantle beneath the southern part of the zone of Ivrea. Proc. 12th Gen. Ass. Europ. Seismol. Comm. (Luxembourg 1970), Obs. Royal de Belgique, Comm. Sér. A-N° 13, Sér. Geophys. 101, 182-183 (1971).

GIESE, P., MORELLI, C., STEINMETZ, L.: Main features of crustal structure in western and southern Europe based on data of explosion seismology. In: The structure of the earth's crust based on seismic data (S. MUELLER, ed.). Tectonophysics 20, 367-379 (1973).

GIESE, P., PAVLENKOVA, N.I.: A comparison of crustal structure in the Rhinegraben area and Dniepr-Donetz depression based on generalized DSS-data. Abstract, 14th Gen. Ass. Europ. Seismol. Comm., Trieste, ESC 12, 1974.

GIESE, P., PRODEHL, C., BEHNKE, C.: Ergebnisse refraktionsseismischer Messungen 1965 zwischen dem Französichen Zentralmassiv und den Westalpen. Z. Geophys. 33, 215-261 (1967).

GIESE, P., STEIN, A.: Versuch einer einheitlichen Auswertung tiefenseismischer Messungen aus dem Bereich zwischen der Nordsee und den Alpen. Z. Geophys. 37, 237-272 (1971).

GLOCKE, A.: Refraktions- und reflexionsseismische Untersuchungen zur Erforschung der Struktur der Erdkruste im Rhein-Main-Gebiet, Versuch einer Gesamtdarstellung. Dipl.-Arbeit, Univ. Frankfurt 1970.

GRAHAM, E.K., BARSCH, G.R.: Elastic constants of single crystal forsterite as a function of temperature and pressure. J. Geophys. Res. 74, 5949-5960 (1969).

GRANT, F., WEST, G.: Interpretation Theory in Applied Geophysics, 583 p. New York-Toronto-London: McGraw Hill 1965.

GREEN, D.A., RINGWOOD, A.E.: The stability fields of aluminous pyroxene peridotite and garnet peridotite and their relevance in upper mantle structure. Earth Planet. Sci. Letters 3, 151-160 (1967).

GRIPP, K.: Erdgeschichte von Schleswig-Holstein, 411 p. Neumünster: Wachholtz 1964.

GRUBBE, K.: Refraktionsseismische Messungen im Jahr 1967 auf einer Linie zwischen Braunschweig und dem Rothaar-Gebirge (Norddeutschland), 66 p. Diplomarbeit, Univ. München 1969.

GUTENBERG, B.: Energy ratio of reflected and refracted waves. Bull. Seism. Soc. Am. 34, 85-102 (1944).

GUTENBERG, B.: Channel waves in the earth's crust. Geophysics 20, 283-294 (1955).

HAAK, V.: Das zeitlich sich ändernde, erdelektrische Feld, beobachtet auf einem Profil über den Rheingraben; eine hiervon abgeleitete Methode der Auswertung mit dem Ziel, die elektrische Leitfähigkeit im Untergrund zu bestimmen, 142 p. Diss., Univ. München 1970.

HADJEBI, B.: Die statistische Auswertung von Reflexionen großer Laufzeiten aus dem norddeutschen Raum und ihre Zuordnung zu den bekannten Diskontinuitäten in der Erdkruste, 29 p. Diplomarbeit, Univ. Hamburg 1966.

HAGEDOORN, J.G.: The plus-minus method of interpreting seismic-refraction sections. Geophys. Prospect. 7, 158-182 (1959).

HÄGELE, U., WOHLENBERG, J.: Recent investigations on the seismicity of the Rhinegraben rift system. In: Graben Problems (H. ILLIES, S. MUELLER, eds.), pp. 167-170. Stuttgart: Schweizerbart 1970.

HAHN, A., KIND, E.G.: Eine Interpretation der magnetischen Anomalie von Bramsche. Fortschr. Geol. Rheinld. u. Westf. 18, 387-394 (1971).

HAHN, A., ZITZMANN, A.: The relation of magnetic anomalies to topography and geologic features in Europe. In: The earth's crust and upper mantle (P.J. HART, ed.), pp. 399-404. Geophys. Monogr. 13, Am. Geophys. Un., Washington, D.C. 1969.

HALES, A.L.: A seismic discontinuity in the lithosphere. Earth and Planet. Sci. Letters 7, 44-46 (1969).

HALES, A.L., HELSLEY, C.E., NATION, J.B.: P-traveltimes for an oceanic path. J. Geophys. Res. 75, 7362-7381 (1970).

HAMILTON, R.M.: Temperature variation at constant pressures of the electrical conductivity of periclase and olivine. J. Geophys. Res. 70, 5679-5692 (1965).

HAMILTON, W., MYERS, W.B.: Cenozoic tectonics of the western United States. Rev. Geophys. 4, 509-549 (1966).

HÄNEL, R.: Ergänzungsmessungen auf dem Profil zwischen Rhön und Hunsrück. 2. Stuttgarter DFG-Kolloquium, 22V5, 2 p. Univ. Stuttgart 1963.

HÄNEL, R.: Ergänzungsmessungen auf einem Profil zwischen Rhön und Ahrgebirge. Bad Kreuznacher DFG-Kolloquium, 9N5, 2 p., Univ. Mainz 1964.

HÄNEL, R.: Interpretation of the terrestrial heat flow in the Rhinegraben. In: Graben Problems (H. ILLIES, S. MUELLER, eds.), pp. 116-120. Stuttgart: Schweizerbart 1970.

HÄNEL, R.: Bestimmungen der terrestrischen Wärmestromdichte in Deutschland. Z. Geophys. 37, 119-134 (1971a).

HÄNEL, R.: Heat flow measurements and a first heat flow map of Germany. Z. Geophys. 37, 975-992 (1971b).

HÄNEL, R., ZOTH, G.: Heat flow measurements in Austria and heat flow maps of central Europe. Z. Geophys. 39, 425-439 (1973).

HARCKE, H.: Die Struktur der Erdkruste im nördlichen Alpenvorland - eine Synthese aus seismischen und gravimetrischen Daten, 68 p. Diss., Univ. Karlsruhe 1972.

HART, P.J. (ed.): The earth's crust and upper mantle, 735 p. Geophys. Monogr. 13, Am. Geophys. Un., Washington, D.C. 1969.

HEACOCK, J.G. (ed.): The structure and physical properties of the earth's crust, 348 p. Geophys. Monogr. 14, Am., Geophys. Un., Washington, D.C. 1971.

HEALY, J.H., WARREN, D.H.: Explosion seismic studies in North America. In: The earth's crust and upper mantle (P.J. HART, ed.), pp. 208-220. Geophys. Monogr. 13, Am. Geophys. Un., Washington, D.C. 1969.

HEHN, K.: Die statistische Auswertung von Reflexionen mit langen Laufzeiten aus dem nordwestdeutschen Raum und ihre Zuordnung zu den bekannten Unstetigkeitsflächen in der Erdkruste, 54 p. Diplomarbeit, Univ. Clausthal 1964.

HELMBERGER, D.V.: Head waves from the oceanic Mohorovičić discontinuity. Bull. Seism. Soc. Am. 58, 179-214 (1968).

HELMBERGER, D.V., MORRIS, G.B.: A traveltime and amplitude interpretation of a marine refraction profile: primary waves. J. Geophys. Res. 74, 483-494 (1969).

HELMBERGER, D.V., MORRIS, G.B.: A traveltime and amplitude interpretation of a marine refraction profile: transformed shear waves. Bull. Seism. Soc. Am. 60, 593-600 (1970).

HERGLOTZ, G.: Über das Benndorfsche Problem der Fortpflanzungsgeschwindigkeit der Erdbebenstrahlen. Phys. Z. 8, 145-147 (1907).

HESS, H.H.: History of ocean basins. In: Petrologic studies, a volume to honor A.F. BUDDINGTON (A.E. ENGEL et al., eds.), pp. 599-620. Geol. Soc. Am., New York 1962.

HILLER, W., ROTHÉ, J.P., SCHNEIDER, G.: La seismicité du fossé Rhénan. In: The Rhinegraben Progress Report 1967 (J.P. ROTHÉ, K. SAUER, eds.). Abh. geol. Landesamt Baden-Württemberg 6, 98-100 (1967).

HIRN, A., STEINMETZ, L., KIND, R., FUCHS, K.: Long range profiles in western Europe: II. Fine structure of the lower lithosphere in France (southern Bretagne). Z. Geophys. 39, 363-384 (1973).

HIRSCHLEBER, H., HJELME, J., SELLEVOLL, M.: A refraction profile through the northern Jutland, 26 p. Geodaetisk Institut Meddelelse no. 41, Copenhagen 1966.

HIRSCHLEBER, H.B., LUND, C.E., MEISSNER, R., VOGEL, A., WEINREBE, W.: Seismic investigations along the Scandinavian "Blue Road" Traverse. Z. Geophys. 41, 135-148 (1975).

HOLUB, K.: Dependence of seismic wave amplitudes on the size of shot-hole explosions. Studia geoph. et geod. 16, 297-302 (1972).

HOLUB, K.: Velocity-depth function of P-waves in the region of the central Bohemian massif. Studia geoph. et geod. 18, 390-393 (1974).

HOMILIUS, J., BLOHM, E.-K.: Modell zur Interpretation der geoelektrischen Tiefensondierung im Rheingraben 1967. Z. Geophys. 39, 441-459 (1973).

HOYER, P., TEICHMÜLLER, R., WOLBURG, J.: Die tektonische Entwicklung des Steinkohlengebirges in Münsterland und Ruhrgebiet. Z. dt. geol. Ges. 119, 549-552 (1967), (1969).

HRISTOW, Wl.K. (CHRISTOV, V.K.): Die Gauß-Krügerschen Koordinaten auf dem Ellipsoid, 80 p., 8th ed. Leipzig-Berlin: Teubner 1943.

HUGHES, D.S., MAURETTE, C.: Variation of elastic wave velocities in granites with pressure and temperature. Geophysics 21, 277-284 (1956).

HUGHES, D.S., MAURETTE, C.: Variation of elastic wave velocities in basic igneous rocks with pressure and temperature. Geophysics 22, 23-31 (1957a).

HUGHES, D.S., MAURETTE, C.: Détermination des vitesses d'onde élastique dans diverses roches en fonction de la pression et de la température. Rev. Inst. Franc. Pétrole 12, 730-752 (1957b).

ILLIES, J.H.: Bauplan und Baugeschichte des Oberrheingrabens. Oberrhein. geol. Abh. 14, 1-54 (1965).

ILLIES, J.H.: Graben tectonics as related to crust-mantle interaction. In: Graben Problems (J.H. ILLIES, S. MUELLER, eds.), pp. 4-27. Stuttgart: Schweizerbart 1970.

ILLIES, J.H.: Taphrogenesis and plate tectonics. In: Approaches to Taphrogenesis (J.H. ILLIES, K. FUCHS, eds.), pp. 433-460. Stuttgart: Schweizerbart 1974.

ILLIES, J.H., FUCHS, K. (eds.): Approaches to Taphrogenesis, 460 p. Stuttgart: Schweizerbart 1974.

ILLIES, J.H., MUELLER, S. (eds.): Graben Problems, 316 p. Stuttgart: Schweizerbart 1970.

Instituto Geografico "Augustin Codazzo": Mapa gravimetrico, Republica Colombia 1959.

International geological map of Europe and the Mediterranean region 1:5 000 000, Western sheet. UNESCO/Bundesanst. f. Bodenforsch., Hannover 1971.

JACKSON, W.H., STEWART, S.W., PAKISER, L.C.: Crustal structure in eastern Colorado from seismic-refraction measurements. J. Geophys. Res. 68, 5767-5776 (1963).

JOHN, H.: Die Gliederung der deutschen Alpenvorlandsmolasse mit Hilfe seismischer Geschwindigkeiten. Erdöl u. Kohle 10, 493-496, 570-573, 661-664 (1957).

JUNG, K., SCHAAF, H., KAHLE, H.G.: Ergebnisse gravimetrischer Messungen im Ries. Geologica Bavarica 61, 337-342 (1969).

KAHLE, H.G.: Deutung der Schwereanomalien im Nördlinger Ries. Z. Geophys. 36, 601-606 (1970).

KAMINSKI, W., FUCHS, K., MENZEL, H.: Crustal investigation along a seismic refraction line from the Harz mountains to the Alps, 1 p., Paper IASPEI - 166, presented at the 14th General Assembly IUGG, Zürich 1967.

KAMMER, E.W., PARDUE, T.E., FRISSEL, H.F.: A determination of the elastic constants for beta-quartz. J. Appl. Phys. 19, 265-270 (1948).

KAPPELMEYER, O.: Vorläufiger Bericht über die Auswertung refraktionsseismischer Messungen am NE-Profil Eschenlohe, 9 p. Frankfurter DFG-Kolloquium, Hannover 1961.

KAPPELMEYER, O.: The geothermal field of the upper Rhinegraben. In: The Rhinegraben progress report 1967 (J.P. ROTHÉ, K. SAUER, eds.). Abh. Geol. Landesamt Baden-Württemberg 6, 101-103 (1967).

KEILIS-BOROK, V.J.: The inverse problem of seismology. In: Mantle and Core in Planetary Physics. New York: Academic Press 1971.

KEMMERLE, K.: Magnetotellurik am Nordrand der Bayerischen Alpen entlang eines Profiles vom Chiemsee bis Reit im Winkl, 104 p. Diplomarbeit, Univ. München 1973.

KERTZ, W., GEHLEN, K.v., GOERLICH, F., KNETSCH, G., WOLF, H.: Das Unternehmen Erdmantel, 376 p. Wiesbaden: Steiner 1972.

KHITAROV, N.I., SLUTSKY, A.B., PUGIN, V.A.: Electrical conductivity of basalts at high

T-P and phase transitions under upper mantle conditions. Phys. Earth Planet. Interiors 3, 334-342 (1970).

KNETSCH, G.: Geologie von Deutschland, 386 p. Stuttgart: Enke 1963.

KNETSCH, G.: Über Funktionswechsel des Rheinischen Lineaments und die Entstehung des Oberrhein-Grabens. Z. dt. geol. Ges. 118, 222-235, 1966 (1969).

KNOTHE, C., WALTHER, K.F.: Vorbereitung und Durchführung einer seismischen Tiefensondierung im Grenzgebiet DDR-ČSSR. Freiberger Forschungshefte C 239, 5-47 (1968).

KNOTHE, H., SCHRÖDER, E.: The German Democratic Republic. In: The Crustal Structure of Central and Southeastern Europe Based on the Results of Explosion Seismology (V.B. SOLLOGUB, D. PROSEN, H. MILITZER, eds.), pp. 80-86. Budapest: Müszaki Könyvkioadó 1972.

KOCHANOWSKY, B.J.: Anlage und Berechnung von Kammersprengungen als Beitrag zur Ermittlung des Sprengstoffbedarfs in der Hartsteingewinnung. Diss., Bergakademie Clausthal 1955.

KONDRATIEV, O.K.: Velocity determination from first breaks. AN SSSR F., Moscow: DFZ 1960.

KONDRATIEV, O.K., GAMBURTSEV, A.G.: Seismic studies in the coastal part of the Eastern Antarctic area. Moscow: Publishing House "Nauka" 1963.

KOSCHYK, K.: Beobachtungen zur Erforschung der Erdkruste mit der Methode der Refraktions-Seismik längs der beiden Profile Eschenlohe-SE und Eschenlohe-E in den Ostalpen 1965-1969, 48 p. Diplomarbeit, München 1969.

KOSCHYK, K.: Seismische Untersuchungen der Erdstöße der Jahre 1962-1971 in Peissenberg, 94 p. Diss., Univ. München 1973.

KOSMINSKAYA, I.P.: Deep Seismic Sounding of the Earth's Crust and Upper Mantle, 184 p. New York-London: Consultants Bureau 1971.

KOSMINSKAYA, I.P., BELYAEVSKY, N.A., VOLVOSKY, I.S.: Explosion seismology in the USSR. In: The earth's crust and upper mantle (P.J. HART ed.), pp. 195-208. Geophys. Monogr. 13, Am. Geophys. Un., Washington, D.C. 1969.

KOSMINSKAYA, I.P., PUZIREV, N.N., ALEKSEYEV, A.S.: Explosion seismology: Its past, present, and future. Tectonophysics 13, 309-323 (1972).

KOSMINSKAYA, I.P., RIZNICHENKO, Y.V.: Seismic studies of the earth's crust in Eurasia. Research in Geophysics, 2, pp. 81-122. Cambridge/Mass.: M.I.T. Press 1964.

KREY, Th., NODOP, I.: Application of refraction seismics to subsalt tectonic problems in a deep saltdome basin. Geophys. Prospect. 18, 364-379 (1970).

KRÖLL, A., WESSELY, G.: Neue Erkenntnisse über Molasse, Flysch und Kalkalpen auf Grund der Ergebnisse der Bohrung Urmannsau I. Erdöl-Erdgas-Ztschr. 10, 342-353 (1967).

LABROUSTE, H.Y.: Rapport preliminaire sur les grands profils sismiques en France. Proc. 10th Ass. Europ. Seismol. Comm. (Leningrad 1968), Vol. I, pp. 405-443. Moscow: Acad. Sciences USSR 1970.

LAMPE, H.: Kammersprengungen im Hartgestein. Nobel-Hefte 1, 17-49 (1959).

LANDISMAN, M., MUELLER, S.: Seismic studies of the earth's crust in continents; part II: Analysis of wave propagation in continents and adjacent shelf areas. Geophys. J.R. Astr. Soc. 10, 539-554 (1966).

LANDISMAN, M., MUELLER, S., MITCHELL, B.J.: Review of evidence for velocity inversions in the continental crust. In: The structure and physical properties of the earth's crust (J.G. HEACOCK, ed.), pp. 11-34. Geophys. Monogr. 14, Am. Geophys. Un., Washington, D.C. 1971.

LAUER, J.P., PETERSCHMITT, E.: Ondes réfléchies des explosions des Bagenelles et de St. Nabor (Vosges, France). Proc. 10th Gen. Ass. Europ. Seismol. Comm. (Leningrad 1968), Vol. I, pp. 444-453. Moscow: Acad. Sciences USSR 1970.

LEMCKE, K.: Zur nachpermischen Geschichte des nördlichen Alpenvorlandes. Geologica Bavarica 69, 5-48 (1973).

LIEBSCHER, H.J.: Reflexionshorizonte der tieferen Erdkruste im bayerischen Alpenvorland, abgeleitet aus Ergebnissen der Reflexionsseismik. Z. Geophys. 28, 162-184 (1962).

LIEBSCHER, H.J.: Deutungsversuche für die Struktur der tieferen Erdkruste nach reflexionsseismischen und gravimetrischen Messungen im deutschen Alpenvorland. Z. Geophys. 30, 51-96, 115-126 (1964).

LITVINENKO, I.V.: DSS on the Baltic Shield. In: Glubinn. seysm. Sondier., pp. 187-206. Leningrad: Gostoptechizdat 1962.

LITVINENKO, I.V.: Besonderheiten des Aufbaus der Erdkruste im Nordwesten der Kola-Halbinsel und Süden der Barents-See (russ.). In: Geologie und Aufbau der Erdkruste im Osten des Baltischen Schildes, pp. 90-95. Leningrad: AN SSSR "Nauka" 1968.

LOSECKE, W.: Bericht über eine erste magnetotellurische Messung (MT) der Bundesanstalt für Bodenforschung und des niedersächsischen Landesamtes für Bodenforschung bei Marwede bei Celle und Vorschläge für den Ausbau der MT. Archiv-Nr. 1157 (Geoelektrik), Bundesamt f. Bodenf. Hannover 1968.

LOTZE, F.: Geologie Mitteleuropas, 4th ed., 491 p. Stuttgart: Enke 1971.

LUBIMOVA, E.A.: Theory of thermal state of the earth's mantle. In: The Earth's Mantle, (T.F. GASKELL, ed.), pp. 231-323. New York: Academic Press 1967.

LUBIMOVA, E.A., POLYAK, B.G.: Heat flow map of Eurasia. In: The earth's crust and upper mantle (P.J. HART, ed.), pp. 82-88. Geophys. Monogr. 13, Am. Geophys. Un., Washington, D.C. 1969.

MAKRIS, J.: Aufbau der Erdkruste in den Ostalpen aus Schweremessungen und die Ergebnisse der Refraktionsseismik. Hamb. Geophys. Einzelschr., Heft 15, 65 p., 1971.

MAYER-ROSA, D.: Die Geschwindigkeitsverteilung seismischer Wellen im oberen Erdmantel Europas, 79 p. Diss., Univ. Stuttgart 1969.

MAYNARD, G.L.: Crustal layer with seismic velocity 6.9-7.6 km/s under deep oceans. Science 168, 120-121 (1970).

McKENZIE, D.P.: Some remarks on heat flow and gravity anomalies. J. Geophys. Res. 72, 6261-6273 (1967).

McMECHAN, G.A., WIGGINS, R.A.: Depth limits in body wave inversions. Geophys. J.R. Astr. Soc. 28, 459-473 (1972).

MEISSNER, R.: An interpretation of the wide angle measurements in the Bavarian Molasse Basin. Geophys. Prospect. 14, 7-16 (1966).

MEISSNER, R.: Zum Aufbau der Erdkruste, Ergebnisse der Weitwinkelmessungen im bayerischen Molassebecken, Teil 1 und 2. Gerl. Beitr. Geophys. 76, 211-254, 295-314 (1967a).

MEISSNER, R.: Exploring deep interfaces by seismic wide angle measurements. Geophys. Prospect. 15, 598-617 (1967b).

MEISSNER, R.: Seismic interpretation methods for structures of the earth's crust. Proc. 10th Gen. Ass. Europ. Seismol. Comm. (Leningrad 1968), pp. 352-385. Moscow: Acad. Sci. USSR 1970.

MEISSNER, R.: The "Moho" as a transition zone. Geophys. Surveys 1, 195-216 (1973).

MEISSNER, R., BERCKHEMER, H.: Seismic refraction measurements in the northern Rhinegraben. In: The Rhinegraben Progress Report 1967 (J.P. ROTHÉ, K. SAUER, eds.). Abh. Geol. Landesamt Baden-Württemberg 6, 105-108 (1967).

MEISSNER, R., BERCKHEMER, H., WILDE, R., POURSADEG, M.: Interpretation of seismic refraction measurement in the northern part of the Rhinegraben. In: Graben Problems (H. ILLIES, S. MUELLER, eds.), pp. 184-190. Stuttgart: Schweizerbart 1970.

MEISSNER, R., VETTER, U.: The northern end of the Rhinegraben due to some geophysical measurements. In: Approaches to Taphrogenesis (H. ILLIES, K. FUCHS, eds.), pp. 236-243. Stuttgart: Schweizerbart 1974.

MENARD, H.W.: The deep ocean floor. Sci. Am. 221(3), 126-142 (1969).

MEREU, R.F.: An iterative method for solving the time-term equations. In: The earth beneath the continents (J.S. STEINHART, T.J. SMITH, eds.). Geophys. Monogr. 10, 495-497 (1966).

MEREU, R.F., HUNTER, J.A.: Crustal and upper mantle structure under the Canadian Shield from project Early Rise data. Bull. Seism. Soc. Am. 59, 147-165 (1969).

MIGAUX, L., ASTIER, J.L., REVOL, P.H.: Un essai de détermination expérimentale de la résistivité électrique des couches profondes de l'écorce terrestre. Ann. Géophys. 16, 555-561 (1960).

MITOFF, S.P.: Bulk versus surface conductivity of MgO crystals. J. Chem. Phys. 41, 2561-2562 (1964).

MÜHLEN, W. von zur: Ergebnisse der "Steinbruch-Seismik" im Siegerland, Kraichgau und in Hessen/Unterfranken. Geol. Jb. 71, 569-594 (1956).

MÜHLEN, W. von zur, TUCHEL, G.: A study of well velocity data in North West Germany. Geophys. Prospect. 1, 159-170 (1953).

MÜLLER, G.: Theoretical seismograms for some types of point sources in layered media. Part II: Numerical calculations. Z. Geophys. 34, 147-162 (1968).

MÜLLER, G.: Exact ray theory and its application to the reflection of elastic waves from vertically inhomogeneous media. Geophys. J. R. Astr. Soc. 21, 261-283 (1970).

MÜLLER, G.: Approximate treatment of elastic body waves in media with spherical symmetry. Geophys. J.R. Astr. Soc. 23, 435-449 (1971).

MUELLER, S.: Geophysical aspects of graben formation in continental rift systems. In: Graben Problems (J.H. ILLIES, S. MUELLER, eds.), pp. 27-37. Stuttgart: Schweizerbart 1970.

MUELLER, S., LANDISMAN, M.: Seismic studies of the earth's crust in continents. Part I: Evidence for a low-velocity zone in the upper part of the lithosphere. Geophys. J. R. Astr. Soc. 10, 525-538 (1966).

MUELLER, S., PETERSCHMITT, E.: Die Geschwindigkeitsverteilung seismischer Wellen im tieferen Untergrund um den Oberrheingraben, 3 p. Wiesloch: DFG-Kolloquium Oberrheingraben 1966.

MUELLER, S., PETERSCHMITT, E., FUCHS, K., ANSORGE, J.: The rift structure of the crust and the upper mantle beneath the Rhinegraben. In: The Rhinegraben Progress Report 1967 (J.P. ROTHÉ, K. SAUER, eds.). Abh. Geol. Landesamt Baden-Württemberg 6, 108-113 (1967).

MUELLER, S., PETERSCHMITT, E., FUCHS, K., ANSORGE, J.: Crustal structure beneath the Rhinegraben from seismic refraction and reflection measurements. Tectonophysics 8, 529-542 (1969).

MUELLER, S., PETERSCHMITT, E., FUCHS, K., EMTER, D., ANSORGE, J.: Crustal structure of the Rhinegraben area. In: The structure of the earth's crust based on seismic data (S. MUELLER, ed.). Tectonophysics 20, 381-392 (1973).

MUELLER, S., RYBACH, L.: Crustal dynamics in the central part of the Rhinegraben. In: Approaches to Taphrogenesis (H. ILLIES, K. FUCHS, eds.), pp. 379-388. Stuttgart: Schweizerbart 1974.

MUELLER, S., SCHICK, R., JENSCH, A.: Beobachtungen auf dem Refraktionsprofil Eschenlohe-Birkenau und Untersuchungen im Steinheimer Becken. 2. Stuttgarter DFG-Kolloquium, 22N7, 4 p. Univ. Stuttgart 1963.

MUELLER, S., STEIN, A., VEES, R.: Seismic scaling laws for explosions on a lake bottom. Z. Geophys. 28, 258-280 (1962).

MUELLER, S., TALWANI, M.: A crustal section across the Eastern Alps based on gravity and seismic refraction data. Pure Appl. Geophys. 85, 226-239 (1971).

MUNDRY, E.: The boundary value problem of geoelectrical exploration for an elliptic cylinder. Z. angew. Math. Mechan. 51, 60-62 (1971).

MUNDRY, E., HOMILIUS, J.: Resistivity measurements in valleys of elliptic cross section. Geophys. Prospect. in press 1975.

NAFE, J.E., DRAKE, D.C.: Society of Exploration Geophysicists, Annual Meeting 1957. Unpublished paper. The graph referred to is published in TALWANI, SUTTON and WORZEL (1959) and STEINHART and MEYER (1961).

NIGGLI, E.: Über den Zusammenhang zwischen der positiven Schwereanomalie am Südfuß der Westalpen und der Gesteinszone von Ivrea. Eclogae geol. Helv. 39, 211-220 (1946).

NITSAN, U.: Electrical conductivity and temperature in the earth's upper mantle. EOS, Trans. AGU 54, 1207 (1973).

NODOP, I.: Vorweisung einer Karte des präkretazischen Untergrundes Nordwestdeutschlands. Z. dt. geol. Ges. 114, 423-426, 1962 (1963).

O'BRIEN, P.N.S.: Seismic observations 20 km from explosions in a lake. Boll. Geofis. Teor. ed Appl. 7, 144-164 (1965).

PAKISER, L.C.: The basalt-eclogite transformation and crustal structure in the western United States. U.S. Geol. Survey Prof. Paper 525B, B1-B8 (1965).

PAKISER, L.C., ROBINSON, R.: Composition and evolution of the continental crust as suggested by seismic observations. Tectonophysics 3, 547-557 (1966).

PAPKE, K.H.: Die Mohorovičić-Diskontinuität. Geologie 16, Beiheft 57, 127 p. (1967).

PARKHOMENKO, E.I.: Electrical Properties of Rocks, 314 p. New York: Plenum Press 1967.

PASECHNIK, J.P.: The determination of the frequency dependence of the absorption coefficient for longitudinal waves which propagate the earth's mantle. Akad. Nauk SSR Doklady 166, 6, 1338-1341 (1966).

PAVLENKOVA, N.I.: Methods of velocity determination from seismic crustal studies. Transactions Internat. Conference of Experts on Explosion Seismology (Leningrad 1968), pp. 104-117. Kiev: Naukova dumka 1969.

PAVLENKOVA, N.I.: Correlation of velocity-depth function of the earth's crust in Ukraina, methods and results, part I and II. Geophys. sbornik, Kiev 39, 12-22, and 42, 46-56 (1971).

PAVLENKOVA, N.I.: Wavefields and Models of the Earth's crust, 219 p. Kiev: NAUKA press 1973a.

PAVLENKOVA, N.I.: Interpretation of refracted waves by the reduced traveltime curve method. Physics of the Solid Earth 8, 544-550 (1973b).

PAVLENKOVA, N.I.: Seismic models of the earth's crust and methods of their determination, 15 p. Report at the meeting of Controlled Sources Seismology, Univ. Paris 1975.

PETERS, K.: Ergebnisse der Gravimetrie im Bereich der Münchberger Gneismasse und der Refraktionsseismik längs eines Profils über die Gneismasse, 82 p. Diss., Univ. München 1974.

PETERSCHMITT, E., MENZEL, H., FUCHS, K.: Seismische Messungen in den Alpen. Die Beobachtungen auf dem NE-Profil Lago Lagorai 1962 und ihre vorläufige Auswertung. Z. Geophys. 31, 41-49 (1965).

PLAUMANN, S.: Bericht über die Auswertung des refraktionsseismischen Profils Adelebsen, 5 p. Frankfurter DFG-Kolloquium. Hannover: Niedersächs. Landesamt Bodenf. 1961a.

PLAUMANN, S.: Bericht über die Auswertung des seismischen Profils Kirchheimbolanden, 9 p. Frankfurter DFG-Kolloquium. Hannover: Niedersächs. Landesamt Bodenf. 1961b.

PLUSCHKELL, W., ENGELL, H.J.: Ionen und Elektronenleitung in Magnesiumorthosilikat. Ber. dt. keram. Ges. 45, 388 (1968).

PRESS, F.: Seismic velocities. In: Handbook of Physical Constants (S.P. CLARK, Jr., ed.). Geol. Soc. Am. Memoir 97, 195-218 (1966).

PRODEHL, C.: Die Kristallinoberfläche zwischen Donau und Inn abgeleitet aus refraktionsseismischen Messungen, 67 p. Diplomarbeit Univ. München 1962.

PRODEHL, C.: Auswertung von Refraktionsbeobachtungen im bayrischen Alpenvorland (Steinbruchsprengungen bei Eschenlohe 1958-1961) im Hinblick auf die Tiefenlage des Grundgebirges. Z. Geophys. 30, 161-181 (1964).

PRODEHL, C.: Struktur der tieferen Erdkruste in Südbayern und längs eines Querprofiles durch die Ostalpen, abgeleitet aus refraktionsseismischen Messungen bis 1964. Boll. Geofis. Teor. ed Appl. 7, 35-88 (1965).

PRODEHL, C.: Seismic refraction study of crustal structure in the western United States. Geol. Soc. Am. Bull. 81, 2629-2646 (1970a).

PRODEHL, C.: Crustal structure of the western United States from seismic-refraction measurements in comparison with central European results. Z. Geophys. 36, 477-500 (1970b).

PRODEHL, C., PAKISER, L.C.: Crustal structure of the southern Rocky Mountains from seismic measurements. Geol. Soc. Am. Bull. 86 (in press, 1976).

PUZIREV, N.N.: Observation schemes in deep seismic sounding, 7 p. Novosibirsk: USSR Acad. of Science 1968.

RAITT, R.W., SHOR, G.G., FRANCIS, T.J.G., MORRIS, G.B.: Anisotropy of the Pacific upper mantle. J. Geophys. Res. 74, 3095-3109 (1969).

REICH, H.: Geophysikalische Karte von Nordwestdeutschland 1:500 000, Bl. 1: Magnetik. Reichsamt f. Bodenforsch. 1948.

REICH, H.: Geologische Ergebnisse der seismischen Beobachtungen der Sprengung auf Helgoland. Geol. Jb. 64, 245-266 (1950).

REICH, H.: Seismische Beobachtungen bei großen Steinbruchsprengungen und deren Ergebnisse. Z. dt. geol. Ges. 104, 174-175 (1952), (1953a).

REICH, H.: In Süddeutschland ermittelte tiefe Grenzflächen. Geol. Rundschau 46, 1-16 (1957).

REICH, H.: Die geologischen Ergebnisse seismischer Registrierungen großer Sprengungen in Deutschland. Geofisica Pura e Applicata 40, 1-46 (1958a).

REICH, H.: Seismische und geologische Ergebnisse der 2 to-Sprengung im Tiefbohrloch Tölz am 11.12.54. Geol. Jb. 75, 1-46 (1958b).

REICH, H.: Seismische Untersuchung des Flyschtroges bei Lenggries westlich und östlich der Isar. Nachr. Akad. Wiss. Göttingen II, Math. Phys. Kl. 11, 205-255 (1960a).

REICH, H.: Zur Frage der geologischen Deutung seismischer Grenzflächen in den Alpen. Geol. Rdsch. 50, 465-473 (1960b).

REICH, H.: Grundlagen der angewandten Geophysik für Geologen. Leipzig: Akademische Verlagsgesellschaft 1960c.

REICH, H., HORRIX, W.: Geophysikalische Untersuchungen im Ries und Vorries und deren geologische Deutung. Beih. Geol. Jb. 19, 119p. (1955).

REICH, H., SCHULZE, G.A., FÖRTSCH, O.: Das geophysikalische Ergebnis der Sprengung von Haslach im südlichen Schwarzwald. Geol. Rdsch. 36, 85-96 (1948).

REICH, H., SCHULZE, G.A., FÖRTSCH, O.: Results of seismic observations in Germany on the Heligoland explosion of April 18, 1947. J. Geophys. Res. 56, 147-156 (1951).

REINHARDT, H.-G.: Steinbruchsprengungen zur Erforschung des tieferen Untergrundes. Freiberger Forschungshefte, C 15, 91 p. (1954).

Rhinegraben Research Group for Explosion Seismology: The 1972 seismic refraction experiment in the Rhinegraben. - First results. In: Approaches to Taphrogenesis (H. ILLIES, K. FUCHS, eds.), pp. 122-137. Stuttgart: Schweizerbart 1974.

RICHTER, M.: Molasse und Alpen. Z. dt. geol. Ges. 102, 177-180 (1951).

RIZNICHENKO, Y.V.: Die Anwendung der Methode der Zeitfelder in der Praxis. Angew. Geophys. 1 (1945).

ROBINSON, E.A.: Statistical Communication and Detection with Special Reference to Digital Data Processing of Radar and Seismic Signals. London: Griffin 1967.

ROTHÉ, J.P.: Quelques expériences sur la structure de la croûte terrestre en Europe occidentale. In: Contributions in Geophysics in Honor of Beno Gutenberg (E. INGERSON, ed.), pp. 135-151. London: Pergamon Press 1958.

ROTHÉ, J.P., PETERSCHMITT, E.: Etude sismique des explosions d'Haslach. Ann. Inst. Phys. Globe Strasbourg 5, 3, 3-28 (1950).

ROTHÉ, J.P., SAUER, K. (eds.): The Rhinegraben Progress Report 1967. Abh. Geol. Landesamt Baden-Württemberg 6, 146 p. (1967).

RÖWER, P., STROBACH, K.: Variationen der P_n-Geschwindigkeiten von Erdbebenwellen im Gebiet des Rheingrabens. In: The Rhinegraben Progress Report 1967 (J.P. ROTHÉ, K. SAUER, eds.). Abh. Geol. Landesamt Baden-Württemberg 6, 121-122 (1967).

RÜLKE, O.: Specific resistivity of the graben fill. In: The Rhinegraben Progress Report 1967 (J.P. ROTHÉ, K. SAUER, eds.). Abh. Geol. Landesamt Baden-Württemberg 6, 66-67 (1967).

RUTTEN, M.G.: The Geology of Western Europe, 520 p. Amsterdam-London-New York: Elsevier 1969.

RYABOI, V.Z.: Kinematic and dynamic characteristics of deep waves associated with boundaries in the crust and upper mantle. Bull. (Izvestiya) Acad. of Sci. USSR (Geophys. Ser.) 3, 177-184 (1966).

RYALL, A.: Improvement of array seismic recordings by digital processing. Bull. Seism. Soc. Am. 54, 277-294 (1964).

RYALL, A., STUART, D.J.: Traveltimes and amplitudes from nuclear explosions, Nevada test site to Ordway, Colorado. J. Geophys. Res. 68, 5821-5835 (1963).

SAPIN, M., HIRN, A.: Results of explosion seismology in the southern Rhône valley. Ann. Géophys. 30, 181-202 (1974).

SATTLEGGER, I.: A method of computing true interval velocities from expanding spread data in the case of arbitrary long spreads and arbitrarily dipping plane interfaces. Geophys. Prospect. 13, 306-318 (1965).

SAUER, H.D.: Seismik Ries 1968, I. Auswertung der Reflexionsseismik für Laufzeiten bis zu einer Sekunde. Diplomarbeit, Univ. München 1969.

SCHANKLAND, T.J.: Transport properties of olivines. In: The Application of Modern Physics to the Earth and Planetary Interiors (S.K. RUNCORN, ed.), pp. 175-190. New York: Wiley Interscience 1969.

SCHEIDEGGER, A.E., WILLMORE, P.L.: The use of a least squares method for interpretation of data from seismic surveys. Geophysics 22, 9-22 (1957).

SCHELIGA, G.: Ergebnisse seismischer Messungen (1965-1970) im Gebiet des Engadiner Fensters, 83 p. Diss., Univ. München 1971.

SCHICK, R.: Untersuchungen über die Bruchausdehnung und Bruchgeschwindigkeit bei Erdbeben mit kleinen Magnituden (M < 4). Z. Geophys. 34, 267-286 (1968a).

SCHICK, R.: Die Tiefenlage der Mohorovičić- und Conrad-Diskontinuität im Bereich des Schwäbischen Juras, 5 p. Veröff. Landeserdbebendienst Baden-Württemberg, Stuttgart 1968b.

SCHICK, R.: A method for determining source parameters of small magnitude earthquake. Z. Geophys. 36, 205-224 (1970).

SCHMUCKER, U.: An introduction to induction anomalies. J. Geomagn. Geoelectr. 22, 9-33 (1970).

SCHNEIDER, G.: Erdbeben und Tektonik in Süd-west-Deutschland. Tectonophysics 5, 459-511 (1967).

SCHNEIDER, G.: Seismizität und Tektonik der Schwäbischen Alb, 79 p. Stuttgart: Enke 1971.

SCHNEIDER, G., MUELLER, S., KNOPOFF, L.: Gruppengeschwindigkeitsmessungen an kurzperiodischen Oberflächenwellen in Mitteleuropa. Z. Geophys. 32, 33-60 (1966).

SCHNEIDER, G., SCHICK, R., BERCKHEMER, H.: Faultplane solutions of earthquakes in Baden-Württemberg. Z. Geophys. 32, 383-393 (1966).

SCHOBER, M.: Messung der elektrischen Leitfähigkeit an einigen Proben natürlichen Olivins bei hohen Drucken und Temperaturen, 75 p. Diss., Univ. München 1970.

SCHOBER, M.: The electrical conductivity of some samples of natural olivine at high temperatures and pressures. Z. Geophys. 37, 283-292 (1971).

SCHULT, A.: The electrical conductivity of minerals and the temperature distribution in the upper mantle. In: Approaches to Taphrogenesis (H. ILLIES, K. FUCHS, eds.), pp. 376-378. Stuttgart: Schweizerbart 1974.

SCHULT, A., SCHOBER, M.: Measurement of electrical conductivity of natural olivine at temperatures up to 950°C and pressures up to 42 kbar. Z. Geophys. 35, 105-112 (1969).

SCHULZ, G.: Reflexionen aus dem kristallinen Untergrund des Pfälzer Berglandes. Z. Geophys. 23, 225-235 (1957).

SCHULZE, G.A.: Anfänge der Krustenseismik. In: Zur Geschichte der Geophysik (H. BIRETT, K. HELBIG, W. KERTZ, U. SCHMUCKER, eds.), pp. 89-98. Berlin-Heidelberg-New York: Springer 1974.

SCHULZE, G.A., FÖRTSCH, O.: Die seismischen Beobachtungen bei der Sprengung auf Helgoland am 18. April 1947 zur Erforschung des tieferen Untergrundes. Geol. Jb. 64, 205-242 (1950).

SCHWEYDAR, W., REICH, H.: Künstliche elastische Bodenwellen als Hilfsmittel geologischer Forschung. Gerl. Beitr. z. Geophys. 17, 121-147 (1927).

SEIDL, D., MUELLER, S., KNOPOFF, L.: Dispersion von Rayleigh-Wellen in Südwestdeutschland und in den Alpen. Z. Geophys. 32, 472-481 (1966).

SEIDL, D., MUELLER, S., REICHENBACH, H.: Dispersion and absorption of seismic surface waves and the structure of the upper mantle based on observations in Europe. Proc. 12th Gen. Ass. Europ. Seismol. Comm. (Luxembourg 1970), Obs. Royal de Belgique, Comm. Sér. A-N° 13, Sér. Géophys. 101, 198-199 (1971).

SHIMSHONI, M., SMITH, S.W.: Seismic signal enhancement with three-component detectors. Geophysics 29, 664-671 (1964).

SIMMONS, G.: Velocity of compressional waves in various minerals at pressures to 10 kb. J. Geophys. Res. 69, 1117-1122 (1964a).

SIMMONS, G.: Velocity of shear waves in various minerals at pressures to 10 kb. J. Geophys. Res. 69, 1123-1130 (1964b).

SIMMONS, G.: Interpretation of heat flow anomalies, 1, contrast in heat production. Rev. Geophys. 5, 42-52 (1967).

SLICHTER, L.B.: The theory of the interpretation of traveltime curves in horizontal structures. Physics 3, 273-295 (1932).

SMITH, T.J., STEINHART, J.S., ALDRICH, L.T.: Lake Superior crustal structure. J. Geophys. Res. 71, 1141-1172 (1966).

SOLLOGUB, V.B.: Seismic studies on deep crustal structure. Transactions Internat. Conference of Experts on Explosion Seismology (Leningrad 1968), pp. 89-103. Kiev: Naukova dumka 1969a.

SOLLOGUB, V.B.: Seismic crustal studies in southeastern Europe. In: The earth's crust and upper mantle (P.J. HART, ed.), pp. 189-195. Geophys. Monogr. 13, Am. Geophys. Un., Washington, D.C. 1969b.

SOLLOGUB, V.B., PROSEN, D., MILITZER, H. (eds.): The crustal structure of central and southeastern Europe based on the results of explosion seismology, 172 p. Budapest: Müszaki Könyvkiadó 1972.

STAUDACHER, W.: Messung der elektrischen Leitfähigkeit von natürlichem Augit, Enstatit und Granat bei Temperaturen bis 1150°K und Drucken bis 48 kbar, 81 p. Diplomarbeit, Univ. München 1968.

STEIN, A.: Ein Gegenschußprofil Kellerwald-Bayrischer Wald. 2. Stuttgarter DFG-Kolloquium, 22V4, 2 p. Univ. Stuttgart 1963.

STEIN, A.: Considerations about DSS. Transactions Internat. Conference of Experts on Explosion Seismology (Leningrad 1968), pp. 67-78. Kiev: Naukova dumka 1969.

STEIN, A.: Sprengseismik; das Unternehmen Erdmantel. Forschungsbericht der Deutschen Forschungsgemeinschaft, pp. 149-159. Wiesbaden: Steiner 1972.

STEIN, A., DRIUVENGA, G.: Refraktionsseismische Untersuchungen 1967 auf dem Profil R16 zwischen Versmold und Geesthacht, Teil 1. NLfB Archiv-Nr. DS 170, Hannover 1970.

Steinbruchs-Berufsgenossenschaft: Die Steinbrüche in der Bundesrepublik Deutschland 1971. Mitteilungsblatt der Steinbruchs-Berufsgenossenschaft, Special Edition, Hannover, Juni 1972.

STEINHART, J.S.: Mohorovičić-discontinuity. In: Intern. Dictionary of Geophysics (K. RUNCORN, ed.) 2, 991-994 (1967).

STEINHART, J.S., MEYER, R.P.: Explosion studies of continental structure. Carnegie Inst. of Washington Publ. 622, 409 p. (1961).

STEINHART, J.S., SMITH, T.J. (eds.): The earth beneath the continents. Geophys. Monogr. 10, 663 p. Am. Geophys. Un., Washington, D.C. 1966.

STEINWACHS, M.: Systematische Untersuchung der kurzperiodischen seismischen Bodenunruhe in der Bundesrepublik Deutschland. Geol. Jb. Series E (Geophysik), Vol. 3, 59 p. Hannover 1974.

STEWART, S.W.: Seismic ray theory applied to refraction surveys of the earth's crust in Missouri, 189 p. Ph.D. thesis, St. Louis University, St. Louis, Miss. 1966.

STEWART, S.W.: Crustal structure in Missouri by seismic-refraction methods. Bull. Seism. Soc. Am. 58, 291-323 (1968).

STILLE, H.: Grundfragen der vergleichenden Tektonik, 443 p. Berlin: Bornträger 1924.

STREICHER, P.: L'influence du remplissage sédimentaire sur la propagation des ondes sismiques dans le fossé Rhénan, 156 p. Mémoire de diplôme, Univ. Strasbourg 1974.

STROBACH, K.: Ein Gegenschußprofil von der Rhön zum Odenwald. 2. Stuttgarter DFG-Kolloquium, 22V6, 4 p. Univ. Stuttgart 1963.

TALWANI, M., Le PICHON, X., EWING, M.: Crustal structure of mid ocean ridges, 2. computed model from gravity and seismic refraction data. J. Geophys. Res. 70, 341-352 (1965).

TALWANI, M., SUTTON, G.H., WORZEL, J.L.: A crustal section across the Puerto Rico Trench. J. Geophys. Res. 64, 1545-1555 (1959).

THORNBURGH, H.R.: Wave front diagrams in seismic interpretation. Bull. Am. Ass. Petr. Geologists 14, 185-200 (1930).

THYSSEN, F.: Ein Beitrag zu den seismischen Untersuchungen im unteren Teufenbereich der Bohrung Münsterland I. Bad Kreuznacher DFG-Kolloquium 11V26, 5 p., Univ. Mainz 1964.

THYSSEN, F., ALLNOCH, H.G., LÜTKEBOHMERT, G.: Einige Ergebnisse geophysikalischer Arbeiten im Bereich der Bramscher Anomalie. Fortschr. Geol. Rheinld. u. Westf. 18, 395-410 (1971).

TOZER, D.C.: Temperature, conductivity composition and heat flow. J. Geomagn. Geoelectric. 22, 35-51 (1970).

UHRI, D.C.: The electrical properties of iron-rich silicates. Ph.D. Thesis, Dept. of Geol. and Geophys., Mass. Inst. Techn. 1961.

VEES, R.: Der seismische Impuls bei Unterwassersprengungen. Diss., Bergakademie Clausthal, Technische Hochschule 1965.

VEIT, E.: Der Bau der südlichen Molasse Oberbayerns auf Grund der Deutung seismischer Profile. Bull. Ver. Schweiz. Petrol. Geol. u. Ing. 30, 15-52 (1963).

VETTER, U., MEISSNER, R.: Überprüfung der Isostasie durch tiefenseismische Sondierungen. Z. Geophys. 36, 225-228 (1970).

VOGEL, A.: Deep-seismic sounding in northern Europe, 98 p. Stockholm: Swedish Natural Sci. Res. Council 1971.

VOLAROVICH, M.P.: The investigation of elastic and absorption properties of rocks at high pressures and temperatures. Tectonophysics 2, 211-217 (1965).

WANGEMANN, E.-K.: Die Geschwindigkeitsverteilung in der Erdkruste im Gebiet des Süddeutschen Dreiecks, abgeleitet aus refraktionsseismischen Messungen auf einem gestaffelten Nord-Süd-Profil, 73 p. Diplomarbeit, Univ. Hamburg 1970.

WARREN, D.H., HEALY, J.H., BOHN, J., MARSHALL, P.A.: Crustal structure under Lasa from seismic-refraction measurements. J. Geophys. Res. 78, 8721-8734 (1973).

WHITTEN, E.H.T.: Trends in computer applications in structural geology. In: Computer Applications in the Earth Sciences (D.F. MERRIAM, ed.), 281 p. New York: Plenum Press 1969.

WIECHERT, E.: Bestimmung des Weges der Erdbebenwellen im Erdinnern, 1. Theoretisches. Phys. Z. 11, 294-304 (1910).

WIECHERT, E.: Untersuchungen der Erdrinde mit dem Seismometer unter Benutzung künstlicher Erdbeben. Nachr. Ges. Wiss. Göttingen, Math.-Phys. Klasse, 57-70 (1923).

WIECHERT, E.: Untersuchung der Erdrinde mit Hilfe von Sprengungen. Geol. Rdsch. 17, 339-346 (1926).

WIECHERT, E.: Seismische Beobachtung von Steinbruchsprengungen. Z. Geophys. 5, 159-162 (1929).

WIGGINS, R.A., McMECHAN, G.A.: Range of earth structure nonuniqueness implied by body wave observations. Rev. Geophys. 11, 87-113(1973).

WILDE, R.: Refraktionsseismische Untersuchungen auf dem Kraichgauprofil. Diplomarbeit, Univ. Frankfurt/Main 1969.

WILL, M.: Seismik Ries 1968, II. Auswertung der Refraktionsmessungen, 92 p. Diplomarbeit, Univ. München 1970.

WILL, M.: Refraktionsseismik im Nordteil der Ostalpen zwischen Salzach und Inn, 1970-1974; Messungen und deren Interpretation, 145 p. Diss., Univ. München 1975.

WILL, M.: Ergebnisse refraktions-seismischer Messungen im Nordteil der Geotraverse Ia. Geol. Rdsch. 65, (in press, 1976).

WILLMORE, P.L.: Seismic experiments on the North German explosions, 1946 to 1947. Phil. Trans. Roy. Soc. London A, 242, 123-151 (1949a).

WILLMORE, P.L.: Seismic aspects of the Heligoland explosion. Nature 160, 350 (1949b).

WILLMORE, P.L., BANCROFT, A.M.: The time-term approach to refraction seismology. Geophys. J.R. Astr. Soc. 3, 419-432 (1960).

WINKLER, H.G.F.: Die Genese der metamorphen Gesteine, 237 p. Berlin-Heidelberg-New York: Springer 1967.

WOLBER, G.: Energieverluste elastischer Wellen in der Erdkruste, 56 p. Zulassungsarbeit, Univ. Karlsruhe 1968.

WOOLLARD, G.P.: Crustal structure from gravity and seismic measurements. J. Geophys. Res. 64, 1521-1544 (1959).

WOOLLARD, G.P.: The relation of gravity anomalies to surface elevation, crustal structure, and geology. Geophys. Polar Res. Center, Univ. Wisconsin, Res. Rept. 62-9, 356 p. 1962.

WOOLLARD, G.P.: The inter-relationship of the crust, the upper mantle, and isostatic gravity anomalies in the United States. In: The crust and upper mantle of the Pacific area (L. KNOPOFF, C.L. DRAKE, P.J. HART, eds.), pp. 312-341. Geophys. Monogr. 12, Am. Geophys. Un., Washington, D.C. 1968.

WOOLLARD, G.P.: Regional variations in gravity. In: The earth's crust and upper mantle (P.J. HART, ed.), pp. 320-341. Geophys. Monogr. 13, Am. Geophys. Un., Washington, D.C. 1969a.

WOOLLARD, G.P.: A study of the problems associated with the prediction of gravity in Europe, 45 p. Hawaii Inst. of Geophys., 69-12, 1969b.

WUNDERLICH, H.G.: Wesen und Ursachen der Gebirgsbildung, 367 p. Mannheim: Bibliograph. Inst. 1966.

YODER, H.S., Jr., TILLEY, C.E.: Origin of basalt magmas: an experiment study of natural and synthetic rock systems. J. Petrology 3, 342-532 (1962).

ZIJL, J.S.V. van: A deep Schlumberger sounding to investigate the electrical structure of the crust and upper mantle in South Africa. Geophysics 34, 450-462 (1969).

ZIJL, J.S.V. van, HUGO, P.L.V., de BELLOCO, J.H.: Ultra deep Schlumberger sounding and crustal conductivity structure in South Africa. Geophys. Prospect. 18, 615-634(1970).

ZOHDY, A.: The auxiliary point method of electrical sounding interpretation, and its relationship to the Dar Zarrouk parameters. Geophysics 30, 644-660 (1965).

ZSCHAU, J.: Bearbeitung digitaler Seismogramme des Profils Eschenlohe-E (Registrierungen bis März 1969), eine Vorbereitung zur Interpretation der Einsätze von Transversalwellen. Diplomarbeit, Univ. München 1969.

ZÜRN, W.: Analyse der Dispersion kurzperiodischer Oberflächenwellen im Alpenvorland aus Magnetbandregistrierungen von Nahbeben, 110 p. Diss., Univ. Stuttgart 1970.

ZVEREV, S.M.: Problems in seismic studies of the oceanic crust. Physics of the Solid Earth. Izvestiya 4, 237-246 (1970).

ZWERGER, R. von: Der tiefere Untergrund des westlichen Peribaltikums. Abh. Geol. Landesamt Berlin, N.F. 210, 74 p. (1948).

Subject Index

Geographical and Geological Index

Appendix: Record Sections

Explanatory notes are given in section 3.4. For details see also Figs. 1 (3.3) and 2 (3.3), Tables 1 (3.3) and 2 (3.3).

The different record sections were compiled by various authors as follows: AICHELE, ANSORGE, BAIER, BARTELSEN, BEHNKE, BERCKHEMER, BRAM, DEGUTSCH, EDEL, EMTER, FÖRTSCH, FUCHS, GEBRANDE, GELBKE, GIESE, GLOCKE, GUTHEIL, HÄNEL, HEEP, E. HINZ, KAMINSKI, KAPPELMEYER, KATZLER, KOSCHYK, MEISSNER, MILLER, S. MUELLER, PETERS, PETERSCHMITT, PLAUMANN, POURSADEG, PRODEHL, SCHICK, SCHRÖDER, STEIN, STROBACH, THYSSEN, VEES, VETTER, VOSS, WANGEMANN, WIEHLE, WILDE, WOLBER.

Map	Number of Record Section	Profile	Map	Number of Record Section	Profile
1	1	17-230-V	1	34	02-325
1	2	06-350	1	35	01-175-F
1	3	14-010	1	36	08-195
1	4	10-305	1	37	01-130-F
1	5	03-250	1	38	14-195-V
1	6a	06-260-V	2	39	02-220-05
1	6b	06-280-V	2	40	09-275
1	7	06-300	2	41	05-040-02
1	8	17-350-06	2	42	15-240-08
1	9	02-350-06	2	43a,b	13-240-20
1	10	02-265	2	44	02-165-01
1	11	24-170	2	45	16-080-02
1	12	01-220-F	2	46	19-110
1	13	13-120-09	2	47	02-215
1	14	12-115-F	2	48	02-240-19
1	15	03-118-F	2	49	02-140
1	16	02-300	2	50a,b	09-240
1	17	09-300-13	2	51	01-210-F
1	18	11-120-09	2	52	16-195
1	19	02-125-09	2	53	01-150-F
1	20	06-080	2	54	01-040
1	21	04-045	2	55	01-290
1	22	17-170-01	2	56	10-135
1	23	08-000	2	57	01-020-09
1	24	17-230	2	58	HE-160
1	25	06-170-02	3	59	16-130
1	26	09-325-27	3	60	HA-120-01
1	27	27-145-09	3	61	09-200-01
1	28	14-090-02	3	62	240-LO-60
1	29	12-260	3	63	01-315-05
1	30	01-005	3	64	01-090
1	31	01-345-02	3	65a,b	X-350-WE
1	32	06-260	3	66a,b,c, d,e	BO-215
1	33	24-090			

Journal of Geophysics
Zeitschrift für Geophysik

A basic information source

Managing Editors
W. Dieminger, J. Untiedt

This journal is edited for the Deutsche Geophysikalische Gesellschaft. In order to increase its international range, the editorial board had been enlarged to include leading scientists from various countries. The journal will continue to publish original papers and reviews from all fields of geophysics, including applied geophysics and related disciplines. Once accepted, papers will be published very quickly, especially Short Communications and Letters to the Editor.

Subscription and back volumes information as well as sample copies available upon request.

Please address:
Springer-Verlag
Abteilung wissenschaftliche Information
Neuenheimer Landstraße 28/30
D-6900 Heidelberg

Springer-Verlag
Berlin
Heidelberg
New York

C. B. Officer
Introduction to Theoretical Geophysics

118 figures. X, 385 pages. 1974
Cloth DM 48,60; US $ 20.00
ISBN 3-540-06485-0

Contents: Introduction: Mathematical Considerations. - Thermodynamics and Hydrodynamics: Thermodynamics of the Earth. Hydrodynamics. Physical Oceanography - Circulation. Physical Oceanography - Waves and Tides. - Seismology, Gravity, and Magnetism: Seismology-Ray Theory. Seismology-Wave Theory. - Gravity. Geomagnetism. - Dynamics of the Earth: Earth-Motion, Rotation, and Deformation. Earth Crustal and Mantle Deformation.

Coordinated treatment of the whole of theoretical geophysics in a basic and elementary manner, including thermodynamics, physical oceanography, seismology, geodesy, geomagnetism, and earth rotation and deformation.

Springer-Verlag
Berlin
Heidelberg
New York

Prices are subject to change without notice.

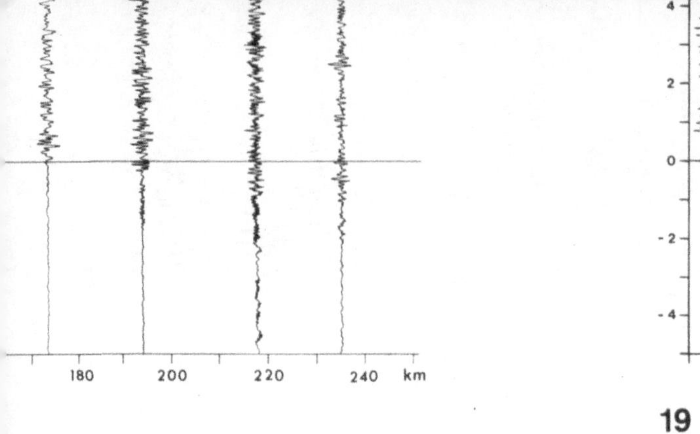

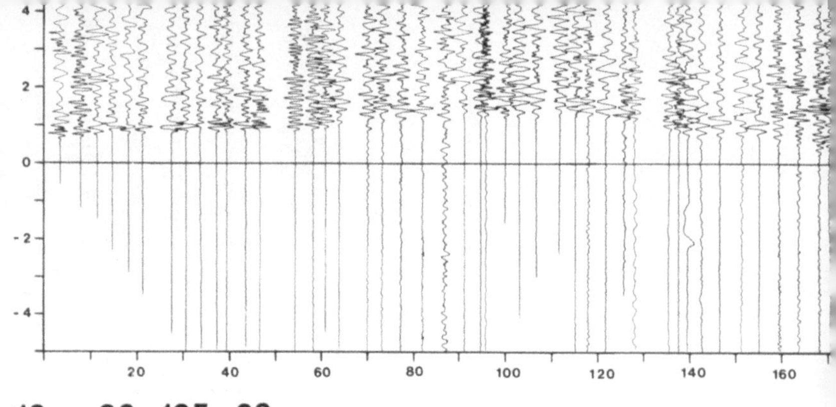

19 02-125-09

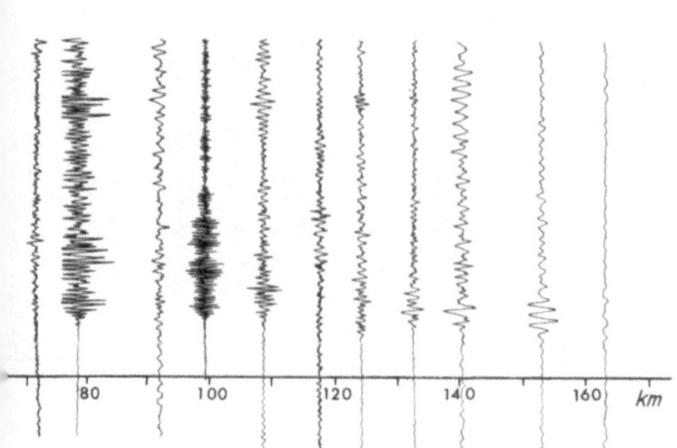

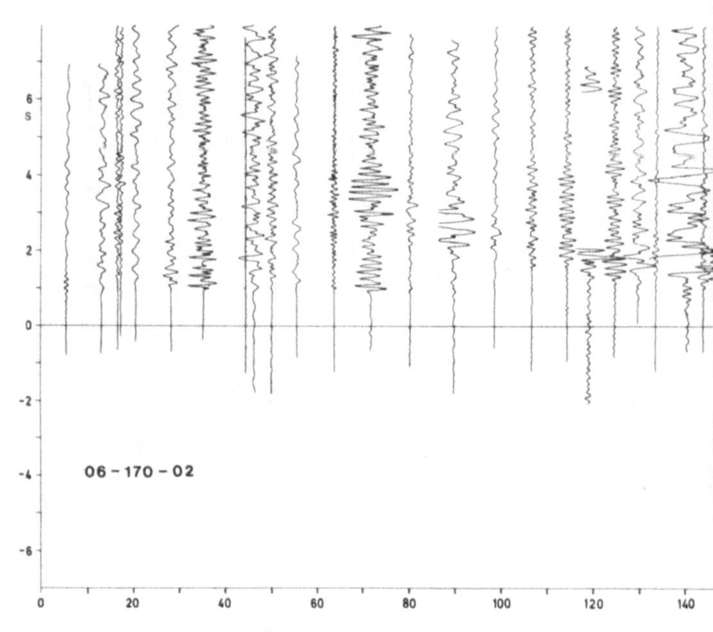

25 06-170-02

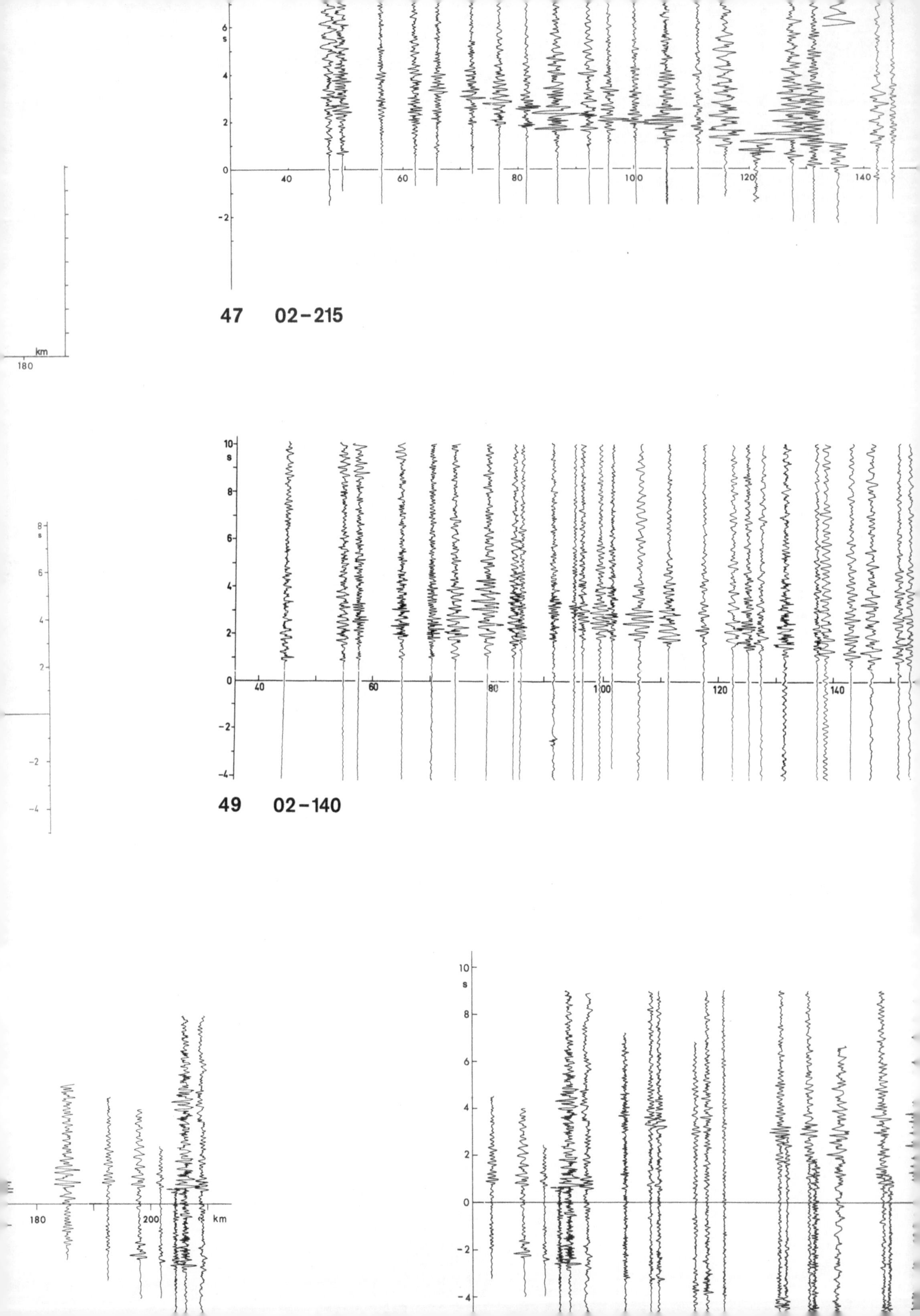

47 02-215
49 02-140
s
km

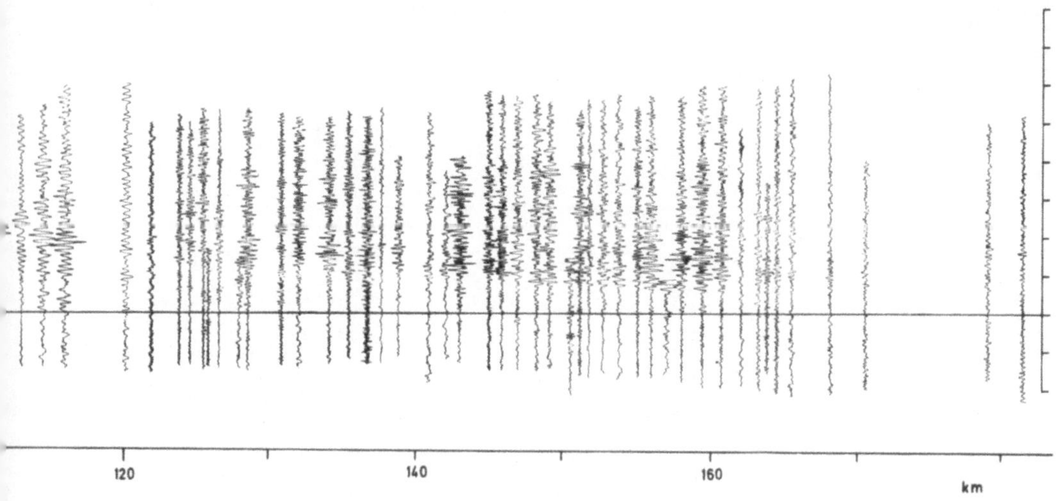

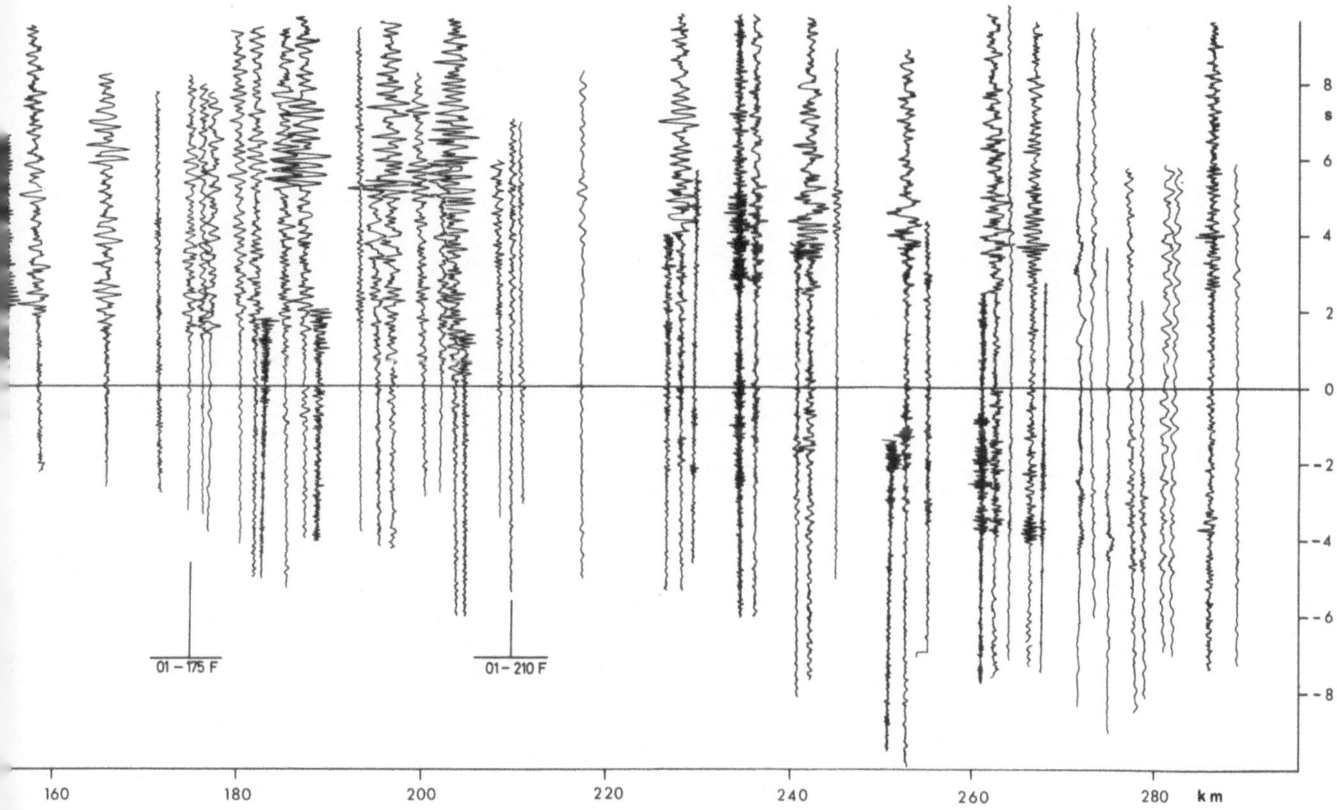

01 – 175 F
01 – 210 F

8
s
6
4
2
0
- 2
- 4
- 6
- 8

160 180 200 220 240 260 280 km